Developments in Agricultural Engineering 11

Soil Compaction in Crop Production

Developments in Agricultural Engineering 11

Soil Compaction in Crop Production

Editors

B.D. Soane

Scottish Centre of Agricultural Engineering, SAC, Bush Estate, Penicuik, Midlothian, EH26 OPH, U.K

C. van Ouwerkerk

Institute for Soil Fertility Research (IB-DLO), P.O. Box 30003, 9750 RA Haren, The Netherlands

ELSEVIER
Amsterdam – London – New York – Tokyo   1994

ELSEVIER SCIENCE B.V.
Molenwerf 1
P.O. Box 211, 1000 AE Amsterdam, The Netherlands

ISBN 0-444-88286-3

Transferred to digital printing 2006

FOREWORD

"Soil Compaction in Crop Production" attempts to provide a global review of the mechanisms, incidence and control measures related to the problems of soil compaction in agriculture, forestry and other cropping systems. Among the disciplines which relate to this subject are soil physics, soil mechanics, vehicle mechanics, agricultural engineering, plant physiology, agronomy, pedology, climatology and economics. While a multi-disciplinary approach is indispensable, the many R & D programmes concerned with this subject are rarely, if ever, sufficiently staffed and funded to cover all relevant aspects. For this reason, it is of vital importance to encourage the fullest possible exchange of information on current work throughout the world. We hope that this book will facilitate such exchange, promote scientific understanding and stimulate the development, evaluation and adoption of practical solutions to these widespread and urgent problems.

During the planning and preparation of the book we received most helpful assistance from the staff at Elsevier Science B.V., Amsterdam, Netherlands.

We acknowledge with thanks the encouragement and assistance which we have received from our employers, the Scottish Centre of Agricultural Engineering, at Penicuik, U.K., and the Institute for Soil Fertility Research, at Haren Gn, Netherlands.

We are greatly indebted to the numerous and widely distributed authors, who cooperated so fully in the onerous work of preparing their contributions. The book would never have been possible without their enthusiastic participation.

Finally, we wish to express our sincere thanks to our wives, Christine and Gera, who, in giving us their loyal support, have greatly eased the burden of our tasks as Editors.

Brennan D. Soane   Cees van Ouwerkerk
Penicuik   Haren Gn
U.K.   Netherlands

This book is dedicated to

Professor Walter Söhne

whose pioneering work laid the foundations

of the scientific study of the compaction

of agricultural soils and its remedies

NOTE ON USAGE OF TERMS, SYMBOLS AND UNITS

In any scientific publication it is highly desirable to adopt a unified system of terminology, symbols and units which is consistent with the SI system and with the numerous attempts by international organizations to standardise usage. However, in considering the chapters submitted by different authors, we became aware of a considerable diversity of usage. This is attributable to the differing traditions which have become entrenched within the various disciplines from which the contributing authors have derived their experience.

We have attempted to reduce the amount of variation in usage but have not found it possible to be entirely consistent, largely because of the regrettable and widespread divergence of usage between the important theoretical and practical application aspects of the subject with which this book is concerned. For the guidance of readers, we have set out on page viii details of those properties which are most commonly used in the book, together with the symbols and units which we have adopted. In some cases these do not comply strictly with scientific recommendations but rather reflect common usage. We have adopted units of negative pressure (kPa, MPa) for both "water potential" and "matric potential". Similarly, we have used the term "air permeability" rather than the standard term "intrinsic permeability".

There were two subjects which were the source of particular variation. Firstly, the use of the terms "mass" and "weight", usually with respect to vehicles, wheels, etc. Many official "standards" concerned with vehicles and tyres use these terms as if they were synonymous, in contrast to the basic scientific distinction between "weight" (e.g., in kN) and "mass" (e.g., in kg). While we prefer that the terms "weight" and "mass" should be used strictly according to their scientific definitions, we have recognised that practical usage does not usually follow this rule. Therefore, we have accepted that, in respect to usage concerning practical applications of vehicles and running gear, the colloquial term "weight" may be used instead of the scientifically correct term "mass", provided that a unit of mass is specified or clearly implied.

Secondly, we encountered much variation in the usage of the terms "pressure" and "stress", which are sometimes used synonymously and sometimes not. The meaning of these terms in physics appears to differ from that in certain branches of engineering. We decided to use the term "stress" for: (1) any application below the soil surface; (2) measured stresses within the soil-wheel or soil-track interface. We have restricted the unqualified term "pressure" to any application related to liquids or gases but, according to common usage, we have adopted the qualified terms "ground contact pressure" and "ground pressure" to denote the average value of the vertical component of the stress in the soil-wheel or soil-track contact area, as calculated from the vehicle weight and the contact area.

The Editors

LIST OF TERMS, SYMBOLS AND UNITS

Property	Symbols	Commonly used unit(s)	Alternative unit(s)
Mass		μg, g, kg, Mg	t (tonne)
Load, weight, force		kN	
Length, depth		mm, cm, m, km	
Area		m^2, ha	
Pressure, stress, resistance		kPa, MPa	bar (for inflation pressure only)
Velocity	v	m s^{-1}	km h^{-1}
Water content (weight)	w	%(w/w), kg kg^{-1}	g g^{-1}
Water content (volume)	θ	%(v/v), m^3 m^{-3}	
Water potential	ψ	kPa, MPa	
Matric potential	ψ	kPa, MPa	
Hydraulic conductivity	K	m s^{-1}	
Total porosity	n, n$_t$	%(v/v), m^3 m^{-3}	
Void ratio	e	m^3 m^{-3}	
Specific volume	v	m^3 m^{-3}	
Dry bulk density	ρ_d	Mg m^{-3}	
Wet bulk density	ρ_w	Mg m^{-3}	
Particle density	ρ_s	Mg m^{-3}	
Air-filled porosity (air content)	n$_a$	%(v/v), m^3 m^{-3}	
Gas concentration (content)		%(v/v), m^3 m^{-3}	
Air permeability (intrinsic permeability)	k	m^2	
Oxygen diffusion rate	ODR	μg m^{-2} s^{-1}	
Gas diffusion coefficient	D	m^2 s^{-1}	
Relative gas diffusion coefficient	D/D$_o$	-	
Redox potential	Eh	mV	
Crop yield		Mg ha^{-1}	kg ha^{-1}, t ha^{-1}

List of Contributors

R.R. Allmaras
University of Minnesota, Department of Soil Science, 1991 Upper Buford Circle, St. Paul, MN 55108, U.S.A.

M.D. Ankeny
USDA-ARS, National Soil Tilth Laboratory, 2150 Pammel Drive, Ames, IA 50011, U.S.A.

E. Audsley
Silsoe Research Institute, Wrest Park, Silsoe, Bedfordshire, MK45 4HS, U.K.

B.C. Ball
Scottish Agricultural College (SAC), Soil Science Department, West Mains Road, Edinburgh, EH9 3JG, U.K.

F.R. Boone
Wageningen Agricultural University, Department of Soil Tillage, Diedenweg 20, 6703 GW Wageningen, Netherlands

L. Brussaard
Institute for Soil Fertility Research (IB-DLO), Department of Soil Biology, P.O. Box 30003, 9750 RA Haren Gn, Netherlands

D.J. Campbell
Scottish Centre of Agricultural Engineering (SAC), Soil Engineering Department, Bush Estate, Penicuik, Midlothian, EH26 0PH, U.K.

W.C.T. Chamen
Silsoe Research Institute, Soil Science Group, Wrest Park, Silsoe, Bedfordshire, MK45 4HS, U.K.

J.T. Douglas
Scottish Centre of Agricultural Engineering (SAC), Soil Engineering Department, Bush Estate, Penicuik, Midlothian, EH26 0PH, U.K.

K. Eradat Oskoui
USDA-ARS, North Central Soil Conservation Research Laboratory, Morris, MN 56267, U.S.A.

D.C. Erbach
USDA-ARS, National Soil Tilth Laboratory, 2150 Pammel Drive, Ames, IA 50011, U.S.A.

Anna Eynard
University of Minnesota, Department of Soil Science, 1991 Upper Buford Circle, St. Paul, MN 55108, U.S.A.

H.G. van Faassen
Institute for Soil Fertility Research (IB-DLO), Department of Soil Biology, P.O. Box 30003, 9750 RA Haren Gn, Netherlands

J. Gliński
Polish Academy of Sciences, Institute of Agrophysics, ul. Doświadczalna 4, 20-280 Lublin, Poland

J. Guérif
INRA, Agronomy Unit of Laon-Péronne, B.P. 101, 02004 Laon Cedex, France

S.C. Gupta
University of Minnesota, Department of Soil Science, 1991 Upper Buford Circle, St. Paul, MN 55108, U.S.A.

A. Hadas
Agricultural Research Organization, Division of Soils and Water, The Volcani Center, P.O. Box 6, Bet Dagan 50-250, Israel

I. Håkansson
Swedish University of Agricultural Sciences, Department of Soil Sciences, P.O. Box 7014, 750 07 Uppsala, Sweden

J.B. Holt
84 Putnoe Lane, Bedford, MK41 9AG, U.K.

R. Horn
Christian-Albrechts-University of Kiel, Institute of Plant Nutrition and Soil Science, Olshausenstrasse 40, 2300 Kiel 1, Germany

R. Horton
Iowa State University, Department of Agronomy, Ames, IA 50011, U.S.A.

B. Kayombo
Sokoine University of Agriculture, Department of Agricultural Engineering and Land Planning, P.O. Box 3003, Chuo Kikuu, Morogoro, Tanzania

Maja J. Kooistra
The Winand Staring Centre for Integrated Land, Soil and Water Research (SC-DLO), Department of Micromorphology and Soil Structure, P.O. Box 125, 6700 AC Wageningen, Netherlands

A.J. Koolen
Wageningen Agricultural University, Department of Soil Tillage, Diedenweg 20, 6703 GW Wageningen, Netherlands

H. Kuipers
Nassaulaan 2, 6721 DZ Bennekom, Netherlands

R. Lal
Ohio State University, Department of Agronomy, 2021 Coffey Road, Columbus, OH 43210-1086, U.S.A.

W.E. Larson
University of Minnesota, Department of Soil Science, 1991 Upper Buford Circle, St. Paul, MN 55108, U.S.A.

M. Lebert
Christian-Albrechts-University of Kiel, Institute of Plant Nutrition and Soil Science, Olshausenstrasse 40, 2300 Kiel 1, Germany

M.J. Lindstrom
USDA-ARS, North Central Soil Conservation Research Laboratory, Morris, MN 56267, U.S.A.

J. Lipiec
Polish Academy of Sciences, Institute of Agrophysics, ul. Doświadczalna 4, 20-280 Lublin, Poland

M.J. McGregor
Scottish Agricultural College (SAC), Rural Resource Management Department, West Mains Road, Edinburgh, EH9 3JG, U.K.

G. Murphy
Forest Research Institute, P.B. 3020, Rotorua, New Zealand

C. van Ouwerkerk
Institute for Soil Fertility Research (IB-DLO), Department of Soil Physics, P.O. Box 30003, 9750 RA Haren Gn, Netherlands

U.D. Perdok
Wageningen Agricultural University, Department of Soil Tillage, Diedenweg 20, 6703 GW Wageningen, Netherlands

H. Petelkau
Friedrich Engels Strasse 13, 1240 Fürstenwalde, Germany

R.L. Raper
USDA-ARS, National Soil Dynamics Laboratory, P.O. Box 792, Auburn, AL 36831-0792, U.S.A.

C. Simota
Research Institute of Soil Science and Agricultural Chemistry, Bld. Mărăşti 61, Bucharest 71331, Romania

B.D. Soane
Scottish Centre of Agricultural Engineering (SAC), Bush Estate, Penicuik, Midlothian, EH26 0PH, U.K.

W. Stępniewski
Polish Academy of Sciences, Institute of Agrophysics, ul. Doświadczalna 4, 20-280 Lublin, Poland

J.H. Taylor
USDA-ARS, National Soil Dynamics Laboratory, P.O. Box 792, Auburn, AL 36831-0792, U.S.A.

F.G.J. Tijink
Institute of Agricultural Engineering (IMAG-DLO), Mechanization Division, P.O. Box 43, 6700 AA Wageningen, Netherlands

N. K. Tovey
University of East Anglia, School of Environmental Sciences, Norwich, NR4 7TJ, U.K.

B.W. Veen
Centre for Agrobiological Research (CABO-DLO), Department of Plant Physiology, P.O. Box 14, 6700 AA Wageningen, Netherlands

G.D. Vermeulen
Institute of Agricultural Engineering (IMAG-DLO), Department of Tillage and Traction, P.O. Box 43, 6700 AA Wageningen, Netherlands

W.B. Voorhees
USDA-ARS, North Central Soil Conservation Research Laboratory, Morris, MN 56267, U.S.A.

E.B. Wronski
Dr. Ed Wronski and Associates, GPO Box 2875, Canberra, ACT 2601, Australia

J.C. van de Zande
Research Station for Arable Farming and Field Production of Vegetables (PAGV), P.O. Box 430, 8200 AK Lelystad, Netherlands

CONTENTS

Foreword v

Note on Usage of Terms, Symbols and Units vii

List of Terms, Symbols and Units viii

List of Contributors ix

PART A: INTRODUCTION

1   Soil Compaction Problems in World Agriculture
     B.D. Soane and C. van Ouwerkerk 1

PART B: SOIL - VEHICLE MECHANICS

2   Mechanics of Soil Compaction
     A.J. Koolen 23

3   Soil Compactability and Compressibility
     R. Horn and M. Lebert 45

4   Prediction of Soil Compaction under Vehicles
     S.C. Gupta and R.L. Raper 71

PART C: EFFECTS OF COMPACTION ON SOIL PROPERTIES

5   Effects of Compaction on Soil Microstructure
     Maja J. Kooistra and N.K. Tovey 91

6   Determination and Use of Soil Bulk Density in Relation to Soil
     Compaction
     D.J. Campbell 113

7   Effects of Compaction on Soil Hydraulic Properties
     R. Horton, M.D. Ankeny and R.R. Allmaras 141

8   Effects of Compaction on Soil Aeration Properties
     W. Stępniewski, J. Gliński and B.C. Ball 167

9   Effects of Compaction on Soil Strength Parameters
     J. Guérif 191

10   Effects of Compaction on Soil Biota and Soil Biological
     Processes
     L. Brussaard and H.G. van Faassen 215

PART D: MECHANISMS AND INCIDENCE OF CROP RESPONSES TO SOIL
        COMPACTION

11   Mechanisms of Crop Responses to Soil Compaction
     F.R. Boone and B.W. Veen 237

12   Responses of Temperate Crops in North America to Soil
     Compaction
     M.J. Lindstrom and W.B. Voorhees 265

13   Responses of Tropical Crops to Soil Compaction
     B. Kayombo and R. Lal 287

14   Responses of Forest Crops to Soil Compaction
     E.B. Wronski and G. Murphy 317

15   Responses of Perennial Forage Crops to Soil Compaction
     J.T. Douglas 343

16   Role of Soil and Climate Factors in Influencing Crop Responses
     to Soil Compaction in Central and Eastern Europe
     J. Lipiec and C. Simota 365

PART E: VEHICLE AND TRAFFIC SYSTEMS IN CROP PRODUCTION

17   Quantification of Vehicle Running Gear
     F.G.J. Tijink 391

18   Quantification of Traffic Systems in Crop Production
     H. Kuipers and J.C. van de Zande 417

19   Benefits of Low Ground Pressure Tyre Equipment
     G.D. Vermeulen and U.D. Perdok 447

20   Benefits of Limited Axle Load
     I. Håkansson and H. Petelkau 479

21   Benefits of Tracked Vehicles in Crop Production
     D.C. Erbach 501

22   Development and Benefits of Vehicle Gantries and
     Controlled-Traffic Systems
     J.H. Taylor 521

PART F: ECONOMIC ASPECTS OF SOIL COMPACTION AND ITS CONTROL

23   Economics of Modifying Conventional Vehicles and Running
     Gear to Minimize Soil Compaction
     K. Eradat Oskoui, D.J. Campbell, B.D. Soane and M.J. McGregor ... 539

24   Economics of Gantry- and Tractor-based Zero-Traffic Systems
     W.C.T. Chamen, E. Audsley and J.B. Holt 569

25   Control and Avoidance of Soil Compaction in Practice
     W.E. Larson, Anna Eynard, A. Hadas and J. Lipiec 597

PART G: CONCLUSION

26   Conclusions and Recommendations for Further Research on
     Soil Compaction in Crop Production
     C. van Ouwerkerk and B.D. Soane 627

INDEX 643

PART A

==========================:

INTRODUCTION

Soil Compaction in Crop Production
B.D. Soane and C. van Ouwerkerk (Eds.)
1

CHAPTER 1

Soil Compaction Problems in World Agriculture

B.D. SOANE[1] and C. van OUWERKERK[2]

[1]Scottish Centre of Agricultural Engineering, SAC, Penicuik, U.K.
[2]Institute for Soil Fertility Research (IB-DLO), Haren Gn, Netherlands

SUMMARY

There is widespread evidence for the prevalence of problems in crop production which are attributable to soil compaction caused by the passage of vehicles, implements and draft animals. Agricultural, horticultural and forestry crops are known to experience these problems in both temperate and tropical regions.

Soil compaction problems were experienced in commercial production long before any coherent research was undertaken on this subject. During the early part of the 19th century, draft animals were observed to cause soil compaction during cultivation, while during the second half of the 19th and early years of the 20th centuries the use of steam engines for cultivation was accompanied by excessive compaction, unless cable traction was employed or soils were extremely dry. The introduction of the internal combustion engine for small tractors did not initially lead to widespread compaction problems but by the middle of the 20th century, and particularly during the past 30 years, mechanization has advanced to such a scale and intensity that compaction problems have become of worldwide importance.

Soil compaction is now considered to be a multi-disciplinary problem in which machine/soil/crop/weather interactions play an important role and which may have dramatic economic and environmental consequences in world agriculture. However, recent progress in scientific understanding of the soil compaction process and its implications, improved insight into proper vehicle use and soil management, and the development of mechanization systems and novel running gear, provide new perspectives for reducing soil compaction problems in crop production.

INTRODUCTION

In this introductory chapter to "Soil Compaction in Crop Production", it seems fitting, first, to define and delineate the subject matter and to put it in its historical context. Second, to review current soil compaction problems in world agriculture and to indicate the economic and environmental importance of the soil compaction problem, which is only one aspect of the much wider problems of physical, chemical and biological soil degradation. Third, to emphasize that the understanding of the soil compaction phenomenon requires an assessment of the many interacting factors which play a role in the soil compaction process. Fourth, to stress that the solution to the soil compaction problem can be found only by interdisciplinary research and development. Fifth, to indicate that this approach may provide technically acceptable and economically feasible perspectives to reduce soil compaction.

Definitions and boundaries of the subject

Soil compaction
A process of densification in which porosity and permeability are reduced, strength is increased and many changes are induced in the soil fabric and in various behaviour characteristics.

Compactness
The state which indicates the extent to which compaction processes have influenced the packing of the constituent solid parts of the soil fabric. It can be measured or assessed by a wide range of soil properties, such as bulk density, porosity, void ratio and pore size distribution. It can also be expressed in terms of the value of such properties as a ratio to the value of that property for the same soil when in a specified reference state (see Chapter 6).

Crop production
The term "crop production" is used in its widest sense and is intended to cover virtually all types of agricultural, horticultural and forestry crops. Although most compaction research has been undertaken with arable field crops, important areas of research now concern forest crops (see Chapter 14) and perennial forage crops (see Chapter 15). Horticultural crops, soft fruit (see Chapter 16), orchard fruit, fibre crops, protected crops (in glasshouses or tunnel houses) and biomass and other non-food crops are all likely to be equally prone to compaction problems, which may be of considerable local importance. Irrigated crops are especially prone to compaction problems and even irrigated desert soils in the central Sahara, with a sand content of 85-95% (w/w), have degraded under the use of heavy machinery and suffer from marked compaction problems (Allan, 1980).

Compaction problems show similarities in widely different cropping systems. It is frequently possible to trace these similarities to the types of vehicle traffic which are employed for specific types of field operations, e.g., land preparation, soil cultivation, post-planting pesticide and fertilizer applications, and harvesting and transport operations. For different crops, these operations follow a common cyclical relationship (Fig. 1), with the duration of the cycle varying from a few months for vegetable crops to nearly a century for some forest crops.

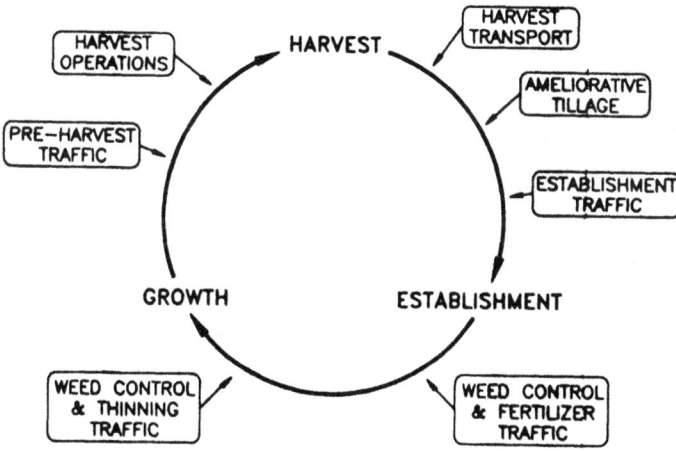

Fig. 1. The cyclical nature of traffic and tillage operations during the establishment, growth and harvest phases of crop production.

Geographical aspects and need for a global approach

Compaction problems have world-wide importance, especially where advanced mechanization forms a significant part of the production techniques. However, research information tends to be more readily available from those areas in which compaction problems have been recognised for a longer period, such as North America (see Chapter 12) and northern, central and eastern Europe (see Chapter 16). More recently, compaction problems have been widely reported in tropical countries (see Chapter 13). It is now apparent that compaction problems in crop production must be considered within a global context because: (1) trade in agricultural and forestry machinery is now world-wide; (2) dissemination of the results of research is increasingly on a world-wide basis. Fortunately, there are now exciting new opportunities for rapid global exchange of ideas and results of mutual interest among research and extension workers in the many different branches of agriculture, horticulture and forestry where previously there was little cooperative effort.

Soil degradation, in which the productivity of the soil suffers a severe decline, largely as a result of mis-management by man, is now recognised as a widespread condition, with physical, chemical and biological processes being identified (Boels et al., 1982). Compaction is seen as an important component of such problems. This is particularly relevant where heavy machinery, used in forest clearing, land levelling and land reclamation, may subsequently result in severe erosion and loss of soil fertility (Lal et al., 1986). Soil degradation may also be asssociated with compaction during treading by working and grazing animals.

SOIL COMPACTION PROBLEMS IN THE PAST

Problems with draft animals

Until about the first third of the 20th century, draft animals were the primary source of power for most field operations throughout the world and even today it is estimated that there are about 400 million working draft animals, mainly in less developed countries. Their use is likely to continue for the foreseeable future, in view of widespread economic and technical difficulties in introducing tractor-based systems in such areas.

The hoofs of draft animals exert high but somewhat variable ground contact pressures, depending on the mass of the animal and the area of the hoof. Typical values for a 750-kg horse are 75 kPa standing and 150 kPa walking (see Chapter 18). The corresponding values for a 530-kg ox are 130 kPa standing and 250 kPa walking. The use of draft animals for mouldboard ploughing and other field operations in Europe was widely recognised as a serious source of compaction (Bonnett, 1965). The "traffic" intensity for two horses pulling a single-furrow plough amounts to approximately 750,000 hoof marks per hectare. In particular, the compaction caused by horses walking in the plough furrow was recognised as a cause of impeded drainage, restricted root growth and reduced yields (Scott, 1908; J. Fowler writing in 1857, quoted by Lane, 1980).

Problems with steam engines

Steam engines for cultivation were introduced into agriculture in Europe and North America during the second half of the 19th century, although the first patents and early trials of such machinery were already recorded at the end of the 18th century. The mass/power ratio of such machines was very high, due to their construction in iron rather than steel, and some early steam-powered vehicles had a mass as high as 30 Mg although most were in the range of 5-20 Mg. It soon became apparent that soil compaction was a serious problem when such vehicles were driven over agricultural fields (Bonnett, 1965). Therefore, the use of steam engines in Europe was almost entirely restricted to cable traction systems in which the engine was stationed at the perimeter of the field, with the cultivation equipment being drawn across the field by cable, a system of traffic control which contributed to the increased crop yields which were widely claimed for this technique.

The use of steam engines for direct traction was rarely adopted in commercial practice in north-western Europe, where soils tended to be wet and soft during the cultivation periods. However, this was not the case in North America, where very large fields and dry soils after harvest permitted the use of large steam engines with towed ploughs and cultivators.

Many alternative designs of running gear were tested to reduce the ground contact pressure and hence limit soil damage. In some cases the wheel dimensions were increased to mammoth proportions, with diameters up to 2.70 m and widths up to 4.50 m (Freitag, 1979). Numerous types of slats attached to wheels and early types of tracks were also tested. Although such modifications, in some cases, reduced the ground contact pressure to acceptable levels, the extremely cumbersome nature of such vehicles and running gear was generally considered unacceptable for within-field traffic.

A more radical approach to controlling compaction problems with steam engines was the attempt by Halkett (1858) to introduce a zero-traffic system in which fixed rails were installed across fields on which steam engines would propel platforms of up to 15 m width, bearing a variety of cultivation and harvesting implements.

Problems during the first half of the 20th century

The introduction of the internal combustion engine and the use of steel led to traction vehicles having a mass/power ratio much lower than steam engines (Fig. 2) and presented an opportunity for direct traction with trailed equipment without the excessive compaction caused by steam engines. Nevertheless, their introduction was followed by widespread concern about posssible soil damage. By the end of the first quarter of the 20th century, Bacon (1929) reported that "ever since the advent of the tractor there has been more or less complaint among farmers relative to the packing of the ground with tractor wheels". However, this concern was, to some extent, muted by the greatly improved productivity of cropping systems in many developed countries during the 1930s and 1940s, a period when many improvements were being made in the power and performance of tractors. Appreciable further reductions in mass/power ratio have been achieved through continuing increases in the power of both 2- and 4-wheel drive tractors (Fig. 3).

One of the earliest field experiments on compaction was that of Huberty (1944) in irrigated citrus orchards in California, in which he measured soil bulk density, penetration resistance, permeability and compactability. By 1947, the effect of organic matter in reducing the compactability of soils had been established (Free et al., 1947), while a year later Veihmeyer and Hendrickson (1948) reported that the threshold soil bulk density for the penetration of sunflower roots was about 1.75 Mg m^{-3} for sand soils and 1.46-1.63 Mg m^{-3} for clay soils.

Of outstanding interest and importance was the work of Gliemeroth (1948) who, in many field experiments in Germany, found evidence of compaction under tractor wheels which he compared to that resulting from horse traffic. He was able to detect compaction from wheels to a depth of 80 cm. In England, Fountaine and Payne (1952) found evidence of compaction under tractor wheels running in the plough furrow. The same effect was confirmed by the Swiss worker

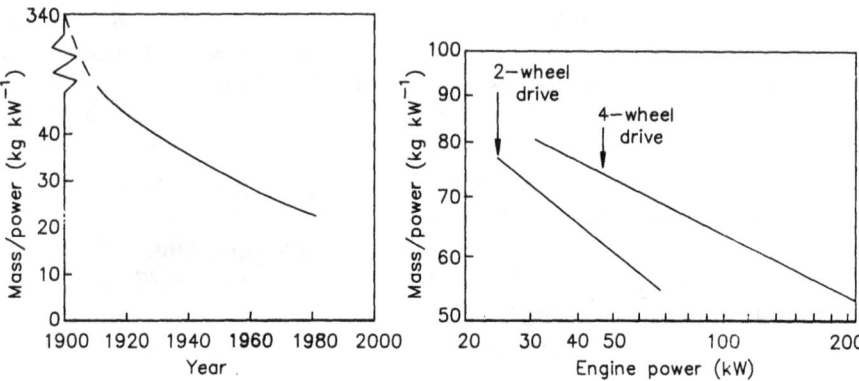

Fig. 2 (left). The reduction in mass/power ratio of tractors, resulting from improvements in their design during the period 1900-1980 (after Reece, 1969-1970).

Fig. 3 (right). With increasing engine power, the mass/power ratio of 2- and 4-wheel drive tractors is improved (after Göhlich, 1984).

Buess (1950) using air permeability and bulk density tests. The extension of compaction below the depth of normal cultivation (about 20 cm) was further confirmed by studies in soil bins at the National Tillage Machinery Laboratory at Auburn, AL, U.S.A., by Weaver (1950) and Weaver and Jamison (1951), who showed that the Proctor test was useful in indicating the compactability of soils under tractor wheels. The relative importance of soil water content and soil looseness in influencing both the intensity and depth of compaction in Cecil clay was identified by Jamison et al. (1950). They reported that compaction was slight for dry soil under all conditions of looseness but extended to 30 cm depth on firm wet soil and even 43 cm depth when the soil was both wet and loose.

Studies by Free (1953) showed the importance of compaction from tractor wheels on the growth of potatoes at Long Island, NY, U.S.A., but he considered that this "does not mean that one must either turn back to horse-drawn equipment or jump ahead at once to performing all operations with airplanes". He considered that the solution lay in the maintenance of a high content of soil organic matter. Outstanding basic research on the mechanics of compaction was undertaken by Söhne (1953, 1958), who was able to relate the behaviour of soil under wheels to the classical theories of stress distribution in elastic media. Although this work was instrumental in awakening widespread interest in compaction in the U.S.A., surprisingly, as late as 1969 he felt able to state that "...the fears of permanent damage to agricultural soils by compaction have not yet been realised in Germany" (Söhne, 1969).

By 1955, the effects of tractor logging in causing deterioration in the physical properties of forest soils was reported by Steinbrenner (1955) and subsequent studies confirmed compaction problems in many types of forests (Beekman, 1987).

The first review of American work on compaction in agricultural soils was that of Raney et al. (1955) with 43 references. This was followed in 1960 by a bibliography issued by the American Society of Agricultural Engineers (ASAE), containing 600 references. In 1961, the ASAE organised a symposium on agronomic and engineering aspects of soil compaction (Bekker et al., 1961). Perhaps the most realistic viewpoint of the 1950's was held by Nichols (1957) who stated: "Unfortunately, we have not gone far enough with these studies to give the farmer or the machine designer specific instructions or to set forth limitations for avoidance of compaction ...".

By the beginning of the 1960s, there was a tendency towards appreciable increases in the power and mass of tractors, which were not accompanied by proportional increases in the size of tyres fitted, and hence the compaction risks were increased. This effect is illustrated (Fig. 4) by changes in the "load index", which is defined as the load per wheel (W) divided by the product of the tyre diameter (d) and section width (b). Although considerably over-estimating the ground contact area, the load index (kPa) "has meaning as it has been shown to be related to vehicle performance on soft soils by a number of investigators" (Freitag, 1979). Having fallen during earlier years, the load index started to rise again after about 1960 (Freitag, 1979), with an accompanying increase in tyre inflation pressures.

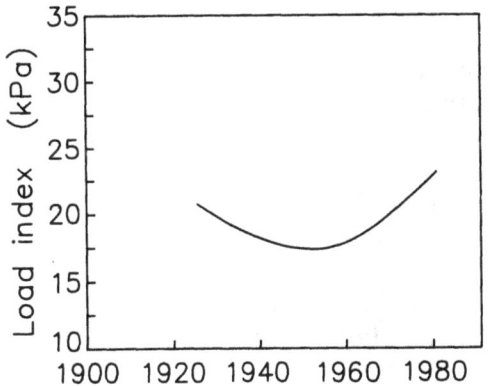

Fig. 4. Changes in the load index (for definition see text) for International Harvester farm tractors from 1925 to 1980 (after Freitag, 1979).

CURRENT SOIL COMPACTION PROBLEMS

At present, farmers, foresters and others concerned with commercial crop production, recognize that the quality and yield of crops is frequently less than expected for the level of inputs applied (e.g., intensity of tillage, quantities of seed, pesticides, fertilizer, fungicides, etc.). Soil degradation as a result of wheel traffic is often clearly visually evident in commercial fields in the form of reduced crop establishment, growth, yield or quality. In some cases this evidence may persist for several years after the traffic event to which it may be attributed, indicating that severe compaction may not be ameliorated by subsequent cycles of cultivation or weather action. The general concern about soil degradation has caused a revival of interest in soil compaction research.

Present-day research interest

There is no difficulty in identifying the evidence for the current world-wide research interest in soil compaction problems, as exemplified by the great number of recently held conferences and workshops, and by numerous books, review papers and monographs on the subject, some of which are listed below.

Conferences

17-19 June 1985. International Conference on "Soil Dynamics", Auburn, AL, U.S.A. (NMTL, 1985).

8-12 July 1985. 10th International Conference of ISTRO on "Reduced Tillage - Rational Use in Sustained Production", Guelph, Ont., Canada (Van Ouwerkerk, 1985).

31 August-4 September 1987. 9th International Conference of ISTVS on "Terramechanics, Mobility and Unconventional Vehicles", Barcelona, Spain (ISTVS, 1987).

11-15 July 1988. 11th International Conference of ISTRO on "Tillage and Traffic in Crop Production", Edinburgh, U.K. (ISTRO, 1988; Van Ouwerkerk, 1988).

5-9 June 1989. International Conference on "Soil Compaction as a Factor Determining Plant Productivity", Lublin, Poland (Gliński, 1990; Van Ouwerkerk, 1991).

24-27 June 1990. ASAE Summer Meeting on "Soil Compaction", Columbus, OH, U.S.A. (ASAE, 1990).

20-24 August 1990. 10th International Conference of ISTVS on "Breaking New Ground", Kobe, Japan (ISTVS, 1990).

8-12 July 1991. 12th International Conference of ISTRO on "Tillage for Sustainable Crop Production", Ibadan, Nigeria (ISTRO, 1991; Lal, 1991, 1993).

8-12 June 1992. International Conference on "Soil Compaction and Soil Management", Tallinn, Estonia (Ecofiller, 1992).

31 August 1992. International Conference on "Problems in Modern Soil Management", Brno, Czechoslovakia (Herman, 1992).

Seminars, Workshops and Symposia

13-15 October 1980. CEC sponsored Land Use Seminar on "Soil Degradation", Wageningen, Netherlands (Boels et al., 1982).

17-18 September 1985. CEC sponsored Workshop on "Soil Compaction: Consequences and Structural Regeneration Processes", Avignon, France (Monnier and Goss, 1987).

11-13 August 1986. NATO sponsored 1st Workshop on "Soil Physics and Soil Mechanics", Hannover, Germany (Drescher et al., 1988).

13-16 September 1988. NATO Advanced Research Workshop on "Mechanics and Related Processes in Structured Agricultural Soils", St. Paul, MN, U.S.A. (Larson et al., 1989).

15 August 1990. ISSS Working Group PT (Pedotechnique) Symposium on "Pedotechnical Approach to Present-Day Soil Tillage and Field Traffic Problems", Kyoto, Japan (ISSS, 1990).

Books

Compaction of Agricultural Soils (Barnes et al., 1971).

Compaction of Arable Soils (Kovda, 1987).

Review papers and monographs

Effect of soil compaction on soil structure and crop yields (Eriksson et al., 1974).

Compaction of soil by agricultural equipment (Chancellor, 1976).

Compaction of forest soil: A review (Greacen and Sands, 1980).

Compaction by agricultural vehicles: A review (Soane et al., 1981a,b, 1982).

Increased mechanisation and soil damage in forests: A review (Wingate-Hill and Jakobson, 1982).

Soil compaction: State-of-the-art report (Taylor and Gill, 1984).

Traction and transport systems as related to cropping systems (Soane, 1985).

Soil compaction in agriculture: A view towards managing the problem (Raghavan et al., 1990).

Reduction of traffic-induced soil compaction (Taylor, 1992).

SOIL COMPACTION: A MULTI-DISCIPLINARY PROBLEM

Wheel/soil interaction

Classical soil mechanics was developed primarily to predict soil behaviour in the context of civil engineering, in which both the rate of loading and the degree of soil failure are much less than usually encountered when agricultural soils in the field are subjected to traffic from vehicles. The development of a science of the mechanics of agricultural soils ("soil dynamics") has therefore required emphasis on different concepts (see Chapter 2).

Satisfactory characterization of all relevant soil properties, both before and after the passage of vehicles, has not yet been fully standardised. This is partly attributable to the very labour-intensive nature of many soil test procedures and partly to the considerable spatial variability of those soil properties which are thought to be of importance. This has restricted the quantification of the relationships between vehicle running gear and the soil (Fig. 5).

While the study of soil reaction to compactive loading is still largely conducted in laboratories (see Chapter 3), there is an urgent need to express the properties of field soils in terms of compactability and compressibility and to use the results, together with appropriate wheel or track characteristics, to predict soil compaction effects in terms of plant sensitive properties, over a wide range of conditions (see Chapter 4).

Soil compactness, as expressed by bulk density (or one of its derived quantities, such as total porosity, void ratio, etc.), is generally considered to be

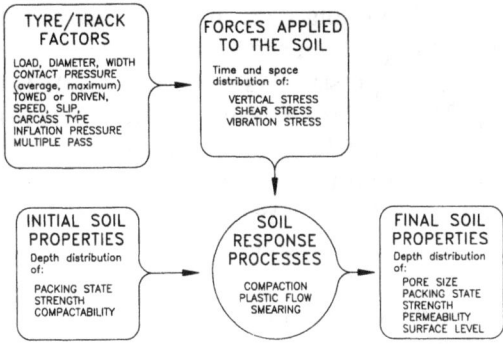

Fig. 5. The relationships between tyre/track factors, initial soil properties and final soil properties during the passage of a vehicle over field soil.

the fundamental criterion in compaction studies (see Chapter 6). However, it is now recognised that crop responses are likely to be only indirectly related to absolute bulk density values and that derived properties, such as "relative compaction" and "degree of compactness" (relative to a standard condition), are likely to be more useful. Similarly, other properties, such as structure and pore size distribution (see Chapter 5), permeability (see Chapter 7), aeration characteristics (see Chapter 8) and strength (see Chapter 9), which are also strongly influenced by compaction, may have a greater influence on subsequent crop growth than compactness *per se*. It is necessary to express the mechanical characteristics of the running gear of field vehicles (see Chapter 17) in terms which are both relevant to their compaction capability and simple enough for universal application by manufacturers, extension workers, etc. The intensity and distribution of vehicular traffic in the field must also be expressed in simple but quantified terms if the compaction hazard is to be evaluated (see Chapter 18).

In qualitative terms, the relationship to be studied, evaluated and validated, may be expressed in the following conceptual relationship:

$$F = f(W,T,S) \tag{1}$$

where F = field compaction incidence; W = wheel or track compaction capability; T = field traffic intensity; S = soil compactability at the time of traffic.

There are a number of alternative types of running gear and vehicles which provide the opportunity to reduce compaction problems. These options can be grouped into: (1) reduction of ground contact pressure (see Chapter 19); (2) reduction of axle load (see Chapter 20); (3) use of tracks (see Chapter 21); (4) use of zero-traffic systems, such as gantries (see Chapter 22). Because of the variety of cropping systems, soil and weather conditions, it is unlikely that any one of these options will be found to have universal relevance. It seems likely, however, that each will find application in certain specific conditions. While each option has relevance independently, it is also possible that, in certain situations, they could be utilised jointly in soil management systems, e.g., a zero-traffic system might be employed during cultivation and crop establishment, while vehicles having low ground contact pressure might be used for harvesting and transport.

Soil/plant interaction

The physical, chemical and biological aspects of the root environment have a profound influence on crop growth and yield characteristics. Compaction imposes considerable changes, both within and below the root zone, and it is clearly important to be able to understand the complex mechanisms of resulting crop responses (see Chapter 11).

Early studies of crop responses to compaction tended to involve entirely empirical attempts to correlate a single soil property (e.g., penetration resistance) with a single aspect of plant response (usually yield). While, in some cases, close correlations could be demonstrated, it soon became apparent that they were often of limited applicability, perhaps only to one season at one location. In certain cases, the variation in the type (positive or negative) and the closeness of fit of observed correlations, was found to be so variable as to discredit the relevance of the soil property. Although not yet fully evaluated, there is now conclusive evidence for the concept of an optimum level of soil compactness for crop production. The optimum level will be influenced by soil type, crop type, weather conditions and other factors. This optimum represents the dynamic balance between interacting soil/plant mechanisms which tend to restrict plant growth or function at either high or low levels of compactness (Table 1).

The root distribution of crop plants can be changed markedly as a result of compaction in both the topsoil and the subsoil (Tardieu, 1988; Van Ouwerkerk and Van Noordwijk, 1991; Fig. 6). Three-dimensional mapping of the root distribution and statistical analysis of the results (Tardieu, 1988), have shown that obstacles to root penetration, such as compacted zones caused by wheel tracks, can cause a reduction in root density, not only in the compacted zone itself, but also in underlying or adjacent soil which has not been compacted ("shadow" effect). A study of root distribution is therefore a vital component in studies on crop responses to compaction. Different crops, and even different varieties, show

TABLE 1

Soil/plant interactions tending to induce adverse crop responses at low and high levels of compactness (after Soane, 1985)

Low compactness	High compactness
a. Restricted germination and emergence due to poor seed/soil contact	a. Anaerobiosis of the soil leading to ethylene accumulation, poor O_2 supply and N loss by denitrification
b. Unfavourable hydraulic properties, reduced water transport and capillary rise	b. Restricted root penetration and clustered root distribution, leading to reduced uptake of water and nutrients
c. Restricted uptake of water and nutrients per unit root length due to poor root/soil contact	c. High root/soil contact, leading to restricted O_2 uptake
d. Trace element deficiencies, e.g., manganese in barley	d. Diminished nitrogen fixation, nitrification and activity of soil fauna

fundamentally different sensitivity to soil compactness. Cowpeas, for example, have been shown to be capable of rooting well at a level of compactness which inhibits more sensitive crops, such as maize or soybean. This variation in sensitivity to soil compactness has already given added importance to the use of crop rotations in the tropics (see Chapter 13).

Negative effects of soil compaction on crop production due to poor root penetration can often be compensated to a large extent by increased supply of water and nutrients (Schuurman, 1971; Van Noordwijk and De Willigen, 1987). Reduced root development, e.g., due to soil compaction, may affect the efficiency of the use of water and nutrients, may force the farmer to use more fertilizer and irrigation water and may thus enhance negative environmental impacts of agriculture (Van Noordwijk and De Willigen, 1991). Nutrient or water uptake efficiency depend on: (1) the required uptake rate per unit root (which is related to shoot/root ratio and actual growth rate of the plant; see Chapter 11); (2) the degree of synlocation of roots and nutrients (depending on nutrient mobility and soil water content); (3) the degree of synchronization of nutrient demand and supply (De Willigen and Van Noordwijk, 1987); (4) spatial variability in fields which are managed as if they were homogeneous units (Van Noordwijk and Wadman, 1992).

It is now recognised that successful crop production is dependent on the presence and vigorous functioning of a wide range of soil-borne micro- and macro-flora and fauna. The habitat of these organisms will be strongly influenced by the level of compactness and in particular by the pore size distribution (see Chapter 10).

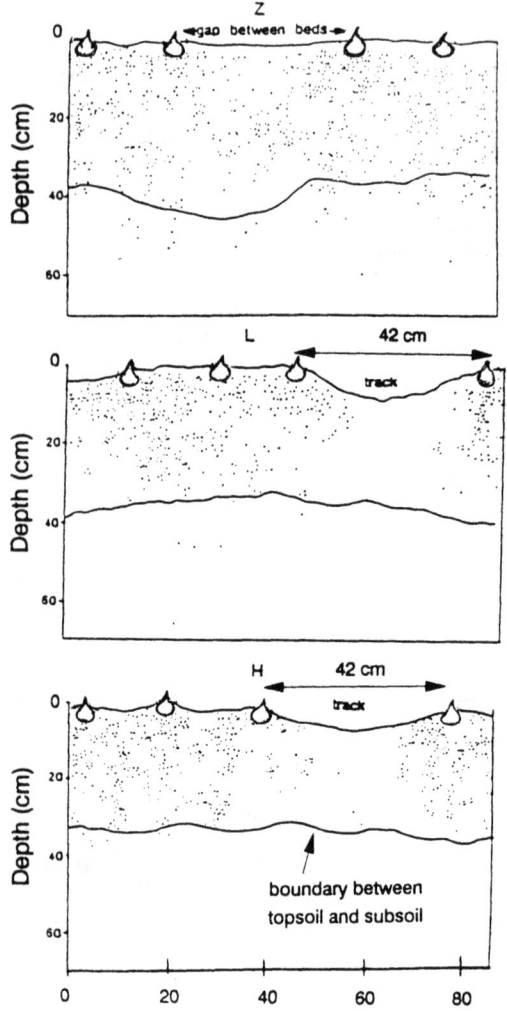

Fig. 6. Distribution of onion roots visible on vertical profile walls of plots under three field traffic systems: zero-traffic (Z), low ground contact pressure, 40-80 kPa (L) and high ground contact pressure, 80-240 kPa (H) (after Van Ouwerkerk and Van Noordwijk, 1991).

Machine/soil/crop/weather interactions

Progress in the understanding of the scientific principles of soil compaction in crop production has been inhibited by the complex nature of the mechanisms involved. Only rarely have the research teams working on this subject included the full range of necessary disciplines. The role of the weather has frequently been found to be paramount in influencing the effect of changes in soil compactness on crop responses. When the effects of seasonal weather are taken into account, particularly rainfall at critical stages of crop growth, it is posssible to explain variations in crop responses to compaction which were previously inexplicable. The dominant influence of weather factors raises important questions concerning the validity of relationships between soil conditions and crop responses which lack a weather component. Experiments undertaken in periods of particularly dry or wet weather conditions may lack general applicability. The role of weather cycles, which are already well documented in some countries, as well as changes in rainfall anticipated from global warming, must therefore assume a major significance. Consequently, the statistical probability of encountering seasonal weather of specified types must accompany any analysis concerning the influence of weather conditions on crop responses to soil compaction.

Tillage and traffic interaction

In practical crop production there is a very close interactive relationship between tillage and traffic systems which, however, has often been overlooked in field experimentation. Attempts to reduce the depth or intensity of tillage have, in some areas, been found to result in crop failure due to unrelieved compaction from vehicles (Westmaas Research Group, 1984). However, there are many soil processes, such as the stabilization of plant root channels, growth of soil-fauna populations and accumulation of organic matter, which are encouraged under no-till cropping and, under certain conditions, these effects may reduce or even reverse the compaction effects induced by machinery.

The adoption of a zero-traffic regime within a conventional tillage system of mouldboard ploughing may result in soil conditions which are below the optimum level of compactness for crop growth (Lamers et al., 1986; Chamen et al., 1992) and in this case the full potential will not be achieved without a suitable modification to the tillage system. From these examples it may be concluded that, in any attempt to improve systems of soil management, the components of both tillage and traffic systems should be considered.

Economic aspects of compaction problems

In the past, field experiments involving different levels of traffic have had limited impact on commercial crop production practices because the economic consequences of the observed crop responses to compaction-limiting techniques had not been fully evaluated. In some cases, the treatments employed in such experiments have been performed with machinery which was not commercially available or have involved traffic distribution treatments which do not correspond to those found in whole-field situations. It has also proved difficult to quantify all the primary and secondary costs of different traffic treatments involving different types of running gear, or even different types of vehicles, when applied to a whole-farm situation (see Chapter 23).

The economic benefits of the control of compaction problems are likely to be much more complex than a simple increase of crop yield. Aspects of crop quality likely to show responses to soil compactness, such as: (1) time of maturity; (2) marketable yield; (3) disease resistance; (4) uniformity of shape; (5) composition, e.g., sugar and protein content, may be much more important than total yield. Furthermore, tillage costs will be reduced as less compacted soil requires less deep and less intensive tillage. However, attempts to establish a relationship between the costs of techniques to limit the incidence of compaction on commercial farms and the benefits in terms of reduced tillage costs and increased value of the produce, are still in an elementary stage. While the modification of running gear on conventional vehicles may be achieved at comparitively modest cost (see Chapter 23), the resulting changes in the operational characteristics of such vehicles may be more difficult to establish. Fundamental approaches, in which all traffic is eliminated from the cropped area, may entail the deployment of complex new vehicles, such as gantries. However, as yet little information exists concerning the economics of their operation in whole-farm situations (see Chapter 24).

PROSPECTS FOR OVERCOMING SOIL COMPACTION PROBLEMS

Improved insight into proper vehicle use and soil management

There are several opportunities to avoid compaction in commercial enterprises (see Chapter 25). In many countries, farmers are already aware of the need for improved methods of avoiding compaction from vehicles. For example, in northern Europe the use of tramlines is widespread in crops of wheat and barley and this has led to much greater uniformity of growth (Fig. 7). These wheel lanes for conventional tractors at normal wheel track spacing are established at the time of drilling, usually with 12, 18 or 24 m distance between the tramlines, and are used for all subsequent traffic involving the application of fertilizers, herbicides and fungicides. The same tramlines may also be used after cereal harvest when applying nitrogen fertilizer to undersown grass and for the subsequent application of herbicides. In future, through the development of suitable sensors for position-finding in motion, farmers are likely to be able to exert a similar close control over the distribution of vehicle traffic over fields for all crops. However, the economic implications of taking remedial action, e.g., mouldboard ploughing and subsoiling, rather than adopting techniques to avoidsoil compaction, will continue to be important and need to be examined for different soil/climate zones.

Fig. 7. Improved uniformity of barley and wheat growth has been widely achieved on farmers' fields as a result of using tramlines at drilling and for all subsequent traffic prior to harvest. The same tramlines may also be used subsequently in undersown grass or green manure.

New concepts in running gear and vehicle design

In view of the economic importance of machinery costs in crop production, it is likely that those concerned with the design and development of running gear and vehicles will pay increasing attention to the commercial opportunities to limit the incidence of compaction. This process has already commenced to a limited extent. Tyre manufacturers have adjusted downwards the recommended minimum inflation pressure of many standard tyres, and have introduced new types of tyres which are designed to run at inflation presssures well below that of standard tyres. Manufacturers of wheeled and tracked tractors and other vehicles are now aware of the need to consider opportunities to reduce compaction when developing new designs.

There are numerous ways in which reductions in vehicle mass and ground contact pressures might be obtained. There is already limited use of light-weight materials, particularly in specialised spraying equipment and spraying vehicles. Further progress in the innovation of similar new concepts will be enhanced by the ability and enthusiasm of scientists and economists to provide conclusive evidence of their technical feasibility and economic advantages.

Progress in scientific understanding

The concept of an optimum level of compaction, to suit the requirements of particular crops growing under particular soil and weather conditions, has allowed a much more rational approach to the opportunities of overcoming compaction problems. It is now possible to foresee the application of computer models to predict the most appropriate soil and machinery management systems, in economic terms, for particular crops under different weather conditions (see Chapter 23).

CONCLUSIONS

(1) Soil compaction problems, in various degree, are found in virtually all cropping systems throughout the world. They are of particular significance where intensive mechanization has been adopted on soils subject to high rainfall or irrigation.

(2) There is urgent need to relate the technical and economic benefits of overcoming compaction problems to the costs of alternative systems of mechanization on a whole-farm scale.

(3) Numerous opportunities exist for reducing compaction problems through the use of new mechanization systems and novel running gear.

REFERENCES

Allan, J.A., 1980. Resource degradation on agricultural schemes in the central Sahara. In: W. Meckelein (Editor), Desertification in Extremely Arid Environments. Univ. Stuttgart, Geogr. Inst., Stuttgart, Germany, pp. 145-156.

ASAE, 1990. Soil Compaction. ASAE Summer Meeting, 24-27 June 1990, Columbus, OH, U.S.A. Am Soc. Agric. Eng., St. Joseph, MI, U.S.A., ASAE Pap. Nos. 901072-901108.

Bacon, C.A., 1929. Some physical aspects of organic matter. Agric. Eng., 10: 83-85.

Barnes, K.K., Carleton, W.M., Taylor, H.M., Throckmorton, R.I. and Vanden Berg, G.E. (Editors), 1971. Compaction of Agricultural Soils. Am. Soc. Agric. Eng., St. Joseph, MI, U.S.A., 471 pp.

Beekman, F., 1987. Soil strength and forest operations. Ph.D. thesis, Wageningen Agricultural University, Wageningen, Netherlands, 168 pp.

Bekker, M.G., Vanden Berg, G.E., Gill, W.R., Vomocil, J.A., Flocker, W.J., Raney, W.A. and Edminster, T.W., 1961. Soil compaction: agronomic and engineering approaches (a symposium). Trans. ASAE, 4: 231-248.

Boels, D., Davies, D.B. and Johnston, A.E. (Editors), 1982. Soil Degradation. Balkema, Rotterdam, Netherlands, 280 pp.

Bonnett, H., 1965. Saga of the Steam Plough. Allen and Unwin, London, U.K., 208 pp.

Buess, O., 1950. Beitrag zur Methodik der Diagnostisierung verdichteter Bodenhorizonte und Ergebnisse von Untergrundlockerungsversuche auf Schweizerischen Ackerböden. (Contribution to the methodology of diagnosing compacted soil horizons and results of subsoiling experiments on Swiss arable fields). Landw. Jb. Schweiz, 64, pp. 1-68 (in German).

Chamen, W.C.T., Watts, C.W., Leede, P.R. and Longstaff, D.J., 1992. Assessment of a wide span vehicle (gantry), and soil and cereal crop responses to its use in a zero traffic regime. Soil Tillage Res., 24: 359-380.

Chancellor, W.J., 1976. Compaction of soil by agricultural equipment. Univ. California, Davis, CA, U.S.A., Bull. 1881, 53 pp.

De Willigen, P. and Van Noordwijk, M., 1987. Roots, plant production and nutrient use efficiency. Ph.D. thesis, Wageningen Agric. Univ., Wageningen, Netherlands, 282 pp.

Drescher, J., Horn, R. and De Boodt, M. (Editors), 1988. Impact of Water and External Forces on Soil Structure. Catena, Cremlingen, Germany, Catena Suppl. 11, 175 pp.

Ecofiller, 1992. Soil Compaction and Soil Management, Proc. Int. Conf., 8-12 June 1992, Tallinn, Estonia. Scientific-Creative Association "Ecofiller", Tallinn, Estonia, Part I (Proc.): 242 pp.; Part II (Suppl.): 50 pp.

Eriksson, J., Håkansson, I. and Danfors, B., 1974. Effect of soil compaction on soil structure and crop yields. Swedish Inst. Agric. Eng., Uppsala, Sweden, Bull. 544, 101 pp.

Fountaine, E.R. and Payne, P.C.J., 1952. The effect of tractors on volume weight and other soil properties. Nat. Inst. Agric. Eng., Silsoe, U.K., Case Study 17, 34 pp.

Free, G.R., 1953. Traffic soles. Agric. Eng., 34: 528-531.

Free, G.R., Lamb, J. and Carleton, E.A., 1947. Compactibility of certain soils as related to organic matter and erosion. J. Am. Soc. Agron., 39: 1068-1076.

Freitag, D.R., 1979. History of wheels for off-road transport. J. Terramech., 16: 49-68.

Gliemeroth, G., 1948. Selbstverschuldete Strukturstörungen des Bodens unter besonderer Berücksichtigung des Schlepperraddrucks. (Man-induced disturbances of soil structure with special reference to tractor wheel ground pressure). Ber. Landtech., 2: 19-54 (in German).

Gliński, J. (Editor), 1990. Soil Compaction Control. Selected papers, Int. Conf. "Soil Compaction as a Factor Determining Plant Productivity", 5-9 June 1989, Lublin, Poland. Soil Technol., Vol. 3, No. 3 (Special Issue), pp. 255-298.

Göhlich, H., 1984. The development of tractors and other agricultural vehicles. J. Agric. Eng. Res., 29: 3-16.

Greacen, E.L. and Sands, R., 1980. Compaction of forest soil: A review. Aust. J. Soil Res., 18: 163-189.

Halkett, P.A., 1858. On guideway agriculture: being a system enabling all the operations of the farm to be performed by steam power. J. Soc. Arts, 7: 41-56.

Herman, M. (Editor), 1992. Problems in Modern Soil Management, Proc. Int. Conf., 31 August-5 September 1992, Brno, Czechoslovakia. Res. Inst. Agroecol. Soil Manage., Hrušovany u Brna, Czechoslovakia, 307 pp.

Huberty, M.R., 1944. Compaction in cultivated soils. Trans. Am. Geophys. Union, 25: 896-899.

ISSS, 1990. Pedotechnical Approach to Present-Day Soil Tillage and Field Traffic Problems. Symposium VI-2: organized by ISSS Working Group PT (Pedotechnique). Trans. 14th Int. Congr. Soil Science, 12-18 August 1990, Kyoto, Japan, Vol. VI, pp. 37-85.

ISTRO, 1988. Tillage and Traffic in Crop Production. Proc. 11th Conf. Int. Soil Tillage Res. Org. (ISTRO), 11-15 July 1988, Edinburgh, U.K. Scot. Centre Agric. Eng., Penicuik, U.K., Vol. 1, pp. 1-444; Vol. 2, pp. 445-938.

ISTRO, 1991. Soil Tillage and Agricultural Sustainability. Proc. 12th Conf. Int. Soil Tillage Res. Org. (ISTRO), 8-12 July 1991, Ibadan, Nigeria. Ohio State Univ., Columbus, OH, U.S.A., 687 pp.

ISTVS, 1987. Terramechanics, Mobility and Unconventional Vehicles. Proc. 9th Int. Conf. Int. Soc. Terrain-Vehicle Systems (ISTVS), 31 August-4 September 1987, Barcelona, Spain, Vol. 1, pp. 1-438; Vol. 2, pp. 439-959.

ISTVS, 1990. Breaking New Ground. Proc. 10th Int. Conf. Int. Soc. Terrain-Vehicle Systems (ISTVS), 20-24 August 1990, Kobe, Japan, Vol. 1, pp. 1-334; Vol. 2, pp. 335-600; Vol. 3, pp. 601-949; Vol. 4, pp. 951-1202.

Jamison, V.C., Weaver, H.A. and Reed, I.F., 1950. The distribution of tractor tire compaction effects in Cecil clay. Soil Sci. Soc. Am. Proc., 15: 34-37.

Kovda, V.A. (Editor), 1987. Pereuplotneniye Pakhotnykh Pochv. (Compaction of Arable Soils). Akad. Nauk U.S.S.R., "Nauka", Moscow, U.S.S.R., 216 pp. (in Russian).

Lal, R. (Editor), 1991. Soil Tillage for Agricultural Sustainability. Proc. 12th Conf. Int. Soil Tillage Res. Org. (ISTRO), 8-12 July 1991, Ibadan, Nigeria, Part I. Soil Tillage Res., Vol. 20, Nos. 2-4 (Special Issue), pp. 133-382.

Lal, R. (Editor), 1993. Soil Tillage for Agricultural Sustainability. Proc. 12th Conf. Int. Soil Tillage Res. Org. (ISTRO), 8-12 July 1991, Ibadan, Nigeria, Part II. Soil Tillage Res., Vol. 27, Nos. 1-4 (Special Issue), pp. 1-386.

Lal, R., Sanchez, P.A. and Cummings, R.W. (Editors), 1986. Land Clearing and Development in the Tropics. Balkema, Rotterdam, Netherlands, 450 pp.

Lamers, J.G., Perdok, U.D., Lumkes, L.M. and Klooster, J.J., 1986. Controlled traffic farming systems in the Netherlands. Soil Tillage Res., 8: 65-76.

Lane, M.R., 1980. The Story of the Steam Plough Works. Northgate, London, U.K., 409 pp.

Larson, W.E., Blake, G., Allmaras, R.R., Voorhees, W.B. and Gupta, S. (Editors), 1989. Mechanics and Related Processes in Structured Agricultural Soils. Kluwer, Dordrecht, Netherlands, NATO ASI Series E: Applied Sciences 172, 273 pp.

Monnier, G. and Goss, M.J. (Editors), 1987. Soil Compaction and Regeneration. Balkema, Rotterdam, Netherlands, 167 pp.

Nichols, M.L., 1957. Soil compaction by farm machinery. Soil Conserv. (Dec): 95-98.

NMTL, 1985. Proc. Int. Conf. Soil Dynamics, 17-19 June 1985, Auburn, AL, U.S.A. Nat. Tillage Machinery Lab., Auburn, AL, U.S.A. Vol. 1: Anniversary Volume, pp. 1-178; Vol. 2: Soil Dynamics as Related to Tillage Machinery Systems, pp. 79-442; Vol. 3: Tillage Machinery Systems as Related to Cropping Systems, pp. 443-604; Vol. 4: Soil Dynamics as Related to Traction and Transport Systems, pp. 605-862; Vol. 5: Traction and Transport as Related to Cropping Systems, pp. 863-1157.

Raghavan, G.S.V., Alvo, P. and McKyes, E., 1990. Soil compaction in agriculture: A view towards managing the problem. Adv. Soil Sci., 11: 1-36.

Raney, W.A., Edminster, T.W. and Allaway, W.H., 1955. Current status of research in soil compaction. Soil Sci. Soc. Am. Proc., 19: 423-428.

Reece, A.R., 1969-1970. The shape of the farm tractor. In: Agricultural and Allied Industrial Tractors. Proc. Inst. Mech. Eng., 184 (Part 3Q): 125-131.

Schuurman, J.J., 1971. Effect of supplemental fertilization of oats with restricted root development. Z. Acker- Pflanzenb., 133: 315-320.

Scott, J., 1908. Text-book of Farm Engineering, Part VI. Crosby Lockwood, London, U.K., 181 pp.

Soane, B.D., 1985. Traction and transport systems as related to cropping systems. Proc. Int. Conf. Soil Dynamics, Auburn, AL, U.S.A., Vol. 5, pp. 863-935.

Soane, B. D., Blackwell, P.S., Dickson, J.W. and Painter, D.J., 1981a. Compaction by agricultural vehicles: A review. I. Soil and wheel characteristics. Soil Tillage Res., 1: 207-237.

Soane, B. D., Blackwell, P.S., Dickson, J.W. and Painter, D.J., 1981b. Compaction by agricultural vehicles: A review. II. Compaction under tyres and other running gear. Soil Tillage Res., 1: 373-400.

Soane, B. D., Dickson, J.W. and Campbell, D.J., 1982. Compaction by agricultural vehicles: A review. III. Incidence and control of compaction in crop production. Soil Tillage Res., 2: 3-36.

Söhne, W., 1953. Druckverteilung im Boden und Bodenverformung unter Schlepperreifen.

(Pressure distribution in the soil and soil deformation under tractor tires). Grundl. Landtech., 5: 49-63 (in German).

Söhne, W., 1958. Fundamentals of pressure distribution and soil compaction under tractor tires. Agric. Eng., 39: 279-281, 290.

Söhne, W., 1969. Agricultural engineering and terramechanics. J. Terramech., 6: 9-30.

Steinbrenner, E.C., 1955. The effect of tractor logging on physical properties of some forest soils in South Western Washington. Soil Sci. Soc. Am. Proc., 19: 372-376.

Tardieu, F., 1988. Effect of the structure of the ploughed layer on the spatial distribution of root density. Proc. 11th Conf. Int. Soil Tillage Res. Org. (ISTRO), Edinburgh, U.K., Vol. 1, pp. 153-157.

Taylor, J.H. (Editor), 1992. Reduction of Traffic-Induced Soil Compaction. Soil Tillage Res., Vol. 24, No. 4 (Special Issue), pp. 301-439.

Taylor, J.H. and Gill, W.R., 1984. Soil compaction: state-of-the-art report. J. Terramech. 21: 195-213.

Van Noordwijk, M. and de Willigen, P., 1987. Agricultural concepts of roots: from morphogenetic to functional equilibrium between root and shoot growth. Neth. J. Agric. Sci., 35: 487-496.

Van Noordwijk, M. and De Willigen, P., 1991. Root functions in agricultural systems. In: B.L. McMichael and H. Persson (Editors), Plant Roots and Their Environment. Elsevier, Amsterdam, Netherlands, Devel. Agric. Managed-Forest Ecol. 24, pp. 381-395.

Van Noordwijk, M. and Wadman, W.P., 1992. Effects of spatial variability of nitrogen supply on environmentally acceptable nitrogen fertilizer application rates to arable crops. Neth. J. Agric. Sci., 40: 51-72.

Van Ouwerkerk, C. (Editor), 1985. Reduced Tillage - Rational Use in Sustained Production. Proc. 10th Conf. Int. Soil Tillage Res. Org. (ISTRO), 8-12 July 1985, Guelph, Ont., Canada. Part I: Soil Tillage Res., Vol. 5, No. 2 (Special Issue), pp. 103-222; Part II: Soil Tillage Res., Vol. 5, Nos. 1-2 (Special Issue), pp. 1-182; Part III: Soil Tillage Res., Vol. 8 (Special Volume), pp. 1-376.

Van Ouwerkerk, C. (Editor), 1988. Tillage and Traffic in Crop Production. Proc. 11th Conf. Int. Soil Tillage Res. Org. (ISTRO), 11-15 July 1988, Edinburgh, U.K. Soil Tillage Res., Vol. 11, Nos. 3-4 (Special Issue), pp. 197-372.

Van Ouwerkerk, C. (Editor), 1991. Soil Compaction and Plant Productivity. Selected papers, Int. Conf. Soil Compaction as a Factor Determining Plant Productivity, 5-9 June 1989, Lublin, Poland. Soil Tillage Res., Vol. 19, Nos. 2-3 (Special Issue), pp. 95-362.

Van Ouwerkerk, C. and Van Noordwijk, M., 1991. Effect of traffic intensity on soil structure and root development in a field experiment on a sandy clay loam soil in the Netherlands. Proc. 11th Conf. Int. Soil Tillage Res. Org. (ISTRO), Ibadan, Nigeria, pp. 253-262.

Veihmeyer, F.J. and Hendrickson, A.H., 1948. Soil density and root penetration. Soil Sci., 65: 487-493.

Weaver, H.A., 1950. Tractor use effects on volume weight of Davidson loam. Agric. Eng., 31: 182-183.

Weaver, H.A. and Jamison, V.C., 1951. Effects of moisture on tractor tire compaction of soil. Soil Sci., 71: 15-23.

Westmaas Research Group on New Tillage Systems, 1984. Experiences with three tillage systems on a marine loam soil. II: 1976-1979. Pudoc, Wageningen, Netherlands, Agric. Res. Rep. 925, 263 pp.

Wingate-Hill, R. and Jakobson, B.F., 1982. Increased mechanisation and soil damage in forests: A review. N.Z. J. For. Sci., 12: 380-393.

PART B

==

SOIL - VEHICLE MECHANICS

Soil Compaction in Crop Production
B.D. Soane and C. van Ouwerkerk (Eds.)
©1994 Elsevier Science B.V. All rights reserved.   23

CHAPTER 2

Mechanics of Soil Compaction

A.J. KOOLEN

Wageningen Agricultural University, Department of Soil Tillage, Wageningen, Netherlands

SUMMARY

An introduction to the mechanics of structured soil is followed by stress and strain theory of relevance to soil compaction, analyses of deformation fields in the soil under a tyre and of deformation paths of soil elements. Tests to measure soil behaviour under loading are described, typical soil reactions and related types of soil-wheel systems are discussed, and an outlook is given on the development of accurate prediction methods.

INTRODUCTION

In agriculture, soil compaction generally refers to the negative aspects of volume decrease and deformation of soil by anthropogenic causes. Among these, the most significant is field traffic, which often is not properly adapted to soil type, structure and water content. Volume decrease is primarily at the cost of soil air, which will be expelled or, sometimes, may be considerably compressed. During compaction, soil water is usually not displaced over macro-distances but water potential is changed. Solid particles do not change in volume, but are subject to rearrangement, which may be accompanied by bending of clay platelets, changes in the shape of organic matter, and breaking of bonds. Immediate compaction effects may change with time, and interact with natural soil structure changes caused by flora, fauna, weather and gravity.

Generally, mechanics may be defined as the knowledge of forces and movements of material points. In civil engineering, soil mechanics is defined as the use of soil mechanical properties in relation to problems of settlement, stability and groundwater flow. Agricultural soil mechanics (mechanics of structured soil) relates to technical activities in the pedon (the structured top layer of the earth's crust), such as soil tillage and field traffic.

At the present state of the mechanics of structured soil, it is assumed that soil behaves as a continuum, i.e., matter in a soil volume is assumed to be continuously distributed rather than to consist of a packing of grains interspaced with water- and air-filled voids. If properties in a continuous soil volume do not change from place to place, the material in the volume is called homogeneous. If properties are independent of the direction in which they are measured, the material is called isotropic. Stresses and strains are called homogeneous if they do not change from place to place. A volume of homogeneous material in a larger soil mass which is subjected to homogeneous stress and strain, is called an elemental volume. In loaded topsoil, such elemental volumes cannot be very large because of spatial stress variation, and also not very small because of the continuum assumption. Continuity is also assumed where the soil condition before and after compaction is expressed in quantities such as permeability and storage characteristics for water, air and heat, strength parameters and resistance against penetration by roots and metal probes. Such quantities (and also quantities not based on assumed continuity) will be called soil qualities if they are of direct importance to the soil user.

Although the continuum approach has proven to be very powerful in a number of situations, a more detailed approach is needed. Especially in explaining phenomena, and where it is tried to find relationships between compaction events and soil processes induced by weather, plant roots, soil fauna, etc., a detailed study is indispensable. In these situations it is appropriate to start from the so-called strength determining factors (strength factors, or microfactors). For a given soil with given intrinsic properties, the following strength factors may be formulated (Koolen and Kuipers, 1983): (1) the proportion of soil particles in a unit volume, which is the complement of porosity; (2) the spatial distribution of soil particles, which is related to the pore size distribution; (3) the volumetric water content (m^3 m^{-3}); (4) the soil water distribution within the soil; (5) the bonds at points of contact between solid particles not arising from soil water potential; (6) the distribution of these bonds.

STRESS THEORY

The main cause of soil compaction is the running gear of tractors and machinery, which usually consists of pneumatic tyres. Initially, it was thought that the pressure in the soil-tyre contact area roughly equals tyre inflation pressure. However, it was soon realized that a tyre has a certain carcass stiffness, which contributes to its bearing capacity. Later, the situation appeared to be even more complex. When a pneumatic tyre moves over a rigid surface and deflects, contacting tyre parts move relative to the rigid surface (Gill and Vanden Berg, 1967). Later, it has been observed that this also holds true for non-rigid surfaces, such as soil. Because both tyres and soils usually are rough, these relative movements must be accompanied by shear stresses in the contact surface. Shear stresses as well as the pressures acting perpendicular to the contact surface (normal stresses) are transmitted to places in the soil more remote from the contact area. These stresses can amplify (or counteract) each other in such a way that normal stresses within the soil can, locally, be larger (or smaller) than the normal stresses in the contact surface. To understand these interactions we need the concept of three-dimensional stress (the stress tensor concept).

Consider a small, imaginary plane with area A, located in a small region (a "physical" point) of a soil mass (Fig. 1a). If the soil is loaded, a force F will act on the plane. Assuming continuity in the region, the force intensity or stress, p, in the physical point can be defined as:

$$\lim_{A \to 0} \frac{F}{A} = p \tag{1}$$

Because usually p is not perpendicular to the plane, p is resolved into components that are perpendicular (normal stress σ) and tangential (shear stress τ) to the plane.

Obviously, p will depend on the position of the plane. Therefore, the above concept is extended to a three-dimensional stress concept in which we assume a cube rather than a plane in the region considered (Fig. 1b). The position of the cube is chosen such that the edges of the cube are parallel to the axes of an orthogonal (x,y,z) co-ordinate system. Due to the load acting on the soil, a stress

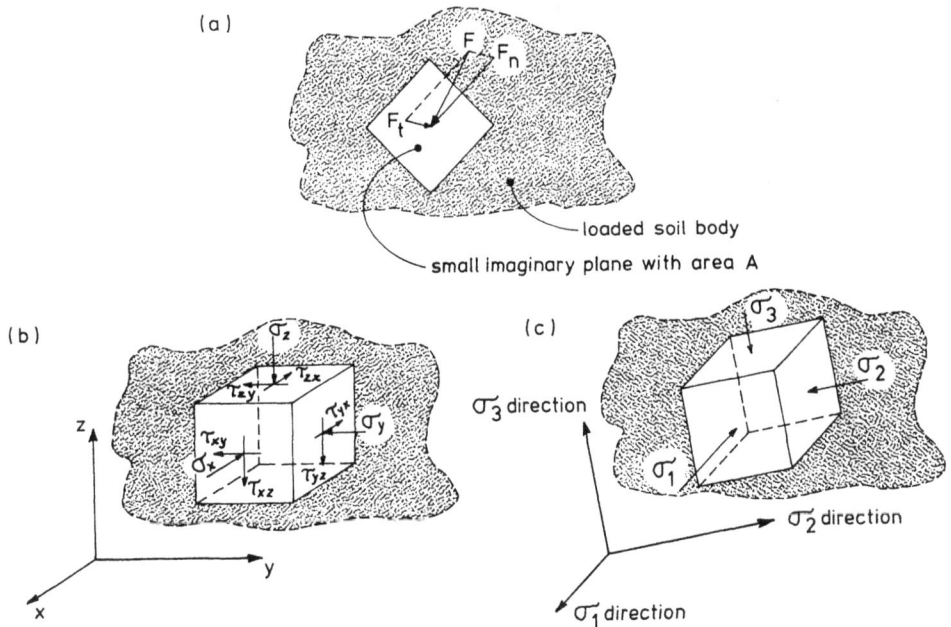

Fig. 1. Stress on an imaginary plane (a) and a cube (b) in a continuous body. Principal stresses after cube rotation are shown in (c).

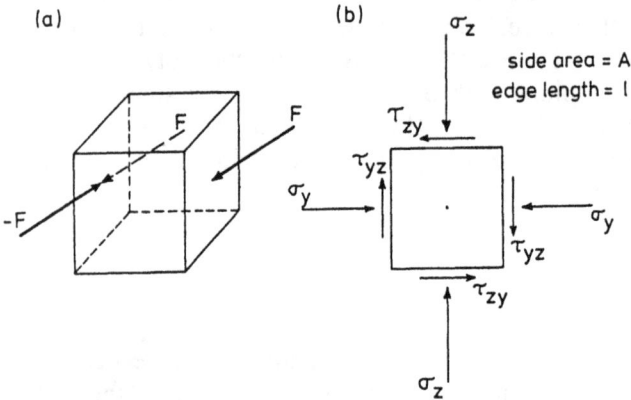

Fig. 2. (a) Forces on opposite sides of a small cube have the same absolute values but opposite directions. (b) All shear stresses shown are equal (see text).

acts on each side of the cube.

Each of these stresses can be resolved into a perpendicular (normal) and a tangential component. The latter is further resolved into tangential components parallel to the co-ordinate axes of the system. Fig. 1b shows these components for the front, upper and right-hand sides of the cube. The components for each of the other sides, which are not shown, have the same absolute values as the components for their opposite side. However, their direction is opposite, because a small parallel shift of a side does not change its state of stress, and a force acting on one face of a plane is equal but opposite to the force on the other face (Fig. 2a). By convention, the index of a normal stress refers to the co-ordinate axis having the same direction as the indicated stress (Fig. 1b). The first and second index of a shear stress refer, respectively, to the normal stress to which it belongs, and to the axis to which it is parallel.

The 9 components shown in Fig. 1b can be put together in a matrix:

$$\begin{pmatrix} \sigma_x & \tau_{xy} & \tau_{xz} \\ \tau_{yx} & \sigma_y & \tau_{yz} \\ \tau_{zx} & \tau_{zy} & \sigma_z \end{pmatrix} \tag{2}$$

Each component has its own fixed place in the matrix. Later on it will be shown that this matrix, called the matrix of the stress tensor at the physical point considered, completely describes the state of stress at that point. The matrix is always symmetrical, i.e.:

$$\tau_{xy} = \tau_{yx}, \quad \tau_{yz} = \tau_{zy}, \quad \tau_{zx} = \tau_{xz} \tag{3}$$

This can be proven by considering the forces on the infinitely small cube in Fig. 2b, where, for simplicity, all stress components in the x-direction have been put at zero. Equilibrium of all force couples, calculated relative to the cube centre, requires:

$$-\tfrac{1}{2}l\tau_{yz}A + \tfrac{1}{2}l\tau_{zy}A - \tfrac{1}{2}l\tau_{yz}A + \tfrac{1}{2}l\tau_{zy}A = 0 \tag{4}$$

or:

$$\tau_{yz} = \tau_{zy} \tag{5}$$

The values of the components in the matrix depend on the choice of the co-ordinate system. If the imaginary cube and co-ordinate axes are slightly rotated while the position of the soil mass and the external load are not changed, the positions of the cube sides change and, generally, the values of the stress components on the cube sides also change. It is possible to calculate the stress components after a rotation if the original components are known (Fig. 3). Again, for simplicity, all stress components in the x-direction have been put at zero. The values of σ_y, σ_z, τ_{yz} and τ_{zy} are known. The cube is rotated counter-clockwise in the y-z plane over an angle θ, and is given a convenient size. The new, still unknown values $\sigma_{z'}$ and $\tau_{z'y'}$ are to be expressed in θ, σ_y, σ_z, τ_{yz} and τ_{zy}.

For this, consider the condition of equilibrium of forces on the hatched prism. If the inclined edge has a length L and a thickness of one unit, then force equilibrium in the $\sigma_{z'}$ direction requires:

$$L \cdot \sigma_{z'} = L \cos\theta \cdot \sigma_z \cos\theta - L \cos\theta \cdot \tau_{zy} \sin\theta + L \sin\theta \cdot \sigma_y \sin\theta - L \sin\theta \cdot \tau_{yz} \cos\theta \tag{6}$$

or:

$$\sigma_{z'} = \sigma_z \cos^2\theta + \sigma_y \sin^2\theta - \tau_{yz} \sin^2 2\theta \tag{7}$$

Similarly, for the $\tau_{z'y'}$ direction:

$$L \cdot \tau_{z'y'} = L \cos\theta \cdot \sigma_z \sin\theta + L \cos\theta \cdot \tau_{zy} \cos\theta - L \sin\theta \cdot \sigma_y \cos\theta - L \sin\theta \cdot \tau_{yz} \sin\theta \tag{8}$$

or:

$$\tau_{z'y'} = \tfrac{1}{2} \cdot (\sigma_z - \sigma_y) \cdot \sin 2\theta + \tau_{yz} \cdot \cos 2\theta \tag{9}$$

For any $(\sigma_z, \sigma_y, \tau_{yz})$ combination a θ-value can always be found at which $\tau_{z'y'}$

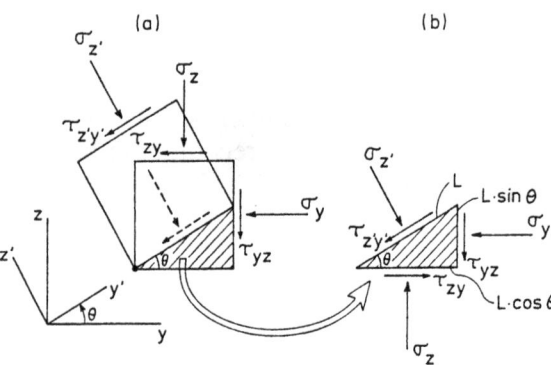

Fig. 3. Calculation of changes in the magnitude of stress components after a change in the reference co-ordinate system (see text).

is zero. Then:

$$\frac{1}{2} \cdot (\sigma_z - \sigma_y) \cdot \sin 2\theta + \tau_{yz} \cdot \cos 2\theta = 0 \qquad (10)$$

or:

$$\theta = \frac{1}{2} \cdot \arctan \frac{2 \cdot \tau_{yz}}{\sigma_y - \sigma_z} \qquad (11)$$

For any state of stress, such as in Fig. 1b, it is possible to rotate the imaginary cube into a position at which all shear stresses acting on its sides have simultaneously become zero. Co-ordinate axes for that position are called principal axes and their directions principal directions. The normal stresses for that position are called principal stresses. Further rotation of the cube appears never to result in larger or smaller normal stresses than the extreme values of the principal stresses. The largest principal stress is called the first (or major) principal stress, σ_1, the smallest the third (or minor) principal stress, σ_3, and the remaining principal stress is called the second (or intermediate) principal stress, σ_2. Consequently, a complete description of the state of stress in a physical point needs at least six quantities, being σ_1, σ_2, σ_3 and their directions (Fig. 1c). If a wheel passes over a physical point, σ_1, σ_2, σ_3 will increase, reach maximum values, and decrease. At the same time, the principal directions will rotate.

The complexity of a complete description of such a loading event is a real problem. Therefore, many simplifying approaches and assumptions have been suggested and more or less accepted. For example, often only maximum stress values are considered, or, even more simple, only the maximum value reached by σ_1. Directional aspects of the stresses are seldom taken into account. However, as yet the effect of the rotation of axes on compaction behaviour cannot be measured adequately. A rational approach starts from the full stress description (including six quantities) and tries to deduce significant characteristics from that description, which are less complex. Classic results of this approach are the so-called stress-invariants. These are quantities, derived from the matrix of the stress tensor, that do not depend on the position of the coordinate system. For example, if we calculate the sum $\sigma_x + \sigma_y + \sigma_z$ for Fig. 1b, and the sum $\sigma_1 + \sigma_2 + \sigma_3$ for Fig. 1c, we will find the same result. This sum is called the first invariant, I_1, which is normally expressed as the sum of the principal stresses:

$$I_1 = \sigma_1 + \sigma_2 + \sigma_3 \tag{12}$$

It means that the mean normal stress $\sigma_m = (\sigma_1 + \sigma_2 + \sigma_3)/3$ is also invariant. Other invariants are the octahedral normal stress σ_{oct} and octahedral shear stress τ_{oct} (the normal and tangential components acting on an octahedral plane, i.e., a plane perpendicular to a space diagonal of the space formed by the σ_1, σ_2, σ_3 directions). It can be derived that:

$$\sigma_{oct} = \frac{1}{3}(\sigma_1 + \sigma_2 + \sigma_3) \tag{13}$$

and

$$\tau_{oct} = \frac{1}{3}\sqrt{(\sigma_1 - \sigma_2)^2 + (\sigma_2 - \sigma_3)^2 + (\sigma_1 - \sigma_3)^2} \tag{14}$$

The maximum shear stress τ_{max} of a state of stress may be a useful quantity:

$$\tau_{max} = \frac{1}{2}(\sigma_1 - \sigma_3) \tag{15}$$

It acts on planes that are parallel to the σ_2-direction and are at an angle of 45° with the σ_1 and σ_3 directions.

The stresses near the tyre-soil contact surface are transmitted to regions in the soil profile that are more remote from the tyre. When a relatively large soil volume under the tyre is considered, the transmitted stresses σ_1, σ_2, σ_3 vary significantly throughout the volume. Therefore, it may not be assumed that a homogeneous state of stress exists that applies to the entire volume. However, it is always possible to divide the volume into smaller volumes (elements), in each of which the state of stress may be considered to be homogeneous. Fig. 4 shows calculated σ_1 values under tyre centres. It is obvious that σ_1 variation in a region depends on the size, as well as on the depth of that region.

STRAIN THEORY

From a soil physical point of view, it is useful to consider deformation as a path function. This is demonstrated in Fig. 5, which shows two different deformation paths to obtain shape B from shape A. The effect of deformation on soil structure in the second deformation process generally differs from the effect in the first and is therefore considered to be path-dependent.

An important building block in descriptions of deformation is the theory of small, homogeneous strain in three dimensions. Small strain means that changes in lengths and angles do not exceed 1-10% of the initial values. It is very characteristic for homogeneous strain that line elements that were parallel and straight initially, are still parallel and straight after the deformation.

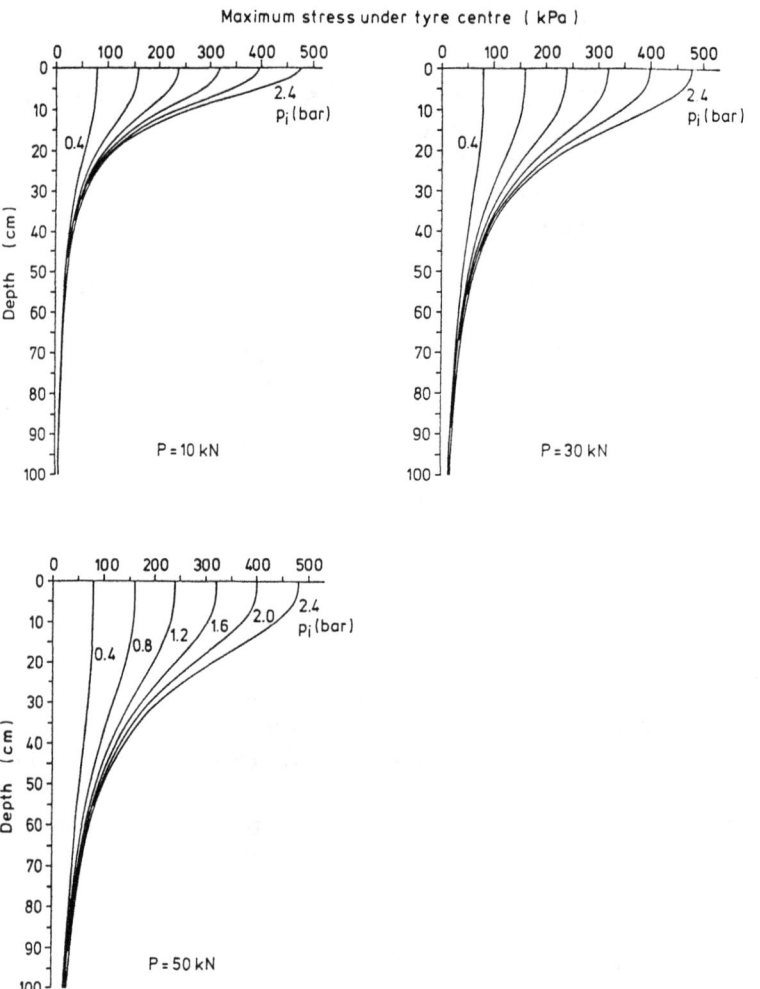

Fig. 4. Calculated maximum stress under tyres as a function of depth, for different wheel loads and tyre inflation pressures. P = vertical wheel load; p_i = tyre inflation pressure (1 bar = 100 kPa = 0.1 MPa).

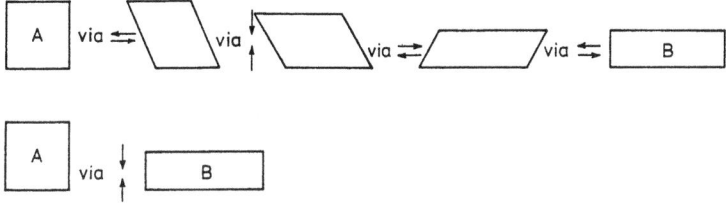

Fig. 5. Shape A can be changed into shape B along different paths.

Small, homogeneous, three-dimensional strain

Analogous to stress, strain can be designated as normal strain and shear strain (Fig. 6a). If a small line element has a length l before deformation and a length $l + \Delta l$ after deformation, the normal strain of that line element is defined as $\Delta l / l$. The shear strain definition starts with a right angle with infinitely small sides (Fig. 6a). If the vertical side is rotated clockwise over the small angle α and the horizontal side counter-clockwise over the small angle ß (both expressed in radials), the shear strain of the initially right angle is $(\alpha + ß)/2$.

Consider a small region in a soil (physical point A in Fig. 6b) and place an imaginary cube in that region, with sides parallel to the axes of a co-ordinate system x, y, z. When the soil is loaded and deformed, the cube will also be deformed and changed into a parallelepiped if deformation in the region is homogeneous. If, in addition, the deformation is small, meaningful quantities can be derived by comparing the parallelepiped with the cube:

e_x = normal strain of the side initially in the x-direction;
e_y = normal strain of the side initially in the y-direction;
e_z = normal strain of the side initially in the z-direction;
e_{xy} = shear strain of the angle of the sides initially in the x-y plane;
e_{xz} = shear strain of the angle of the sides initially in the x-z plane;
e_{yz} = shear strain of the angle of the sides initially in the y-z plane.

Like the stress components, these strain components are placed in a symmetrical matrix (Fig. 6b) with important properties. The co-ordinate system can always be positioned such that all shear strains are zero. For this position, the normal strains are referred to as principal strains, e_1, e_2 and e_3, and the co-ordinate directions as principal directions. With respect to the principal directions, the strain tensor matrix equals:

$$\begin{pmatrix} e_1 & 0 & 0 \\ 0 & e_2 & 0 \\ 0 & 0 & e_3 \end{pmatrix} \tag{16}$$

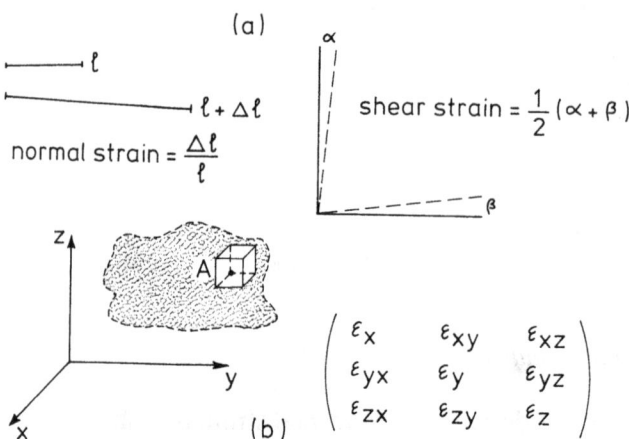

Fig. 6. (a) Normal strains and shear strains on an imaginary cube located at physical point A. (b) Strain tensor.

The shear strain has a maximum value $(e_1 - e_3)/2$, which will be attained when one axis of the co-ordinate system is parallel to the e_2 direction and the other two axes are at an angle of 45° with the e_1 and e_3 directions. The first invariant of the strain tensor,

$$J_1 = e_x + e_y + e_z = e_1 + e_2 + e_3 \tag{17}$$

equals the relative volume change of the cube. The octahedral normal strain and octahedral shear strain are, respectively,

$$e_{oct} = \frac{1}{3}(e_1 + e_2 + e_3) \tag{18}$$

$$\gamma_{oct} = \frac{2}{3}\sqrt{(e_1 - e_2)^2 + (e_2 - e_3)^2 + (e_1 - e_3)^2} \tag{19}$$

The above concepts for small strains are also applicable to a large deformation by dividing the large deformation into a number of successive small deformations and describing each deformation step by the procedure presented above. Usually, strain components in such a stepwise approach are referred to as incremental strains Δe_x,..., etc., rather than as e_x,..., etc. Dividing by the duration t or Δt of the deformation or deformation step, respectively, provides strain rates e_x/t,..., etc., or $\Delta e_x/\Delta t$,..., etc. If the step is infinitely small, the notations dt and de are used instead of Δt and Δe.

Although the description of small strain presented above is an important aid, it is still far away from the solution of compaction problems.

Deformation paths and deformation fields

Consider the soil near the tyre as a continuum that flows with respect to a spatial co-ordinate system. If properties and flow characteristics at each position in space remain invariant with time, the flow is called steady flow. A time-dependent flow, on the other hand, is called unsteady flow. Often, a steady flow may be derived from an unsteady flow field by changing the space reference. For example, the unsteady flow pattern created by a tyre moving over initially undisturbed soil at constant speed V_0 relative to a reference xy (Fig. 7a) can be

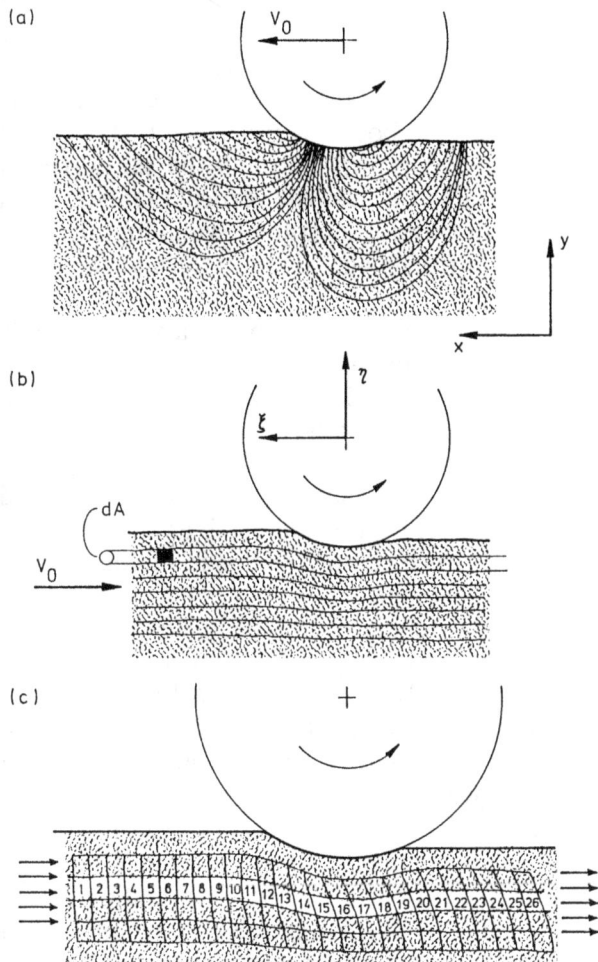

Fig. 7. Streamlines in a soil-wheel system. Unsteady flow in (a) is changed into steady flow in (b) by changing from xy- to $\xi\eta$-reference. A stream tube is indicated in (b). Successive positions of a volume element moving through a stream tube are numbered in (c).

transformed to a steady flow in the space reference co-ordinates ξ and η (Fig. 7b) with their origin in the wheel axis, by superimposing a velocity V_o on the entire flow field of Fig. 7a.

Flows can be presented graphically by means of streamlines. These lines are drawn so as to be always tangent to the velocity vectors of soil particles in a flow (Fig. 7a). For steady flow, the orientation of the streamlines will be fixed, and soil particles will proceed along stream paths coincident with the streamlines. Streamlines proceeding through the periphery of a very small area dA at some time t will form a tube, called stream tube (Fig. 7b). According to the definition of the streamline, there can be no flow through the walls of the stream tube. As in steady flows the position of stream tubes is fixed, the small black volume element fixed to the undisturbed soil in Fig. 7b will just flow through the indicated stream tube, its lateral sides coinciding with the stream tube wall.

Here, the element will move initially as a rigid body, deform in the curved part of the stream tube, and again move as a rigid body when it has left the sphere of influence of the wheel. Volume elements flowing through the same stream tube in a steady flow field behave identically. Because the vertical plane through the centre of the wheel in the direction of travel is a plane of symmetry, the flow pattern on one side of this plane is the mirror image of the pattern at the other side of the plane.

Fig. 7c shows successive, numbered stages of deformation of a small initially cubic soil element passing through a stream tube in a steady flow field. The time interval between two successive stages has been selected such that it equals the original soil element length divided by the velocity of the untouched soil. Because the picture in Fig. 7c presents a steady flow, its determination is relatively easy. Although the picture as such is not an instantaneous picture, it can be obtained from an instantaneous picture of the soil interior, provided with an internal grid, taken at the moment the wheel is passing. The reason for this is, of course, that all successive volume elements behave identically.

The soil physical effect of the deformation subjected to the volume element in Fig. 7c, cannot be determined uniquely by merely comparing its final shape (Position 21) with its initial shape (Position 9), because the final shape can be obtained from the initial shape via an infinite number of possible, different "deformation paths". However, theoretically, there is always a unique "shortest path" to arrive at a given shape from a given initial shape. When two shapes are compared between which only little deformation has occurred (for example, Nos. 15 and 16 in Fig. 7c), the "true" path will resemble the "shortest" path. Therefore, a large deformation path can be analysed by dividing the deformation into a number of small steps of small deformation, calculating relevant quantities for each step on the basis of the "shortest"-path assumption and, finally, adding up the relevant quantities. It appears that the sum found is independent of the chosen number of steps, provided the steps are small. An important question is, what are the relevant quantities that must be added up to obtain the most useful relationship between deformation and soil physical effect.

In volume-element deformation studies, it is convenient to choose the elemental volume so small that the flow in a small domain containing the element may be assumed to be homogeneous. Straightness and parallelism of small line elements in the domain are then conserved during deformation. When the volume element is cubic initially, homogeneity means that the element transforms to a non-cubic rectangular block or a parallelepiped.

Fig. 8a shows a very small volume element moving through a stream tube. A small domain containing the initially cubic element is supposed to undergo homogeneous deformation and the stream tube is straight, so that in the plane of the drawing the element deforms from a square (labelled i) to a rectangle (labelled f), its height changing from l_i to l_f. Two intermediate positions, close to each other, are indicated, one occurring at time t, the other an infinitesimally small time interval dt later. If decrease in length is taken as positive, then the relative vertical compression, de_1, in time dt is:

$$de_1 = \frac{l_t - l_{t+dt}}{l_t} \tag{20}$$

The total relative vertical compression (true or natural vertical strain) occurring between positions i and f is:

$$\bar{e}_1 = \int_{l_i}^{l_f} de_1 \tag{21}$$

Obviously,

$$\bar{e}_1 = \int_{l_i}^{l_f} \frac{dl}{l} = \ln\left(\frac{l_i}{l_f}\right) \tag{22}$$

The so-called large, or technical, vertical strain between positions i and f is defined as:

$$e_1 = \frac{l_i - l_f}{l_i} \tag{23}$$

Consequently:

$$\bar{e}_1 = -\ln (1 - e_1) \tag{24}$$

When the deformation path is divided into n small but finite steps (Fig. 8b), with vertical strain $(\Delta e_1)_j$ in step j, then (Tijink et al., 1988):

$$\bar{e}_1 \approx \sum_{j=1}^{n} (\Delta e_1)_j \tag{25}$$

The above approach can be applied to any line element of the body, e.g., side h, and to volume changes.

Usually in soil, a flow as in Fig. 8a involves compaction and lateral expansion (perpendicular to the drawing plane). It means that the largest change in length occurs in the vertical direction. The important question, which deformation quantity is most related to the effect on physical soil properties, is still not completely answered. It may be volume change, in unstructured loose sands, or $\bar{e}_1$, in incompressible, plastic clays.

Generally, a stream tube is not straight, but curved, as in Fig. 8c. An initially cubic volume element will deform to a parallelepiped, which changes continuously when travelling through the stream tube. The shapes at time t and a short time interval Δt later, are presented in Fig. 8c. It may be assumed that the deformation between t and Δt can be described by the theory of small, homogeneous strain, resulting in incremental principal strains Δe_1, Δe_2, Δe_3. Calculation procedures for 2-dimensional homogeneous deformation have been given in detail by Koolen and Kuipers (1983), and for 3-dimensional homogeneous deformation by Tijink et al. (1988). These authors also discuss

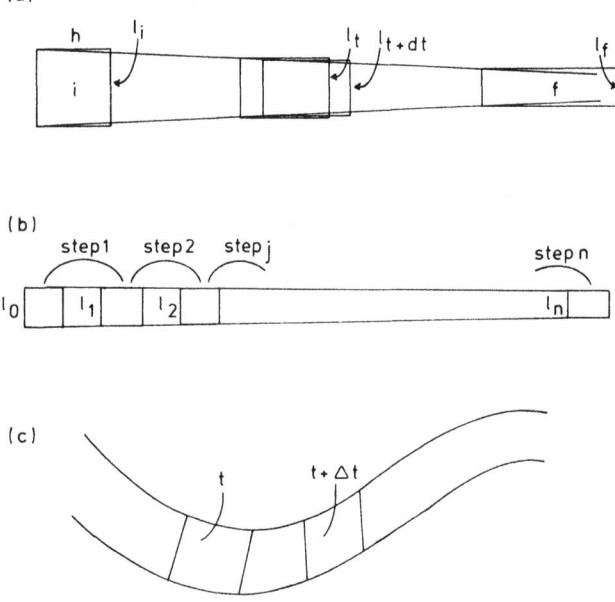

Fig. 8. Illustration of the definitions of incremental and natural strain of a volume element moving through a stream tube. (a) Changes in the shape of a volume element in a straight tube in an infinitesimally small time interval. (b) Changes in large time intervals. (c) Change in a curved stream tube.

rotation of principal directions during the course of loading. Just as $\bar{e}_1$ may be a good measure for the danger to soil structure from the flow in Fig. 8a, the natural largest principal strain, $\Sigma\Delta e_1$, may relate strongly to the soil physical changes of flows in curved stream tubes.

MEASURING STRESS-STRAIN RELATIONSHIPS

Soil deformations in a soil-wheel system depend on the wheel load, wheel characteristics, soil mechanical properties (relationships between stress and strain for a volume element) and system characteristics, such as speed. At least three types of typical systems may be distinguished, although they hardly occur in isolation: (1) the non-deforming type; (2) the hardening type; (3) the flow type, the names referring to the typical soil reactions in the systems. The soil mechanical properties primarily determine which type will occur under specific conditions. Well-known tests to measure these properties are the uniaxial and the triaxial tests.

Uniaxial and triaxial tests

The relationships that may occur between stresses on a soil volume element and the corresponding element strains, can be assessed more or less completely by means of uniaxial or triaxial tests.

In a uniaxial test (Fig. 9a) soil is compacted "quickly" in a rigid cylinder under a steadily downward-moving plate until a certain stress is achieved, after which the load is removed. During the test, the load, F, required to maintain the pre-set downward speed, is measured continuously as a function of sinkage. From these measurements a σ_1-void ratio (or porosity) relationship may be derived.

σ_1-void ratio curves measured on disturbed, precompacted samples or on undisturbed samples of settled field soil will exhibit a "knee" as in the curve in Fig. 9b. From such curves the (equivalent) precompaction stress (pre-consolidation stress, compaction resistance) σ_c can be derived. When the load is removed, the soil volume recovers slightly. When the loading cycle is repeated

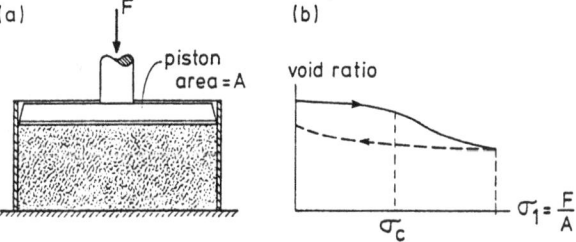

Fig. 9. (a) Uniaxial compression test. (b) General shape of stress-void ratio curves measured on field samples.

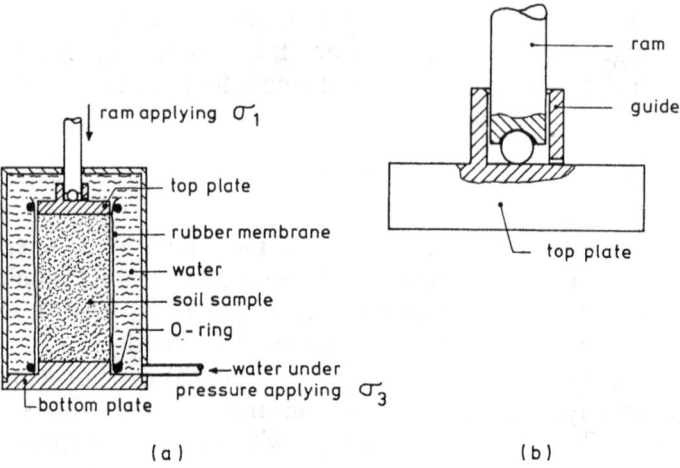

Fig. 10. Triaxial test.

repeated, bulk density increases further. It should be noted that any loading repetition increases bulk density somewhat, although the increase diminishes with an increasing number of loading cycles. Because the test is "quick", water will not be expelled from the sample and the volume decrease is fully at the cost of the soil air volume. If, during the compression water content and decrease of volume are large enough to reach near-saturation conditions, volume decrease will be hampered and the required load will rise dramatically in that stage of the test. Compaction events in which such hampering occurs are said to involve "wet" compaction, as opposed to the dry compaction discussed above. After dry compaction, there still exist locations which have retained the initial soil structure. However, after wet compaction, finer soil particles are found as an amorphous mass between larger particles. The latter case, in particular, worsens many soil qualities.

In the triaxial test, an apparatus is used in which a cylindrical soil sample with a height of 10 cm and a diameter of 5 cm is enclosed by rigid top and bottom plates and a cylindrical rubber membrane (Fig. 10). A rigid transparant cylinder (the cell) is placed over the enclosed sample and filled with water. The water in the cell is pressurized to a constant cell pressure, σ_3. A loading ram is then moved downward at constant speed to deform the sample. The degree of deformation at any moment can be expressed by the vertical strain $e = \Delta l/l_o$ where l_o = initial sample height, and Δl = decrease in sample height due to the ram movement.

During the test, the force exerted by the ram on the sample, F, is continuously measured as a function of Δl. The vertical normal stress σ_1 on the sample is due partly to F and partly to σ_3. σ_1 can be calculated at any time as:

$$\sigma_1 = \frac{F}{A} + \sigma_3 \tag{26}$$

where A = surface area of the horizontal cross section of the sample. The surface area A is derived assuming a constant soil volume when anticipating volume changes are small:

$$A = A_0 \, l_0 / (l_0 - \Delta l) \tag{27}$$

where A_0 = initial value of A.

When significant volume changes are likely to occur, A may be derived by measuring either the sample diameter, D, directly using sensors, or by determining the sample volume change, ΔV, through measuring the volume of the water that is expelled from the cell:

$$A = (A_0 \, l_0 - \Delta V) / (l_0 - \Delta l) \tag{28}$$

In the latter case, care should be taken that the cell does not contain air. If sample volume change is measured through direct measurement of the sample diameter D with sensors, the following relationship applies:

$$\Delta V = V_0 - \pi D^2 (l_0 - \Delta l) / 4 \tag{29}$$

where V_0 = initial sample volume. At any stage of the test, e_1 (or de_1, or Δe_1) and e_3 (or de_3, or Δe_3) can be derived from $\epsilon = \Delta l / l$ and ΔV. The triaxial apparatus may be used to measure three typical soil reactions, each of significance in one of the three typical soil-wheel systems.

Compaction resistance

Similar to σ_c derived from a uniaxial test, compaction resistance of a soil may be represented by the Mohr-Coulomb failure criterion, which uses cohesion, c , and angle of internal soil friction, ϕ, as soil properties. Koolen and Kuipers (1983) presented procedures to determine c and ϕ by means of triaxial tests.

Compaction

Currently, the most complete description of stress-volume strain relationships is probably the Auburn soil compaction model (Bailey and Johnson, 1989):

$$\ln (V/V_0) = (A + B\sigma_{oct}) (1 - e^{-C\sigma_{oct}}) + D(\tau_{oct}/\sigma_{oct}) \tag{30}$$

where σ_{oct} = octahedral normal stress; τ_{oct} = octahedral shear stress; A,B,C = compactability coefficients; D = coefficient for the component of natural volumetric strain due to shearing stress.

Flow (viscoplasticity)

When a sample of wet, dense, clayey soil is loaded in a triaxial apparatus at a given σ_3 and a given ram speed, σ_1 initially rises to a certain level and then remains near constant as deformation continues. If the test is run at a higher ram speed, the σ_1-level is higher. If the test is run at a higher σ_3, the σ_1-level is also higher but σ_1 minus σ_3 is independent of σ_3. The loading appears to induce soil flow without volume change rather than compaction in the form of volume decrease. Flow involves the widely used concept of viscosity. The coefficient of viscosity (η), expressing the fluid flow-ability, may be defined as indicated in Fig. 11a. In a time-interval dt, due to shear stress (τ), a small square fluid element adjacent to a wall deforms to a parallelogram. When, in time-interval dt, $d\gamma$ is the angular change for an element side that initially was perpendicular to the wall, then, for many fluids,

$$\tau = \eta \, (d\gamma/dt) \tag{31}$$

In materials such as tooth-paste, and also in wet, dense soil, the shear stress must exceed a threshold value before movement is initiated. Therefore, a threshold value (ζ) is defined and the previous equation changes into:

$$\tau = \zeta + \eta \, (d\gamma/dt) \tag{32}$$

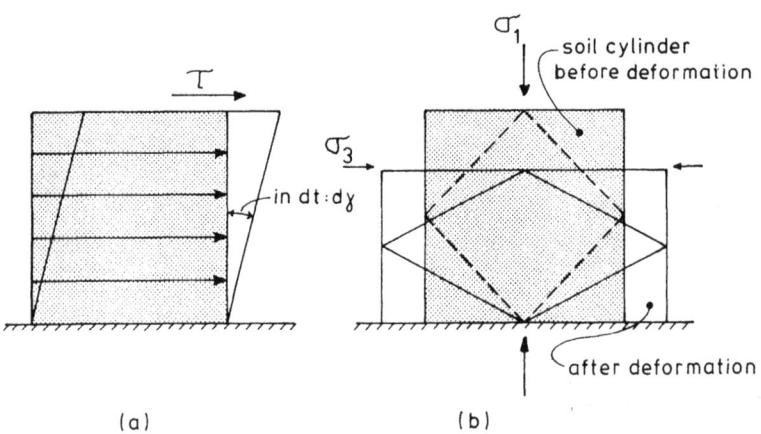

(a)   (b)

Fig. 11. Illustration of the usual definitions of fluid viscosity (a) and of soil viscosity as measured in a triaxial apparatus (b).

Materials which behave according to this formula are called Bingham materials. The above definitions can be applied to a soil sample in a triaxial apparatus, according to Fig. 11b. The square bounded by broken lines resembles the square fluid element, and the rhombus resembles the fluid element after time-interval dt. According to stress and strain theories, if volume change is negligible, $\tau = (\sigma_1\text{-}\sigma_3)/2$ and $d\gamma/dt = 1.5\ de_1/dt$.

TYPES OF SOIL BEHAVIOUR UNDER WHEELS

Non-deforming type

If wheel-induced soil stresses are low relative to soil strength, these stresses cause only very small soil strains, and rut depth will virtually be zero. This situation may be assumed to occur if σ_1 under the tyre does not exceed the compaction resistance σ_c, or, similarly, if σ_1 and σ_3 under the tyre do not give rise to soil failure according to the Mohr-Coulomb failure criterion. In the latter assumption, c and ϕ are the important soil properties. This type, involving only very small strain, allows the calculation of the soil stresses with methods from the mechanics of elastic bodies. To account for the relatively small decrease of stress level with depth in the case of soil-wheel systems, these calculations often use a so-called concentration factor υ. Usually, the value of υ is put at 4, while $\upsilon = 3$ if soil behaviour would be perfectly elastic. Many analytical as well as numerical methods to determine stress fields have been published. Perhaps the most realistic and simple method is the one in which it is assumed that the soil-wheel contact area is a circle, that stresses acting on that circle are vertical and uniformly distributed and amount to twice the tyre inflation pressure, and that $\upsilon = 4$ (Koolen et al., 1992). Results of calculations according to this method were presented in Fig. 4. When using this method, only the variables vertical wheel load and tyre inflation pressure are taken into account, and no variables such as the soil-tyre contact stress distribution. It may be judged from such calculated results whether a particular soil-wheel system is of the non-deforming type or not, e.g., by comparing the calculated stress level with σ_c. As there is now widespread interest concerning the care for soil structure, it may be anticipated that the importance of soil-wheel systems that hardly deform soil will increase.

Hardening type

If wheel-induced soil stresses exceed soil strength, and the soil condition is such that the soil becomes stronger by compaction (compaction hardening), soil will deform and compact until a new state of soil strength has been reached which is able to counteract the wheel-induced stress field. The soil stresses corresponding to that new state of soil strength may be calculated according to elastic-body mechanics just as for the non-deforming type and "translated" into final strain values using stress-strain relationships measured on soil samples. Unfortunately, such translations are not always satisfactory, especially in the case of wet compaction.

Flow type

Very few attempts have been made to calculate stresses under tyres where soil behaves as a Bingham material because little is known about the strain field of the flow type. The following relationships may be hypothesized.

The effect of a wheeling of the flow type on soil qualities is strongly related to the largest natural principal strain $\bar{e}_1$. For given tyre dimensions, wheel slip and depth of the rigid subsoil, $\bar{e}_1$ is almost linearly related to rut depth. The stress level influences the rate of soil deformation and, thus, the rate at which tyre sinkage (rut formation) is realized. Strain at a given point increases at the stress-level dependent strain rate as long as the stress prevails at that point.

There is no direct relationship between stress level and effects on soil qualities. In contrast to the non-deforming type and the hardening type, quantitative relationships still hardly exist for the flow type. However, probably, there are two important chains of causes and effects: (1) if stress level increases → strain rate increases → rut depth and $\bar{e}_1$ increase → effect on soil qualities increases; (2) if tyre travelling speed increases → time of influence (flow duration) decreases → $\bar{e}_1$ and rut depth decrease → effect on soil qualities decreases. Note that these chains do not prevail in the first two types of soil-wheel systems.

STRESS AND STRAIN EFFECTS ON SOIL QUALITIES

When the stresses and strains induced by a wheel are known, the effects on soil qualities of the wheel loading can be derived if, for each soil volume element influenced by the wheel, the relationships between stresses and strains on the one hand, and the changes in soil qualities on the other hand are known. Such relationships may be more or less completely established by loading soil samples in a uniaxial and/or triaxial apparatus and measuring soil qualities on these samples after load removal. From these tests it appears that the effect of loading is determined mainly by soil type and wetness. This may explain why these effects are usually presented for a specified soil type as a function of gravimetric water content at loading in the case of uniaxial compression tests on unsaturated samples (Dawidowski and Lerink, 1990) and as a function of initial soil water potential in the case of triaxial tests on near-saturated samples (Dawidowski and Koolen, 1987).

The accuracy of the prediction of soil qualities following the procedure outlined above is not always high, as could be expected from the discrepancy between uniaxial and triaxial test capabilities and deformation paths in real soil-wheel systems. Therefore, Lerink (1990), following the principles of the prediction theory (Koolen and Kuipers, 1983), measured soil qualities under wheelings immediately after different field operations on one field during a period of three years. The field operations included fertilizing, seedbed preparation, sowing, harvesting and ploughing, each at a number of field moisture contents. Lerink plotted the results for each type of field operation in diagrams as a function of gravimetric field moisture content. In this way, he was able to predict accurately the direct effects of field traffic from the actual soil water content.

CONCLUSIONS

(1) Soil behaviour under wheels may be one of three types: non-deforming, hardening or plastic flow. Hardening type behaviour may occur during both dry and wet compaction. Soil qualities are readily and adversely affected, particularly by wet compaction. Plastic flow, involving deformation at near-constant volume, also reduces soil qualities. The relative importance of non-deforming soil behaviour will probably grow as concern for good soil structure encourages farmers to adopt recent developments in agricultural engineeering. In those situations where compaction cannot be avoided, farmers should aim at restricting wheel traffic to periods when the soil is dry. However, increasing use of low tyre inflation pressures gives farmers easier access to soft terrains, so that in future plastic flow type soil behaviour may occur more frequently.

(2) Wheel traffic is not the only cause of deterioration in soil structural conditions; climate, fertilizers, and biological factors may also contribute to these effects. Where climatic and biological influences are strong, the adoption of a system of mechanization which minimizes soil structural damage will reduce the likelihood that field traffic is the primary cause of structural deterioration. In areas where climate has little or no influence on soil structure, activities such as tillage and traffic will remain decisive for maintaining the structural state of the soil.

REFERENCES

Bailey, A.C. and Johnson, C.E., 1989. A soil compaction model for cylindrical stress states. Trans. ASAE, 32: 822-825.

Dawidowski, J.B. and Koolen, A.J., 1987. Changes of soil water suction, conductivity and dry strength during deformation of wet undisturbed samples. Soil Tillage Res., 9: 169-180.

Dawidowski, J.B. and Lerink, P., 1990. Laboratory simulation of the effects of traffic during seedbed preparation on soil physical properties using a quick uni-axial compression test. Soil Tillage Res., 17: 31-45.

Gill, W.R. and Vanden Berg, G.E., 1967. Soil Dynamics in Tillage and Traction. USDA-ARS, Washington, DC, U.S.A., Agric. Handbook 316, 511 pp.

Koolen, A.J. and Kuipers, H., 1983. Agricultural Soil Mechanics. Springer, Heidelberg, Germany, Advanced Series in Agricultural Sciences 13, 241 pp.

Koolen, A.J., Lerink, P., Kurstjens, D.A.G., Van den Akker, J.J.H. and Arts, W.B.M., 1992. Prediction of aspects of soil-wheel systems. Soil Tillage Res., 24: 381-396.

Lerink, P., 1990. Prediction of the immediate effects of traffic on field soil qualities. Soil Tillage Res., 16: 153-166.

Tijink, F.G.J., Lerink, P. and Koolen, A.J., 1988. Summation of shear deformation in stream tubes in soil under a moving tyre. Soil Tillage Res., 12: 323-345.

Soil Compaction in Crop Production
B.D. Soane and C. van Ouwerkerk (Eds.)
©1994 Elsevier Science B.V. All rights reserved.   45

CHAPTER 3

Soil Compactability and Compressibility *

R. HORN and M. LEBERT

Christian-Albrechts-University Kiel, Institute of Plant Nutrition and Soil Science, Kiel, Germany

SUMMARY

The strength of structured soils during loading depends on both effective stresses and neutral stresses. Thus, soil compaction affects the inter- as well as the intra-aggregate pore size distribution and results in changes of several parameters of the effective-stress equation. Furthermore, differences in the hydraulic conductivity of the bulk soil and in single aggregates, and also differences in the pore continuity must be considered. Therefore, determination of compressibility and compactability requires physically and mechanically well-defined measurements and methods and a detailed analysis on both macro- and micro-scale, in order to deal adequately with the complex relationships between the requirements of growing plants and the soil physical characteristics as affected by loading.

INTRODUCTION

During the last five decades, positive effects of favourable soil structure and negative effects of soil compaction on crop growth and/or yield have often been described. At present, besides *in-situ* and laboratory experiments, modelling of soil compaction is performed in order to predict the compressibility of arable soils and the effects on plant growth (e.g., Rynasiewicz, 1945; Page and Willard, 1947; Barnes et al., 1971; Eriksson, 1975; Emerson et al., 1978; Kral et al., 1982; Gupta and Larson, 1982; Gliński and Stępniewski, 1985; Drescher et al., 1988; Proc. 11th Conf. ISTRO, Edinburgh, 1988; Håkansson et al., 1988; Gupta and Allmaras, 1989; Larson et al., 1989; Horn et al., 1989b; Lebert, 1989).

The application of heavier agricultural machinery and the intensification of plant and soil treatments during the growing season, make the question of maximum acceptable mechanical compressibility or trafficability of arable soils more cogent. Anthropogenic soil compaction induces higher drought sensitivity, inadequate soil aeration, reduced nutrient uptake efficiency and reduced nutrient accessibility by plant roots. Soil compaction furthermore induces a reduction in root penetration due to higher soil strength (Håkansson et al., 1988; Horn, 1990). Therefore it is very important to establish cause-effect relationships between soil structure and crop growth (e.g., Gupta et al., 1985; Håkansson et al., 1987). To explain differences in soil strength between sites under well-defined climatic and hydraulic conditions, the influence of soil type, soil internal strength, load-dependent variations in soil structure and external factors have to be determined.

DEFINITIONS

In silty, loamy and clayey soils, due to biological activity and to swelling and shrinkage processes, the single mineral particles tend to form structured units known as aggregates. To characterize soil stucture in these aggregated soils, a distinction has to be made between inter- and intra-aggregate pore systems. While the inter-aggregate pore system (e.g., cracks, earthworm or root channels) is coarser and more continuous, the intra-aggregate pore system has a smaller diameter, and is less continuous, partly due to a higher aggregate bulk density (Horn, 1990). Fig. 1 informs about the scale of particles, aggregates, pore functions and biota.

It is appropriate to differentiate between several terms used to define soil compaction. These definitions have been taken from Kézdi (1969), Bradford and Gupta (1986) and Hartge and Horn (1991).

Forces applied to the soil have to be related to an area to provide information concerning soil strength or soil deformation. The force per unit area is defined as stress. Stresses applied parallel to the soil surface will also induce stresses in the soil, which may result in a three-dimensional deformation of the soil volume or will be transmitted as a rigid body. Stresses perpendicular to a plane are called normal stresses, denoted by σ_x, σ_y, σ_z, while the tangential components are called shear stresses, denoted by τ_{xz}, τ_{yz} and τ_{xy}.

In saturated soils, total normal stresses (σ) are divided into effective stresses (σ') and neutral stresses (u) (= pore water pressure):

$$\sigma' = \sigma \pm u \tag{1}$$

The effective stress σ' is transmitted via the solid particles, the neutral stress (u) via the liquid phase.

In unsaturated soils, stresses are transmitted via the solid, liquid and gaseous phase. Thus, eqn. (1) becomes:

$$\sigma' = (\sigma - u_a) + X (u_a - u_w) \tag{2}$$

where u_a = pore air pressure, u_w = pore water pressure, and X = a factor which depends on the degree of saturation of the soil. At saturation, X = 1, while in

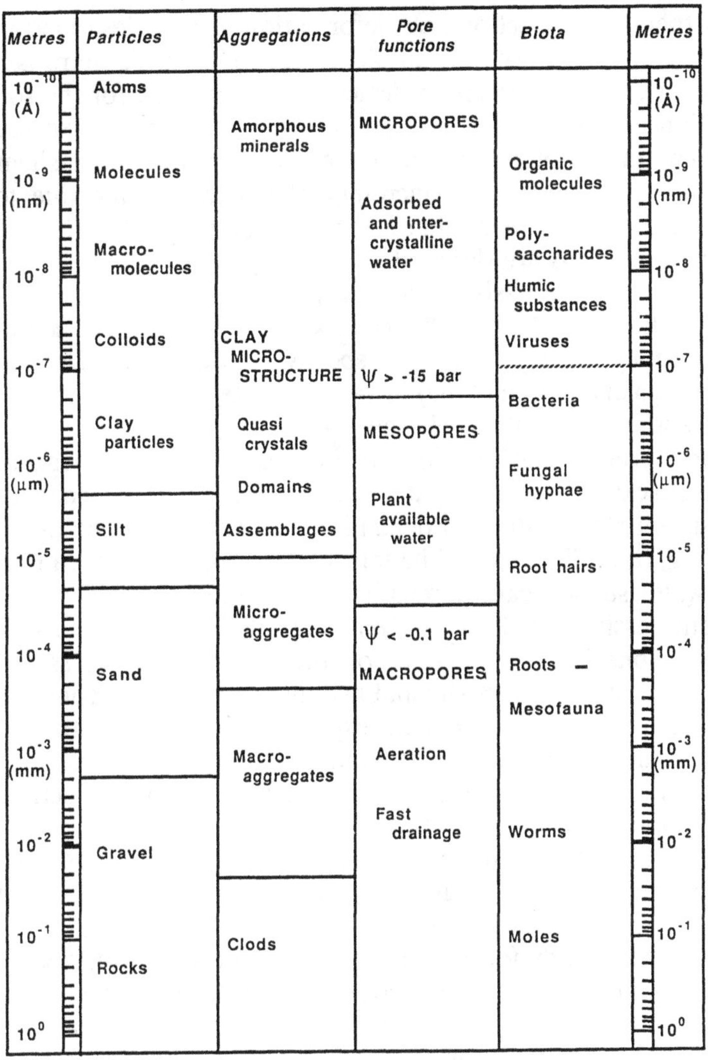

Metres	Particles	Aggregations	Pore functions	Biota	Metres
10^{-10} (Å)	Atoms	Amorphous minerals	MICROPORES	Organic molecules	10^{-10} (Å)
10^{-9} (nm)	Molecules		Adsorbed and inter-crystalline water	Poly-saccharides	10^{-9} (nm)
10^{-8}	Macro-molecules			Humic substances	10^{-8}
10^{-7}	Colloids	CLAY MICRO-STRUCTURE	$\psi > -15$ bar	Viruses	10^{-7}
				Bacteria	
10^{-6} (μm)	Clay particles	Quasi crystals	MESOPORES	Fungal hyphae	10^{-6} (μm)
		Domains	Plant available water		
10^{-5}	Silt	Assemblages		Root hairs	10^{-5}
10^{-4}	Sand	Micro-aggregates	$\psi < -0.1$ bar	Roots —	10^{-4}
			MACROPORES	Mesofauna	
10^{-3} (mm)		Macro-aggregates	Aeration		10^{-3} (mm)
10^{-2}	Gravel		Fast drainage	Worms	10^{-2}
10^{-1}		Clods		Moles	10^{-1}
	Rocks				
10^{0}					10^{0}

Fig. 1. Variation in size of soil components (from Kay, 1990).

air-dry soil X = 0.

For sandy, less compressible and non-aggregated soils, X can be calculated by

$$X = 0.22 + 0.78 \, S \qquad\qquad (3)$$

where S = degree of saturation. For silty and clayey soils, the values of the parameters in eqns. (1) and (2) mainly depend on soil aggregation, soil strength and hydraulic properties.

According to strain theory, compaction and deformation can be described in terms comparable to the stress theory. For an elemental soil volume there is a certain relationship between stress and strain, which is characteristic for the soil considered. Such relationships are known as stress/strain relationships. Likewise as for stresses, the components of deformation are normal strains and shear strains. A normal strain is defined as the incremental length increase due to deformation $(l + \Delta l)$. Then, the normal strain of that line element is $\Delta l/l$. Normal strains in the x, y and z directions are denoted as ϵ_x, ϵ_y, ϵ_z.

A shear strain is defined on the basis of changes in three directions initially at right angles to each other. Shear strains between x and y, y and z, and x and z directions are denoted by ϵ_{xy}, ϵ_{yz}, and ϵ_{xz}, respectively. For more detailed information, see e.g. Larson and Gupta (1980) or various textbooks (Terzaghi and Jelinek, 1954; Koolen and Kuipers, 1983; Hartge and Horn, 1991).

Compression refers to a process that describes the increase in soil mass per unit volume (= increase in bulk density) due to externally applied load or to changes in internal water pressure. External static or dynamic loads may be applied in the form of vibration, rolling, trampling, etc., while internal forces per unit area may be water pressure or water suction caused by a hydraulic gradient.

In saturated soils, compression is called *consolidation*, while in unsaturated soils it is called *compaction*. Consolidation depends on the drainage of excess soil water according to the hydraulic conductivity and hydraulic gradient. In contrast, during compaction part of the soil air will be expelled, according to the air permeability, pore continuity and water saturation in the whole profile. Consolidation tests are mainly used in civil engineering, e.g. in road construction, and have only limited application to agricultural soils.

Compressibility is defined as the resistance against volume decrease when the soil is subjected to a mechanical load. Thus, it is described by the shape of the stress-strain curve.

Compaction tests are used both for laboratory and field soil compression characterizations. In the laboratory, soil compaction refers to the compression of small soil samples, whereas in the field it refers to the three-dimensional increase in bulk density, the change in pore size distribution, or the changes in strength parameters of an elemental soil volume down to deeper depths and farther away from the loaded area. Compaction tests in the laboratory are performed with either homogenized or undisturbed bulk soil samples in steel cylinders at different soil water potentials. To minimize friction between the cylinder wall and the sample, the ratio of the diameter to the height of the cylinder should be ≥ 5. Additionally, even single naturally formed aggregates may be compacted in order to determine their strength (Horn, 1990).

Compactability is the difference between the initial bulk density and the maximum bulk density to which a soil can be compacted by a given amount of energy at a defined water content.

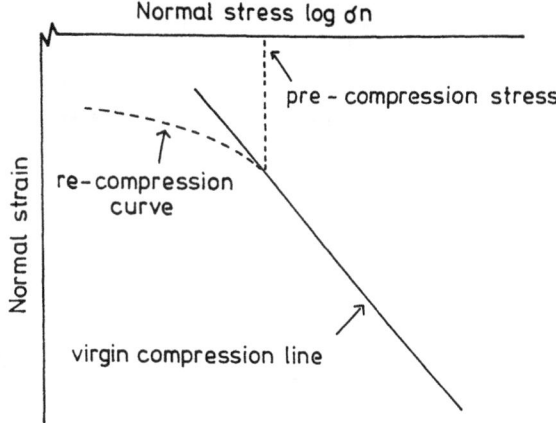

Fig. 2. Stress/strain curve as a means to determine the pre-compression stress value (from Lebert and Horn, 1991).

Soil strength of undisturbed (= aggregated) bulk soils or of completely homogenized soil material at a given matric water potential is quantified by the *pre-compression stress* value, which is determined by the stress-strain characteristics of the soil. The pre-compression stress is defined as the stress value at the transition of the less declined re-compression curve to the virgin compression line (Fig. 2).

There are several methods available to determine the pre-compression stress. One of the most frequently used is that according to Casagrande. For more detailed information, see Bölling (1971).

METHODS OF MEASUREMENT

The wide variety of methods used to determine soil strength, can be divided into those which only give an indirect (= mostly qualitative or relative) result and those which give a direct value with mechanically well defined dimensions. The advantage of methods of the latter category is their direct applicability, e.g. for the prediction of soil compressibility or trafficability of arable soils.

Some of the most common methods used to quantify soil mechanical parameters under *in-situ* and laboratory conditions will be described (see Tables 1 and 2).

Stress measurements under in-situ conditions

Stress distribution measurements in undisturbed soil profiles during trafficking at different speeds, with different loads and contact areas, reveal the kind and intensity of soil strength, stress attenuation, or soil deformation. In unsaturated soils, the water potential at the time of loading further affects these parameters.

One of the major problems with respect to the validation of compaction models is the installation of pressure sensors in the soil. The determination of soil stresses requires installation at different depths and distances from the vertical line through the centre of the wheeling. If the original soil structure is disturbed during the excavation, installation of the sensors and the backfilling of the soil, the obtained data only describe the stress pattern for more or less homogenized or artificially mixed soil material. Thus, only in non-aggregated sandy soils are no great differences to be expected between the values obtained after various kinds of installation. In aggregated soils, however, the influence of soil structure is so dominant that very precise sensor installation is required from a lateral position into completely undisturbed soil in well-defined directions.

Generally, the pressure sensor is a foreign body with deformation properties which differ from those of the soil material. If the pressure sensor itself is weaker than the soil, the registered stresses will underestimate the real stresses at that depth. If, however, the stiffness of the pressure sensor exceeds that of the surrounding soil, stresses will concentrate at the more rigid transducer body and, therefore, overestimate the real soil stresses. Generally, the deformation of the sensors can be either plastic or elastic. Table 1 informs about different types of stress transducers.

Plastic bodies, such as pneumatic or hydraulic cylinders, balls or discs made from silicone or rubber (Kögler, 1933; Blackwell and Soane, 1978; Bolling, 1986), change their volume according to the applied stresses. Before the measurement, the pressure cells have to be filled with water or air and pre-stressed up to 80 kPa. However, the pre-pressure influences the stress-strain modulus of the sensor and, therefore, the measured pressure value. Theoretically, the sensor elasticity should be the same as that of the surrounding undisturbed soil, which in practice is nearly impossible to obtain. Generally, plastic sensors tend to behave weaker than the soil and stresses will be underestimated.

A plastic stress transducer measures an average normal stress. Shear stresses cannot be determined. If cylindrical or spherical transducers are used, the direction of stresses cannot be identified.

Rigid bodies which are used as stress transducers, consist of piezoelectric materials (Hesse, 1983) or strain gauges (Horn, 1980; Nichols et al., 1987) applied to an aluminium or steel diaphragm. In contrast to the plastic stress transducers, it is not possible to match the stress-strain modulus of the rigid transducer with the surrounding soil. According to Peattie and Sparrow (1954), the optimum ratio of the stress-strain moduli of the transducer and the soil should be ≥ 10. Depending on the size of the stress state transducer, stresses from different, well-defined directions can be measured. With a stress state transducer as developed by Nichols et al. (1987), six normal stresses on three mutually orthogonal planes and three other, non-orthogonal planes, can be determined. Based on continuum mechanics theory, octahedral normal and shear stresses can be partly measured

TABLE 1

Types of stress state transducers (after Bolling, 1986)

Principle	Material	Shape	Deformation	Measured values	Authors
Pneumatic	rubber	cylinder	plastic	soil stiffness	Kögler (1933)
Hydraulic	rubber	disk	plastic	1 defined stress	Söhne (1951)
Hydraulic	steel	disk	elastic	1 defined stress	Franz (1958)
Hydraulic	rubber	sphere	plastic	mean normal stress	Blackwell and Soane (1978)
Hydraulic	silicon	cylinder	plastic	mean normal stress	Bolling (1986)
Strain gauge	silicon	sphere	plastic	mean normal stress	Verma et al. (1976)
Strain gauge	steel	disk	elastic	1 defined normal stress	Cooper et al. (1975)
Strain gauge	aluminum	disk	elastic	1 defined normal stress	Horn (1980)
Strain gauge	steel	cube	elastic	3 defined normal stresses	Prange (1960)
Strain gauge	aluminum	quarter sphere	elastic	6 defined normal stresses[1]	Nichols et al. (1987); Horn et al. (1992)

[1]From these, $\sigma_{1,2,3}$ and $\tau_{xy,yz,xz}$ may be determined.

and partly calculated for a cube, isolated from the continuum. Therefore, only with this device can the state of stress at a certain point in the continuum be defined.

Measurements of soil strength under laboratory conditions

The determination of soil strength parameters requires measurements under well-defined soil conditions. Therefore, these measurements are mainly performed in the laboratory. Generally, soil strength measurements are divided into indirect and direct stability tests (Table 2).

TABLE 2

Methods for determining soil strength

Method	Parameter	Derived	Soil condition
Water stability test[1]	length (cm)	-	single aggregate
Atterberg test[1]	water content (%,w/w)	-	homogenized soil
Proctor test[1]	water content (%,w/w)	-	homogenized soil
	bulk density (Mg m^{-3})	-	single aggregate
Uniaxial unconfined compression test[2]	stress (Pa)	-	homogenized soil single aggregate structured bulk soil
Confined compression test[2]	stress (Pa)	pre-compression stress (Pa)	homogenized soil structured bulk soil
Triaxial test[2]	stress (Pa)	cohesion (Pa) angle of internal friction (°)	homogenized soil single aggregate structured bulk soil
Direct shear test[2]	stress (Pa)	cohesion (Pa) angle of internal friction (°)	homogenized soil single aggregate structured bulk soil

[1]Indirect determination.
[2]Direct determination.

Indirect stability tests

Indirect stability tests have been described in detail by Burke et al. (1986).

Water stability
Aggregate stability is often determined by wet sieving and percolating or irrigating packages of aggregates with water, alcohol or benzene under defined conditions. In each case, the volume reduction of the various aggregate fractions after the treatment is taken as a characteristic of aggregate strength. The larger the average diameter after a defined time of sieving under water, the higher the stability of the aggregates (De Leenheer and De Boodt, 1959).

Atterberg test
Consistency limits of homogeneous soil material as a function of water content relate to soil strength properties and are applied to predict the soil workability.

The plastic limit is the water content at which the soil begins to crumble when rolled between the hand and a glass plate into a thread of about 5 mm diameter. The liquid limit is defined as the water content at which the soil starts to flow after a certain amount of energy has been applied to the soil (25 blows) in the Casagrande apparatus. The plasticity index defines the difference in the water content at the liquid and plastic limits. The higher the plasticity index, the smaller the angle of internal friction for sandy soils (Kézdi, 1969). Generally, this test has to be performed under well-defined conditions, using homogenized soil. Extrapolation to strength properties of structured bulk soil is impossible.

Proctor test

The Proctor test is recommended as a standard test mainly for homogenized soil material in order to define the effect of the water content and of the organic and mineral composition on soil compactability. After a series of soil tests have been performed with different water contents under a constant energy application by means of a falling weight, the maximum attainable bulk density and the optimum water content for maximum compactability of the soil sample can be determined.

In these stability tests, neither the homogenized soil material nor the way of loading the soil are comparable with *in-situ* soil conditions or with agricultural treatments. For that reason, in soil physics and soil mechanics, direct stability tests have to be preferred to quantify the strength of soil structure elements and of the bulk soil at a given soil water potential under defined loading conditions.

Direct stability tests

The dimension of soil strength is force per unit area, i.e., a stress (Pa). Thus, stability tests which result in a stress value are defined as direct stability tests. For the following descriptions of direct stability tests, reference is made to Kézdi (1969).

Uniaxial unconfined compression test

The uniaxial unconfined compression test is used to define the stress at which the homogenized or structured soil sample starts to fail at a given water content. One defined vertical normal stress (σ_1) is applied to the soil sample, while the stresses on the planes which are mutually perpendicular to the σ_1-direction ($\sigma_2 = \sigma_3$) are zero. The uniaxial unconfined compression test is also used to determine the crushing strength of single aggregates (crushing test).

Confined compression test

The stress-strain relationships of undisturbed or homogenized soils and of single aggregates can be quantified in the confined compression test. In this test, in contrast to the uniaxial unconfined compression test, stresses in the σ_2 and σ_3

direction are undefined because of the rigid wall of the test cylinder. Both the time- and the load-dependent alteration of soil deformation are quantified, and the slope of the virgin compression line (i.e., the compression index) and the transition from the re-compression curve to the virgin compression line (i.e., the pre-compression stress) are determined (cf. Fig. 2).

Triaxial test

In this test, undisturbed cylindrical soil samples are loaded with increasing vertical principal stress σ_1, while the horizontal principal stresses $\sigma_2 = \sigma_3$ are defined and kept constant throughout the test. Shear stresses occur in any other plane than in the planes of the principal stresses. The shear parameters cohesion and angle of internal friction, can be determined from the slope of the envelope of the Mohr's circles. However, the triaxial test results are affected by the number of contact points, strength per contact point and the pore geometry.

Various kinds of triaxial tests can be distinguished:

In the *consolidated drained test* (CD), prior to the increase in the vertical stress σ_1, the soil sample is equilibrated with the mean normal stresses and the pore water can be drained from the soil sample when the volume reduction exceeds the volume of the air-filled pore space. Therefore, the applied stresses are assumed to be transmitted as effective stresses via the solid phase. However, measurements made by Baumgartl and Horn (1991) showed an additional change in the pore water pressure, even during very long-lasting triaxial tests under "drained and consolidated" conditions, depending on the hydraulic properties of the soil. Also the shear speed, the degree of tortuosity of the pores and the magnitude of the hydraulic gradients affect the drainage of excess soil water and the effective stresses.

In the *consolidated undrained test* (CU), pore water cannot be drained from the soil during vertical stress increase. Therefore, high hydraulic gradients occur and the pore water acts as a lubricant with a low surface tension value. Thus, in the CU test, the shear parameters are much smaller and the pore water pressure values are much higher compared to those in the CD test.

The highest neutral stresses and therefore the lowest shear stresses can be measured in the *unconsolidated undrained test* (UU), where at the start of the test neither the effective stresses nor the neutral stresses are equilibrated with the applied principal stresses.

It may be concluded that the shear parameters cohesion and angle of internal friction, are strongly influenced by the compression/drainage conditions during the triaxial test. Thus, the characterization of the potential strength properties requires the *consolidated drained test*. If, however, the initial soil strength (= *in-situ* strength under running gear) has to be quantified, the *consolidated undrained triaxial test* or even the *unconsolidated undrained test* should be applied. The last two tests quantify transitional stages of soil deformation which mainly depend on the hydraulic properties of the soil.

Direct shear test

In the direct shear test, the kind and the direction of the shear plane are fixed. The shear plane is assumed to be affected only by normal and shear stresses. The normal stress is applied in the vertical direction and the shear stress in the horizontal direction. For a given soil, similar to the triaxial test, the values of the shear parameters cohesion and angle of internal friction are influenced by the shear speed and the drainage conditions.

FACTORS INFLUENCING SOIL STRENGTH

Internal parameters

Soil strength mainly depends on: (1) grain size distribution; (2) kind of clay minerals, and kind and amount of exchangeable cations; (3) content and kind of organic substances; (4) aggregation induced by swelling and shrinking, root proliferation and organic substances; (5) bulk density, pore size distribution and pore continuity of the bulk soil and of single aggregates; (6) water content and/or water potential.

Soil compressibility is less pronounced the coarser the structure and the less aggregated the soil. Thus, generally, gravelly soils are less compressible than sandy or finer-textured soils (Horn, 1988). However, any deviation of the particles from the spherical form results in an increase in the shearing resistance and, thus, in soil strength (Gudehus, 1981; Hartge and Horn, 1991).

The cohesive forces between illite, smectite or vermiculite particles are higher than for kaolinite. However, because of the size and the shape of the single particles, kaolinite has a larger angle of internal friction (Sommer, 1978). Additionally, the higher the valency of the adsorbed cations, and the higher the ionic strength, the higher the shearing resistance (Yong and Warkentin, 1966; Mitchell, 1976).

With an increasing amount of organic substances, the shearing resistance becomes higher compared with the parent material (Horn, 1981). However, the kind of the organic substance also has to be considered (Bachmann, 1988). In agricultural soils with similar physical and chemical properties, the pre-compression stress value, the angle of internal friction and the cohesion seem to be the larger, the higher the proportion of carbohydrates, condensed lignin sub-units, bound fatty acids and aliphatic polymers to the total organic material (Hempfling et al., 1990).

Owing to the formation of water menisci during swelling and shrinkage and the formation of organic-mineral and chemical bonds, aggregated soils are always stronger than homogenized material. However, increased strength can result either from an increase in the total number of contact points between single particles (i.e., an increase in effective stresses) or from an increase in the shear resistance per contact point (Hartge and Horn, 1984). Therefore, even if the bulk density of the bulk soil is similar, soil strength may be quite different. This apparent anomaly occurs because the bulk density of single aggregates (which is closely related to the number of contact points per unit volume) and the proportion of finer intra-aggregate pores are always higher than in the bulk soil (Horn, 1986; Gunzelmann, 1990; Becher, 1991).

However, aggregate strength also depends on the drying intensity and the number of drying events. Semmel et al. (1990) made an experiment in which lysimeters were filled with loess soil to an initial bulk density of 1.34 Mg m^{-3}. After wetting and subsequent drainage to -6 kPa, the soil was repeatedly dried to water potentials of -30, -60 and -1600 kPa, respectively. Fig. 3 shows that, at the same initial bulk density of the homogenized soil, the final aggregate bulk density increased, the wetter the soil was kept and the larger the number of wetting-drying cycles imposed.

At the same bulk density of the bulk soil, the final aggregate tensile strength also varied, even when the aggregates were re-wetted to the same water content. The explanation is that soil strength not only depends on bulk soil and aggregate density but also on the free entropy, which is related to the arrangement of the single particles in the aggregate. During repeated wetting and drying, normal and residual shrinkage may cause considerable changes in this arrangement (Horn and Dexter, 1989). Additionally, in aggregated soils, the grain size distribution influences the shear parameters. When the clay content exceeds 40% (w/w), the soil strength at a given water potential becomes increasingly smaller because of

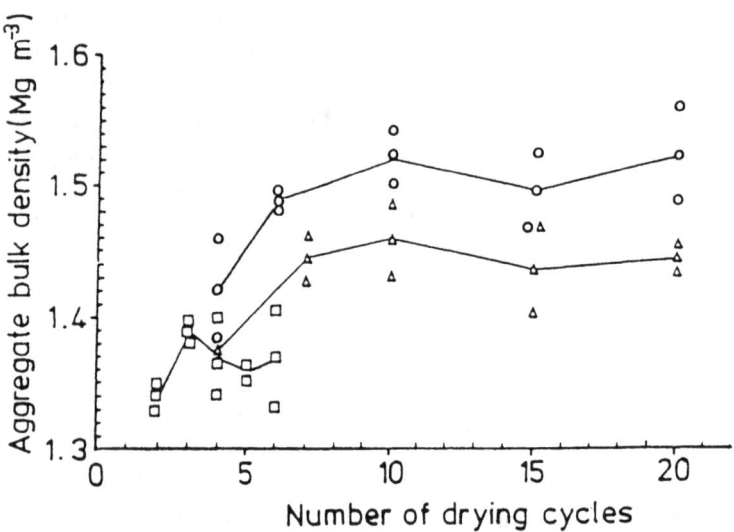

Fig. 3. Effect of drying intensity ($\circ$ = -6 → -30 kPa; $\triangle$ = -6 → -60 kPa; $\square$ = -6 → -1600 kPa) and number of drying cycles on aggregate bulk density. Real values as well as the corresponding mean values are shown (from Semmel et al., 1990).

changes in the ratio of effective and neutral stresses (Horn, 1990).

External forces

Soil strength generated by internal forces must counteract external forces, the impact of which may be very different, according to factors such as: (1) type of loading (static or dynamic); (2) duration of loading and number of loading events; (3) water potential during loading.

Dynamic as well as static loading generally induce a load-dependent increase in internal stresses in the soil. When these stresses exceed the soil strength, the pore volume decreases, because the mineral particles are incompressible.

Stress-strain behaviour

With increasing applied stresses, the magnitude of the changes in void ratio and/or bulk density depends on the grain size distribution, bulk density, water potential, organic matter content and soil aggregation (= internal factors).

At the same bulk density of the homogenized bulk soil and at the same water potential, soil samples are the more compressible the higher the clay content and the smaller the amount of organic substances. At the same clay content, soil samples are more compressible the lower the bulk density and the lower the water potential.

However, arable or forest soils are never homogeneous, nor do they have the same bulk density or water potential throughout the whole profile. With increasing aggregate formation, soil strength is increased compared to structureless, homogenized soil material. With increasing aggregation as well as with increasing water potential, the pre-compression stress value increases (Fig. 4).

Therefore, at the same grain size distribution, bulk density, and water potential, less aggregated, i.e., coherent soil horizons, are more compressible than those with prismatic or polyhedral structure. However, at the same water potential, the more the clay content exceeds 40% (w/w), the weaker the aggregates become. Generally, strength increase or decrease always depends on the changes in σ', u, and X (eqn. 2), which are induced by changes in the water potential. As long as the decrease in water potential exceeds the decrease in the X factor, soil strength increases. Thus, every soil has a maximum strength at a certain water potential that corresponds to the pore size distribution. In sandy soils, the highest strength is obtained at higher water potential values than in silty or clayey soils, due to the smaller amount of finer pores. Therefore, the overconsolidated load range in homogenized soil material corresponds to the soil water potential as the value of effective stress σ' (Horn, 1981).

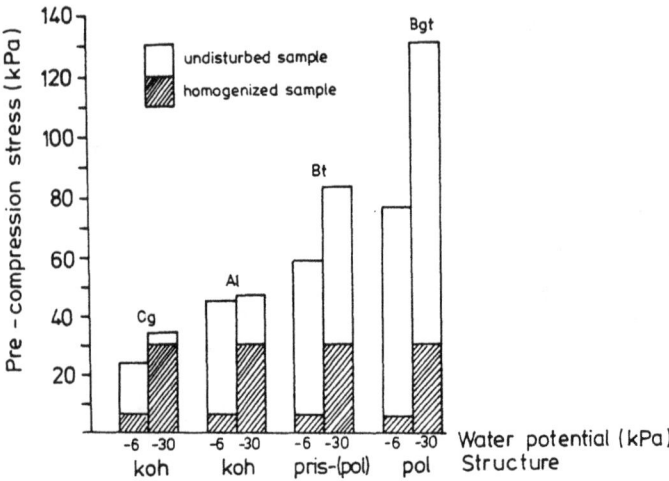

Fig. 4. Influence of soil structure and water potential in the Cg, Al, Bt and Bgt horizons on the pre-compression stress of a Luvisol. Koh = coherent; pris = prismatic; pol = polyhedral structure (from Horn, 1988).

Soil strength is further affected by freezing and thawing, because due to ice-lens formation during freezing, aggregates become either more dense or, if they are highly compacted, they are crushed by the growing ice-lenses. Both effects are called soil curing, but they result in completely different strength values and physical properties of the bulk soil (Horn, 1985; Kay, 1990).

Soils with a pronounced vertical pore system are stronger than those with a randomized or extremely horizontal pore system, because these vertical (bio) pores are opposing the maximum principal stress σ_1 (Bohne and Hartge, 1982). Thus, untilled or minimum-tilled soils are stronger than those which are conventionally tilled (Horn, 1986; Ehlers, 1982).

However, the pre-compression stress value depends to a great extent not only on the maximum pre-desiccation (Hartge, 1986) and actual water potential, but also on the hydraulic properties of the soil. The smaller the hydraulic conductivity, the hydraulic gradient and the pore continuity of the sample, the more stable are soils during short-term loading, because the changes in the proportions of the solid, liquid and gaseous phases during loading are time-dependent. In sandy soils, this effect is small, because the initial settlement (= timeless settlement) equals the total strain (Fig. 5).

With increasing clay content, the time-dependent soil settlement, i.e., the proportion of the initial to the primary and secondary settlement, is enlarged. This results in an increase in the pre-compression stress value for short-term loading. Again, these differences are smaller in strongly aggregated loamy and clayey soils than in sandy soils (Lebert et al., 1989).

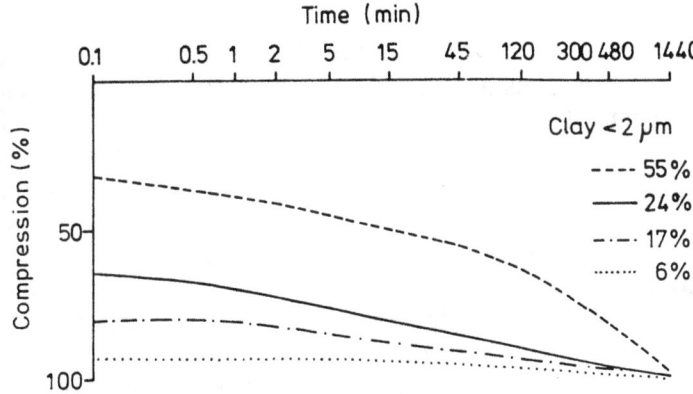

Fig. 5. Time-dependent settlement of soils with different clay content. Load: 30 kPa, water potential: -6 kPa (from Horn, 1988).

The factor X (eqn. 2) depends not only on the load applied to the bulk soil and on bulk density, aggregation and soil strength, but it is also different for the bulk soil and for a single aggregate. The kind of aggregate also affects the slope of the X vs. water potential curve. Generally, the slope of this curve is steeper for aggregates than for bulk soil samples, especially at high water potential. The higher the aggregate strength, the more pronounced is the load-dependent alteration of the X vs. water potential curve for the bulk soil than for single aggregates (Horn, 1989).

To obtain direct strength values and reproducible correlations with soil properties, the pre-compression stress value is most useful. As long as the pre-compression stress value is not exceeded, all deformations are elastic, i.e., reversible, while only in the virgin compression load range will plastic, i.e., irreversible deformation occur. For differently textured and aggregated arable soils, the pre-compression stress value can be predicted, especially for silty, loamy and clayey soils, if the shear parameters and some general physical properties are known. Lebert (1989) and Lebert and Horn (1991) have made such predictions for 37 soils with a very high degree of significance.

Shear strength

Under an applied stress, deformation will occur at the weakest points in the soil matrix. Further increases in stress result in the formation of failure zones. The strength of the failure zone equals the energy required to create a new unit of surface area or to propagate a crack (Skidmore and Powers, 1982), the so-called apparent surface energy (Hadas, 1987). Consequently, the stability of the soil relates to the distribution of strength in the failure zones. If the applied stress is less than the strength in the failure zone, soil structure will remain completely stable.

The strength required to deform the soil depends on the grain size distribution, organic matter content, soil bulk density and soil water potential, as well as on soil aggregation and aggregate strength. An increase in the volume fraction of solids in an elemental volume of soil, resulting in an increase in strength, is associated with shrinkage caused by drying or freezing, or with compression around roots. Loss of strength occurs in other elemental volumes, in which the volume fraction of solids decreases by the formation of root channels and cracks. The strength and spatial distribution of the failure zones depend on the coherence between points of low strength. In general, the higher the strength of single aggregates and the more stable the arrangement of single aggregates in the bulk soil, the larger the angle of internal friction. At the same bulk density of the bulk soil and at the same water potential, the angle of internal friction is smaller for prisms than for polyhedrons or subangular blocks. For a given aggregate type, the higher the bulk density, the larger the angle of internal friction. On the other hand, the more compressed, drier and finer-textured soils are, the higher their cohesion (Horn, 1981; 1988; Horn et al., 1989a; Lebert, 1989; Zhang and Hartge, 1989; Baumgartl and Horn, 1991).

At a given normal load, the shear strength of single aggregates is much higher than the shear strength of the bulk soil or the completely homogenized soil material. As long as the load applied during a shear test does not exceed the aggregate strength, shear results in a decrease in the number of contact points. However, if the load exceeds the aggregate strength, the aggregates will be destroyed and the number of contact points between the single particles in the

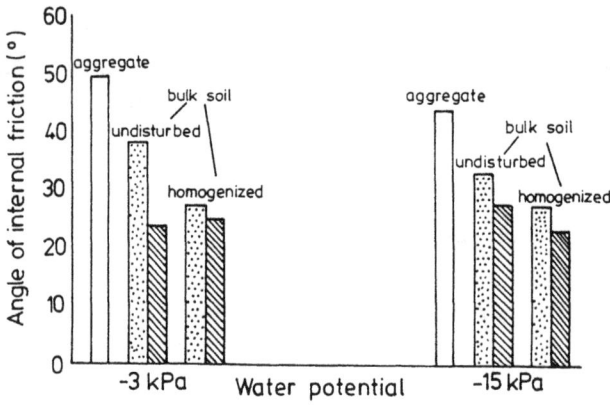

Fig. 6. Variation of the angle of internal friction with applied load for undisturbed (structured) and homogenized bulk soil, and for single aggregates. Pre-desiccation to matric water potentials of -3 and -15 kPa (from Baumgartl and Horn, 1991).

bulk soil increases (Lebert et al., 1987).

Fig. 6 shows the value of the angle of internal friction of single aggregates and of undisturbed and homogenized bulk soil samples. The angle of internal friction is highest for the single aggregates, owing to the larger number of contact points per unit area within the dense aggregates, compared to the undisturbed or homogenized bulk soil. In the normal stress range of 0-50 kPa, the aggregates remain stable during the shear test. Therefore, the angle of internal friction of the undisturbed, aggregated bulk soil is higher, compared to the homogenized bulk soil. However, the angle of internal friction of the homogenized soil depends on soil texture. In the normal stress load range of 150-400 kPa, the aggregates in the bulk soil are disturbed. This process results in an angle of internal friction similar to that in the homogenized bulk soil.

When the shear speed is increased, the Mohr-Coulomb failure line of structured soils more and more resembles an envelope of Mohr's circles rather than a straight line. Fig. 7 shows the Mohr-Coulomb failure lines of undisturbed and homogenized bulk soil samples as a function of shear speed. With increasing shear speed, the angle of internal friction decreases but soil cohesion increases.

If, during shearing, the expelled pore water cannot be immediately drained, the slope of the Mohr-Coulomb failure line for structured soils or single aggregates at a given initial water potential, can be either steeper or flatter. Thus, shearing can induce an increase in either neutral stress or effective stress. When the pore water pressure increases, the shearing resistance is reduced as a result of the very small surface tension of water. However, if, during shearing, the displacement of fine particles into coarser and air-filled pores reduces the diameter of the pores

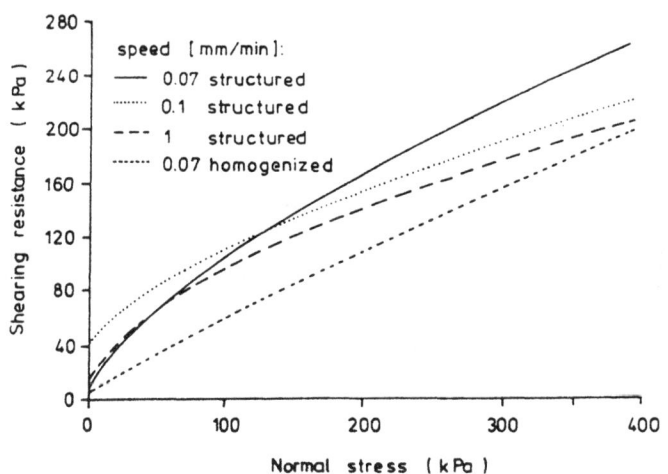

Fig. 7. Mohr-Coulomb failure lines of homogenized and structured bulk soil samples as a function of shear speed. Pre-desiccation to a matric water potential of -6 kPa (from Horn et al., 1991).

which theoretically should be water-saturated, the water potential will be further increased. Consequently, the strength will be increased (Horn, 1981; Gupta and Larson, 1982; Bohne, 1983).

Spatial stress transmission

Any load applied to the soil surface is three-dimensionally transmitted to the soil via the solid, liquid and gaseous phases. If we assume that air permeability is sufficiently high to allow immediate deformation of the air-filled pores, soil settlement is mainly affected by liquid flow. However, liquid flow may be delayed because changes in water content or water potential depend on the hydraulic conductivity, hydraulic gradient and pore continuity. Thus, the intensity and form of the stress transmission are again affected by soil strength.

Based on the theory of Boussinesq, Fröhlich (1934) has described the form of the equipotential lines by merely texture-dependent concentration factors. Under

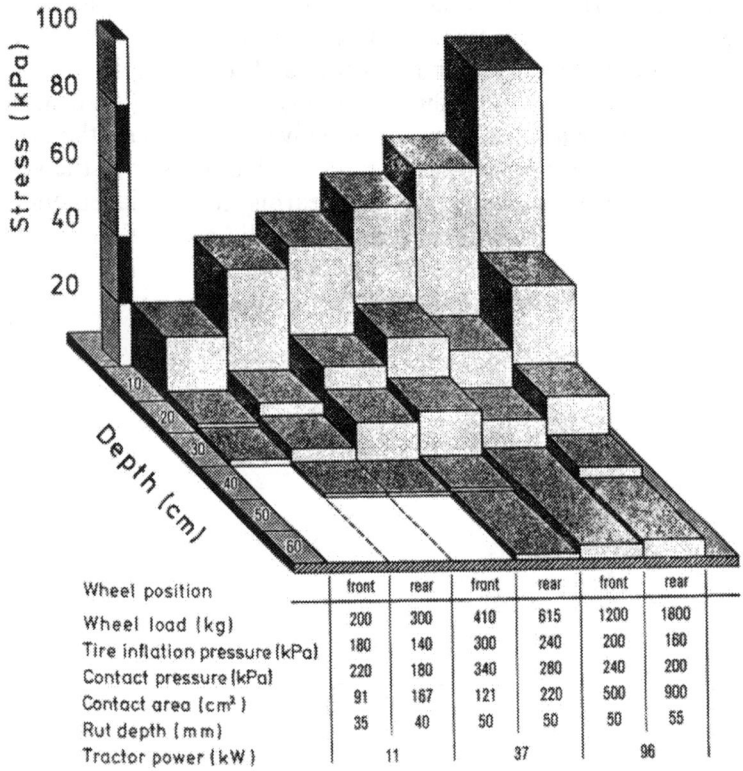

Wheel position		front	rear	front	rear	front	rear
Wheel load (kg)		200	300	410	615	1200	1800
Tire inflation pressure (kPa)		180	140	300	240	200	160
Contact pressure (kPa)		220	180	340	280	240	200
Contact area (cm²)		91	167	121	220	500	900
Rut depth (mm)		35	40	50	50	50	55
Tractor power (kW)		11		37		96	

Fig. 8. Stress distribution in a loamy sand as a function of wheel load and contact area (from Burger et al., 1988).

saturated conditions, their values range between 3 for very strong material and 9 for soft soil, and they increase with increases in clay content. In weak soils with high concentration factor values, stress is transmitted to deeper depths but the stresses remain closer to the perpendicular centre-line through the contact area. On the other hand, in strong soils, with low values of the concentration factor, the stress is transmitted more horizontally, in a shallower soil layer.

Under *in-situ* conditions in undisturbed soils with the same internal parameters, it is obvious that the stronger the soils are aggregated, the more pronounced is the stress attenuation. Thus, the concentration factor values are the smaller, the stronger and the drier the soils (Burger et al., 1987). It is also well known that not only the contact stress but also the size and the shape of the contact area affect the stress distribution in structured, unsaturated soils. Fig. 8 shows results of an experiment on a loamy sand, which was loaded by 3 tractors, differing in power, wheel load and wheel contact area but exerting similar contact pressures. It is clear that at the same contact pressure, stresses are transmitted to deeper depths, the larger the wheel load and the larger the contact area. However, the stress distribution pattern in the soil is not only different for the areas beneath lugs and between the lugs, but it is also affected by the stiffness of the tyre carcass (Horn et al., 1987). Thus, in undisturbed bulk soils, there are no well-defined unique stress equipotential lines (Horn et al., 1989b).

The stress distribution and compaction also depend on the forward speed of the tractor. In Fig. 9, a Luvisol was wheeled by a tractor at 2 different speeds. At higher speed (2.18 m s^{-1}), the vertical stresses are slightly increased, especially in the weak, ploughed topsoil (0-30 cm depth). Thus, during short-time loading, the soil may react as a rigid body, where the stresses propagate mainly in the vertical direction. However, the horizontal stresses in the soil are higher at a low speed of wheeling (0.83 m s^{-1}). Due to the time-dependency of soil settlement, pore

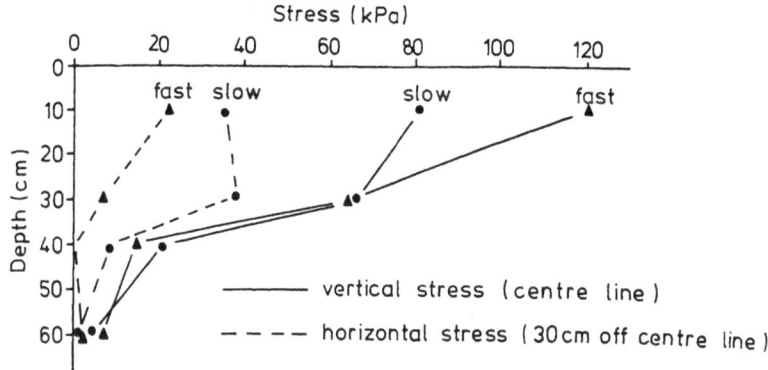

Fig. 9. Horizontal and vertical stress distribution (kPa) in a Luvisol during wheeling at low (0.83 m s^{-1}) and high (2.18 m s^{-1}) speed. Water content: 30% (w/w).

water will be expelled during loading, which results in more intense compaction and, therefore, in an increase in the horizontal stresses in the soil.

As a further consequence of the time-dependent soil settlement (=densification), the change in soil strength due to loading does not increase the pre-compression stress value to its theoretical value, which is defined by the contact area pressure of the wheeling tractor (Horn and Hartge, 1990).

According to Newmark (1942), irrespective of these interrelationships among the various parameters and the stress distribution pattern, mean values for the concentration factor, v, in a particular soil can be correlated with the pre-consolidation value and applied load at the top of each distinct soil horizon (Table 3). At the same pre-compression stress value and comparable values for total stress applied at the soil surface (σ_o), the concentration factor decreases with increasing contact area and with increasing pre-compression stress.

TABLE 3

Mean values of the concentration factor, v, for silty soils at -30 kPa water potential, in relation to the contact area radius, the contact area stress, and the pre-compression stress (from Horn et al., 1989b)

Contact area radius (cm)	Contact area pressure (kPa)	Pre-compression stress (kPa)				
		60	100	140	180	>180
0-10	100	4.1	3.7	2.1	2.0	1.1
	150	4.3	3.8	3.1	2.3	1.7
	200	4.5	3.9	3.4	2.5	1.8
	>200	4.7	4.2	3.8	2.9	2.0
10.1-15	100	3.4	3.3	2.6	2.6	1.9
	150	3.6	3.0	2.8	2.5	1.9
	200	3.7	3.0	2.9	2.8	2.0
	>200	3.9	3.7	-	-	-
15.1-20	150	3.2	-	2.5	-	-
	200	3.5	-	2.6	-	-
	>200	3.7	-	2.8	-	-
20.1-25	100	2.5	-	2.1	-	-
	150	2.9	-	2.5	-	-
	200	3.4	-	2.7	-	-
	>200	3.6	-	2.9	-	-

Dynamic effects

Dynamic stresses, such as caused by the slip of pulling tractor tyres, trampling or kneading by animals, or by the vibration of agricultural machinery, may induce a further alteration of physical as well as chemical properties of the soil (Horn et al., 1991). If soils are kneaded in the presence of excess soil water, a considerable deterioration of the structure takes place. The duration of kneading as well as the puddling intensity (load and velocity of kneading) have to be considered, because soil structure deterioration during shearing always coincides with an improved accessibility of the mineral surface areas for water. Thus, swelling potential is also increased. Consequently, the angle of internal friction and the cohesion become smaller, the concentration factor values increase, soil aeration becomes less, and both water permeability and penetration resistance decrease. Furthermore, the availability of nutrients (e.g. Fe, Mn) as well as their mobility are increased (Koenigs, 1963; Horn, 1976, 1985).

CONCLUSIONS

(1) The strength of structured agricultural soils differs to a great extent from that of homogenized soil samples.

(2) Single aggregates in structured soils are stronger, i.e., they have a larger angle of internal friction and a larger cohesion than the bulk soil or the homogenized soil material.

(3) Soil strength is strongly influenced by spatial differences in soil water potential. Thus, the stress distribution under running wheels (intensity and spatial direction) varies greatly, in dependence of soil and machinery parameters.

(4) Soil compressibility and compactability are strongly affected by the internal and external parameters of arable and forest soils and can seldom be predicted by values obtained from homogenized soil material.

(5) The determination of compressibility requires multifunctional methods in order to deal adequately with the complex relationships between the soil physical characteristics as affected by loading and the subsequent response of growing plants to differences in soil structure.

REFERENCES

Bachmann, J., 1988. Auswirkungen der organischen Substanz verschiedenen Zersetzungsgrades auf physikalische Bodeneigenschaften. (The effects of organic matter at various stages of decomposition on soil physical properties). Ph.D. thesis, Univ. Hannover, Germany (in German).

Barnes, K.K., Carleton, W.M., Taylor, H.M., Throckmorton, R.I. and Vanden Berg, G.E. (Editors), 1971. Compaction of Agricultural Soils. Am. Soc. Agric. Eng., St. Joseph, MI, U.S.A., ASAE Monograph, 471 pp.

Baumgartl, T. and Horn, R., 1991. Effect of aggregate stability on soil compaction. Soil Tillage Res., 19: 203-213.

Becher, H.H., 1991. Über die Aggregatdichte und deren mögliche Auswirkung auf den

Bodenlösungstransport. (Aggregate density and its possible effects on the transport of the soil solution). Z. Pflanzenern. Bodenkd., 154: 3-8.

Blackwell, P. S. and Soane, B. D., 1978. Deformable spherical devices to measure stresses within field soils. J. Terramech., 15: 207-222.

Bohne, H., 1983. Mechanismen bei der Stabilisierung von Aggregaten aus Tonen mit Calciumoxid. (Mechanisms during the stabilization of clay aggregates with calciumoxide). Ph.D. thesis, Univ. Hannover, Germany (in German).

Bohne, H. and Hartge, K.H., 1982. Auswirkungen der Gefügegeometrie auf den Wuchs von Roggenkeimlingen. (Effect of pore geometry on growth of rye seedlings). Mitt. Dtsch. Bodenkd. Ges., 34: 141-145 (in German).

Bolling, I., 1986. Beanspruchung des Bodens beim Schlepper- und Maschineneinsatz. (Stresses in soils under tractors and agricultural machines). Landwirtschaftsverlag, Münster-Hiltrup, Germany, KTBL-Schrift 308, Bodenverdichtungen, pp. 49-72 (in German).

Bölling, W. H., 1971. Zusammendrückung und Scherfestigkeit von Böden. (Compressibility and shear strength of soils). Springer, Wien, Austria, 194 pp. (in German).

Bradford, J.M. and Gupta, S., 1986. Soil compressibility. In: A. Klute (Editor), Methods of Soil Analysis. Part I: Agronomy. Am. Soc. Agron., Madison, WI, U.S.A., pp. 479-492.

Burger, N., Lebert, M. and Horn, R., 1987. Druckausbreitung unter fahrenden Traktoren im natürlich gelagerten Boden. (Pressure distribution in naturally structured soils under travelling tractors). Mitt. Dtsch. Bodenkd. Ges., 55: 135-141 (in German).

Burger, N., Lebert, M. and Horn, R., 1988. Prediction of the compressibility of arable land. In: J. Drescher, R. Horn and M. De Boodt (Editors), Impact of Water and External Forces on Soil Structure. Catena, Cremlingen, Germany, Catena Supplement 11, pp. 141-151.

Burke, W., Gabriels, D. and Bouma, J., 1986. Soil Structure Assessment. Balkema, Rotterdam, Netherlands, 92 pp.

Cooper, A.W., Vanden Berg, G.E., McColly, H.F. and Erickson, A.E., 1975. Strain gage cell measures soil pressure. Agric. Eng., 38: 232-235.

De Leenheer, L. and De Boodt, M., 1959. Determination of aggregate stability by the change in the mean weight diameter. Proc. Int. Symp. Soil Structure, Ghent, 1958. Meded. Landbouwhogeschool Ghent, Belgium, 24, pp. 290-300.

Drescher, J., Horn, R. and De Boodt, M. (Editors), 1988. Impact of Water and External Forces on Soil Structure. Catena, Cremlingen, Germany, Catena Supplement 11, 175 pp.

Ehlers, W., 1982. Die Bedeutung des Bodengefüges für das Pflanzenwachstum bei moderner Landbewirtschaftung. (The importance of soil structure for plant growth under modern farm management). Mitt. Dtsch. Bodenkd. Ges., 34: 115-128 (in German).

Emerson, W.W., Bond, R.D. and Dexter, A.R. (Editors), 1978. Modification of Soil Structure. Wiley, Chichester, U.K., 438 pp.

Eriksson, J., 1975. Influence of extremely heavy traffic on clay soil. Grundförbättring, 27: 33-51.

Fröhlich, O.K., 1934. Druckverteilung im Baugrund. (Pressure distribution in soil foundation). Springer, Berlin, Germany, 178 pp. (in German).

Franz, G., 1958. Unmittelbare Spannungsmessung in Beton und Baugrund. (Direct stress measurement in concrete and foundation). Der Bauingenieur, 33: 190-195 (in German).

Gliński, J. and Stępniewski, W., 1985. Soil Aeration and its Role for Plants. CRC Press, Boca Raton, FL, U.S.A., 229 pp.

Gudehus, G., 1981. Bodenmechanik. (Soil Mechanics). Enke, Stuttgart, Germany, 281 pp. (in German).

Gunzelmann, M., 1990. Die Quantifizierung und Simulation des Wasserhaushalts von Einzelaggregaten und strukturierten Gesamtböden unter besonderer Berücksichtigung der Wasserspannungs-/Wasserleitfähigkeitsbeziehung von Einzelaggregaten. (The quantification and simulation of the water regime of single aggregates and structured bulk soils with special regard to the relationship between water tension and water conductivity). Bayer. Bodenkd. Ber., 15, 130 pp. (in German).

Gupta, S. and Allmaras, R.R., 1989. Models to assess the susceptibility of soils to excessive compaction. Adv. Soil Sci., 6: 65-100.

Gupta, S. and Larson, W.E., 1982. Modeling soil mechanical behavior during tillage. In: D.M. Kral, S. Hawkins, P.W. Unger and D.M. Van Doren Jr. (Editors), Predicting Tillage Effects on Soil Physical Properties and Processes. Am. Soc. Agron., Madison, WI, U.S.A., Spec. Publ. 44, pp. 151-178.

Gupta, S., Hadas, A., Voorhees, W.B., Wolf, D., Larson, W.E. and Schneider, E.C., 1985. Development of guides for estimating the ease of compaction of world soils. Res. Report BARD, No. US 337-80, 131 pp.

Hadas, A., 1987. Dependence of true surface energy of soils on air entry pore size and chemical constituents. Soil Sci. Soc. Am. J., 51: 187-191.

Håkansson, I., Voorhees, W.B., Elonen, P., Raghavan, G.S.V., Lowery, B., Van Wijk, A.L.M., Rasmussen, K. and Riley, H., 1987. Effect of high axle load traffic on subsoil compaction and crop yield in humid regions with annual freezing. Soil Tillage Res., 10: 259-268.

Håkansson, I., Voorhees, W.B. and Riley, H., 1988. Weather and other environmental factors influencing crop responses to tillage and traffic. Soil Tillage Res., 11: 239-282.

Hartge, K.H., 1986. Ein Konzept des Verdichtungszustandes. (A concept of the compacted condition of soil). Z. Pflanzenern. Bodenk., 149: 361-370 (in German).

Hartge, K.H. and Horn, R., 1984. Untersuchungen zur Gültigkeit des Hooke'schen Gesetzes bei der Setzung von Böden bei wiederholter Belastung. (Investigations on the validity of Hooke's Law with regard to the compaction of soil under repeated loading). Z. Acker-Pflanzenbau, 153: 200-207 (in German).

Hartge, K.H. and Horn, R., 1991. Einführung in die Bodenphysik. (Introduction to Soil Physics). Enke, Stuttgart, Germany, 303 pp. (in German).

Hempfling, A., Schulten, H. and Horn, R., 1990. Relevance of humus composition for the physical/mechanical stability of agricultural soils: A study by direct pyrolysis mass spectrometry. J. Anal. Appl. Pyrolysis, 17: 275-281.

Hesse, Th., 1983. Druckspannungsmessdosen für körnige Haufwerke. (Pressure sensors for granular materials). Grundl. Landtechnik, 33: 121-131 (in German).

Horn, R., 1976. Festigkeitsänderungen infolge von Quellung und Schrumpfungsprozesses eines mesozoischen Tones. (Changes in the strength of a mesozoic clay as a result of swelling and shrinking processes). Ph.D. thesis, Techn. Univ. Hannover, Germany, 96 pp. (in German).

Horn, R., 1980. Die Ermittlung der vertikalen Druckfortpflanzung im Boden mit Hilfe von Dehnungsmessstreifen. (The determination of vertical stress propagation in the soil with strain gauges). Z. Kulturtech. Flurber., 21: 343-349 (in German).

Horn, R., 1981. Die Bedeutung der Aggregierung von Böden für die mechanische Belastbarkeit in dem für Tritt relevanten Auflastbereich und deren Auswirkungen auf physikalische Bodenkenngrössen. (The significance of soil aggregation for soil strength and its effects on soil physical properties). Schriftenreihe des FB 14, Techn. Univ. Berlin, Germany, 10, 200 pp. (in German).

Horn, R., 1985. Die Bedeutung der Trittverdichtung durch Tiere auf physikalische Eigenschaften alpiner Böden. (The importance of trampling by animals on the physical properties of alpine soils). Z. Kulturtech. Flurber., 26: 42-51 (in German).

Horn, R., 1986. Auswirkung unterschiedlicher Bodenbearbeitung auf die mechanische Belastbarkeit von Ackerböden. (The effect of different types of soil tillage on the mechanical loading of agricultural soils). Z. Pflanzenern. Bodenkd., 149: 9-18 (in German).

Horn, R., 1988. Compressibility of arable land. In: J. Drescher, R. Horn and M. De Boodt

(Editors): Impact of Water and External Forces on Soil Structure. Catena, Cremlingen, Germany, Catena Supplement 11, pp. 53-71.

Horn, R., 1989. Strength of structured soils due to loading - a review of the processes on macro- and microscale; European aspects. In: W.E. Larson, G.R. Blake, R.R. Allmaras, W.B. Voorhees and S. Gupta (Editors), Mechanics and Related Processes in Structured Agricultural Soils. Kluwer, Dordrecht, Netherlands, NATO ASI Series E: Applied Science 172, pp. 9-22.

Horn, R., 1990. Aggregate characterization as compared to soil bulk properties. Soil Tillage Res., 17: 265-289.

Horn, R. and Dexter, A.R., 1989. Dynamics of soil aggregation in a homogenized desert loess. Soil Tillage Res., 13: 254-266.

Horn, R. and Hartge, K.H., 1990. Effect of short time loading on soil strength and some physical properties. Soil Tillage Res., 15: 247-256.

Horn, R., Burger, N., Lebert, M. and Badewitz, G., 1987. Druckfortpflanzung in Böden unter fahrenden Traktoren. (Pressure transmission in soils under travelling tractors). Z. Kulturtech. Flurber., 28: 94-102 (in German).

Horn, R., Blackwell, P.S. and White, R., 1989a. The effect of speed of wheeling on soil stresses, rut depth and soil physical properties in an ameliorated transitional red brown earth. Soil Tillage Res., 13: 353-364.

Horn, R., Lebert, M. and Burger, N., 1989b. Vorhersage der mechanischen Belastbarkeit von Böden als Pflanzenstandort auf der Grundlage von Labor- und in situ-Messungen. (Prediction of the mechanical compressibility of soils, based on laboratory and in-situ measurements). Abschlussbericht Bayer. StMLU, Germany, Bewilligungsnr. 6333-972-57238, 178 pp. (in German).

Horn, R., Baumgartl, T., Kühner, S., Lebert, M. and Kayser, R., 1991. Zur Bedeutung des Aggregierungsgrades für die Spannungsverteilung in strukturierten Böden. (The effect of aggregation on stress distribution in structured soils). Z. Pflanzenern. Bodenkd., 154: 21-26 (in German).

Horn, R., Johnson, C., Semmel, H., Schafer, R. and Lebert, M., 1992. Räumliche Spannungsmessungen mit dem Stress State Transducer (SST) in ungesättigten aggregierten Böden - theoretische Betrachtungen und erste Ergebnisse. (Spatial stress measurements with the Stress State Transducer (SST) in unsaturated aggregated soils - theoretical considerations and first results). Z. Pflanzenern. Bodenkd., 155: 269-274 (in German).

Kay, B.D., 1990. Rates of change of soil structure under different cropping systems. Adv. Soil Sci., 12: 1-41.

Kézdi, A., 1969. Handbuch der Bodenmechanik. (Soil Mechanics Handbook). VEB Verlag, Berlin, Germany, Vol. 1: 259 pp., Vol. 2: 309 pp., Vol. 3: 274 pp. (in German).

Koenigs, F.R., 1963. The puddling of clay soils. Neth. J. Agric. Sci., 11: 145-156.

Kögler, F., 1933. Baugrundprüfung im Bohrloch. (Soil foundation testing). Der Bauingenieur, 8: 266-270 (in German).

Koolen, A. and Kuipers, H., 1983. Agricultural Soil Mechanics. Springer, Berlin, Germany, 241 pp.

Kral, M., Hawkins, S., Unger P.W. and Van Doren Jr., D.M. (Editors), 1982. Predicting Tillage Effects on Soil Physical Properties and Processes. Am. Soc. Agron., Madison, WI, U.S.A., Spec. Publ. 44, 198 pp.

Larson, W.E. and Gupta, S., 1980. Estimating critical stresses in unsaturated soils from changes in pore water pressure during confined compression. Soil Sci. Soc. Am. J., 44: 1127-1132.

Larson, W.E., Blake, G.R., Allmaras, R.R., Voorhees, W.B. and Gupta, S. (Editors), 1989. Mechanics and Related Processes in Structured Agricultural Soils. Kluwer, Dordrecht, Netherlands, NATO ASI Series E: Applied Science 172, 273 pp.

Lebert, M., 1989. Beurteilung und Vorhersage der mechanischen Belastbarkeit von Ackerböden. (Determination and and prediction of the mechanical compressibility of agricultural soils). Ph.D. thesis, Bayreuther Bodenkd. Ber., 12, 131 pp. (in German).

Lebert, M. and Horn, R., 1991. A method to predict the mechanical strength of agricultural soils. Soil Tillage Res., 19: 275-286.

Lebert, M., Burger, N. and Horn R., 1987. Welche Bedeutung kommt der Aggregatstabilität während des Schervorganges zu? (The effect of aggregate stability on shear strength). Mitt. Dtsch. Bodenkd. Ges., 53: 427-432 (in German).

Lebert, M., Burger, N. and Horn, R., 1989. Effects of dynamic and static loading on compaction of structured soils. In: W.E. Larson, G.R. Blake, R.R. Allmaras, W.B. Voorhees and S. Gupta (Editors), Mechanics and Related Processes in Structured Agricultural Soils. Kluwer, Dordrecht, Netherlands, NATO ASI Series E: Applied Science 172, pp. 73-82.

Mitchell, J.K., 1976. Fundamentals of Soil Behaviour. Wiley, New York, NY, U.S.A., 421 pp.

Newmark, N.U., 1942. Influence charts for computation of stress in elastic foundations. Univ. Illinois, Urbana, IL, U.S.A., Bull. 338, 28 pp.

Nichols, T.A., Bailey, A.C., Johnson, C.E. and Grisso, R.D., 1987. A stress state transducer for soil. Trans. ASAE, 30: 1237-1241.

Page, J.B. and Willard, C.J., 1947. Cropping systems and soil properties. Soil Sci. Soc. Am. Proc., 11: 81-88.

Peattie, K.R. and Sparrow, R.W., 1954. The fundamental action of earth pressure cells. J. Mech. Phys. Solids, 2: 141-155.

Prange, B., 1960. Neuartiges Verfahren zur Messung bodenmechanischer Zustandsgrössen. (New procedure for the measurement of soil mechanical parameters). Der Bauingenieur, 35: 389-396 (in German).

Proc. 11th Conf. ISTRO, Edinburgh, 1988. Tillage and Traffic in Crop Production. Scot. Centre Agric. Eng., Penicuik, Midlothian, U.K., 939 pp.

Rynasiewicz, J., 1945. Soil aggregation and cotton yield. Soil Sci., 60: 387-396.

Semmel, H., Horn, R., Hell, U., Dexter, A.R. and Schulze, E.D., 1990. The dynamics of soil aggregate formation and the effect on soil physical properties. Soil Technol., 3: 113-129.

Skidmore, E.L. and Powers, D.H., 1982. Dry soil aggregate stability: Energy based index. Soil Sci. Soc. Am. J., 46: 1274-1279.

Söhne, W., 1951. Das mechanische Verhalten des Ackerbodens bei Belastungen unter rollenden Rädern sowie bei der Bodenbearbeitung. (The mechanical behaviour of soils under stresses of rolling wheels and during tillage). Grundl. Landtechnik, 9. Konstrukteur Heft 1, pp. 87-94 (in German).

Sommer, C., 1978. Befahrbarkeit und Bearbeitbarkeit. (Trafficability and workability). Mitt. Dtsch. Bodenkd. Ges., 29: 1090-1094 (in German).

Terzaghi, J. and Jelinek, P., 1954. Theoretische Bodenmechanik. (Theoretical Soil Mechanics). Springer, Berlin, Germany, 456 pp. (in German).

Verma, P.B., Bailey, A.C., Schafer, R.L. and Futral, J.G., 1976. A pressure transducer in soil compaction study. Trans. ASAE, 19: 442-447.

Yong, R.N. and Warkentin, B.P., 1966. Introduction to Soil Behaviour. Macmillan, New York, NY, U.S.A., 345 pp.

Zhang, G. and Hartge, K.H., 1989. Die Wechselwirkung zwischen Winkel der inneren Reibung, organischen Substanzen und Wasserspannung. (The interaction between the angle of internal friction, organic matter and water tension). Mitt. Dtsch. Bodenk. Ges., 59: 279-284 (in German).

Soil Compaction in Crop Production
B.D. Soane and C. van Ouwerkerk (Eds.)
©1994 Elsevier Science B.V. All rights reserved.

CHAPTER 4

Prediction of Soil Compaction under Vehicles

S.C. GUPTA[1] and R.L. RAPER[2]

[1]University of Minnesota Department of Soil Science, St. Paul, MN, U.S.A.
[2]USDA-ARS, National Soil Dynamics Laboratory, Auburn, AL, U.S.A.

SUMMARY

Modeling of soil compaction processes in agriculture provides a means to evaluate harmful vs. desired soil compaction. The steps involved in the development of a basic soil compaction simulation model include: (1) prediction of surface applied forces from farm vehicles; (2) representation of soil stress-strain relationships; (3) modeling the propagation of stresses in the soil profile. Simple relationships are available that predict the surface stress from easily measurable parameters, such as inflation pressure, applied load and tire characteristics. Because of the presence of lugs on tires, prediction of surface stress is much more complicated and needs additional research. Stress-strain relationships obtained from both uniaxial or triaxial laboratory tests have been shown to be adequate for describing soil compaction from farm vehicles. In general, uniaxial tests are easy and inexpensive, whereas triaxial tests are more accurate and apply to many different soil conditions. All existing models of stress propagation in soil are based on the theory of elastic deformation. These models can be categorized into two groups: finite element models and analytical models. Both groups of models have been shown to be adequate for a limited set of experimental conditions. Much more research is needed to make these models acceptable in everyday use by researchers and practitioners. The procedures to link soil compaction models to crop growth models and the research needed both in methods to estimate parameters of the model and in describing the missing processes are also discussed.

INTRODUCTION

Considerable field and laboratory research has taken place in the study of soil compaction in agricultural soils and of the consequences of soil compaction on crop production. The following review articles and books summarize some of this research to date: Barnes et al. (1971), Soane et al. (1981a, 1981b, 1982), Gupta and Larson (1982), International Conference on Soil Dynamics (1985), Monnier and Goss (1987), Gupta and Allmaras (1987), Håkansson et al. (1988), Drescher et al. (1988), Larson et al. (1989), International Conference on Soil Compaction

(1989) and Schafer et al. (1990). This research has been useful in providing a broad understanding of principles and factors affecting soil compaction. For example, Horn (1988) concluded that mechanical compressibility of soil depends on both internal and external factors. The internal factors include parameters (grain size distribution, type of clay mineral, organic matter content, bulk density, pore size distribution, pore continuity and soil wetness) that influence the ability of soil to withstand mechanical loading, whereas the external factors include those factors (kind of loading, load intensity, and time and number of compaction events) that are acting against the internal factors.

Apart from this broad understanding of principles and factors, there is a scarcity of reliable information (Gupta and Allmaras, 1987) concerning soil compaction that can be widely used to develop guidelines for: (1) the maximum pressure a specific soil can withstand over a range of water contents; (2) the range of applied stresses and soil water contents that are conducive to excessive compaction; (3) the soil types and areas in a geographical region that are susceptible to excessive compaction; (4) the dominant soil and plant limiting conditions in a given area. This lack of reliable information is partially due to the many possible ways a soil can respond to externally-applied stresses, the uncertainties in identifying stresses that cause compaction, the variability in soil physical and chemical properties across the landscape, and the complexity of identifying the response of plants to soil compaction. The modeling of soil compaction processes and its linkage to crop growth models offers an inexpensive way to integrate existing knowledge of soil compaction processes and provides a means to evaluate soil conditions and management alternatives that are not conducive to harmful compaction. Our goal in this chapter is to review the existing soil compaction models and to show how they can be linked to crop growth models in order to predict soil compaction effects on crop production.

The process of building a soil compaction model could be divided into four broad steps: (1) modeling surface-applied mechanical forces from farm vehicles; (2) identifying stresses that cause compaction; (3) defining stress-strain relationships of a given soil; (4) modeling the propagation of forces in soil and the effect of these forces on soil compaction. The following sections review the details of each of the above steps in the construction of soil compaction models and show how some of these models are linked to crop growth models.

MODELING SURFACE-APPLIED FORCES FROM FARM VEHICLES

Many studies have been conducted to develop relationships between farm vehicle load and the stress at the soil surface. Because of the complex geometry of tires on farm vehicles, it has been difficult to develop a relationship that will apply to a range of soil and tire types and vehicle operating conditions, such as vehicle speed, wheel slip, etc. Koolen and Kuipers (1983) suggested a simple method to estimate mean contact pressure, p_m, from the inflation pressure, p_i:

$$p_m = k \, p_{iz} \tag{1}$$

where k = a factor determined by the ply rating of the tire. The k value varies from 1.1 for 4- or 6-ply tires to 1.25 for 16-ply tires. Since eqn. (1) was developed from measurements of a deflected tire on a rigid surface under static load conditions, it does not account for variation in soil and the dynamic load conditions usually encountered in the field.

One of the reasons for simplistic approaches is the complexity of measuring the contact area between soil and tire. Generally, the contact area of a tire is measured under static load conditions by pressing the inked tire against paper on a rigid surface. Another approach, commonly used under static load conditions, is to fill the footprint of a tire in soil with plaster of Paris and then determine its area. To estimate the contact area under a moving tire, the most common method is to multiply the tire contact length by the tire section width. Once the contact area is determined, the average contact pressure can be estimated by dividing the load by the contact area. Söhne (1958) estimated that the maximum contact pressure for a tractor tire without high lugs was 1.4-2.0 times the average pressure measured with the above technique. Based on stress measurements with transducers placed on tires and in the soil, Steiner and Söhne (1979) developed the following relationship for estimating the average contact pressure on the soil surface, P_{avg}:

$$P_{avg} = 2.677 + 0.575 \, IPR + 0.011 \, DL - 0.016 \, d \tag{2}$$

where IPR = inflation pressure (bar); DL = dynamic load (kN); d = outside diameter of the tire (cm).

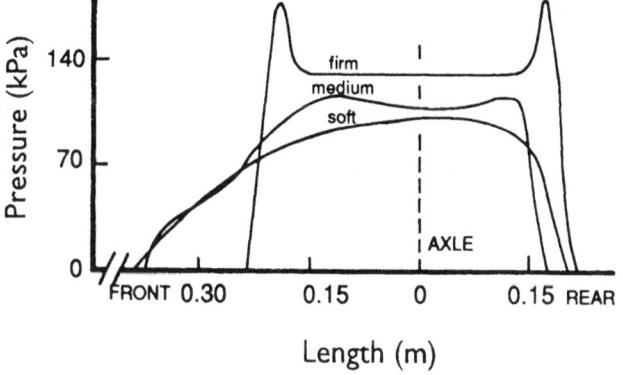

Fig. 1. Longitudinal ground contact pressure distribution under the center of an 11-38 smooth tire, inflated to 97 kPa, in sandy soil having three different degrees of firmness (from Vanden Berg and Gill, 1962).

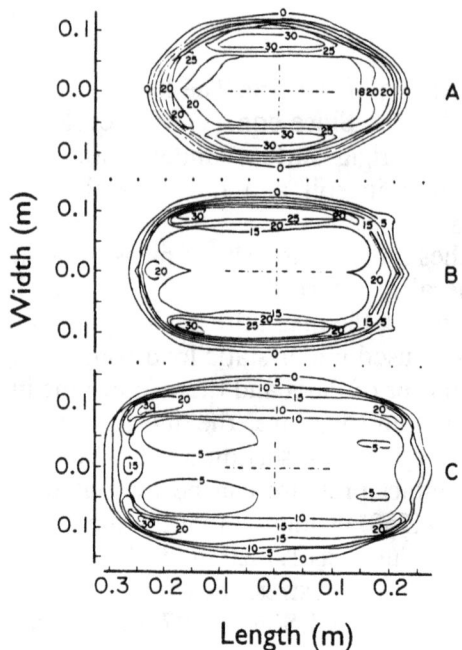

Fig. 2. Contour maps of the distribution of the ground contact pressure (lb in² = 0.145 kPa) under an 11-38 smooth tire on firm sand. (A) inflated to 97 kPa; (B) inflated to 69 kPa; (C) inflated to 41 kPa. The direction of travel of the tire was from right to left (from Vanden Berg and Gill, 1962).

Lugs on tires are another complicating factor in determining the stresses at the soil-tire interface. To simplify the problem, Vanden Berg and Gill (1962) determined the pressure distribution under the center of a smooth tire by embedding transducers in the tire surface. They found that higher pressure was generated at the soil-tire interface in firm soils than in soft soils (Fig. 1) and that the interface stresses also varied with the inflation pressure (Fig. 2). Higher inflation pressure tended to reduce the tire footprint and to concentrate the loads near the side of the tire. As the inflation pressure decreased, the footprint increased and thus the interface stress decreased, except for a small area near the tire edge.

Under lugs, interface stresses are larger than between the lugs or on a smooth tire. Trabbic et al. (1959) determined that the maximum value of normal stress measured between the lugs was only about 50% of the interface normal stress measured on the lugged area. Horn et al. (1987) showed that stresses between lugs are lower in the center of the tire than at the edge. Because rubber tires are flexible, and the exact location of the transducer is no longer known once the tire enters the soil, there is some uncertainty in accounting for all interface stresses from the transducer measurements.

Burt et al. (1987) installed a 3-dimensional sonic digitizing system inside a pneumatic tire to determine the geometric orientation of pressure transducers. Using this system, Burt et al. (1989) measured normal and tangential interface stresses with bi-directional transducers on the lug and between the lug for a combination of two soils and two soil conditions. From this study, these authors concluded that soil condition rather than soil type influenced the pressure distribution in soils.

In a related study, Burt et al. (1990) determined that the average contact pressure was close to the tire inflation pressure. Examination of the data from this research leads to the conclusion that, in general, the inflation pressure is a fair indicator of the average soil-tire interface pressure for a loose soil condition (cone index <0.20 MPa), even in the lugged area (Fig. 3). For firm soil conditions (cone index 0.20-0.78 MPa), the inflation pressure gave a poor estimate of the average soil-tire interface pressure. A better estimate of the average interface pressure in firm soils is three times the inflation pressure (Fig. 3). Also, these estimates are constant across soil types and dynamic loads. These conclusions are similar to the observations of Söhne (1958), that in soft soils the pressure under lugs may vary from 1-2 times the pressure under a smooth tire, whereas in firm soils this pressure may correspond to 3-4 times the pressure under a smooth tire. The increased magnitude of the peak resultant stress over the peak normal stress in Fig. 3 is due to the tangential stress. This would suggest

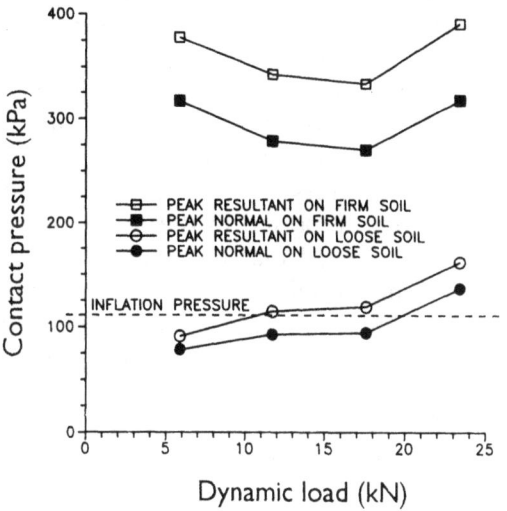

Fig. 3. Comparison of peak resultant and peak ground contact pressure on firm and loose soils (from Burt et al., 1990).

that the peak resultant stress rather than the peak normal stress should be used for prediction of soil compaction because the former term includes the applied tangential stress. This observation is similar to the conclusion of Gill and Vanden Berg (1968) that even in rigid wheels some of the applied load is transformed into tangential stress.

In summary, interface stresses greatly in excess of the inflation pressure have been documented near the edges of smooth tires and underneath the lugs of tractor tires. Lugs on agricultural tires cover more than 20% of the total contact area (Monroe et al., 1990). Assuming 15-20% slip for high draft conditions, a significant proportion of the contact area will be affected by the lugs and should be considered in modeling of soil compaction.

MODELING THE RELATIONSHIP BETWEEN APPLIED STRESS AND SOIL STRAIN

There has been a debate in the literature concerning the stresses which are appropriate for describing soil compaction in the profile during the passage of farm vehicles over the soil surface. Basically there are two approaches that are currently being used in modeling soil compaction. The first approach was suggested by Söhne (1953), who assumed that the major principal stress on a volume element in the soil profile was adequate to predict the soil strain under a tractor tire. Based on this concept, Gupta and Larson (1982) used the measured relationship of normal stress to soil bulk density in a uniaxial compression test to estimate iso-bulk density lines from iso-principal stress lines that could be predicted using Boussinesq's (1885) equations. A concern has been expressed whether a uniaxial compression test (which provides a change in linear strain caused by normal stress) is adequate in describing the changes in soil bulk density (volumetric strain) due to surface-applied loads from farm vehicles. In other words, triaxial test data which gives a change in volumetric strain due to both major (σ_1) and minor principal stresses (σ_3), may be needed to describe the soil bulk density changes due to surface-applied stress. Using the principles of critical state soil mechanics and triaxial test data reported in the literature on agricultural soils, Koolen and Kuipers (1983) showed that bulk density varied mainly according to σ_1 and only little with σ_3. This would suggest that the uniaxial compression relationship of normal stress to bulk density may be adequate in estimating the iso-bulk density lines from the iso-stress lines predicted by Söhne (1953).

Dexter and Tanner (1973) suggested the following relationship to describe the change in packing density, D, during hydrostatic compression in a triaxial apparatus:

$$D = D_0 + B \exp (-K \sigma_a) - C \exp (-L \sigma_a) \tag{3}$$

where D_0 = maximum limiting packing density; $(D_0 + B + C)$ = initial packing density; σ_a = normal applied pressure; K and L are measures of how rapidly the maximum density is attained by increasing σ_a.

To account for the effects of both soil wetness and applied load, Amir et al. (1976) suggested eqn. (4) to describe the compression of thin laboratory samples:

$$N = A_n - B_n \ln (\sigma_r - \sigma_a) - C_n \ln \theta \tag{4}$$

where N = porosity of the soil; σ_r = residual pressure; σ_a = normal applied pressure; θ = volumetric water content; A_n, B_n, C_n = constants.

Larson et al. (1980) showed that for an initial pore water pressure of -5 to -60 kPa, bulk density, ρ, along the virgin compression curve in a uniaxial compression test, was related to normal applied stress, σ_a, according to the following relationship:

$$\rho = \rho_k + \Delta_T (S_1 - S_k) + C \log (\sigma_a/\sigma_k) \tag{5}$$

where ρ_k = bulk density at a reference degree of water saturation, S_k, and a reference applied stress, σ_k; Δ_T = slope of ρ_k vs. degree of water saturation curve; S_1 = desired degree of water saturation; C = compression index.

Based on the laboratory data of 54 soils from various countries covering nine soil orders, Larson et al. (1980) provided regression relationships of compression parameters (eqn. (3)) to easily measurable soil properties, such as particle size distribution. Gupta and Larson (1982) and Gupta et al. (1985) updated these relationships to cover over 100 soils. This database, when linked to a model of propagation of stress, provides a means to estimate iso-bulk density lines in a given soil from easily measurable soil properties.

The second approach to modeling the stress-strain relationship of soils deals with the description of volumetric strain (Grisso et al., 1987). Bailey and Johnson (1989) improved earlier stress-strain models to include shear stress in the relationship. With bulk density as a dependent variable, Bailey and Johnson's relationship can be expressed as:

$$\ln\rho = \ln\rho_i - [(a + b \, \sigma_{oct}) \{1\text{-exp} (-c \, \sigma_{oct}) + d(\gamma_{oct}/\sigma_{oct})\}] \tag{6}$$

$$\sigma_{oct} = 1/3 \, (\sigma_1 + \sigma_2 + \sigma_3) \tag{7}$$

or

$$\sigma_{oct} = 1/3 \, (\sigma_x + \sigma_y + \sigma_z) \tag{8}$$

$$\tau_{oct} = 1/3 \, [(\sigma_1 - \sigma_2)^2 + (\sigma_2 - \sigma_3)^2 + (\sigma_1 - \sigma_3)^2]^{0.5} \tag{9}$$

where ρ_i = initial bulk density; σ_{oct} = octahedral normal stress or mean normal stress; τ_{oct} = octahedral shear stress; a, b, c = compactability coefficients; d = coefficient of natural volumetric strain due to shearing stress ratio; σ_1, σ_2 and σ_3 are the principal stresses; σ_x, σ_y and σ_z are the normal stresses in x, y and z directions. The last term in eqn. (6) accounts for the changes in volumetric strain due to shear stress. In contrast to eqn. (5), this relationship (eqn. (6)) can only be determined from a triaxial test, which allows accurate control of σ_2 and σ_3 and enables more exact determination of compression behavior. The triaxial test, however, is rather time consuming and expensive to run.

MODELING THE PROPAGATION OF FORCES IN THE SOIL

Most models which describe the propagation and distribution of stresses reported in the literature, are based on the elastic stress-strain theory (Gupta et al., 1989). This theory states that, at equilibrium, the displacement within a body due to an applied load is directly proportional to the load. Existing soil compaction models of propagation and distribution of stresses can be categorized into two groups: (a) finite element models; (b) analytical models.

Finite element models

The basis of a finite element soil compaction model is the division of the soil continuum into an assemblage of a finite number of elements that are connected at the nodal points (Fig. 4). The displacement at each nodal point due to externally applied stress is calculated using a displacement function that satisfies the compactability conditions at all times. In most finite element soil compaction models, the stress-strain relationship is taken to be similar to that of a linear-elastic material:

$$\{\sigma\} = [C] \{\varepsilon\} \tag{10}$$

where $\{\sigma\}$ = stress vector; $\{\varepsilon\}$ = strain vector; [C] = constitutive matrix that is dependent upon the material properties, i. e., Young's modulus (E) and Poisson's ratio (υ). These parameters are defined as follows:

$$E = \sigma/\varepsilon \tag{11}$$

$$\upsilon = - \text{(lateral strain / axial strain)} \tag{12}$$

where σ = stress and ε = strain in linear elastic materials in one dimension.

Equilibrium equations for potential energy and the unknown nodal displacements are developed for each element. A matrix containing a set of equations of all elements of the continuum is solved simultaneously for all nodal

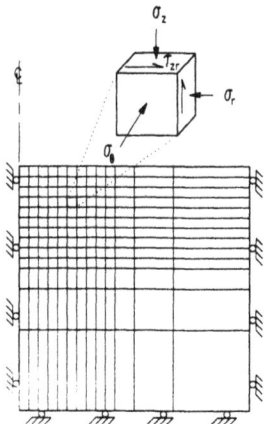

Fig. 4. Finite element mesh of an axisymmetric soil profile. The exploded element shows the appropriate stresses from the axisymmetric assumption.

displacements (Perumpral et al., 1971).

Since agricultural soils behave as non-linear plastic material, the assumption of a single value of E (stress-strain slope) over a range of applied stress is not valid. During simulation, adjustments are continuously needed in Young's modulus at each increment of load or at each iteration for a total applied load.

Several methods of incorporating Young's modulus in finite element soil compaction models have been suggested in the literature. Perumpral et al. (1971) assumed a constant Poisson's ratio and estimated the tangential Young's modulus from a relationship of octahedral normal stress (σ_{oct}) vs. the octahedral shearing stress (τ_{oct}) at various octahedral shearing strains (γ_{oct}). Duncan and Chang (1970), Pollock et al. (1986) and Chi and Kushwaha (1989) used the following hyperbolic model to estimate a tangential Young's modulus:

$$E_t = [1-\{R_f(1-\sin \phi)(\sigma_1-\sigma_3)\}/\{2 \, c \cos \phi + 2(\sigma_3+p_a) \sin \phi\}]^2 \, K \, p_a[(\sigma_3+p_a)/p_a]^n \quad (13)$$

where R_f = failure ratio; c and ϕ = Mohr-Coulomb strength parameters; K = modulus number; n = exponent determining the rate of variation of the initial E with σ_3; p_a = atmospheric pressure.

Duncan and Chang (1970) also found that Poisson's ratio, determined from the triaxial tests, increased from 0.11 to 0.65 with an increase in the deviatoric stress. Therefore, these authors also stressed the need for an incremental change in Poisson's ratio during simulation with a finite element model. In their model, Poisson's ratio was obtained from the relationship:

$$\upsilon = (\Delta e_1 - \Delta e_v)/2 \, \Delta e_1 \quad (14)$$

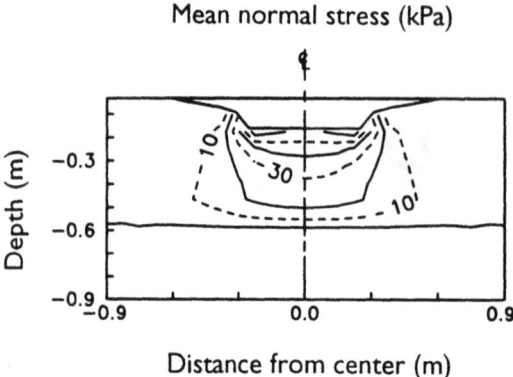

Distance from center (m)

Fig. 5. Mean normal stress distribution beneath a flat circular plate when a load of 25 kN is applied and a hard pan is assumed at a depth of 0.54 m (from Raper and Erbach, 1990b).

where Δe_1 = incremental axial strain; Δe_v = incremental volumetric strain.

Raper and Erbach (1990a) concluded from a finite element soil compaction model that soil stress predictions were dependent upon Poisson's ratio and almost independent of Young's modulus. In a simulation with a high constant value of Poisson's ratio, the initially loose soil transferred too much force laterally, creating excessive and artificial soil strength and thereby restricted soil compaction. They suggested the need to vary both parameters throughout the simulation. Raper and Erbach (1990b), using eqn. (10) and a derivative of an earlier version of the soil compaction relationship given in eqn. (6) to estimate Poisson's ratio and Young's modulus, successfully predicted the stresses beneath a flat loaded circular steel plate (Fig. 5). Most of the predicted stress values were within the 95%-confidence interval of the measured values. However, recent testing (Raper et al., 1990) of the finite element soil compaction model with the stress-strain relations described by eqn. (6) showed that the finite element model over-predicted the stresses in the soil profile and most values exceeded the 95%-confidence interval.

Analytical models

Most analytical models of propagation and distribution of stresses in the soil are based on the analytical solution of stress distribution in a homogeneous, elastic, isotropic, semi-infinite medium under a point load on the ground surface. The solution was originally proposed by Boussinesq (1885) and later modified by Fröhlich (1934) to account for an increase of Young's modulus with soil depth due to overburden pressure. Söhne (1953) was the first who used the modified Boussinesq's equations to predict the stress distribution in the soil from an agricultural vehicle. Modified Boussinesq's equations that describe various stresses on a soil element (Fig. 6) as a result of point load, P, at the soil surface are as follows (Söhne, 1953):

1. Vertical normal stress, σ_z:

$$\sigma_z = (vP/2\pi r^2) \cos^v \beta = (vP/2\pi z^2) \cos^{v+2} \beta \tag{15}$$

2. Horizontal normal stress, σ_h:

$$\sigma_h = (vP/2\pi r^2)\cos^{v-2} \beta (\sin^2 \beta) = (vP/2\pi z^2)\cos \beta \sin^2 \beta \tag{16}$$

3. Shear stress (τ) belonging to σ_z and σ_h:

$$\tau = (vP/2\pi r^2)\cos^{v-1} \beta (\sin \beta) = (vP/2\pi z^2)\cos^{v+1}\beta \sin^2\beta \tag{17}$$

where r = radial distance from the point load; z = vertical distance from the point load; β = angle bisected by a vertical line from point O with a line to the center of gravity of the volume element in question (Fig. 6); P = point load; v = Fröhlich's concentration factor.

Stresses calculated from eqns. (15)-(17) correspond to those in an elastic body whose Poisson ratio equals 0.5 and an isotropic body if v is equal to 3. An increase in v leads to deeper penetration of given iso-stress lines. Söhne (1953) used eqns. (15)-(17) to predict iso-stress lines under a tractor tire. The procedure

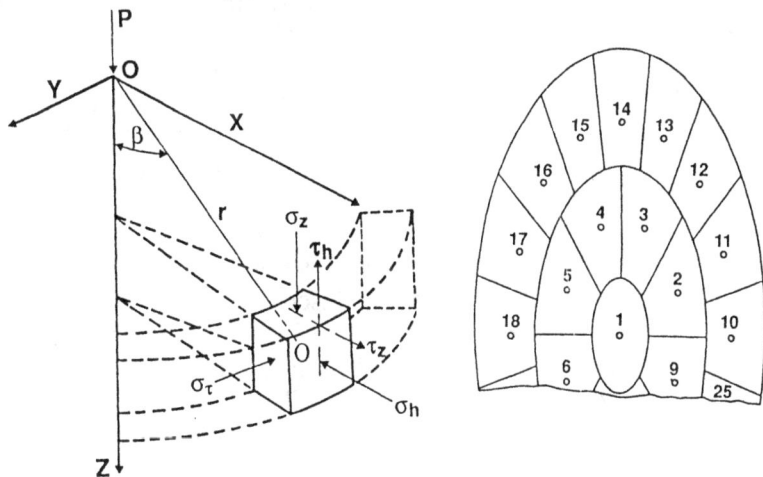

Fig. 6. (left). Stresses on a soil element caused by a point load, P (from Söhne, 1953).

Fig. 7. (right). A surface load broken up into 25 load elements. Open circles represent the center of gravity with a pressure distribution according to a parabola of 5th potency (from Söhne, 1953).

involved dividing the contact area between the tire and the ground into 25 load elements (Fig. 7) and applying a fraction of the total load in the center of gravity of a load element. Söhne (1953) assumed that soil compaction (vertical strain) at a point in the soil profile was mostly caused by the major principal stress, σ_1, calculated from the summation of stresses from the 25 load elements using eqns. (18)-(24). Fig. 8 defines the variables used in eqns. (18)-(24).

$$\sigma_1 = \sigma_z \cos^2 \alpha + \sigma_x \sin^2 \alpha + \tau_{xz} \sin 2\alpha \tag{18}$$

$$\sigma_z = \Sigma(\nu P_i/2\pi z^2)\cos^{\nu+2} \theta \tag{19}$$

$$\sigma_x = \Sigma(\nu P_i/2\pi z^2)\cos\theta \sin^2\theta \cos^2\phi \tag{20}$$

$$\tau_{xz} = \Sigma(\nu P_i/2\pi z^2)\cos^{\nu+1} \theta \sin \theta \cos^2 \phi \tag{21}$$

where

$$\cos \phi = (x_i + x) / [(x_i + x)^2 + y_i^2]^{0.5} \tag{22}$$

$$\cos \theta = z/[(x + x_i)^2 + y_i^2 + z^2] \tag{23}$$

$$\tan 2\alpha = 2\tau_{xz} / (\sigma_z - \sigma_x) \tag{24}$$

Söhne (1953) used the laboratory measured stress-strain relationship in a uniaxial compression setup to predict changes in vertical soil strain or soil bulk density distributions from a predicted σ_1-profile. He suggested ν values of 4, 5, and 6 for hard, firm and soft soil conditions, respectively. Söhne's description of firmness applied to combinations of both bulk density and water status of soil.

Blackwell and Soane (1981) showed that the modified Boussinesq's equations required different ν values with soil depth in order to match predicted bulk density with measured values. In a laboratory test, Gupta and Larson (1982) found that ν values that gave the best predictions of iso-bulk density lines were small near the surface and increased with depth. Based on a field experiment, Ram (1984) concluded that: (1) ν varied from 3.28 to 5.84 for sandy and loamy soils in normal field conditions and increased to 7.74 in low-strength, powdered silt loam soil; (2) ν increased with an increase in soil moisture content at a constant bulk density and decreased with an increase in bulk density of soil at the same moisture level.

Using the procedure of Newmark (1942), Horn (1988) calculated the value of the concentration factors for various soil horizons at matric potentials of -60 and -300 kPa. From the calculated values, Horn (1988) concluded that a decrease in soil wetness stabilized the soil and led to a decrease in the value of the concentration factor. Upon drying, even a weak soil stabilized and gave a lower

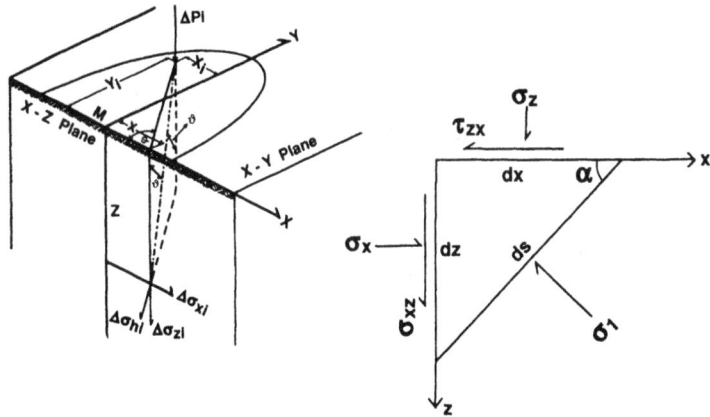

Fig. 8. Left: stresses at a point on the XZ plane as a result of a point load at x_i, y_i on the soil surface. Right: equilibrium of the stresses on a triangular element whose hypotenuse is vertical to the direction of the main stress (from Söhne, 1953).

value of the concentration factor. Also, as the soil horizons became more strongly aggregated, the concentration factor decreased. The author also observed that, because the degree of settlement depended upon the time after application of load, the concentration factor also varied as a function of the time. In his experiment, the value of the concentration factor increased over time, with the majority of the increase occurring in the first minute after application of load. At a given time, the value of the concentration factor also increased with an increase in the surface-applied load.

Although several methods are available to estimate the concentration factors, these methods are time-consuming and expensive. Databases are needed that can provide estimates of the concentration factor from easily measurable soil and tire characteristics.

Based on the concepts of Söhne (1953), Gupta and Larson (1982) developed a model to predict the distribution of applied stress after passage of an agricultural vehicle. Predicted iso-stress lines were converted to iso-bulk density lines, using the laboratory-measured uniaxial stress-strain relationship of Larson et al. (1980). This procedure assumed that the major principal stress (σ_1) from eqn. (18) and the vertical stress in a uniaxial compression test were equivalent. The model was tested on Waukegan silt loam (Typic Hapludoll) and Doland loam (Udic Haploboroll) in above-ground soil bins at Rosemount and Morris, Minnesota, respectively, and *in situ* on Nahal Oz silt loam (Calcic Malic Haploxeralf) in Israel (Gupta et al., 1985). Field tests in Israel covered several moisture contents, and two passes of three tractors operating at two speeds. At the Rosemount site, the soil compaction model under-predicted the mechanical stress profile (Fig. 9), but the predicted bulk density profiles (Fig.10) were close

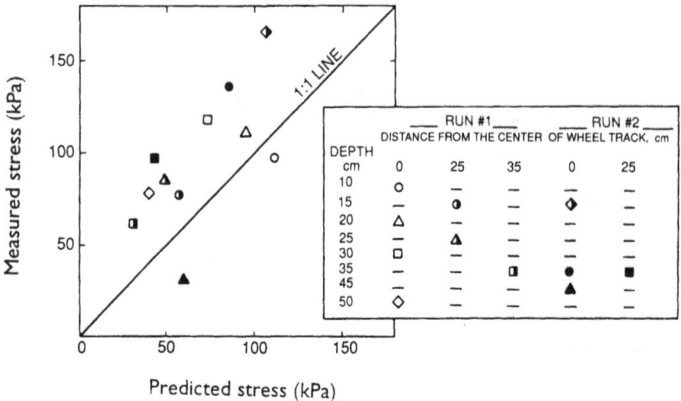

Fig. 9. Comparison between measured and predicted mechanical stresses in an artificial soil profile of Waukegan silt loam during the passage of a 2-wheel drive tractor (117kW; 5.8 Mg) on the soil surface (from Gupta et al., 1985).

to the measured values. Maximum deviations between predicted and measured bulk densities were 0.12, 0.13, and 0.33 Mg m⁻³ for Waukegan silt loam, Doland loam, and Nahal Oz loam, respectively. Some of the differences between measured and predicted stress profiles were related to uncertainties of soil-transducer contact and the soil variability in the soil bins. A sensitivity analysis of the soil compaction model to changes in v value did not significantly alter the differences between measured and predicted bulk densities on a Waukegan silt loam and Doland loam (Gupta et al., 1985). Bulk densities predicted on the assumption that v values are between 3 and 6, were within the range of errors encountered in the measurements of field bulk densities.

Van den Akker and Van Wijk (1987) also developed a soil compaction model using the procedure of Söhne (1953). Their model differed from that of Gupta and Larson (1982) in that it also included horizontal load exerted by the tire at the tire-soil interface. A test of the model in a laboratory soil bin showed that measured increases in bulk density were greater than predicted values based on main stress (σ_1) calculations, but less than the values calculated based on mean normal stress or octahedral normal stress (σ_m). With some additional analysis, these authors concluded that differences in the predicted and measured increases in bulk density may indicate a lack of consideration of plastic deformation in the soil compaction process. They estimated that, to a depth of 70 cm, 75% of the soil compaction can be described by elastic deformation. These authors also concluded that, due to smaller changes in soil volume during compaction, the

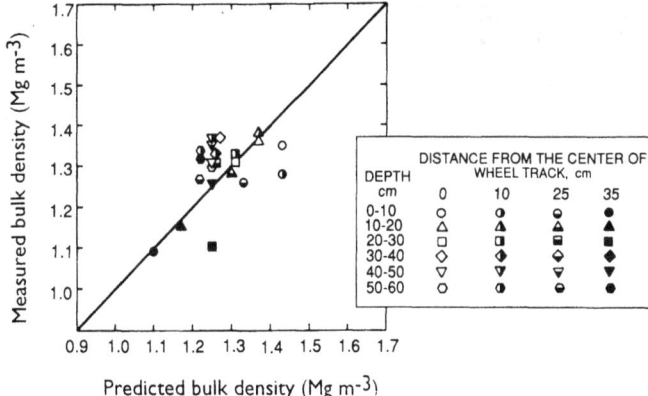

Fig. 10. Comparison between measured and predicted bulk densities of Waukegan silt loam after one pass of a 2-wheel drive tractor (117 kW; 5.8 Mg) at Rosemount, MN, USA (from Gupta et al., 1985)

contribution of plastic deformation decreases with depth. Although they assumed a constant value of $v = 4$ in their calculation, they implied that some of the plastic deformation may be accounted for by varying v with soil depth.

LINKING SOIL COMPACTION MODELS TO CROP GROWTH MODELS

In most crop growth models, bulk density has been used as the surrogate variable to account for the effects of compaction on crop growth. Except for Reddy et al. (1987), the authors are unaware of any crop growth model that simulates the soil compaction process starting with the weight of the tractor, contact area between the tire and the soil surface, and soil firmness properties such as wetness and initial bulk density. Generally, a compaction event in the crop growth models is simulated by physically inputting the new bulk density of the soil. There is no description as to how this bulk density was determined, what surface applied loads were used, or what was the soil wetness at the time of compaction. Since most crop growth models are one-dimensional, it is further assumed that the bulk density is constant at a given depth. This assumption is incorrect because bulk density varies with depth and in the transverse direction in the plane below the tire.

In most crop growth simulation models, the effect of increased bulk density due to a compaction event is simulated by a decrease in root growth and, in some cases, a decrease in soil hydraulic conductivity. Decreased root growth is mostly due to an increase in soil penetration resistance (soil strength), assumed to be caused by an increase in bulk density. Statistical relationships of root growth

parameters (extension, branching, death) as a function of penetration resistance
and relationships of penetration resistance to bulk density, provide a mechanism
to adjust root growth in response to a soil compaction event. For example,
Shaffer and Clapp (1987) described soil strength (f) as a function of bulk density
(ρ_b) with the following relationship:

$$f = -1.095 + 0.1084 \times 10^{-1}\ \Psi + 0.9735 \times 10^{5}\ \Psi^2 + 4.636\ \rho_b - 2.044\ \rho_b^2 \qquad (25)$$

where Ψ = matric potential varying from 0 to -70 kPa. Eqn. (25) was derived from
the data of Taylor and Gardner (1963) and applies to variation in soil bulk
density from 1.0 to 2.0 Mg m^{-3}. Fig. 11 gives the graphical representation of eqn.
(25).

Other effects of soil compaction that have been incorporated in crop growth
models (Shaffer and Clapp, 1987; Jones et al., 1991) include the aeration effect.
Shaffer and Clapp (1987) assumed that root growth was equal to the optimum
value in the range of soil wetness between 0 and 85% of saturation and
decreased linearly from its optimum value to zero when soil wetness increased
from 85 to 100% saturation. Thus, compaction effects on soil aeration will vary
with degree of saturation. Similar approaches have been used by others (Baker
et al., 1983; Williams et al., 1989; Jones et al., 1991) to incorporate the effects of
soil compaction on root development in crop growth models.

Compaction effects on root growth are then simulated by multiplying the
minimum value of f (from soil compaction factors, such as strength, aeration)
with the value of root growth at a reference state (i.e., when no soil physical and

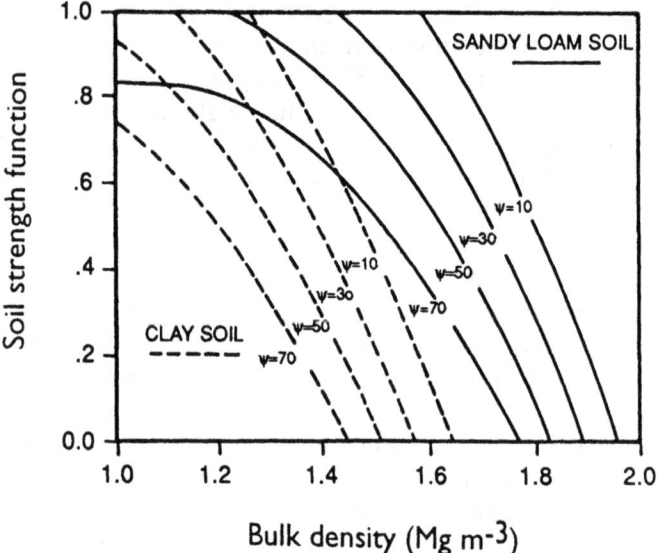

Fig. 11. Soil strength as a function of bulk density at various matric potentials, ψ (-kPa), for
sandy loam and clay soils (from Shaffer and Clapp, 1987).

chemical factors are limiting root growth). Since other factors, such as soil temperature, soil salinity, water and nutrient availability also effect root growth, overall effects of all factors in this model are simulated by selecting a minimum f value over all factors.

Reddy et al. (1987) incorporated the Gupta and Larson (1982) soil compaction model in the cotton growth model GOSSYM (Baker et al., 1983), to study long-term soil compaction effects on cotton yield in the U.S.A. Cotton Belt. Input for simulation of soil compaction included tractor weights, tire sizes, soil texture, and water contents. Predicted soil bulk density profiles from the soil compaction sub-model were then used in GOSSYM to alter soil hydraulic and impedance properties within the wheel tracks zone. These altered properties provided a means to evaluate long-term effects of soil compaction on cotton yield.

The effect of soil compaction on root growth in the GOSSYM model is assessed through a calculation of soil impedance from soil water content and bulk density. Soil impedance was related to root growth with the relationship:

$$RG = 104.6 - 35.3 \, PR \tag{26}$$

where RG = root growth relative to unimpeded growth (%); PR = penetration resistance (Pa). Reddy et al. (1987) suggested that this relationship would hold for cotton over a wide range of soil types.

Using the combined GOSSYM and the soil compaction model, Reddy et al. (1987) evaluated the effects of soil compaction on cotton lint yield from 1963 through 1983 at five sites in the U.S.A. Cotton Belt. At Florence, SC, the average predicted cotton lint yield response to compaction from 2-row wheel traffic over this period was 5.0% (w/w), although in individual years the response was often negligible, sometimes positive and sometimes negative. Similar comparisons at Stoneville, MS, College Station, TX, and Phoenix, AZ, showed changes in cotton lint yield of -53.3, +22.2 and -12.9%, respectively.

CONCLUSIONS

(1) Soil compaction models are useful tools for evaluating harmful vs. desired compaction in agricultural fields. The steps involved in the construction of a soil compaction model include: (1) modeling surface-applied forces from farm vehicles; (2) representation of soil stress-strain relationships; (3) modeling the propagation of stresses in the soil profile.

(2) Simple relationships are available that predict the surface stress from easily measurable tire and vehicle characteristics. Because of the presence of lugs on tires, additional techniques are needed that can predict stress distribution between lugged and non-lugged areas for a wide range of soil and tire conditions.

(3) Both uniaxial or triaxial laboratory measurements of stress-strain relationships on soil samples seem to be adequate for predicting soil compaction,

either from finite element or analytical models of stress propagation. There is an extensive database on laboratory measured compressibility indices using uniaxial compression equipment. Similar data bases for Young's modulus and Poisson's ratio are needed to facilitate a greater use of finite element models. Since both the finite element and analytical models are based on the theory of elastic deformation, additional models are also needed to describe stress propagation for conditions where the soil deformation is plastic. As wheel slip is common on the farm, existing models of stress propagation also need to include the horizontal stress at the soil surface due to wheel slip.

(4) The challenge to make the existing soil compaction models more useful is to link them with crop growth models. Soil mechanics research is also needed to develop a 3-dimensional soil compaction model that considers the total stress state of the soil under an applied load.

(5) Continuous testing of all soil compaction models is necessary to extend the range of conditions where these models can be applied.

REFERENCES

Amir, I., Raghavan, G.S.V., McKyes, E. and Broughton, R.S., 1976. Soil compaction as a function of contact pressure and soil moisture content. Can. Agric. Eng., 18: 54-57.

Bailey, A.C. and Johnson, C.E., 1989. A soil compaction model for cylindrical stress states. Trans. ASAE, 32: 822-825.

Baker, D.N., Lambert, J.R. and McKinion, J.M., 1983. GOSSYM: A simulator of cotton growth and yield. South Carolina Agric. Exp. Stn., Clemson, SC, U.S.A., Tech. Bull. 1089, 183 pp.

Barnes, K.K., Carleton, W. M., Taylor, H.M., Throckmorton, R.I. and Vanden Berg, G.E. (Editors), 1971. Compaction of Agricultural Soils. Am. Soc. Agric. Eng., St. Joseph, MI, U.S.A., ASAE Mono. 1, 471 pp.

Blackwell, P.S. and Soane, B.D., 1981. A method of predicting bulk density changes in field soils resulting from compaction by agricultural traffic. J. Soil Sci., 32: 51-65.

Boussinesq, J., 1885. Application des Potentials à l'Etude de l'Equilibre et du Mouvement des Solides Elastiques. (Application of Potentials in the Study of the Equilibrium and the Movement of Elastic Solids). Gauthier-Villais, Paris, France, 30 pp. (in French).

Burt, E.C., Wood, R.K. and Bailey, A.C., 1987. A three-dimensional system for measuring tire deformation and contact stresses. Trans. ASAE, 30: 324-327.

Burt, E.C., Wood, R.K. and Bailey, A.C., 1989. Effects of dynamic load on normal soil-tire interface stresses. Trans. ASAE, 32: 1843-1846.

Burt, E.C., Wood, R.K. and Bailey, A.C., 1990. Some comparisons of average to peak soil-tire contact pressures. Am. Soc. Agric. Eng., St. Joseph, MI, U.S.A., ASAE Pap. 90-1094.

Chi, L. and Kushwaha, R.L., 1989. Finite element analysis of soil forces on two shapes of tillage tool. Am. Soc. Agric. Eng., St. Joseph, MI, U.S.A., ASAE Pap. 89-1103.

Dexter, A.R. and Tanner, D.W., 1973. The response of unsaturated soils to isotropic stress. J. Soil Sci., 24: 491-502.

Drescher, J., Horn, R. and De Boodt, M. (Editors), 1988. Impact of Water and External Forces on Soil Structure. Catena, Cremlingen, Germany, Catena Suppl. 11, 175 pp.

Duncan, J.M. and Chang, C.Y., 1970. Nonlinear analysis of stress and strain in soils. J. Soil Mech. Foun. Div., Proc. Am. Soc. Civil Eng., 96: 1629-1653.

Fröhlich, O.K., 1934. Druckverteilung im Baugrunde. (Pressure distribution in soil foundation).

Springer, Wien, Austria, 178 pp. (in German).

Gill, W.R. and Vanden Berg G.E., 1968. Soil Dynamics in Tillage and Traction. USDA-ARS, Washington, DC, U.S.A., Agric. Handbook 316, pp. 355-362.

Grisso, R.D., Johnson, C.E., and Bailey, A.C., 1987. Soil compaction by continuous deviatoric stress. Trans. ASAE, 30: 1293-1301.

Gupta, S.C. and Larson, W.E., 1982. Modeling soil mechanical behavior during tillage. In: P. Unger, D.M. Van Doren Jr, F.D. Whisler and E.L. Skidmore (Editors), Predicting Tillage Effects on Soil Physical Properties and Processes. Am. Soc. Agron., Madison, WI, U.S.A., Spec. Publ. 44, pp. 151-178.

Gupta, S.C. and Allmaras, R.R., 1987. Models to assess the susceptibility of soils to excessive compaction. Adv. Soil Sci., 6: 65-100.

Gupta, S.C., Hadas, A., Voorhees, W.B., Wolf, D., Larson, W.E. and Schneider, E.C., 1985. Field testing of a soil compaction model. Proc. Int. Conf. Soil Dynamics, Auburn, AL, U.S.A., Vol. 5, pp. 979-994.

Gupta, S.C., Hadas, A. and Schafer, R.L., 1989. Modeling soil mechanical behavior during compaction. In: W.E. Larson, G.R. Blake, R.R. Allmaras, W.B. Voorhees and S.C. Gupta. (Editors), Mechanics and Related Processes in Structured Agricultural Soils. NATO ASI Series E, Applied Science 172, Kluwer, Dordrecht, Netherlands, pp. 137-152.

Håkansson, I., Voorhees, W.B. and Riley, H. 1988. Vehicle and wheel factors influencing soil compaction and crop response in different traffic regimes. Soil Tillage Res., 11: 239-282.

Horn, R. 1988. Compressibility of arable land. In: J. Drescher, R. Horn and M. de Boodt (Editors), Impact of Water and External Forces on Soil Structure. Catena, Cremlingen, Germany, Catena Suppl. 11, pp. 53-71.

Horn, R., Burger, N., Lebert, M. and Badewitz, G., 1987. Druckfortpflanzung im Boden unter fahrenden Traktoren. (Pressure transmission in soils under travelling tractors). Z. Kulturtech. Flurber., 28: 94-102 (in German).

International Conference on "Soil Compaction as a Factor Determining Plant Productivity", 1989, Lublin, Poland. Soil Tillage Res., 19: 95-362.

International Conference on Soil Dynamics, 1985. Nat. Tillage Machinery Lab./ Auburn Univ., Auburn, AL, U.S.A., Vol. 1-5, 1157 pp.

Jones, C.A., Bland, W.L., Ritchie, J.T. and Williams, J.R., 1991. Simulation of root growth. In: R.J. Hanks, and J.T. Ritchie (Editors), Modeling Plant and Soil Systems. Am. Soc. Agron., Madison, WI, U.S.A., ASA Mono. 31, pp. 92-120.

Koolen, A.J. and Kuipers, H., 1983. Agricultural Soil Mechanics. Springer, New York, NY, U.S.A., 241 pp.

Larson, W.E., Gupta, S.C. and Useche, R.A., 1980. Compression of agricultural soils from eight soil orders. Soil Sci. Soc. Am. J., 44: 450-457.

Larson, W.E., Blake, G.R., Allmaras, R.R., Voorhees, W.B. and Gupta, S.C. (Editors), 1989. Mechanics and Related Processes in Structured Agricultural Soils. Kluwer, Dordrecht, NATO ASI Series E, Applied Science 172, 273 pp.

Monroe, G.E., Burt, E.C. and Bailey, A.C., 1990. Tire performance using different treads on traffic lanes. Trans. ASAE, 33: 51-55.

Monnier, G. and Goss, M. J. (Editors), 1987. Soil Compaction and Regeneration. Balkema, Rotterdam, Netherlands, 167 pp.

Newmark, N.U., 1942. Influence charts for computation of stress in elastic foundations. Univ. Illinois, Urbana, IL, U.S.A., Bull. 338, 28 pp.

Perumpral, J.V., Liljedahl, J.B. and Perloff, W.H., 1971. The finite element method for predicting stress distribution and soil deformation under tractive devices. Trans. ASAE, 14: 1184-1188.

Pollock, D. Jr., Perumpral, J.V. and Kuppusamy, T., 1986. Finite element analysis of multipass effects of vehicles on soil compaction. Trans. ASAE, 29: 45-50.

Ram, R.B., 1984. Pressure measurement in the soil under the load. Soil Tillage Res., 4: 137-145.

Raper, R.L. and Erbach, D.C., 1990a. Effect of variable linear elastic parameters on finite element prediction of soil compaction. Trans. ASAE, 33: 731-736.

Raper, R.L. and Erbach, D.C., 1990b. Prediction of soil stresses using finite element analysis. Trans. ASAE, 33: 725-730.

Raper, R.L., Johnson, C.E. and Bailey, A.C., 1990. Coupling normal and shearing stresses to use in finite element analysis of soil compaction. Am. Soc. Agric. Eng., St. Joseph, MI, U.S.A., ASAE Pap. 90-1086.

Reddy, V.R., Baker, D.N., Whisler, F.D., Wanjura, D.F., Barker, G.L. and McKinion, J.M., 1987. Yield and productivity in cotton - system analysis of factors affecting crop yields. USDA-ARS Crop Sci. Res. Lab., Mississippi State, MS, U.S.A./Clemson Univ., Clemson, SC, U.S.A./Mississippi State Univ., Mississippi State, MS, U.S.A., Final Rep., 142 pp.

Schafer, R.L., Johnson, C.E., Koolen, A.J., Gupta, S.C. and Horn, R., 1990. Future research needs. Am. Soc. Agric. Eng., St. Joseph, MI, U.S.A., ASAE Pap. 90-1078.

Shaffer, M.J. and Clapp, C.E., 1987. Root growth submodel. In: M.J. Shaffer and W.E. Larson (Editors), NTRM, Soil-Crop Simulation Model for Nitrogen, Tillage, and Crop-Residue Management. USDA-ARS, Conserv. Res. Rep. 34-1, pp. 63-72.

Soane, B.D., Blackwell, P.S., Dickson, J.W. and Painter, D.J., 1981a. Compaction by agricultural vehicles: A review. I. Soil and wheel characteristics. Soil Tillage Res., 1: 207-237.

Soane, B.D., Blackwell, P.S., Dickson, J.W. and Painter, D.J., 1981b. Compaction by agricultural vehicles: A review. II. Compaction under tyres and other running gear. Soil Tillage Res., 1: 373-400.

Soane, B.D., Dickson, J.W. and Campbell, D.J., 1982. Compaction by agricultural vehicles: A review. III. Incidence and control of compaction in crop production. Soil Tillage Res., 2: 3-36.

Söhne, W., 1953. Druckverteilung im Boden und Bodenverformung unter Schlepperreifen. (Pressure distribution in the soil and soil deformation under tractor tires). Grundl. Landtech., 5: 49-63 (in German).

Söhne, W., 1958. Fundamentals of pressure distribution and soil compaction under tractor tires. Agric. Eng., 39: 276-281, 290.

Steiner, M. and Söhne, W., 1979. Berechnung der Tragfähigkeit von Ackerschlepperreifen sowie des Kontaktflächenmitteldruckes und des Rollwiderstandes auf starrer Fahrbahn. (Calculation of the load capacity of tractor tires as well as the medium pressure in the contact area and the rolling resistance on a rigid surface). Grundl. Landtech., 29: 145-152 (in German).

Taylor, H. M. and Gardner, H. R., 1963. Penetration of cotton seedling taproots as influenced by bulk density, moisture content and strength of soil. Soil Sci., 96: 153-156.

Trabbic, G.W., Lask, K.V. and Buchele, W.F., 1959. Measurement of soil-tire interface pressures. Agric. Eng., 40: 678-681.

Van den Akker, J.J.H. and Van Wijk, A.L.M., 1987. A model to predict subsoil compaction due to field traffic. In: G. Monnier and M.J. Goss (Editors), Soil Structure and Regeneration. Balkema, Rotterdam, Netherlands, pp. 69-84.

Vanden Berg, G.E. and Gill, W.R., 1962. Pressure distribution between a smooth tire and the soil. Trans. ASAE, 5: 126-129, 132.

Williams, J.R., Jones, C.A. and Kiniry, D. A., 1989. The EPIC crop growth model. Trans. ASAE, 32: 497-511.

PART C

====================================

EFFECTS OF COMPACTION ON SOIL PROPERTIES

Soil Compaction in Crop Production
B.D. Soane and C. van Ouwerkerk (Eds.)

CHAPTER 5

Effects of Compaction on Soil Microstructure

Maja J. KOOISTRA[1] and N.K. TOVEY[2]

[1]The Winand Staring Centre for Integrated Land, Soil and Water Research (SC-DLO), Wageningen, Netherlands
[2]University of East Anglia, School of Environmental Sciences, Norwich, U.K.

SUMMARY

Microstructure encompasses all aspects of soil structure that are revealed when soil structure is examined at a magnification of x5 or greater. The microstructure plays an important part in determining the properties and behaviour of soils as it regulates and influences many processes in the soil, such as physical processes, decomposition and transport processes, and the functioning of soil organisms and plant root systems.

The architecture of the microstructure can be divided into several levels: (1) the macroaggregation; (2) the macroporosity in aggregates and non-aggregated materials; (3) the packing of primary particles in the groundmass; (4) the arrangement of clay particles. Changes in microstructure due to compaction occur at all these levels. They can be subdivided into changes in microporosity and changes in particle arrangement in the groundmass. During compaction, either by cultivation activities or by field traffic, voids are reduced in size and changed in shape, they may become deformed or disrupted, or disappear completely. Generally, the larger voids are reduced in size and more smaller voids, lacking interconnections, result. Consequently, soil properties and soil behaviour change considerably. Soil fauna and plant rooting are mainly responsible for the establishment of continuous void systems.

Compacted conditions in the subsoil can be removed temporarily, but often recompaction occurs within a few years, resulting in reduced permeability to water and air, due to changes in the pore size distribution and pore continuity. Compacted soil conditions can also be due to disintegration of soil aggregates at the surface, which can lead to the formation of compact, laminated crusts which hamper infiltration and germination, or can result in filling and blocking voids below the surface. Consequently, total porosity is reduced and most of the continuous voids disappear.

When the soil is subject to a load, such as by farm machinery, changes occur also within the groundmass. The arrangement of the primary particles collapses, whereby fine material will be squeezed between larger sand and silt grains. Mechanical deformation, such as that occurring during tillage, causes shear failure characterized by realignment of particles. Changes in the arrangement of particles and degree of realignment can be quantified and linked with the exerted load or shear stress and/or with changes in soil properties, such as, for example, permeability.

Microscopic studies provide considerable insight into how soils compact, and they explain why and when soil properties and behaviour change. Already a number of processes can be analysed but more research is needed before all processes involved will be fully understood.

INTRODUCTION

An understanding of the formation, dynamics and stability of soil structure is important as it determines and influences many soil processes: physical processes, such as water, air and temperature regimes, decomposition and transport processes, and the occurrence, survival and functioning of soil organisms and plant root systems. In various disciplines dealing with soil structure, e.g., soil science, engineering and geology, different definitions are used (McKeague and Wang, 1982; Jastrow and Miller, 1991). A recent definition, accommodating the many different aspects of soil structure manifest at different size scales, was given by Dexter (1988): "Soil structure is the spatial heterogeneity of the different components or properties of soils". This definition is adopted for this chapter. It has long been recognized that the microstructure of soils and sediments plays an important part in determining the behaviour of these materials when they are subjected to external loading.

Microstructure encompasses all aspects of soil structure that are revealed when soil material is examined at a magnification of x5 or greater. However, macroscopic structural features, examined with the naked eye, are also important in the study of soil compaction (Bullock et al., 1985). In the literature, besides microstructure, the term microfabric is also used, sometimes with a slightly different meaning. To avoid confusion, the term microstructure is used here as defined above.

Soil structure is the dynamic result of many abiotic and biotic factors and processes. In natural environments, the main structure-forming factors are: texture, organic matter, soil organisms, depth of the water table and weather conditions (Kooistra, 1991). In non-cultivated land, the stability of the soil structure is relatively high and the dynamics are mainly governed by the weather, regulating physical aggregate formation and biological processes. In cultivated land, however, the natural development of soil structure is disturbed and the human impact, i.e., tillage, traffic and machinery use, application of pesticides, fertilizers and manures, regulates the soil structure throughout the year. Here, natural physical aggregate formation and biological processes play only a secondary role. Besides the planned effects of cultivation practices on soil structure, a number of side-effects occur, such as soil compaction and deformation, surface and/or internal slaking, and crust formation due to transport of soil material along and through the surface. Most of these side-effects lead to some kind of compacted conditions and largely determine soil properties and behaviour and the processes occurring in the soil.

To follow the changes in microstructure, an understanding of the architecture of soil structure is essential. On the field scale, the pedality or aggregation can be observed with the naked eye (Fig. 1). Pedality is defined as the occurrence of individual, natural soil aggregates or peds (Soil Survey Staff, 1975). The individual peds can be classified according to their shape and arrangement into prisms, columns, plates, blocky peds, granules and crumbs, delineated by planar voids. Natural soil aggregates are formed by physical and/or biological processes. Non-natural aggregates are mainly due to cultivation or civil engineering activities. They can be classified in the same way as peds and are delineated by compound packing voids.

Groundmass is a general term used for the coarse and fine materials which form the basic constituents of the soil. In the groundmass of aggregates and peds or in apedal soil material, several kinds of voids other than primary tillage voids may occur, formed by physical (natural or mechanical) and biological processes or combinations of these. The groundmass itself consists of primary particles which can also be arranged in different ways (packing). Separate grains may occur as an entity in sands and gravels. However, finer material is often present as bridges between the larger grains, as coatings around them or as smaller aggregates between them. When soil material contains > 15-18 % (w/w) of clay, the larger grains generally are completely surrounded by small clay particles.

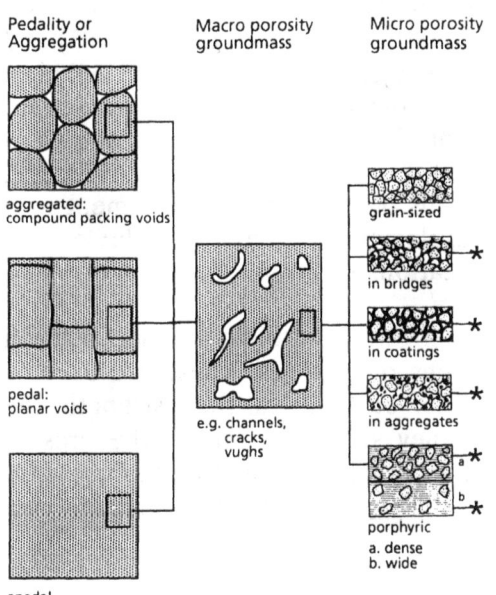

Fig. 1. Schematic presentation of types of voids present in soils at three different scales of observation: (left) pedality or aggregation (cm); (centre) macro-porosity (mm); (right) microporosity (μm). Microstructures of clay-sized particles (*) are illustrated in Fig. 10.

Changes in microstructure due to compaction occur at all scales described above. However, many voids and most soil particles are small, and beyond the resolution of the naked eye. Therefore, changes in microstructure can be observed and changes in soil properties can be understood only by the use of techniques such as optical microscopy, electron microscopy and X-ray diffraction.

METHODS TO OBSERVE CHANGES IN MICROSTRUCTURE

Introduction

Microscopic analyses are performed by means of visible light microscopy using a polarizing light microscope, and by submicroscopic techniques. Submicroscopy involves the use of instruments that analyse emitted radiation of wavelengths shorter than that of the visible light; it is used when morphological or chemical information is required that cannot be obtained using light microscopy. Many techniques can be used (e.g., Bisdom et al., 1990), of which scanning electron microscopy (SEM), transmission electron microscopy (TEM) and energy dispersive X-ray analysis (EDXRA) are the most common. For microstructure studies, light microscopy and SEM are mainly used.

Soil microstructure may be studied qualitatively as well as quantitatively. In the past, quantitative analyses were performed using a microscope with a calibrated eye piece graticule or other types of grids. The most common techniques are estimation on the basis of charts (e.g., Bullock et al., 1985) and point counting (e.g., Brewer, 1964; Weibel, 1979). Nowadays, provided technical know-how and advanced equipment are available, most quantitative studies are performed using digitized images, which may be acquired by means of a CCD (Charge Coupled Device) camera attached to the tube of a light microscope, or directly from the output of a SEM. This permits the establishment of a number of characteristics of microstructural elements, such as area/volume ratio, perimeter, number, length, width, orientation, position, shape factors and neighbour distances.

Both visible light and submicroscopic investigations result in data sets that can be presented as descriptions, tables, graphs or pictures. These are used to interpret processes, which caused the observed features, their strength, the interrelation of processes and whether they still occur under the present conditions or not. This information together with data from other sources leads to a synthesis, from which extrapolations, models or predictions can be derived (Kooistra, 1990).

Sample preparation

To study compacted and non-compacted soil material *in situ*, specific sample preparation is required. The basic techniques used in soil micromorphology are

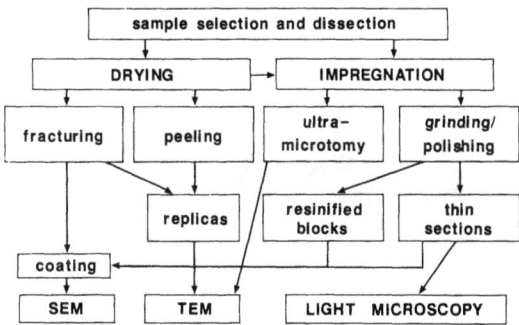

Fig. 2. Flow chart showing methods of sample preparation for the most common microscopic studies. SEM = Scanning Electron Microscopy; TEM = Transmission Electron Microscopy.

well documented in the literature (e.g., Jongerius and Heintzberger, 1975; FitzPatrick, 1984; Murphy, 1986). They mainly concern the standard techniques to produce thin sections for light microscopy. Preparation techniques for SEM studies were described by Pusch (1973), Whalley (1979), Bennett et al. (1990) and in the December 1980 issue of the Journal of Microscopy. An overview of preparation techniques for soil microstructure and soil biota interrelationships was given by Kooistra (1991).

Fig. 2 summarizes the methods available for preparing samples for microscopic studies, including the preparation of thin sections. It is important to choose a method which minimizes damage to the structure. The standard technique of air-drying causes shrinkage of wet inorganic and organic soil materials and, therefore, other methods are required. The method chosen will depend on the *in-situ* water content of the soil, on its swelling potential and whether the sample needs to be impregnated or not. When samples need to be impregnated, the whole preparation takes about three months. However, samples selected for SEM or TEM studies can be dried and analysed within a few days.

For submicroscopic studies of saturated soils which show limited swelling and need not to be impregnated, the best drying method is critical point drying, with freeze-drying as a good second choice. For partially saturated soils, and those showing significant swelling, careful freeze-drying should be practiced. However, dry soils (at water contents below the shrinkage limit) may be successfully air-dried. For wet samples to be impregnated, two methods are used: (1) acetone replacement (Miedema et al., 1974); (2) freeze-drying, usually for high-swelling soils (Jongerius and Heintzberger, 1975). Unimpregnated, dried samples for SEM studies require a clean surface, which is achieved by fracturing or peeling (Smart and Tovey, 1982). Thin sections (25-30 μm thick) are prepared from resinified blocks for light microscopy, while polished resinified blocks as well as thin sections may be observed in the back-scattered electron (BSE) mode of SEM. All samples for observation by SEM must be adequately coated to

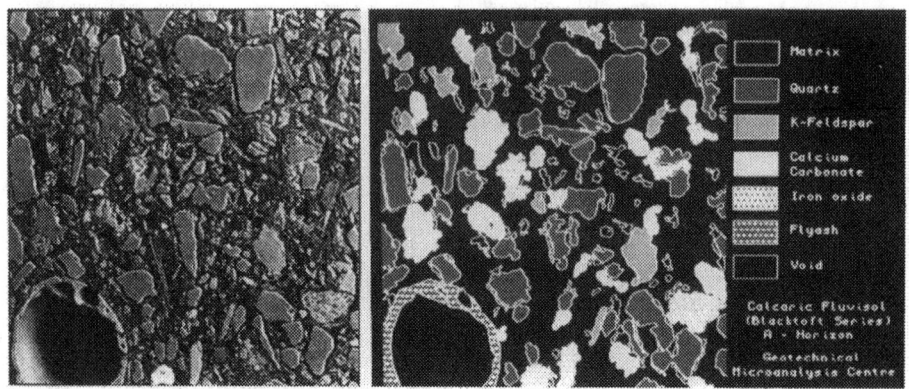

Fig. 3. (left). Back-scattered image of the tilled layer of a sandy loam soil (magnification 200x).

Fig. 4. (right). Example of mineral mapping of the image in Fig. 3.

render them conducting. In recent years, the resolution in the BSE mode has improved dramatically. Contrast arises from small chemical differences in the material present (atomic number contrast), with areas having high atomic numbers appearing brighter. An example of a back-scattered image is given in Fig. 3. For all observations of the microstructure it is important to recognize the difference between resolution and magnification, particularly when image analysis is used (Smart and Tovey, 1982).

Quantification

The reliability of quantitative analyses of images depends on the quality of the images and how well the image analyzer can define an image without degrading it. Methods to reconstruct the image without degradation are now becoming more widespread, such as Wiener filtering (Gonzalez and Wintz, 1987).

Quantitative measurements of porosity have been attempted using both visible light microscopy and back-scattered electron microscopy. A basic requirement for such quantifications is that the image is correctly segmented into the two components representing the solid features and the voids. Several methods are mentioned in the literature, of which those developed by Jongerius et al. (1972) and Ismail (1975), with some later modifications, give best results for most soils (Kooistra, 1991). Voids with diameters $< 30 \, \mu$m can only be examined accurately on images produced by SEM when using the BSE mode (Jongerius and Bisdom, 1981).

Most image analysers have facilities for evaluating the size, shape, orientation, etc., of particles within an image. The distribution of individual elements across the area covered by an imag⌐ can be mapped using the X-ray mode of operation. Several different elemental maps may be stacked with the original BSE image. Many common minerals of silt or sand size can be identified by their chemical composition, e.g., quartz, feldspar, pyrite, etc. Standard classification techniques used in remote sensing can be used to delineate areas of different minerals (Tovey et al., 1992a). Fig. 4 illustrates the application of this technique on the image in Fig. 3.

The fine-grained particles within the groundmass can only be adequately resolved using SEM in the BSE mode with a resolution of 0.1 μm or better. Specific techniques, such as the one developed from the intensity gradient method (Tovey, 1980; Smart and Tovey, 1988; Tovey et al., 1989), are particularly promising in that an index of anisotropy and the direction of preferred orientation are readily computed. The orientation values can be aggregated into rosette diagrams, while areas with preferred orientation of fine particles can be delineated. A newly developed technique (Tovey et al., 1992b) permits automatic delineation of such regions or domains (domain segmentation). Fig. 5 presents the complete analysis of the soil material shown in Fig. 3. The shaded areas show the direction of predominant orientation as determined by domain segmentation, and it is possible to study domains which form pronounced coatings around the larger mineral grains.

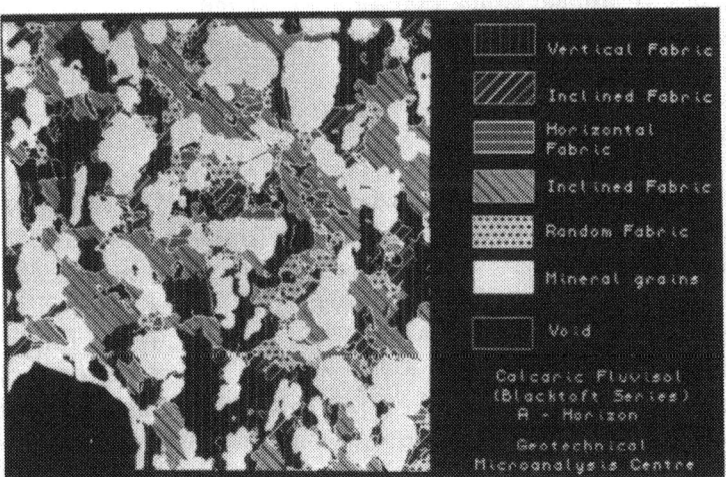

Fig. 5. Example of domain segmentation (shaded areas) combined with mineral segmentation of the image in Fig. 3.

CHANGES IN MICROPOROSITY

The first publications on microscopic studies of soil structure which focused on agriculture, appeared in the second half of the 1950s (e.g., Altemüller, 1957). Afterwards, an increasing number of publications followed. These publications concerned the effects of different kinds of tillage implements, field traffic, different management systems, changes in land use, and the application of fertilizers, manures and slurries. The introduction of image analysis for microscopic research (Jongerius et al., 1972) led to a more accurate characterization and quantification of the microporosity. However, the quantification of void systems does not in itself add information on the function and properties of soils. Therefore, soil microstructure research was soon linked with soil physical analysis (Bouma et al., 1977; Kooistra et al., 1985; Norton and Schreuder, 1987). In most of these studies, continuous voids, determining primarily the permeability and matric potential, were quantified separately and correlated with hydraulic properties. In the last decade, environmental aspects of agriculture have become important, with added interest in the biotic impact on microstructure (Jaillard and Callot, 1987; Kooistra et al., 1989a; Van Noordwijk et al., 1992). The results of microstructure studies, linked with soil physical data and/or biological analyses, could be translated into land qualities, such as the number of workable days and water deficit (Van Lanen et al., 1987), which can be applied in recommendations for specific kinds of integrated arable farming (Brussaard et al., 1988) and used as input data in models of C and N cycling in soils (De Willigen and Van Noordwijk, 1987).

The effects of cultivation practices on microstructure may be found in the cultivated layer as well as in the zone beneath it. These effects involve: (1) stability and dynamics of tillage voids (created by tillage practices); (2) compaction as a result of traffic and machinery use; (3) disintegration of soil coherence; slaking.

Tillage voids, their stability and dynamics

Soil tillage results in loosening of the soil, whereby aggregates or clods are formed, separated by rough-walled, irregular tillage voids (open spaces between aggregates, clods and soil particles). These voids are not stable as the soil becomes denser by natural settlement, other cultivation activities and traffic. The aggregates merge, the continuity of tillage voids is phased out and finally only isolated irregular voids remain. After the harvest, the more or less compacted soil material with mainly isolated voids is cultivated again to loosen the soil for a new cycle. Tillage voids may become modified after formation. In soils with a high clay content, shrinkage cracks are formed, starting from the tillage voids. In all soils, effects of soil organisms, plant roots and soil meso- and macro-fauna can be found. Plant roots predominantly follow the irregular voids between the aggregates and their impact on soil structure is low as long as the soil material has not strongly settled. With increasing bulk density their impact increases. The meso-fauna enlarges the necks between voids or locally ingests fine-grained soil material. Larger soil fauna produces channel-like tracks on the surface of aggregates or produces channel systems through aggregates, often to or from the soil surface.

In Fig. 6 an example is given of a quantification of voids in the same sandy loam soil under conventional farming and ley farming with reduced tillage.

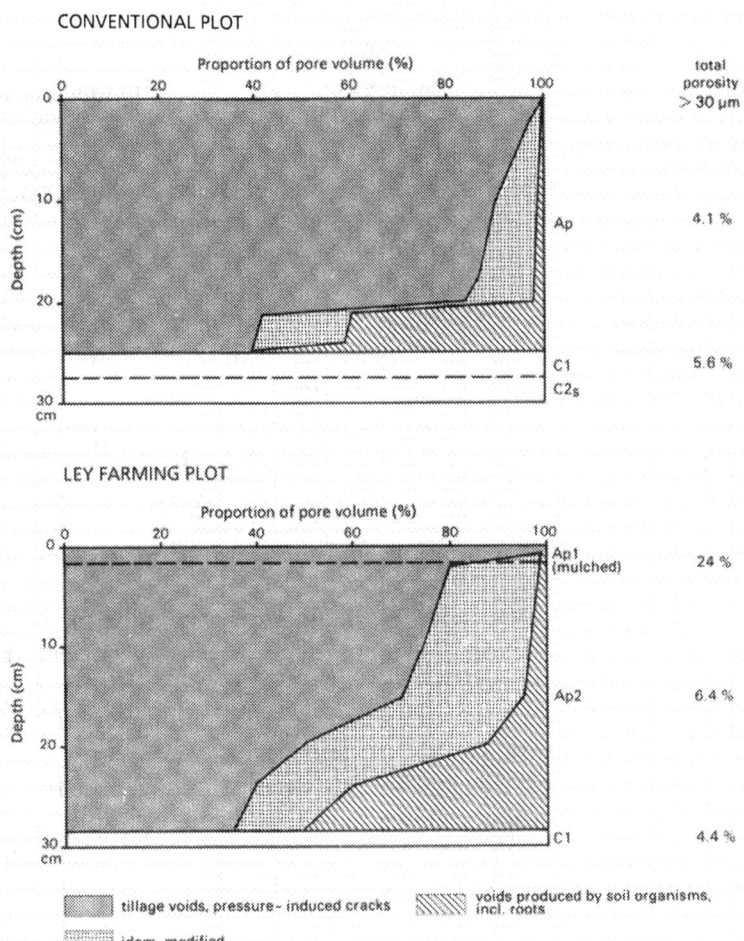

Fig. 6. Analysis and quantification of the microporosity in the upper 30 cm of the soil profile in plots under conventional farming (top) and under ley farming with reduced tillage (bottom) on the same sandy loam soil, obtained by visible light microscopy and image analysis (after Kooistra et al., 1989a).

Tillage voids constitute about 90% of the voids present in the conventional system and 80% in the ley farming system. In the conventional system, primary biological voids constitute <2% (v/v), which are mainly faunal channels. In the ley farming system, primary biological voids occupy twice this volume, which also are mainly faunal channels. In this system, the group of biologically modified tillage voids has increased considerably compared to the conventional system. In both cases, roots as well as soil fauna are the modifying agents (Kooistra et al., 1989a). The impact of soil organisms on soil structure formation may seem small, but effects on soil physical properties can be large as they produce continuous voids.

Compaction as a result of traffic and machinery

Over the last decades, trends in agricultural engineering have resulted in machines of increased size and weight, often with more tyres. In soils, wheel loads lead to compaction, deformation of the groundmass and rut formation. During the passage of wheeled machines, existing voids are reduced in size and changed in shape, and may become deformed, disrupted, closed or disappear completely. During compaction, at first large voids are reduced in size. A reduction of the macroporosity (diameter of the voids $> 100 \, \mu m$) by 3% (v/v) or more, is common (Kooistra, 1987). Total porosity is also reduced, but often less than the macroporosity, due to increased microporosity (diameter of the voids $< 100 \, \mu m$).

Fig. 7 illlustrates the difference in porosity of an undisturbed, uncompacted subsoil and the massive plough pan above. In this case, the macroporosity $> 100 \, \mu m$ and the the microporosity $< 100 \, \mu m$ in the plough pan are lower than in the undisturbed subsoil by 2.0 and 0.5 % (v/v), respectively (Jager et al., 1983). Due to the changes in porosity during compaction, void continuity is strongly affected. This factor determines the actual permeability and aeration, and the transport processes in the soil (Kooistra, 1987).

After passage of a loaded wheel, rather small horizontal cracks can be formed just below the rut as a de-loading effect on the soil particles, which were oriented by the external force. Since during ploughing one set of wheels drives over the furrow bottom, these cracks also occur below the tilled zone and often give a platy appearance to the plough pan. Słowińska-Jurkiewicz and Domżał (1991) observed that where these cracks occurred, most tillage voids had disappeared. Traffic from wheeled vehicles accompanies many operations other than tillage, e.g., spraying, slurry spreading and harvesting. Therefore, these horizontal, pressure-induced cracks are common phenomena in arable fields. In the example of Fig. 6, pressure-induced cracks constitute >50% of the voids present in the upper 4 cm of the profile in both farming systems.

The compacted zones with horizontal cracks near the surface can easily be loosened by shallow working cultivation implements. Such zones below the

Fig. 7. Back-scattered image of a sandy loam. Soil material: greys and white; voids: black. Left: undisturbed subsoil; right: massive plough pan. In the right-hand image the black voids are smaller and the total void space is less than in the left-hand image.

cultivated layer need to be loosened when they severely hamper aeration and plant rooting. This is often the case in soils with little natural cracking, generally containing < 15% (w/w) clay. However, it has been observed that, after loosening, aeration and root development were improved for a few years only. After 3 or more years, often an even worse situation was found when the soil management system and machinery use were not suitable (Kooistra et al., 1985; Kooistra, 1987).

Fig. 8 shows the different soil physical properties of the same subsoil zone under varying agricultural histories. Permanent pasture land had better physical properties than did the arable land. The microstructure in the subsoil of the never-compacted arable land forms the reference for the undisturbed plough pan and for the loosened and recompacted plough pan. The temporary character of the improvement by subsoil loosening is evident. Microscopic studies showed that in permanent pasture land soil fauna produced many continuous channel systems, resulting in satisfactory porosity and other soil physical properties (Fig. 9A). In the arable land these faunal channels were missing and vertical root channels were responsible for the hydraulic conductivity. In the undisturbed plough pan these channels were severely compacted, which resulted in reduction in size and number of root channels, but still some conductivity remained. The voids present in the

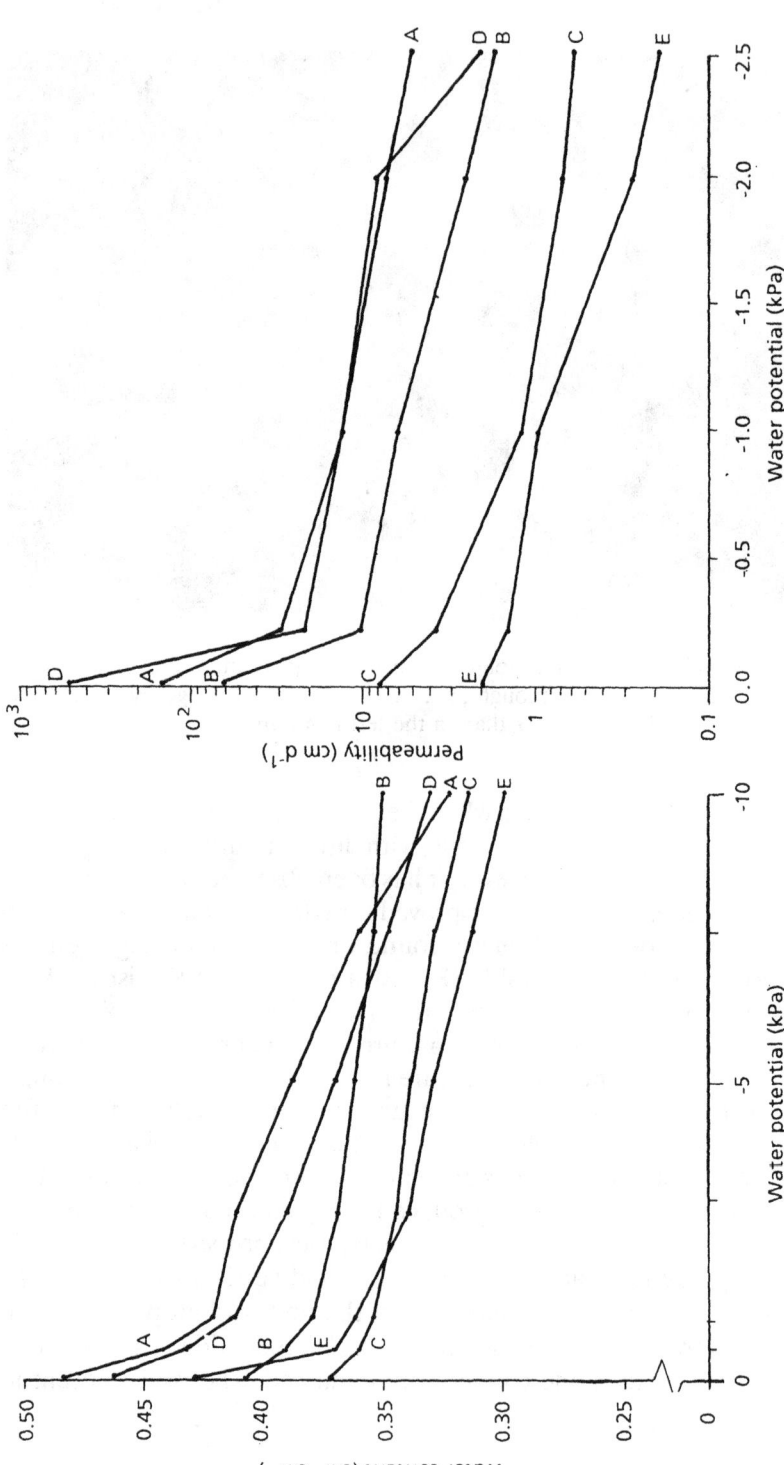

Fig. 8. Water retention curves (left) and permeability (= hydraulic conductivity) curves (right) of the zone between 25 and 45 cm below the soil surface. A = permanent pasture land; B = never-compacted arable land; C = undisturbed plough pan; D = loosened subsoil after 1 year; E = loosened subsoil after 3 years (loosened and recompacted plough pan) (from Kooistra et al., 1989b).

loosened and recompacted zone were mainly isolated tillage voids with hardly any conductivity (Fig. 9B,C). These microscopic research data could be extrapolated and used in simulation models for improvement of land qualities (Van Lanen et al., 1987).

Subsoil compaction, when severely hampering aeration and root development, can best be treated by shallow subsoiling with plough-mounted subsoilers every few years, if conditions are favourable. This procedure ensures that before serious recompaction occurs, enough vertical root channels will have been formed (Kooistra, 1987). Results of computer simulation indicate that all changes in soil management which lead to a higher degree of biological activity, result in improved soil physical properties for growing crops.

Disintegration of soil coherence; slaking

When a soil is brought under cultivation, the surface soil is repeatedly tilled and the land lies bare during large parts of the year. Soil structure is not in equilibrium with natural physical forces and wind and rain have a larger impact than in vegetation-covered, stabilized areas. At the surface, larger aggregates may disintegrate, often accompanied by separation of particles into their size categories ("shifting" or sorting of soil material). When no transport of particles occurs, the disintegrated soil material forms compact crusts at the surface. However, very often transport takes place, either along the surface as soil erosion or downward into the soil as internal slaking. Microscopic studies revealed that there are many types of erosional and depositional crusts, occurring in all climates (Courty, 1986; Kooistra and Siderius, 1986; Valentin and Ruiz Figueroa, 1987). Such crusts all cause restricted infiltration and limited germination due to their sorted, often laminated composition, with hardly any continuous voids.

In arable land, internal slaking occurs on a large scale. Consequently, in the international classification "Soil Taxonomy" (Soil Survey Staff, 1975), a special diagnostic horizon has been described: the agric horizon. It is defined as an illuvial horizon, formed under cultivation, that contains considerable amounts of illuvial silt, clay and humus. The illuviated material, derived from the surface, is deposited in continuous voids as coatings (Fig. 10) or as complete fills, accumulating in the B-horizon. Internal slaking results in a decrease of total porosity and reduction of continuous voids, which are essential for maintenance of satisfactory water and air regimes and transport processes. Moreover, the most fertile soil material is illuviated and lost for crop growth. This process occurs in all climates. In the two farming systems mentioned in Fig. 6, about 5% of the finer-grained soil material illuviates each year from the surface into the soil. In this case, most of it is caught in the cultivated layer, but soil physical processes and biological activity are adversely affected (Kooistra et al., 1989a). To reduce slaking, the fields should be covered as long as possible by, for example, the growing of green manure crops and intercropping.

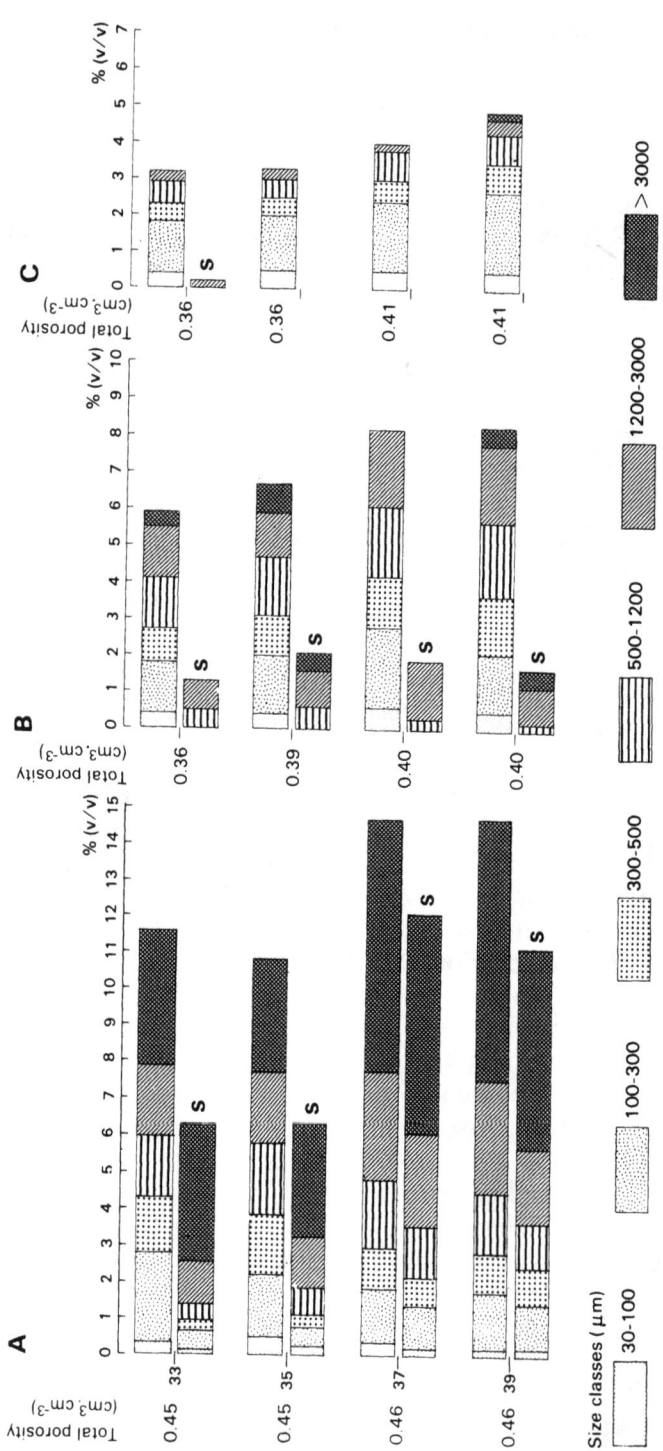

Fig. 9. Total porosity and pore-size distribution by image analysis of horizontal thin sections of soil columns stained with methylene-blue. A = permanent pasture land; B = arable land with undisturbed plough pan; C = arable land with a loosened and recompacted plough pan, three years after deep tillage; S = proportion of conductive voids (stained walls). The sampling depth (33, 35, 37, 39 cm) is indicated to the left of the bars in part A (from Kooistra et al., 1989b).

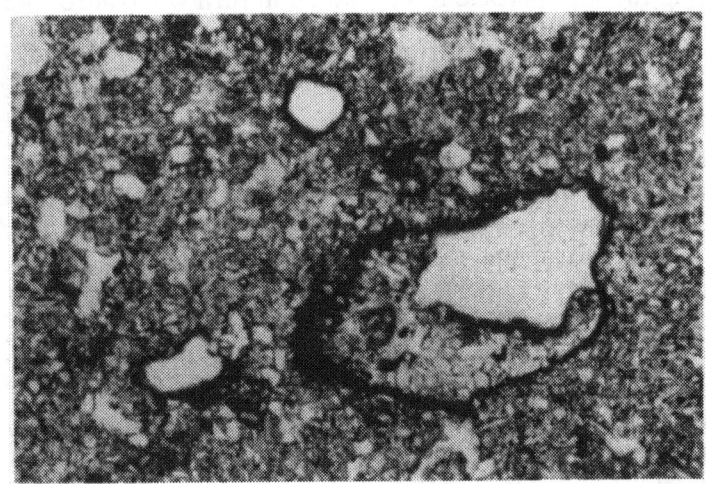

Fig. 10. Example of illuviated coatings of silt- and clay-sized material in continuous root and faunal channels below the cultivated layer (Ganges Plain, India) (from Kooistra, 1982).

CHANGES IN THE GROUNDMASS

Soil tends to be in equilibrium or near-equilibrium with the natural forces exerted upon it. However, since both the nature and magnitude of these forces change with time, this equilibrium can be disturbed, causing the soil to deform slowly. If the changes in external forces occur rapidly, as by mechanical tillage, massive deformation can take place, and this may be associated with significant changes in particle alignment, particle aggregation and, in extreme cases, particle breakage. The three main physical factors which influence the compaction of the groundmass are water regime, loading and shear.

Effects of water

In most instances, a soil just saturated with water will be in equilibrium with the overburden, the interparticle forces, and the pore water potential. The ability to resist deformation depends on the shear resistance of the material. In the absence of other effects, such as machinery loading, tillage, etc., the magnitude of the pore water potential is of considerable importance. During a dry period, decreasing pore water potential will increase the interparticle forces and the ability of the soil to resist deformation. Conversely, a build-up of excess pore water can reduce the interparticle forces to the extent that the equilibrium is upset and the soil deforms. This process is evident from tillage aggregates, which become solid and hard when dry, and soft and weak when wet. However, during this process the water will migrate, thereby changing the interparticle forces, which in turn will control the subsequent rate of deformation of the fabric.

Effects of loading

When a saturated soil is subjected to a load, such as from farm machinery, the load will initially be absorbed by increases in the pore water potential and water will migrate upwards (and downwards if there are permeable strata below). This leads to a collapse of the microstructure as a greater proportion of the load is absorbed by the particles themselves. Orderly collapse of microstructure occurs if the initial porosity and permeability are relatively low. In all other cases, loading is likely to result in rapid collapse, during which the fine clay particles will flow and be squeezed between larger sand and silt particles, causing the formation of orientated coatings around these grains. An example of this flow was noted by Tovey (1990) in marine soils from Hong Kong. Compaction of soil also occurs if the water table is considerably lowered by drainage. Soil which was previously below the water table no longer experiences buoyancy effects and will compact by collapse of the microstructure. In some areas, such as young polders, this kind of compaction may continue for several hundred years.

Effects of shear

If, during loading, the soil is also subjected to shear stresses, as may be the case on a slope or during tillage, increased pore water potential may cause shear failure. Such failure is characterized by a general realignment of particles in a direction approximately parallel to the induced deformation.

The rate at which these deformation processes occur is controlled to a large extent by the rate at which water flows through the soil, i.e., its permeability, and this in turn depends on the size, nature, orientation and packing of the particles. Since most clay particles are platy or rod-like, then, if the particles are highly aligned, the resistance to shear deformation parallel to that alignment will be less than that at right angles to that direction. Similarly, significant differences in permeability can be expected into the two directions.

Microstructure studies of soils under load are now providing considerable insight into the soil deformation process, although much more work is needed before we will fully understand all processes involved.

SOIL MICROSTRUCTURE-SOIL LOADING RELATIONSHIPS

Microstructure of clays in unloaded samples

Many hypotheses about the deformation of soil microstructure, and clay micro-

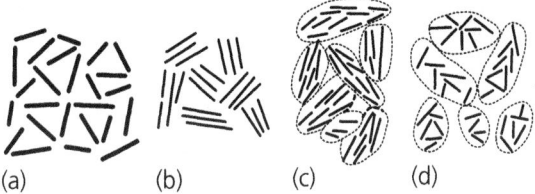

(a) (b) (c) (d)

Fig. 11. Different microstructures of clay-sized particles. (a) card house; (b) book house; (c) domains; (d) micro-peds (from Kooistra et al., 1989b).

structure in particular, have been advanced. In some hypotheses, it has been suggested that individual plate-like particles form an open network or card house, which progressively collapses as load is applied. Others have suggested a variation in the form of a book house structure, in which the individual components comprise packets of nearly-parallel, equi-dimensional particles which behave as integral units during deformation. These packets often have an overall shape similar to an approximate rectangle or parallelogram. Yet another hypothesis considers domains of sub-parallel, irregularly-sized particles as a dominant unit.

These domains themselves are of irregular size and shape, and may grow or deform during deformation. Finally, another hypothesis suggests that the basic associations do not have to consist of sub-parallel particles, but that micro-peds would form a series of basic units which remain largely intact during deformation, although once again their shape and/or overall size may vary. Fig. 11 illustrates the various hypotheses.

Changes in microstructure during compaction

There is some evidence to suggest that during compaction the particles become aligned in a direction perpendicular to the applied load. Initially, the large voids will tend to collapse, causing a reduction in volume. Subsequently, the particles may slide over each other into a more dense arrangement. However, although a few micrographs have shown such effects, recent quantitative studies have indicated that alignment becomes significant only at relatively high stress levels (>1 MPa). At present there is no clear evidence to support any of the above hypotheses. Current work by Tovey et al. should provide some information within the near future. Fig. 12 shows that the index of anisotropy (a measure of alignment ranging from 0 for a random microstructure to 1.0 for one which is perfectly aligned) consistently increases as the stress level rises. The plotted points represent the mean values obtained from approximately 60 separate images of soil microstructure at each stress level (Tovey and Martinez, 1991).

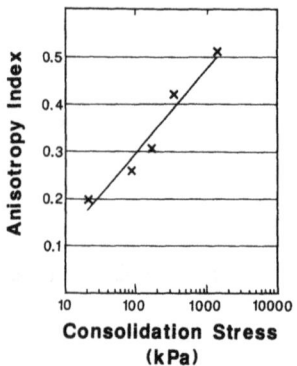
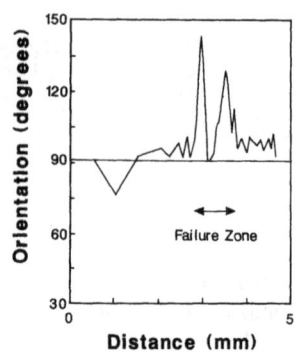

Fig. 12. (left). Relationship between the increase in stress level and the alignment of clay-sized particles (from Tovey and Martinez, 1991).

Fig. 13. (right). Relationship between the degree of orientation of clay-sized particles in and near failure zones due to shearing (from Tovey and Martinez, 1991).

Changes in microstructure during shearing

Mechanical deformation, such as by tillage, is characterized by the development of shear failure zones. Fig. 13 shows how the orientation of particles varies as such a zone is traversed. The original near-horizontal orientation (90°) is replaced by alignments at 135° to the upward vertical in precisely the expected direction. At the same time the index of anisotropy drops at the edge of the zone before regaining a high value within the zone itself. It would appear that the initial high degree of alignment becomes more random initially along the incipient failure zone, and then shows increased alignment again as failure is reached.

CONCLUSIONS

(1) Microscopic studies of the microstructure reveal how and when soils are compacted under different external mechanical forces. The changes observed can be quantified and related, either to the exerted forces, or to the properties of the soil materials concerned. The insights obtained also explain why soil properties and behaviour differ before and after subjection to external forces, such as loading and shear stress.

(2) Total porosity, bulk density or soil strength in themselves do not determine properties and behaviour of soils. To explain the processes involved, knowledge concerning the kind of voids present and their interconnections, and the alignment of soil particles, including clay, is needed.

(3) It is necessary to gain an understanding of the processes of soil modification under applied loads before data can be extrapolated and used in simulation models.

In the past, because of lack of necessary quantitative methods, it was not possible to test adequately the theories concerning the ways in which soil structure behaves when it is subjected to external forces. However, in view of the recent developments described in this chapter, many of these aspects should be solved within the next few years.

ACKNOWLEDGEMENTS

The authors wish to thank their colleagues, who over the years have given many constructive comments and advice on the techniques described in this chapter, especially Dr. M.W. Hounslow (Univ. of East Anglia, School of Env. Sc., Norwich, U.K.), Dr. P. Smart (Univ. of Glasgow, Dept. of Civil Eng., Glasgow, U.K.) and Dr. E.B.A. Bisdom, Mr. O.H. Boersma and Mr. D. Schoonderbeek (The Winand Staring Centre for Integrated Land, Soil and Water Research, Wageningen, Netherlands).

REFERENCES

Altemüller, H.-J., 1957. Bodentypen aus Löss in Raume Braunschweig und ihre Veränderungen unter dem Einfluss des Ackerbaues. (Soil types originated from loess in the Braunschweig region and their changes under the influence of arable farming). Ph.D. thesis, Univ. Bonn, Germany, 250 pp. (in German).

Bennett, R.H., Bryant, W.R. and Hulbert, M.H. (Editors), 1990. Microstructure of Fine-Grained Sediments. Springer, New York, U.S.A., 582 pp.

Bisdom, E.B.A., Tessier, D. and Schoute, J.F.Th., 1990. Micromorphological techniques in research and teaching (submicroscopy). In: L.A. Douglas (Editor), Soil Micromorphology. Elsevier, Amsterdam, Netherlands, Developments in Soil Science 19, pp. 581-605.

Bouma, J., Jongerius, A., Boersma, O.H., Jager, A. and Schoonderbeek, D., 1977. The function of different types of macropores during saturated flow through four swelling soil horizons. Soil Sci. Soc. Am. J., 41: 945-950.

Brewer, R., 1964. Fabric and Mineral Analysis of Soils. Wiley, New York, U.S.A., 470 pp.

Brussaard, L., Kooistra, M.J., Lebbink, G. and Van Veen, J.A., 1988. The Dutch Programme on Soil Ecology of Arable Farming Systems I. Objectives, approach and preliminary results. In: H. Eijsackers and A. Quispel (Editors), Ecological Implications of Contempory Agriculture. Ecol. Bull., 39: 35-40.

Bullock, P., Fedoroff, N., Jongerius, A., Stoops, G. and Tursina, T., 1985. Handbook for Soil Thin Section Description. Waine Research Publ., Wolverhampton, U.K., 152 pp.

Courty, M.A., 1986. Morphology and genesis of soil surface crusts in semi-arid conditions (Hissar region, Northwest India). In: F. Callebaut, D. Gabriels and M. De Boodt (Editors), Assessment of Soil Surface Sealing and Crusting. State Univ. Ghent, Belgium, pp. 32-40.

De Willigen. P. and Van Noordwijk, M., 1987. Roots, plant production and nutrient use efficiency. Ph. D. thesis, Agric. Univ. Wageningen, Netherlands, 282 pp.

Dexter, A.R., 1988. Advances in characterization of soil structure. Soil Tillage Res., 11: 199-239.

FitzPatrick, E.A., 1984. Micromorphology of Soils. Chapman and Hall, London, U.K., 433 pp.

Gonzalez, R.C. and Wintz, P., 1987. Digital Image Processing. Addison-Wesley Publ. Comp., Reading, MS, U.S.A., 503 pp.

Ismail, S.N.A., 1975. Micromorphometric soil-porosity characterization by means of electro-optical image analysis (Quantimet 720). Stiboka, Wageningen, Netherlands, Soil Survey Papers 9, 104 pp.

Jager, A., Boersma, O. and Bisdom, E.B.A., 1983. The characterization of microporosity in a ploughpan by submicroscopic and Quantimet techniques. In: E.B.A. Bisdom, and J. Ducloux (Editors), Submicroscopic Studies of Soils. Elsevier, Amsterdam, Netherlands, Developments in Soil Science 12, pp. 227-285.

Jaillard, B. and Callot, G., 1987. Action des racines sur la ségrégation minéralogique des constituants du sol. (Effect of roots on the mineral disintegration of soil constituents). In: N. Fedoroff, L.M. Bresson and M.A. Courty (Editors), Soil Micromorphology. AFES, Plaisir, France, pp. 371-376 (in French).

Jastrow, J.D. and Miller, R.M., 1991. Methods for assessing the effects of biota on soil structure. Agric. Ecosystems Environm., 34: 279-303.

Jongerius, A. and Heintzberger, G., 1975. Methods in soil micromorphology; a technique for the preparation of large thin sections. Netherlands Soil Survey Inst., Wageningen, Netherlands, Soil Survey Papers 10, 48 pp.

Jongerius, A. and Bisdom, E.B.A., 1981. Porosity measurements using the Quantimet 720 on backscattered electron images of thin sections of soil. In: E.B.A. Bisdom (Editor), Submicroscopy of Soils and Weathered Rocks. Pudoc, Wageningen, Netherlands, pp. 207-216.

Jongerius, A., Schoonderbeek, D. and Jager, A., 1972. The application of the Quantimet 720 in soil micromorphometry. Microscope, 20: 243-254.

Kooistra, M.J., 1982. Micromorphological Analysis and Characterization of 70 Benchmark Soils of India. Netherlands Soil Survey Institute, Wageningen, Netherlands, 778 pp.

Kooistra, M.J., 1987. The effects of compaction and deep tillage on soil structure in a Dutch sandy loam soil. In: N. Fedoroff, L.M. Bresson and M.A. Courty (Editors), Soil Micromorphology, AFES, Plaisir, France, pp. 445-450.

Kooistra, M.J., 1990. The future of soil micromorphology. In: L.A. Douglas (Editor), Soil Micromorphology. Elsevier, Amsterdam, Netherlands, Developments in Soil Science 19, pp. 1-8.

Kooistra, 1991. A micromorphological approach to the interactions between soil structure and soil biota. Agric. Ecosystems Environm., 34: 315-328.

Kooistra, M.J. and Siderius, W., 1986. Micromorphological aspects of crust formation in Savanna climate under rainfed subsistence agriculture. In: F. Callebaut, D. Gabriels and M. de Boodt (Editors), Assessment of Soil Surface Sealing and Crusting. State Univ. Ghent, Belgium, pp. 9-17.

Kooistra, M.J., Bouma, J., Boersma, O.H. and Jager, A., 1985. Soil-structure differences and associated physical properties of some loamy Typic Fluvaquents in The Netherlands. Geoderma, 36: 215-228.

Kooistra, M.J., Lebbink, G. and Brussaard, L., 1989a. The Dutch Programme on Soil Ecology of Arable Farming Systems II. Geogenesis, agricultural history, field site characteristics and present farming systems at the Lovinkhoeve experimental farm. Agric. Ecosystems Environm., 27: 361-387.

Kooistra, M.J., Tovey, N.K. and Pagliai, M., 1989b. Agricultural Applications. In: M.J. Kooistra (Editor), European Training Course on Soil Micromorphology II. Applications to Soil Micromorphology. Agric. Univ., Wageningen, Netherlands, pp. 1-48.

McKeague, J.A. and Wang, C., 1982. Soil structure: concepts, description and interpretation. Land Res. Res. Inst., Ottawa, Ont., Canada, LRRI-Contribution 82-15, 52 pp.

Miedema, R., Pape, T. and Van der Waal, G.J., 1974. A method to impregnate wet soil samples, producing high quality thin sections. Neth. J. Agric. Sci., 22: 37-39.

Murphy, C.P., 1986. Thin section preparation of soils and sediments. A.B. Academic Publishers, Berkhamsted, U.K., 149 pp.

Norton, L.D. and Schreuder, S.L., 1987. The effects of various cultivation methods on soil loss: a micromorphological approach. In: N. Fedoroff, L.M. Bresson and M.A. Courty (Editors), Soil Micromorphology. AFES, Plaisir, France, pp. 431-437.

Pusch, R. (Editor), 1973. Proceedings of the International Symposium on Soil Structure. Swedish Geotech. Soc., Stockholm, Sweden, 251 pp.

Słowińska-Jurkiewicz, A. and Domżał, H., 1991. The structure of the cultivated horizon of soil compacted by the wheels of agricultural tractors. Soil Tillage Res., 19: 215-226.

Smart, P. and Tovey, N.K., 1982. Electron Microscopy of Soils and Sediments: Techniques. Oxford University Press, Oxford, U.K., 264 pp.

Smart, P. and Tovey, N.K., 1988. Theoretical aspects of intensity gradient analysis. Scanning, 10: 115-121.

Soil Survey Staff, 1975. Soil Taxonomy. A Basic System of Soil Classification for Making and Interpreting Soil Surveys. US Government Printing Office, Washington, DC, U.S.A., 754 pp.

Tovey, N.K., 1980. A digital computer technique for orientation analysis of micrographs of soil fabric. J. Microscopy, 120: 303-315.

Tovey, N.K., 1990. The microfabric of some Hong Kong marine soils. In: R.H. Bennett, W.R. Bryant and M.H. Hulbert (Editors), Microstructure of Fine-grained Sediments. Springer, New York, U.S.A., pp. 519-530.

Tovey, N.K. and Martinez, M.D., 1991. A comparison of different intensity gradient formulae for orientation analysis of electron micrographs. Scanning, 13: 289-298.

Tovey, N.K., Smart, P., Hounslow, M.W. and Leng, X.L., 1989. Practical aspects of automatic orientation analysis. Scanning Microscopy, 3: 771-784.

Tovey, N.K., Krinsley, D.H., Dent, D.L. and Corbett, W.C., 1992a. Techniques to quantitatively study the microfabric of soils. Geoderma, 53: 217-237.

Tovey, N.K., Smart, P., Hounslow, M.W. and Leng, X.L., 1992b. Automatic domain mapping of certain types of soil fabric. Geoderma, 53: 179-201.

Valentin, C. and Ruiz Figueroa, J.F., 1987. Effect of kinetic energy and water application rate on the development of crusts in a fine sandy loam soil using sprinkling irrigation and rainfall simulation. In: N. Fedoroff, L.M. Bresson and M.A. Courty (Editors), Soil Micromorphology. AFES, Plaisir, France, pp. 401-409.

Van Lanen, H.A.J., Bannink, M.H. and Bouma, J., 1987. Use of simulation to assess the effects of different tillage practices on land qualities of a sandy loam soil. Soil Tillage Res., 10: 347-361.

Van Noordwijk, M., Kooistra, M.J., Boone, F.R., Veen, B.W. and Schoonderbeek, D., 1992. Root-soil contact of maize, as measured by thin section technique. I. Validity of the method. Plant Soil, 139: 109-118.

Weibel, E.R., 1979. Stereological Methods, Vol. 1. Academic Press, London, U.K., 334 pp.

Whalley, W.B. (Editor), 1979. Scanning Electron Microscopy in the Study of Sediments. Geoabstracts, Norwich, U.K., 414 pp.

Soil Compaction in Crop Production
B.D. Soane and C. van Ouwerkerk (Eds.)
113

CHAPTER 6

Determination and Use of Soil Bulk Density in Relation to Soil Compaction

D.J. CAMPBELL

Scottish Centre of Agricultural Engineering, SAC, Penicuik, U.K.

SUMMARY

Bulk density and some related terms which quantify soil compactness are defined and discussed. Methods for measuring bulk density are considered in two groups. Direct methods involve the measurement of sample mass and volume, while indirect methods are based on the effect of soil on nuclear radiation. Methods outlined and discussed include core sampling, frame sampling, sand replacement, rubber balloon and clod direct methods and both backscatter and transmission radiation methods. Methods are compared in relation to the selection of a method for a particular purpose. Consideration is given to statistical problems arising from equipment calibration, the number of test samples, and the presence of stones, and to problems associated with the use of gamma radiation. Attention is given to the alternative ways in which test results can be presented. In conclusion, some unresolved problems of bulk density measurement are discussed.

INTRODUCTION

Quantifying changes in compactness

The process of compaction or densification of soils, which is induced by the application of mechanical loads, is accompanied by an increase of compactness. It is most important to quantify such changes accurately by measuring a soil property which is relevant both to the process of compaction and to the interpretation of the resulting soil conditions. The most fundamental and widely used property for this purpose is bulk density.

Soil bulk density or, more precisely, soil wet bulk density, is defined as the mass of soil particles plus the mass of soil water in a unit volume of soil. In order to allow soils at different water contents to be compared, soil compaction is usually described in terms of the soil dry bulk density, which is derived from the wet bulk density by subtracting the mass of water present in unit volume.

Changes in soil dry bulk density provide a quantitative evaluation of both the loosening effect of tillage implements and the compacting effect of wheels. A number of other properties, closely associated with bulk density, are frequently employed by engineers and scientists to indicate changes in compactness. These include void ratio, total porosity, specific volume and unit weight. The inter-relationships of these properties are given in Table 1.

TABLE 1

Some properties related to soil compactness and their interrelationships

Soil property	Symbol	Formula	Interrelation-ship
Particle specific gravity	G_s	$\dfrac{\text{weight of any volume of soil particles}}{\text{weight of an equal volume of water}}$	
Gravimetric water content	w	$\dfrac{\text{mass of water}}{\text{mass of solids}}$	$w = (v - 1)/G_s$ (for saturated soil only)
Volumetric water content[1]	θ	$\dfrac{\text{volume of water}}{\text{total volume}}$	$\theta = w\, \rho_d / \rho_w$
Void ratio	e	$\dfrac{\text{volume of voids}}{\text{volume of solids}}$	$e = n/(1 - n)$
Total porosity	n	$\dfrac{\text{volume of voids}}{\text{total volume}}$	$n = e/(1+e)$
Air-filled porosity	n_a	$\dfrac{\text{volume of air}}{\text{total volume}}$	$n_a = n - \theta$
Dry bulk density[1]	ρ_d	$\dfrac{\text{mass of dry soil}}{\text{total volume}}$	$\rho_d = \rho_w\, G_s/(1+e)$
Dry unit weight[2]	γ_d	$\dfrac{\text{weight of dry soil}}{\text{total volume}}$	$\gamma_d = \gamma_w\, G_s/(1+e)$
Specific volume	v	total volume of soil containing unit volume of solids	$v = 1 + e$
Degree of saturation	S	$\dfrac{\text{volume of water}}{\text{volume of voids}}$	$S = w\, G_s/e$ $S = \theta/n$

[1] ρ_w = density of water.
[2] γ_w = unit weight of water.

The choice of a property as the primary indicator of compactness from among those listed in Table 1, is dependent on the perceived advantage for particular applications. Total porosity and void ratio can be calculated from dry bulk density and have the advantages of being dimensionless and independent of particle density. Void ratio is widely used by civil engineers and is considered to be easier to handle than total porosity as the volume of the solids remains essentially constant for a given initial total volume of soil, whereas with porosity, both the volume of voids and the total volume change (Karafiath and Nowatzki, 1978).

Relative expressions of compactness

Bulk density measurements may show limited correlation with plant growth because of the significance of particle- and pore-size distribution and the presence of other soil components, such as organic matter. There have been a number of attempts to obtain more useful properties by expressing the measured bulk density, or a property derived from it, such as void ratio, as a ratio of the value of that property for the same soil when it is in a defined reference state.

Relative density is a term used by civil engineers for expressing the void ratio of soil *in situ* in relation to the void ratio of the same soil when it has been subjected to arbitrary packing techniques in the laboratory to simulate two reference states termed "densest" and "loosest". Relative density values of 0-0.15 and 0.85-1.0 are broadly classed as very loose and very dense, respectively, with three intermediate classes (Karafiath and Nowatzki, 1978). Relative density can similarly be defined in terms of bulk density as follows:

$$D_r = (\rho_d - \rho_{min})/(\rho_{max} - \rho_{min}) \tag{1}$$

where D_r = relative density; ρ_d = dry bulk density; ρ_{max} and ρ_{min} = maximum and minimum dry bulk density, respectively.

Another dimensionless ratio is the relative compaction (Anon., 1964) in which the observed bulk density is expressed as a proportion of the maximum bulk density obtained in the Proctor compaction test. This ratio has found application in describing the properties of field soils (Pidgeon and Soane, 1977). The bulk density of field soils has also been compared with another standard laboratory test, which measures the equilibrium bulk density reached when a large sample of loose wet soil (350 mm diameter, 120 mm depth) is subject to an axial pressure of 200 kPa within a confined cylinder (Eriksson et al., 1974; Håkansson, 1988). The bulk density in this reference state may be used to calculate the degree of compactness which has found application for comparing the growth of different varieties of crops (Håkansson, 1973) and for the comparison of the compacting effects of different wheel treatments over a range of soil types (Ljungars, 1977).

The compactness of field soils may be related to a reference state associated with some specified management practice. The concept of an equilibrium bulk density (Heinonen, 1977), reached in response to annual tillage, has proved useful in studying soil structure in relation to tillage (Pidgeon and Soane, 1977; Carter, 1990).

In the following sections, dealing with measurement methods and the presentation of results, the emphasis has been placed on bulk density as the primary property of interest. However, the techniques described are generally equally relevant to the study of the related properties shown in Table 1.

METHODS OF MEASUREMENT

A wide range of methods is available for the measurement of soil bulk density, but no single method is the most suitable for all circumstances. A very crude but quick method may be appropriate when all that is required is to characterise soil conditions. In contrast, when comparing the compaction produced by different vehicles or the loosening effects of different tillage implements, the use of a much slower method, which requires expensive apparatus, may be justified if the details of the treatment differences could not otherwise be detected.

Methods for measuring soil bulk density generally fall into one of two groups. In the first group, both the mass and the volume of a sample are determined. Most such methods are long-established, require relatively inexpensive apparatus and have been used to determine the bulk density of soils by both agricultural soil scientists (Freitag, 1971) and civil engineers (Anon., 1964). Basically, these direct methods differ from each other only in the way in which the sample volume is determined and it is in this respect that an assessment has to be made as to the suitability of any method for a given purpose since all the methods have some limitations. In an attempt to overcome these limitations, a second group of methods has evolved, in which the attenuation or scattering of nuclear radiation by the soil sample is used to give an indirect measurement of bulk density. This is achieved by calibrating the apparatus by means of samples of known bulk density. However, although the use of radiation methods generally results in more accurate and precise measurements of bulk density, they too have limitations, including the complexity and cost of the necessary apparatus.

Generally, it is the change in bulk density following application of an experimental treatment which is of interest and this is obtained by measuring bulk density both before and after the treatment is applied. In this context, soil variability can give rise to statistical problems when the initial and final measurements are compared and so there can be some advantage in using a method which makes both measurements on the same sample. However, the necessary access for the apparatus in such circumstances raises the question of whether the provision of access interferes with the treatment effect.

A review of methods, especially the direct methods, is given by Blake and Hartge (1986) while Campbell and Henshall (1991) describe both direct and indirect methods, explain their theoretical basis and discuss their use in traffic and tillage research.

Direct measurement methods

Core sampling

In this method, an open-ended, metal cylinder is either pressed or hammered into the soil. The cylinder is then excavated and weighed together with the soil core after the latter has been trimmed flush with the ends of the cylinder. As the volume of the cylinder is known, the bulk density can be calculated. The method has been widely used over several decades (Lutz, 1947; Jamison et al., 1950; Tessier and Steppuhn, 1990) and has the advantage that the water release characteristic may be determined for the same sample (Ball and Hunter, 1988).

Sampling is easiest in cohesive soils with water contents close to field capacity, while sands and gravels do not give satisfactory samples. The method is susceptible to error arising from compression or shattering of the core while the cylinder is inserted. Baver et al. (1972) have suggested that if the cylinder is inserted by steady pressure, compression of the core may occur, while insertion by hammering may cause shattering. Freitag (1971) refers to a survey of core sampling undertaken for civil engineering purposes by Hvorslev, who considered that sample distortion was minimised when the cylinder was pressed steadily rather than hammered into the soil. As the risk of sample compression increases with decreasing core diameter, Freitag (1971) suggested that the diameter should be selected to give an adequate sample size and that the length should be no more than three times the diameter, while Baver et al. (1972) regarded a diameter of 75-100 mm as a satisfactory compromise. Sample disturbance can also be minimised by making the cylinder wall thickness as small as possible, consistent with it remaining rigid during insertion (Anon., 1964), by relieving the inner and outer diameters of the cylinder immediately behind the cutting edge (Buchele, 1961) and by lightly greasing the inner wall of the cylinder (Veihmeyer, 1929).

Nevertheless, despite such precautions against sample disturbance, compression and shattering may occur. While compression is readily observed as any difference between the length of the core and that of the excavation, shattering may be less easily detected. Ball and Hunter (1988) found that most sample disturbance arose from the cylinder striking stones and took the form of cracking, twisting and loosening of the core. This occurred particularly at soil bulk densities >1.2 Mg m^{-3}, when stones are not readily displaced. Clearly, undetected changes in structure and hence bulk density may occur during sampling. Campbell and Hunter (1986) found marked differences in drop-cone penetration on cores and *in situ* and their term "minimally disturbed" is probably a more appropriate description of carefully sampled cores than the frequently encountered term "undisturbed".

Although conventional core sampling is unsuitable for hard, brittle soils since they may shatter during sampling, rotary core samplers have been developed to overcome the problem. As the cylinder of a rotary sampler is inserted, an auger attached to the outside excavates an annulus of soil from around the sample. Raper and Erbach (1987) found that use of such an auger gave better, less variable results than conventional core sampling. There is a risk of torsional shear of the sample if the auger is allowed to corkscrew rather than cut its way into the soil. This may be avoided by using engine power to drive the auger to ensure that the annulus of soil is removed ahead of the cutting edge of the sample cylinder (Buchele, 1961).

Erbach (1982) has briefly reviewed core samplers, including some which are tractor-mounted. While the latter may be convenient, their use may be inappropriate on experimental plots where the necessary tractor traffic is unacceptable.

Frame sampling

Where field soils are highly stratified, recently cultivated, very loose or coarse-textured, frame sampling may be more satisfactory than core sampling (Andersson and Håkansson, 1967; Håkansson, 1988). A steel sampling frame of 0.707 x 0.707 m (equivalent to 0.5 m^2) is hammered through the layer of interest (Fig. 1). Using the upper edge of the frame as a reference plane, the elevation of the soil surface is measured, either with a hand or an electrical relief meter, using approximately 200 elevation readings within the area of the frame. A layer of soil of appropriate thickness is removed using hand tools and weighed, a sample being retained for water content measurement. The elevation of the new soil surface is then measured as before and the process repeated as required.

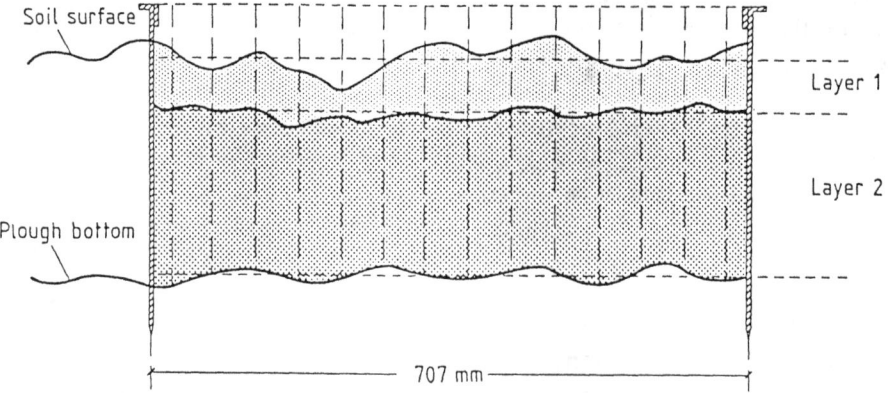

Fig. 1. Section through the steel frame used in the frame sampling method and details of samples taken from two layers within the plough layer (after Håkansson, 1988).

Discontinuities, such as the depths of primary and secondary cultivation, are useful boundary surfaces. The average thickness, bulk density, total porosity, water content and air content of each sampling layer can be calculated.

Sand replacement

In this method an approximately cylindrical sample of soil is excavated and its mass and water content determined. The sample volume is then determined with the aid of a device commonly referred to as a sand bottle (Fig. 2), which consists of a vertical metal cylinder containing sand which is placed over the excavation. A tap in the base of the cylinder is opened and sand allowed to fill the excavation. The difference in weight of the sand bottle before and after filling the excavation is recorded. The bulk density of the sand in the bottle is determined from a calibration test in which sand from the bottle is used to fill cylindrical containers of known dimensions. Thus, the volume of the excavated hole can be calculated and hence the *in-situ* bulk density of the soil. It is, of course, necessary to allow for the sand between the tap and the soil surface level and this is done by opening the tap while the bottle rests on a flat plate. The sand bottle test takes about 30 min to perform and has the advantage that it can be used on virtually all types of soil (Anon., 1964; Blake and Hartge, 1986).

There are several potential sources of error in the method but most of them can be avoided. It is crucial that the sand used is dry and particular care has to be taken if sand is recovered from the excavation at the end of a test to ensure that it is dry and uncontaminated by the test soil. Regular calibration checks are the quickest way of ensuring that such errors are avoided but if sufficient sand is available it is a good policy not to use the sand after recovery until it has been dried and also sieved since grading is also important. Typically 0.2-2.0 mm sand is used although the need for closely graded material is more important than the

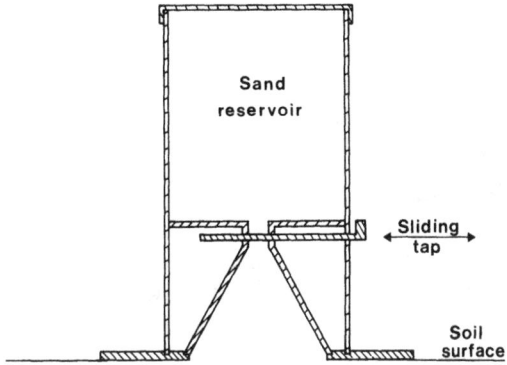

Fig. 2. Schematic section through a typical sand bottle used in the sand replacement method, showing the sliding tap in the closed position (from Campbell and Henshall, 1991).

actual size used if segregation and hence variation in sand bulk density is to be avoided. A 1-% reduction in sand bulk density is produced by a 50-mm reduction in the initial height of the sand in the bottle and it is also important that the dimensions of the calibration container should be similar to those of the excavated hole since a 25-mm decrease in the depth of the container gives a 1-% decrease in sand bulk density (Anon., 1964).

Among recent improvements to the method, Cernica (1980) employed a sand bottle which is calibrated in terms of volume, as in a measuring cylinder. In this way the difference in sand volume before and after filling the excavation corresponds to the volume of the hole and so a knowledge of the sand bulk density is not required. Further, the new method is reported to give smaller errors than the conventional method. Freitag (1971) described a variation in the method, in which oil rather than sand is used as the replacement material. The oil readily fills all irregularities in the excavation and is not sensitive to handling techniques. Disadvantages include the possible effect of temperature on oil bulk density, the need to excavate the soil from an initially horizontal surface and the necessity, for most soils, to spray the inside of the excavation with a material which will prevent penetration of the soil pores by the oil. Howard and Singer (1981) used plastic film to line the excavation and water rather than oil to fill it. They considered the method satisfactory in terms of accuracy, speed and simplicity. The need to line the excavation was avoided by Laundré (1989) who used a hard-setting foam to fill it. The volume of the cast was subsequently found by water displacement and the method gave results which were not significantly different from those obtained with the core and sand bottle methods.

Rubber balloon

A soil sample is excavated from the layer under test as for the sand replacement method. The sample mass is obtained by weighing the excavated soil and its water content is then determined. The volume of the sample is found by inserting a thin rubber balloon into the excavation and then filling the balloon with a measured volume of water. Ideally, the excavation should have a regular shape to ensure that the balloon fills it completely (Anon., 1964; Freitag, 1971; Blake and Hartge, 1987). In a modification of the basic apparatus described by Freitag (1971), the balloon is clamped to the base of a device designed to force water into the balloon from a calibrated water container. Pressures of between 21 and 48 kPa have been proposed as a suitable compromise between ensuring that the excavation is completely filled and avoiding compression of the walls of the excavation. Anon. (1964) described a further modification, in which water is forced from the device into the balloon until the device is raised 20 mm above ground level. The water supply to the balloon is then closed off and the operator stands on the base plate of the device to force the balloon into any irregularities in the walls of the excavation. Cooper and Fleming (1990) describe a version of the water-filled balloon test which they developed specifically to test samples up to 200 mm in diameter and 1 m deep in trench backfill material. Such large samples in presumably reasonably uniform soil resulted in excavated volume errors of < ± 1%.

Although the method is suitable for a range of soil types, it is generally considered to suffer from sources of error which are not readily dealt with. These include entrapped air between the balloon and the walls of the excavation and the dependence of the derived sample volume on the pressure used to insert the balloon.

Clod

Rather than make measurements on the bulk soil, it is sometimes appropriate to make measurements on individual clods. Russell and Balcerek (1944) have produced a useful review of many of such methods. In the basic method, the clod is weighed and its volume is determined by coating it in paraffin wax and immersing it in a volumenometer (Anon., 1964). Alternatively, the volume may be determined by weighing the waxed clod in air and in water. In both versions of the method, it is necessary to remove and then weigh the wax coating. Although the method can give reliable results, it is limited to cohesive soils and is inevitably rather slow since care has to be taken to ensure that the wax coats the clod but does not penetrate the pore system. This is best achieved by dipping the clod, suspended by a fine wire which passes around it, into molten paraffin wax which is maintained at a temperature just above its melting point.

Several alternatives to paraffin wax have been reported. Brasher et al. (1966) used a commercially available resin, Saran F-220. As the resin is permeable to water vapour but not to water, and as it does not melt when the coated clod is oven-dried at 105°C, it is possible to use the clods thus coated to study their drying and shrinkage behaviour. Although the resin coating is flexible, sandy soils may puncture the coating. Abrol and Palta (1968) substituted rubber solution for paraffin wax and claimed improved accuracy and convenience. A flotation technique was used by Campbell (1973) in which the clod is sprayed with a resin solution and then immersed sequentially in liquids of different specific gravity. The specific gravities of the liquids in which the clod just floated and just sank gave the range of values within which that of the clod bulk density lay (Fig. 3). As the method does not require the determination of either clod mass or clod volume, it was found to be ten times as rapid as the wax coating method, while giving bulk density values closely similar to those obtained by the wax coating method (Table 2). The method has been used to demonstrate that there is no effect of clod size on clod wet bulk density (Campbell, 1973).

It is not necessary to coat the clod if the fluid in which the clod is immersed will not penetrate the clod pore system. Various oils and mercury have been used but the technique is probably restricted to soils with very small pores and Gill (1959) used it successfully to study puddled soils. Olson and Zobeck (1989) proposed a method in which pore intrusion by mercury was controlled by reducing

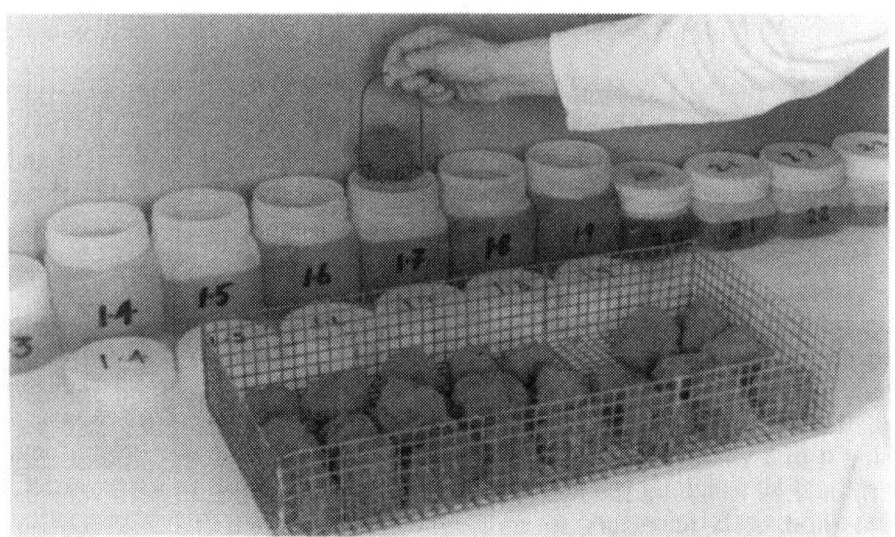

Fig. 3. Determination of the specific gravity of the liquids in which a clod just sinks and just floats in the flotation method (Campbell, 1973).

the pressure in a chamber containing both the mercury and the aggregates under test.

TABLE 2

Comparison of flotation and wax-coating methods (from Campbell, 1973)

Soil texture	Mean water content (%, w/w)	Wet bulk density (Mg m⁻³)				Significance of differences[1]	
		Flotation method		Wax method			
		Mean	S.E.	Mean	S.E.	Means	Variances
Loam	17	1.70	0.022	1.73	0.024	n.s.	n.s.
Loam	0	1.95	0.006	1.88	0.012	**	**
Sandy loam	15	1.71	0.017	1.76	0.081	n.s.	**
Sandy clay loam	18	2.11	0.014	2.06	0.017	*	n.s.
Peat	550	1.19	0.007	1.08	0.015	**	**

[1]n.s. = not significant; * = significant at $P_{0.05}$; ** = significant at $P_{0.01}$.

Alternatively, the immersion fluid may be allowed to enter the pores of a clod provided this is allowed for in subsequent calculation of clod wet bulk density. Both kerosene and, more recently, hexane have been used for this purpose (Ross and Prebble, 1989).

Other methods of measuring clod bulk density which, although not widely used, may have advantages in particular circumstances, include immersion in a bed of glass beads (Voorhees et al., 1966), elutriation in a vertical air stream (Chepil, 1950) and the use of X-rays to determine the variation in bulk density within a clod (Greacen et al., 1967).

Radiation methods

Gamma-rays emitted by a radioactive source may either pass through or be

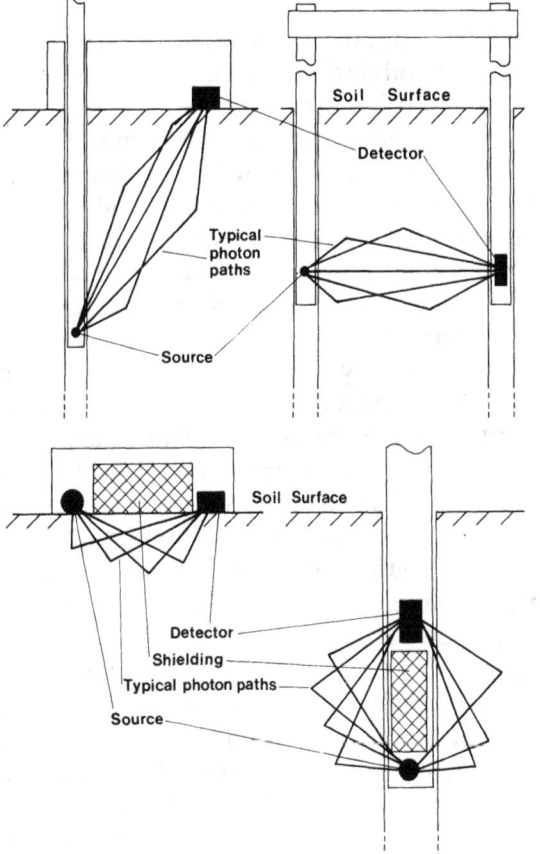

Fig. 4. Schematic diagrams of typical gamma-ray transmission gauges (top) and backscatter gauges (bottom) (from Campbell and Henshall, 1991).

absorbed or scattered by a soil sample. Both absorption and scattering increase with soil bulk density. Thus, a bulk density gauge comprises a radioactive source and a facility for detecting and counting either transmitted or scattered gamma-rays (Fig. 4). High-resolution transmission gauges count only those photons with an energy which is characteristic of the source, while backscatter gauges count photons of lower energy. Some simpler transmission gauges, which do not include energy discrimination in the detector system, will also count scattered photons. All gauges give only an empirical measurement of soil bulk density and therefore must be calibrated by making measurements in samples of known bulk density. Campbell and Henshall (1991) detailed the physical principles on which such gauges are based and discussed the implications for gauge design and calibration, sample resolution, sources of error and the accuracy of bulk density measurements.

Gauge design and calibration

In backscatter gauges, the source and detector are shielded from each other in an assembly which may either rest on the soil surface or be lowered into an access hole prepared in the soil (Fig. 4). In both cases, Campbell and Henshall (1991) reported that, typically, only soil lying within about 75 mm of the source/detector axis is found to influence the count rate. Further, surface gauges indicate the mean bulk density for the top 75 mm of the soil profile only and so their main application is in civil engineering, where bulk densities are often uniform with depth. Carlton (1961) gives a typical accuracy for surface gauges of ± 0.16 Mg m^{-3}. Backscatter gauges which are lowered into lined access holes have the advantage that they may be combined with neutron moisture probes. However, a major disadvantage is that, since they are biased toward photons scattered by soil close to the gauge, both the liner tube and any air gap between soil and tube will influence results unduly. All forms of backscatter gauge are clearly of limited value in the layered soils common in agriculture, in which it is usually desirable to be able to detect abrupt changes in bulk density which can be important in relation to root penetration and gas and water movement.

In transmission gauges, the radioactive source and the detector are each contained in a separate tubular probe, the two probes being held rigidly at a fixed distance apart. The test sample is the soil between the two access holes which are provided for the two probes either by augering or by hammering spikes into the soil via an alignment jig. Augered holes are usually fitted with liners and have the attraction that water content samples may be obtained from the extracted soil. In one type of equipment, the detector is housed in the assembly on the soil surface, with the advantage that only one access hole is required, but with the considerable disadvantage that the test result is the mean for all soil layers between the source depth and the surface. Most of the transmission gauges currently available have Geiger-Müller detectors which are incapable of energy discrimination and so detect both transmitted and scattered photons. A consequence of this is the error introduced when these gauges are used to make measurements close to either the soil surface or the transition between layers of markedly different bulk density, such as the upper face of a plough pan. Henshall and Campbell (1983) found that while the error in bulk density was about 5% at a depth of 90 mm below an air/water interface, the error exceeded 35% at 20 mm depth. In contrast, a high resolution gauge (Fig. 5) in which only transmitted photons were detected gave an error in bulk density of <5% at 10 mm depth.

Most radiation gauges have provision for making a test count in a reference material of fixed density, such as a steel plate, and all counts made in soil samples are expressed as a fraction of the reference count. This is necessary not only from theoretical considerations (Campbell and Henshall, 1991) but also to allow for any long-term changes in detector efficiency or source activity, which will affect the calibration of the gauge.

It is essential to calibrate each individual gauge. Even if the manufacturer supplies a calibration, this may be affected by detailed differences in the mode of operation, the most likely source of such errors being the method whereby probe access is provided (Rawitz et al., 1982; Bertuzzi et al., 1987; Rousseva et al., 1988; Culley and McGovern, 1990; Campbell and Henshall, 1991). The access method must be exactly the same for both calibration and subsequent test samples. Calibration in terms of field soils may be *in situ*, but this method

Fig. 5. Prototype high-resolution gamma-ray transmission gauge developed at the Scottish Centre of Agricultural Engineering (from Campbell and Henshall, 1991).

assumes, unreasonably, that the alternative method used to measure soil bulk density, which is usually core sampling, is absolutely accurate. Careful repacking of field soils into containers and calculating their bulk density from sample mass and volume has been shown to be more satisfactory (Soane et al., 1971; Henshall and Campbell, 1983; Van Bavel et al., 1985). Coppola and Reiniger (1974) calculated that, for photon energies >0.3 MeV, soil composition had little effect on transmission gauge calibration and the error in bulk density due to variation among 9 soils was shown by Reginato (1974) to be about 0.5%. Theoretical considerations led Campbell and Henshall (1991) to suggest, however, that soil composition effects could be important in relation to backscatter gauges. Morris and Williams (1989) considered it impractical to obtain calibration samples of the soft, layered coal tailings with which they were working and they derived a satisfactory calibration relation for their backscatter gauge from first principles according to the elemental composition of the tailings. Hodgson (1988) used a backscatter gauge in a swelling clay and found wet bulk densities to be poorly related to those from core sampling. This was thought to be due to the negative correlation between dry bulk density and water content. Culley and McGovern (1990) suggested that wet rather than dry bulk density should be used in comparing the effects of tillage treatments on soil bulk density in shrink-swell soils.

A further development of the transmission gauge is one in which two radioactive sources are employed. In addition to the usual ^{137}Cs source, a second source is employed, usually ^{241}Am, which emits low-energy photons, the attenuation of which is affected by chemical composition, particularly hydrogen content, and hence water content. Thus, by calibrating the gauge for each energy level, both bulk density and water content may be determined simultaneously (Soane, 1967). In practice, however, the complexity and weight of the required equipment and the effect of soil chemical composition renders the method suitable primarily for detailed examination of a single soil under carefully controlled laboratory conditions (Hopmans and Dane, 1986; Fahad, 1989).

A more recent development is the use of X-ray transmission computed tomography which was originally developed to give non-destructive scans of the human body. Petrovic et al. (1982), Bertuzzi et al. (1987) and Tollner and Ramseur (1988) tested the method and found that, in general, measurements related well to bulk densities and that the method was a promising technique for laboratory tests. Phogat and Aylmore (1989) used gamma-rays rather than X-rays and considered the method to have considerable potential in the study of changes in macroporosity during wetting and drying of soil columns.

Statistical considerations

When several measurements are made to characterise the bulk density of a soil known to be variable, no increase in accuracy in the mean value obtained from measurements at several positions is gained by counting large numbers of photons for each measurement. Instead, less precise measurements should be made on a larger number of samples. In replicated field experiments, counting between 2,000 and 3,000 photons in two or three positions per plot will typically give coefficients of variation of about 10% (Soane et al., 1971). However, if the soil is known to be uniform, or if the volume of interest is small enough for virtually the whole soil to be tested, longer, more precise measurements of count rate should be made. Typically, between 2,000 and 10,000 photons are counted corresponding to precision levels of 2.5 and 1.0%, respectively (Campbell and Henshall, 1991). Errors in the measurement of the reference count rate in the standard material are particularly important since the reference count subsequently affects all deduced bulk densities. Accordingly, the reference count period is usually up to five times longer than that for the subsequent test counts.

If detailed information on the variation of bulk density with depth is sought with a single-probe transmission gauge, the number of depths at which measurements can be made is limited to those for which calibrations have been established. In addition, for twin-probe gauges, there is no advantage in making measurements at depth intervals which are much smaller than the spatial resolution of the gauge. The spatial resolution is primarily dependent on the dimensions of the detector when energy discrimination is employed (Campbell and Henshall, 1991).

Stones may prevent the provision of access holes. If the problem is avoided by moving to a new test position, bulk density may be over-estimated since there may be a bias toward samples in which stones lie within the sample under test rather than in the access hole. In the case of twin-probe gauges, stones may deflect the probes during insertion, causing variation in the source/detector separation and hence introducing calibration errors. Soane (1968) overcame the problem by measuring the separation at each depth for which counts were made, while Bertuzzi et al. (1987) showed that, for a probe separation of 200 mm and measurements at 400 mm depth, a probe deflection of ± 10 mm could give an error in bulk density of ± 0.15 Mg m^{-3}. The statistical problems caused by stones and soil variability have been discussed by O'Sullivan et al. (1987) in relation to the measurement of cone resistance, while Glasbey and O'Sullivan (1988) proposed a method for dealing with the resulting bias in cone resistance measurements when stones prevent penetration.

Generally, the closer together sampling positions are, the more soil properties tend to be similar (Burgess and Webster, 1980) and if this is taken into account, the number of sampling positions can sometimes be reduced without loss of precision (McBratney and Webster, 1983). Dane et al. (1986) discuss the statistical problems arising when practical considerations limit the number of positions at which bulk density may be measured.

SELECTING A MEASURING METHOD

In making a decision as to which method to use in a given set of circumstances, a major factor is often the cost. Campbell and Henshall (1991) have summarised the specifications and costs of a selection of backscatter and transmission gauges, all of which are very much more expensive than the equipment required in, for example, the sand bottle or core sampling methods. Thus, although the radiation methods, especially the transmission method employing energy discrimination, may be the more accurate and have the ability to detect thin layers of soil of high density, cost may preclude their use. The need to comply with radiation safety regulations (Anon., 1985) may also be a consideration in relation to radiation methods, although this does not usually constitute a problem.

Possibly because the equipment required for the core sampling method is probably the cheapest and simplest to use, it is both widely used and is usually the method with which radiation methods are compared when the latter are being assessed. It has been pointed out by Raper and Erbach (1985), however, that the core sampling method has many inherent errors and that it is therefore questionable whether it should be used as a reference method. It certainly seems unreasonable to assume that the results of the core test are totally accurate. In the many comparisons which have been made of the core sampling and radiation methods, the two methods have been found to be in general agreement. Soane et al. (1971) tested three mineral soils and found results to agree to within 3%, although in tests on a peat soil sample the core sampling

TABLE 3

Comparison of dry bulk density ($Mg\ m^{-3}$) determined by low-resolution transmission gamma-ray gauge (GRG) and core sampling (CS) methods for different soils (after Soane et al., 1971)

	Darvel topsoil (sandy loam, 3% gravel)		Winton topsoil (sandy clay loam, 17% gravel)		Winton subsoil (clay loam, 13% gravel)		Peat (0% gravel)	
	GRG	CS	GRG	CS	GRG	CS	GRG	CS
Mean[1]	1.371	1.355	1.321	1.350	1.760	1.710	0.223	0.283
S.E.	0.014	0.007	0.022	0.026	0.028	0.013	0.007	0.009
C.V. (%)	3.2	1.6	5.2	6.2	5.1	2.4	10.4	9.6
Sign. of differences[2]:								
in means (paired data)	n.s.		n.s.		n.s.		**	
in variances	*		n.s.		*		n.s.	

[1]Number of comparisons: 10.
[2]n.s. = not significant; * = significant at $P_{0.05}$; ** = significant at $P_{0.01}$.

method gave significantly higher densities, which could not be accounted for. Variability in the results for both methods was similar (Table 3). They found the radiation method to be up to three times faster than core sampling, over which it also had the advantage of being able to make more detailed measurements of the variation of bulk density with depth. Swinford and Meyer (1985) reported that the transmission method was up to 5 times quicker than core sampling. Schafer et al. (1984) avoided the statistical problems associated with testing the two methods on different samples by using a laboratory gamma-ray gauge to test soil cores from the field. For 80% of the 236 cores tested, the difference in the results for the two methods was <1% and for only two samples was it >2%.

Core sampling was compared with both a single- and a twin-probe transmission gamma-ray gauge by Gameda et al. (1983). Although they found good agreement among methods for a clay and a sand, poorer agreement for a loam was attributed to stones. The core results were more closely related to those of the twin-probe method than those of the single-probe method.

King and Parsons (1959) compared a single-probe backscatter gauge with the sand bottle and found the methods to agree within 3% in sand and clay soils. An unacceptably large difference of 11% was found for a gravelly soil and several possible explanations for the disagreement were discussed, although, in this early comparison, suspicion tended to fall on the gamma-ray gauge. Erbach (1983) considered the sand bottle method to be good for use in a gravelly soil.

Comparisons of the various direct methods have also been made. Generally, the rubber balloon method has been relatively unreliable. Cernica (1980) found it to give systematic errors of 4.85%. This was in contrast to 2.95% for the sand bottle method or 0.53% when the latter was based on the measurement of sand volume rather than mass. Although the bulk density of clods might be expected to be greater than that for the bulk soil from which they are taken, bulk densities by the clod method have generally been similar to those obtained by core sampling, while sand bottle values have been about 2% lower. Howard and Singer (1981) compared the clod method with an excavation method in which a film-lined hole was filled with water. Results for the two methods were linearly correlated and equally precise, but, for 14 soils, the clod bulk densities were consistently higher by up to about 2%. Håkansson (1990) compared the results obtained by both frame and core sampling. Within the ploughed layer, core sampling gave consistently lower porosity (0-6%, v/v) than frame sampling, the difference being attributed to compaction of loose soil within the core. There was no comparable effect in the firm subsoil.

Despite the general agreement usually found between direct and radiation methods, the greater accuracy of radiation methods, together with the better spatial resolution of the transmission gauges, make the latter the most suitable method for the layered soils usually found in agriculture. Campbell and Henshall (1991) gave examples of the use of transmission systems in tillage and field traffic experiments.

PRESENTATION OF RESULTS

Selection of a method of presentation of the test results will generally be determined by the objective of the presentation. For example, if the soil loosening produced by alternative tillage treatments at sowing depth is the only interest, the bulk densities may simply be tabulated. More often, however, the variation of bulk density with depth is of interest since different tillage or traffic treatments give different results at each depth. In such situations, graphical presentation is the preferred method. While this can be a satisfactory method for comparing the variation of bulk density with depth before and after the passage of a vehicle, problems can arise, especially when comparisons are to be made between vehicles which produce ruts of different depth. In such circumstances, soil which was present at a given depth in the uncompacted profile may, for different vehicles, end up at different depths below the original surface and/or the bottoms of the wheel ruts produced.

In response to this problem, Henshall and Smith (1989) developed a method of presenting results in which comparisons of bulk density could be made on elements of soil in the compacted profiles which originated from the same depth in the uncompacted soil profile. Fig. 6 shows the variation of dry bulk

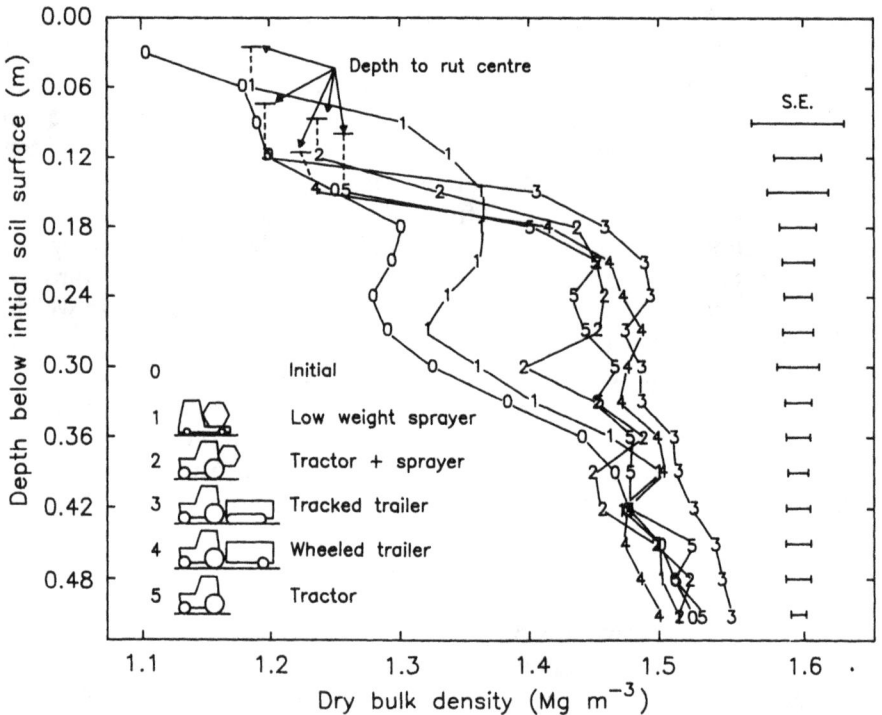

Fig. 6. Variation of soil bulk density with depth below the initial soil surface for five experimental treatments in comparison with the initial dry bulk density profile. Treatment numbers indicate depths of measurements (from Henshall and Smith, 1989).

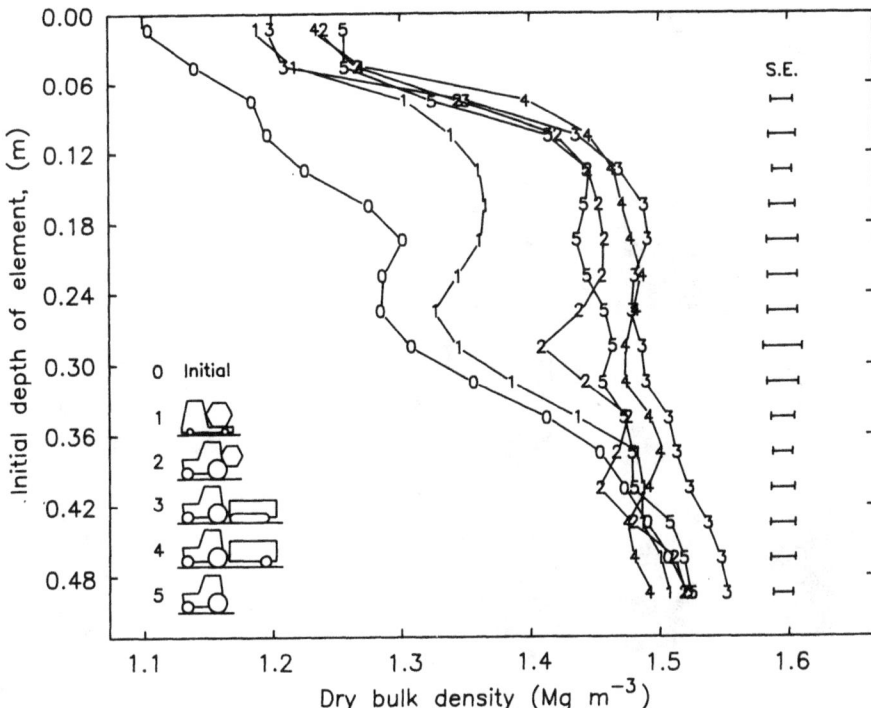

Fig. 7. Data in Fig. 6 replotted so that the mean final density of each soil element is plotted according to its depth in the initial profile (from Henshall and Smith, 1989).

density with depth for five experimental treatments, in comparison with the initial bulk density profile. Comparisons between treatments are difficult, mainly as a result of differences in rut depth. Fig. 7 shows the same data, replotted according to the technique described by Henshall and Smith (1989). In this presentation, it is now clear that treatment number one produces smaller increases in bulk density than all the other treatments among which there are essentially no differences.

When comparing several compaction treatment effects, especially when they are complex, it is often desirable to have an overall qualitative impression of the effects which can be easily assimilated. In order to achieve this, Henshall and Smith (1989) suggested that the soil profile should be represented by a stack of elements, each element being shaded according to its bulk density and varying in width, according to the horizontal soil displacement occurring within that element. Fig. 8 shows the data of Fig. 6 displayed in this way.

Generally, soil compaction by wheels results in changes in bulk density not only below the centre line of the wheel rut but also for some distance either side of the centre line. This can be an important consideration, particularly in relation to row crops (Campbell, 1982), but also in relation to tillage generally since the

Initial

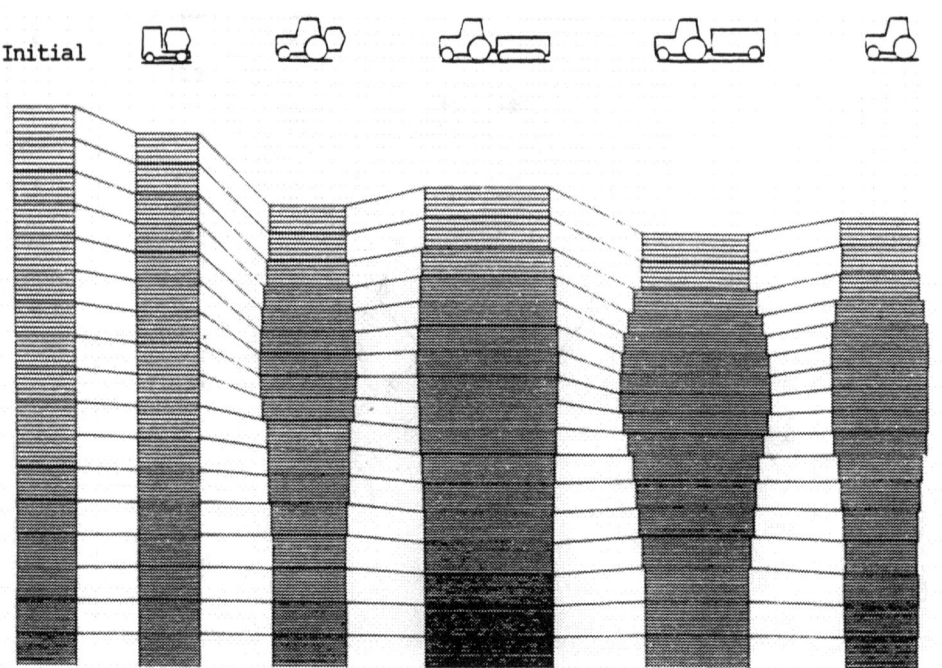

Fig. 8. Data in Fig. 6 presented as stacks of soil elements, each element being shaded according to its bulk density and varying in width according to horizontal soil displacement (from Henshall and Smith, 1989).

efficiency of any loosening operation will be affected. The problem may be overcome by making bulk density measurements at a series of positions in a line at right angles to the wheel rut. This has been done with both a hand-held transmission gauge (Dickson and Smith, 1986) and a version of the method in which the source and detector probes were mounted on an electrically propelled carriage which scanned a soil sample in the form of a rectangular block at right angles to the wheel rut (Soane, 1973). Measurements at 825 positions on a 20 mm x 20 mm grid resulted in diagrams showing contours of dry bulk density. These diagrams (Fig. 9) clearly show differences in the variation of bulk density beneath a tyre, a tyre fitted with a cage wheel of slightly smaller diameter, and a steel crawler track. It can be seen that, had measurements been restricted to positions beneath the centre line of each rut, a false impression of the differences in compaction produced by the different sets of running gear would have been gained, particularly in relation to the volume of soil compacted.

Although both the qualitative and quantitative results obtained with scanning equipment are undoubtedly useful, problems can arise in quantifying the comparison of contour diagrams for different treatments. Soane et al. (1976), faced with the same problem in relation to cone resistance contours, found that

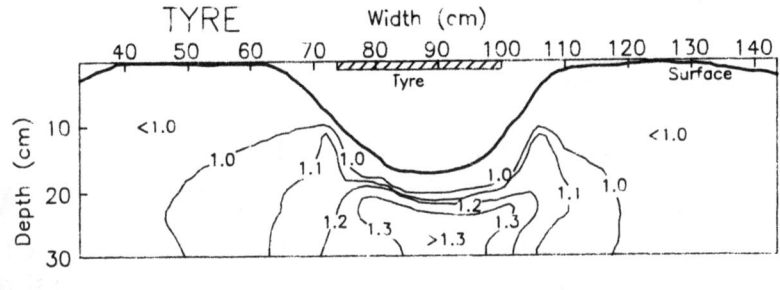

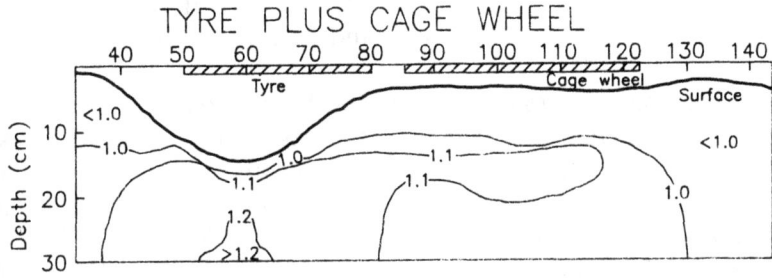

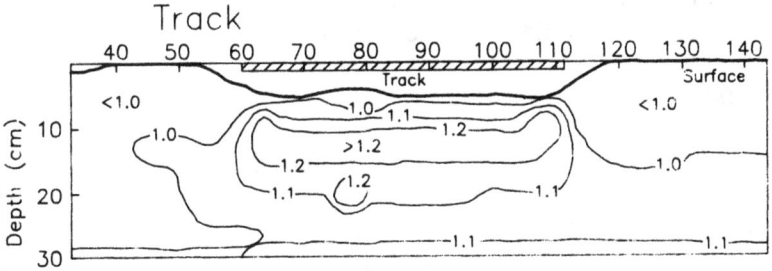

Fig. 9. Two-dimensional variation of soil dry bulk density (Mg m⁻³) below the ruts produced by alternative running gear. Top and centre: 45-kW, 2-wheel tractor; bottom: 48-kW tracked tractor (from Soane, 1973).

the sum, mean and standard deviation of the differences in cone resistance at each point in the scan before and after the treatment was applied, were useful statistics. In particular, they indicated not only the volume of soil compacted but the extent to which the peak increase in cone resistance occurred within the volume compacted. Dickson and Campbell (1991) also produced cone resistance scans. In assessing the volume of soil suitable for crop growth in both zero and conventional traffic systems, they found that the variation across the scan of the depth at which a given contour was found, was a useful indicator.

UNRESOLVED PROBLEMS

Problems with stones

Stones present several problems, in addition to those already discussed, which often apply to more than one method. For example, in calculating soil porosity from bulk density values, it is necessary to use the soil particle density. Anon. (1964) showed that the particle density of stones >5 mm in diameter ranged from 2.54 to 2.98 Mg m^{-3}, while Soane et al. (1972) found that the particle density for material <2 mm ranged from 2.37 to 2.68 Mg m^{-3} for 58 Scottish soils. The use of a single value for particle density may therefore result in erroneous porosity values, especially for different soil types and different depths in layered profiles.

During core sampling, some stones will lie partly within and partly outwith the ends of the core sample. Usually the trimming process results in such stones being removed. The cavities so created are often packed with soil to an unknown bulk density. Where the cavities are not filled, the calculated volume of the core will be in error. Similar errors can arise in cutting rectangular blocks of field soils for testing by the scanning gamma-ray transmission method (Soane, 1973). In such circumstances, stones are not only a possible source of error, but can make preparation of the sample a very slow operation thereby detracting from the appeal of a method which can give very detailed information on the horizontal and vertical variation of soil bulk density below wheel ruts.

Problems with radiation methods

With twin-probe transmission gauges, accurate results depend on the source and detector probes being maintained at their designed separation. Probe deflection by stones has already been mentioned, but there is also the risk of error with lined access holes in that the liners may be deflected although the slightly smaller diameter probes may not be, thereby introducing calibration errors.

The advent of high-resolution transmission gauges has enabled acceptably accurate measurements to be made at much shallower depths below the soil surface than previously. This ability has emphasised the need to define zero depth accurately but this is not always easy because of the surface roughness of cultivated soils. Even when measurements are made relative to the bottom of a nominally smooth wheel rut, the indentations caused by tyre ribs and lugs present difficulties. Absorbers placed on the soil surface have been used with low resolution transmission probes to allow effective measurements closer to the surface, but these have always been rigid and could not accommodate variations in surface level. A flexible absorber such as a bag of sand is one possible solution to the problem although this appears not to have been tested.

High-resolution transmission gauges based on energy discrimination usually have a very low-activity source fixed to the detector crystal to provide a reference count rate to allow the stabiliser, which corrects for temperature fluctuations, to operate when the source is shielded from the detector. Calibration relations usually assume that the relative count rates of the main- and very low-activity sources remain unchanged. In the long term, this may not be the case. A further possible source of error is that in dense soils the proportion of the total count rate attributable to the main source may become too small for satisfactory gauge performance.

Henshall and Campbell (1983) suggested that the stability of a high-resolution system can be monitored by comparing the variation in reference counts over a period of use with the predicted variation associated with random radioactive emissions. However, there may also be a trend in the variation in reference counts due to, for example, a trend in the difference in temperature of the soil under test and the material used to make reference counts. It is possible that, if a rolling mean reference count rather than an overall mean was used, bulk densities could be more accurate. However, the possibility appears not to have been investigated.

Where a high-resolution system allows measurements to be made in layers as thin as, for example, 10 mm, it will often be appropriate to make measurements at 10-mm depth intervals at shallow depths, where compaction effects are greatest and possibly most important, while increasing the depth increment at greater depths. In these conditions it may well be appropriate, where the range of photon energies detected can be easily altered, to reduce the resolution of the gauge to correspond to the increased depth increment. In this way, the count rate would be increased and the test procedure speeded up by an amount which could be important when multiple measurements are being made in replicated field experiments.

CONCLUSIONS

(1) The measurement of bulk density and related properties which quantify changes in compactness, is an essential requirement in studies on soil compaction by vehicles.

(2) Interpretation of crop responses to soil bulk density may be facilitated by expressing bulk density in relation to the bulk density of the same soil in a specified reference condition. Further examination of this method of presenting results is needed for a wider range of soils and crops.

(3) While direct methods of measuring bulk density are cheap and simple to use, they are not without errors and often prove to be very tedious, especially when measurements are undertaken in large numbers at depth in field soils.

(4) The greater expense and complexity of radiation methods, together with the need for higher technical skills on the part of the operator, may be perceived as

serious drawbacks. However, experience of their use over extended periods has confirmed their accuracy, reliability, safety, speed of measurement and ability to measure with high spatial resolution close to the surface and at depth.

ACKNOWLEDGEMENT

The author is grateful to his colleague J.K. Henshall (Scottish Centre of Agricultural Engineering, Penicuik, U.K.) for helpful discussion during the preparation of this chapter.

REFERENCES

Abrol, I.P. and Palta, J.P., 1968. Bulk density determination of soil clods using rubber solution as a coating material. Soil Sci., 106: 465-468.

Andersson, S. and Håkansson, I., 1967. Determination of the volume relationship in the different layers of the topsoil. In: M. de Boodt, L. de Leenheer, H. Frese, A.J. Low and P.K. Peerlkamp (Editors), West-European Methods for Soil Structure Determinations. The State Faculty of Agric. Sci., Ghent, Belgium, VII, pp. 25-29.

Anon., 1964. Soil Mechanics for Road Engineers. HMSO, London, U.K., 541 pp.

Anon., 1985. The Ionising Radiations Regulations. HMSO, London, U.K., 83 pp.

Ball, B.C. and Hunter, R., 1988. The determination of water release characteristics of soil cores at low suctions. Geoderma, 43: 195-212.

Baver, L.D., Gardner, W.H. and Gardner, W.R., 1972. Soil Physics. Wiley, New York, NY, U.S.A., 498 pp.

Bertuzzi, P., Bruckler, L., Gabilly, Y. and Gaudu, J.C., 1987. Calibration, field-testing and error analysis of a gamma-ray probe for in situ measurement of dry bulk density. Soil Sci., 144: 425-436.

Blake, G.R. and Hartge, K.H., 1986. Bulk density. In: A. Klute (Editor), Methods of Soil Analysis, 1. Physical and Mineralogical Methods. Am. Soc. Agron./Soil Sci. Soc. Am., Madison, WI, U.S.A., pp. 363-375.

Brasher, B.R., Franzmeir, D.P., Valassis, V. and Davidson, S.E., 1966. Use of Saran resin to coat natural soil clods for bulk density and moisture retention measurements. Soil Sci., 101: 108.

Buchele, W.F., 1961. A power sampler of undisturbed soil. Trans. ASAE, 4: 185-187, 191.

Burgess, T.M. and Webster, R., 1980. Optimal interpolation and isorithmic mapping of soil properties, I. The semi-variogram and punctual kriging. J. Soil Sci., 31: 315-331.

Campbell, D.J., 1973. A flotation method for the rapid measurement of the wet bulk density of soil clods. J. Soil Sci., 24: 239-243.

Campbell, D.J., 1982. A review of the clod problem in potato production. J. Agric. Eng. Res., 27: 373-395.

Campbell, D.J. and Henshall, J.K., 1991. Bulk density. In: K.A. Smith and C.E. Mullins (Editors), Soil Analysis: Physical Methods. Marcel Dekker, New York, NY, U.S.A., pp. 329-366.

Campbell, D.J. and Hunter, R., 1986. Drop-cone penetration *in situ* and on minimally disturbed soil cores. J. Soil Sci., 37: 153-163.

Carlton, P.F., 1961. Application of nuclear soil meters to compaction control for airfield pavement construction. In: Symposium on Nuclear Methods of Measuring Soil Density and Moisture. Am. Soc. Testing Mater., Philadelphia, PA, U.S.A., Spec. Tech. Publ. 293, pp.

27-35.

Carter, M.R., 1990. Relative measures of soil bulk density to characterize compaction in tillage studies on fine sandy loams. Can. J. Soil Sci., 70: 425-433.

Cernica, J.N., 1980. Proposed new method for the determination of density of soil in place. Geotech. Testing J., 3: 120-123.

Chepil, W.S., 1950. Methods of estimating apparent density of discrete soil grains and aggregates. Soil Sci., 70: 351-362.

Cooper, M.R. and Fleming, P.R., 1990. A large-capacity batch filling water balloon apparatus for deep in-situ density tests. Geotech. Testing J., 12: 222-226.

Coppola, M. and Reiniger, P., 1974. Influence of the chemical composition on the gamma-ray attenuation by soils. Soil Sci., 177: 331-335.

Culley, J.L.B. and McGovern, M.A., 1990. Single and dual probe nuclear instruments for determining water contents and bulk densities of a clay loam soil. Soil Tillage Res., 16: 245-256.

Dane, J.H., Reed, R.B. and Hopmans, J.W., 1986. Estimating soil parameters and sample size by bootstrapping. Soil Sci. Soc. Am. J., 50: 283-287.

Dickson, J.W. and Campbell, D.J., 1991. Soil and crop responses to zero- and conventional-traffic systems for winter barley in Scotland. Soil Tillage Res., 18: 1-26.

Dickson, J.W. and Smith, D.L.O., 1986. Compaction of a sandy loam by a single wheel supporting one of two masses each at two ground pressures. Scot. Inst. Agric. Eng., Penicuik, U.K., Dep. Note SIN/479, 21 pp.

Erbach, D.C., 1982. State of the art of soil density measurement. Am. Soc. Agric. Eng., St. Joseph, MI, U.S.A., ASAE Pap. 82-1541, 11 pp.

Erbach, D.C., 1983. Measurement of soil moisture and bulk density. Am. Soc. Agric. Eng., St. Joseph, MI, U.S.A., ASAE Pap. 83-1553, 16 pp.

Eriksson, J., Håkansson, I. and Danfors, B., 1974. The effect of soil compaction on soil structure and crop yields. Swed. Inst. Agric. Eng., Uppsala, Sweden, Bull. 354, 101 pp. (English translation by J.K. Aase).

Fahad, A.A., 1989. A computer-controlled dual-gamma scanner for measurement of soil water content and bulk density. Int. J. Radiation Appl. Instrum., A, Applied Rad. Isotopes, 40: 340-342.

Freitag, D.R., 1971. Methods of measuring soil compaction. In: K.K. Barnes, W.M. Carleton, H.M. Taylor, R.I. Throckmorton and G.E. Vanden Berg (Editors), Compaction of Agricultural Soils. Am. Soc. Agric. Eng., St. Joseph, MI, U.S.A., pp. 47-103.

Gameda, S., Raghavan, G.S.V., McKyes, E. and Thériault, R., 1983. Single and dual probes for soil density measurement. Am. Soc. Agric. Eng., St. Joseph, MI, U.S.A., ASAE Pap. 83-1550, 9 pp.

Gill, W.R., 1959. Soil bulk density changes due to moisture changes in soil. Trans. ASAE, 2: 104-105.

Glasbey, C.A. and O'Sullivan, M.F., 1988. Analysis of cone resistance data with missing observations below stones. J. Soil Sci., 39: 587-592.

Greacen, E.L., Farrel, D.A. and Forrest, J.A., 1967. Measurement of density patterns in soil. J. Agric. Eng. Res., 12: 311-313.

Håkansson, I., 1973. The sensitivity of different crops to soil compaction. Proc. 6th Conf. Int. Soil Tillage Res. Org. (ISTRO), Wageningen, Netherlands, Pap. 14, pp. 1-3.

Håkansson, I., 1988. A method of characterizing the state of compaction of an arable soil. In: J. Drescher, R. Horn and M. de Boodt (Editors), Impact of Water and External Forces on Soil Structure. Catena, Cremlingen, Germany, Catena Supplement 11, pp. 101-105.

Håkansson, I., 1990. A method for characterizing the state of compactness of the plough layer. Soil Tillage Res., 16: 105-120.

Heinonen, R., 1977. Toward "normal" soil bulk density. Soil Sci. Soc. Am. J., 41: 1214-1215.

Henshall, J.K. and Campbell, D.J., 1983. The calibration of a high resolution gamma-ray transmission system for measuring soil bulk density and an assessment of its field performance. J. Soil Sci., 34: 453-463.

Henshall, J.K. and Smith, D.L.O., 1989. An improved method of presenting comparisons of soil compaction effects below wheels. J. Agric. Eng. Res., 42: 1-13.

Hodgson, A.S., 1988. Use of neutron and gamma radiation meters to estimate bulk density and correct for bias of sampling for water content in a swelling clay soil. Aust. J. Soil Res., 26: 261-268.

Hopmans, J.W. and Dane, J.H., 1986. Calibration of a dual energy gamma-radiation system for multiple point measurements in soil. Water Resour. Res., 22: 1109-1114.

Howard, R.F. and Singer, M.J., 1981. Measuring forest soil bulk density using irregular hole, paraffin clod, and air permeability. For. Sci., 27: 316-322.

Jamison, V.C., Weaver, H.H. and Reed, I.F., 1950. A hammer-driven soil core sampler. Soil Sci., 69: 487-496.

Karafiath, L.L. and Nowatzki, E.A., 1978. Soil Mechanics for Off-Road Vehicle Engineering. Trans Tech., Clausthal, Germany, 515 pp.

King, F.G. and Parsons, A.W., 1959. Portable radioactive equipment for measuring soil density. Road Res. Lab., Crowthorne, U.K., Res. Note RN/3628/FGK.AWP, 20 pp.

Laundré, J.W., 1989. Estimating soil bulk density with expanding polyurethane foam. Soil Sci., 147: 223-224.

Ljungars, A., 1977. Olika faktorers betydelse för traktorernas jordpackningsverkan. Mätningar 1974-1976. (Importance of different factors of soil compaction by tractors. Measurements in 1974-1976). Agric. College Sweden, Uppsala, Sweden, Dep. Soil Sci., Div. Soil Manage., Rep. 52, 43 pp. (in Swedish with English summary).

Lutz, J.F., 1947. Apparatus for collecting undisturbed soil samples. Soil Sci., 64: 399-401.

McBratney, A.B. and Webster, R., 1983. How many observations are needed for regional estimation of soil properties? Soil Sci., 135: 177-183.

Morris, P.H. and Williams, D.J., 1989. Generalized calibration of a nuclear moisture/density depth gauge. Geotech. Testing J., 13: 24-35.

O'Sullivan, M.F., Dickson, J.W. and Campbell, D.J., 1987. Interpretation and presentation of cone resistance data in tillage and traffic studies. J. Soil Sci., 38: 137-148.

Olson, K.R. and Zobeck, T.D., 1989. Improved mercury-displacement method to measure the density of soil aggregates. Soil Sci., 147: 71-75.

Petrovic, A.M., Siebert, J.E. and Rieke, P.E., 1982. Soil bulk density analysis in three dimensions by computed tomographic scanning. Soil Sci. Soc. Am. J., 46: 445-450.

Phogat, V.K. and Aylmore, L.A.G., 1989. Evaluation of soil structure by using computer assisted tomography. Aust. J. Soil Res., 27: 313-323.

Pidgeon, J.D. and Soane, B.D., 1977. Effects of tillage and direct drilling on soil properties during the growing season in a long-term barley mono-culture system. J. Agric. Sci., Camb., 88: 431-442.

Raper, R.L. and Erbach, D.C., 1985. Accurate bulk density measurements using a core sampler. Am. Soc. Agric. Eng., St. Joseph, MI, U.S.A., ASAE Pap. 85-1542, 24 pp.

Raper, R.L. and Erbach, D.C., 1987. Bulk density measurement variability with core samplers. Trans. ASAE, 30: 878-881.

Rawitz, E., Etkin, H. and Hazan, A., 1982. Calibration and field testing of a two-probe gamma gauge. Soil Sci. Soc. Am. J., 46: 461-465.

Reginato, R.J., 1974. Gamma radiation measurements of bulk density changes in the soil pedon following irrigation. Soil Sci. Soc. Am. Proc., 38: 24-29.

Ross, P.J. and Prebble, R.E., 1989. Non-destructive measurement of soil clod volume using hexane displacement. Aust. J. Soil Res., 27: 39-44.

Rousseva, S.S., Ahuja, L.R. and Heathman, G.C., 1988. Use of a surface gamma-neutron gauge for in situ measurement of changes in bulk density of the tilled zone. Soil Tillage Res., 12: 235-251.

Russell, E.W. and Balcerek, W., 1944. The determination of the volume and airspace of soil clods. J. Agric. Sci., Camb., 34: 123-132.

Schafer, G.J., Barker, P.R. and Northey, R.D., 1984. Density of undisturbed soil cores by gamma-ray attenuation. New Zealand Soil Bureau, Lower Hutt, N.Z., Rep. 67, 11 pp.

Soane, B.D., 1967. Dual energy gamma-ray transmission for coincident measurement of water content and dry bulk density of soil. Nature, 214: 1273.

Soane, B.D., 1968. A gamma-ray transmission method for the measurement of soil density in field tillage studies. J. Agric. Eng. Res., 13: 340-349.

Soane, B.D., 1973. Techniques for measuring changes in the packing state and cone resistance of soil after the passage of wheels and tracks. J. Soil Sci., 24: 311-323.

Soane, B.D., Campbell, D.J. and Herkes, S.M., 1971. Hand-held gamma-ray transmission equipment for the measurement of bulk density of field soils. J. Agric. Eng. Res., 16: 146-156.

Soane, B.D., Campbell, D.J. and Herkes, S.M., 1972. The characterisation of some Scottish topsoils by agricultural and engineering methods. J. Soil Sci., 23: 93-104.

Soane, B.D., Kenworthy, G. and Pidgeon, J.D., 1976. Soil tank and field studies of compaction under wheels. 7th Int. Conf. Int. Soil Tillage Res. Org. (ISTRO), Uppsala, Sweden, Agric. College of Sweden, Dep. Soil Sci., Div. Soil Manage., Rep. 45, Pap. 48, pp. 1-6.

Swinford, J.M. and Meyer, J.H., 1985. An evaluation of a nuclear density gauge for measuring infield compaction in soils of the South African sugar industry. Proc. 59th Ann. Conf., South African Sugar Technol. Ass., Durban and Mount Edgecombe, South Africa, pp. 218-224.

Tessier, S. and Steppuhn, H., 1990. Quick-mount soil core sampler for measuring bulk density. Can. J. Soil Sci., 70: 115-118.

Tollner, E.W. and Ramseur, E.L., 1988. Using computed tomography to measure soil moisture and bulk density in the presence of a growing plant. Am. Soc. Agric. Eng., St. Joseph, MI, U.S.A., ASAE Pap. 88-1625, 13 pp.

Van Bavel, C.H.M., Lascano, R.J. and Baker, J.M., 1985. Calibrating two-probe, gamma-gauge densitometers. Soil Sci., 140: 393-395.

Veihmeyer, F.J., 1929. An improved soil sampling tube. Soil Sci., 27: 147-152.

Voorhees, W.B., Allmaras, R.R. and Larson, W.E., 1966. Porosity of surface soil aggregates at various moisture contents. Soil Sci. Soc. Am. Proc., 30: 163-167.

Soil Compaction in Crop Production
B.D. Soane and C. van Ouwerkerk (Eds.)
1994 Elsevier Science B.V. All rights reserved.   141

CHAPTER 7

Effects of Compaction on Soil Hydraulic Properties

R. HORTON[1], M.D. ANKENY[2] and R.R. ALLMARAS[3]

[1]Iowa State University, Department of Agronomy, Ames, IA, U.S.A.
[2]National Soil Tilth Laboratory, Ames, IA, U.S.A.
[3]USDA-ARS and University of Minnesota, Department of Soil Science, St. Paul, MN, U.S.A.

SUMMARY

Compactive processes affect soil hydraulic properties and associated soil water flow. Soil water retention and transport properties are altered in response to changes in pore space geometry. Soil water flow is affected not only by soil hydraulic properties but additionally by the distribution of sources and sinks of water in the soil system. Compaction can alter soil pore geometry, and can also affect sources and sinks of water by changing surface configuration, and crop rooting distribution. This paper reviews the literature and presents data and relationships showing the effects of compaction on soil hydraulic properties and water flow, presents numerically modeled water flow for some management systems, and identifies future directions for research.

INTRODUCTION

Compaction significantly influences soil hydraulic properties, infiltration, soil water retention, soil water flow, and hydrologic response (Klute, 1982; Onstad and Voorhees, 1987). Unfortunately, comparative soil hydraulic properties or soil hydrologic components for various management systems are often not consistent (Culley et al., 1987b). Given specific soil management operations, few general statements can be made to describe the effects of compaction on soil hydraulic properties and soil hydrology (Baker, 1987). Measurement and interpretation are often difficult because compaction and soil loosening action usually take place spatially within the same unit implement width. Zonal loosening often requires zonal compaction in a horizontal/vertical arrangement. Thus, inconsistencies are often attributed to variance in climate, topography and spatial/temporal variance of the soil, because the spatial character of tillage/traffic (Cassel, 1983) was not recognized. Measured tillage effects on soil hydraulic properties often disagree because investigations are not consistent as to how, when, and where in the soil profile measurements were made. Results obtained in loosened and compacted zones may have been wrongly lumped together. A better understanding of the effects of management on soil hydraulic properties and hydrology is badly needed; it requires joint efforts in theory development, field measurement and modeling of systems with validation. Field traffic has a fundamental influence on soil hydraulic properties, although its adverse effect on soil permeability is not always as obvious as demonstrated in the accumulation of surface water in wheel ruts (Fig. 1).

The objectives of this paper are to: (1) review some fundamental principles of water retention and transmission; (2) highlight observations of soil hydraulic properties and hydrologic components for various compactive and tillage situations; (3) report results of modeling efforts for predicting the hydrology of soils; (4) suggest future research directions.

Water flow and associated water content of the soil have many indirect effects on plant rooting and growth, aeration, and nutrient availability. Our approach will be to concentrate on those hydraulic factors needed to predict water flow and associated water contents in a soil affected by compaction.

Fig. 1. Ponded water in wheel ruts illustrates the greatly reduced permeability of compacted soil (foreground) when heavy rain followed the mechanical harvesting of potatoes, whereas there was no ponded water in the unharvested area (background).

PORE SPACE, WATER RETENTION AND WATER TRANSPORT

Water retention and transport in soil via the non-solid or the pore spaces, underscores the importance of porosity (see Chapter 5). Total porosity, n, is defined as the ratio of non-solid volume to total volume, and can be calculated using the following equation:

$$n = 1 - (\rho_d/\rho_s) \tag{1}$$

where ρ_d = dry soil bulk density and ρ_s = density of soil solids. Bulk density and total porosity, which are affected by compactive and tillage operations, are commonly measured (Allmaras et al., 1977; Gantzer and Blake, 1978; Akram and Kemper, 1979; Bauder et al., 1981; Reicosky et al., 1981; Hill and Cruse, 1985; Potter et al., 1985; Voorhees et al., 1985; Culley et al., 1987b; Allmaras et al., 1988). Intra- and inter-aggregate porosity were determined in relation to tillage-induced soil structure (Allmaras et al., 1977); yet additional information is needed to estimate soil water retention and transport functions.

The size, shape, continuity and tortuosity of pores in structured soil all contribute to the water retention and transport characteristics (Hill et al., 1985). Therefore, total porosity alone should not be expected to correlate with either the water retention or hydraulic conductivity function (McBride et al., 1987; Kluitenberg et al., 1988).

The height of water rise in a capillary tube, h, is described by Jurin's equation:

$$h = (2 \gamma \cos \psi)/r \rho g \tag{2}$$

where γ = surface tension of water; ψ = contact angle between liquid and solid; r = pore radius; ρ = density of water; g = gravitational acceleration. The Jurin equation can be applied to soil by assuming that a given value of h is the matric potential at which all pores greater than the associated radius, r, must drain. Based on this analogy, soil water retention at a given matric potential is dependent upon soil pore size and geometry. The soil water retention curve is expressed as a continuous functional relationship between volumetric water content and negative pressure required to remove water from a soil (Hillel, 1980).

Numerous investigators have theoretically estimated a relative soil hydraulic conductivity, given a measured water retention function (Marshall, 1958; Millington and Quirk, 1961; Brutsaert, 1967; Green and Corey, 1971; Campbell, 1974; Mualem, 1976; Van Genuchten, 1980). A measured saturated or unsaturated hydraulic conductivity, K, is used to convert relative to absolute estimated values. Others have reported empirical equations using texture and bulk density to estimate soil water retention and hydraulic conductivity (Clapp and Hornberger, 1978; Gupta and Larson, 1979; Arya and Paris, 1981; Rawls and Brakensiek, 1982; Saxton et al., 1986). Wu et al. (1990) developed an empirical technique to account for both particle size and aggregation effects on the water retention function. Unfortunately, these techniques need more evaluation for application to a wider range of soils, states of aggregation and compaction.

Van Genuchten (1980) presented the following equations to define mathematically the soil water retention relationship (eqn. (3)) and unsaturated hydraulic conductivity (eqn. (4)):

$$\theta = \theta_r + (\theta_s - \theta_r)[\frac{1}{1+(\alpha h)^n}]^{(1-1/n)} \tag{3}$$

$$K(h) = K_s \frac{1-(\alpha h)^{n-1}[1+(\alpha h)^n]^{(1-n)/n^2}}{[1+(\alpha h)^n]^{n-(1/2n)}} \tag{4}$$

where θ = actual water content; θ_s and θ_r = saturated and residual water content, respectively; $K(h)$ = unsaturated hydraulic conductivity; h = matric potential; K_s = saturated hydraulic conductivity; α and n = parameters describing the shape of the soil water retention function. The parameter n is closely related to pore size distribution (Horton et al., 1987). This mathematical form (Van Genuchten, 1980) is the most general formulation available; others are restrictive in that the fit is best at some special range of the water retention function.

Often the water retention function and/or the hydraulic conductivity function (of water content) are not available for a test soil over the whole water content range. Wösten and Van Genuchten (1988) suggested that simultaneous fitting of the functions in eqns. (3) and (4), using the available incomplete data, can provide good estimates of the water retention and conductivity functions for simulations of the water flux.

SOIL WATER DIFFUSIVITY RESPONSE TO BULK DENSITY

Libardi et al. (1982) estimated soil water diffusivity, D, of various soils for a range of bulk density values, provided D is known for at least one bulk density. The following equation was used to represent D (cm^2 s^{-1}) for soil i:

$$D_i(w, \rho_d) = A\ m_i^2\ \exp(Bw) \tag{5}$$

where w = a dimensionless water content equal to $(\theta-\theta_o)/(\theta_s-\theta_o)$; ρ_d = dry bulk density; A and B = constants studied in detail by Brutsaert (1979); m_i ($cm\ s^{-0.5}$) = the slope in a plot of distance (cm) from the water source to the wetting front in horizontal infiltration (Bruce and Klute, 1956), as a function of the square root of the time (s). The volumetric water content, θ, has a value θ_o for the initial air-dry condition and a value θ_s near the water source. The following equation can be used to relate m_i and ρ_d:

$$m_i = a_i + c_i \, \rho_d \tag{6}$$

where a_i and c_i = empirical coefficients. Rearranging eqn. (6) and substituting c as the average value of c_i gives:

$$m_i - a_i = c \, \rho_d \tag{7}$$

where c (= -0.464) and a_i are values which translate the observations to a common ordinate. Libardi et al. (1982) presented the combined results obtained from 13 soil types ranging in textural classification from sand to clay.

If m_i is obtained for one soil at one bulk density value, a_i can be estimated using eqn. (6) and the value of c from eqn. (7). Hence, eqn. (5), written as a function of w, becomes:

$$D_i(w,\rho_d) = A \, (a_i - 0.464\rho_d)^2 \exp \, (Bw) \tag{8}$$

The derivative of eqn. (8) with respect to ρ_d shows how changes in bulk density affect soil water diffusivity:

$$(\partial D_i/\partial \rho_d) = 0.928 \, A \, (a_i - 0.464\rho_d) \exp \, (Bw) \tag{9}$$

FIELD AND LABORATORY MEASUREMENT OF HYDRAULIC PROPERTIES

Field and laboratory techniques for measurement of the unsaturated hydraulic properties of soil are described, respectively, by Green et al. (1986) and Klute and Dirksen (1986). The solution of unsaturated flow problems generally has required experimental determination of the relationship between hydraulic conductivity and water potential or water content. Field methods used to obtain these relationships include the instantaneous profile method, steady-flux methods (with sprinkler irrigation or artificial crusts), sorptivity measurements, and use of tension infiltrometers (Clothier and White, 1981; Ankeny et al., 1988; Elrick et al., 1988; White and Perroux, 1987, 1989). Soil profile and steady-flux techniques often require installation of tensiometers or neutron probe access tubes, which may limit sample numbers. Internal drainage rates in the subsoil often limit the range of applicable θ. When rapid field techniques and straightforward calculations are needed for measuring unsaturated hydraulic properties of the soil, especially at a number of sites and soil depths, tension and positive head infiltrometers have proven successful. They are especially useful for comparing soil management treatments.

Saturated and near-saturated hydraulic properties, especially those in the tilled layer, are of particular interest but are difficult to predict without *in-situ* measurements. Spatial variability encountered in compaction studies due to both intrinsic soil properties and management effects (e.g., compaction or tillage), often necessitates intensive sampling to reach experimental objectives. Description of field-scale water or solute movement also requires that the distribution of hydraulic properties be known. Increasing interest in near-saturated hydraulic properties has prompted improved methods of measuring infiltration under negative water potential (Elrick et al., 1988; Ankeny et al., 1988; Perroux and White, 1988). An experimental approach to quantify compaction effects on soil hydraulic properties is to measure steady-state unconfined infiltration rates with ponded and tension infiltrometers (Ankeny et al., 1990b).

Unconfined, saturated infiltration rates are measured by ponding water in a ring pressed a short distance into the soil. The sharpened ring defines the infiltration surface area and prevents lateral surface flow of ponded water. After a steady-state saturated rate is measured, sand is applied to the infiltration surface to establish hydraulic continuity between the tension infiltrometer and the soil surface. Both saturated and unsaturated water flow are three-dimensional and usually reach steady-state rates rapidly (typically within 30 min). Dense and dry soils require the most time to reach a steady-state rate. At the exact location of the infiltration measurements, soil cores for laboratory measurements of desorption or hydraulic conductivity may conveniently be taken afterwards because the soil water content is then much higher. Confined laboratory measurements of unsaturated hydraulic conductivity can be made by a method of steady-state head control using a device suggested by Klute and Dirksen (1986) or as modified by Ankeny et al. (1991). Tension infiltrometers with a larger contact area (Perroux and White, 1988) may be used with a membrane in direct contact with the soil.

Steady-state infiltration rates can be used directly to compare compaction treatments (Ankeny et al., 1990b). In turn, these rates can be used to estimate saturated and unsaturated hydraulic conductivities (Ankeny et al., 1991). There are several practical advantages in estimating conductivities from unconfined measurements: (1) the estimates are independent of antecedent water potential or content; (2) only steady-state infiltration rate measurements are needed; (3) both ponded and tension measurements are taken on the same soil surface area; (4) flow through longer macropores is not interrupted by driving a ring or isolating a monolith; (5) calculations are straightforward. Capillary lengths as well as hydraulic conductivities can be calculated (White and Sully, 1987; Ankeny et al., 1991).

Available instrumentation also allows fast, accurate field determination of sorptivity. For example, an automated tension infiltrometer (Ankeny et al., 1988) can be used in the field or laboratory to measure sorptivity, a measurement of water uptake by soil in the absence of gravitational forces (Philip, 1957), at different surface-applied water potentials. Improved precision of a transducer-equipped tension infiltrometer allows measurement of sorptivity even at low infiltration rates. An alternative method is to use two tension infiltrometers that have a large difference in contact area (White and Sully, 1987; Sauer et al., 1990). Macropore contributions to sorptivity can also be determined using a tension infiltrometer. Sorptivity measurements can be used to model the effect of antecedent water content on soil erosion, and water flux into soil peds from cracks and other macropores in a clay soil.

COMPACTIVE EFFECTS ON SOIL HYDRAULIC PROPERTIES

The soil compaction process increases soil bulk density and decreases total pore space; as a result, water-related soil properties are significantly altered. Another effect of loading on a soil is shear without a significant change in volume. This type of soil strain, associated with traffic and implement use on wet soils, most significantly alters water-related properties. According to Koolen and Kuipers (1983), compaction and/or shear may be produced by traction, transport, and depth control devices on machinery (tires, wheels, tracks, and sliding plates). There are therefore many possibilities for changing water relations in soils as related to tillage and traffic. The number of soil water-related properties chosen for measurement should be sufficient to simulate the water flux process; a single water-related characteristic rarely suffices to explain the full impact of compaction on water relations of an arable soil.

Measured water retention curves and predicted unsaturated hydraulic conductivity for Barnes loam packed at several bulk densities, demonstrated that compaction decreases total porosity (Reicosky et al., 1981). However, unsaturated water contents were larger for a wide range of matric potentials in compacted versus non-compacted soil (Fig. 2). Gupta et al. (1989) have shown a similar response to soil compaction when volumetric water content is related to water potential. Moreover, their relationship was based upon model computation. When the gravimetric water content is related to water potential, the effect of compaction is characteristically related to soil texture (Gupta et al., 1989). The original work of Hill and Sumner (1967) explains that the characteristic influence of compaction on the water retention curve is based on three combined effects: (1) prominence of large pores; (2) distribution of smaller pores; (3) overall reduction in pore volume caused by the compaction. Hill et al. (1985) characterized and explained water retention curves related to tillage systems without traffic-induced compaction, but this technique can also be used to describe compaction effects on soil water retention.

Hydraulic conductivity as a function of soil water content generally decreases with compaction (Fig. 2); however, over some of the compactive range, hydraulic

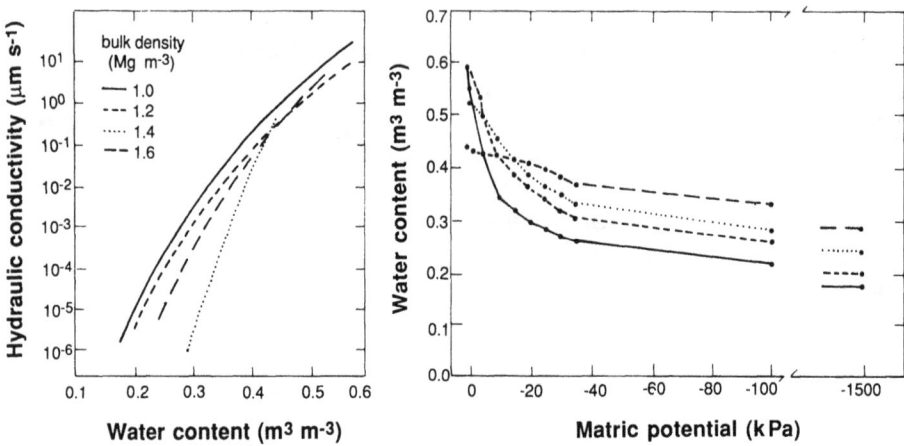

Fig. 2. Unsaturated hydraulic conductivity and soil water characteristics for Barnes loam packed to known bulk densities (from Reicosky et al., 1981).

conductivity as a function of matric potential may increase with compaction (Mapa et al., 1986; Horton et al., 1989). Saturated hydraulic conductivity is especially sensitive to soil compaction caused by field traffic (Table 1). Dawidowski and Lerink (1990) showed that stress produced during traffic has a relatively major influence on saturated hydraulic conductivity compared to the influence on total pore space, while soil water content at the beginning of compaction had a relatively stronger influence on total pore space than on hydraulic conductivity (Fig. 3).

Changes of these two hydraulic properties during uniaxial compression in undisturbed cores were sensitive to both applied stress and soil water content at the time of compaction. The LGP and HGP curves describe the soil response to field traffic at planting (and during traffic for harvest of root crops) using low- and high-pressure tractor tires, respectively (Fig. 3). Differences between uniaxial stress and field traffic were related to soil structure. In a triaxial experiment, where shear was involved during deformation with very small volume changes, the saturated hydraulic conductivity was reduced from 2.6 to 0.06 μm s^{-1} with an axial deformation of only 10% (Dawidowski and Koolen, 1987).

A maximum bulk density of laboratory-compacted soil samples generally occurred when soils were compacted at water contents near field capacity (Akram and Kemper, 1979). Soils containing a wide texture range were compacted with loads equivalent to 340 kPa. The soil compacted at field capacity had infiltration rates <1% of those compacted at air-dry water contents. Dawidowski and Lerink (1990) and Lerink (1990) made similar comparisons. Walker and Chong (1986) reported that sorptivity responded to soil compaction in a manner similar to void ratio (Fig. 4). Sorptivity was more sensitive to changes in soil structure than in

TABLE 1

Summary of means and coefficients of variation for hydraulic conductivities of a Webster silty clay loam after different treatments, measured near the soil surface (after M.D. Ankeny, 1990, unpublished)

Tillage	Position	Traffic	Matric potential (-kPa)	Hydraulic conductivity (μm s^{-1})	CV[1] (%)
No-till	Interrow	No	0.00	166.4	52
			0.30	8.4	42
			0.60	3.0	57
			1.50	1.1	46
		Yes	0.00	26.6	96
			0.30	2.7	116
			0.60	0.9	83
			1.50	0.3	4
	In-row	No	0.00	257.4	70
			0.30	8.8	50
			0.60	3.5	60
			1.50	1.1	44
Chisel	Interrow	No	0.00	219.3	44
			0.30	28.5	42
			0.60	11.6	42
			1.50	2.0	57
		Yes	0.00	33.7	98
			0.30	2.8	116
			0.60	1.1	147
			1.50	0.4	116
	In-row	No	0.00	168.0	88
			0.30	13.2	68
			0.60	5.0	69
			1.50	1.9	67
Plow	In-row	No	0.00	593.2	84
			0.30	12.7	33
			0.60	4.5	27
			1.50	1.2	46

[1]n = 16 for trafficked and non-trafficked; n = 8 for in-row measurements.

soil water content. Mapa et al. (1986) report sorptivity to be sensitive to temporal changes in soil hydraulic properties.

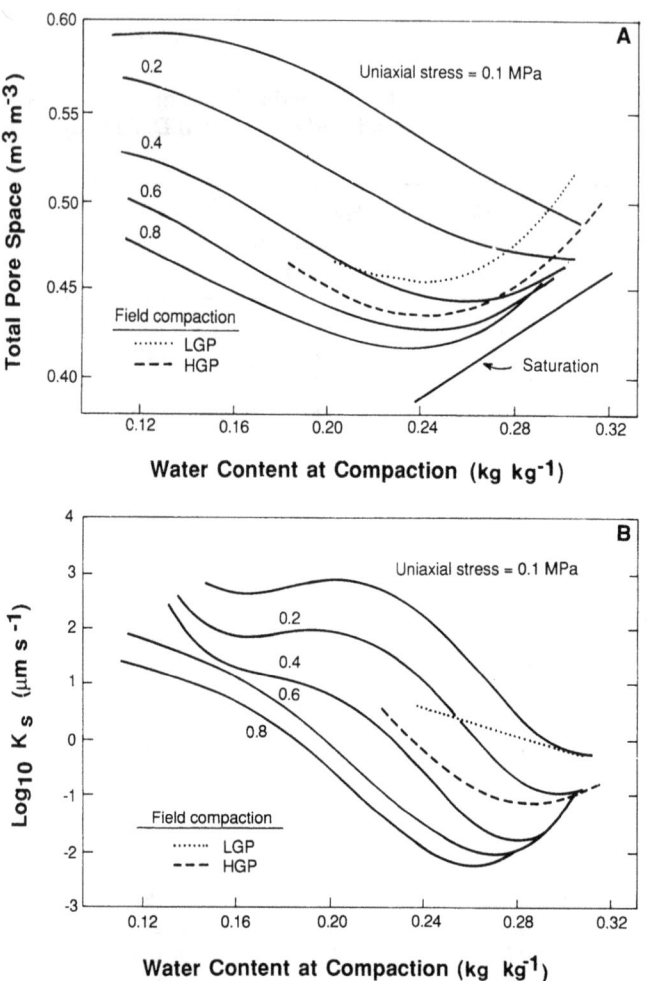

Fig. 3. Changes in total pore space (A) and saturated hydraulic conductivity (B) when a sandy clay loam, after field traffic with low ground pressure (LPG) or high ground pressure (HGP) tires, was subjected to different uniaxial stresses at various water contents. Solid lines refer to compression of aggregate mixtures (from Dawidowski and Lerink, 1990).

Fig. 5 shows the effect of tillage and wheel traffic on infiltration rates at different nominal pore diameters (diameter is inversely proportional to negative water potential) in a silty clay loam. As sequentially smaller macropores emptied, infiltration rates decreased. Compaction reduced infiltration rates in both no-till and chisel treatments. Compaction also changed the slope of the lines in Fig. 5. A decrease of the intercept with constant slope shows proportional reduction in both larger and smaller macropores in which there is water flow. The observed decrease in slope shows that flow in larger pores is more affected by compaction

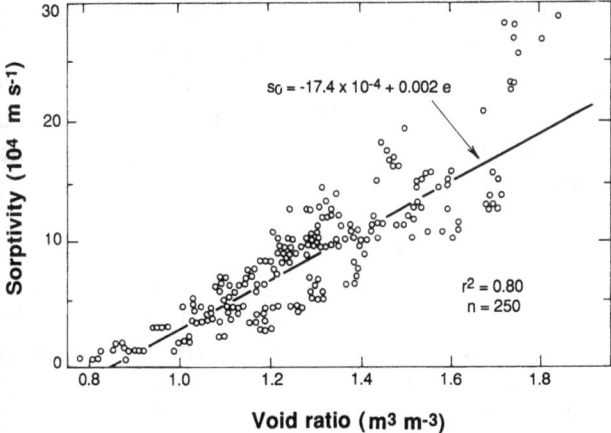

Fig 4. Relationship between sorptivity (S_o) and void ratio (e) in a silt loam soil with different water contents (from Walker and Chong, 1986).

than flow in smaller pores. This observation suggests that compaction destroys more large than small macropores, a phenomenon often measured in desorption curves. Because measured infiltration rates under negative water potential are quite sensitive to compaction, the simplicity and rapidity of these field techniques proves useful in quantifying compaction and other management effects

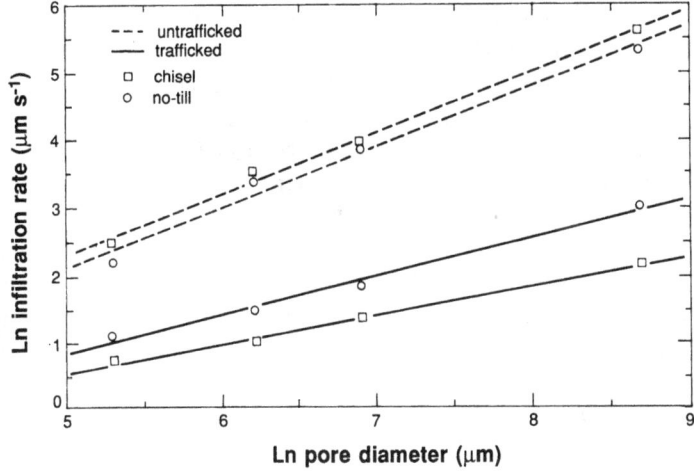

Fig. 5. Influence of compaction on the relationship between unconfined infiltration rate and largest nominal water-filled pore diameter for two tillage treatments on trafficked and untrafficked soil (after Ankeny et al., 1990a).

on hydraulic properties. When there was traffic, the increased separation of datapoints between chisel and no-till treatments (Fig. 5) suggests that infiltration may have responded to the degree of rutting related to antecedent water content or soil strength. Sorptivity also shows great sensitivity to compaction (Walker and Chong, 1986) and can be readily measured in the field. Sorptivity and conductivity measurements might therefore be useful in predicting, as well as measuring, compaction.

The high variability of unconfined hydraulic conductivity measured in a tillage study on a Webster silty clay loam is typical (Table 1). Means within each treatment shown in Table 1 display a response to matric potential. The coefficients of variation are typical for controlled wheel traffic plots. Higher variability when there was traffic is at least partially due to increased measurement error associated with measuring low flow rates. Because traffic has such a profound influence on infiltration properties (Table 1), variability can be higher where measurements are taken without knowledge of the compaction history of the site.

Soil compaction dramatically influences solute transfer, particularly because of the influence on water flow. In fact, solute measurements can infer water flow responses to compaction. Kluitenberg et al. (1988) compared solute breakthrough curves for undisturbed and compacted soil samples. Similarly to Hill et al. (1985), they found that compaction reduces the number of large pores, which is shown in breakthrough curves by reductions in both mean pore-water velocity and hydrodynamic dispersion.

Kiuchi et al. (1990) and Kirkham and Horton (1990) have proposed to use soil compaction as a means to retard nitrate leaching. Where fertilizer is banded, a compacted layer either above or below the band, will direct water flow away from the band into less compacted soil on both sides. In fact, banding fertilizer without formation of a compacted layer would leave a macroporous pathway for enhanced water flux through the banded fertilizer, which is analogous to the flow observed through chisel marks (Pikul et al., 1990). Both laboratory and field studies show that such water flow barriers from selected compaction of soil zones are effective retardants to fertilizer-nitrate leaching. Further investigation of this use of soil compaction to control water and solute fluxes is warranted.

Compaction of soil generally, but not always, increases bulk density, whereas changes in porosity and pore geometry are always produced. Changes of these soil properties must be considered in the context of compaction in a soil layered and variably structured by tillage systems.

LOOSENING EFFECTS OF TILLAGE ON SOIL HYDRAULIC PROPERTIES

Tillage deforms (strains) soil by applying tensile, shear, and compressive stresses. Every tillage tool is unique in its spatial application of stress and therefore, the strain or deformation would also be spatially unique (Koolen and

Kuipers, 1983). When tillage tools are used to loosen soil, various traction, transport, and depth-control devices (tires, wheels, tracks, sliding plates) may produce a zone of soil compaction just below the depth of tillage tool action, alter strain in the soil in response to the tillage tool, and/or recompact bands of soil. Soil structure may also change in response to subsequent wheel traffic without tillage, as well as to biological activity and weather-related inputs of energy. Currently no theoretical basis exists for predicting soil water properties from use of a tillage tool alone or in combination with similar or different tillage tools (Hadas et al., 1988).

For purposes of water flow, a generalized one-dimensional model consists of three soil layers: (1) a tilled and packed A_p layer; (2) an intermediate sublayer rarely tilled but subject to packing; (3) a subsoil unaffected directly by tillage and traffic (Allmaras and Logsdon, 1990). A fourth layer is the surface of the tilled layer subject to sealing and crusting. Such a model systematizes the frequency, spatially and temporally, with which water flow-related parameters must be measured/remeasured. Layering not only affects water flow rates but can also alter water retention of a soil profile after subsequent drainage and redistribution of water (Miller and Bunger, 1963; Miller, 1964; Allmaras et al., 1982).

After tillage and before water has infiltrated, the tilled layer is an admixture of clods, incorporated residue, and tillage voids (unstable voids created by tillage; see Chapter 5) (Soane, 1990; Staricka et al., 1991); the incorporated materials, including broadcast agrichemicals and weed seeds, are located in the tillage voids. The clods may or may not have a wide variation in density depending on soil management prior to the most recent tillage. Vertical uniformity is rare. After some infiltration and biological activity, these tillage voids should resemble flow paths around clods or aggregates.

Rawls et al. (1983) predicted hydraulic properties in the tilled layer and their change during subsequent slaking and natural recompaction. They combined the usual prediction of retention and hydraulic conductivity functions (using organic carbon content, soil texture, effective saturation and pore size distribution) with field-measured porosity changes produced by moldboard plowing, which were also related to soil texture. After first accounting for total porosity produced immediately after moldboard plowing, they then accounted for both seasonal decrease of total porosity (in the absence of wheel traffic) and changes in total porosity for other tillage systems (i.e. chisel, plow-disk-harrow, rotary, plow-pack using traffic) relative to that produced by moldboard plowing.

Brakensiek and Rawls (1983) modeled hydraulic conductivity of the surface crust of the tilled soil, which was made responsive to inputs of rainfall energy (original K_s estimated by Rawls et al., 1983), random roughness of the surface, and a texture-based steady-state matric potential drop across the surface seal. Recent refinements in interpreting the influence of random roughness on depressional storage and soil deposition can improve the Brakensiek and Rawls (1983) model. Moore and Larson (1979) developed a system of routing water over the soil surface to estimate surface storage by use of point data originally measured to estimate random roughness produced by tillage or compaction. Onstad (1984) demonstrated that measurement of precipitation excess (to fill field-estimated depressional storage) was also required because runoff was initiated before maximum depressional storage volume was attained. Linden et al. (1988) presented a theoretical model to relate ponded area, depressional storage volume, and runoff volume to random roughness and its decline during the application of rainfall energy. They included the concept that soil detachment began after infiltration rate lagged behind rainfall rate and a ponded area was initiated; as roughness declined, the surface seal formed in the depressional areas.

INFLUENCES OF SOIL STRUCTURE ON HYDRAULIC PROPERTIES OF SUBSOILS

The pedal nature of soil structure can be used to predict hydraulic properties in soil layers below the tilled layer. Pedal properties are routinely described by pedologists and may be changed directly by soil tillage and traffic in the upper 30 cm of a soil profile (Bouma et al., 1975; Wang et al., 1985; McKeague et al., 1987). However, pedal properties are more obvious factors in the structure and hydraulic properties of soil layers below the tilled layer. When the subsoil is compacted by heavy axle loads, particularly during harvest (Håkansson et al., 1988), changes in the hydraulic properties of a subsoil may adversely influence hydrologic responses in the tilled layer.

McKeague et al. (1987) define pedality as "the natural organization of soil particles into units (peds) which are separated by surfaces of weakness that persist through more than one cycle of wetting and drying in place." McKeague et al. (1982) estimated K_s from soil morphological observations. Eight classes of K_s, ranging from <0.05 to >139 μm s^{-1}, were described on the basis of combinations of texture and morphology. Soils with high sand or clay contents had a high K_s, but biopores and a blocky structure had the largest influence on K_s. The textural relation is not the same as that given by Rawls et al. (1982), in which K_s predictions decreased as sand content decreased and clay content increased. The predictions agree with field measurements of K_s by Topp et al. (1980). Horizons with clay texture had massive and compressed structure, few or no macropores and low K_s; some of these horizons were in the tilled layer and were often at the base of the A_p or upper part of the B horizon and were associated with tillage pans. The degree of compression and associated K_s estimate were difficult to assess visually.

Micromorphometric information concerning planar voids, tillage voids, and tubular pores in both aggregated and non-aggregated structure was used to predict unsaturated hydraulic conductivity and a water retention characteristic (Bouma and Anderson, 1973). However, it was concluded that direct physical measurement is much easier.

The estimation of K_s based on observed structure and texture in the field, has most application in soil layers deeper than the depth of tillage tool action but it can also be used in the tilled layer, especially if severe compaction is involved. Both measured K_s and K_s estimated from soil morphology (structure) detected a tillage pan (produced by moldboard plowing and furrow traffic) at 10-25 cm under continuous corn, but not in adjacent hayland or in first year corn after hayland on clayey soils (Wang et al., 1985). These K_s changes (as much as 100-fold reduction) were much greater than any changes of dry bulk density or total porosity; thus, the massive and compressed structure without macropores or a blocky structure could be detected only by observing structure or measuring K_s. In tillage- and traffic-affected layers, a structure that is more pedal than angular blocky or subangular blocky is rarely found; rather the structure may be platy (compressed), granular or massive.

Water movement in these tillage-affected zones (usually not deeper than 45 cm) may be affected by pedal structures in B horizons below 45 cm depth. Compared to coarse prismatic structure, medium subangular blocky structure (both with the same texture) had a much higher K_s and a much larger hydrodynamic dispersion coefficient during saturated flow. Although hydrodynamic dispersion was much reduced in drained soil columns with pulse applications of water, the subangular blocky structure had a much larger dispersion coefficient than the prismatic structure (Anderson and Bouma, 1977).

MULTI-DIMENSIONAL ASPECTS OF COMPACTION

Compaction is often associated with two-dimensional hydrologic effects because some form of tool action or trafficking occurs on or within the tilled layer (Lindstrom et al., 1981; Reicosky et al., 1981; Voorhees and Lindstrom, 1984; Voorhees et al., 1985). Local slope configuration and orientation, such as in ridge tillage (Hamlett, 1987; Van Es et al., 1988), and strips of heavier residue cover, accentuate these hydrologic effects. They are usually most intense just after tillage; their degree of influence is often tempered with time.

A wheel track may cause significant non-uniform drying during an evaporation period (Reicosky et al., 1981). Matric potential profiles at a selected time after initiation of surface evaporation for both "dry" and "wet" wheel track compaction methods, show clearly that matric potential varies in a two-dimensional fashion, with depth and horizontal location (Fig. 6). Both the soil surface and the soil beneath the wheel track are sinks for water because water potential gradients decrease toward the surface and laterally towards the wheel track region. Vertical and lateral fluxes of water were not determined experimentally, but this experiment shows clearly the need for further two-dimensional experimental as well as two-dimensional numerical studies to characterize traffic compaction effects on soil water flow.

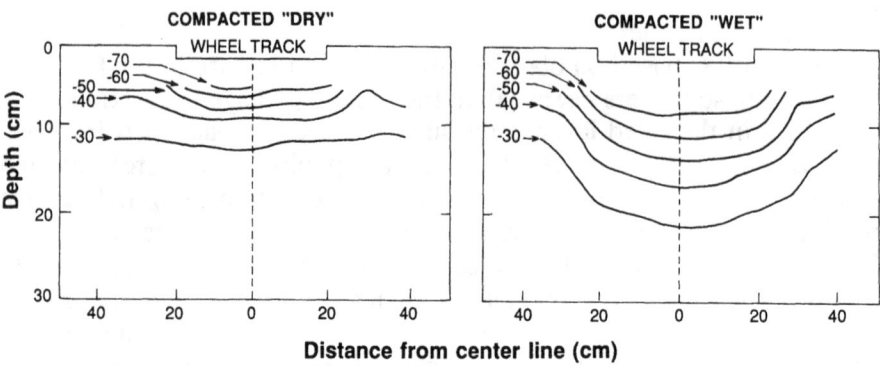

Fig. 6. Matric potential (kPa) distributions in "dry" and "wet" compacted wheel tracks, after 14 days of drying (from Reicosky et al., 1981).

Variability of soil water properties in tilled soils has been studied spatially (Hamlett et al., 1986; Cressie and Horton, 1987) and temporally (Mapa et al., 1986). Matric potential and infiltration of soil water showed spatial variability in both tilled and no-till systems. Analyses indicated that tillage with larger soil disturbance levels may provide more spatial correlation in physical condition of the soil surface. Water retention in freshly tilled soils changed dramatically with successive wetting and drying cycles (Mapa et al., 1986); associated hydraulic conductivity curves also changed with time as a function of wetting and drying cycles. Wetting and drying cycles and freezing and thawing cycles were also reported to increase infiltration rates into previously compacted soil (Akram and Kemper, 1979).

MODELING SOIL WATER FLOW

Numerous models of soil water flow have been developed, some of which have potential to characterize field compaction effects on soil water flow. Both one-dimensional and two-dimensional models are included.

Van Genuchten (1978) modeled flow in layered soil profiles using several numerical schemes; both finite-difference and finite-element simulations offered advantages depending on initial and boundary conditions. Soil water flow and distributions of water content and potential were predicted for soil profiles with both gradual and abrupt changes in properties with depth.

Tillage effects on soil water flow were predicted using one-dimensional numerical models (Hammel et al., 1981; Mapa et al., 1986; Culley et al., 1987a). Hammel et al. (1981) predicted and measured seed zone volumetric water contents to be higher in conventionally tilled soil than in no-till under fallow conditions with no mechanical disturbance; the no-till condition was too dry for successful establishment of winter wheat. Culley et al. (1987a) used two models, a simple water budget and an integrated soil-plant-atmosphere model (NTRM), to predict soil water regimes under no-till and conventional tillage of a highly structured Mollisol. Output from the NTRM model agreed better than the simple water budget model with experimental data, showing water contents to be lower under conventional tillage than no-till. Mapa et al. (1986) predicted the impact of temporal changes in hydraulic functions on water movement in the soil by comparing water content profiles predicted with a simulation model using hydraulic input functions obtained before and after irrigation. Water content profiles after infiltration and redistribution differed substantially, depending upon hydraulic properties of the soil. Temporal changes in soil hydraulic properties after tillage can therefore influence soil water flow and soil water content.

For water flow simulation in fine-textured soils with cracks and macroporous channels sensitive to shrink-swell, Jarvis and Leeds-Harrison (1990) have developed a dual-region model that treats water flow in large continuous cracks in dynamic equilibrium with water content changes in aggregates (or blocks of swelling/ shrinking soil). Some unusual soil parameters were used, e.g., stable crack porosity, slope of the shrinkage characteristic, aggregate diameter distribution and sorptivity at the wilting point in each of the soil horizons specified.

Two-dimensional models have been used to study water flow. Chung and Horton (1987) predicted that a partial surface mulch changes surface evaporative water flux, water contents, temperature, and water potential variations compared to bare soil. The mulch, however, was shown to have almost no effect on drainage. Hamlett (1987) used a finite-element model to predict infiltration and redistribution of water and anions in flat and ridged soil surfaces. He showed that water entered uniformly into the flat profile, while more water infiltrated from the base of the ridge than from the ridge top. Thus, a tracer anion in the ridge did not move as deeply as it did under the flat surface condition. Whisler et al. (1982) modified the GOSSYM cotton growth simulation model to account for changes in hydraulic properties produced by cultivation and wheel traffic. The model predicted rooting patterns to vary according to external processes, and the authors state that the predicted root patterns agreed with observations.

Benjamin et al. (1990a,b) described a two-dimensional heat and water flow model that includes variable soil properties and surface configurations. The model allows for compacted and loosened zones of soil in a ridge tillage system (Fig. 7). Soil physical properties in the planted row, the untracked interrow, and the wheel-tracked interrow positions, were used to model the effects of soil variability (also compaction) on subsurface water and heat transport. Soil below the dotted lines in Fig. 7 was assumed to have properties of the untracked interrow. Simulated water and heat flow over a period of 134 h, produced isolines of soil water potential and soil temperature that were sensitive to ridge configuration and traffic compaction in the furrow (Fig. 8). Isopotential response to traffic compaction was noted as deep as 50 cm. Isopotentials under the non-

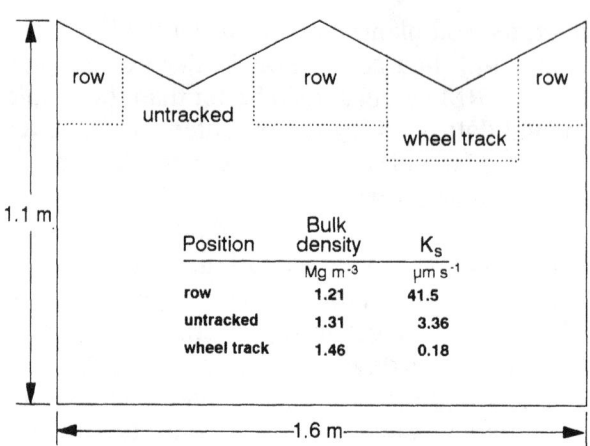

Fig. 7. Soil compaction-related soil properties measured in a ridge tillage system on a Monona silt loam (from Benjamin et al., 1990a).

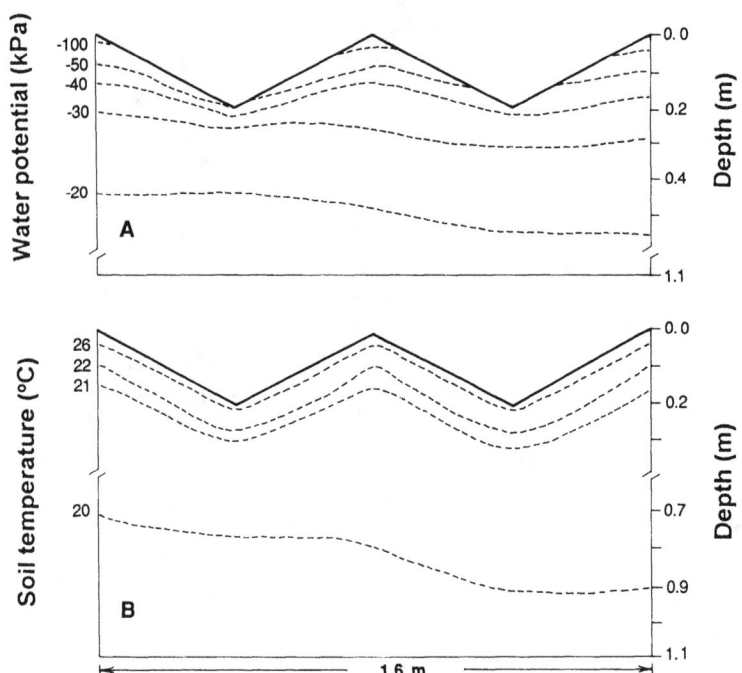

Fig. 8. Predicted soil water potential (A) and soil temperature (B) after 134 h in a ridge tillage system with variable degrees of compaction (cf. Fig. 7) and exposed to a typical diurnal boundary condition at the surface (after Benjamin et al., 1990a).

tracked furrow responded to changes in coupled heat and water flow produced under the tracked furrow. Isothermals more clearly followed the ridge shape and did not show asymmetrical trends due to compaction until deeper than 70 cm. Future research efforts should include studies of variable surface configuration and variable soil properties on soil heat, water and chemical dynamics because it is virtually impossible to maintain constant soil properties horizontally in an arable field.

Most literature on plant response to compaction does not separate the direct from the indirect effects. Research must first identify the direct soil water responses to compaction and tillage and then use simulation with validation to identify the indirect effects. Several especially helpful discussions on this matter are given by Richter (1987), Larson et al. (1989), Rendig and Taylor (1989), and Gliński and Lipiec (1990).

CONCLUSIONS

(1) Among soil properties affected by field compaction and related tillage, perhaps least is known about soil hydraulic properties and processes in spite of their importance.

(2) Although compaction usually does not directly change soil water properties below 30 cm depth, changes of soil water properties above 30 cm depth may dramatically affect water regimes in the subsoil. Rarely is matric flow the dominant mechanism for water flux in these subsoil layers, which contain biopores and planar voids.

(3) Changes of soil water properties in the tilled layer as related to traffic, temporal changes after tillage and extreme layered arrangements, all make it difficult to measure and account for changed soil hydraulic properties and processes in the tilled layer. Matric flow does not dominate because tillage, planar, and biopore voids are usually present in the tilled layer.

(4) A variety of methods are available for *in-situ* measurements and subsequent laboratory measurements on undisturbed samples. Positive/negative pressure infiltrometers are especially useful for studying soil hydraulic properties *in situ* or in the laboratory on undisturbed samples. The proportion of total water flow which passes either through the macropores or through the soil matrix, is influenced by structural changes by tillage and compaction. There is no direct method to estimate soil water properties from a given tillage system and specified soil properties before tillage. However, soil water properties can be estimated, at least generally, from a knowledge of soil structural properties after tillage is completed.

(5) Numerical models are available to predict field compaction effects on soil water regimes in soils that are layered or consist of a matric flux into and out of macro-porous paths. Two-dimensional models utilize surface configuration features related to wheel tracks and residue placement in rows and between rows.

(6) Matrix and macro-porous flow occur simultaneously in all but very sandy soils. Such awareness will improve the flow predictions in the upper soil layers formed by tillage and compaction. Coupled heat and water flow simulations demonstrate relatively greater influences of compaction on the soil water regime relative to the thermal regime close to the soil surface. Soil water regimes can have strong influences on the dynamics of the movement of mineral plant nutrients and pesticide chemicals.

(7) Direct influences of compaction on hydraulic properties may control infiltration and redistribution of water in the soil profile. This water flow process has many indirect and important influences on plant rooting and growth, such as aeration status, soil mechanical resistance to rooting, root extension and uptake of water and plant nutrients. These effects should be included in models predicting the influence of soil compaction on soil hydraulic properties.

ACKNOWLEDGEMENT

Journal Paper No. J-14729 of the Iowa Agriculture and Home Economics Experiment Station, Ames, IA, U.S.A., Projects No. 2556 and 2715.

REFERENCES

Akram, M. and Kemper, W.D., 1979. Infiltration of soils as affected by the pressure and water content at the time of compaction. Soil Sci. Soc. Am. J., 3: 1080-1086.

Allmaras, R.R. and Logsdon, S.D., 1990. Soil structural influences on the root zone and rhizosphere. In: J.E. Box, Jr. and L.C. Hammond (Editors), Rhizosphere Dynamics. Westview, Boulder, CO, U.S.A., pp. 8-54.

Allmaras, R.R., Hallauer, E.A., Nelson, W.W. and Evans, S.E., 1977. Surface energy balance and soil-thermal property modifications by tillage-induced soil structure. Minn. Agric. Expt. Stn., St. Paul, MN, U.S.A., Bull. 306, 44 pp.

Allmaras, R.R., Ward, K., Douglas Jr., C.L. and Ekin, L.G., 1982. Long-term cultivation effects on hydraulic properties of Walla Walla silt loam. Soil Tillage Res., 2: 265-279.

Allmaras, R.R., Pikul Jr., J.L., Kraft, J.M. and Wilkins, D.E., 1988. A method for measuring incorporated crop residue and associated soil properties. Soil Sci. Soc. Am. J., 52: 1128-1133.

Anderson, J.L. and Bouma, J., 1977. Water movement through pedal soils: I. Saturated flow; II. Unsaturated flow. Soil Sci. Soc. Am. J., 41: 413-423.

Ankeny, M.D., Kaspar, T.C. and Horton, R., 1988. Design for an automated tension infiltrometer. Soil Sci. Soc. Am. J., 52: 893-896.

Ankeny, M.D., Horton, R. and Kaspar, T.C., 1990a. Field estimates of hydraulic conductivity from unconfined infiltration measurements. In: K. Roth, H. Fluhler, W.A. Jury and J.C. Parker (Editors), Field-Scale Solute and Water Transport Through Soil. Birkhauser, Basel, Switzerland, pp. 95-100.

Ankeny, M.D., Kaspar, T.C. and Horton, R., 1990b. Characterization of tillage and traffic effects on unconfined infiltration measurements. Soil Sci. Soc. Am. J., 54: 837-840.

Ankeny, M.D., Ahmed, M., Kaspar, T.C. and Horton, R., 1991. Simple field method for determining unsaturated hydraulic conductivity. Soil Sci. Soc Am. J., 55: 467-470.

Arya, L.M. and Paris, J.F., 1981. A physioempirical model to predict the soil moisture characteristics from particle size distribution and bulk density data. Soil Sci. Soc. Am. J., 45: 1023-1030.

Baker, J.L., 1987. Hydrologic effects of conservation tillage and their importance relative to water quality. In: T.J. Logan, J.M. Davidson, J. L. Baker and M.R. Overcash (Editors), Effects of Conservation Tillage on Groundwater Quality: Nitrates and Pesticides. Lewis, Chelsea, MI, U.S.A., pp. 113-124.

Bauder, J.W., Randall, G.W. and Swan, J.B., 1981. Effect of four continuous tillage systems on mechanical impedance of a clay loam soil. Soil Sci. Soc. Am. J., 45: 802-806.

Benjamin, J.G., Blaylock, A.D., Brown, H.J. and Cruse, R.M., 1990a. Ridge tillage effects on simulated water and heat transport. Soil Tillage Res., 18: 167-180.

Benjamin, J.G., Ghaffarzadeh, M.R. and Cruse, R.M., 1990b. Coupled water and heat transport in ridged soils. Soil Sci. Soc. Am. J., 54: 963-969.

Bouma, J. and Anderson, J.L., 1973. Relationships between soil structure characteristics and hydraulic conductivity. In: R.R. Bruce, K.W. Flach and H.M. Taylor (Editors), Field Soil Water Regime. Soil Sci. Soc. Am., Madison, WI, U.S.A., Spec. Publ. 5, pp. 77-105.

Bouma, J., Van Rooyen, D.J. and Hole, F.D., 1975. Estimation of comparative water transmission into two pairs of adjacent virgin and cultivated pedons in Wisconsin. Geoderma, 13: 73-78.

Brakensiek, D.L. and Rawls, W.J., 1983. Agricultural management effects on soil water processes. Part II: Green and Ampt parameters for crusting soils. Trans. ASAE, 26: 1753-1757.

Bruce, R.R. and Klute, A., 1956. The measurement of soil moisture diffusivity. Soil Sci. Soc. Am. J., 20: 458-462.

Brutsaert, W., 1967. Some methods of calculating unsaturated permeability. Trans. ASAE, 10: 400-404.

Brutsaert, W., 1979. Universal constants for scaling the exponential soil water diffusivity. Water Resour. Res., 15: 481-483.

Campbell, G.S., 1974. A simple method for determining unsaturated conductivity from moisture retention data. Soil Sci., 117: 311-314.

Cassel, D.K., 1983. Spatial and temporal variability of soil physical properties following tillage of Norfolk loamy sand. Soil Sci. Soc. Am. J., 47: 196-201.

Chung, S.O. and Horton, R., 1987. Soil heat and water flow with a partial surface mulch. Water Resour. Res., 23: 2175-2186.

Clapp, R.B. and Hornberger, G.M., 1978. Empirical equations for some soil hydraulic properties. Water Resour. Res., 14: 601-604.

Clothier, B.E. and White, I., 1981. Measurement of sorptivity and soil water diffusivity in the field. Soil Sci. Soc. Am. J., 45: 241-245.

Cressie, N.A.C. and Horton, R., 1987. A robust, resistant spatial analysis of soil water infiltration. Water Resour. Res., 23: 911-917.

Culley, J.L.B., Larson, W.E., Allmaras, R.R. and Shaffer, M.J., 1987a. Soil-water regimes of a Typic Haplaquoll under conventional and no-tillage. Soil Sci. Soc. Am. J., 51: 1604-1610.

Culley, J.L.B., Larson, W.E. and Randall, G.W., 1987b. Physical properties of a Typic Haplaquoll under conventional and no-tillage. Soil Sci. Soc. Am. J., 51: 1587-1593.

Dawidowski, J.B. and Koolen, A.J., 1987. Changes in soil water suction, conductivity and dry strength during deformation of wet undisturbed samples. Soil Tillage Res., 9: 169-180.

Dawidowski, J.B. and Lerink, P., 1990. Laboratory simulation of the effects of traffic during seedbed preparation on soil physical properties using a quick uni-axial compression test. Soil Tillage Res., 17: 31-45.

Elrick, D.E., Reynolds, W.D., Baumgartner, N., Tan, K.A. and Bradshaw, K.L., 1988. In situ measurements of hydraulic properties of soils using the Guelph permeameter and the

Guelph infiltrometer. Proc. 3rd Int. Workshop on Land Drainage, Columbus, OH, U.S.A., Dec. 7-11, 1987, pp. G13-G24.

Gantzer, C.J. and Blake, G.R., 1978. Physical characteristics of Le Sueur clay loam following no-till and conventional tillage. Agron. J., 70: 853-857.

Gliński, J. and Lipiec, J., 1990. Soil Physical Conditions and Plant Roots. CRC Press, Boca Raton, FL, U.S.A., 250 pp.

Green, R.E. and Corey, J.C., 1971. Calculation of hydraulic conductivity: A further evaluation of some predictive methods. Soil Sci. Soc. Am. J., 35: 3-8.

Green, R. E., Ahuja, L.R. and Chong, S.K., 1986. Hydraulic conductivity, diffusivity, and sorptivity of unsaturated soils: Field methods. In: A. Klute (Editor), Methods of Soil Analysis. Part 1. Physical and Mineralogical Methods, Am. Soc. Agron./Soil Sci. Soc. Am., Madison, WI, U.S.A., pp. 771-798.

Gupta, S.C. and Larson, W.E., 1979. Estimating soil water retention characteristics from particle size distribution, organic matter percent, and bulk density. Water Resour. Res., 15: 1633-1635.

Gupta, S.C., Sharma, P.P. and De Franchi, S.A., 1989. Compaction effects on soil structure. Adv. Agron., 42: 311-338.

Hadas, A., Larson, W.E. and Allmaras, R.R., 1988. Advances in modeling machine-soil-plant interactions. Soil Tillage Res., 11: 239-282.

Håkansson, I., Voorhees, W.B. and Riley, H., 1988. Vehicle and wheel factors influencing soil compaction and crop response in different traffic regimes. Soil Tillage Res., 11: 239-282.

Hamlett, J.M., 1987. Nitrate movement under a ridge configuration: A field and model investigation. Ph.D. thesis, Iowa State Univ., Ames, IA, U.S.A.

Hamlett, J.M., Horton, R. and Cressie, N.A.C., 1986. A resistant analysis of spatially related data. Soil Sci. Soc. Am. J., 50: 868-875.

Hammel, J.E., Papendick, R.I. and Campbell, G.S., 1981. Fallow tillage effects on evaporation and seedzone water content in a dry summer climate. Soil Sci. Soc. Am. J., 45: 1016-1022.

Hill, J.N.S. and Sumner, M.E., 1967. Effect of bulk density on moisture characteristics of soils. Soil Sci., 103: 234-238.

Hill, R.L. and Cruse, R.M., 1985. Tillage effects on bulk density and soil strength of two Mollisols. Soil Sci. Soc. Am. J., 49: 1270-1273.

Hill, R.L., Horton, R. and Cruse, R.M., 1985. Tillage effects on soil water retention and pore-size distribution of two Mollisols. Soil Sci. Soc. Am. J., 49: 1264-1270.

Hillel, D., 1980. Fundamentals of Soil Physics. Academic Press, New York, NY, U.S.A., 413 pp.

Horton, R., Thompson, M.L. and McBride, J.F., 1987. Method of estimating the travel time of noninteracting solutes through compacted soil material. Soil Sci. Soc. Am. J., 51: 48-53.

Horton, R., Allmaras, R.R. and Cruse, R.M., 1989. Tillage and compactive effects on soil hydraulic properties and water flow. In: W.E. Larson, G.R. Blake, R.R. Allmaras, W.B. Voorhees and S.C. Gupta (Editors), Mechanics and Related Processes in Structured Agricultural Soils. Kluwer, Dordrecht, Netherlands, pp. 187-203.

Jarvis, N.J. and Leeds-Harrison, P.B., 1990. Field test of a water balance model of cracking clay soils. J. Hydrol., 112: 203-218.

Kirkham, D. and Horton, R., 1990. Managing soil-water and chemical transport with subsurface flow barriers. II. Theoretical. Agron. Abstr., 82: 213.

Kiuchi, M., Horton, R. and Kaspar, T.C., 1990. Managing soil-water and chemical transport with subsurface flow barriers. I. Experimental. Agron. Abstr., 82: 214.

Kluitenberg, G.J., Horton, R., Thompson, M.L. and McBride, J.F., 1988. Recompact Iowa soil materials before use as liners for waste containment. Iowa Acad. Sci. J., 95: 114-116.

Klute, A., 1982. Tillage effects on the hydraulic properties of soils: A review. In: D.M. Kral and

S. Hawkins (Editors), Predicting Tillage Effects on Soil Physical Properties and Processes. Am. Soc. Agron., Madison, WI, U.S.A., Spec. Publ. 44, pp. 29-41.

Klute, A. and Dirksen, C., 1986. Hydraulic conductivity and diffusivity: Laboratory methods. In: A. Klute (Editor), Methods of Soil Analysis. Part 1. Physical and Mineralogical Methods. Am. Soc. Agron./Soil Sci. Soc. Am., Madison, WI, U.S.A., pp. 687-734.

Koolen, A.J. and Kuipers, H., 1983. Agricultural Soil Mechanics. Springer, Berlin, Germany, 241 pp.

Larson, W.E., Blake, G.R., Allmaras, R.R., Voorhees, W.B. and Gupta, S.G. (Editors), 1989. Mechanics and Related Processes in Structured Agricultural Soils. Kluwer, Dordrecht, Netherlands, 273 pp.

Lerink, P., 1990. Prediction of the immediate effects of traffic on field soil qualities. Soil Tillage Res., 17: 153-166.

Libardi, P.L., Reichardt, K., Jose, C., Bazza, M. and Nielsen, D.R., 1982. An approximate method of estimating soil water diffusivity for different soil bulk densities. Water Resour. Res., 18: 177-181.

Linden, D.R., Van Doren Jr., D.M. and Allmaras, R.R., 1988. A model of the effect of tillage-induced soil surface roughness on erosion. Proc. 11th Conf. Int. Soil Tillage Res. Org., (ISTRO), Edinburgh, U.K., Vol. 1, pp. 373-378.

Lindstrom, M.J., Voorhees, W.B. and Randall, G.W., 1981. Long-term tillage effects on interrow runoff and infiltration. Soil Sci. Soc. Am. J., 45: 945-948.

Mapa, R.B., Green, R.E. and Santo, L., 1986. Temporal variability of soil hydraulic properties with wetting and drying subsequent to tillage. Soil Sci. Soc. Am. J., 50: 1133-1138.

Marshall, T.J., 1958. A relation between permeability and size distribution of pores. J. Soil Sci., 9: 1-8.

McBride, J.F., Horton, R. and Thompson, M.L., 1987. Evaluation of three Iowa soil materials as liners for hazardous-waste landfills. Iowa Acad. Sci. J., 94: 73-77.

McKeague, J.A., Wang, C. and Topp, G.C., 1982. Estimating saturated hydraulic conductivity from soil morphology. Soil Sci. Soc. Am. J., 46: 1239-1244.

McKeague, J.A., Fox, C.A., Stone, J.A. and Protz, R., 1987. Effects of cropping system and structure of Brookston clay loam in long term experimental plots at Woodslee, Ontario. Can. J. Soil Sci., 67: 571-584.

Miller, D.E., 1964. Estimating moisture retained by layered soils. J. Soil Water Conserv., 19: 235-237.

Miller, D.E. and Bunger, W.C., 1963. Moisture retention by soil with coarse layers in the profile. Soil Sci. Soc. Am. Proc., 27: 586-589.

Millington, R.J. and Quirk, J.P., 1961. Permeability of porous solids. Trans. Faraday Soc., 57: 1200-1207.

Moore, I.D. and Larson, C.L., 1979. Estimating microrelief surface storage from point data. Trans. ASAE, 20: 1073-1077, 1079.

Mualem, Y., 1976. A new model for predicting the hydraulic conductivity of unsaturated porous media. Water Resour. Res., 12: 513-522.

Onstad, C.A., 1984. Depressional storage on tilled surfaces. Trans. ASAE, 27: 729-732.

Onstad, C.A. and Voorhees, W.B., 1987. Hydrologic soil parameters affected by tillage. In: T.J. Logan, J.M. Davidson, J.L. Baker and M.R. Overcash (Editors), Effects of Conservation Tillage on Groundwater Quality: Nitrates and Pesticides. Lewis, Chelsea, MI, U.S.A., pp. 95-112.

Perroux, K. M. and White, I., 1988. Designs for disc permeameters. Soil Sci. Soc. Am. J., 52: 1205-1215.

Philip, J.R., 1957. The theory of infiltration: 4. Sorptivity and algebraic infiltration equations. Soil Sci., 84: 257-264.

Pikul Jr., J.L., Zuzel, J.F. and Ramig, R.E., 1990. Effect of tillage-induced soil macroporosity on water infiltration. Soil Tillage Res., 17: 153-165.

Potter, K.N., Cruse, R.M. and Horton, R., 1985. Tillage effects on soil thermal properties. Soil Sci. Soc. Am. J., 49: 968-973.

Rawls, W.J. and Brakensiek, D.L., 1982. Estimating soil water retention characteristics from soil properties. J. Irrig. Drain., ASCE, 108, No. IR2: 166-171.

Rawls, W.J., Brakensiek, D.L. and Saxton, K.E., 1982. Estimation of soil water properties. Trans. ASAE, 19: 1316-1320, 1328.

Rawls, W.J., Brakensiek, D.L. and Soni, B., 1983. Agricultural management effects on soil water processes. Part I: Soil water retention and Green and Ampt infiltration parameters. Trans. ASAE, 26: 1747-1752.

Reicosky, D.C., Voorhees, W.B. and Radke, J.K., 1981. Unsaturated water flow through a simulated wheel track. Soil Sci Soc. Am. J., 45: 3-8.

Rendig, V.V. and Taylor, H.M., 1989. Principles of Soil-Plant Interrelationships. McGraw-Hill, New York, NY, U.S.A., 275 pp.

Richter, J., 1987. The Soil as a Reactor. Catena, Cremlingen, Germany, 192 pp.

Sauer, T.J., Clothier, B.E. and Daniel, T.C., 1990. Surface measurements of the hydraulic properties of a tilled and untilled soil. Soil Tillage Res., 15: 359-369.

Saxton, K.E., Rawls, W.J., Romberger, J.S. and Papendick, R.I., 1986. Estimating generalized soil-water characteristics from texture. Soil Sci. Soc. Am. J., 50: 1031-1036.

Soane, B.D., 1990. The role of organic matter in soil compactibility: A review of some practical aspects. Soil Tillage Res., 16: 179-201.

Staricka, J.A., Allmaras, R.R. and Nelson, W.W., 1991. Spatial variation of crop residue incorporated by tillage. Soil Sci. Soc. Am. J., 55: 1668-1674.

Topp, G.C., Zebchuk, W.D. and Dumanski, J., 1980. The variation of in situ measured soil water properties within soil map units. Can. J. Soil Sci., 60: 497-509.

Van Es, H.M., Thompson, M.L., Henning, S.J. and Horton, R., 1988. Effect of deep tillage and microtopography on corn yield on reclaimed surface-mined lands. Soil Sci., 145: 173-179.

Van Genuchten, M. Th., 1978. Numerical solutions of the one-dimensional saturated-unsaturated flow equation. Princeton Univ., Princeton, NJ, U.S.A., Water Resour. Program, Dept. Civil Eng., Res. Rep. 78-WR-09.

Van Genuchten, M. Th., 1980. A closed-form equation for predicting hydraulic conductivity of unsaturated soils. Soil Sci. Soc. Am. J., 44: 892-898.

Voorhees, W.B. and Lindstrom, M.J., 1984. Long-term effects of tillage method on soil tilth independent of wheel traffic compaction. Soil Sci. Soc. Am. J., 48: 152-156.

Voorhees, W.B., Evans, S.D. and Warnes, D.D., 1985. Effect of preplant wheel traffic on soil compaction, water use, and growth of spring wheat. Soil Sci. Soc. Am. J., 49: 215-220.

Walker, J. and Chong, S.K., 1986. Characterization of compacted soil using sorptivity measurements. Soil Sci. Soc. Am. J., 50: 288-291.

Wang, C., McKeague, J.A. and Switzer-Howse, K.D., 1985. Saturated hydraulic conductivity as an indicator of structural degradation in clayey soils of Ottawa area, Canada. Soil Tillage Res., 5: 19-31.

Whisler, F.D., Lambert, J.R. and Landivar, J.A., 1982. Predicting tillage effects on cotton growth and yield. In: D.M. Kral and S. Hawkins (Editors), Predicting Tillage Effects on Soil Physical Properties and Processes. Am. Soc. Agron., Madison, WI, U.S.A., Spec. Publ. 44, pp. 179-198.

White, I. and Perroux, K.M., 1987. Use of sorptivity to determine field soil hydraulic properties. Soil Sci. Soc. Am. J., 51: 1093-1101.

White, I. and Perroux, K.M., 1989. Estimation of unsaturated hydraulic conductivity from field sorptivity measurements. Soil Sci. Soc. Am. J., 53: 324-329.

White, I. and Sully, M.J., 1987. Macroscopic and microscopic capillary length and time scales from field infiltration. Water Resour. Res., 23: 1514-1522.

Wösten, J.H.M. and Van Genuchten, M. Th., 1988. Using texture and other soil properties to predict the unsaturated soil hydraulic function. Soil Sci. Soc. Am. J., 52: 1762-1770.

Wu, L., Vomocil, J.A. and Childs, S.W., 1990. Pore size, particle size, aggregate size, and water retention. Soil Sci. Soc. Am. J., 54: 952-956.

Soil Compaction in Crop Production
B.D. Soane and C. van Ouwerkerk (Eds.)
©1994 Elsevier Science B.V. All rights reserved.

CHAPTER 8

Effects of Compaction on Soil Aeration Properties

W. STĘPNIEWSKI[1], J. GLIŃSKI[1] and B.C. BALL[2]

[1]Polish Academy of Sciences, Institute of Agrophysics, Lublin, Poland
[2]Soil Science Department, SAC, Edinburgh, U.K.

SUMMARY

Important indicators of soil aeration status, such as air-filled porosity, air permeability, relative gas diffusion coefficient, soil air composition, oxygen availability (ODR) and redox potential (Eh) are presented. Relationships among these indicators and techniques of measurement are discussed. Field and laboratory experimentation is described. Compaction is shown to restrict aeration and thereby impair crop growth. Examples are given of field experiments, generally involving cereals, with compaction treatments applied either before sowing to simulate unalleviated previous compacton by harvest machinery or during seedbed preparation. Crops, particularly winter crops, are most vulnerable in wet, warm periods shortly after sowing or fertilizer application. Impaired crop growth resulting from compaction is attributed to the interacting effects of poor aeration and mechanical impedance. Poor aeration can also result in gaseous losses of plant-available nitrogen. Experimental evidence of recovery from compaction in undisturbed soils is also presented. Subject areas requiring further research efforts are identified.

INTRODUCTION

Soil aeration is a dynamic soil process which concerns the exchange of gases between the atmosphere and the soil air to support aerobic respiration by plant roots and microorganisms. Aeration is dynamic because soil respiration depends on temperature, the availability of organic matter for oxidation, the composition of the soil air and the volume of air-filled pores. The aeration status of a soil is influenced by both the biological uptake and production of gases and the physical transport of gases between their sites of production or absorption and the free atmosphere. Gas transport occurs in the air-filled pores by a process of diffusion. This permits the two-way exchange of gases involved in respiration, with carbon dioxide passing upward from the soil to the atmosphere and oxygen moving in the opposite direction. The flow rate depends on the volume of air-filled pores and on their number, size distribution and continuity. These properties are closely related to the matric potential of the soil water, the soil bulk density and soil structure (Grable, 1971).

Soil aeration is, in general, a problem for crop growth only in wet soils where the oxygen supply fails to meet the demand and anaerobic zones may develop. This can retard or kill growing roots, can cause the production of physiologically harmful compounds, such as ethylene, or the loss of nitrogen by denitrification (Smith, 1977). Monitoring of oxygen diffusion rates (ODR), which relate to the movement of oxygen to the root under wet conditions, was considered by Erickson (1982) and Gliński et al. (1990) to be the best method to describe the soil-plant interactions associated with aeration status. Restricted aeration impairs plant growth by several mechanisms (Gliński and Stępniewski, 1985) and ultimately reduces crop yields. Henderson and Patrick (1982) reviewed the influence of soil aeration on plant productivity. Since soil wetness and temperature depend on the weather, aeration problems are often of short duration and are difficult to detect and to relate to crop yields and to soil compaction. Smith (1977) concluded that climatic factors were more important than compaction or tillage in determining soil aeration status.

Compaction is generally accepted as being a problem in soils prone to wetness. Consequently, the most vulnerable soils tend to be in the northern temperate regions where the climate is moist (Soane et al., 1981). High soil wetness thus provides conditions both for restriction of soil aeration and for disruption of soil structure by compaction.

Our objective in this review is to indicate the most suitable methods to measure relevant soil properties, to suggest limiting values, to identify the extent of the aeration problem under compaction and to identify future research priorities.

SOIL AERATION PROPERTIES

Since soil aeration depends on a range of both plant and soil properties, these may be used individually as indicators of soil aeration status (Gliński et al., 1990). However, no single property can describe soil aeration completely. Several indicators (Table 1) and their importance for soil processes and plant performance have been presented by Gliński and Stępniewski (1985) and by Gliński and Lipiec (1990).

Air-filled porosity (n_a) is the volume fraction of the soil occupied by gas and relates to the ability of the soil both to store and to transport gas. Relative gas diffusion coefficient and air permeability are measures of the gas transport capability of the soil. Soil air composition and oxygen diffusion rate (ODR) concern the immediate aeration status of the soil. Redox potential, respiratory quotient, Fe^{2+} content, enzymatic activity and the plant mineral composition give evidence of the chemical and biological soil processes associated with the soil

TABLE 1

Indicators of the soil aeration status

Feature	Character
Air-filled porosity (n_a)	Physical
Water-filled porosity (WFP)	Physical
Air permeability (k)	Physical
Relative gas diffusion coefficient (D/D_o)	Physical
O_2 content in soil air	Chemical
CO_2 content in soil air	Chemical
C_2H_4 content in soil air	Chemical
N_2O content in soil air	Chemical
Soil oxygen availability (ODR)	Chemical
Soil redox potential (Eh)	Chemical
Fe^{2+} content in soil	Chemical
Respiratory quotient (RQ)	Biological
Soil enzymatic activity	Biological
Plant mineral composition	Biological

aeration status.

Air-filled porosity

Air-filled porosity is the simplest and probably the oldest indicator of soil aeration status but it is still widely used and relevant. Values of air-filled porosity are often standardised at a matric potential corresponding to field capacity, i.e. the air capacity. Aeration is maintained by gas movement occurring principally within macropores (Grable, 1971). Lower limits of air-filled porosity for maintaining satisfactory aeration are frequently quoted. These depend on the respiratory demand of the soil biological components in addition to soil structure. The influence of compaction on such limits will be discussed later but it can be assumed that an air-filled porosity of 25% (v/v) provides good aeration, that in the range 10-25% (v/v) there may be a limitation to gas exchange under certain conditions and that air-filled porosities < 10% (v/v) are characteristic of deficient aeration (Grable, 1971). An alternative, albeit complementary, parameter of aeration is water-filled porosity. Linn and Doran (1984) indicated that aerobic microbial activity in soil decreased as a result of poor aeration when >60 % of the soil pore space was water-filled. In order to relate air-filled porosity to the capability of the soil for gas exchange, the relationship between air-filled porosity and gas diffusion coefficient is required. Relationships were reviewed by Ball et al. (1988).

Relative gas diffusion coefficient

The gas diffusion coefficient, D, is the main parameter characterising the capability of the soil for gas exchange. Diffusion is the movement of individual gases within the soil air from zones of higher concentration to zones of lower concentration. Gases are principally exchanged by diffusion. D depends not only on the kind of gas, temperature and pressure conditions, and air-filled porosity,

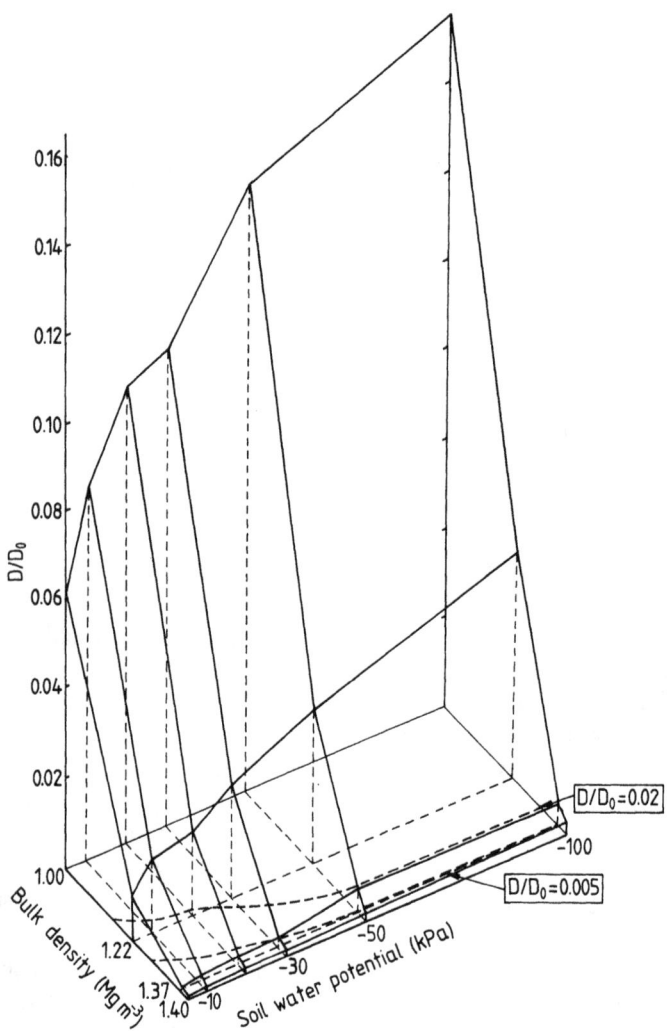

Fig. 1. Dependence of the relative gas diffusion coefficient (D/D_0) of heavy alluvial soil (30% clay, 2.2% organic C) on soil water potential and bulk density (from Stępniewski, 1981).

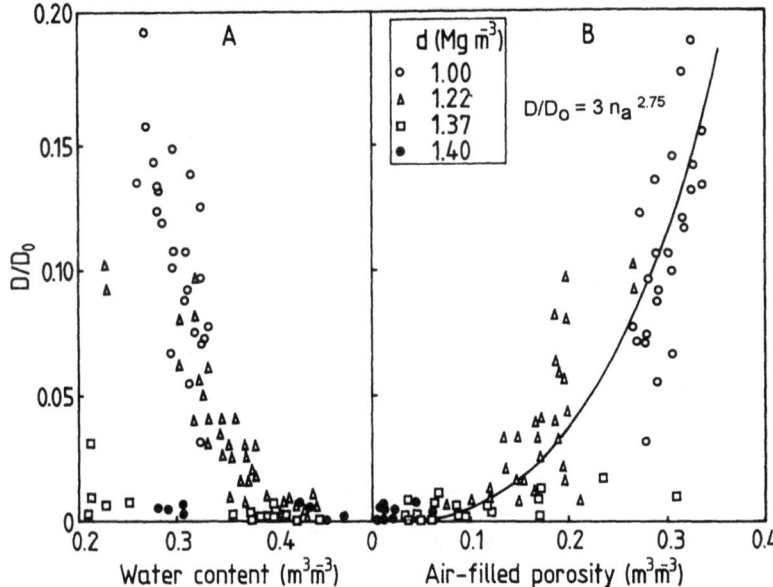

Fig. 2. Dependence of D/D_o of alluvial soil (the same as in Fig. 1) on volumetric water content (A) and air-filled porosity (B), at different bulk densities (A: W. Stępniewski, unpublished data; B: from Stępniewski, 1981).

but also on the continuity and tortuosity of the pores. These are determined by the spatial arrangement of the soil particles, aggregates and water films, viz., soil structure and water potential. Soil gas diffusivity is usually expressed as the relative gas diffusion coefficient, D/D_o, where D is the gas diffusion coefficient in soil and D_o is the diffusion coefficient of the same gas in atmospheric air, under the same conditions of pressure and temperature. An advantage of the use of the relative gas diffusion coefficient is that it does not depend on the temperature, pressure or the type of diffusing gas. Values of D/D_o for air-dry soils range from 0.02 to 0.50. Since D/D_o is insensitive to pore diameter, it can be used with measurements of air-filled porosities to derive indices of pore continuity (Gliński and Stępniewski, 1985; Ball et al., 1988).

The relationship of D/D_o to soil water potential at different levels of compaction (Stępniewski, 1981) is shown in Fig. 1. D/D_o increases, in a curvilinear manner, with a decrease in the soil water potential and decreases rapidly with an increase in the soil bulk density. Soil compaction strongly influences the relationship between D/D_o and moisture content (Fig. 2A). However, the dependence of D/D_o on the air-filled porosity, n_a, of soil is little changed by compaction (Fig. 2B) as the D value does not depend on pore diameter.

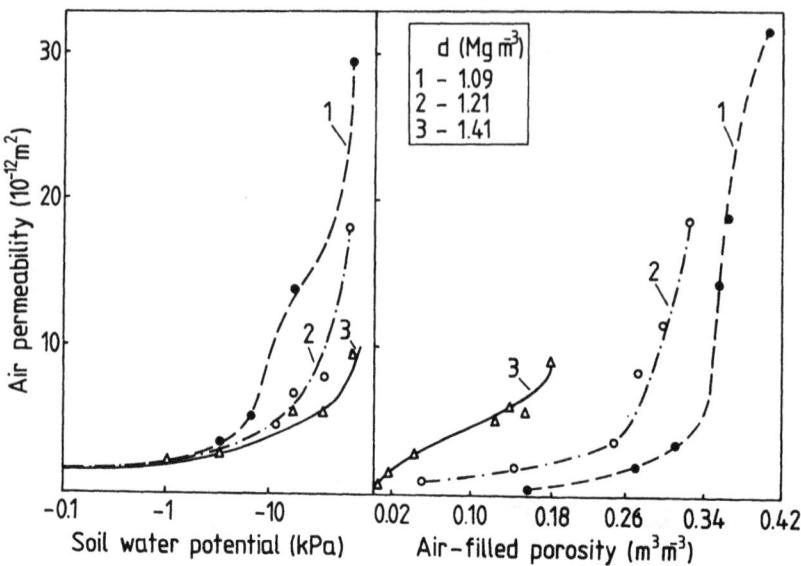

Fig. 3. Air permeability of sandy loam soil as a function of soil water potential and air-filled porosity at different bulk densities (from Turski et al., 1978).

Air permeability[1]

Air permeability characterises the soil's ability to conduct mass flow of soil gas in response to pressure gradients. Mass flow only makes a minor contribution to gas exchange, e.g., in response to changes in barometric pressure, temperature or soil water content. Air permeability of soil is usually within the range from 0.01 to 500.10^{-12} m^2 and depends directly on the number, continuity, tortuosity and the square of the diameter of the air-filled pores. This dependence on pore diameter makes air permeability highly sensitive to the presence of continuous macropores. Air permeability thus depends on water potential, bulk density and soil structure. An example of the relationship between water potential, air-filled porosity and air permeability at different bulk densities is shown in Fig. 3. The large dependence of air permeability, k, on pore diameter makes the measurement highly variable and establishing an average value in structured soil requires high replication (Ball and Smith, 1991). Since the relative diffusion coefficient has no such dependence on the pore diameter, k and D/D_o are not related, though both decrease with increasing compaction as shown in Fig. 4.

---

[1]To comply with common usage, the term "air permeability (m^2)" is used here although the standard term of this property is "intrinsic permeability (m^2)".

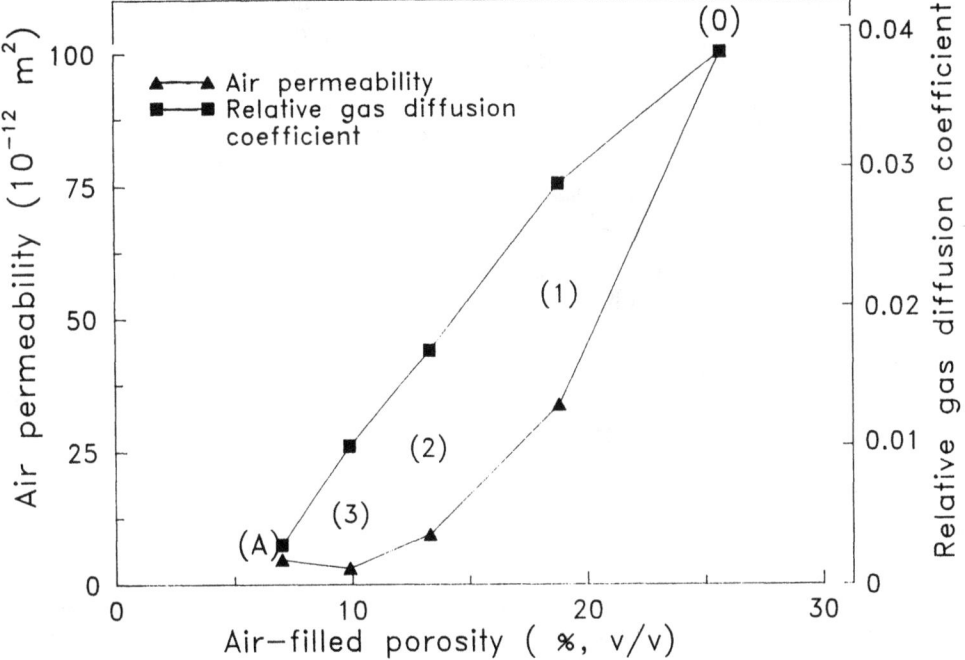

Fig. 4. Relative gas diffusion coefficient, air permeability and air-filled porosity measured on intact cores during progressive uniaxial compaction. Results were normalised to the starting or zero compaction level (0), where bulk density was 1.174 Mg m^{-3}. At compaction levels (1), (2) and (3), stresses applied and resultant bulk densities were: (1) 78 kPa, 1.287 Mg m^{-3}; (2) 155 kPa, 1.375 Mg m^{-3} and (3) 258 kPa, 1.43 Mg m^{-3}. Measurements made on cores from poorly drained zero-tilled areas subject to normal traffic and equilibrated to -6 kPa water potential are located at A (B.C. Ball, unpublished data).

Soil air composition

The soil air composition, unlike the transport parameters which reflect only the potential for gas exchange in the soil, provides a direct indicator of the aeration status. Soil air composition is a measure of the lack of equilibrium between the respired soil gases and the free atmosphere due to impeded gas transport.

The basic components of the soil air are nitrogen, oxygen, carbon dioxide and water vapour. The two most important components are CO_2 and O_2. Almost all of the soil oxygen is contained in air-filled pores because oxygen, like nitrogen, has a very low solubility in water. However, CO_2 is more soluble within water and is distributed in similar quantities within the air- and water-filled pores. Further, the concentration of carbon dioxide in the soil does not change significantly with a change in the proportion of the liquid and gaseous phases at temperatures around 15°C.

In comparison to the atmospheric air, the soil air contains less oxygen (usually 0.1-0.2 m^3 m^{-3}, although it can drop to zero) and more carbon dioxide (usually 0.001-0.05 m^3 m^{-3}, but occasionally >0.1 m^3 m^{-3}). Under conditions of oxic metabolism, the sum of the contents of oxygen and carbon dioxide is about 0.2 m^3 m^{-3}. When this total amount changes, the content of argon and other inert gases remains virtually unchanged, while the content of nitrogen varies within a range of several percent. Campbell (1985) provided simple computer programs to calculate oxygen concentrations and fluxes as a function of depth within a soil profile. In certain cases, when oxygen supply is insufficient, other gases may be produced in the soil air, such as CH_4, H_2S, N_2O, C_2H_4 and H_2. The presence of any of these gases, despite their low concentrations, is an important indicator of restricted aeration status of the soil (Stolzy et al., 1981). Soil air composition is difficult to monitor in the field. Samples need to be extracted periodically from porous chambers buried in the soil and taken to the laboratory for analysis (Smith and Arah, 1991).

Oxygen diffusion rate

As the availability of soil oxygen to plants depends not only on its concentration in the soil air but also on soil physical conditions around the roots,

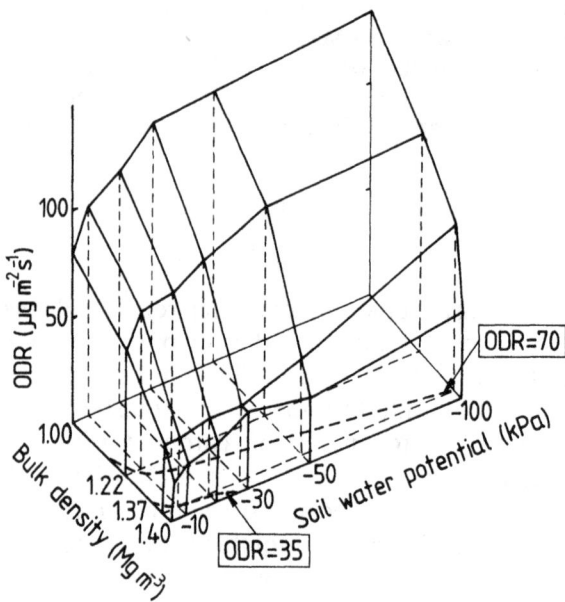

Fig. 5. Dependence of oxygen diffusion rate (ODR) on bulk density and soil water potential of alluvial soil (the same soil as in Fig. 1) (from Stępniewski, 1980).

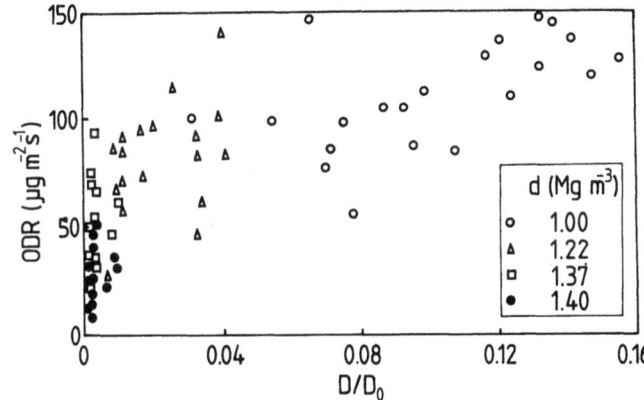

Fig. 6. Dependence of ODR on D/D_0 for alluvial soil (the same soil as in Fig. 1) (from Stępniewski, 1980).

an index called oxygen diffusion rate (ODR) was proposed by Lemon and Erickson (1955). This parameter is measured with a platinum microelectrode on which oxygen is reduced, thereby simulating its uptake by the root. Thus, ODR is a measure of the rate of supply of oxygen from the soil air through the surrounding soil via a film of water adjacent to the root.

The ODR value decreases rapidly with increasing soil bulk density, and increases with decreasing soil water potential (Fig. 5), which is associated with the resulting increase in air-filled porosity and thus with the increase in D/D_0 (Fig. 6). At present, ODR is the best indicator of the potential oxygen availability for plant roots in the soil and it is well correlated with plant emergence, growth and yield (Gliński and Stępniewski, 1985). ODR also provides a good method of identifying the influence of anaerobic microsites on soil aeration (Stolzy et al., 1981). Since soil aeration effects are transient, ODR must be measured punctually during periods of aeration stress (Erickson, 1982). The applicability of the test is limited to wet soils where water potentials are between 0 and -100 kPa.

Redox potential

Another characteristic of soil aeration status is the redox potential, Eh. Redox potential is a measure of the intensity of reduction in soils containing no molecular oxygen. It is a very important indicator of soil reduction processes in water-logged and anoxic soils when other indicators such as ODR and gaseous oxygen concentration are insensitive and of little value. The range of Eh in soils is from +800 to -400 mV. In normal, moist, non-flooded soils, it is governed by the oxygen concentration. Flooding of the soil causes the redox potential to decrease gradually, the rate of decrease depending on soil type, e.g., in various

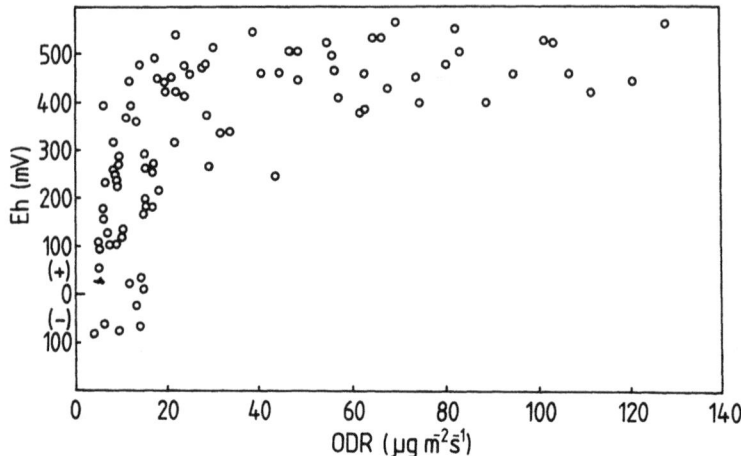

Fig. 7. Relationship between Eh and ODR of a loess soil (1% organic C) (W. Stępniewski, unpublished data).

flooded Polish soils the time for Eh to decrease to 300 mV (t_{300} indicator) ranges between 0.2 and 18 days at 18°C (Gliński and Stępniewska, 1986). The relationship between Eh and ODR for a loess soil is shown in Fig. 7.

MEASUREMENT TECHNIQUES

Each indicator of the soil aeration status generally needs a specific measuring technique. Measuring techniques were reviewed by Gliński and Stępniewski (1985) and Ball and Smith (1991).

Air-filled porosity is usually calculated from samples of known volume using volumetric moisture content and total porosity, or using air pycnometers (Danielson and Sutherland, 1986). In order to monitor air-filled porosity in the field, the relationship between air-filled porosity and soil water potential (water retention characteristic) needs to be identified and soil water potential monitored in the soil layer of interest. Methods of measurement or estimation of the water retention characteristic were reviewed by Klute (1986) and Reeve and Carter (1991) and for monitoring soil water potential in the field by Mullins (1991).

The relative gas diffusion coefficient in soil can be measured using steady-state or nonsteady-state methods (Rolston, 1986). In steady-state methods, diffusion of a gas or vapour is measured at a constant gradient of concentration or partial pressure. These methods have been used under laboratory conditions only and diffusing substances used include carbon dioxide, carbon disulphide, acetone, methanol, ethylene dibromide and oxygen. In nonsteady-state methods, the gradient of concentration or partial pressure of the diffusing agent varies with time. These methods are quicker and, therefore, are more frequently employed, primarily under laboratory conditions. However, some of these methods can also be adapted to field conditions. Various diffusing substances, such as acetone, carbon disulphide, carbon dioxide, nitrogen, hydrogen, oxygen, labelled krypton and labelled xenon have been used. In order to identify conditions where aeration is limiting from relative gas diffusion coefficients, it may be necessary to measure respiration rates.

Air permeability measurement using several types of permeameters was described by Corey (1986).

Soil air composition is analysed using paramagnetic oxygen analyzers, polarographic membrane-covered sensors, gas chromatographs, mass and infrared spectrometers, in addition to older volumetric absorption methods. Smith and Arah (1991) reviewed analysis and sampling techniques.

Redox potential and oxygen diffusion rate can be measured continuously, using multiple probes sampled by one detector. Blackwell (1983) presented methods to measure ODR and Eh, which include the use of cathodes which can be left in the soil and remain functional for several months without cleaning. For microscale measurements, microelectrode techniques have been developed (Greenwood and Goodman, 1967; Revsbech and Ward, 1983; Sextone et al., 1985; Stępniewski et al., 1991). Eh and ODR suffer the limitations of small sampling volume and difficulty of keeping electrodes clean and in working order in the field.

CHANGES IN AERATION IN FIELD SOILS SUBJECT TO COMPACTION

An optimum level of soil compactness is generally accepted as providing the best soil structure for crop growth (Boone, 1986). Reductions in crop yield as a result of compaction beyond this optimum are associated with either poor soil aeration or restricted root growth due to mechanical impedance (Eavis, 1972; Boone et al., 1986).

Field experimentation on compaction has increased in the past ten years because of the growing perception of the importance of soil compaction in crop production. However, more experiments have been made in the laboratory than in the field. Our discussion below divides compaction experiments into two types. First, experiments where aeration status was inferred from air porosity and transport and water status measurements in terms of soil structure, such that values limiting soil aeration are produced. Second, experiments where aeration was monitored directly as oxygen diffusion rate or soil air composition in order to relate crop performance to these parameters.

Compaction, aeration and soil structure

Compaction of loose seedbeds by wheel traffic prior to sowing on a clay loam in a moist climate in Scotland reduced winter barley yields dramatically in some seasons (Campbell et al., 1986). This effect was associated mainly with reductions in plant populations caused by transient waterlogging in early winter. Compaction generally decreased the content of air-filled macropores. The extent of decrease depends on local soil and climate conditions and is usually between 3 and 10% (v/v) (Hodgson and MacLeod, 1989; McAfee et al., 1989; see also Chapters 6 and 7). In aggregated soils, the loss of porosity on compaction was attributed to the collapse of interaggregate pores (Angers et al., 1987).

Uniaxial compaction in the laboratory of intact samples of field soil, with applied stresses in the same order of magnitude as the average ground contact pressure under tractor tyres, resulted in a more rapid decrease in air permeability than in relative gas diffusion coefficient as macropores closed (Fig. 4). More extreme stresses (representative of laden trailer wheels) reduced the relative gas diffusion coefficient more than air permeability. This was attributed to a reduction in macropore continuity and tortuosity. Measurements made on intact samples taken from the same experiment and equilibrated to -6 kPa water potential from poorly drained zero-tilled areas subject to normal traffic (shown at A in Fig. 4), corresponded to very high levels of uniaxial compaction in the laboratory. Thus, Ball concluded that repeated, unrelieved compaction under wet conditions may have a net effect similar to extreme pressures applied in the laboratory.

In Sweden, compaction by a tractor with a soil-tyre contact pressure of 200 kPa

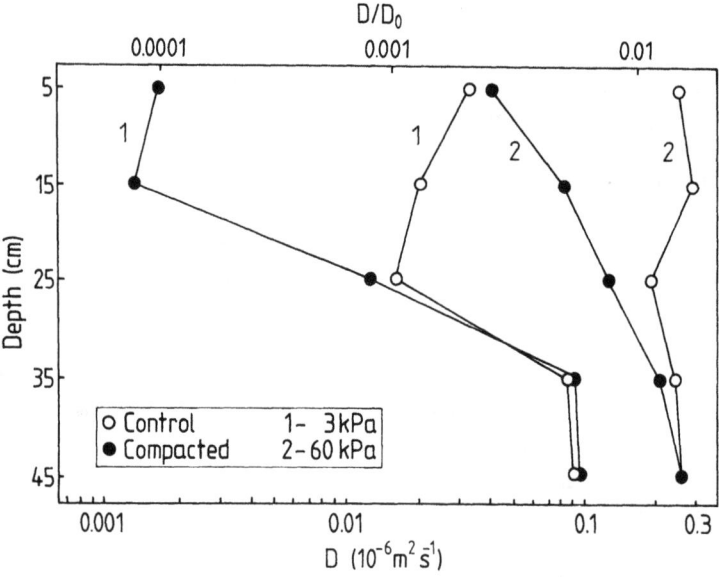

Fig. 8. Variation with depth of the gas diffusion coefficient (D) and relative gas diffusion coefficient (D/Do) in compacted clay soil (49% clay) and in the non-compacted control in Central Sweden at soil water potentials of -3 and -60 kPa (after McAfee et al., 1989).

on clay soil (McAfee et al., 1989), resulted in a decrease in air-filled porosity, greatly reduced air permeability and significantly reduced the relative gas diffusion coefficient (Fig. 8). According to Ball (1987), the assessment of soil aeration status from average values of air permeability is restricted by large variability among replicate samples. In his experiments in Scotland, made on clay loam after one tractor pass, the permeability of seedbed samples at -1 kPa soil water potential ranged from zero to 700×10^{-12} m^{-2}. At this water potential, about half of the samples from the compacted soil were impermeable to gas flow. This impermeability was accompanied by zero or very low relative gas diffusion coefficients. Thus, the establishment of the relationship between both relative diffusion coefficients and air permeabilities and water potential is important. Mean values of diffusion coefficient and air permeability should be determined at water potentials < -2 kPa, since at greater potentials gas movement may be blocked completely.

Although permeability and relative gas diffusion coefficient are poorly related because of the dependence of permeability on macropore diameter, very low permeabilities are almost always accompanied by very low relative gas diffusion coefficients (B.C. Ball, unpublished data). This contrasts with the observation recorded by Grable (1971) that aeration is likely to be adequate provided that mass flow can occur.

The sensitivity of air permeability to compaction in soils subjected to the range of normal traffic pressures may explain why Dickson and Campbell (1990) found that air permeability related well to differences in winter barley yields between zero and normal traffic-compacted plots. Their experiment was on a clay loam in Scotland in a moist climate. They found that although air-filled porosities were frequently <5% (v/v) in both zero and normally compacted soils, such low values limited aeration, and hence crop growth, in the normally compacted soil only. Hence they revised their limiting air-filled porosity to 5% (v/v) under zero compaction and to 10% (v/v) under normal compaction. Similarly, Blackwell et al. (1985, 1986) found that in soils subject to wheel traffic, soil aeration was poor enough to impair crop establishment, growth and yield when air-filled porosity was ≤5% (v/v). Perdok and Lamers (1985) found that, on a marine sandy clay-loam soil, zero-traffic systems maintained air-filled porosities >15% (v/v) and suggested a limiting value of 15% (v/v).

In the Swedish experiment of McAfee et al. (1989), compaction reduced the relative gas diffusion coefficient at -60 kPa soil water potential at 0-15 cm depth from 0.01 to <0.005. Such a low value would be considered to limit gas exchange under Polish soil conditions (Stępniewski, 1981). Soil water potential had a large effect. At -3 kPa, compaction reduced D/D_o by one order of magnitude, although in both treatments the D/D_o value was <0.005 throughout the 0-45-cm depth.

The importance of layers restricting aeration was identified by Boone et al. (1986). Although compaction of loose soil in seedbeds is perhaps the most damaging in terms of soil structure, the depth of the layer where aeration is most affected may vary. Thin layers or crusts near the surface and plough pans were considered by Grable (1971) to have only a minor influence on soil aeration. However, both Dickson and Campbell (1990) and J.T. Douglas (personal communication, 1991) stress the importance of a layer near the surface acting as a throttle controlling gas movement. Douglas and Goss (1987) found that the relative gas diffusion coefficient and air permeability were smaller at -1 kPa water potential at 25-30 cm depth in soil ploughed to 23-25 cm depth than in undisturbed soil. They attributed this to the presence of a plough pan and to the disruption of the continuity of channel-like macropores. Aura (1983) identified the importance of compaction in reducing porosity below the seedbed.

Zero tillage can result in the development of compact layers as a result of wheelings and weathering (Fig. 4). However, biological activity and weathering also may change the structure such that the effects of compaction are moderated (Campbell et al., 1986). This is shown by the relative gas diffusion coefficient in intact soil samples at a given air-filled porosity being greater in zero-tilled than in ploughed soil (Boone et al., 1976), indicating greater pore continuity and less tortuosity in the zero-tilled soil.

In an attempt to describe the range of soil structure suitable for maize growth in the Netherlands, Boone et al. (1986) recognised the importance of water potential in determining limits on aeration critical to crop growth. Their work is perhaps the most penetrating attempt so far to relate soil aeration and compaction to crop growth. They studied treatments of moderate and severe tractor compaction on seedbeds prepared from ploughed soil and derived critical aeration limits for root growth in sandy soils in terms of soil water potential. They realised that the most important effect of compaction on oxygen diffusion coefficient was through its effect on air-filled porosity at a given water potential. Two critical limits for oxygen diffusion coefficients were derived (Table 2). The upper (UCAL) and lower (LCAL) critical aeration limits corresponded to very high and very low respiration (oxygen consumption) rates, respectively. The oxygen content of the soil atmosphere needs to be maintained at a higher level

TABLE 2

Lower and upper critical aeration limits for 0-30 cm depth (after Boone et al., 1986)

Soil aeration characteristic	Lower limit (LCAL)	Upper limit (UCAL)
Oxygen consumption rate (mg m^{-3} s^{-1})	0.093	0.93
Minimum oxygen concentration (%, v/v)	1	10
Oxygen diffusion coefficient (m^2 s^{-1})	7.5×10^{-8}	1.5×10^{-6}
Range of water potentials (kPa)	-2 to -8	-6 to -20
Range of air-filled porosities (%, v/v)	5-8	15-20

by a greater oxygen diffusion coefficient and air-filled porosity for the UCAL than for the LCAL. Correspondingly, the water potential for UCAL is lower than for LCAL.

Compaction can increase the rate of soil respiration. On Dutch soils, Boone et al. (1986) observed that in the most severe compaction treatment oxygen consumption was doubled and they concluded that any mechanical disturbance of the soil can influence oxygen consumption. This effect may, however, result from the generally greater soil water content of compacted soil.

Compaction and soil air composition

As a result of soil compaction, the composition of the soil air changes significantly. It is difficult, however, to assess to what extent these changes affect plant performance. McAfee et al. (1989), in the earlier mentioned experiment on clay soil compacted with a normal four-wheeled tractor, showed that the oxygen concentration of the soil atmosphere was significantly reduced in the compacted treatment, especially at high moisture contents. Fluctuations in O_2 concentration with time were greater in the compacted soil than in the control. An example of O_2 distribution with depth for 3 sampling dates is presented in Fig. 9. Simojoki et al. (1991) also confirmed that the change in soil air composition as a result of compaction is more pronounced during wet periods. In a compacted treatment on clay soil, the O_2 concentration covered a wider range of values at 50 cm depth than in the upper layers (Fig. 10). In a wet period (June) O_2 concentration dropped to 6-7% (v/v), while in the non-compacted soil it was always above 10% (v/v). Klimanek and Greilich (1981) in a 2-year field experiment on loess soil with five bulk densities between 1.11 and 1.49 Mg m^{-3}, and on sandy loam soil with five bulk densities between 1.35 and 1.71 Mg m^{-3}, found a much smaller

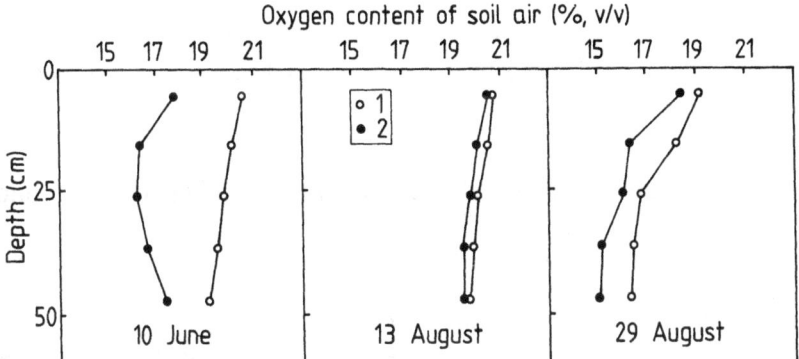

Fig. 9. Changes in oxygen content in the soil air with depth for non-compacted (1) and compacted (2) clay soil (soil and conditions are the same as in Fig. 8) (after McAfee et al., 1989).

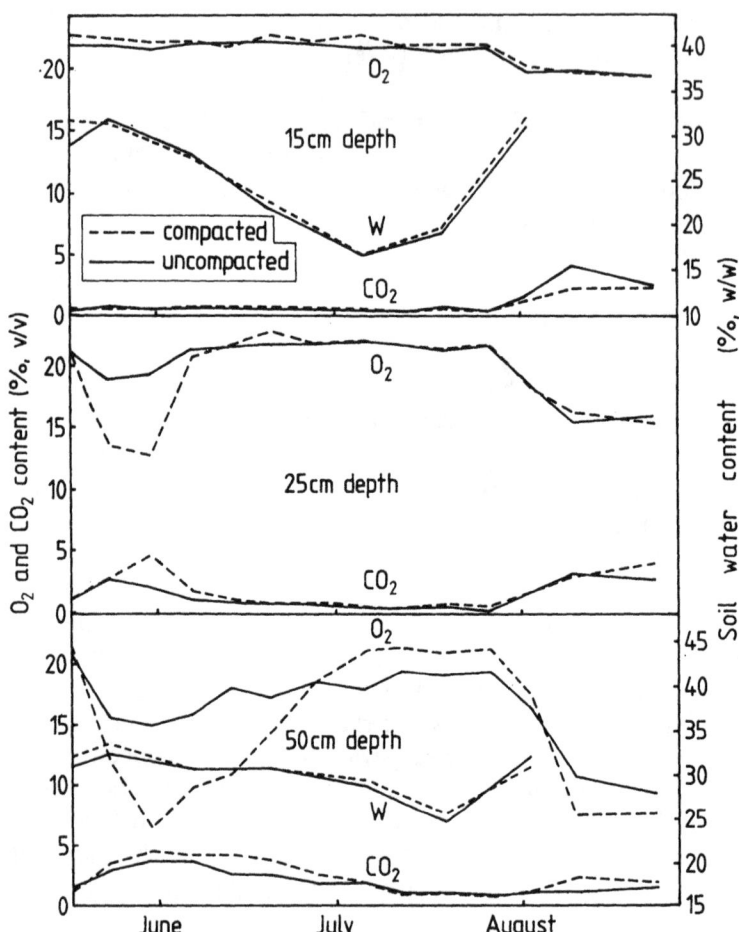

Fig. 10. Average contents of O_2 and CO_2 at 15, 25 and 50 cm depth and water content (W) in a field experiment on Finnish clay soil (48% clay and 2.8% organic C for the 0-30-cm layer and 65% clay and 0.5% organic C below 30 cm) compacted by heavy tractor traffic (after Simojoki et al., 1991).

range in oxygen concentration. Usually the O_2 content varied between 19.5 and 20.8% (v/v) and did not drop below 17.5% (v/v).

In the experiment of Simojoki et al. (1991), the CO_2 concentration apparently changed less with compaction and depth than the O_2 concentration. The maximum concentration of CO_2 was about 5% (v/v) in the compacted treatment and about 4% (v/v) in non-compacted soil and was similar at 25 and 50 cm depth. In the experiment of Klimanek and Greilich (1981) on loess and sandy loam soils, the increase in CO_2 content due to compaction was significant but did not exceed 1% (v/v) between the extreme compaction treatments. For both soils,

CO_2 contents at depths of 15 and 45 cm usually did not exceed 2% (v/v), with only one exception on the loess soil where it reached 2.9% (v/v).

An important consequence of poor soil aeration is the production of anaerobic conditions which enhance loss of fertilizer nitrogen as nitrous oxide. In a field experiment on a Norwegian loam, Bakken et al. (1987) found that tractor traffic on wet soil (matric potential > -5 kPa) reduced total porosity, increased nitrogen loss by denitrification by a factor of 3-4 and reduced wheat yield by about 25 %. In a 75-day period in summer, 15-25 kg ha^{-1} N were lost from this wet compacted soil.

Incorporation of organic material in the soil can impede aeration by increasing microbial respiration and consumption of oxygen, thereby adding to any compaction effects. Ball and Robertson (1990) found that incorporation of straw in a shallow (10-cm deep) layer overlying compact untilled soil, resulted in oxygen concentrations <5 % (v/v) and significant losses of nitrous oxide from the topsoil in early autumn.

Compaction and oxygen diffusion rate

Several field studies have revealed significant effects of compaction on oxygen diffusion rate and redox potential. Redox potential is influenced by changes in oxygen content of the soil air. The extent of the influence depends on soil biological activity and redox buffering capacity.

Soil compaction applied by a smooth powered roller on irrigated soil in a two-year field experiment (Wiecko, 1990), impaired soil physical properties and subsequent shoot growth of Tifway bermudagrass. The oxygen deficit, deduced from ODR values, was greater in the compacted surface layer (5 cm) than deeper in the soil profile. Values of ODR <25 μg m^{-2} s^{-1}, regarded as critical for bermudagrass, were found 27 h and, in some cases, 48 h after irrigation. Further, during drying over a 48-h period, ODR was substantially lower at 5 cm depth in compacted plots than in non-compacted plots. Deeper in the soil profile (at 13 cm depth), differences between treatments were not significant. Large-scale field traffic experiments in England included detailed measurements of soil aeration, both on silt loam (Blackwell et al., 1986) and on a swelling clay (Blackwell et al., 1985). Compaction treatments included zero traffic, tractor traffic and combine harvester traffic applied before sowing winter wheat on plots where all subsequent traffic was localised to lanes between plots. In early spring, redox potential at 15 cm depth decreased periodically to below 250 mV in the silt loam subjected to one pass of a large combine harvester simulator plus one pass of a large tractor. Redox potentials <250 mV are generally associated with exhaustion of oxygen. The effect of wheeling was greater in the clay, where the redox potential in wheeled soil at 15 cm depth was more frequently below 250 mV (Fig. 11). Such low redox potentials occurred after periods of very heavy rainfall. At 5 cm depth the effects of wheeling were slightly less. Aeration conditions were

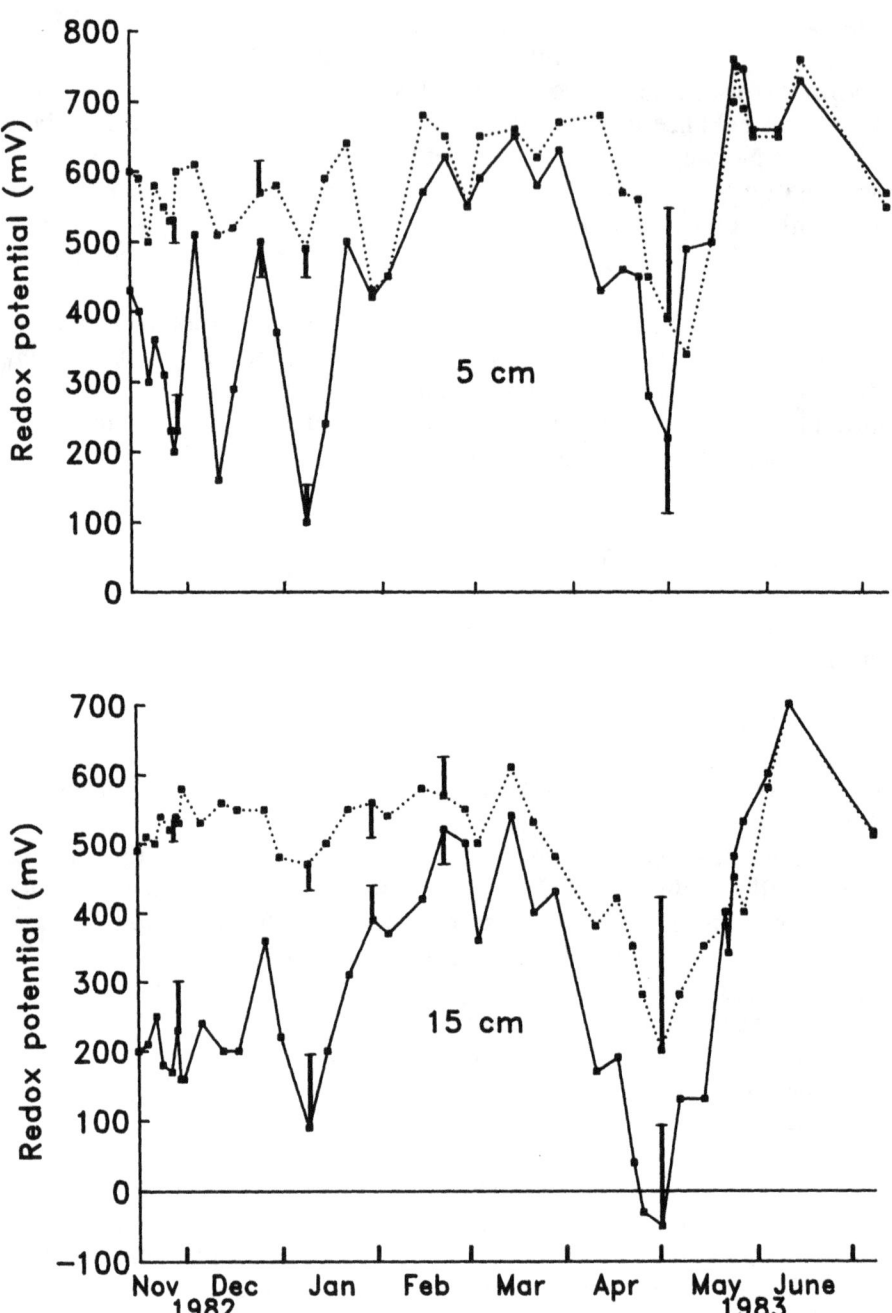

Fig. 11. Redox potential in unwheeled soil (dotted line) and in wheeled soil (solid line) at 5 and 15 cm depth. Standard errors of the differences are shown as bars attached to some points (after Blackwell et al., 1985).

generally satisfactory in the unwheeled soils. Blackwell et al. (1985) considered that the oxygen flux in the wheeled clay was often small enough ($< {\sim}8\,\mu g\,m^{-2}\,s^{-1}$) to have slowed or stopped root growth in the spring in the absence of any other restriction. Redox potentials of <250 mV were also considered by Blackwell et al. (1985) to indicate suitable conditions for denitrification. Although amounts of nitrogen lost may be small in terms of total fertilizer requirements (<10 kg ha^{-1} year^{-1}), these may have been sufficient to cause localised restriction of nitrate supply to the roots in wheeled soil. Once again, however, Blackwell et al. (1985) identified that restriction of root growth was related to a combination of poor soil aeration (restricted oxygen supply) and mechanical impedance.

Cultivation under unsuitable conditions can give problems related to compaction and loss of soil structure. For example, cultivation of wet clay soil was related by Hodgson and MacLeod (1989) to an increase in bulk density of ~0.1 Mg m^{-3}. After wet cultivation, the oxygen flux density (and thus ODR) was 1.2-6 times lower than after dry cultivation. Further, in comparison to dry cultivation, wet cultivation decreased air-filled porosity by up to 11% (v/v).

Although soil aeration is restricted by compaction, there is evidence for recovery of soil structure in undisturbed soil, which should improve soil aeration. Campbell et al. (1986) found that after several years of treatment, moderately compacted soil under zero-tillage allowed faster infiltration of rainfall than under annual chisel ploughing. This improvement in structure was not accompanied by any reduction in bulk density. In England, recovery of topsoil after compaction was associated with a gain in macroporosity of 3% (v/v) at 15 cm depth (Blackwell et al., 1986). Blackwell et al. (1985) reported a similar recovery in compacted clay soil which was restricted to the top 15 cm and associated with wider, deeper cracks during summer drying than in the uncompacted soil. In regenerating soil structure, wetting and drying were considered to be more important than freezing and thawing.

Finally, an interesting interaction between compaction and root growth was identifed by Asady and Smucker (1989) in a pot experiment. They found that plugging by roots of 5% or more of the macropores responsible for aeration, significantly reduced gas movement. The effect increased with increasing soil compactness.

CONCLUSIONS

(1) Soil aeration is a function of soil structure, water potential, bulk density and soil respiration rate, all of which are influenced by soil compaction. Air-filled porosity and oxygen diffusion rate are perhaps the most suitable indicators of soil aeration status in the field. Air permeability is also useful as it indicates the presence of macropores which are vulnerable to compaction.

(2) Compaction can restrict aeration, which impairs root growth. The problem is apparently greatest in northern temperate zones, where unalleviated

compaction caused by traffic during the previous harvest or compaction occurring during seedbed preparation, can severely affect soil aeration during the growth of the subsequent crop. Soil aeration seems to be most impaired by compaction near the surface, although compaction of layers deeper in the soil profile can restrict drainage and increase the likelihood of waterlogging.

(3) Aeration is most likely to be limiting after periods of unusually heavy rainfall or irrigation, when soil temperatures are also relatively high. The influence of restricted aeration is hard to distinguish from that of restriction to root growth caused by mechanical impedance. Impeded aeration causes gaseous losses of nitrogen which are generally small in terms of fertilizer loss but may starve growing roots during vulnerable periods.

(4) There are several areas where further research could lead to a further understanding of the effect of compaction on aeration and subsequent plant growth. These include: (1) more long-term field experimentation; (2) more extensive measurement of air permeability, preferably in relation to other aeration indicators; (3) identification of soil layers where aeration is limiting; (4) more precise experiments, both laboratory and field, to determine the oxygen requirements of individual plant species and to determine plant responses to oxygen stress due to compaction and their interaction with mechanical stress.

REFERENCES

Angers, D.A., Kay, B.D. and Groenevelt, P.H., 1987. Compaction characteristics of a soil cropped to corn and bromegrass. Soil Sci. Soc. Am. J., 51: 779-783.

Asady, G.H. and Smucker, A.J.M., 1989. Compaction and root modification of soil aeration. Soil Sci. Soc. Am. J., 53: 251-254.

Aura, E., 1983. Soil compaction by the tractor in spring and its effect on soil porosity. J. Sci. Agric. Soc. Finland, 55: 91-107.

Bakken, L.R., Børresen, T. and Njøs, A., 1987. Effect of soil compaction by tractor traffic on soil structure, denitrification and yield of wheat (*Triticum aestivum* L.). J. Soil Sci., 38: 541-552.

Ball, B.C., 1987. Air permeability and gas diffusion measurements to quantify soil compaction. In: G. Monnier and M.J. Goss (Editors), Soil Compaction and Regeneration. Balkema, Rotterdam, Netherlands, pp. 15-24.

Ball, B.C. and Robertson, E.A.G., 1990. Straw incorporation and tillage methods: straw decomposition, denitrification and growth and yield of winter barley. J. Agric. Eng. Res., 46: 223-243.

Ball, B.C. and Smith, K.A., 1991. Gas movement. In: K.A. Smith and C.E. Mullins (Editors), Soil Analysis: Physical Methods. Marcel Dekker, New York, U.S.A., pp. 511-549.

Ball, B.C., O'Sullivan, M.F. and Hunter, R., 1988. Gas diffusion, fluid flow and derived pore continuity indices in relation to vehicle traffic and tillage. J. Soil Sci., 39: 327-339.

Blackwell, P.S., 1983. Measurements of aeration in waterlogged soils: some improvements in techniques and their application to experiments using lysimeters. J. Soil Sci., 34: 271-285.

Blackwell, P.S., Ward, M.A., Lefevre, R.N. and Cowan, D.J., 1985. Compaction of a swelling clay soil by agricultural traffic; effects upon conditions for growth of winter cereals and evidence for some recovery of structure. J. Soil Sci., 36: 633-650.

Blackwell, P.S., Graham, J.P., Armstrong, J.V., Ward, M.A., Howse, K.R., Dawson, K.J. and Butler, A.R., 1986. Compaction of a silt loam soil by wheeled agricultural vehicles. I. Effects upon soil conditions. Soil Tillage Res., 7: 97-116.

Boone, F.R., 1986. Towards compaction limits for crop growth. Neth. J. Agric. Sci., 34: 349-360.

Boone, F.R., Slager, S., Miedema, R. and Eleveld, R., 1976. Some influences of zero-tillage on the structure and stability of a fine textured river levee soil. Neth. J. Agric. Sci., 24: 105-119.

Boone, F.R., Van der Werf, H.M.G., Kroesbergen, B., Ten Hag, B.A. and Boers, A., 1986. The effect of compaction of the arable layer in sandy soils on the growth of maize for silage. 1. Critical matric water potentials in relation to soil aeration and mechanical impedance. Neth. J. Agric. Sci., 34: 155-171.

Campbell, D.J., Dickson, J.W., Ball, B.C. and Hunter, R., 1986. Controlled seedbed traffic after ploughing or direct drilling under winter barley in Scotland, 1980-1984. Soil Tillage Res., 8: 3-28.

Campbell, G.S., 1985. Soil Physics with Basic: Transport Models for Soil-Plant Systems. Elsevier, Amsterdam, Netherlands, Developments in Soil Science 14, 150 pp.

Corey, A.T., 1986. Air permeability. In: A. Klute (Editor), Methods of Soil Analysis. Part 1. Physical and Mineralogical Methods. Am. Soc. Agron./Soil Sci. Soc. Am., Madison, WI, U.S.A., pp. 1121-1136.

Danielson, R.E. and Sutherland, P.L., 1986. Porosity. In: A. Klute (Editor), Methods of Soil Analysis. Part 1. Physical and Mineralogical Methods. Am. Soc. Agron./Soil Sci. Soc. Am., Madison, WI, U.S.A., pp. 443-461.

Dickson, J.W. and Campbell, D.J., 1990. Soil and crop responses to zero- and conventional-traffic systems for winter barley in Scotland, 1982-1986. Soil Tillage Res., 18: 1-26.

Douglas, J.T. and Goss, M.J., 1987. Modification of pore space by tillage in two stagnogley soils with contrasting management histories. Soil Tillage Res., 10: 303-317.

Eavis, B.W., 1972. Soil physical conditions affecting seedling growth. I. Mechanical impedance, aeration and moisture availability as influenced by bulk density and moisture levels in a sandy loam soil. Plant Soil, 36: 613-622.

Erickson, A.E., 1982. Tillage effects on soil aeration. In: D.M. Kral, S. Hawkins, P.W. Unger and D.M. Van Doren Jr. (Editors), Predicting Tillage Effects on Soil Physical Properties and Processes. Am. Soc. Agron./Soil Sci. Soc. Am., Madison, WI, U.S.A., Spec. Publ. 44, pp. 91-104.

Gliński, J. and Stępniewski, W., 1985. Soil Aeration and Its Role for Plants. CRC Press, Boca Raton, FL, U.S.A., 229 pp.

Gliński, J. and Stępniewska, Z., 1986. An evaluation of soil resistance to reduction processes. Polish J. Soil Sci., 19: 15-19.

Gliński, J. and Lipiec, J., 1990. Soil Physical Conditions and Plant Roots. CRC Press, Boca Raton, FL, U.S.A., 250 pp.

Gliński, J., Stępniewska, Z. and Stępniewski, W., 1990. Indicators of soil aeration. Ernst-Schlichting-Gedächtnis-Kolloquium, Hohenheim, Germany, Tagungsband S. 1, pp. 1-11.

Greenwood, D.J. and Goodman, D., 1967. Direct measurement of the distribution of oxygen in soil aggregates and in columns of fine soil crumbs. J. Soil Sci., 18: 182-196.

Grable, A.R., 1971. Effects of compaction on content and transmission of air in soils. In: K.K. Barnes, W.M. Carleton, H.M. Taylor, R.I. Throckmorton, G.E. Vanden Berg (Editors), Compaction of Agricultural Soils. Am. Soc. Agric. Eng., St. Joseph, MI, U.S.A., pp. 154-164.

Henderson, R.E. and Patrick, W.H., 1982. Soil aeration and plant productivity. In: CRC Handbook of Agricultural Productivity 1. CRC Press, Boca Raton, FL, U.S.A., pp. 51-69.

Hodgson, A.S. and MacLeod, D.A., 1989. Oxygen flux, air-filled porosity, and bulk density as indices of Vertisol structure. Soil Sci. Soc. Am. J., 53: 540-543.

Klimanek, E.M. and Greilich, J., 1981. Untersuchungen zum CO_2- und O_2-Gehalt in der Bodenluft bei unterschiedlichen Lagerungsdichten der Ackerkrume. (Investigation of CO_2 and O_2 content of differentiated density of the plow layer). Arch. Acker- Pflanzenb. Bodenkd., Berlin, Germany, 25: 525-530 (in German).

Klute, E., 1986. Water retention: laboratory methods. In: A. Klute (Editor), Methods of Soil Analysis. Part 1. Physical and Mineralogical Methods. Am. Soc. Agron./Soil Sci. Soc. Am., Madison, WI, U.S.A., pp. 635-662.

Lemon, E.R. and Erickson, A.E., 1955. Principles of the platinum microelectrode as a method of characterizing soil aeration. Soil Sci., 79: 383-392.

Linn, D.M. and Doran, J.W., 1984. Effect of water-filled pore space on carbon dioxide and nitrous oxide production in tilled and nontilled soils. Soil Sci. Soc. Am. J., 48: 1267-1272.

McAfee, M., Lindstrom, J. and Johansson, W., 1989. Effects of pre-sowing compaction on soil physical properties, soil atmosphere and growth of oats on a clay soil. J. Soil Sci., 40: 707-717.

Mullins, C. E., 1991. Matric potential. In: K.A. Smith and C.E. Mullins (Editors), Soil Analysis: Physical Methods. Marcel Dekker, New York, U.S.A., pp. 75-109.

Perdok, U.D. and Lamers, J.G., 1985. Studies of controlled agricultural traffic in the Netherlands. Proc. Int. Conf. Soil Dynamics, Auburn, AL, U.S.A., Vol. 5, pp. 1070-1085.

Reeve, M.J. and Carter, A.D., 1991. Water release characteristic. In: K.A. Smith and C.E. Mullins (Editors), Soil Analysis: Physical Methods. Marcel Dekker, New York, NY, U.S.A., pp. 111-160.

Revsbech, N.P. and Ward, D.M., 1983. Oxygen microelectrode that is insensitive to medium chemical composition: use in an acid microbial mat dominated by *Cyanidium caldarium*. Appl. Environ. Microbiol., 45: 755-759.

Rolston, D.E., 1986. Gas diffusivity, gas flux. In: A. Klute (Editor), Methods of Soil Analysis. Part 1. Physical and Mineralogical Methods. Am. Soc. Agron./Soil Sci. Soc. Am., Madison, WI, U.S.A., pp. 1089-1119.

Sextone, A.J., Revsbech, N.P., Parkin, T.B. and Tiedje, J.M., 1985. Direct measurement of oxygen profiles and denitrification rates in soil aggregates. Soil Sci. Soc. Am. J., 49: 645-651.

Simojoki, A., Jaakkola, A. and Alakukku, L., 1991. Effect of compaction on soil air in a pot experiment and in the field. Soil Tillage Res., 19: 175-186.

Smith, K.A., 1977. Soil aeration. Soil Sci., 123: 284-291.

Smith, K.A. and Arah, J.R.M., 1991. Gas chromatographic analysis of the soil atmosphere. In: K.A. Smith (Editor), Soil Analysis: Modern Instrumental Techniques. Marcel Dekker, New York, U.S.A., pp. 505-546.

Soane, B.D., Blackwell, P.S., Dickson, J.W. and Painter, D.J., 1981. Compaction by agricultural vehicles: a review. I. Soil and wheel characteristics. Soil Tillage Res., 1: 207-237.

Stępniewski, W., 1980. Oxygen diffusion and strength as related to soil compaction. 1. ODR. Polish J. Soil Sci., 13: 3-12.

Stępniewski, W., 1981. Oxygen diffusion and strength as related to soil compaction. 2. Oxygen diffusion coefficient. Polish J. Soil Sci., 14: 3-13.

Stępniewski, W., Zausig, J., Niggemann, S. and Horn, R., 1991. A dynamic method to determine the O_2-partial pressure distribution within soil aggregates. Z. Pflanzenern. Bodenkd., 153: 1-3.

Stolzy, L.H., Focht, D.D. and Fluhler, H., 1981. Indicators of soil aeration status. Flora, 171: 236-265.

Turski, R., Domżał, H. and Słowińska-Jurkiewicz, A., 1978. Przepuszczalność powietrzna jako wskaźnik stanu fizycznego gleby. (Air permeability as an index of the physical state of soil). Roczn. Glebozn., 29: 3-25 (in Polish, with English summary).

Wiecko, G., 1990. Effects of cultivation and wetting agents on turfgrass growth and on alleviation of soil compaction. Ph.D. thesis, Univ. Georgia, Athens, GA, U.S.A., 125 pp.

Soil Compaction in Crop Production
B.D. Soane and C. van Ouwerkerk (Eds.)
©1994 Elsevier Science B.V. All rights reserved.   191

CHAPTER 9

Effects of Compaction on Soil Strength Parameters

J. GUERIF

INRA, Agronomy Unit of Laon-Péronne, Laon Cedex, France

SUMMARY

Physical and chemical origins as well as basic theoretical aspects of soil strength are presented. The effects of soil structure and organisation, soil constitution (mineral and organic) and water content, are outlined. It is concluded that different scales of observation and measurement are necessary to analyse soil strength origins and to determine changes in field soils subject to traffic, tillage and climate. The variety in measurement techniques, due to the wide field of research where soil strength is involved, is outlined. The different methods are briefly reviewed as well as their fields of application and their limitations. The spatial distribution and variability of soil strength in field soils subject to compaction and tillage is demonstrated. Factors influencing soil strength, as induced by compaction, are investigated. Practical consequences for trafficability and workability are briefly discussed. The unresolved problems in this field of research are discussed in terms of theoretical gaps and methodological needs.

INTRODUCTION

Soil strength is normally defined as the resistance which has to be overcome to obtain a given soil deformation. When submitted to external forces, soils react in different modes, according to the characteristics of the internal stress tensor, which accounts for the distribution, orientation and magnitude of the generated internal stresses. Each mode corresponds to a given stress-strain relationship, which may involve compaction, brittle failure with dilation, and/or plastic flow without volumetric change. The theoretical background to such processes in a material as complex as soil (mainly heterogeneous and discontinuous) remains incomplete or inadequate. As a consequence, agronomists, soil scientists and agricultural engineers commonly use a wide variety of empirical or semi-empirical methods to assess soil strength, which they select according to their field of research, e.g., root growth and seedling emergence studies, assessment of aggregate stability and erodibility, modelling of bearing capacity and rolling resistance, prediction of compaction, analysis of draft force requirements, etc.

The mechanical resistance of the soil may be either a favourable or an unfavourable factor for a particular objective. This depends on its magnitude and spatial distribution, but also on the agricultural practices concerned, the required level of plant productivity and/or on environmental requirements. Increased soil strength, due to drying and/or compaction, may be favourable to trafficability by improving soil bearing capacity but unfavourable to soil tillage by increasing draught force requirements. It may be favourable for efficient subsoiling, but may cause difficulties in obtaining a desired soil structure by tillage, increase unfavourable cloddiness, and restrict root growth and establishment (Boone et al., 1985; Boone et al., 1986; Masle and Passioura, 1987; Tardieu and Manichon, 1987; Tardieu, 1988).

Among the main factors affecting soil strength, compactness, water status and conditions at the time of compaction, are commonly considered (Chancellor, 1971).

SOIL PROPERTIES CONCERNED

Origins of soil strength

Soil deformation can occur when elementary particles are able to separate and to move with respect to each other. Such movements are restricted by particle-to-particle friction and inter-particle bonds. The denser the soil, and the more intricately arranged the particles, the higher the friction forces, which, in conjunction with the overburden pressure (Bradford et al., 1971), are mainly responsible for the strength of dry granular material.

The types of bonds involved in soil strength include cohesion forces due to pore water menisci (i.e., surface tension forces at the air-water interface and negative pore water pressure) and solid phase bonds at particle-to-particle contacts (i.e., mineral-mineral, mineral-organic-mineral). Any of these bonds is enhanced by a large number of contacts per particle. Clay minerals and organic compounds, as a result of their size, specific area, and surface properties, act as cementing agents between coarser particles (skeleton particles), such as silt and sand (Guérif, 1988c). This cementation includes both types of bonds, i.e., capillary forces and solid-phase bonds. At the meso (cm) scale, soil compaction can lead to a closer contact between aggregates and clods and be responsible for the increase in cohesion forces.

Theoretical aspects of soil strength, limitations

Soil mechanics theories are derived from the mechanics of continuous, homogeneous solid bodies. However, in soils, and particularly in agricultural soils, most of the common assumptions are only valid at a given scale of observation or measurement and for given conditions. Except for sample shape and size factors, the bulk behaviour of a continuous, homogeneous and saturated soil volume subjected to external loading, would be similar to the behaviour of an elementary volume (see Chapter 2). However, if the soil volume is heterogeneous, discontinuous (i.e., particulate) and unsaturated, its bulk mechanical behaviour can differ from the behaviour of elementary structural units, clods and aggregates, and will depend on their interactions. The rupture mode may be qualified differently, according to the internal structure of the stressed volume of soil, in conjunction with the scale at which the process is described. For example, the compaction of aggregate beds results, according to their consistency, from the local tensile failure and/or the shear failure of aggregates, in combination with the rearrangement of the fragments of the fractured aggregates (Dexter, 1975; Braunack et al., 1979; Guérif, 1982; Angers et al., 1987; Dexter, 1988).

Organisation, structure, volume concerned, heterogeneity

Soil strength intensity varies widely within the soil profile. Independent of the effect of the distribution of water content, the strength distribution in the profile depends on the distribution of structural elements, such as clods or aggregates, and of compacted or fragmented zones (O'Sullivan et al., 1987; Billot and Marionneau, 1988).

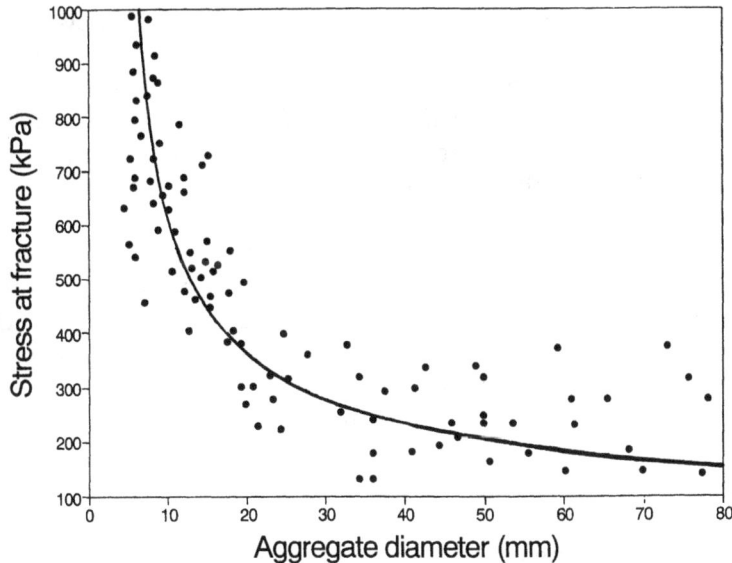

Fig. 1. Stress at fracture under compression as a function of aggregate diameter at constant water potential (from Hadas and Wolf, 1984a).

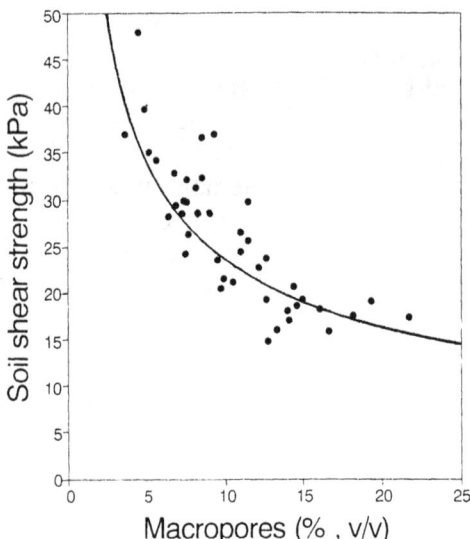

Fig. 2. Relationship between vane shear strength and macropore volume at constant soil water potential (from Carter, 1990).

When sampling structural units within the profile, several authors (Rogowski and Kirkham, 1976; Braunack et al., 1979; Hadas and Wolf, 1984a; Lipiec and Tarkiewicz, 1986; Willatt, 1987; Angers et al., 1987; Carter, 1990) have shown that the measured soil strength and its distribution were dependent on the size (Fig. 1) and on the compactness of the soil volume concerned in the test. However, as bulk density varies with water content according to swelling and shrinkage properties, the effects of bulk density and water content on soil strength are generally confounded. Experimental procedures are often necessary to study these factors separately. Soil strength seems to be largely dependent on compactness, as it increases the number of particle-to-particle contacts and, thus, enlarges the binding forces between elementary particles.

For both relationships, i.e., strength-volume and strength-porosity, the trend observed is independent of the mode of rupture, tension, shearing or compression. It is generally observed that, as the volume concerned increases, the number of cracks or other types of flaws or structural weaknesses increases and, consequently, the strength decreases (Fig. 2).

Braunack et al. (1979) related the strength of aggregates, S, to their volume, V:

$$\ln S = -k \ln V + A \tag{1}$$

where A = an adjustment parameter; k = an indication of how readily large clods will break down into a range of smaller aggregates (Fig. 3). Utomo and Dexter (1981) identified this parameter k with "soil friability", as defined by "the

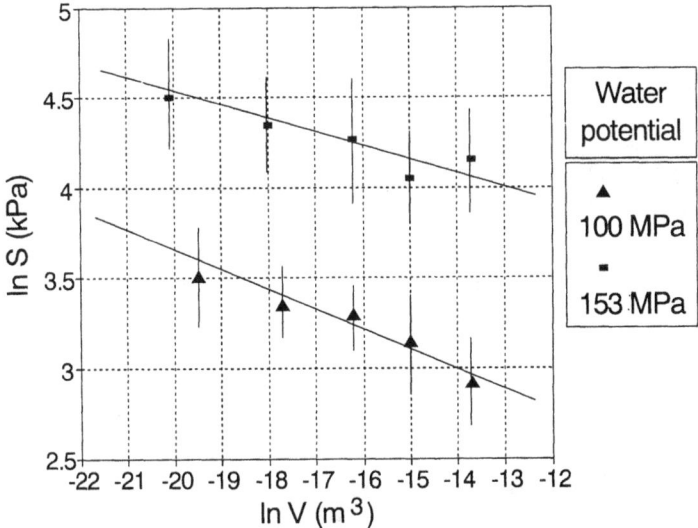

Fig. 3. Mean aggregate tensile strength, S, as a function of aggregate volume, V (from Braunack et al., 1979).

tendency of a mass of unconfined soil to break down and crumble under applied stress into a particular size range of smaller fragments".

Constitution: particle size distribution, clay mineralogy, organic matter

The scattered distribution of strength within a given range of aggregate sizes (Fig. 4), suggests the presence of aggregates of different internal strength due to different internal structure, which may be correlated to different conditions of compaction, or to tillage or weathering effects (Hadas and Wolf, 1984a). This variability of internal structure has often obscured the search for a relationship between soil strength and soil constitution (Spivey et al., 1986).

Various experimental procedures have been used to study the influence of mineral and organic constitution (soil particle distribution, mineralogy) on soil strength, and how it may be modified by natural (organic manure) or chemical additives (conditioners). One technique has been to eliminate internal structure variability by compacting (Spivey et al., 1986), puddling or remoulding natural soils (Middleton, 1924; Towner, 1973; Aylmore and Sills, 1978) and by making artificial mixtures of known proportions of clay, silt and sand fractions (Middleton, 1924; Mullins and Panayiotopoulos, 1984).

Guérif (1988b,c) postulated that natural soil strength could be analysed according to the partition of pore space into two types (Fig. 5), as proposed by Monnier et al. (1973) and by Fiès and Stengel (1981): (1) the "textural" pore space, which is due mainly to the packing state of the elementary particles; (2)

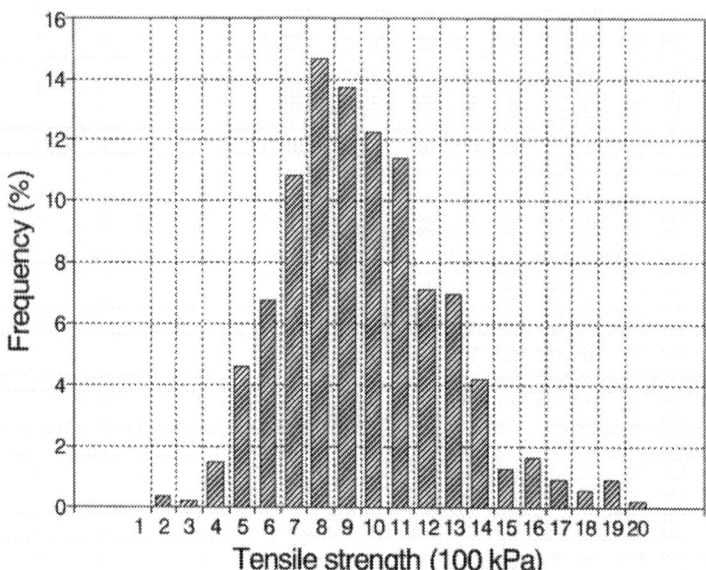

Fig. 4. Tensile strength distribution of 546 dry (over silica gel), 2-3-mm spherical aggregates (from Guérif, 1988b).

the "structural" pore space, which results from the arrangement of structural elements created by tillage and/or weathering (Stengel, 1990). For a given soil, the textural void ratio varies with the water content and is not affected by traffic-

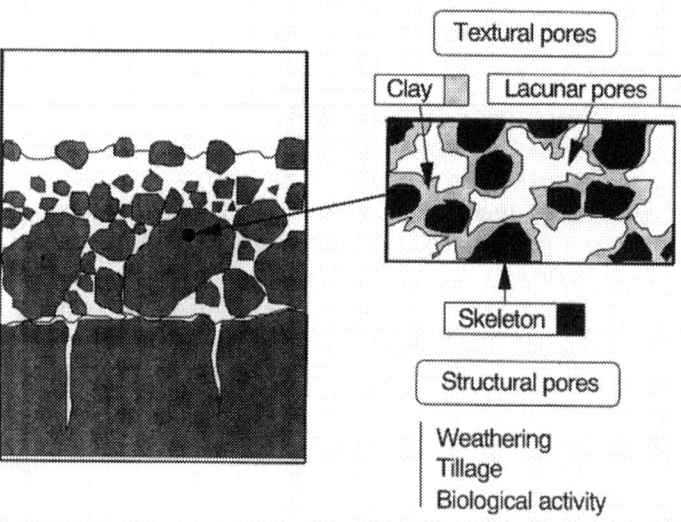

Fig. 5. Pore space partition within a tilled layer.

induced compaction (Guérif, 1988b).

The textural soil strength should be measured on a sufficiently small volume of unremoulded soil (in practice 2-3-mm spherical aggregates) to avoid the presence of cracks and planes of weakness caused by tillage and/or weathering, and to focus on the strength which is due mainly to the arrangement of the skeleton (sand, silt and organic fragments) and the cementing effect of clay, possibly modified by humus (Guérif, 1988a-c, 1990). The overall strength, to be measured on larger volumes of soil, results from the textural strength as well as from the structural strength, due to inter-aggregate bonds and inter-aggregate friction. Most authors consider that soil strength is influenced mainly by the colloid fraction. Sometimes a statistical correlation between soil strength and clay content is encountered, the data being scattered by the variability of organic matter content, exchangeable sodium percentage, or the presence of sesquioxides (Middleton, 1924; Aylmore and Sills, 1978). Guérif (1988c) has shown that the textural tensile strength of calcic soils from temperate regions was highly correlated with their clay content (Fig. 6). A statistical relationship could be established:

$$^aT_o = p \, CLAY + q \qquad (2)$$

where ^aT_o = mean textural tensile strength of dry spherical aggregates of 2-3 mm diameter; CLAY = clay content (g g^{-1}); q = constant. The coefficient p can be interpreted as the mean tensile strength of an ideal clay, representative of the

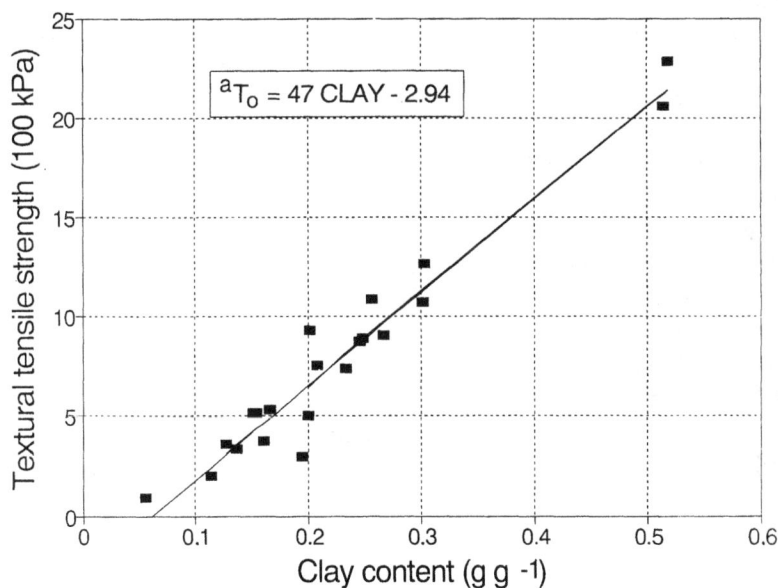

Fig. 6. Relationship between textural tensile strength and clay content (from Guérif, 1990).

different soils involved in the regression analysis.

The arrangement of the skeleton material should have no effect *per se* on tensile strength, even if particle size and grading could modify the location and the thickness of the clay bridges which are responsible for strength. However, when a shear failure was involved in the rupture, as in unconfined compressive rupture of cylindrical cores, the intricate arrangement of the skeleton particles increased the friction forces and thus the strength intensity (Middleton, 1924).

The effects of organic matter on soil compressibility and strength, as reported in the literature, often appear to be contradictory (Soane, 1990). Organic matter includes a great number of very different compounds which, for a better understanding of their specific influence, are best considered in different groups of similar properties.

The first group can be defined at a molecular level, and is included in the clay-humus complex. This complex is generally considered as an aggregating agent, creating bonds between particles and, thus, increasing soil strength. Experimental evidence of such action on natural soils is always difficult to obtain directly, although the importance of increased organic matter content in increasing aggregate stability is often advocated (Aylmore and Sills, 1978). Artificial mixtures of pure clay and organic compounds have shown that clay strength was increased by natural organic polymers, such as polysaccharides (Chenu and Guérif, 1991), or artificial organic polymers, such as soil conditioners (Williams et al., 1967).

The second group concerns organic fragments of undecomposed crop residues, the size of which ranges from a few microns to a few millimetres; when they are included in the textural organisation, they act as a particular part of the skeleton. They are generally supposed to induce weak sites, as well as to increase porosity (Guérif, 1979,1988a). In natural soils the content of such fragments is generally correlated with other organic fractions and, therefore, statistical evidence of their effect is difficult to achieve. However, their experimental incorporation in remoulded and compacted samples induces the same effects (Guérif, 1979; Ohu et al., 1985; Ekwue, 1990).

The last group concerns the larger organic fragments, mostly elongated, such as roots (Willatt and Sulistyaningsih, 1990) and straw, the fibrous structure of which increases the overall soil strength (shear and tension) and generally increases the bearing capacity of soils.

These considerations lead to the conclusion that part of the strength properties of soils are related to their textural characteristics. In most cases, textural tensile strength may be considered an intrinsic property and, thus, for given types of soils, can be predicted. This textural strength, being defined at the scale of the smallest significant elementary volume of cohesive material, may be considered as the upper limit of the strength that a given soil may exhibit after severe compaction.

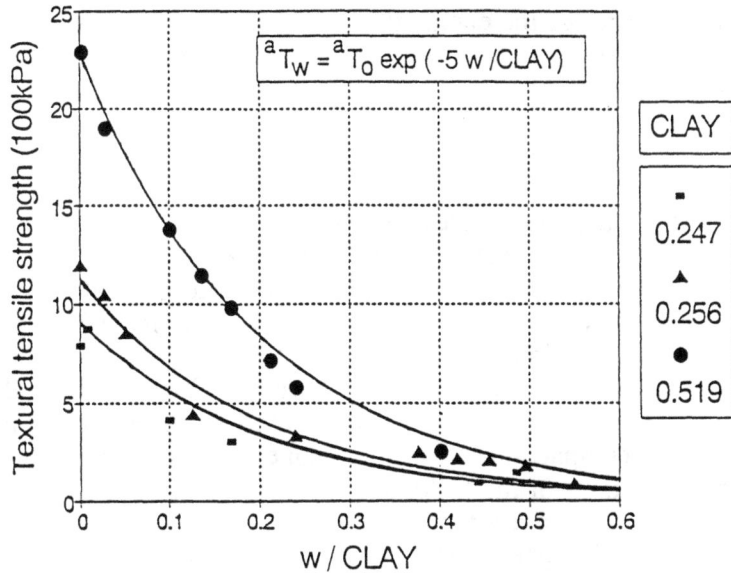

Fig. 7. Relationship between textural tensile strength, ^aT_w, and relative water content of the clay fabric, for three clay contents (from Guérif, 1990).

Water content, water potential

For any mode of rupture (tension, shear or compression) and for most soils, soil strength is increased by drying (Farrell et al., 1967; Spoor and Godwin, 1979; Guérif, 1988c, 1990; Chan, 1989; Carter, 1990). In this phenomenon, two effects interact: (1) drying increases the capillary cohesion as it increases the effective stress; (2) drying increases the compactness by shrinkage.

The mean textural tensile strength of a soil, ^aT_w, at a given water content, w (g g^{-1}), may be fairly well predicted by an exponential relationship (Fig. 7):

$$^aT_w = {}^aT_0 \exp(-kw/CLAY) \tag{3}$$

where $^aT_0$ = dry textural tensile strength (kPa); CLAY = clay content (g g^{-1}); k = adjustment parameter.

Wetting also induces modification of the internal structure (Stengel, 1988), such as microcracks, which decreases strength. These phenomena are part of the weathering effect (Utomo and Dexter, 1981; McKenzie and Dexter, 1985).

MEASUREMENT TECHNIQUES

Procedures for assessing soil mechanical resistance usually involve the measurement of either the force or the energy that is required to obtain a given strain, including sudden collapse, or the measurement of the strain which is induced by a given force.

Empirical and semi-empirical assessments

Certain methods determine a strength index, i.e., a compound parameter which includes components of shear, compressive and tensile strength, the relative contributions of which are not known. Their main interest is that generally they are easy to handle; they are therefore used to compare soil management treatments in rather similar conditions.

Measurements of soil resistance to penetrometer probes of different shape and size, driven into the soil at constant rate, are widely used to investigate soil strength, and particularly compaction effects, as cone resistance is easily measured. Despite the limitations of the technique, stressed by Mulqueen et al. (1977), different applications of cone resistance determinations can be made, such as crushing strength of clods (Campbell, 1976), cone resistance distribution with depth within tilled layers, and analysis of the spatial variability of soil strength (Perfect et al., 1990). Cone resistance is a point measurement. The coordinates of points within a vertical plane through the soil profile can be registered and the contours of iso-resistance can be identified by two-dimensional plotting. The effect of compaction and loosening can therefore be localized and the volume concerned inferred from a transect perpendicular to the direction of traffic (Soane et al., 1986; O'Sullivan et al., 1987; Billot and Marionneau, 1988).

Various tests are carried out to determine the energy requirements of tillage operations for dry soil fragmentation. Instrumented tines are used in the field by various authors (Owen et al., 1987). The analysis of forces required to fracture clods, is performed with different impact loading devices (Bateman et al., 1965, Campbell, 1976; Boyd et al., 1983).

Elementary strength components and types of failure

Most procedures for measuring elementary strength components, as partly reviewed by Snyder and Miller (1985), involve either direct application of stress to a sample (direct methods), or indirect failure induced by applying external compressive forces or bending moments that generate tensile or shear stresses within the sample (indirect methods). Knowledge of elementary strength components is useful when they are to be included into analytical models of soil mechanical behaviour or when strengths measured at different scales and/or with different methods are to be compared.

Direct methods

In the direct shear test, the soil is forced to fail along a certain predetermined plane. In the simple shear box (Casagrande test), the shear strain is induced by horizontal displacement at the bottom of the sample which is submitted to a given normal load. The horizontal dimensions are supposed to remain unchanged. However, as the test proceeds, the area of the shear plane decreases, and the density may increase locally and no longer reflect the original properties of the soil. A ring shear box was developed (Hvorslev apparatus) to avoid these problems (Hartge, 1974). Preparing ring samples, either remoulded or undisturbed, but at a given structural state, remains the major difficulty.

Various torsional shear vane testers have been designed. Motorized equipment was developed for laboratory purposes by Schjønning (1986) and for field utilization by Collis-George and Lloyd (1978), Wong (1980) and Olsen (1984), the principle of which is torque measurement during a rotational shearing action of a grousered shear annulus. Normal load, torque and displacements are now generally measured by electronic transducers and controlled by microcomputer. The soil must be wet enough to allow vane or grouser penetration, as driving the instrument into dry soil often results in appreciable loss of strength by loosening the soil.

Most direct tension tests require either grasping a sample of known geometry at the ends by means of an adhesive (Farrell et al., 1967), or along its sides, and pulling until failure occurs (Gill, 1959). Compressive and shear stresses are often induced and may influence the results (Al-Hussaini and Townsend, 1973).

Indirect methods

Indirect methods for shear strength assessment, such as the drop-cone test (Towner, 1973; Campbell and Hunter, 1986) and the flat-tip or cone penetrometer (Carter, 1990), are generally based on empirical equations obtained by correlation. The normal stress during shear is not controlled. The measurement of tensile strength by indirect methods involves applying external compressive stresses or bending moments to samples of geometrical configurations such that they will fail in tension.

Compression tests were adapted to soils by Kirkham et al. (1959) for cylindrical core samples (brazilian test), and by Rogowski et al. (1968) and Dexter (1975) for aggregates which were assumed to approximate spheres. The tensile strength was estimated from the applied compressive force at rupture and from the geometry of the samples. The assumptions underlying these tests are: (1) continuity of the sample; (2) elasticity of the deformation; (3) uniqueness of the modulus of elasticity in tension and in compression. These conditions prevail only at a relatively low soil water content. Frydman (1964) determined a criterion of applicability of the brazilian test from the amount of flattening at load points of the sample. Guérif (1988b,c) found this range to correspond roughly with a water potential < -500 kPa and suggested the assessment of textural tensile strength by crushing 2-3-mm aggregates which were previously rounded off by mechanical abrasion.

General or complex rupture analysis

The most rigorous analytical procedure to study soil strength is achieved on cylindrical samples in the triaxial test (see Chapter 2). As two parameters (cohesion and angle of internal friction) are estimated by the Mohr-Coulomb equation, the test should be performed at least twice. This experimental procedure assumes that the samples are identical from one test to another, which is most improbable for undisturbed cores. Another limitation is due to the difficulty of collecting, in tilled layers, homogeneous core cylinders, the height of which should be twice their diameter.

Despite the uncontrolled effect of the confinement, the uniaxial compression test under drained conditions is widely used to assess soil compactability (Dexter, 1975; Koolen, 1978; Braunack et al. 1979; Guérif, 1982; Willatt, 1987; Dexter, 1988; McBride, 1989; see Chapter 3).

In order to test the bearing capacity in the field, plate sinkage equipment is used to assess overall soil strength (Soane et al., 1976; Wong, 1980) and the data are often analysed by means of Bekker's approximation (Bekker, 1956). Combining the plate sinkage test and transmission gamma-densitometry allows study of the compactability of tilled soils (Guérif, 1984).

Morphological evaluation of soil structure

As the structure of tilled and wheeled soils and, consequently, their strength distribution, is mainly heterogeneous and discontinuous, it is recommended for field studies that soil strength assessments and morphological observation of soil structure should be combined (Manichon, 1987; Tardieu and Manichon, 1987). The Ap horizon can be partitioned vertically and horizontally into rather homogeneous zones as determined by operating tools and traffic-induced compaction. Each zone can be classified and mapped according to the internal morphology of the structural units and their strength, and also according to the way they are packed.

As a quantification of the result of a given tillage operation, the number, the mass or the volume of a given type of structural unit per unit volume of Ap horizon can be registered (Campbell, 1976; Manichon, 1988). The clod size distribution can be analysed or summarized into the mean-weight diameter.

CHANGES IN STRENGTH PROPERTIES OF FIELD SOILS SUBJECT TO COMPACTION

Strength of compacted layers

Experimental evidence of strength variations under field conditions, due to soil management or wetting and drying, have been reported by many authors

(Pidgeon and Soane, 1978; Douglas et al., 1986). Cone penetration resistance profiles are greatly influenced by soil wetness, which interacts with bulk density. The effect of treatments (cultivation techniques or compaction intensities) on the distribution with depth of cone penetration resistance, generally follows a pattern similar to that in profiles of bulk density (Blackwell et al., 1986; Douglas et al., 1986).

The variations in bulk density are due either to compaction treatments or to natural settlement during wetting and drying. Blackwell et al. (1986), Lipiec and Tarkiewicz (1986), Culley and Larson (1987), Hammel (1988), and Lowery and Schuller (1988) demonstrated in field compaction experiments that soil strength increased with the number of passes and/or axle load, and that the strength increase could persist for at least 3 years.

The stress-strain relationship in soil compaction is time-dependent (Stafford and De Carvalho Mattos, 1981) and Horn and Hartge (1990) demonstrated that soil strength resulting from compaction due to a given load, was dependent on the duration of loading.

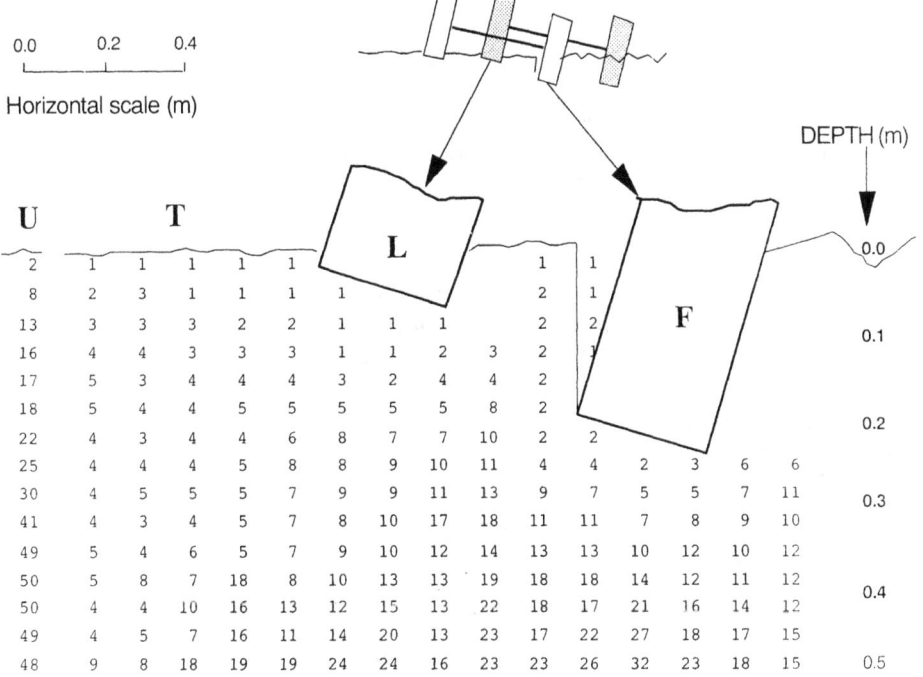

Fig. 8. Penetration resistance transect showing the initial, undisturbed condition (U), the loosened profile (T), and tractor land wheels (L) and furrow wheels (F) effects for mouldboard ploughing. Units: 10^{-1} MPa (after Soane et al., 1986).

Douglas (1986) reported good correlations between bulk density and shear vane strength, as well as between water content and shear vane strength, measured during the growing season from November to May, but he emphasized the weathering effect on shear vane strength in the tilled layer. Blackwell et al. (1986) showed that cone penetration resistance increased proportionally more by drying in wheeled than in unwheeled soil. But he showed also, as did Stengel and Bourlet (1987), that during the growing season the densest soil exhibited the largest variations in total porosity during wetting/drying cycles and, consequently, lost strength. On the other hand, the unwheeled soil lost cohesion (shear vane strength) at water contents below the plastic limit.

Soane et al. (1986) studied the recompaction effect by land and furrow wheels of a tractor during mouldboard ploughing subsequent to deep loosening (Fig. 8). They showed that under wheelings the benefit of the loosening effect would be lost in one ploughing operation unless particular techniques are used. Under wheelings the bulk density was as high as in the undisturbed profile. Soil strength was not as high as in the undisturbed profile, but they argued that insufficient time had passed for the cohesive bonds to reform. These results point out how important on the resulting soil strength can be the effects of compaction and subsequent variation in water content.

Spatial distribution and variability of soil strength

Origin of spatial variability across the field and within the soil profile

Cultivation techniques involve an annual cycle of both loosening and compaction. Loosening is achieved by primary tillage and/or subsoiling, compaction is due to several operations, including seedbed preparation, weed, pest and disease control, fertilizer application and harvest.

The distribution of wheel tracks over the field and the soil water content at the time of operations will be primarily responsible for the distribution of soil strength within the field (Manichon, 1988). Soane et al. (1981) reviewed the variability of traffic distribution for different cultivation systems and reported that seedbed preparations for barley can involve nearly 90% coverage, while 10% of the area of the field may receive from 4 to 9 passes.

At a secondary level, tillage practices and climatic conditions will influence the distribution of soil strength within the soil profile (Manichon, 1987, 1988). The two-dimensional plotting of cone resistance measured along a transect perpendicular to the direction of traffic (Soane et al., 1976; Soane et al., 1986; O'Sullivan et al., 1987; Billot and Marionneau, 1988) is of particular value for assessing the volume concerned by compaction, and the strength distribution within the profile (Fig. 8). At a meso (cm) scale, Perfect et al. (1990) have reported that the spatial variability of resistance to penetration decreased when the density increased, from which they deduced that the distribution of aggregate strength should be narrower in compacted zones.

Wheel tracks
Compaction under wheels causes localised increases in bulk density and, thus, regardless of the type of test used, increases in soil strength. The soil volume concerned and the final density are determined by the initial depth and strength of the soil horizons (Guérif, 1984), as well as by wheel and vehicle characteristics, such as tyre inflation pressure and axle load (see Chapters 3 and 4). The maxima of soil strength increases are generally not encountered just beneath the wheel-soil boundary but deeper (Soane et al., 1976; Soane et al., 1986). On the one hand, this is probably due to the stress concentration at a given depth and to the overburden pressure (Blackwell et al., 1986). On the other hand, a superficial soil loosening can occur which is due to the shear stresses induced near the surface by wheel lugs and is enhanced by wheel slip.

Tillage pan
Under rather wet conditions, most tillage implements (discs, mouldboard ploughs, powered implements), while loosening the surface layer, are liable to create a dense tillage pan. This type of pan is due to smearing rather than to compaction. On the other hand, a plough pan, which is induced by mouldboard ploughing, combines smearing in wet conditions and compaction by the passage of the furrow wheel of the tractor. A plough pan is located at the boundary of the tilled layer and the subsoil and may exhibit a high dry bulk density, which corresponds to maxima in cone resistance profiles (Douglas et al., 1986; Daniel et al., 1988). The compactness of this plough pan will be increased by the passage of the furrow wheel of the tractor.

Several authors have demonstrated, as did Boone et al. (1985) for potatoes, that in a relatively dry growing season severe plough pans can act as a barrier for roots. Generally, roots of winter crops are not limited by a plough pan, while spring crops are more likely to encounter such problems in a rather dry growing season.

Structural units
After cultivation, compacted zones can result in rather large and stable clods, which can persist for several years (Voorhees, 1983; Manichon, 1988). When they have no or only little structural porosity, they may have a rather high strength, nearly reaching textural strength, depending on the condition of compaction (Lyles and Woodruff, 1961). Håkanson et al. (1985) showed that seedbeds, prepared after compaction, contained >50% of aggregates with diameters >5 mm, while the uncompacted treatment contained only 25% of these aggregates. Campbell (1976) reported that compaction induced by tractor wheels in the sides of potato ridges, will increase the draught forces during harvesting and increase the amount of clods that have to be separated from potatoes on the harvester. Tardieu and Manichon (1987) and Tardieu (1988) have shown that these massive clods were not penetrated by maize roots, which were concentrated in the

remaining zones of the tilled layer. This heterogeneous distribution of the root system in the tilled layer markedly influenced the spatial distribution of maize roots in the subsoil.

Experimental analysis of factors influencing soil strength

Influence of the compaction procedure
At least three factors interact in compaction-induced increases in soil strength: the load and the water content during compaction, and the resulting compactness. According to Guérif (1990), the overall tensile strength of tilth (aggregates of 2-3 mm diameter), which was compacted at a given water content, increased exponentially with decreasing structural void ratio (Fig. 9). Apparently, the tensile strength varied with the area of the flat interfaces developing between aggregates as the stress was increased. Under the chosen conditions of compaction (water content $w = 0.18$ g g^{-1}), even when the structural void ratio was zero, the overall tensile strength did not reach the textural tensile strength. The tensile strength of tilth compacted at equal porosity (Koolen, 1976) was shown to be dependent on the water content at precompaction.

Guérif (1990) showed that the overall tensile strength, ^cT_w, of aggregate beds, compacted at equal structural void ratio, varied according to the water content at precompaction and was modified by subsequent drying (Fig. 10). Although the textural tensile strength was high in dry conditions, ^cT_w remained zero until the water content was sufficient for aggregates to adhere at the given pressure. As the water content increased, the overall tensile strength increased up to a maximum, where the inter- and intra-aggregate bonds happened to be rather similar. From this point the overall tensile strength decreased simultaneously with textural tensile strength, as water content increased. Drying of the samples increased their overall tensile strength, and, moreover, the higher the initial water content at compaction, the higher was the resulting tensile strength after drying.

In practice, the tensile strength of clods originating from a compacted zone will depend on their compactness but also on the quality of the bonds between adhering aggregates, which has been shown to depend on the water content at compaction. Moreover, after drying, the quality of these bonds was enhanced when sodium ions were applied and reduced when organic matter was applied (Ohu et al., 1985; Guérif, 1990).

Natural reversibility
Climatic phenomena such as wetting and drying cycles, or freezing and thawing cycles are known to reduce soil strength (Voorhees, 1983; Hadas, 1990), and increase friability (Utomo and Dexter, 1981). Compacted soils may thus exhibit natural reversibility. During the wetting phase, fragmentation occurs (McKenzie and Dexter, 1985; Stengel and Bourlet, 1987; Stengel, 1988), which is induced by

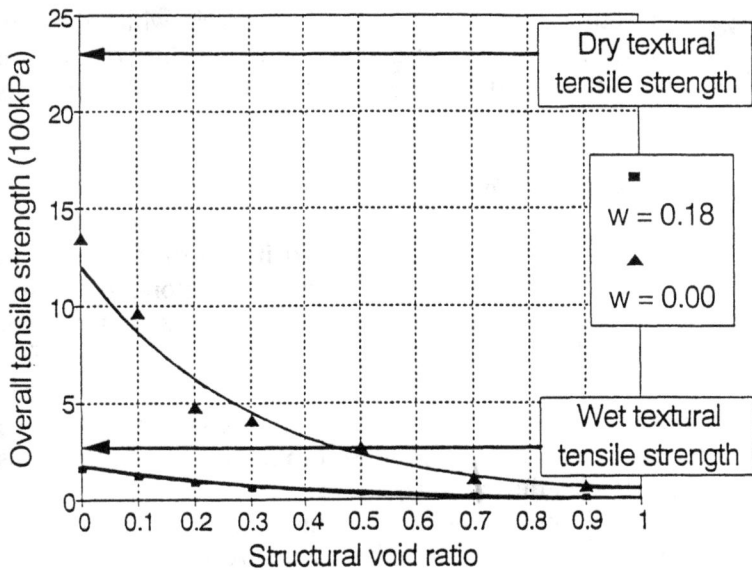

Fig. 9. Influence of the structural void ratio on the overall tensile strength, at the moisture content during compaction (w = 0.18 g g⁻¹) and after drying (w = 0 g g⁻¹), in comparison with the dry and wet textural tensile strength (from Guérif, 1990).

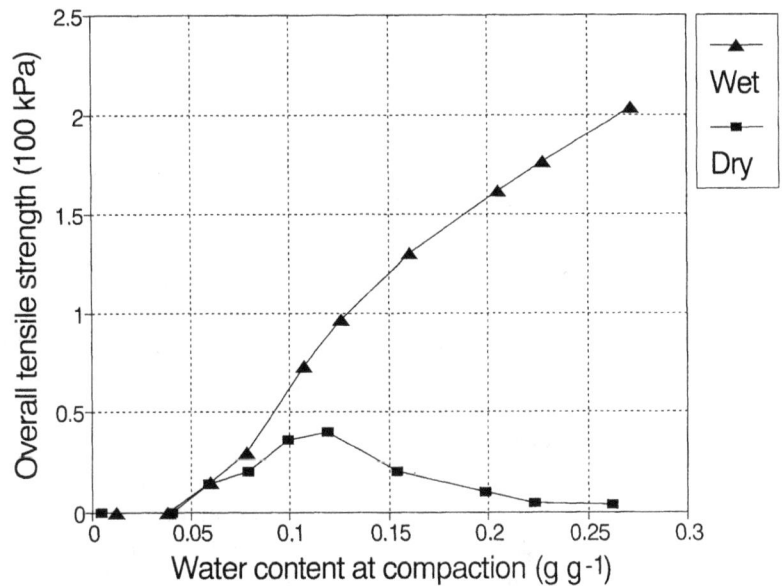

Fig. 10. Overall tensile strength, in relation to the water content at compaction ("wet") and after drying ("dry") (from Guérif, 1990).

differential swelling at the wetting front. The length of the cracks which develop during wetting, was shown to depend on the magnitude of swelling of the soil material, as determined by initial and final water content or potential, and by its intrinsic swelling properties (Stengel, 1988).

Practical aspects (trafficability, workability)

Trafficability results from soil-wheel interactions and it determines wheeled vehicle performance in given field situations.   Various components of soil strength are responsible for soil sinkage beneath the wheel and sometimes for horizontal deformation or bulldozing resistance in front of the wheel. Soil compaction, by increasing soil strength, generally improves wheeled vehicle performance. Soil management techniques, such as traffic-lane or tramline systems and zero-tillage practices, are often proposed to improve the conditions of access to the fields for vehicles during wet periods.

Workability results from soil-implement interactions, and it determines the ability of a soil layer under given conditions of initial structure and water content to react satisfactorily to a given implement, i.e., to assume a desired soil structure. Soil compaction increases the amount of energy required to fragment soils (Bateman et al., 1965; Koolen and Kuipers, 1983). Apparently, the higher the compactness, the higher the dilation when submitted to shear (Olsen, 1986), but the resulting clods are generally bigger and stronger (Håkansson et al., 1985; Manichon, 1987).

UNRESOLVED PROBLEMS

Theoretical aspects

Among the peculiarities and difficulties of applying soil mechanics theory to agricultural practices are the limitations to the traditional assumptions of classical mechanics, due to the structure of tilled soil layers. In most cases, agricultural soils are heterogenous because of the variability in size, shape and constitution of the structural elements (clods and aggregates) and their distribution within the profile. This heterogeneity is enhanced by the water gradients within the profile or within the clods. Agricultural soils are discontinuous and their properties often vary abruptly along a given direction. Moreover, the strains during wheel compaction or tillage cannot be considered as small.

In the comprehension and modelling of the mechanical behaviour of heterogeneous soil material, a number of questions have to be answered, such as: (1) how to model the stress transmission in heterogeneous soils; (2) how to deal with the spatial distribution of soil strength, stress transmission and subsequent local failure of structural elements; (3) how to combine into a dynamic process the spatial rearrangement of fragments of failed structural elements and the subsequent redistribution of stress?

In a theoretical application of soil mechanics, the concept of effective stress in unsaturated conditions remains the most problematic aspect. The need to take into account simultaneously the effect of water potential on effective stress and the changes in the water potential under loading imposes difficulties.

Methods

As stated above, it is not always possible to assess soil strength with the same technique throughout the whole range of possible water contents. This technical problem requires that checks be made on the compatibility of the different methods which are supposed to measure the same mechanical variable.

In view of the spatial variability of soil strength within the profile, soil strength should be assessed on soil volumes (size and geometry) which are adapted to the scale at which the studied phenomenon takes place. Particularly, efforts should be made to assess soil impedance according to biomechanics requirements such as root growth and establishment or seedling emergence.

Generally, soil strength measurements are destructive methods. Non-destructive methods should be developed, that would allow, on the same site or on the same sample, assessment of strength variations related to time-dependent factors.

CONCLUSIONS

(1) Soil strength parameters are involved in processes concerning wheel-soil, soil-implement, soil-climate and soil-plant interactions. These parameters must be evaluated as both input and output variables for models to describe and predict these phenomena. Further studies, dedicated to both theoretical and methodological aspects, are required.

(2) Soil strength measurements are a means for assessing soil structure "quality", following compaction, tillage and weathering. They should be linked through models to other behaviour characteristics of cultivated soils, such as mass and heat transfer processes, to gain a better general understanding and management of agricultural practices and their environmental consequences and/or requirements.

REFERENCES

Al-Hussaini, M.M. and Townsend, F.C., 1973. Tensile testing of soils: a literature review. U.S. Army Eng. Waterways Exp. Station, Misc. Pap. S-7324, 96 pp.

Angers, D.A., Kay, B.D. and Groenevelt, P.H., 1987. Compaction characteristics of a soil cropped to corn and bromegrass. Soil Sci. Soc. Am. J., 51: 779-783.

Aylmore, L.A.G. and Sills, I.D., 1978. Pore structure and mechanical strength of soils in relation to their constitution. In: W.W. Emerson, R.D. Bond and A.R. Dexter (Editors), Modification of Soil Structure. Wiley, Chicester, U.K., pp. 69-77.

Bateman, H.P., Naik, M.P. and Yoerger, R.R., 1965. Energy required to pulverize soil at different degrees of compaction. J. Agric. Eng. Res., 10: 132-144.

Bekker, M.G., 1956. Theory of Land Locomotion. Univ. Mich. Press, Ann Arbor, MI, U.S.A., 210 pp.

Billot, J. F. and Marionneau, A., 1988. Soil compaction : Analysis of field and laboratory experiments. Proc. 11th Conf. Int. Soil Tillage Res. Org. (ISTRO), Edinburgh, U.K., Vol. 1, pp. 197-202.

Blackwell, P.S., Graham, J.P., Armstong, J.V., Ward, M.A., Howse, K.R., Dawson, C.J. and Butler, C.J., 1986. Compaction of a silt loam soil by agricultural vehicles. I. Effects upon soil conditions. Soil Tillage Res., 7: 97-116.

Boone, F.R., De Smet, L.A.H. and Van Loon, C.D., 1985. The effect of a ploughpan in marine loam soils on potato growth. I. Physical properties and rooting patterns. Potato Res., 28: 295-314.

Boone, F.R., Van der Werf, H.M.G., Kroesbergen, B., Ten Hag, B.A. and Boers, A., 1986. The effect of compaction of the arable layer in sandy soils on the growth of maize for silage. I. Critical matric water potential in relation to soil aeration and mechanical impedance. Neth. J. Agric. Sci., 34: 155-171.

Boyd, D.W., Skidmore, E.L. and Thomson, J.G., 1983. A soil-aggregate crushing-energy meter. Soil Sci. Soc. Am. J., 47: 313-316.

Bradford, J.M., Farrel, D.A. and Larson, W.E., 1971. Effect of soil overburden pressure on penetration of fine metal probes. Soil Sci. Soc. Am. J., 35: 12-15.

Braunack, M.V., Hewitt, J.S. and Dexter, A.R., 1979. Brittle fracture of soil aggregates and the compaction of aggregate beds. J. Soil Sci., 30: 653-667.

Campbell, D.J., 1976. The occurence and prediction of clods in potato ridges in relation to soil physical properties. J. Soil Sci., 27: 1-9.

Campbell, D.J. and Hunter, R., 1986. Drop-cone penetration in situ and on minimally disturbed soil cores. J. Soil Sci., 37: 153-163.

Carter, M.R., 1990. Relationship of strength properties to bulk density and macroporosity in cultivated loamy sand to loam soils. Soil Tillage Res., 15: 257-268.

Chan, K.Y., 1989. Friability of a hardsetting soil under different tillage and land use practices. Soil Tillage Res., 13: 287-298.

Chancellor, W.J., 1971. Effects of compaction on soil strength. In: K.K. Barnes, W.M. Carleton, H.M. Taylor, R.I. Throckmorton and G.E. Vanden Berg (Editors), Compaction of Agricultural Soils. Am. Soc. Agric. Eng., St. Joseph, MI, U.S.A., ASAE Monograph, 471 pp.

Chenu, C. and Guérif, J., 1991. Mechanical strength of clay minerals as influenced by an adsorbed polysaccharide. Soil Sci. Soc. Am. J., 55: 1076-1080.

Collis-George, N. and Lloyd, J.E., 1978. Description of seedbeds in terms of shear strength. In: W.W. Emerson, R.D. Bond and A.R. Dexter (Editors), Modification of Soil Structure. Wiley, Chichester, U.K., pp. 371-378.

Culley, J.L.B. and Larson, W.E., 1987. Susceptibility to compression of a clay loam Haplaquoll. Soil Sci. Soc. Am. J., 51: 562-567.

Daniel, H., Jarvis, R.J. and Aylmore, L.A.G., 1988. Hardpan development in loamy sand and its effects upon soil conditions and crop growth. Proc. 11th Conf. Int. Soil Tillage Res. Org. (ISTRO), Edinburgh, U.K., Vol. 1, pp. 233-238.

Dexter, A.R., 1975. Uniaxial compression of ideal brittle tilths. J. Terramech. 12: 3-14.

Dexter, A.R., 1988. Strength of soil aggregates and aggregate beds. In: J. Drescher, R. Horn and M. de Boodt (Editors), Impact of Water and External Forces on Soil Structure. Catena, Cremlingen, Germany, Catena Suppl. 11, pp. 35-52.

Douglas, J.T., 1986. Effect of season and management on the vane shear strength of clay top-soil. J. Soil Sci., 37: 669-679.

Douglas, J.T., Jarvis, M.G., Howse, K.R. and Goss, M.J., 1986. Structure of a silty soil in relation to management. J. Soil Sci., 37: 137-151.

Ekwue, E.I., 1990. Organic matter effects on soil strength properties. Soil Tillage Res., 16: 289-297.

Farrell, D.A., Greacen, E.L. and Larson, W.E., 1967. The effect of water content on axial strain in a loam soil under tension and compression. Soil Sci. Soc. Am. Proc., 31: 445-450.

Fiès, J.C. and Stengel, P., 1981. Densité texturale des sols naturels. I. Méthode de mesure. (Textural bulk density of natural soils. I. Measurement method). Agronomie, 1: 651-658 (in French).

Frydman, S., 1964. The applicability of the brazilian (indirect tension) test to soils. Aust. J. Appl. Sci., 15: 335-343.

Gill, W.R., 1959. The effect of drying on the mechanical strength of Lloyd clay. Soil Sci. Soc. Am. Proc., 23: 255-257.

Guérif, J., 1979. Mechanical properties of straw. The effects on the soil. In: E. Grossbard (Editor), Straw Decay and its Effect on Disposal and Utilisation. Wiley, Chichester, U.K., pp. 73-91.

Guérif, J., 1982. Compactage d'un massif d'agrégats: effet de la teneur en eau et de la pression appliquée. (Compaction of aggregate beds: effect of water content and applied pressure). Agronomie, 2: 287-294 (in French).

Guérif, J., 1984. The influence of water-content gradient and structure anisotropy on soil compressibility. J. Agric. Eng. Res., 29: 367-374.

Guérif, J., 1988a. Effects of changing straw disposal on soil physical properties. In: Energy Saving by Reduced Soil Tillage. Proc. CEC Workshop, Göttingen, 10-11 June 1987. C.E.C Luxembourg, 1989, EUR-11258, pp. 117-126.

Guérif, J., 1988b. Détermination de la résistance en traction des agrégats terreux: revue bibliographique et mise au point technique. (Assessment of soil aggregate tensile strength: review and technical set up). Agronomie, 8: 281-288 (in French).

Guérif, J., 1988c. Résistance en traction des agrégats terreux : Influence de la texture, de la matière organique et de la teneur en eau. (Soil aggregate tensile strength: Effect of texture, organic matter content and water). Agronomie, 8: 379-386 (in French).

Guérif, J., 1990. Factors influencing compaction-induced increases in soil strength. Soil Tillage Res., 16: 167-178.

Hadas, A., 1990. Directional strength in aggregates as affected by aggregate volume and by a wet/dry cycle. J. Soil Sci., 41: 85-93.

Hadas, A. and Wolf, D., 1984a. Soil aggregates and clod strength dependence on clod size, cultivation, and stress load rates. Soil Sci. Soc. Am. J., 48: 1157-1165.

Hadas, A. and Wolf, D., 1984b. Refinement and re-evaluation of the drop-shatter soil fragmentation method. Soil Tillage Res., 4: 237-249.

Håkansson, I., Henriksson, L. and Gustafsson, L., 1985. Experiments on reduced compaction of heavy clay soils and sandy soils in Sweden. Proc. Int. Conf. Soil Dynamics, Auburn, AL, U.S.A., Vol. 5, pp. 995-1009.

Hammel, J.E., 1988. Influence of high axle loads on subsoil physical properties on crop yields in the Pacific northwest USA. Proc. 11th Conf. Int. Soil Tillage Res. Org. (ISTRO), Edinburgh, U.K., Vol. 1, pp. 275-280.

Hartge, K.H., 1974. Der Scherwiderstand von Bodenaggregaten. (Shear strength of soil aggregates). Trans. 10th Int. Congr. Soil Sci., Vol. 1, pp. 194-202 (in German).

Horn, R. and Hartge, K.H., 1990. Effects of short-time loading on soil deformation and strength of an ameliorated Typic Paleustalf. Soil Tillage Res., 15: 247-256.

Kirkham, D., De Boodt, M.F. and De Leenheer, L., 1959. Modulus of rupture determination on undisturbed core samples. Soil Sci., 87: 141-144.

Koolen, A.J., 1976. Mechanical properties of precompacted soil as affected by the moisture content at precompaction. Proc. 7th Conf. Int. Soil Tillage Res. Org. (ISTRO), Uppsala, Sweden, pp. 20: 1-6.

Koolen, A.J., 1978. The influence of a soil compaction process on subsequent soil tillage processes. A new research method. Neth. J. Agric. Sci., 26: 191-199.

Koolen, A.J. and Kuipers, H., 1983. Agricultural Soil Mechanics. Springer, Berlin, Germany, 241 pp.

Lipiec, J. and Tarkiewicz, S., 1986. The effect of moisture on the crushing strength of aggregates of loamy soil of various density levels. Polish J. Soil Sci., 19: 27-31.

Lowery, B. and Schuller, R.T., 1988. Effects and duration of compaction on soil and plant growth. Proc. 11th Conf. Int. Soil Tillage Res. Org. (ISTRO), Edinburgh, U.K., Vol. 1, pp. 293-298.

Lyles, L. and Woodruff, N.P., 1961. Surface soil cloddiness in relation to soil density at time of tillage. Soil Sci., 23: 178-182.

Manichon, H., 1987. Observation morphologique de l'état structural et mise en évidence d'effets de compactage des horizons travaillés. (Morphological observation of the structural state and enhancement of compaction effects in tilled layers). In: G. Monnier and M.J. Goss (Editors), Soil Compaction and Regeneration. Balkema, Rotterdam, Netherlands, pp. 39-52 (in French).

Manichon, H., 1988. Compactage, décompactage du sol et système de culture. (Compaction, decompaction and crop system). C.R. Acad. Agric. France, 74: 43-54 (in French).

Masle, J. and Passioura, J.B., 1987. The effect of soil strength on the growth of young wheat plants. Austr. J. Plant Physiol., 14: 643-656.

McBride, R.A., 1989. Estimation of density-moisture-stress functions from unixial compression of unsaturated, structured soils. Soil Tillage Res., 13: 383-397.

McKenzie, B.M. and Dexter, A.R., 1985. Mellowing and anisotropy induced by wetting of moulded soil samples. Austr. J. Soil Res., 23: 37-47.

Middleton, H.E., 1924. Factors influencing the binding power of soil colloids. J. Agric. Res., 23: 499-513.

Monnier, G., Fiès, J.C. and Stengel, P., 1973. Une méthode de mesure de la densité apparente de petits agglomérats terreux, application à l'analyse de porosité du sol. (A method to measure the bulk density of small aggregates of soil, application to soil porosity analysis). Ann. Agron., 24: 533-545 (in French).

Mullins, C.E. and Panayiotopoulos, K.P., 1984. The strength of unsaturated mixtures of sand and kaolin and the concept of effective stress. J. Soil Sci., 35: 459-468.

Mulqueen, J., Stafford, J.V. and Tanner, D.W., 1977. Evaluation of penetrometers for measuring soil strength. J. Terramech., 14: 137-151.

O'Sullivan, M.F., Dickson, J.W. and Campbell, D.J., 1987. Interpretation and presentation of cone resistance data in tillage and traffic studies. J. Soil Sci., 38: 137-148.

Ohu, J.O., Raghavan, G.S.V., McKyes. E. and Mehuys, G., 1985. The shear strength of compacted soils with varying organic matter contents. Am. Soc. Agric. Eng., St. Joseph, MI, U.S.A., ASAE Pap. 85-1039, 14 pp.

Olsen, H.J., 1984. A torsional shearing device for field tests. Soil Tillage Res., 4: 599-611.

Olsen, H.J., 1986. Soil mechanical behaviour of a heavy clay soil after three long-term compaction treatments. Soil Tillage Res., 7: 145-156.

Owen, G.T., Drummond, H., Cobb, L. and Godwin, R.J., 1987. An instrumentation system for deep tillage research. Trans. ASAE, 30: 1578-1582.

Perfect, E., Groenevelt, P.H., Kay, B.D. and Grant, C.D., 1990. Spatial variability of soil penetrometer measurements at the mesoscopic scale. Soil Tillage Res., 16: 257-271.

Pidgeon, J.D. and Soane, B.D., 1978. Soil structure and strength relations following tillage, zero-tillage and wheel traffic in Scotland. In: W.W. Emerson, R.D. Bond and A.R. Dexter (Editors), Modification of Soil Structure. Wiley, Chichester, U.K., pp. 371-378.

Rogowski, A.S. and Kirkham, D., 1976. Strength of soil aggregates: Influence of size, density and clay and organic matter content. Med. Fac. Landbouww. Rijksuniv. Gent, Belgium, 41: 85-100.

Rogowski, A.S., Moldenhauer, W.C. and Kirkham, D., 1968. Rupture parameters of soil aggregates. Soil Sci. Soc. Am. Proc., 32: 720-724.

Schjønning, P., 1986. Shear strength determination in undisturbed soil at controlled water potential. Soil Tillage Res., 8: 171-179.

Snyder, V.A. and Miller, R.D., 1985. A pneumatic fracture method for measuring the tensile strength of unsaturated soils. Soil Sci. Soc. Am. J., 49: 1369-1374.

Soane, B.D., 1990. The role of organic matter in soil compactibility: a review of some practical aspects. Soil Tillage Res., 16: 179-201.

Soane, B.D., Kenworthy, G. and Pidgeon, J.D., 1976. Soil tank and field studies of compaction under wheels. Proc. 7th Conf. Int. Soil Tillage Res. Org. (ISTRO), Uppsala, Sweden, pp. 48: 1-6.

Soane, B.D., Blackwell, P.S., Dickson, J.W. and Painter, D.J., 1981. Compaction by agricultural vehicles: a review. I. Soil and wheel characteristics. Soil Tillage Res., 1: 207-237.

Soane, G.C., Godwin, R.J. and Spoor, G., 1986. Influence of deep loosening techniques and subsequent wheel traffic on soil structure. Soil Tillage Res., 8: 231-237.

Spivey, L.D., Jr., Busscher, W.J. and Campbell, R.B., 1986. The effect of texture on strength of Southeastern Coastal Plain soils. Soil Tillage Res., 6: 351-363.

Spoor, G. and Godwin, R.J., 1979. Soil deformation and shear strength characteristics of some clay soils at different water contents. J. Soil Sci., 30: 483-498.

Stafford, J.V. and De Carvalho Mattos, P., 1981. The effect of forward speed on wheel-induced soil compaction : laboratory simulation and field experiments. J. Agric. Eng. Res., 26: 333-347.

Stengel, P., 1988. Cracks formation during swelling : Effects on soil structure regeneration after compaction. Proc. 11th Conf. Int. Soil Tillage Res. Org. (ISTRO), Edinburgh, U.K., Vol. 1, pp. 147-152.

Stengel, P., 1990. Caractérisation de l'état structural, objectifs et méthodes. (Characterization of soil structure: objectives and methods). In: J. Boiffin and A. Marin la Flèche (Editors), La Structure du Sol et son Evolution: Conséquences Agronomiques et Maîtrise par l'Agriculteur. (Soil structure and its evolution: agricultural consequences and management). Coll. INRA., pp. 15-36 (in French).

Stengel, P. and Bourlet, M., 1987. Fissuration d'un sol argileux gonflant après compactage: Effet de l'humectation. (Cracking of a swelling clay soil after compaction: effect of wetting). In: G. Monnier and M.J. Goss (Editors), Soil Compaction and Regeneration. Balkema, Rotterdam, Netherlands, pp. 95-110 (in French).

Tardieu, F., 1988. Effect of the structure of the ploughed layer on the spatial distribution of root density. Proc. 11th Conf. Int. Soil Tillage Res. Org. (ISTRO), Edinburgh, U.K., Vol. 1, pp. 153-157.

Tardieu, F. and Manichon, H., 1987. Etat structural, enracinnement et alimentation hydrique du maïs : Modélisation d'états structuraux types de la couche labourée. (Soil structure, maize rooting and water supply: modelling different standards of the structural state of the tilled layer). Agronomie, 7: 123-131 (in French).

Towner, G.D., 1973. An examination of the fall-cone method for the determination of some strength properties of remoulded agricultural soils. J. Soil Sci., 24: 470-479.

Utomo, W.H. and Dexter, A.R., 1981. Soil friability. J. Soil Sci., 32: 203-213.

Voorhees, W.B., 1983. Relative effectiveness of tillage and natural forces in alleviating wheel-induced soil compaction. Soil Sci. Soc. Am. J., 47: 129-133.

Willatt, S.T., 1987. Influence of aggregate size and water content on compactability of soil using short-time static loads. J. Agric. Eng. Res., 37: 107-115.

Willatt, S.T. and Sulistyaningsih, N., 1990. Effect of plant roots on soil strength. Soil Tillage Res., 16: 329-336.

Williams, B.G., Greenland, D.J. and Quirk, J.P., 1967. The tensile strength of soil cores containing polyvinyl alcohol. Austr. J. Soil Res., 5: 85-92.

Wong, J.Y., 1980. Data processing methodology in the characterization of the mechanical properties of terrain. J. Terramech., 17: 13-41.

Soil Compaction in Crop Production
B.D. Soane and C. van Ouwerkerk (Eds.)

215

CHAPTER 10

Effects of Compaction on Soil Biota and Soil Biological Processes

L. BRUSSAARD[1,2] and H.G. van FAASSEN[1]

[1]Institute for Soil Fertility Research (IB-DLO), Haren Gn, Netherlands
[2]Wageningen Agricultural University, Wageningen, Netherlands

SUMMARY

Optimum crop production requires integrated management of interacting soil physical-chemical properties and biological processes. This chapter considers how soil biology aspects can be integrated into agro-ecosystem management, to overcome the problems of conventional high-input, "physical-chemical" agriculture. A good soil structure is a prerequisite for satisfactory crop growth and adequate functioning of soil organisms. Soil structure supplies plant roots and soil organisms with "habitable pore space" and controls many processes, e.g., the transport of water, oxygen and nutrients. In turn, soil organisms can contribute to an optimum soil structure through aggregate formation and the creation of biopores. Soil organisms also play an important role in plant development, e.g., by supplying nutrients and by controlling pests and plant pathogens. Intensive soil tillage and traffic with heavy machinery can damage soil organisms and their habitats, and thus reduce their positive role. In reduced-input management systems, the role of soil organisms in maintaining or improving soil structure and supplying plant nutrients, becomes more important. In-depth studies on the functioning of soil biota (roots, microbes, fauna) and on soil biological processes, as influenced by soil compaction, are reviewed in the context of integrated management in crop production.

INTRODUCTION

In the soil science community we observe increasing acknowledgement of the neglect of the soil biota and of the inadequacy of the implicit replacement of its function by "physical-chemical" agriculture. This trend seems to be part of a re-orientation in agricultural research and, in modern agriculture, a shift from problem solving to systems research and sustainability. In terms of the subject of this book: from solving soil compaction to maintaining or improving soil structure, i.e., from working the soil (opposing to nature) to letting the soil work for us (cooperating with nature) (Elliott and Coleman, 1988).

Conventional high-input, "physical-chemical" agriculture has led to environmental problems (water pollution by leaching of nutrients and pesticide residues; air pollution by ammonia and nitrous oxides) and to deterioration of soil structure (soil erosion and compaction). To solve these problems as well as the economic problems of agriculture (overproduction and low prices), reduced-input ("integrated") soil management systems have been developed.

A reduction in the intensity of soil tillage and in the use of mineral fertilizers and pesticides makes the role of soil organisms in crop production of crucial importance. Integration of soil biology views in farm management forms part of a systems approach to agriculture, which aims at optimizing rather than maximizing crop production and at reducing negative effects on the environment. Lower costs of inputs should compensate for lower crop yields and maintain the net financial result achieved by the farmer. At present, soil tillage and field traffic are the main management practices that influence the degree of soil compaction. The interrelationships between soil tillage, field traffic, soil structure and soil physical, chemical and biological properties are shown in Fig. 1.

Soil tillage, in particular inversion tillage, reduces soil compaction but, generally, it also reduces the soil biota. Conversely, reduced soil tillage is usually associated with an increase in the soil biota, but generally leads to increased compaction. Until a critical soil bulk density is reached, compaction can have a positive effect on crop growth by improving soil-root contact and thereby root

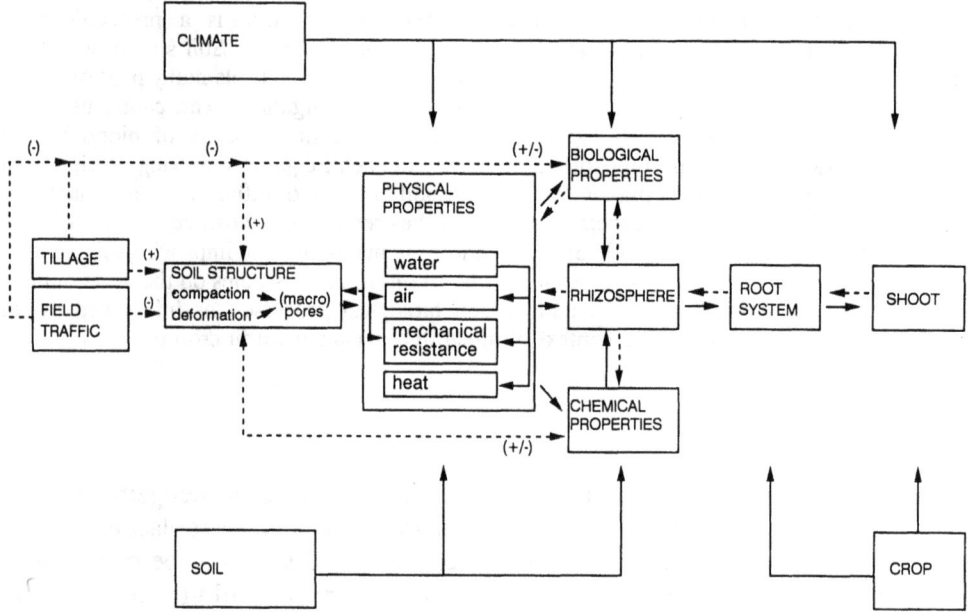

Fig. 1. Simplified scheme of functional relationships between soil structure and soil biota (after Boone, 1986).

functioning. A further small increase in soil density can greatly reduce crop yields by restriction of aeration and root penetration (Boone, 1986).

Soil compaction and tillage affect the soil biota (roots, microbes and fauna) through several direct and indirect mechanisms (Doran and Power, 1983). Soil compaction increases soil bulk density and volumetric soil moisture content and decreases air-filled porosity. By altering the fluxes and storage of air, water, nutrients and heat, soil compaction affects the living conditions (micro-environments) for roots and soil organisms. Thus, in general, soil structure controls the activities of roots and soil organisms. On the other hand, roots and soil organisms can also affect soil structure by creating biopores and by increasing aggregate formation and stability. Well aggregated soils always contain a substantial population of both macro- and meso-fauna. Stimulation rather than destruction of soil fauna may offer economic approaches to structural improvement, e.g., by stubble mulching or improving water infiltration rates by production of biopores (Hole, 1981; Oades, 1984; Fig. 2).

Over the last decades, economic constraints and reduced prospects of financial returns have induced the adoption of reduced tillage in both temperate and tropical areas, although not all climates and soils are compatible with this farming practice. Conservation tillage, i.e., any tillage that reduces soil or water loss compared to mouldboard ploughing (Gebhardt et al., 1985), is conducive to soil compaction, with continuous no-till as the extreme version. Biological activity in the upper 10-cm layer of soil is generally greater under no-till than in ploughed soil, partly owing to a concentration of crop residues. However, total biological activity in the 0-25-cm depth can be almost equal under both management systems (House and Parmelee, 1985).

Chemical and biological effects of conservation tillage have been reviewed by Dick and Daniel (1987). Generally, the micro-environment of the non-tilled soil is less oxidative than in tilled soil, which promotes anaerobic processes such as

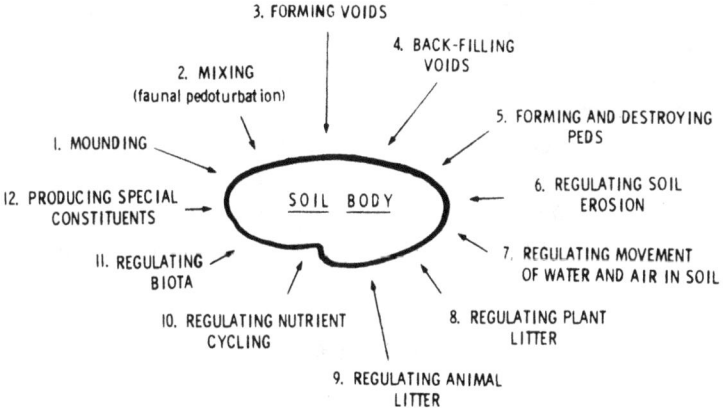

Fig. 2. Diagrammatic classification of the effects of animals on soil (from Hole, 1981).

denitrification and fermentation, and decreases aerobic breakdown of organic matter. By changing the soil micro-environment, conservation tillage can increase the incidence of some soil-borne diseases, while decreasing that of other diseases. Generally, any change in soil management can alter the delicate balance between the different groups of soil organisms and it can take several years before a new level of biological equilibrium is reached.

Biological activity, such as earthworm and root penetration into the soil, creates macropores that, at least under no-till, can remain undisturbed for many years. These macropores will affect water flow and leaching of nutrients and pesticides. Depending on the situation, increased or decreased transport and loss of solutes have been found. Rapid infiltration of water through macropores limits interaction with solutes in micropores and therefore limits their loss from the topsoil, whereas heavy rainfall shortly after nitrogen fertilization may lead to rapid downward transport and loss.

There are many studies in which the increase of soil compaction under reduced tillage is documented (e.g., Westmaas Research Group on New Tillage Systems, 1984) and other studies in which the biological changes following reduced tillage are shown (El Titi, 1984; House and Parmelee, 1985; Hendrix et al., 1986, 1990; Brussaard et al., 1990; El Titi and Landes, 1990). However, there are only few studies in which both soil compaction and soil biological properties are explicitly addressed.

The objective of this chapter is to review in-depth studies on the interactions between soil compaction and soil biota (roots, microbes and fauna) and biological processes. We will also explore the prospects of managing the soil biota to overcome soil compaction.

SOIL COMPACTION AND ROOTS

Soil strength increases with increasing bulk density and decreases with increasing soil water potential (Taylor and Bar, 1991). Since the hydraulic properties of soils may be improved by moderate soil compaction while root development may be hindered, the overall effect of moderate soil compaction, as found under no-till, cannot easily be predicted. In this section, recent literature on the pressures which roots can exert to overcome soil compaction is summarized, and possible agricultural implications are indicated.

Studies on the behaviour of roots at the interface of loose and compacted soil and on their exudation and uptake response, yield some insight into the functional aspects of root-soil interactions. The greater the strength of the soil beneath a seedbed, the less likely it is that roots can penetrate the subsoil. Soil strength can arise in a number of ways: natural pedogenic processes, compaction by agricultural vehicles and implements, hard-setting on drying, and age-hardening or curing after mechanical disturbance. The proportion of roots penetrating a compacted subsoil below a tilled seedbed decreased exponentially with subsoil strength (Dexter, 1986a). Provided the seedbed is deep enough to prevent sideways displacement of aggregates, the penetrometer resistance of the subsoil should not exceed 0.4 MPa for plants with a single seminal axis (dicotyledons) or 3 MPa for plants with four seminal axes (monocotyledons).

Apart from soil strength, subsoil cracks and biopores determine the penetrability of the soil to roots (Dexter, 1986b,c). Biopores (created mainly by roots and earthworms) add significantly to root penetration at depth, even when the content of small, vertical pores is only 0.1% (v/v) (Jakobsen and Dexter, 1988). In the long term, the penetrability of a soil to roots is determined to a large extent by the persistence of biopores, which in turn depends on their resistance to compaction by stresses exerted by agricultural vehicles. The stability of lucerne taproot channels to uniaxial stresses in clay soil was investigated by Blackwell et al. (1990) in soil cores from the B horizon of an ameliorated loam soil, each including one nearly vertical biopore. In uncompressed soil cores, as in soil cores which had been subjected to uniaxial compression at a water potential of -10 kPa with vertical stresses of 50-400 kPa, air-filled porosity at -10 kPa was similar for soil cores with biopores of 3.5, 5.0, 6.0 or 7.5 mm mean diameter and for soil cores without biopores, over the full range of applied stresses. In all cores, air-filled porosity decreased rectilinearly at stresses >50 kPa. However, in uncompressed soil cores, intrinsic permeability at -10 kPa increased in proportion to the biopore diameter. In soil cores without biopores and in cores with biopores of 3.5 mm mean diameter, intrinsic permeability gradually decreased at stresses >50 kPa. In cores with biopores of 5.0, 6.0 or 7.5 mm mean diameter, intrinsic permeability decreased only, but steeply, at stresses >200 kPa. In these larger pores, fluid flow remained efficient at stresses up to 200 kPa, but decreased above that value. The greater stability of the wider biopores was ascribed to the greater thickness of the surrounding cylinder of denser soil, which is due to the fact that thin roots displace less soil laterally and increase the density of the nearby soil less than thick roots. A surface load of 6 Mg and a surface pressure of 300 kPa may be needed to exert a vertical stress of 200 kPa at 40 cm depth in clay soils, which will seldom occur. Hence, biopores from taproots grown into clay subsoils are able to maintain their integrity as a pathway for vertical movement and redistribution of roots, water and solutes, despite large surface stresses.

SOIL COMPACTION, SOIL MICROBES AND SOIL BIOCHEMICAL PROCESSES

Protozoa and nematodes are the predators of soil bacteria, while protozoa are also eaten by nematodes. Because most protozoa are smaller than nematodes, they can enter soil pores with neck-sizes smaller than the diameter of nematodes. Once inside pores containing bacteria, the protozoa population will increase and some may move into larger pores where they can be consumed by nematodes. Consequently, the number of nematodes will be larger in soil with bacteria and

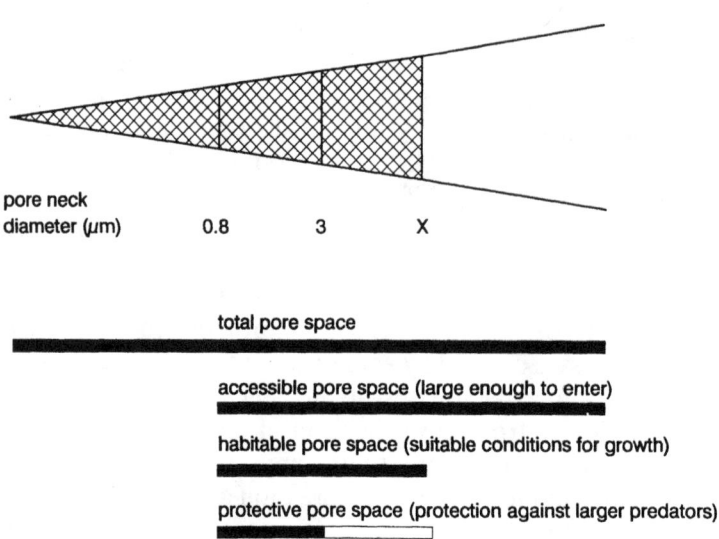

Fig. 3. Schematic presentation of total, accessible, habitable, and protective pore space for rhizobial cells. Pores <0.8 μm in diameter are considered to be inaccessible for rhizobial cells. Hatched areas represent pores filled with water, and x is the pore neck diameter that is still water-filled at the water potential used (at -10 kPa, x = 30 μm). ▬ and ▭ indicate if pore volume or pore surface area, respectively, of a certain pore diameter class are expected to be important (from Postma and Van Veen, 1990).

protozoa than in soil with bacteria only, especially in fine-textured soil (Elliott et al., 1980). Similarly, it may be expected that bacteria will be better protected from predation by protozoa in finer-textured soil (Heijnen et al., 1988; Postma et al., 1990; Heijnen and Van Veen, 1991).

Identification of these phenomena has led to the subdivision of total pore space into accessible and inaccessible pore space and of accessible pore space into protective and habitable pore space (Postma and Van Veen, 1990; Fig. 3). Soil compaction will affect the relative abundance of these pore space categories and, thereby, the abundance of microbial populations. This hypothesis was investigated for introduced rhizobial cells and naturally occurring bacteria by varying the bulk density and water content of repacked samples of the <2 mm fraction of sterilized and natural loamy sand and silt loam soil (Postma and Van Veen, 1990). Bulk densities and water contents compared are given in Table 1. In the loamy sand only the volume of pores >60 μm was diminished by compaction (pressing by hand) and rhizobial cell numbers were only slightly lower as compared to the uncompressed soil. In the silt loam, the volume of pores <15 μm was increased by pressing the soil, but no significant influence on rhizobial cell numbers was found. To explain these results for various pore-neck diameter

TABLE 1

Water content of a loamy sand and a silt loam soil at two bulk densities and various water potentials[1] (after Postma and Van Veen, 1990)

	Bulk density (Mg m³)	Water content (%,w/w) at			
		Saturation	-5 kPa	-10 kPa	-20 kPa
Loamy sand	1.33	41.5	28.0	15.9	11.9
Loamy sand compacted	1.42	39.0	27.7	15.8	12.1
LSD ($P_{0.05}$)	0.02	0.9	1.0	1.0	1.0
Silt loam	0.89	80.1	39.6	35.6	32.9
Silt loam compacted	1.11	60.7	41.7	36.7	34.0
LSD ($P_{0.05}$)	0.02	1.1	1.0	1.0	1.0

[1]Equivalent pore neck diameters: 60 μm (-5 kPa); 30 μm (-10 kPa); 15 μm (-20 kPa).

TABLE 2

Calculated pore volume and surface area of a loamy sand and a silt loam for different pore neck diameters (after Postma and Van Veen, 1990)

Water potential (-kPa)	Pore-neck diameter (μm)	Loamy sand		Silt loam	
		Volume (cm³ kg⁻¹)	Surface area (m² kg⁻¹)	Volume (cm³ kg⁻¹)	Surface area (m² kg⁻¹)
10⁶-400	<0.8	71	> > >[1]	306	> > >[1]
400-100	0.8-3	20	55	26	92
100-10	3-30	67	21	114	27
10-0	>30	201	15	155	11

[1]Extremely large surface area.

classes, the pore space was estimated from the water retention curves of the loamy sand and the silt loam at bulk densities of about 1.40 and 1.00 Mg m⁻³, respectively (Table 2).

Assuming a volume of 1 μm³ for each of the introduced rhizobial cells, the number of cells that could, theoretically, occupy the pore volume, can be calculated from Table 2. For the loamy sand and the silt loam, these numbers are, per g of dry soil, for the accessible pore space 28.8×10^{10} and 29.5×10^{10}, for the habitable pore space 8.7×10^{10} and 14.0×10^{10}, and for the protective pore space

2.0×10^{10} and 2.6×10^{10}, respectively.

In the sterilized samples, at -10kPa, rhizobial cells numbered $2.5-3.2 \times 10^8$ and $5.0-6.3 \times 10^8$ colony-forming units per g of dry soil, for the loamy sand and the silt loam, respectively. This means that, at -10 kPa, at most 1.6 and 2.4% (v/v), respectively, of the protective pore space was occupied in the loamy sand and the silt loam, respectively. Corresponding figures for the maximum occupation by rhizobium cells of the habitable pore space were 0.37 and 0.44% (v/v), respectively, and for the accessible pore space 0.11 and 0.21% (v/v), respectively. Hence, habitable and protective pore spaces are not expected to be limiting for the survival of rhizobial cells. This explains the minor impact of increased bulk density. However, bacteria are not evenly distributed throughout the soil, and it cannot be excluded that locally, where substrate is plentiful, pore space may limit bacterial growth (Postma and Van Veen, 1990).

In the study by Postma and Van Veen (1990), no observations on soil protozoa or nematodes were made. However, soil compaction may reduce the predation efficiency of these organisms through a decrease in accessible pore space. This was probably not the case for protozoa and hardly so for nematodes in a saturated silt loam soil at bulk densities of 0.89 and 1.11 Mg m^{-3} (Table 2, Fig. 4).

Fig. 4 shows a theoretical pore volume distribution, as calculated from the water retention curves for a silt loam at different bulk densities. It demonstrates that the total pore space decreases with increasing bulk density, whereas the protective pore space remains constant over nearly the whole range for both bacteria and protozoa. However, the volume of large pores (>30 μm), the habitat of nematodes, decreases strongly with increasing bulk density. The volume of the accessible pore space cannot be read from this figure, since it is dependent

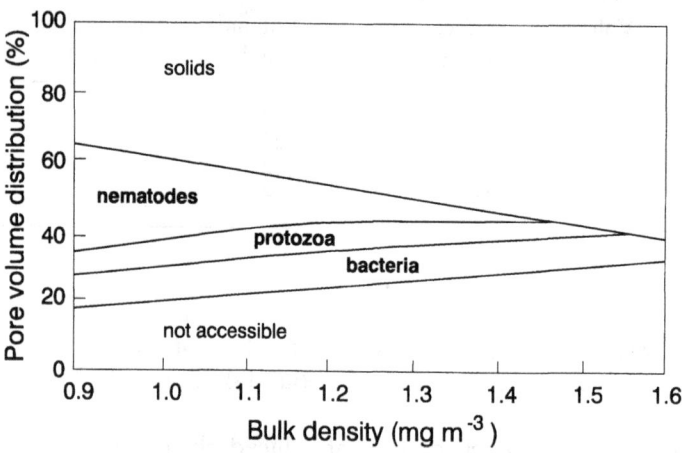

Fig. 4. Pore volume distribution in a silt loam soil at different bulk densities and the assumed protective pore space for some groups of soil biota. Neck diameter (p) of accessible pores: nematodes p >30 μm; protozoa p >5 μm; bacteria p >0.2 μm. Not accessible pores: p <0.2 μm (after Van der Linden et al., 1989).

on the soil water content. At decreasing water potential, the larger pores will be emptied before the smaller ones. Therefore, the accessible pore space will always be greatest for bacteria, followed by protozoa.

Carbon and nitrogen transformations

Due to the resultant changes in substrate, nutrients and oxygen availability, soil compaciton may affect not only microbial cell numbers but also microbial activity. Soil respiration, i.e., the aerobic microbial decomposition of organic residues, is known to increase with increasing soil water content until the supply rate of oxygen by diffusion through gas-filled pores becomes limiting. For soil respiration, a threshold air-filled porosity of about 10% (v/v) has often been found, somewhat depending on soil type and bulk density (Kroeckel and Stolp, 1986; Prade and Trolldenier, 1988). When oxygen becomes limiting, parts of the microbial populations switch to using nitrate instead of oxygen, a process which is called denitrification.

In cropped soils, plant roots also use oxygen for their respiration. The capacity of plant roots to survive periods without oxygen is limited and damage may occur, especially when hydrogen sulphide is produced by anaerobic microbes. However, transport of oxygen by diffusion through channels in their roots into the soil may help plants to survive periods when the soil becomes anaerobic (De Willigen and Van Noordwijk, 1989).

Generally, compacted soils have lower water infiltration rates than uncompacted soils and, therefore, are more prone to problems with stagnant water, anaerobic conditions and nitrate loss through denitrification (Bakken et al., 1987). Besides, in no-till soils crop residues and plant roots are concentrated in a thin layer, thus locally increasing oxygen demand. The same phenomenon may be observed when, following conventional tillage, easily degradable crop residues (e.g., sugar beet leaves) are concentrated at the bottom of the ploughed layer and anaerobic conditions prevail because of limited oxygen supply.

In reduced-input systems, appropriate residue management and adapted fertilization should account for the differences with conventional systems to optimize crop development and minimize environmental pollution (Lopez-Real and Hodges, 1986; Doran and Werner, 1990). Mineral fertilizers are often partly replaced by animal manures, compost or organic wastes. In these materials, plant nutrients become available through microbial mineralization/immobilization turnover. The net outcome and timing of this turnover can make it more difficult to adjust supplementary use of mineral fertilizers. In the long term this problem will be reduced, as gradually a larger buffer of slowly mineralizable nutrients builds up. Application of manures and organic wastes should be such that the nutrients they contain become available only in periods of active crop uptake; in other periods nutrient immobilization in the soil biomass should prevent losses to the environment. Because of their high biological oxygen demand, liquid organic wastes can promote nitrate loss from soil through denitrification, which is further promoted by soil compaction induced by the heavy machinery used to apply these wastes. In this respect, more stabilised, solid organic wastes give fewer problems and, thus, are more suitable to replace mineral fertilizers in reduced-input systems.

SOIL COMPACTION, SOIL ANIMALS AND SOIL BIOPHYSICAL PROCESSES

The interactions between soil compaction and soil biota and biochemical (mainly microbial) and biophysical (mainly "macrobial") processes are part of the relational web within the agro-ecological system, in which the soil biota are an integral part of the soil, not just inhabitants (McGill and Spence, 1985).

There is ample evidence for the important role of the soil mesofauna (mites, springtails, enchytraeids, fly larvae, etc.) in building the microstructure of the soil, and for the substantial contribution of the soil macrofauna (termites, ants, diplopods, beetles, earthworms, etc.) to the formation of the soil macrostructure, and in both humification and accelerated decomposition of organic matter under natural conditions (Hole, 1981; Bal, 1982; McGill and Spence, 1985; Mermut, 1985; Pawluk, 1985, 1987).

It is generally agreed that the soil mesofauna plays only a minor role in the ingestion and resulting transport of mineral soil (Didden, 1990). In an experiment in which springtails were introduced in small cores which contained loose as well as compact soil, the compact soil hindered migration and had a repellent effect on the animals (Didden, 1987). Larger soil animals, however, may burrow down the profile. In moist and freely-drained sandy soil, bulk densities up to 1.7 Mg m^{-3} had no effect on the depth of dung beetle burrows (Brussaard and Slager, 1986), but at soil water potentials < -50 kPa, the survival of eggs and larvae declined.

Among the data available on the soil macrofauna, those on earthworms are most abundant. Earthworms can be divided into three ecological categories (Lavelle, 1981): (1) epigeic earthworms, which feed on and live in the leaf litter; (2) anecic earthworms, which feed on leaf litter that they mix with soil of the upper horizons, but shelter in semi-permanent vertical burrows dug into the soil; (3) endogeic earthworms, which live in the soil and feed on soil organic matter and on dead or live roots. The last two categories are of interest in terms of soil compaction effects. In pasture soil, Rushton (1986) established a significant correlation between the length of the burrows of the anecic earthworm *Lumbricus terrestris* and soil compaction in the bulk density range 1.38-1.66 Mg m^{-3}. However, the explained variation was rather low: $r^2 = 0.40$. There was no correlation with the soil water content, which ranged from 13.9 to 16.1% (w/w). In sandy silt and loamy silt soils, Joschko et al. (1989) found longer burrows of *L. terrestris* at low bulk densities (1.06-1.12 Mg m^{-3}) than at high bulk densities (1.41-1.60 Mg m^{-3}). According to Kemper et al. (1988), *L. terrestris* would not penetrate soil with a bulk density >1.60 Mg m^{-3}. However, these data are difficult to interpret without reference to the soil strength, which is determined by both bulk density and water potential.

Soil strength was used as a parameter in a number of studies on endogeic worms. Tunneling of *Aporrectodea caliginosa* was independent of soil strength over the range of micropenetrometer resistance from 0.3-3 MPa (Dexter, 1978). No compacted zone was observed around the burrows, suggesting that the principal mechanism was ingestion of the soil ahead of the worm. The mean maximum axial pressure generated by *Aporrectodea rosea* was 72.8 kPa at a mean maximum penetrometer resistance of about 1.5 MPa (McKenzie and Dexter, 1988a). This is about half the resistance which can be overcome by monocotyledonous plants and about 4 times the resistance which can be overcome by dicotyledonous plants (Whiteley et al., 1981; Dexter, 1986a). The same worm species was found to exert a mean radial stress of 230 kPa (McKenzie and Dexter, 1988b). This was considered not to be at variance with the absence of compacted zones or cracks around tunnels made by earthworms in compacted soil blocks reported by Dexter (1978), considering that on the soils studied loads >230 kPa will have occurred. The better ability of *A. rosea* to overcome soil strength as compared with *L. terrestris,* may be related to the different ecological strategies of the species: for the endogeic *A. rosea*, penetration of soil has clearly more survival value than for the anecic *L. terrestris* (McKenzie and Dexter, 1988b).

On the assumption that the weight of worm casts on the surface is correlated with mechanical constraints, Kretzschmar (1991) assessed the effects of applied pressures in the range 50-350 kPa (resulting in bulk densities in the range 1.08-1.49 Mg m^{-3}) and of soil water potential in the range between -65 and -7 kPa, on burrowing by the earthworm *Aporrectodea longa* in soil columns. Cast production increased with soil compaction up to an applied pressure of about 250 kPa, which is close to the maximum radial pressure generated by the ecologically similar *A. rosea* (McKenzie and Dexter, 1988b). Beyond this value, soil strength seemed to be a limiting factor for burrowing activity, especially in wet soil columns. At water potentials <-30 kPa, cast production was severely limited. Different degrees of compaction did not change the effects of the water potential, except at the lower compaction levels where cast production was low, independent of the water content. There seems to be no literature on the resistance of earthworm burrows to applied mechanical stress.

The ability of soil animals to overcome soil compaction may have important effects on soil physical properties. In sandy soil, which is too compact for root penetration, plant roots may preferentially follow (back-filled) dung beetle burrows (Brussaard and Hijdra, 1986) and vertical earthworm channels may cause bypass-flow, i.e. preferential flow along continuous macropores through an unsaturated soil matrix (Bouma et al., 1982). Preferential flow along continuous vertical macropores such as root and earthworm channels has often been reported to enhance the risk of groundwater contamination with nitrate and pesticides. Whether or not this occurs may depend on the location of the available nitrogen: in the soil matrix (from nitrogen mineralization) or in the discharge water (from fertilizer application) (Elliott et al., 1986). Even in the latter case the total movement of nitrate through earthworm channels may be small (Edwards et al., 1990). With large continuous voids, such as earthworm channels, penetrometer resistance is not an adequate measure of soil strength (Baeumer, 1981) and large tensiometer cups intercepting water-conducting macropores erroneously suggest saturation of the entire soil matrix (Bouma et al., 1982). For management recommendations in this area the standard theoretical soil physical and chemical framework is currently being modified (Bouma, 1991).

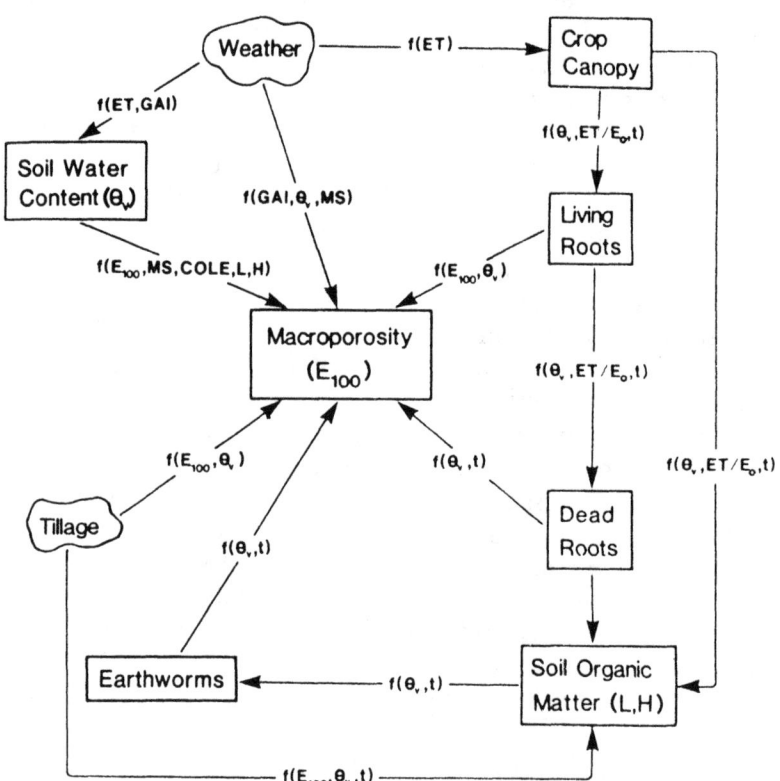

Fig. 5. Conceptual model of changes in soil structure under different cropping systems. f = function; ET = actual evapotranspiration (m day^{-1}); E_o = potential evapotranspiration (m day^{-1}); GAI = green area index (m^2 m^{-2}); θ_v = soil water content (m^3 m^{-3}); COLE = coefficient of linear expansibility (m m^{-1}); E_{100} = pores >100 μm in diameter (m^3 m^{-3} or %, v/v); MS = macropore stability; t = time (usually in days); L = light or litter fraction (kg m^{-3}); H = heavy or humified fraction (kg m^{-3}) (from Gibbs and Reid, 1988).

MODELLING THE INTERACTIONS BETWEEN SOIL STRUCTURE AND SOIL BIOTA

The complex interactions between soil structure and soil biota take place at very different scales: micrometre-scale (soil particles and microaggregates, microflora, microfauna and root hairs); millimetre-scale (macroaggregates, mesofauna and small roots); and centimetre-scale (crumbs and clods, macrofauna, main roots). The scale at which the effects of soil structure-soil biota interactions on soil biochemical and biophysical processes should be studied depends on the objective. For mechanistic understanding, studies should take place at all three scales. If one is interested in plant response, the micrometre and millimetre scales may be too detailed. If the objective is to evaluate effects on trafficability, the plant scale (m^2) is too small; if the aim is to assess the risk of nitrogen losses, only the field scale (ha) may be adequate.

In view of this hierarchy of scales of observation, it is likely that soil structure-soil biota interactions will have to be described by a hierarchy of models, in which soil compaction will be an important parameter. Such models could: (1) describe soil structure-soil biota interactions; (2) generate research hypotheses on the impacts of soil structure-soil biota interactions on biochemical and biophysical processes and plant performance; (3) predict the effects of soil structure-soil biota interactions on next-higher spatial scales.

The literature in this area is very scarce and far from comprehensive. A conceptual model on structure-forming and structure-following physical and biological factors and processes was given by Kooistra (1991) and a conceptual model on changes in macroporosity (Fig. 5) was presented by Gibbs and Reid (1988). Such models may be used as a framework to support modelling efforts at various scales, such as those of Malicki et al. (1991) and Kretzschmar (1988).

MANAGEMENT OF SOIL COMPACTION-SOIL BIOTA INTERACTIONS

The body of primary, quantitative results from studies on compaction - soil biota interrelationships is only small. However, these empirical results can be used to explore the management options for sustained land use. In addition, much can be learnt from the formation of macropores under natural conditions, which are very different for temperate and tropical regions.

Temperate regions

Deep-rooting plant species play an important part in soil formation in originally densely packed sedimentary soils. The vegetation that spontaneously colonized the areas which were reclaimed from the IJssel Lake (Netherlands) during the period 1940-1970, consisted of a few species with taproots penetrating deeply into the anaerobic subsoil, thereby increasing soil macroporosity, soil aeration, and evapotranspiration. Reed (*Phragmites australis*) was sown to enhance these processes and the first agricultural use was the sowing of deep-rooting cereals, such as wheat and barley that were cropped continuously for some years. The vertical root channels have remained largely intact in the soil profile ever since (Kooistra et al., 1989). Subsequent soil formation, along with an increase in macroporosity, was largely effectuated by earthworms where they were present (Van Rhee, 1969; Bal, 1982) or hampered where they were absent, due to a lack of colonization ability (Hoogerkamp et al., 1983), flooding with sea-water (Van Rhee, 1963) or pesticide application (Van de Westeringh, 1972). The activity of introduced worms increased air permeability within the surface horizon and counteracted soil compaction in wheel tracks (Rogaar and Boswinkel, 1978).

A study on morphological features and faunal activity in a loess-derived soil in France with a long history of either forest, pasture or arable land use, revealed a decline in biological activity in that order, as evidenced by, e.g., lower porosity. In the arable land, earthworm activity was smaller than in the forest or in the pasture, which was due to less surface litter, larger temperature fluctuations and longer dry periods (Thompson et al., 1990). Cultivation of the soil destroys much of the environment of epigeic and anecic worms; it also damages them mechanically and brings them to the surface where they are an easy prey for birds (Edwards, 1983).

Such effects are reversible by reduction of tillage. According to Edwards and Lofty (1982), after only eight years of direct drilling the number of anecic earthworms increased thirty-fold as compared to ploughing. Non-inversion tillage usually resulted in earthworm populations intermediate between those in non-tilled and ploughed soils. When normal field populations of deep-burrowing or shallow-working earthworms were introduced into small, sterilized plots on a site that had been direct-drilled with cereals for six successive years, barley plant populations and the weight and depth of roots and height and amount of foliage significantly increased. It was concluded that most of the effects were due to the creation of channels for root growth in the compacted soil by deep-burrowing species (Edwards and Lofty, 1980).

On a heavy clay soil, where during 20 years the fields were either trafficked by tractors, leading to soil compaction, or the implements were hauled by a winch, numbers and biomass of earthworms were significantly higher in the winch treatment than in the tractor treatment and the number of juveniles per adult was significantly lower in trafficked than in untrafficked plots. These effects were attributed to the susceptibility of earthworms to the pressing and crushing effects of the tractor wheels (Boström, 1986).

Another key to prevention and remedy of soil compaction is organic matter. Addition of crop residues and organic manure has a direct beneficial effect on soil bulk density (Avnimelech and Cohen, 1988; Ball et al., 1990). For earthworm activity it comprises the essential source of energy and nutrients. In plots which were continuously under cereals for more than 135 years, many more earthworms were found than in plots with root crops or fallow, probably because the root system of wheat, which forms 50% of its biomass, remains in the soil after harvest (Edwards, 1983).

The adoption of measures for prevention and amelioration of soil compaction, such as mentioned above and summarized by Dexter (1991), may be constrained by economic and technological considerations and a lack of basic knowledge of the biology of the species considered for introduction, or the functioning of agro-ecosystems (Karlen, 1990). However, it seems that where conservation tillage is considered or already practiced, there is increasing scope for application of these measures.

Tropical regions

Biotic and abiotic conditions in the tropics differ in some important respects from those in temperate regions. In large parts of the tropics, especially in the humid regions, earthworms are the most important soil-dwelling invertebrates. However, termites and ants are also notorious burrowers. Quantitative data were given by Bal (1982).

A more important, functional difference is that in most of the tropics, in contrast to the temperate regions, the organic matter is tied up in the above-ground biomass rather than in the soil. Following deforestation, due to the extremely humid conditions and very high temperatures, the relatively small amount of organic matter in the soil rapidly decomposes. The disappearance of organic food sources and the drastic changes in abiotic conditions lead to the demise of the soil fauna, resulting in the decline in structural properties of the soil. This process is enhanced by the use of heavy machinery during deforestation (Lal, 1987). Often, soil compaction is the result (Seubert et al., 1977; Allen, 1985) but even if it is not, a decline in macroporosity, a shift towards smaller macropores, higher susceptibility to puddling and decreased hydraulic conductivity becomes apparent (Spaans et al., 1989).

The micromorphological effects of various management practices on an Alfisol in south-western Nigeria were studied by Kooistra et al. (1990). In a forest pedon, faunal voids (>30 μm in diameter) were abundant. In a 12-year old, manually-cropped field, zero-tillage without crop residue mulch resulted in the formation of a laminated surface layer, poor water infiltration and restricted root growth. However, plots where maize residues were left at the surface as a mulch had a loose surface soil, resulting from high soil faunal activity, and soil erosion did not occur. In a 6-year old mechanized maize field, tractor ploughing without mulch cover resulted in severe crusts, poor water infiltration and severe erosion. In a conventionally tilled field with residue mulch, the mulch cover prevented the formation of a surface crust, but burrowing fauna was absent. Voids available for infiltration were unstable tillage voids. In all soils, voids >30 μm in diameter had drastically decreased as compared to the forest pedon, but much less so in the mulch treatments than in the treatments without mulch (Kooistra et al., 1990). Hence, it would seem that mulching is essential for cultivating tropical soils that are prone to compaction and erosion in order to maintain adequate soil fauna.

In the humid tropics, earthworms adapted to pasture or annual cropping systems often do not occur in forests. Therefore, following deforestation, suitable species may have to be introduced. However, in some forest-derived cropping systems, such as agroforestry, suitable species may be retained or made to colonize spontaneously by addition of prunings and maintenance of a suitable microclimate. Reduced tillage and introduction of alley-cropping and other forms of agroforestry which are now widely investigated as alternatives to the traditional slash-and-burn agriculture (Kang et al., 1990), would appear especially promising for the prevention and cure of soil compaction and the decline in macroporosity if deep-rooting trees can be grown (Van Noordwijk et al., 1991). Studies on an Alfisol in south-western Nigeria showed that in alley-cropping with maize and *Leucaena leucocephala*, two earthworm species are abundant, i.e., the deep-burrowing *Hyperiodrilus africanus* and the shallow-working *Eudrilus eugeniae*, whereas in maize monoculture the former hardly occurs. The total amount of worm castings was 41 Mg ha^{-1} in alley cropping versus 27 Mg ha^{-1} in the monoculture; total amounts of macronutrients in the casts were 2-3 times higher in alley cropping. It was concluded that permanent shading in the vicinity of hedgerow trees in alley cropping systems enhances earthworm activity (Hauser, 1993).

It may be concluded that the options for management of the soil biota in the tropics, in particular where agroforestry can be practiced, are much the same as those listed in the section on temperate regions (cf. Heinonen, 1986; Lavelle et al., 1989). However, greater knowledge of the plant (tree) and animal species concerned and better control of the microclimatic conditions and organic inputs is required to ensure satisfactory results.

CONCLUSIONS

(1) Quantitative data are available on the limits of soil strength to the penetrability of roots and on the limits of vertical stress that taproot biopores can withstand. However, these data apply to a very limited number of soil types and plant species.

(2) Quantitative data are available on the limits of soil strength for penetrability by soil animals and on the effects of transport of water and solutes through the macropores they produce. However, as with roots, these data apply to a very limited number of soils and animal species.

(3) In temperate regions, the options for managing the soil biota for prevention and amelioration of soil compaction are: (1) introduction of deep-rooting plants and/or tunnelling macrofauna in situations where the species already present are unable to colonize naturally in sufficient numbers; (2) reduction of soil tillage and

field traffic to reduce the speed of decomposition of organic matter and to promote the activities of tunnelling soil macrofauna; (3) addition of organic matter in the form of manure, and/or by growing crops that root abundantly and produce sufficient residues which may serve as a source of food for the soil macrofauna.

(4) In the humid tropics, agroforestry may be promising for the prevention and amelioration of soil compaction if deep-rooting trees can be grown. By the addition of their prunings, burrowing soil fauna can be promoted.

REFERENCES

Allen, J., 1985. Soil response to forest clearing in the United States and in the tropics: geological and biological factors. Biotropica, 17: 15-27.
Avnimelech, Y. and Cohen, A., 1988. On the use of organic manures for amendment of compacted clay soils: effects of aerobic and anaerobic conditions. Biol. Wastes, 29: 331-339.
Bakken, L.R., Børresen, T. and Njøs, A., 1987. Effect of compaction by tractor traffic on soil structure, denitrification, and yield of wheat. J. Soil Sci., 38: 541-552.
Bal, L., 1982. Zoological ripening of soils. I. The concept and impact in pedology, forestry and agriculture. II. The process in two Entisols under developing forest in a recently reclaimed Dutch polder. Pudoc, Wageningen, Netherlands, Agric. Res. Rep. 850, 365 pp.
Ball, B.C., Bickerton, D.C. and Robertson, E.A.G., 1990. Straw incorporation and tillage for winter barley: soil structural effects. Soil Tillage Res., 15: 309-327.
Baeumer, K., 1981. Tillage effects on root growth and crop yield. In: Agricultural Yield Potentials in Continental Climates. Proc. 16th Coll. Int. Potash Inst., Bern, Switzerland, pp. 57-75.
Blackwell, P.S., Green, T.W. and Mason, W.K., 1990. Responses of biopore channels from roots to compression by vertical stresses. Soil Sci. Soc. Am. J., 54: 1088-1091.
Boone, F.R., 1986. Towards soil compaction limits for crop growth. Neth. J. Agric. Sci., 34: 349-360.
Boström, U., 1986. The effect of soil compaction on earthworms *(Lumbricidae)* in a heavy clay soil. Swed. J. Agric. Res., 16: 137-141.
Bouma, J., 1991. Influence of soil macroporosity on environmental quality. Adv. Agron., 46: 1-37.
Bouma, J., Belmans, J.F.M. and Dekker L.W., 1982. Water infiltration and redistribution in a silt loam subsoil with vertical worm channels. Soil Sci. Soc. Am. J., 46: 917-921.
Brussaard, L. and Hijdra, R.D.W., 1986. Some effects of scarab beetles in sandy soils of the Netherlands. Geoderma, 37: 325-330.
Brussaard, L. and Slager, S., 1986. The influence of soil bulk density and soil moisture on the habitat selection of the dung beetle *Typhaeus typhoeus* in the Netherlands. Biol. Fertil. Soils, 2: 51-58.
Brussaard, L., Bouwman, L.A., Geurs, M., Hassink, J. and Zwart, K.B., 1990. Biomass, composition and temporal dynamics of soil organisms of a silt loam soil under conventional and integrated management. Neth. J. Agric. Sci., 38: 283-302.
De Willigen, P. and Van Noordwijk, M., 1989. Model calculations on the relative importance of internal longitudinal diffusion for aeration of roots of non-wetland plants. Plant Soil, 113: 111-119.
Dexter, A.R., 1978. Tunnelling in soil by earthworms. Soil Biol. Biochem., 10: 447-449.

Dexter, A.R., 1986a. Model experiments on the behaviour of roots at the interface between a tilled seed-bed and a compacted sub-soil. I. Effects of seed-bed aggregate size and sub-soil strength on wheat roots. Plant Soil, 95: 123-133.

Dexter, A.R., 1986b. Model experiments on the behaviour of roots at the interface between a tilled seed-bed and a compacted sub-soil. II. Entry of pea and wheat roots into sub-soil cracks. Plant Soil, 95: 135-147.

Dexter, A.R., 1986c. Model experiments on the behaviour of roots at the interface between a tilled seed-bed and a compacted sub-soil. III. Entry of pea and wheat roots into cylindrical pores. Plant Soil, 95: 149-161.

Dexter, A.R., 1991. Amelioration of soil by natural processes. Soil Tillage Res., 20: 87-100.

Dick, W.A. and Daniel, T.C., 1987. Soil chemical and biological properties as affected by conservation tillage: environmental implications. In: T.J. Logan, J.M. Davidson and J.L. Baker (Editors), Effects of Conservation Tillage on Groundwater Quality: Nitrates and Pesticides. Lewis, Chelsea, MI, U.S.A., pp. 126-147.

Didden, W.A.M., 1987. Reactions of Onychiurus fimatus (Collembola) to loose and compact soil. Methods and first results. Pedobiologia, 30: 93-100.

Didden, W.A.M., 1990. Involvement of enchytraeidae (Oligochaeta) in soil structure evolution in agricultural fields. Biol. Fertil. Soils, 9: 152-158.

Doran, J.W. and Power, J.F., 1983. The effects of tillage on the nitrogen cycle in corn and wheat. In: R.R. Lowrance, R.L. Todd. L.E. Asmussen and R.A. Leonard (Editors), Nutrient Cycling in Agricultural Ecosystems. Univ. Georgia, Athens, GA, U.S.A., Special Publ. 23, pp. 441-455.

Doran, J.W. and Werner, M.R., 1990. Management and soil biology. In: C.A. Francis, C. Butler Flora and L.D. King (Editors), Sustainable Agriculture in Temperate Zones. Wiley, New York, NY, U.S.A., pp. 205-230.

Edwards, C.A., 1983. Earthworm ecology in cultivated soils. In: J.E. Satchell (Editor), Earthworm Ecology from Darwin to Vermiculture. Chapman and Hall, Londen, U.K., pp. 123-137.

Edwards, C.A. and Lofty, J.R., 1980. Effects of earthworm inoculation upon the root growth of direct drilled cereals. J. Appl. Ecol., 17: 533-543.

Edwards, C.A. and Lofty, J.R., 1982. The effect of direct drilling and minimal cultivation on earthworm populations. J. Appl. Ecol., 19: 723-734.

Edwards, W.M., Shipitalo, M.J. Owens, L.B. and Norton, L.D. 1990. Effect of Lumbricus terrestris L. burrows on hydrology of continuous no-till corn fields. Geoderma, 46: 73-84.

Elliott, E.T. and Coleman, D.C. 1988. Let the soil work for us. Ecol. Bull., 39: 23-32.

Elliott, E.T., Anderson, R.V., Coleman, D.C. and Cole, C.V., 1980. Habitable pore space and microbial trophic interactions. Oikos, 35: 327-335.

Elliott, E.T., Tracy, P., Peterson, P.A. and Cole, C.V., 1986. Leaching of mineralization N is less under no-till cultivation. Trans. 13th Int. Congr. Soil Sci., Hamburg, Germany, Vol. II, pp. 53-54.

El Titi, A., 1984. Auswirkung der Bodenbearbeitungsart auf die edaphischen Raubmilben (Mesostigmata: Acarina). (Effect of type of tillage on the edaphic predatory mites (Mesostigmata: Acarina). Pedobiologia, 27: 79-88 (in German).

El Titi, A. and Landes, H., 1990. Integrated farming system of Lautenbach: a practical contribution toward sustainable agriculture in Europe. In: C.A. Edwards, R. Lal and P. Madden (Editors), Sustainable Agricultural Systems. Soil Water Conserv. Soc., Ankeny, IA, U.S.A., pp. 265-286.

Gebhardt, M.R., Daniel, T.C., Schweizer, E.E. and Allmaras, R.R., 1985. Conservation tillage. Science, 230: 625-630.

Gibbs, R.J. and Reid, J.B., 1988. A conceptual model of changes in soil structure under different cropping systems. Adv. Soil Sci., 8: 123-149.

Hauser, S., 1993. Distribution and activity of earthworms and contribution to nutrient recycling in alley cropping. Biol. Fertil. Soils, 15: 16-20.

Heinonen, R., 1986. Alleviation of soil compaction by natural forces and cultural practices. In: R. Lal, P.A. Sanchez and R.W. Cummings (Editors), Land Clearing and Development in the Tropics. Balkema, Rotterdam, Netherlands, pp. 285-297.

Hendrix, P.F., Parmelee, R.W., Crossley Jr., D.A., Coleman, D.C., Odum, E.P. and Groffmann, P.M., 1986. Detritus food webs in conventional and no-tillage agroecosystems. Biosci., 36: 374-380.

Hendrix, P.F., Crossley Jr., D.A., Blair, J.M. and Coleman, D.C., 1990. Soil biota as components of sustainable agroecosystems. In: C.A. Edwards, R. Lal and P. Madden (Editors), Sustainable Agricultural Systems. Soil Water Conserv. Soc., Ankeny, IA, U.S.A., pp. 637-654.

Heijnen, C.E. and Van Veen, J.A., 1991. A determination of protective microhabitats for bacteria introduced into soil. FEMS Microbiol. Ecol., 85: 73-80.

Heijnen, C.E., Van Elsas, J.D., Kuikman, P.J. and Van Veen, J.A., 1988. Dynamics of *Rhizobium leguminosarum* biovar *trifolii* introduced into soil; the effect of bentonite clay on predation by protozoa. Soil Biol. Biochem., 20: 483-488.

Hole, F.D., 1981. Effects of animals on soil. Geoderma, 25: 75-112.

Hoogerkamp, M., Rogaar, H. and Eijsackers, H.J.P., 1983. Effect of earthworms on grassland on recently reclaimed polder soils in the Netherlands. In: J.E. Satchell (Editor), Earthworm Ecology from Darwin to Vermiculture. Chapman and Hall, London, U.K., pp. 85-105.

House, G.F. and Parmelee, R.W., 1985. Comparison of soil arthropods and earthworms from conventional and no-tillage agroecosystems. Soil Tillage Res., 5: 351-360.

Jakobsen, B.F. and Dexter, A.R., 1988. Influence of biopores on root growth, water uptake and grain yield of wheat *(Triticum aestivum)* based on predictions from a computer model. Biol. Fertil. Soils, 6: 315-321.

Joschko, M., Diestel, H. and Larink, H., 1989. Assessment of earthworm burrowing efficiency in compacted soil with a combination of morphological and soil physical measurements. Biol. Fertil. Soils, 8: 191-196.

Kang, B.T., Reynolds, L. and Atta-Krah, A.N., 1990. Alley farming. Adv. Agron., 43: 315-359.

Karlen, D.L., 1990. Conservation tillage research needs. J. Soil Water Conserv., 45: 365-369.

Kemper, W.D., Jolley, P. and Rosenau, R.C., 1988. Soil management to prevent earthworms from riddling irrigation ditch banks. Irrig. Sci., 9: 79-87.

Kooistra, M.J., 1991. A micromorphological approach to the interactions between soil structure and soil biota. Agric. Ecosyst. Environ., 34: 315-328.

Kooistra, M.J., Lebbink, G. and Brussaard, L., 1989. The Dutch programme on soil ecology of arable farming systems. 2. Geogenesis, agricultural history, field site characteristics and present farming systems at the Lovinkhoeve (experimental farm). Agric. Ecosyst. Environ., 27: 361-387.

Kooistra, M.J., Juo, A.S.R. and Schoonderbeek, D., 1990. Soil degradation in cultivated Alfisols under different management systems in Southwestern Nigeria. In: L.A. Douglas (Editor), Soil Micromorphology. Elsevier, Amsterdam, Netherlands, pp. 61-69.

Kretzschmar, A., 1988. Structural parameters and functional patterns of simulated earthworm burrow systems. Biol. Fertil. Soils, 6: 252-261.

Kretzschmar, A., 1991. Burrowing ability of the earthworm *Aporrectodea longa* limited by soil compaction and water potential. Biol. Fertil. Soils, 11: 48-51.

Kroeckel, L. and Stolp, H., 1986. Influence of the water regime on denitrification and aerobic respiration in soil. Biol. Fertil. Soils, 2: 15-21.

Lal, R., 1987. Tropical Ecology and Physical Edaphology. Wiley, Chichester, U.K., 732 pp.
Lavelle, P., 1981. Stratégies de réproduction chez les vers de terre. (Reproduction strategies of earthworms). Acta Oecol. Gén., 2: 117-133 (in French).
Lavelle, P., Barois, I., Martin, A., Zaidi, Z. and Schaeffer, R., 1989. Management of earthworm populations in agro-ecosystems: a possible way to maintain soil quality? In: M. Clarholm and L. Bergström (Editors), Ecology of Arable Land. Kluwer, Dordrecht, Netherlands, pp. 109-122.
Lopez-Real, J.M. and Hodges, R.D. (Editors), 1986. The Role of Microorganisms in a Sustainable Agriculture. A B Academic Publishers, Berkhamsted, U.K., 246 pp.
Malicki, J., Bieganowski, A. and Dąbek-Szreniawska, M., 1991. Mathematical modeling of biological activity in differently compacted soils. Soil Tillage Res., 19: 357-362.
Mermut, A.R., 1985. Faunal influence on soil microfabrics and other soil properties. Quaest. Entomol., 21: 595-608.
McGill, W.B. and Spence, J.R., 1985. Soil fauna and soil structure: feedback between size and architecture. Quaest. Entomol., 21: 645-654.
McKenzie, B.M. and Dexter, A.R., 1988a. Axial pressures generated by the earthworm Aporrectodea rosea. Biol. Fertil. Soils, 5: 323-327.
McKenzie, B.M. and Dexter, A.R., 1988b. Radial pressures generated by the earthworm Aporrectodea rosea. Biol. Fertil. Soils, 5: 328-332.
Oades, J.M., 1984. Soil organic matter and structural stability: mechanisms and implications for management. In: J. Tinsley and J.F. Darbyshire (Editors), Biological Processes and Soil Fertility. Nijhoff/Junk, The Hague, Netherlands, pp. 319-338.
Pawluk, S., 1985. Soil micromorphology and soil fauna: problems and importance. Quaest. Entomol., 21: 473-496.
Pawluk, S., 1987. Faunal micromorphological features in moder humus of some western Canadian soils. Geoderma, 40: 3-16.
Postma, J. and Van Veen, J.A., 1990. Habitable pore space and survival of Rhizobium leguminosarum biovar trifolii introduced into soil. Microb. Ecol., 19: 149-161.
Postma, J., Hok-a-Hin, C.H. and Van Veen, J.A., 1990. Role of microniches in protecting introduced Rhizobium leguminosarum biovar trifolii against competition and predation in soil. Appl. Environ. Microb., 56: 495-502.
Prade, K. and Trolldenier, G., 1988. Effects of wheat roots on denitrification at varying soil air-filled porosity and organic-carbon content. Biol. Fertil. Soils, 7: 1-6.
Rogaar, H. and Boswinkel, J.A., 1978. Some soil morphological effects of earthworm activity; field data and X-ray radiography. Neth. J. Agric. Sci., 26: 145-160.
Rushton, S.P., 1986. The effects of soil compaction on Lumbricus terrestris and its possible implications for populations on land reclaimed from open-cast coal mining. Pedobiologia, 29: 85-90.
Seubert, C.E., Sanchez, P.A. and Valverde, C., 1977. Effects of landclearing methods on soil properties of an Ultisol and crop performance in the Amazon jungle of Peru. Trop. Agric. Trinidad, 54: 307-321.
Spaans, E.J.A., Baltissen, G.A.M., Bouma, J., Miedema, R., Lansu, A.L.E., Schoonderbeek, D. and Wielemaker, W.G., 1989. Changes in physical properties of young and old volcanic surface soils in Costa Rica after clearing of tropical rain forest. Hydrol. Processes, 3: 383-392.
Taylor, H.M. and Bar, G.S., 1991. Effect of soil compaction on root development. Soil Tillage Res., 19: 111-119.
Thompson, M.L., Fedoroff, N. and Fournier, B., 1990. Morphological features related to agriculture and faunal activity in three loess-derived soils in France. Geoderma, 46: 329-349.

Van der Linden, A.M.A., Jeurissen, L.J.J., Van Veen, J.A. and Schippers, B., 1989. Turnover of the soil microbial biomass as influenced by soil compaction. In: J.A. Hensen and K. Hendriksen (Editors), Nitrogen in Organic Wastes Applied to Soils. Academic Press, Londen, U.K., pp. 25-36.

Van Noordwijk, M., Widianto, Heinen, M. and Hairiah, K., 1991. Old tree root channels in acid soils in the humid tropics: Important for crop root penetration, water infiltration and nitrogen management. Plant Soil, 134: 37-44.

Van Rhee, J.A., 1963. Earthworm activities and the breakdown of organic matter in agricultural soils. In: J. Doeksen and J. van der Drift (Editors), Soil Organisms. Proc. Coll. Soil Fauna, Soil Microflora and their Relationships, Oosterbeek, Netherlands, pp. 55-59.

Van Rhee, J.A., 1969. Development of earthworm populations in polder soils. Pedobiologia, 9: 133-140.

Van de Westeringh, W., 1972. Deterioration of soil structure in worm-free orchard soils. Pedobiologia, 12: 6-15.

Westmaas Research Group on New Tillage Systems, 1984. Experiences with three tillage systems on a marine loam soil. II: 1976-1979. Pudoc, Wageningen, Netherlands, Agric. Res. Rep. 925, 263 pp.

Whiteley, G.M., Utomo, W.H. and Dexter, A.R., 1981. A comparison of penetrometer pressures and the pressures exerted by roots. Plant Soil, 61: 351-363.

PART D

==

MECHANISMS AND INCIDENCE OF CROP RESPONSES TO SOIL COMPACTION

Soil Compaction in Crop Production
B.D. Soane and C. van Ouwerkerk (Eds.)
©1994 Elsevier Science B.V. All rights reserved.

CHAPTER 11

Mechanisms of Crop Responses to Soil Compaction

F.R. BOONE[1] and B.W. VEEN[2]

[1]Wageningen Agricultural University, Department of Soil Tillage, Wageningen, Netherlands
[2]Centre for Agrobiological Research (CABO-DLO), Wageningen, Netherlands

SUMMARY

Crop growth is less than potential when the uptake of water, oxygen or nutrients is less than the demand of the crop. This may be caused by a limited supply from the soil to the root system, a limited activity of the root system, or a limited length of the growing period. In this chapter the influence of soil compaction on each of these factors is analyzed. The limited supply and uptake of water, oxygen or nutrients are assumed to be complementary. They are discussed on three levels of integration: (1) as a rate per unit root length; (2) as a rate per root system; (3) as an amount for a period of crop growth.

For a given field situation, the potential crop demand is determined by the above-ground growth factors. Comparing the demands of a crop with the availability of the soil growth factors, will yield the most limiting soil factor.

Soil water is considered to play a central role in all relationships discussed. Therefore, it is preferred to explain the effects of soil compaction on crop growth in relation to soil water. By defining criteria for soil aeration and mechanical resistance for plant roots, it is possible to establish critical soil water potentials as a function of the degree of soil compaction. The more compacted a soil, the smaller the range of soil water potentials which can be tolerated during crop growth. A soil can also be too loose, i.e., when unsaturated hydraulic conductivity and root-soil contact are too small. Consequently, the relationship between soil compaction and crop yield may be described by an optimum curve, the characteristics of which depend on a large number of crop growth factors.

INTRODUCTION

Potential crop growth rate and, therefore, the demand for water, oxygen and nutrients from the soil, is determined by the stage of crop development, the actual crop size and the prevailing weather conditions. Actual crop growth is considered to be potential when, at any moment, the uptake of water, oxygen and nutrients from the soil equals the demand of the crop. Actual crop growth is less than potential if the uptake of one or more of these elements is smaller than the demand of the crop. This may be caused by a limited supply from the soil to the root system or by a limited root activity. These limitations may be continually or only temporarily present and may occur in the whole or only in a part of the soil-root system.

Crop yield is the cumulative result of crop growth during the growing season. If the length of the growing period is short compared to the crop requirements, crop yield will be less than potential. Therefore, three types of limitations can be distinguished: (1) length of the growing period; (2) supply of water, oxygen and nutrients; (3) root activity.

In humid temperate climates, sowing and planting in spring are carried out as early as possible to maximimize the length of the growing period. However, the earliest possible date of sowing or planting is determined by soil trafficability and workability and, therefore, these operations often have to be delayed, which may shorten the length of the growing period considerably. Germination and emergence are determined mainly by soil temperature and may, therefore, be seriously delayed by unfavourable weather and soil conditions. A (forced) early date of harvesting (e.g., imposed by a delivery contract), may also shorten the length of the growing period. The latest possible date of harvesting is determined by soil trafficability and crop harvestability.

For the supply of water, oxygen and nutrients to the root system, a distinction is made between the potential specific supply rate, the potential supply rate and the potential supply. The potential specific supply rate (kg cm^{-1}day^{-1}) is defined as the potential rate of transport by a soil to a unit root length. The potential supply rate (kg ha^{-1}day^{-1}) is defined as the potential rate of transport by a soil to a whole root system. It is the product of the potential specific supply rate and the total root length per ha. The potential supply (kg ha^{-1}) is defined as the potential total supply to the root system during the growing period, i.e., the integration over time of the potential supply rate of the soil. Similarly, for the uptake of water, oxygen and nutrients by a root system, a distinction is made between the potential specific uptake rate (kg cm^{-1}day^{-1}), the potential uptake rate (kg ha^{-1} day^{-1}) and the potential uptake (kg ha^{-1}). These potential values are related to prevailing environmental conditions.

Crop size and development stage, together with the actual weather conditions, determine potential crop growth and thereby crop demand (CD) for water, oxygen and nutrients (Fig. 1). Potential crop growth as well as soil physical conditions determine root growth rate (RGR) and root size. The potential supply rate of water, oxygen and nutrients by the soil to the crop is dependent on the size (length) of the root system and on soil physical parameters. When the potential supply rate is optimum, the potential uptake rate can still be restricted by root size as well as by crop demand. The actual uptake rate is limited either by the potential supply rate or by the potential uptake rate. In turn, the actual uptake rate determines actual crop growth, which has an ultimate effect on crop size. The potential supply rate and the potential uptake rate are highly variable

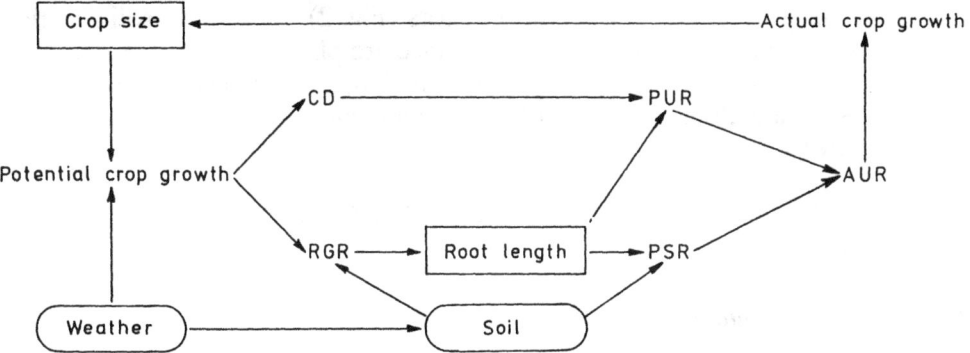

Fig. 1. Scheme of the influence of weather and soil factors on root growth, root activity and crop growth. CD = Crop Demand; RGR = Root Growth Rate; PUR = Potential Uptake Rate; PSR = Potential Supply Rate; AUR = Actual Uptake Rate.

in space and time. This is caused by spatial variation in soil structure and root length density, and by variable weather conditions which induce variations in soil water potential, and in oxygen and ion concentrations. Only when the potential supply rate and actual uptake rate are fully synchronised and synlocalised, are natural resources used most efficiently.

Soil compaction directly affects the system of macropores and, therefore,

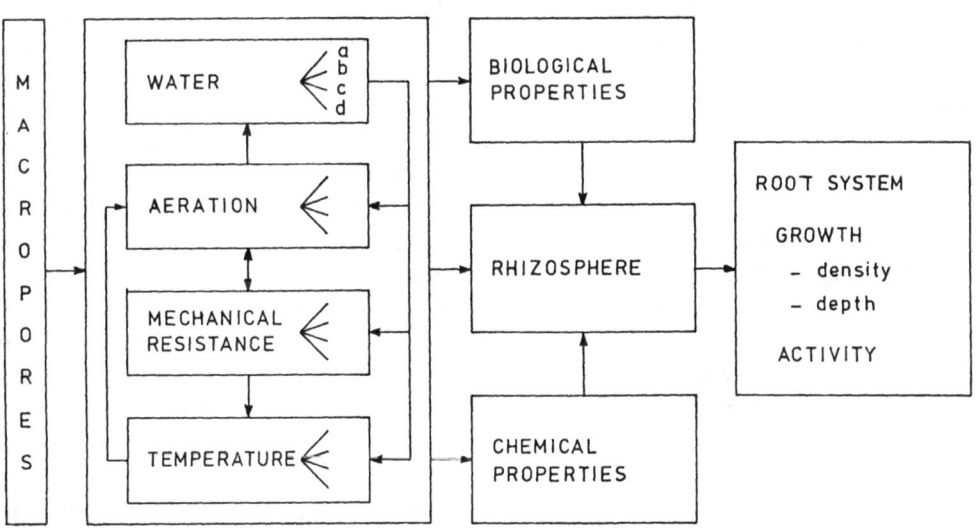

Fig. 2. Scheme of the relationships between soil macropores, soil physical, chemical and biological properties, rhizosphere and root system. a = surface boundary, b = storage, c = transport; d = sink or source aspects (from Boone, 1988).

primarily the soil physical root growth factors (Fig. 2). Soil water, soil aeration, soil temperature and soil mechanical resistance are physical properties which are modified to a variable degree by soil compaction. This change in soil physical environment may affect specific biological and chemical properties, such as the availability of nitrogen (Boone, 1988).

SOIL COMPACTION AND CROP GROWTH LIMITING SITUATIONS IN THE FIELD

Length of the growing period

The actual length of the growing season may be less than potential, either by delayed sowing and planting or by an early harvest. Too early sowing causes a sub-standard seedbed quality and serious damage to soil structure at greater depths. Soil compaction has a direct effect on soil trafficability, soil workability and harvestability of root crops. Soil compaction also affects soil conditions which influence germination and emergence and, therefore, the date of emergence. These effects are discussed below.

Field operations

In general, a compact topsoil has a better trafficability than a loose topsoil. This may make it possible to sow earlier or to harvest root crops later. However, under wet conditions this advantage is totally undone by a worse soil workability or harvestability on a compact than on a loose topsoil. This is especially the case in micro-depressions of the soil surface and in wheel ruts which collect rainwater and thus remain wet for a longer period.

Soil workability is primarily determined by soil water content at sowing or planting depth. The limiting water potential depends on soil type and the depth of the soil layer to be prepared. Soil compaction decreases the saturated hydraulic conductivity and increases the unsaturated hydraulic conductivity (see Chapter 7). Therefore, in wet soil conditions, infiltration may be delayed and capillary rise to the soil surface enhanced to such extent that in the top layer of a compacted soil the water potential is higher than in a loose soil. In temperate humid climates, adequate drainage greatly increases the number of workable days in early spring (Van Wijk and Feddes, 1986). However, on non-tilled, strongly compacted soil, even if well-drained, in some years sowing and planting have to be delayed as compared to a tilled soil (Van Ouwerkerk and Lumkes, 1984).

Through abundant rainfall and enhanced by poor drainage, water potential in the arable layer may return to high values at the end of the growing season, especially in a compacted soil. This adversely affects the possibilities of mechanical harvesting of root crops without seriously damaging soil structure (Buitendijk, 1985).

Crop emergence

Normally, the supply of oxygen to seeds sown into a loose configuration of aggregates at the bottom of the seedbed is satisfactory. The supply of oxygen may become insufficient only if water infiltration into the compacted soil below the seedbed is so slow that water accumulates at the bottom of the seedbed. Moreover, in soils with a low mechanical stability or a high amount of small aggregates, slaking of the soil surface is enhanced by an increased water content. Water ponding on the slaked surface strongly diminishes the availability of oxygen because diffusion of oxygen through water is 10,000 times slower than through air. When the soil surface is sealed over areas >1 m^2, seeds are dependent on the oxygen present at the time of sealing. A 5-cm deep seedbed with an air-filled porosity of 10% (v/v), contains at maximum 1 l m^{-2} of oxygen. This amount will be used within a few hours when soil oxygen consumption is high and within one day when soil oxygen consumption is low. Consequently, germination will be hampered when no oxygen is available from the atmosphere or from the soil below the seedbed. However, effective sealing of the surface is restricted to the period of complete saturation of the slaked surface (Brown et al., 1965). Oxygen stress in wet conditions, possibly followed by a high mechanical impedance through crust formation after drying of the slaked surface, may cause delayed emergence or, in the most affected parts of the field, total failure. This results in irregular distribution of plants, which in many crop species cannot be outweighed by compensatory growth.

The risk of incomplete emergence caused by insufficient aeration is enhanced in a compact, wet topsoil when the sowing coulters smear the surface of the slits in which the seeds are sown. Special coulter designs, e.g., with a "W" shape (Choudhary et al., 1985) or strip seedbeds in crops with a large row spacing, can alleviate this problem if drainage is sufficient.

Evaporation of water has a strong effect on soil temperature. Because a compact soil tends to be wetter, soil temperature at sowing depth may be lower than in a loose soil. Therefore, in temperate humid climates, soil compaction of sub-optimally drained soils may delay emergence. However, in hot climates it may prevent heat stress.

SOIL COMPACTION AND LIMITED SUPPLY OF WATER, OXYGEN AND NUTRIENTS BY THE SOIL

Limited supply of water

The potential specific supply rate of soil water

The rate of water supply from the soil to a root depends on the difference in water potential between the root and the soil, the conductivity of water in the soil around the root and the soil-root contact.

The direct effect of soil compaction on the soil water potential is negligible, unless conditions are extreme. Very severe compaction at high moisture content causes soil deformation and decreases soil water potential. More important are the indirect effects of soil compaction on soil water potential by a change in hydraulic conductivity, which affects infiltration, evaporation, redistribution, capillary rise and downward flow of water. Compaction of a loose soil decreases the saturated hydraulic conductivity. At decreasing soil water potentials the differences in hydraulic conductivity become smaller and, beyond a certain soil water potential, compaction increases the unsaturated hydraulic conductivity. These effects are a function of the changes in pore size distribution and continuity of the pore system. In some soils even small deformations will cause large decreases in the saturated hydraulic conductivity (Dawidowski and Koolen, 1987).

Changes in pore size distribution are attended with changes in the volumetric water content. At every soil water potential there is a specific soil porosity where soil water content is maximum. These maxima shift to higher porosities at higher soil water potentials. Consequently, moderate compaction of a loose subsoil in contact with a groundwater table increases capillary rise and, therefore, the water potential in the arable layer is increased more than by strong compaction (Fig. 3). On the other hand, compaction of the topsoil potentially decreases the water potential in the arable layer by increased evaporation. However, usually this effect is only small because the difference in time with first-stage evaporation is small. When rainfall exceeds the infiltration capacity of the soil matrix, water runs to microdepressions, wheel ruts and large vertical macropores or cracks. By reducing the number, dimensions and continuity of large macropores and vughs, soil compaction increases the occurrence of runoff but may decrease water losses by leaching. Water accumulating in microdepressions and wheel ruts will cause a temporarily higher water potential in the immediate vicinity.

The contact between individual roots and the soil, expressed as the proportion of the root surface which is in direct contact with solid soil particles, is improved by soil compaction. For soil samples composed of sieved aggregates compacted to total porosities of 60, 51 and 44% (v/v), the average degree of root-soil contact for maize roots was 60, 72 and 87%, respectively (Kooistra et al., 1992). These values are larger than may be expected for random positioning of roots in the voids present. The maximum increase in root-soil contact parallelled the increase in volume of solid particles in the soil. The root-soil contact did not depend on root diameter. The extremes, 0% and especially 100%, occurred relatively frequently. The larger root-soil contact caused by soil compaction, implies that for a certain uptake rate of water, the required gradient in matric water potential can be smaller than in loose soil. This effect is additional to the effect of the increased unsaturated hydraulic conductivity of the compacted soil around the roots.

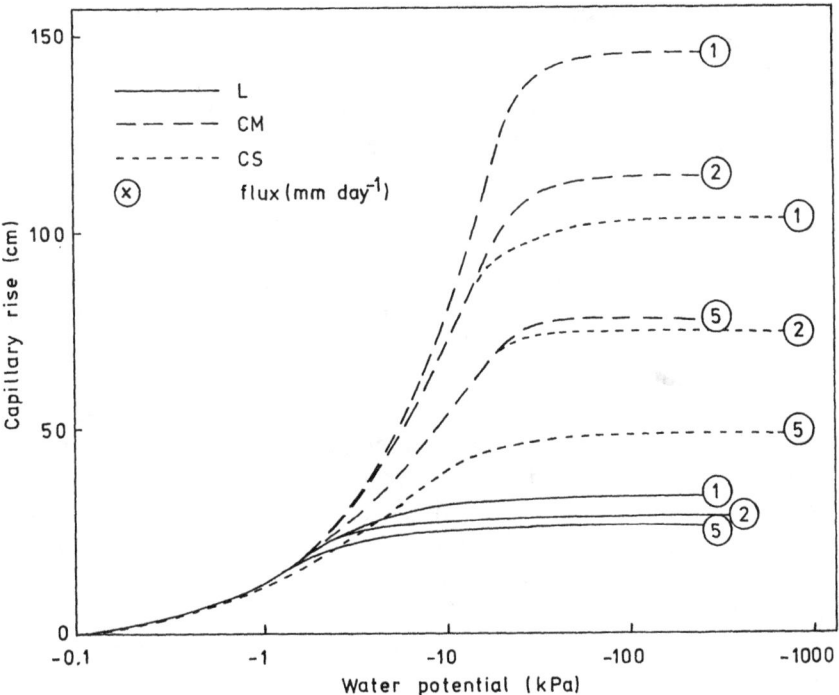

Fig. 3. Theoretical heights (z) of capillary rise, calculated for three steady upward fluxes (1, 2 and 5 mm day⁻¹) at three levels of compaction of the arable layer in a loamy sand soil. L = loose soil; CM = moderately compacted; CS = severely compacted (from Boone et al., 1978).

The potential supply rate of soil water

The potential supply rate of soil water is determined by the potential specific supply rate and the rooted volume and, to a lesser extent, by the root length density because water is very mobile in the soil. Therefore, the most severe limitation to the rate of crop water supply is to be expected when the rootability of the subsoil is strongly diminished by soil compaction, and rainfall and irrigation or capillary rise from a groundwater table are insufficient. However, in many soils maximum rooting depth is determined by the genesis of the soil. For example, in well-drained soils, layered deposits often form a natural mechanical barrier to root development, irrespective of additional compaction (Boone et al., 1985). On the other hand, a superficial root system developing in a poorly drained soil is not a direct result of a high potential supply rate in the topsoil of water but of less favourable aeration conditions at greater depths.

When soil compaction effects are restricted to the arable layer and the rootability of the subsoil is satisfactory, the potential supply rate of water may be only temporarily diminished. Once the roots have reached the moist subsoil, root growth in the subsoil is stimulated, at the expense of root growth in the unfavourable arable layer (Boone et al., 1984). This increases the potential supply rate of water compared to a more uniform root distribution with depth. When sowing depth is 4 cm, and the root growth rate in a loose soil is 3 cm day^{-1} (Veen and Boone, 1990), vertically downwards growing roots will reach the subsoil at 25 cm depth in one week. At a high mechanical impedance, root growth rate can be reduced to 0.3 cm day^{-1} and roots will reach the subsoil only after 10 weeks. By then, crop development has proceeded so far that investment in new roots growing into the subsoil is reduced.

The spatial variation in mechanical resistance, soil aeration and nutrient availability affects the degree of clustering of the roots. Preferential growth of roots occurs in less compacted spots with large macropores, lower mechanical resistance, sufficient aeration, or a high nutrient availability. When the scale of structural heterogeneity is larger than the extension of the root system of individual plants, the position of the plants relative to the most compact parts of the soil controls the possibilities for root growth (Boone, 1988). The least favourable position of the plant is in the middle of a severely compacted wheel track when no escape route through a loose seedbed for at least some of the roots is present. When no compensatory growth of roots in favourable soil parts occurs, water supply is strongly restricted. However, when the slow extraction of water from the compact soil induces planes of weakness or cracks, the restriction is only temporary.

The potential supply of soil water

Usually the rate of water uptake is not limited in the range of soil water potentials between -10 kPa (close to field capacity) and -50 kPa. However, the rate is increasingly limited between -50 kPa and -1600 kPa (permanent wilting

TABLE 1

Effect of different levels of compaction of the arable layer and at plough pan depth of a loamy sand soil on the weight of potato foliage on two dates in 1976 (after Van Loon and Bouma, 1978)

Date	Yield of foliage	Treatment[1]					
		(kg m^{-2}) L	LI	CM	CMI	CSP
14 June	Fresh weight	1.11	1.24	0.96	1.15	0.51	1.31
	Dry weight	0.10	0.11	0.09	0.10	0.05	0.12
16 August	Fresh weight	2.22	2.85	2.41	2.91	2.26	1.42
	Dry weight	0.25	0.34	0.28	0.33	0.28	0.20

[1] L = loose soil; CM = moderately compacted; CS = severely compacted; P = strongly compacted plough pan; I = irrigated.

point). Between -10 kPa and -1600 kPa the gravimetric water content is not influenced by modest compaction, but very severe compaction reduces water content at -10 kPa. Thus, on a volume basis, the amount of available water increases from a loose to a modestly compacted soil and decreases at further compaction. These effects are more pronounced the higher the gravimetric water content at -10 kPa, e.g., for soils with high clay or humus contents. However, expressed in mm water, the compaction effects are small, unless the compacted layer is very deep (e.g., 1 m) and the change in pore space extremely large (e.g., >10%,v/v).

Under favourable conditions, roots at the rooting front grow at a soil water potential close to -10 kPa. The evapotranspiration rate can be, at most, equal to the rate at which water is becoming available at the rooting front. In medium-textured soils, the amount of water available between -10 and -1600 kPa is about 2 mm water per cm depth of soil. Therefore, in situations where evapotranspiration is >2 mm day^{-1}, in combination with a root growth rate <1

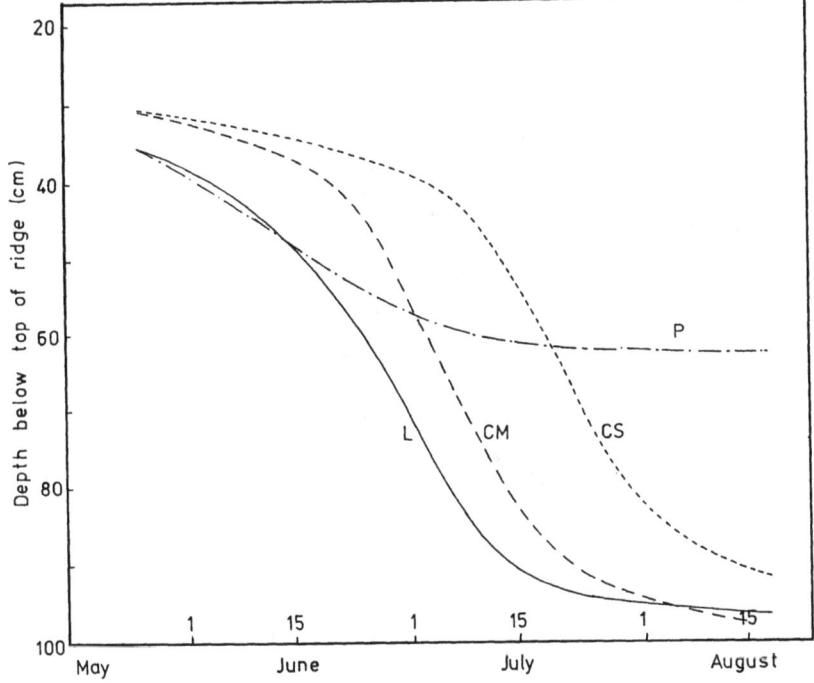

Fig. 4. Rooting depths (100% of the roots) of potatoes during the 1976 growing season for several levels of compaction of the arable layer in a loamy sand soil. L = loose soil; CM = moderately compacted; CS = severely compacted; P = strongly compacted ploughpan (from Boone et al., 1978).

cm day⁻¹, water potential at the rooting front will decrease. Consequently, the mechanical impedance increases and root growth rate decreases. In a homogeneous dense soil without continuous macropores having the size of roots and at high evapotranspiration rates, root growth to depth may ultimately stop and thereby limit the potential supply rate of water.

If, in an uncompacted subsoil, the capillary flux from a shallow groundwater table to the rooting front in the arable layer is smaller than the evapotranspiration rate, subsoil compaction may temporarily favour early crop growth (Van Loon and Bouma, 1978; Table 1). However, if root growth rate in the subsoil is strongly decreased by compaction (Boone et al., 1978; Fig. 4), this

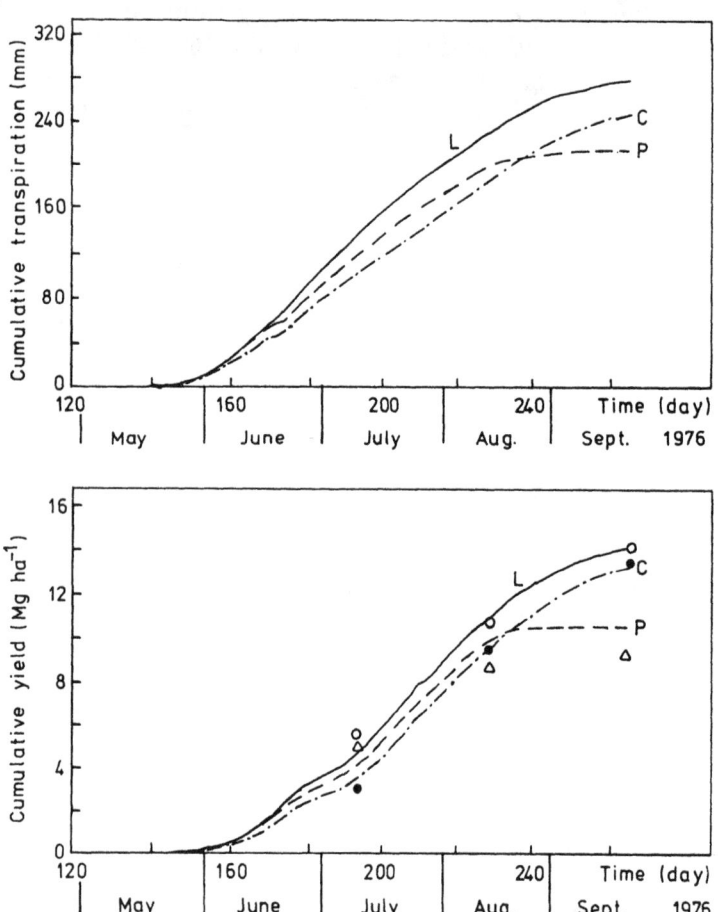

Fig. 5. Simulated cumulative transpiration of potatoes (top), and simulated (curves) and measured (data points) cumulative dry matter production of potato tubers (bottom) during the 1976 growing season for three levels of soil compaction. L = loose soil (○); C = strongly compacted arable layer (●); P = strongly compacted plough pan (▲) (from Feddes et al., 1988).

advantage turns into a disadvantage because root growth to depth lags behind the fall of the groundwater table. Consequently, the net effect of subsoil compaction on the potential supply of water will be strongly negative (Boone et al. 1985; Feddes et al., 1988), which is also reflected in the final yield of potato tubers (Fig. 5).

Limited supply of soil oxygen

The potential specific supply rate of soil oxygen
Soil aeration is determined by the air-filled pore space and is, therefore, strongly influenced by soil drainage and soil compaction (see Chapter 8). Soil compaction decreases the oxygen diffusion rate and increases the root-soil contact. Both aspects have a negative effect on the potential specific supply rate of oxygen.

During compaction, the largest air-filled pores disappear and are replaced by smaller, mainly water-filled pores. The decrease in air-filled porosity is 1.5-2 times larger than the decrease in total pore space. The resulting decrease in the oxygen diffusion coefficient depends on the geometry and stability of the network of air-filled pores and the degree of deformation during compaction. Consequently, this decrease is more pronounced in soils with a low than with a high stability. The oxygen diffusion coefficient asymptotically decreases to zero at air-filled porosities between 2% (v/v) on clay and 10% (v/v) on sand (Bakker et al., 1987).

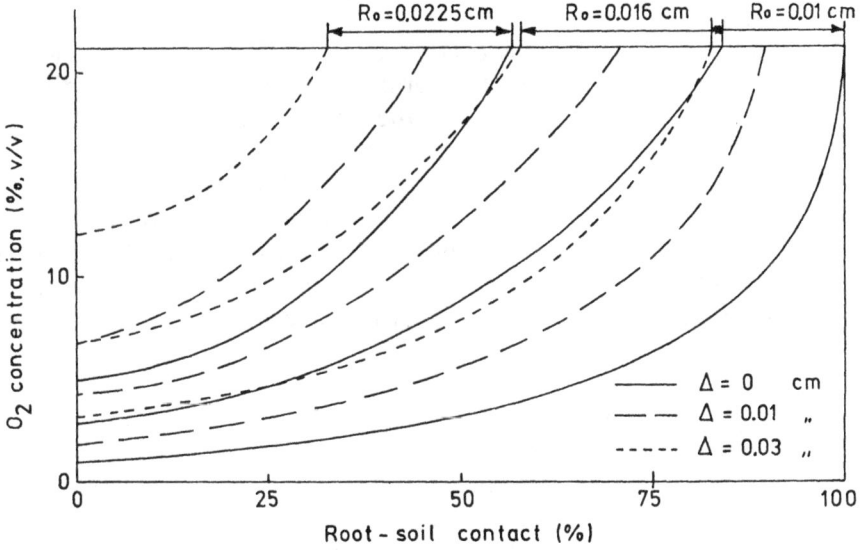

Fig. 6. Oxygen concentration required in soil air for aerobic respiration by all root cells as a function of the root-soil contact, root radius (R_0) and thickness of the water film (Δ) (from Van Noordwijk and De Willigen, 1984).

For root growth, the distribution of the well-aerated parts of the soil volume is more relevant than mean values at a particular depth. Compaction reduces the number of interconnected air-filled pores. Hence, the number and size of poorly aerated volume elements in the soil increase. The network of remaining well-aerated volume elements consists of interconnected large pores, which serve as pathways for preferential root growth. The indirect effect of soil compaction on the potential specific supply rate of oxygen is caused by the potential changes in amount and distribution of water over time.

The fraction of the root surface which is in direct contact with air-filled pores, i.e., the root-air contact, is complementary to the root-soil contact. Therefore, soil compaction causes a decreased transport of oxygen to the root surface (Van Noordwijk and De Willigen 1984; Fig. 6).

The potential supply rate of soil oxygen
The potential supply rate of soil oxygen is determined by the oxygen concentration at the root surface, the root-air contact, the oxygen diffusion coefficient of the soil and the distance to the soil surface. Soil compaction increases the mechanical resistance which causes a more superficial root system and may increase the specific oxygen consumption (Schumacher and Smucker, 1981). This causes a steeper gradient in oxygen concentration with depth, which can be enhanced by wet soil conditions further reducing the potential rooting depth.

Root growth and root activity in the subsoil can be restricted by a topsoil which limits the supply of oxygen to the subsoil. When the topsoil has been strongly compacted in a wet state and the gas diffusion coefficient has been severely reduced, decreased oxygen supply may occur even when the topsoil has a water potential of -10 kPa (Boone et al., 1986). Alternating wet and dry soil conditions may induce aeration stress in the roots in the subsoil, thereby reducing the maximum rooting depth and increasing the risk of water stress in dry periods.

The spatial variation in soil aeration induces clustering of roots in less compacted spots with sufficient aeration. However, insufficient aeration is only temporary when a slight decrease in soil water content increases the gas diffusion coefficient. This is especially the case when small shrinkage cracks are induced. In some plant species aeration stress is relieved to some extent by internal oxygen transport via aerenchyma.

The potential supply of soil oxygen
The amount and distribution of soil water in the soil profile change continuously and, thus, a steady state in the potential supply rate of oxygen is very rare (Leffelaar, 1987). The occurence of periods with a potential supply rate of oxygen lower than crop demand, strongly depends on the amount and distribution of rainfall (Van Lanen and Boersma, 1988). During early crop growth evapotranspiration increases, which increases the probability of a lower soil water

TABLE 2

Typical soil and crop characteristics in two periods after emergence of maize on 15 May 1981, for four levels of compaction of the arable layer in a sandy soil (from Boone et al., 1987)

Period (day)	Characteristic	Depth (cm)	Treatment[1] CS	CM	CL	L
0-37	Matric water potential (-kPa)	15	9-17	-	8-15	-
	Oxygen concentration (%, v/v)	15	11	19	19	-
		35	11	18	18	-
	Cone resistance (MPa)	15	1.7	1.0	0.8	0.8
	Root depth (90%) on day 26 (cm)		13	-	16	-
	Crop development on day 32[2]		7.4	7.2	6.4	5.4
38-77	Matric water potential (-kPa)	15	6-18	-	5-11	-
		35	4-7	-	3-8	-
	Oxygen concentration (%, v/v)	15	8	18	19	-
		35	5	16	16	-
	Root depth (90%) on day 52 (cm)		16	-	28	28
	Functional rooting depth (cm)[3]		12	-	22	26
	Crop development on day 59[2]		5.9	7.2	7.2	6.4

[1]CS = severely compacted; CM = moderately compacted; CL = slightly compacted; L = loose soil.
[2]Visual assessment in a scale of 0-10.
[3]Depth at which root density on day 52 was 0.2 cm cm^{-3}.

content and an increased potential supply rate of oxygen. The oxygen demand of the root-soil system increases as the crop biomass increases, and in temperate regions also as the soil temperature increases. Consequently, the occurrence of insufficient oxygen supply is not limited to emergence and early crop growth, but may also occur later in the growing season when the crop is more vulnerable to oxygen stress (Table 2). Soil compaction increases both the duration and the severity of the periods with oxygen stress (Boone et al., 1986, 1987).

Limited supply of nutrients

The potential specific supply rate of nutrients
There is no direct effect of soil compaction on the nutrient concentration in the soil water. Modest soil compaction increases the volumetric water content, which increases the apparent diffusion coefficient (So and Nye, 1989) and, therefore, the potential specific supply rate of nutrients. Very severe compaction might

decrease the volumetric water content and thus the potential specific supply rate of nutrients.

The potential specific supply rate of nitrate is indirectly affected by a change in the balance between nitrification and denitrification processes. Anaerobic conditions are a prerequisite for denitrification. Soil compaction increases the volume of soil which is vulnerable to denitrification, especially at a high soil water potential. Even short periods with anaerobic conditions cause large losses of nitrate and strongly affect the potential specific supply rate of nitrate (Leffelaar, 1987). Nitrification is slow as long as the supply of oxygen to anaerobic spots is limited.

The potential supply rate of nutrients

In temperate humid climates, root growth to depth usually proceeds with a water potential at the rooting front close to field capacity. Root growth rate at field capacity in strongly compacted soils is less than 20% of the growth rate in uncompacted soils. In soils low in plant nutrients, the potential supply rate of potassium and phosphate thus might become lower than crop demand (Boone and Veen, 1982). When root growth at the rooting front proceeds at lower water potentials, this limitation will be reached at higher levels of nutrient concentration. Moreover, a superficial root system increases the probabilities of leaching of mobile nutrients.

Clustering of roots negatively affects the potential supply rate of ions, especially those with a low mobility, e.g., phosphate, and consequently the number of days that a soil can supply nutrients at a specified rate (De Willigen and Van Noordwijk, 1987; Fig. 7). As the arable layer contains the majority of the nutrients, this effect is more important in that layer than at greater depths.

Parts of the soil with a low rootability may temporarily induce a relatively low root length density in the soil just below these parts ("shade effect"; Tardieu and Manichon, 1987). This effect is more important the larger the horizontal extension of these zones and the higher in the soil profile they are present. Severely compacted wheel tracks combine both aspects.

The potential supply of nutrients

Both the total amount of nutrients the soil can supply and the distribution of this amount over the growing season, are of relevance. The first aspect dominates when soil compaction reduces the root growth rate very strongly. In the course of the growing period, even when water supply is sufficient, the supply of ions, especially those with a low mobility, may become smaller than crop demand. This will occur more frequently at a limited water supply and a low level of chemical soil fertility. Denitrification, on the other hand, may occur in any wet period during crop growth. It suddenly reduces the availability of nitrogen to a very low level, whereas recovery of this availability is only gradual and proceeds at a low rate. Consequently, the potential supply of nitrate is reduced (Leffelaar, 1987).

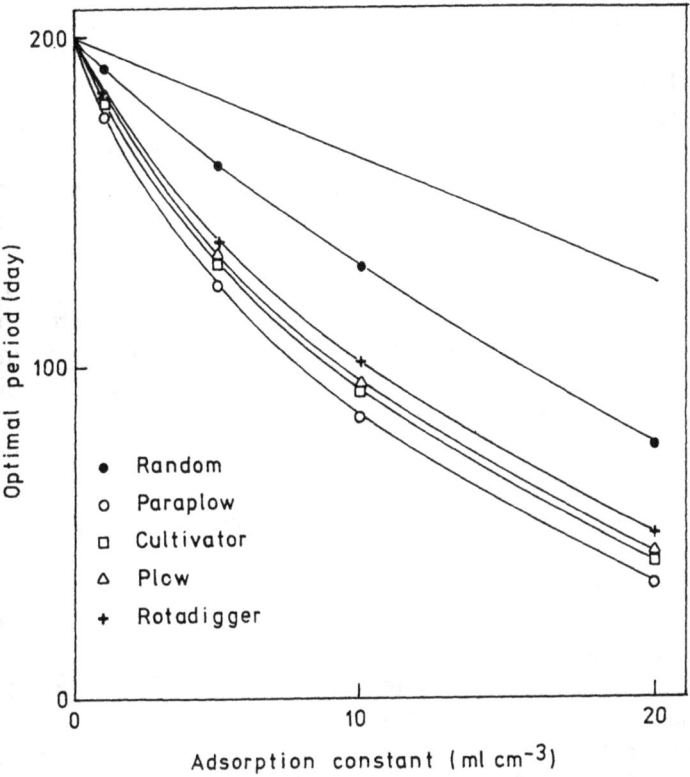

Fig. 7. Optimal period of unrestricted uptake of potassium as a function of the adsorption constant for a regular root distribution (straight line), a random root distribution and four root distributions (curves) after various tillage treatments. The average root density was assumed to be 1 cm cm^{-3} (from De Willigen and Van Noordwijk, 1987).

SOIL COMPACTION AND UPTAKE OF WATER, OXYGEN AND NUTRIENTS LIMITED BY ROOT ACTIVITY

Limited uptake of soil water

The potential specific uptake rate of soil water

The potential specific uptake rate of water is determined by the water potential of the root and the internal resistance of the root to water transport. Although in some cases root resistance is independent of the water flux (Hernandez and Orioli, 1985), root resistance generally decreases with increasing water uptake rate (Black, 1979). The potential specific uptake rate, which depends on the degree of suberization of the endodermis, is largest in that part of the root which is situated between 2 and 10 cm behind the root tip (Sierp and Brewig, 1935). Nevertheless, water uptake in the fully suberized zone is still possible and is more important the greater the water uptake rate (Brouwer, 1954). Root growth rate decreases with soil compaction, whereas the suberization rate is hardly influenced. Therefore, slow growing roots have a higher degree of suberization and an increased average root resistance per unit root length (Kramer, 1959). Root resistance to water transport is also higher at a low soil temperature or at a low oxygen supply, which reduces root respiration severely. Moderate reductions in root respiration do not influence the potential specific uptake rate of water (Trought and Drew, 1980; Veen, 1989). To a certain extent, temperature effects are counteracted by physiological adaptations in the root (Kuiper, 1964).

Soil compaction has only a small negative effect on soil temperature. In situations where soil temperature decreases with depth, the deeper root system in a loose soil experiences a lower mean temperature than the more superficial root system in a compacted soil. This effect may compensate for the effect of a lower seedbed temperature in the compacted soil. Anyway, effects disappear in the course of time by the increasing influence of the shoot on soil temperature. After emergence, the effect of soil compaction on root temperature is negligible.

The potential uptake rate of soil water

The potential uptake rate of water by a root system depends on the potential transpiration of the plant and the resistance of the root system to water transport. Although different parts of a root system can take up water at different rates, these differences are usually not taken into account in calculations of water uptake for whole root systems. On sunny days in temperate climates, evapotranspiration rates of 5 mm day^{-1} can be achieved. Consequently, taking into account measured inflow rates (Lawlor, 1972; Allmaras et al., 1975; Gregory et al., 1978) and assuming a homogeneous root distribution within a depth of 1 m, root length densities of 0.4, 2.5 and 7.5 cm cm^{-3} are required for unrestricted water uptake of soybean, wheat and ryegrass, respectively.

Soil compaction reduces root growth rate and, therefore, increases the resistance to water transport. Only part of the effect on the root growth rate is compensated by osmo-regulation (Greacen and Oh, 1972). At a low plant water potential, stomatal resistance is increased, which results in a decrease in potential uptake rate. It is assumed that plant growth is reduced at a plant water potential < -1 MPa, a value which is reached sooner at a high than at a low atmospheric transpiration demand. De Willigen and Van Noordwijk (1987) suggested that when root development is restricted, the potential uptake rate of water limits plant growth before the potential uptake rate of nutrients becomes limiting.

A strongly compacted wet topsoil may limit the oxygen supply to the subsoil to such an extent that growth and activity of roots in the subsoil are restricted (Boone et al., 1985). In this case, although water supply is not limiting, root resistance may be so much increased that the potential uptake rate of water limits plant growth. On the other hand, when the topsoil is relatively dry, root growth in the subsoil with more easily available water is stimulated (Klepper et al., 1973), compensating for a decrease in potential uptake rate.

The potential uptake of soil water

The potential uptake of soil water depends on the transpirational demand of a crop during the growing season. Only under extreme limitations to root development in any part of the growing season, combined with a high transpirational demand, will root length limit the potential uptake of soil water. This occurs when root resistance is of the same order of magnitude as the stomatal resistance (Veen, 1977).

The potential uptake of water is also affected by compensatory crop responses after a period of limited growth. A possible consequence of water stress is more rapid crop senescence, which limits the potential uptake of water.

Limited uptake of soil oxygen

The potential specific uptake rate of soil oxygen

The potential specific uptake rate of soil oxygen is related to ion uptake, root growth and maintenance processes. Respiration may limit these root activities but can also be determined by these activities and is therefore indirectly dependent on soil temperature. The uptake rate of oxygen per fresh weight of root, about 0.2 mg $g^{-1}h^{-1}$ O_2, depends to a large extent on root activity. Because root growth occurs at the root tip and uptake is mainly in the zone with root hairs, the potential specific uptake rate of oxygen depends on the specific part of the root; it is highest at the root tip (Armstrong and Gaynard, 1976).

Soil compaction at most doubles the root diameter and, thus, the potential specific uptake rate of oxygen may be increased by a factor 4 (Boone and Veen, 1982). In addition, a high mechanical resistance experienced by a root increases the respiration per unit root weight (Schumacher and Smucker, 1981).

The potential uptake rate of soil oxygen

The potential uptake rate of soil oxygen is determined by the potential specific uptake rate of oxygen and the total root length per unit surface area of the soil. Besides the root system, other biota also require oxygen. In temperate climates, the sum of soil and root respiration in the summer amounts to values between 300 and 1500 mg $m^{-2}h^{-1}$ O_2 (Payne and Gregory, 1988). Half of these amounts can be attributed to root respiration (Russell, 1973). Assuming that in a moderately compacted soil the air-filled porosity decreases linearly from 10% (v/v) at the soil surface to 0% (v/v) at the groundwater table at 1 m depth, the amount of air is 50 l m^{-2}. Thus, the amount of oxygen is 10 l m^{-2}, which corresponds to about 7 g m^{-2}. According to Payne and Gregory (1988), when the soil surface is completely sealed, this amount will be sufficient for only 5-24 h. In a strongly compacted soil the period will be even less.

With increasing oxygen concentration in the soil air in contact with the root,

root respiration increases up to the critical oxygen concentration, above which root respiration becomes constant (Berry and Noris, 1949). This critical concentration may vary between <5 and 10% (v/v), due to the variability in the magnitude of the diffusion barriers between the air-filled pores and the root cells which consume the oxygen. Macro-transport in the air-filled pore space as well as the micro-transport from the air-filled pores to the root cells may be hampered by soil compaction to such an extent that the potential uptake rate of oxygen is not reached. Some plant species are able to supply a part of the required amount of oxygen to the root system by continuous air-filled pore spaces in the root system (aerenchyma) which enables diffusion from the shoot (Armstrong, 1979). This internal aeration, which in many arable crops is only of limited significance, decreases the demand of oxygen from the root environment.

The potential uptake of soil oxygen
The root mass reaches its maximum in the generative stage of crop development and, therefore at this stage, root respiration is also maximum. Decomposition of dead roots at the end of the growing period increases oxygen consumption of the soil, especially when soil water and temperature are still favourable. Root crops, such as sugar beet and potatoes, develop storage organs in the soil which constitute a considerable respiring living mass at the end of the growing season.

Limited uptake of nutrients

The potential specific uptake rate of nutrients
The potential specific uptake rate of nutrients depends on the nutritional status of the crop and may vary by a factor 6 (Deane-Drummond, 1982; Jungk et al., 1990). A high mechanical resistance of the soil may reduce root length by a factor 4 without diminishing the volume of the root system (Boone and Veen, 1982). In general, therefore, when root growth is impeded, the demand for nutrients can still be satisfied by an increase in the potential specific uptake rate (Mengel and Steffens, 1985). The potential specific uptake rate of nutrients is also determined by root respiration and therefore by the availability of oxygen. Usually, the effect on the uptake of K is larger than on the uptake of NO_3 (Veen, 1989).

The potential uptake rate of nutrients
The potential uptake rate of nutrients is determined by the crop demand. When the uptake rate of nutrients in part of the root system is restricted by soil compaction, it is increased by compensatory mechanisms in other parts of the root system within the limits indicated for the potential specific uptake rate.

The potential uptake of nutrients
In principle, when soil compaction restricts the potential uptake rate of

nutrients for a considerable length of time, it may cause early senescense of the crop, which further reduces the potential uptake.

SUPPLY OF WATER, OXYGEN AND NUTRIENTS BY THE SOIL RELATIVE TO ROOT ACTIVITY

At optimal supply, the actual uptake rate of water and nutrients is limited by the growth rate of the crop. In the field, however, the actual uptake rate of water, oxygen and nutrients is mainly determined by the transport processes in the soil and, therefore, by the apparent diffusion coefficients and the total root length (De Willigen and Van Noordwijk, 1987). Both factors are directly or indirectly affected by soil compaction.

The potential uptake rate of ions by roots is highly adaptable and dependent on the physiological condition of the plant (Wild et al., 1979; Deane-Drummond, 1982; Jungk et al., 1990). This implies that over a wide range of ion concentrations the demand of the plant can be satisfied (Wild et al., 1979). The influx rates can increase by a factor 6, which means that at unlimited nutrient supply, root length can be decreased by a factor 6 without physiological constraints (Jungk et al., 1990). This is supported by the observation that in a nutrient solution the supply of nutrients to only one quarter of the root system was sufficient to support optimum growth (De Jager, 1985).

However, in soil, the effect of reduced root length on nutrient uptake is completely different. Van Noordwijk et al. (1990) calculated the required inflow of P for seven crops on the basis of crop demand and root length data. For maize the required inflow rate was 2.0×10^{-14} mol $cm^{-1}s^{-1}$, which is far less than the physiological maximum rate (Jungk et al., 1990). It may be concluded that the uptake kinetics of the root did not limit ion uptake rates but rather the transport processes in the soil, determined by apparent diffusion coefficients and root length. This means that a reduction in root length by soil compaction will lead to an equivalent reduction in uptake rate of the growth limiting soil factor. However, when the supply of nutrients is not limiting the uptake rates, soil compaction will influence root development without consequences for shoot growth (Boone and Veen, 1982). In moderate climates a daily dry matter crop production of 200 kg ha^{-1}, with a content of 1.5% N, 0.22% P and 1.0% K, may be assumed (De Willigen and Van Noordwijk, 1987). This implies that 440 g ha^{-1} day^{-1} P would be required. Assuming for maize a maximum influx of 5.4×10^{-7} g $cm^{-1}day^{-1}$ (Jungk et al., 1990; Fig. 8), a minimum root length of 810 m m^{-2} would be required to meet the demand for phosphate. This is far less than the root length required for unrestricted water uptake. In addition, a shortage of water, oxygen or nutrients may induce secondary effects and thereby influence the duration of crop growth. For instance, when soil compaction increases denitrification to such an extent that crop growth is stressed by a shortage of nitrogen, this may cause early senescense.

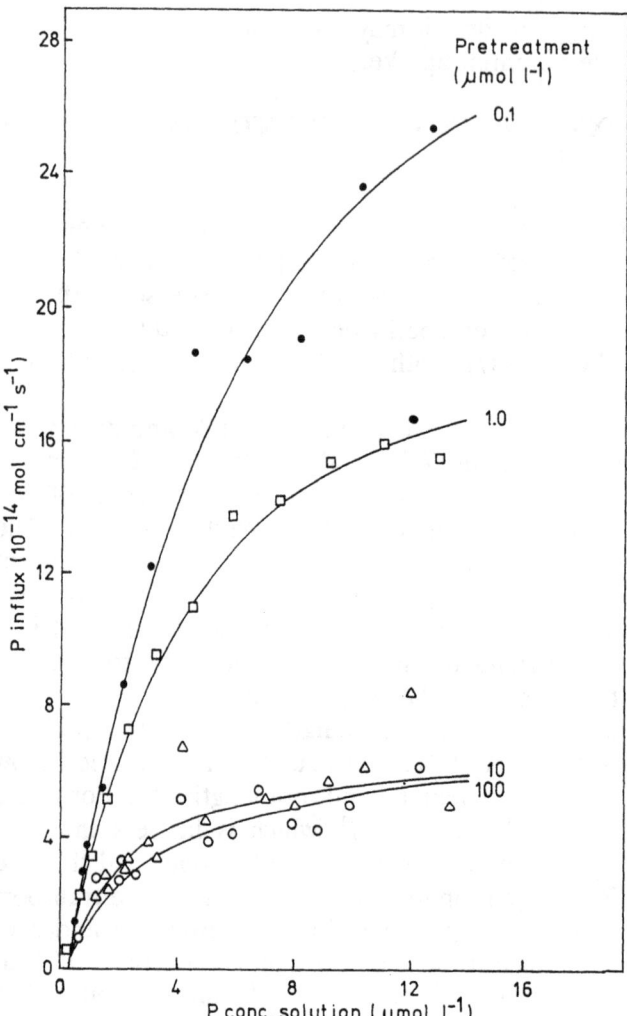

Fig. 8. Kinetics of the uptake of phosphate by maize from solutions with different phosphate concentrations, calculated from depletion curves. Maize plants were pretreated by growing them in solutions with phosphate concentrations as indicated along the curves (from Jungk et al., 1990).

GROWING CONDITIONS AND POSSIBLE LIMITING SOIL FACTORS

Plant shoots and roots react as a physiological entity to environmental conditions (Brouwer and De Wit, 1969). The required supply from the soil of water, oxygen and nutrients strongly depends on the combination of above-ground factors (irradiation, temperature, rainfall), soil-physical properties, chemical soil fertility and specific crop demands. At a high level of photosynthesis, and therefore high crop demand, shoot growth can be limited by a small and superficial root system. At a low light intensity, photosynthesis is the primary shoot growth limiting factor (Boone and Veen, 1982).

Comparing the demands of a specific crop with the level and variability of the soil-borne growth factors, will reveal one or more potential limiting soil factors, which can be classified according to their order of expected importance. It is assumed that soil wetness and soil temperature are independent variables. Soil compaction has no effect on the range of crop growth limiting temperatures. When the ambient temperature is not limiting crop growth, aeration may be limiting in very wet situations, especially at higher temperatures. In temperate humid conditions, both an insufficient aeration as well as water shortage caused by a too superficial root system, induced by a high mechanical resistance, are potential crop growth limiting factors. Although in relative dry situations water is the primary limiting factor, it has been shown that the nutrient supply sometimes is even more limiting than the supply of water (Gales, 1979).

Critical soil water potentials and soil compaction

Water influences crop growth in a complex way. Water is both a direct crop growth factor and a transient medium for the nutrient supply, and indirectly influences photosynthesis by changing stomatal resistance. Moreover, soil aeration, mechanical resistance and soil temperature are strongly dependent on soil water. Therefore, the effects of soil structure on crop growth are best quantified in relation to soil water. Simulation models, such as SWACRO, in which water is regarded as a crop growth limiting factor, are well developed (Feddes et al., 1978; 1988). In principle, submodels which express the relationships between soil structure and soil water, soil aeration, soil temperature or nutrient availability, could be added to these models. In the SWACRO model, the water uptake by the root system governs crop growth and is an empirical function of the soil water potential. Crop growth is limited by deficient soil aeration at high water potential and by deficient water supply at low water potential. The latter soil water potential depends on the evapotranspiration demand and, to some extent, also on the root density. At higher evaporation demand a higher soil water potential is required.

By defining criteria for soil aeration and mechanical resistance, it is possible to establish critical soil water potentials as a function of soil compaction. A lower critical aeration limit has been defined as the value of the oxygen diffusion coefficient necessary to ensure an oxygen concentration in the air-filled pore space of 1% (v/v) at the lower boundary of a soil layer with a low and uniform distribution of oxygen consumption of 10^2 mg m^{-2}h^{-1} O_2 (Boone, 1986). For a 30-cm thick arable soil layer, the required intrinsic gas diffusion coefficient (D/D_0) is then 7.5×10^{-2}. Such a low oxygen consumption may occur in spring in temperate humid climates in soils with a low biological activity, in combination

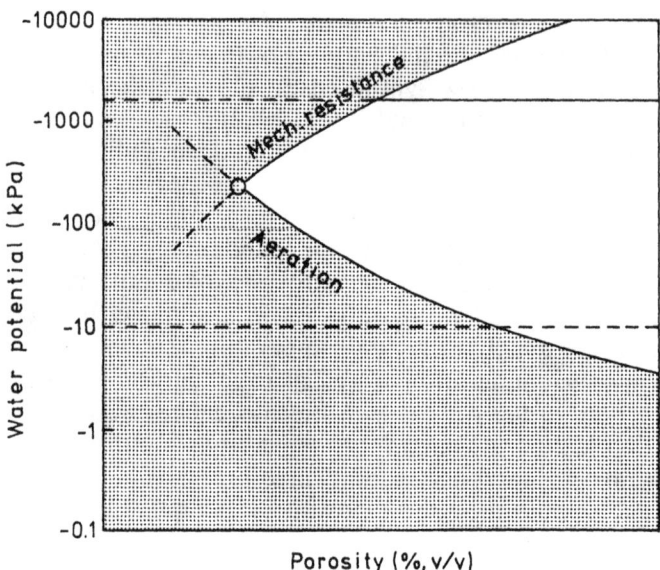

Fig. 9. Conceptual relationship between soil porosity and the soil water potential at which soil aeration and mechanical resistance meet specified root requirements. Root growth is insufficient in the shaded area and impossible beyond a soil water potential of -1600 kPa (from Boone, 1988).

with a low root length density. If the relationships between the oxygen diffusion coefficient, the air-filled pore space and the water retention curve are known for different degrees of compaction, critical soil water potentials can be derived as a function of soil porosity (Boone, 1988; Fig. 9). Because a specific air-filled porosity is attained at lower soil water potentials the more compacted the soil is, the lower critical aeration limit shifts to drier soil conditions with increasing compaction.

Similarly, an upper critical aeration limit has been defined above which no aeration stress is expected. In that case the oxygen concentration in the air-filled pore space does not drop below 10% (v/v) at an oxygen consumption of 10^3 mg $m^{-2}h^{-1}$ O_2. Both the oxygen concentration and the consumption are 10 times higher than at the lower critical aeration limit. To meet these requirements, D/D_0 must be 1.5×10^{-2}, which is reached at lower soil water potentials than the lower critical aeration limit. To predict potential aeration hazards, it is necessary to know at each soil depth the actual oxygen consumption of both soil and root system, and the actual soil oxygen diffusion coefficients. However, all these factors change continually. Therefore, sufficient accuracy can be obtained only by means of dynamic simulation models.

Soil water potentials higher than those at the specified aeration limits are potentially damaging (Fig. 10). Actual yield losses depend on numerous factors, such as duration and intensity of the stress, the proportion of the root system which is stressed, the specific vulnerability for aeration stress and the ability to recover and compensate for aeration stress, as a function of crop development and the level of other environmental factors during and after the stress period.

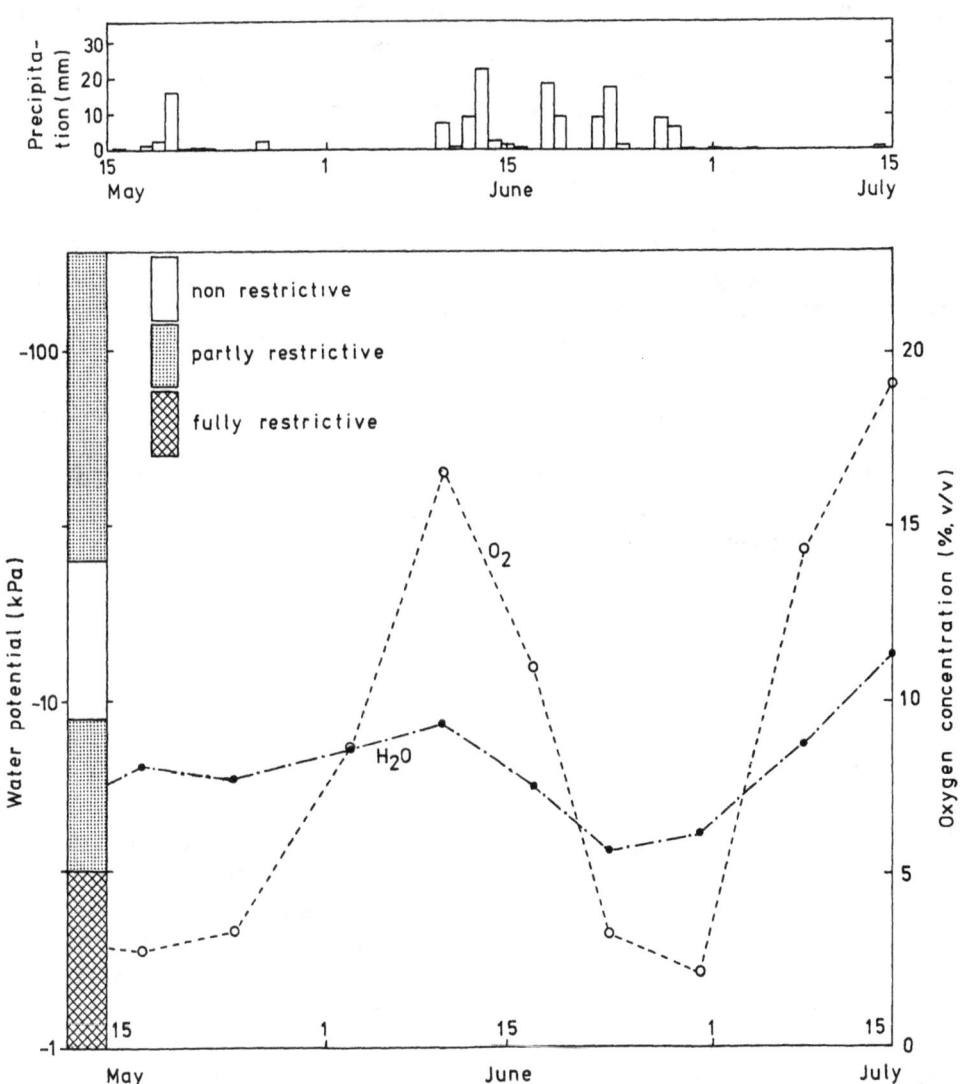

Fig. 10. Measured soil water potentials (scale at left hand side) and oxygen concentrations (scale at right hand side) at a depth of 35 cm in a strongly compacted sandy soil cropped with silage maize. The bar at the left indicates the ranges of non-restrictive, partly and fully restrictive soil water potentials for root growth, based on gas diffusion coefficients and penetration resistances (from Boone et al., 1987).

A potentially damaging situation occurs when the root system is superficial and excessively wet growing conditions are followed by a dry period, especially when harvestable products are being synthesized in this period. Severe soil compaction may enhance soil anaerobiosis in the first period and slow down root growth to depth in the second period by a high mechanical resistance.

Lower and upper critical mechanical limits are reached when root growth rate is significantly reduced, i.e., to 50% and 0%, respectively, of the potential root growth rate. As a first approximation for these critical mechanical limits, cone indices of 1.5 and 3.0 MPa, respectively, were chosen. The latter value should be increased considerably when numerous large and continuous macropores are present. This situation may develop when a compacted soil is not disturbed mechanically for several years as in zero-tillage. It is obvious that both values are reached at higher soil water potentials the more compact is the soil. To predict potentially damaging situations caused by mechanical impedance, the potential evapotranspiration and soil water potential of the soil profile should be known. These objectives can be reached only by dynamic simulation models of root growth in relation to mechanical impedance.

The applicability of the aeration and mechanical impedance submodels depends on the accuracy of the prediction by simulation of the soil and plant water status. These submodels should include soil heterogeneity on at least two specified integration levels: a micro (root) level (mm-cm) for root growth and root activity and a macro (plant) level (dm-m) for integration of different parts of a root system experiencing different soil physical conditions.

COMBINATION OF THE VARIOUS CROP GROWTH LIMITING SOIL FACTORS

If the various soil factors which limit crop growth are combined, it can be predicted that, provided the range of soil porosities is large enough, the relationship between soil porosity and crop yield will follow an optimum curve (Fig. 11). In a very dense, wet soil, soil aeration is sub-optimal, and in relatively dry situations mechanical resistance is too high. On the other hand, in a very loose soil, root-soil contact and hydraulic conductivity are too small to meet crop demands, notwithstanding rapid root growth. In fact, there is a conflicting situation: water and nutrient transport and uptake are favoured by a very dense soil and a large root-soil contact, whereas the transport and uptake of oxygen is favoured by a small root-soil contact.

The range of porosities at which crop growth is potential depends on: (1) the potential yield level: crop requirements are higher at a high than at a low yield level, which narrows the range of optimal soil porosities; (2) the overlap of conflicting requirements: the optimum will be broader when two or more soil factors which limit crop growth act in neighbouring ranges of soil porosities; (3) soil wetness during crop growth: the optimum will shift to higher soil porosities

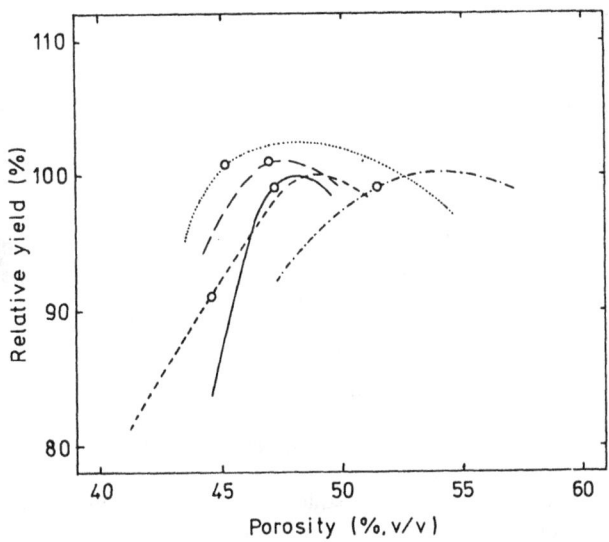

Fig. 11. Relative yields of silage maize in various experiments in the Netherlands as a function of soil porosity in the arable layer of five sandy soils. The porosity obtained by driving once track-to-track over the field (100% coverage) with a 44-kW tractor with twin rear wheels after ploughing and before seedbed preparation, is indicated by a circle (from Boone, 1986).

when the soil is relatively wet and to lower soil porosities when the soil is relatively dry; (4) heterogeneity of soil structure: the optimum will shift to lower soil porosities the higher the number of continuous large pores which are present at low porosities.

CONCLUSIONS

(1) Crop responses to soil compaction may be explained if the appropriate soil and crop growth factors are adequately determined.

(2) The usually large fluctuations in weather conditions and their strong interaction with soil conditions, imply that predictions of the appropriate range of soil porosities have a statistical nature. They quantify the risk of yield losses when soil porosities are outside a specified range.

(3) The greatest challenge is to expand knowledge about the effects of soil structural heterogeneity on transport processes and on root growth and functioning, and to incorporate this knowledge into computer submodels concerning the effects of soil and root aeration and mechanical resistance on root growth and root activity.

REFERENCES

Allmaras, R.R., Nelson, W.W. and Voorhees, W.B., 1975. Soybean and corn rooting in

Southwestern Minnesota. 2. Root distributions and related water inflow. Soil Sci. Soc. Am. Proc., 39: 771-777.

Armstrong, W., 1979. Aeration in higher plants. Adv. Bot. Res., 7: 225-332.

Armstrong, W. and Gaynard, T.J., 1976. The critical oxygen pressure for respiration in intact plants. Physiol. Plant., 37: 200-206.

Bakker, J.W., Boone, F.R. and Boekel, P., 1987. Diffusion of gases into the soil and oxygen diffusion coefficients in Dutch arable soils. Inst. Land Water Manage Res. (ICW), Wageningen, Netherlands, Rep. 20, 44 pp.

Berry, L.J. and Noris, W.E., 1949. Studies on onion root respiration. 1. Velocity of oxygen consumption in different segments of roots at different temperatures as a function of the partial pressure of oxygen. Biochem. Biophys. Acta, 3: 593-606.

Black, C.R., 1979. A quantitative study of the resistances to transpirational water movement in sunflower (*Helianthus annuus* L.). J. Exp. Bot., 30: 947-953.

Boone, F.R., 1986. Towards soil compaction limits for crop growth. Neth. J. Agric. Sci., 34: 349-360.

Boone, F.R., 1988. Weather and other environmental factors influencing crop responses to tillage and traffic. Soil Tillage Res., 11: 283-324.

Boone, F.R. and Veen, B.W., 1982. The influence of mechanical resistance and phosphate supply on morphology and function of maize roots. Neth. J. Agric. Sci., 30: 179-192.

Boone, F.R., Bouma, J. and De Smet, L.A.H., 1978. A case study on the effect of soil compaction on potato growth in a loamy sand soil. 1. Physical measurements and rooting patterns. Neth. J. Agr. Sci., 26: 405-420.

Boone, F.R., Kroesbergen, B. and Boers, A., 1984. Soil conditions and growth of spring barley on a tilled and untilled marine loam soil. In: F.R. Boone (Editor), Experiences with Three Tillage Systems on a Marine Loam Soil. II: 1976-1979. Pudoc, Wageningen, Netherlands, Agric. Res. Rep. 925, pp. 124-166.

Boone, F.R., De Smet, L.A.H. and Van Loon, C.D., 1985. The effect of a ploughpan in marine loam soils on potato growth. 1. Physical properties and rooting patterns. Potato Res., 28: 295-314.

Boone, F.R., Van der Werf, H.M.G., Kroesbergen, B., Ten Hag, B.A. and Boers, A., 1986. The effect of compaction of the arable layer in sandy soils on the growth of maize for silage. 1. Critical matric water potentials in relation to soil aeration and mechanical impedance. Neth. J. Agric. Sci., 34: 155-171.

Boone, F.R., Van der Werf, H.M.G., Kroesbergen, B., Ten Hag, B.A. and Boers, A., 1987. The effect of compaction of the arable layer in sandy soils on the growth of maize for silage. 2. Soil conditions and plant growth. Neth. J. Agric. Sci., 35: 113-128.

Brouwer, R., 1954. Water absorption by the roots of *Vicia faba* at various transpiration strengths. 3. Changes in water conductivity artificially obtained. Proc. Kon. Ned. Acad. Wetensch., C 57: 68-80.

Brouwer, R. and De Wit, C.T., 1969. A simulation model of plant growth with special attention to root growth and its consequences. In: W.J. Whittington (Editor), Root Growth. Butterworths, London, U.K., pp. 224-242.

Brown, N.J., Fountaine, E.R. and Holden, M.R., 1965. The oxygen requirement of crop roots and soil under near field conditions. J. Agric. Sci., Camb., 64: 195-203.

Buitendijk, J., 1985. Effect of workability index, degree of mechanization and degree of certainty on the yield of sugar beet. Soil Tillage Res., 5: 247-257.

Choudhary, M.A., Guo, P.Y. and Baker, C.J., 1985. Seed placement effects on seedling establishment in direct-drilled fields. Soil Tillage Res., 6: 79-93.

Dawidowski, J.B. and Koolen, A.J., 1987. Changes of soil water suction, conductivity and dry strength during deformation of wet undisturbed samples. Soil Tillage Res., 9: 169-180.

De Jager A., 1985. Response of plants to a localized nutrient supply. Ph.D. thesis, Univ.

Utrecht, Netherlands, 137 pp.

Deane-Drummond, C.E., 1982. Mechanisms for nitrate uptake into barley (*Hordeum vulgare* L. cv. Fergus) seedlings grown at controlled nitrate concentrations in the nutrient medium. Plant Sci. Letters, 24: 79-89.

De Willigen, P. and Van Noordwijk, M., 1987. Roots, plant production and nutrient use efficiency. Ph.D. thesis, Wageningen Agric. Univ., Wageningen, Netherlands, 282 pp.

Feddes, R.A., Kowalik, P.J. and Zaradny, H., 1978. Simulation of Field Water Use and Crop Yield. Pudoc, Wageningen, Netherlands, Simulation Mono., 189 pp.

Feddes, R.A., De Graaf, M., Bouma, J. and Van Loon, C.D., 1988. Simulation of water use and production of potatoes as affected by soil compaction. Potato Res., 31: 225-239.

Gales, K., 1979. Effects of water supply on partitioning of dry matter between roots and shoots in *Lolium perenne*. J. Appl. Ecol., 16: 863-877.

Gregory, P.J., McGowan, M., Biscoe, P.V. and Hunter, B., 1978. Water relations of winter wheat. 1. Growth of the root system. J. Agric. Sci., Camb., 91: 91-102.

Greacen, E.L. and Oh, J.S., 1972. Physics of root growth. Nature (New Biol.), 235: 24-25.

Hernandez, L.F. and Orioli, G.A., 1985. Relationships between root permeability to water, leaf conductance and transpiration rate in sunflower (*Helianthus annuus* L.) cultivars. Plant Soil, 85: 229-235.

Jungk, A., Asher, C.J., Edwards, D.G. and Meyer, D., 1990. Influence of phosphate status on phosphate uptake kinetics of maize (*Zea mays*) and soybean (*Glycine max.*). In: M.L. van Beusichem (Editor), Plant Nutrition - Physiology and Applications. Kluwer, Dordrecht, Netherlands, pp. 135-142.

Klepper B., Taylor, H.M., Huck, M.G. and Fiscus, E.L., 1973. Water relations and growth of cotton in drying soil. Agron. J., 65: 307-310.

Kooistra, M.J., Schoonderbeek, D., Boone, F.R., Veen, B.W. and Van Noordwijk, M., 1992. Root-soil contact of maize, as measured by thin-section technique. 2. Effects of soil compaction. Plant Soil, 139: 119-129.

Kramer, P.J., 1959. Transpiration and water economy of plants. In: F.C. Stewart, (Editor), Plant Physiology, a Treatise, Vol. 2: Plants in Relation to Water and Solutes. Acad. Press, New York, NY, U.S.A., 758 pp.

Kuiper, P.J.C., 1964. Water uptake by higher plants as affected by root temperature. Meded. Landb. Hogesch. Wageningen, Wageningen, Netherlands, 63: 1-11.

Lawlor, D.W., 1972. Growth and water use of *Lolium perenne*. 1. Water transport. J. Appl. Ecol., 9: 79-98.

Leffelaar, P.A., 1987. Dynamics of partial anaerobiosis, denitrification and water in soil: experiments and simulation. Ph.D. thesis, Wageningen Agric. Univ., Wageningen, Netherlands, 117 pp.

Mengel, K. and Steffens, D., 1985. Potassium uptake of rye-grass (*Lolium perenne*) and red clover (*Trifolium pratense*) as related to root parameters. Biol. Fert. Soils, 2: 53-58.

Payne, D. and Gregory, P. J., 1988. The soil atmosphere. In: E.W. Russell (Editor), Soil Conditions and Plant Growth. Longman, London, U.K., pp. 298-314.

Russell, E.W., 1973. Soil Conditions and Plant Growth, Longman, London, U.K., 688 pp.

Schumacher, T.E. and Smucker, A.J.M., 1981. Mechanical impedance effects on oxygen uptake and porosity of dry bean roots. Agron. J., 43: 51-55.

Sierp, H. and Brewig, A., 1935. Quantitative Untersuchungen über die Wasserabsorptionszone der Wurzeln. (Quantitative research on the water absorption zone of the roots). Jb. Wiss. Bot., 82: 99-122 (in German).

So, H.B. and Nye, P.H., 1989. The effect of bulk density, water content and soil type on the diffusion of chloride in the soil. J. Soil Sci., 40: 743-749.

Tardieu, E. and Manichon, H., 1987. Conséquences de l'état du profil cultural sur l'enracinement: cas du maïs. (Consequences of the structure of the soil profile for the rootability of maize). In: G. Monnier and M.J. Goss (Editors), Soil Compaction and Regeneration. Balkema, Rotterdam, Netherlands, pp. 131-143 (in French).

Trought, M.C.T. and Drew, M.C., 1980. The development of waterlogging damage in young wheat plants in anaerobic solution cultures. J. Exp. Bot., 31: 1573-1585.

Van Lanen, H.A.J. and Boersma, O.H., 1988. Use of soil-structure type data in a soil-water model to assess the effects of traffic and tillage on moisture deficit, aeration, and workability of sandy loam and clay soils. Proc. 11th Conf. Int. Soil Tillage Res. Org. (ISTRO), Edinburgh, U.K., Vol. 1, pp. 415-420.

Van Loon, C.D. and Bouma, J., 1978. A case study on the effect of soil compaction on potato growth in a loamy sand soil. 2. Potato plant responses. Neth. J. Agric. Sci., 26: 421-429.

Van Noordwijk, M. and De Willigen, P., 1984. Mathematical models on diffusion of oxygen to and within plant roots, with special emphasis on effects of soil-root contact. 2. Applications. Plant Soil, 77: 233-241.

Van Noordwijk, M., De Willigen, P., Ehlert, P.A.I. and Chardon, W.J., 1990. A simple model of P uptake by crops as a possible basis for P fertilizer recommendations. Neth. J. Agric. Sci., 38: 317-332.

Van Ouwerkerk, C. and Lumkes, L.M., 1984. Timing of field work and effect of primary and secondary tillage. In: F.R. Boone (Editor), Experiences with Three Tillage Systems on a Marine Loam Soil, II: 1976-1979. Pudoc, Wageningen, Netherlands, Agric. Res. Rep. 925, pp. 47-57.

Van Wijk, A.L.M. and Feddes, R.A., 1986. Simulating effects of soil type and drainage on arable crop yield. In: A.L.M. van Wijk and J. Wesseling (Editors), Agricultural Water Management. Balkema, Rotterdam, Netherlands, pp. 97-112.

Veen, B.W., 1977. The uptake of potassium, nitrate, water and oxygen by a maize root system in relation to its size. J. Exp. Bot., 28: 1389-1398.

Veen, B.W., 1989. Influence of oxygen deficiency on growth and function of plant roots. Plant Soil, 111: 259-266.

Veen, B.W. and Boone, F.R., 1990. The influence of mechanical resistance and soil water on the growth of seminal roots of maize. Soil Tillage Res., 16: 219-226.

Wild, A., Woodhouse, P.J. and Hopper, J.J., 1979. A comparison between the uptake of potassium by plants from solutions of constant potassium concentration and during depletion. J. Exp. Bot., 30: 697-704.

Soil Compaction in Crop Production
B.D. Soane and C. van Ouwerkerk (Eds.)
1994 Elsevier Science B.V. All rights reserved.

CHAPTER 12

Responses of Temperate Crops in North America to Soil Compaction

M.J. LINDSTROM and W.B. VOORHEES

USDA-ARS, North Central Soil Conservation Research Laboratory, Morris, MN, U.S.A.

SUMMARY

Soil compaction concerns are increasing in the temperate region of North America. Increased size and weight of field equipment, plus changes in farming patterns to a predominate row-crop culture, are contributing to this problem. Soils in the region are annually subjected to a freezing and thawing cycle plus several wetting and drying cycles. However, these natural forces do not necessarily ameliorate soil structural changes resultant from compaction. Yield decreases from soil compaction have been identified for all investigated crops grown in the region, but responses have been variable and in some cases difficult to assess, although compaction problems are magnified when combined with other plant stress situations. Soil compaction effects have been identified by increased bulk density, higher soil strength and reduced pore volumes, which in turn exert a negative influence on soil aeration, root exploration, and water and nutrient uptake. Crop responses to compaction are strongly influenced by soil texture and soil water content at the time of load application. However, crop responses are not always negative, as they are also a function of growing season precipitation. Subsoil compaction, which is mainly due to high axle loads, is causing increasingly serious concerns about future productivity. Subsoil compaction is long-lasting and difficult to correct. In the temperate region of North America, efforts to reduce soil compaction have been limited and compaction will remain a production problem until changes in farm practices are made.

INTRODUCTION

Over thirty years ago, Adams et al. (1960) stated "increased weight of equipment and traffic due to ease of tillage with modern machines have increased the problem of soil compaction". Since that time, equipment weight has increased dramatically, traffic intensity has increased, and soil compaction problems have intensified (Fig. 1). Other reports on early studies on soil compaction in Minnesota were made by Blake et al. (1960a,b). They showed significant yield declines resulting from soil compaction for sugar beet (*Beta*

Fig. 1. Primary tillage operation in corn stover remaining after silage cutting with a modern four-wheel drive tractor (axle loads: front 8 Mg, rear 8 Mg).

vulgaris L.), potatoes (Solanum tuberosum L.), corn (Zea mays L.), and wheat (Triticum aestivum L.). The maximum compactive load applied during these experiments was 6.8 Mg on a single axle applied over the entire soil surface, which at that time was considered extreme. The effects of soil compaction were identified as an increase in soil bulk density and a reduction in air-filled porosity to approximately 9% (v/v), which is clearly below the commonly accepted threshold value for adequate aeration at high water potentials. In addition to observed yield declines, the quality of harvested sugar beet and potatoes was also adversely affected.

Compaction can influence crop production by changing important soil properties, particularly bulk density, soil strength, aggregate size distribution, and the size distribution and continuity of pores. In turn, these changes affect infiltration, drainage, water availability, aeration, root exploration and nutrient uptake, all of which can have a direct bearing on crop production. To describe adequately the effect of soil compaction on crop production, it will be necessary to understand and describe changes in soil conditions caused by machinery traffic, as well as crop response to changes in the root-zone environment. Raghavan et al. (1990) categorized the effects of soil compaction into three main elements: machine, soil and crop. This concept reflects an action-reaction-output approach. The first element is subdivided into compaction and tillage machinery. The second element consists of physical, biological and chemical soil characteristics. The third element reflects the water, air and nutrient needs of the crop. Interactions between these elements and sub-elements make soil compaction a complicated problem.

Soil compaction can also affect crop production in less direct ways. For example, Voorhees et al. (1976) reported reduced nodulation of soybean (*Glycine max* (L.) Merr.), in terms of the number of nodules and individual nodule mass, due to surface compaction associated with planting and cultivation wheel traffic. Campbell and Moreau (1979) reported increased ethylene levels and associated reduced oxygen levels in a compacted field, which resulted in leaf injury, lower potato yields and lower tuber quality. Garcia et al. (1988), in a greenhouse study designed to simulate plow pan formation or wheel tracks on one or both interrows, showed differential nitrogen uptake by corn depending on fertilizer placement and zones of compaction.

The objective of this chapter is to review experimental evidence of wheel-induced soil compaction effects on crop growth and yield in the temperate regions of North America. It is not intended to be a complete review of all literature pertaining to this subject, but rather a discussion of selected experiments to illustrate principles. This discussion will be limited to the areas of the U.S.A. and Canada which are subjected to an annual "hard freeze" to a depth of ≥30 cm, and it will concentrate primarily on the zone receiving >600 mm annual precipitation where corn and soybeans (row crop culture) are the major crops. A wide variety of soil types and climatic conditions are present, but generally the soils can be characterized as having a relatively high clay content (≥25%, w/w) with expanding-lattice type clays and relatively high organic matter content (≥3%, w/w). Precipitation, while on average exceeding 600 mm annually, is often erratic between and within seasons. This is the area where most compaction research has been conducted but this does not imply that compaction is not a problem in the lower rainfall areas of western U.S.A. and Canada producing wheat and barley (*Hordeum vulgare* L.).

CHANGE IN FARMING SYSTEMS

Soil compaction problems are increasing in the temperate regions of North America for a number of reasons. First, farm size is continually becoming larger. Average farm size in the U.S.A. was about 200 ha in 1986 as compared to about 100 ha in 1956, but larger farm sizes are common. Of the cultivated area in the U.S.A., 45% is currently located on farms which are 500 ha in size or greater. Due to the decreasing farm population, farmers have been forced to use larger capacity farm machinery to enable timely field operations. Larger capacity machinery entails heavier farm equipment. In 1989, the average tractor size sold in the U.S.A. was 85 kW, with a gross weight of 6 Mg. Primary and secondary tillage operations are commonly undertaken with larger, four-wheel drive tractors that may weigh up to 12-16 Mg. Eight- and twelve-row planting equipment (6- to 9-m working widths) require tractor power units of 100-115 kW with gross weights of 8.5-9.5 Mg. Large combine harvesters can have a loaded weight of 24 Mg with about 75% of that weight on the front axle. Large sugar beet wagons and grain carts can carry loads of 20-36 Mg on a single axle.

Farmers are also becoming less diversified in their cropping systems. For example, in 1986 corn and soybeans accounted for over 85% of the cultivated land area in Iowa while hay only accounted for about 7%, as compared to less than 60% for corn and soybeans and 15% for hay in 1956. This change in farming system to a preponderance of row crop production results in more tillage activity and field traffic, particularly early in the cropping season when the potential for compacting soils is greatest. Current recommendations for crop establishment are for early planting to take advantage of the full growing season with later maturing varieties. This, in many cases, results in field work being undertaken early in the spring when the soils are wet and most susceptible to compaction.

The effect of continuous row cropping on soil structural properties and compaction have been reported by McKeague et al. (1987) on a Brookston clay loam in Ontario. They showed an increase in topsoil bulk density from 1.38 to 1.50 Mg m^{-3} and a reduction in the content of biopores (>0.5 mm) from >0.2 to <0.02% (v/v) in a continuous corn rotation, as compared to a corn-oats (*Avena sativa* L.)-alfalfa (*Medicago sativa* L.) rotation. Angers et al. (1987) measured an increase in tensile strength of soil aggregates in a bromegrass (*Bromus inermis* Leyss.) culture as compared to a continuous corn rotation. Inter-aggregate porosity also increased under bromegrass and was positively correlated with tensile strength which resulted in a higher compression index. These conditions resulted in a soil structure which was more resistant to compaction from the ground contact pressures commonly encountered in the field (<500 kPa). However, when cropping practices changed, the compression index, tensile strength and inter-aggregate porosity appeared to change rapidly, i.e., within a few cropping seasons (Angers et al., 1987).

CONCEPT OF RELATIVE COMPACTION

In modern agriculture, although a necessary part of it, wheel traffic appears to be the major source of soil compaction. Tramlines and wide spanner-type vehicles that eliminate wheel traffic have proven beneficial in several respects for certain combinations of crops and soils (Taylor, 1985). However, such equipment does not readily lend itself to routine production of row crops on the large non-uniform fields common to the temperate region of North America. Thus, for the immediate future, present levels of wheel traffic will likely remain largely

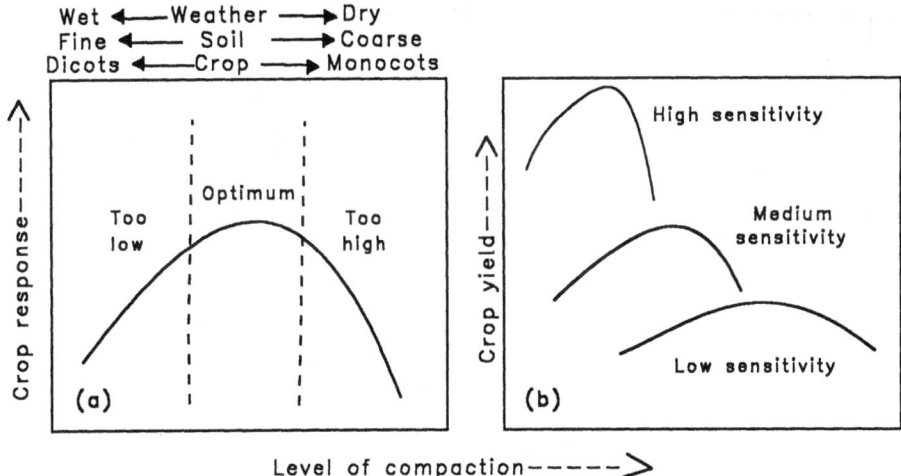

Fig. 2. Conceptual relationship of crop response to the level of compaction showing the interaction between weather, soil structure and crop type (a) and the influence of crop sensitivity on the variation of crop yield to changes in the level of compaction (b) (from Soane, 1985).

unchanged as a consequence of farming practices. Compaction is, however, not always detrimental to crop production but presents a complex interaction between soils, crop type and weather. While discussed in detail in Chapter 11, this interaction is conceptually shown in Fig. 2 to emphasize that plant responses to compaction will not be qualitatively or quantitatively uniform across all regions or crop years.

Bulk density or penetrometer resistance values have not proven to be adequate indicators of how surface soil compaction will influence crop growth and yield. A number of other parameters, such as relative compaction, have therefore been used to obtain better correlations with crop responses. Relative compaction can be obtained by dividing the measured soil bulk density derived from soil cores by the maximum soil bulk density determined from the Proctor test and expressing the results on a percentage basis. For two fine sandy loam soils, Carter (1990) examined the relationship between relative compaction and both volume of macropores (equivalent diameter ≥ 50 μm) (Fig. 3a) and the yield of winter wheat and spring barley (Fig. 3b). A relative compaction range of 83-86% was the optimum for highest yields and this corresponded to a macroporosity range of 12-14% (v/v) which was found to maintain an optimum aerobic environment under humid soil water regime in the two sandy loam soils (Carter et al., 1988). Macropore volume decreased rapidly with compaction or increased bulk density and this will cause significant changes in other soil properties such as hydraulic conductivity, air permeability and soil strength.

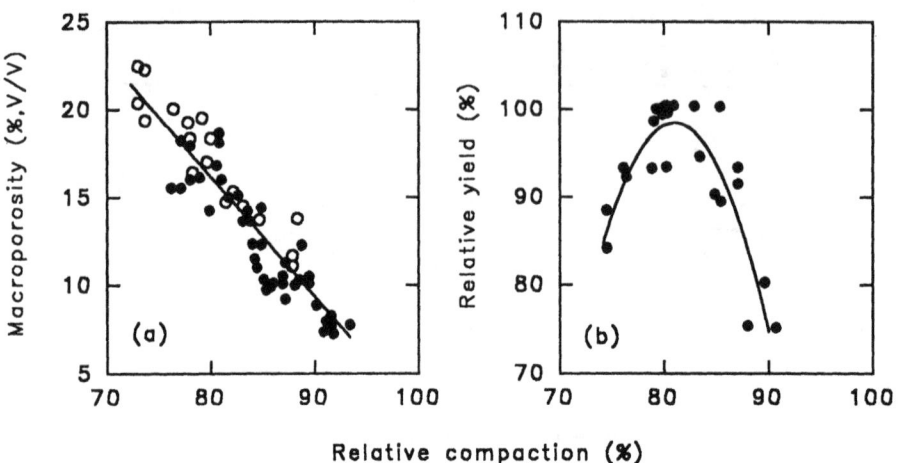

Fig. 3. Relationship between macropore volume (a) and relative yield of small grains (b) and relative compaction on two sandy loam soils (after Carter, 1990).

Carter (1990) stated that the soil behavior properties (soil strength, hydraulic conductivity, pore size distribution and water retention), either individually or collectively, are the main influences (direct or indirect) on plant growth rather than bulk density per se. For example, at low bulk density a large volume of macropores may adversely influence soil water storage capacity, while at high bulk density a reduced macropore volume, which results in lower air or water permeability and increased soil strength, could reduce root growth and crop productivity. Therefore, a certain bulk density may be optimum under a semi-arid moisture regime, but deleterious under a humid moisture regime (Carter et al., 1988). Carter (1990) further stated that, for mineral soils, relative compaction is generally independent of soil particle size distribution, although the soil water regime will influence the optimum relative compaction for wheat and barley. Thus, under similar environments, on different mineral soils, the bulk density effect on crop response may be best evaluated using the concept of relative compaction.

EXTENT AND PERSISTENCE OF COMPACTION

Early field research on soil compaction was concerned primarily with plow pan formation and root impedance or reduced crop emergence. It was generally perceived that soil compaction was not a serious problem in the temperate regions and that such compaction effects that did occur were not long-lasting. Wetting and drying or freezing and thawing were generally thought to ameliorate any compaction problems. The presence of expanding-lattice type clay and relatively high organic matter contents were thought to limit compaction effects.

Phillips and Kirkham (1962), undertaking field research on a clay soil (42% clay) near Ames, IA, U.S.A., showed corn yield reductions up to 41% by a "severe" compaction treatment. Severe compaction was applied by driving over the plow furrow bottom ten times with a tractor when plowing, in addition to a "moderate" compaction treatment applied after plowing. The moderate compaction treatment consisted of driving over the soil surface 20 times before planting, which decreased corn yields by up to 21%. Tractor weight was not mentioned but was likely to have been <3 Mg. There was no significant increase in bulk density deeper than 20 cm. Factors contributing to this crop yield decrease were: reduced seedling emergence, root mass and nutrient uptake. Mechanical impedance, as measured by bulk density and penetrometer resistance, was found to be the physical property most highly correlated with reduction in growth and corn yields. The authors speculated that inadequate oxygen content may have also been a factor, but results of aeration measurements were not conclusive. Even though compaction in the top 20 cm decreased corn yield, this effect was eliminated by normal cultivation after the growing season. Thus, there were no significant residual effects of compaction on corn yield the following year. They concluded that "...one plowing and a winter's freezing and thawing had apparently restored the soil to a nearly normal state". Based on this and other studies with similar results, it was easy to conclude at that time that wheel traffic applied during normal spring field operations in the Corn Belt would not cause any long-term detrimental compaction effects.

Role of freezing and thawing

In a study reported by Blake et al. (1976) in southwestern Minnesota, a packing wheel (static weight 2.25 Mg) was run in the plow furrow (25-cm depth) four times. The applied stress at the bottom of the furrow was calculated to be equivalent to 740 kPa. Soil water potential below the depth of plowing at the time of compaction was about -30 kPa. Plot areas were compacted once in 1960 and then planted to continuous corn or alfalfa for six cropping seasons. In 1966 the entire plot area was seeded to alfalfa. Ten years later, after successive years of freezing and thawing (freezing depth estimated to be approximately 1 m), along with the corn and alfalfa cropping, the restrictive layer generated by the packing was still easily detected by bulk density and penetrometer measurements. Interestingly, no apparent adverse effect on crop yield was observed although different rooting patterns in the alfalfa were measured ten years after the original packing. This study was a definite indication that freezing and thawing may not be as effective in ameliorating soil compaction as originally thought and this agrees with data reported by Akram and Kemper (1979) that several freezing and thawing cycles are required to restore a compacted soil to its original condition. Although soils in this region are subjected to annual freezing and thawing cycles and may freeze to depths of 1 m or more, only the top 5-10 cm will experience more than one freezing and thawing cycle per year.

The common belief that freezing and thawing ameliorates compacted soil under field conditions was perhaps reinforced by two factors. First, the machinery commonly in use when early field studies were conducted weighed much less than that currently in use. Although heavier tractors may not apply more force per unit of soil contact area (because of wider and more tires), the distribution of stress with depth in the soil will be greater (Söhne, 1958). Thus, the degree of compaction experienced in the 1950's was likely to have been much less than under current agriculture, especially as grass and legumes would have been grown in the rotation. The second factor is that there was almost always a primary tillage operation undertaken between the time of harvesting one crop and planting the next crop. The study conducted by Phillips and Kirkham (1962) indicated that, during the 1950's, normal cultivation during the first growing season alleviated the compaction effects on yield, suggesting that a primary tillage operation would almost certainly ameliorate and prevent long-term compaction associated with farming techniques in the temperate regions.

Voorhees (1983) studied the relative effectiveness of fall tillage and freezing and thawing in ameliorating a compacted soil. In his study all wheel traffic was confined to specific interrow areas and the maximum tractor weight was 7.3 Mg. Over the growing season, wheel traffic increased bulk density in the top 10 cm from 1.2 to 1.6 Mg m^{-3}. Freezing and thawing alone decreased bulk density only in the top 10 cm of soil in the wheel-tracked area, but this layer remained more dense than in the non-tracked soil (Fig. 4). Reduced forms of fall tillage, such as

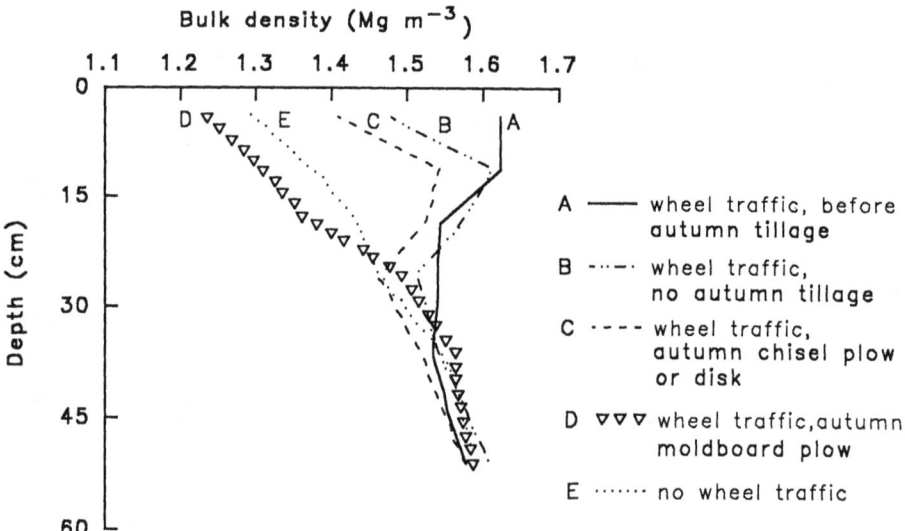

Fig. 4. Relative effectiveness of fall tillage (moldboard plowing, chisel plowing and disking) and freezing and thawing action on ameliorating a compacted soil (after Voorhees, 1983).

chisel plow or disk, were more effective in reducing the bulk density of the wheel-tracked soil than natural forces alone, but did not completely ameliorate the soil. However, fall moldboard plowing, a more intensive and disruptive type of tillage, effectively ameliorated the compacted soil.

Voorhees (1983) also studied residual compaction effects by measuring penetrometer resistance and concluded that natural weathering forces reduced penetrometer resistance by up to 50%, depending on soil water content. Loosening by frost action was more apparent when the soil was wetter at the time of freezing. The geometric mean diameter of soil aggregates was smaller and clod bulk density was lower in the non-wheel-tracked, autumn-tilled areas compared to wheel-tracked, autumn-tilled or non-autumn-tilled treatments. The larger, more dense clods in the wheel-tracked zones presented a less desirable condition for seedbed preparation. This experiment provided more evidence that natural forces alone are not sufficient to ameliorate soil compaction by wheel traffic of modern farm machinery, at least not within a one-year period.

Kay et al. (1985) found that soil bulk densities remained persistently higher with no-till than with conventional tillage systems employing fall moldboard plowing in Ontario. This occurred in spite of the fact that the formation of ice lenses over winter cause upward displacement of the ground surface of several centimeters, which may introduce porosity. The formation of ice lenses appears to create voids which are inherently unstable and collapse as the ice melts and the soil drains. Therefore, Kay et al. (1985) concluded that pores created by tillage were more stable than pores created by ice lenses because of their formation by the vertical and horizontal displacement of soil peds.

CROP RESPONSE TO COMPACTION

Crop growth and yield responses to wheel-induced soil compaction in the temperate region of North America have been quite variable and complex. The deterioration of soil structural properties described above by McKeague et al. (1987), while not resulting in a yield decrease if adequate fertilizer was applied (Stone et al., 1985), definitely indicates specific soil physical problems.

Surface compaction along one side of corn and soybeans rows was studied by Fausey and Dylla (1984) on a silty clay loam soil in Ohio. A total of five passes with a tractor with 3.5 Mg load on the rear axle was made after fall moldboard plowing to a depth of 25 cm and secondary spring tillage to a depth of 12-15 cm with a field cultivator. Soil water content at the time of wheel traffic, for both corn and soybeans, was between the upper and lower plastic limits. Bulk density measurements clearly illustrate the effect of wheel traffic in the 0-15-cm and 15-30-cm depth increments for both the corn and soybean treatments; effects on bulk density were not observed below 30 cm. Differences in penetrometer resistance were measured in the 0-15-cm depth for the corn treatments but not at deeper increments, nor for any depth increments in the soybean plots. Soil water potential at the 30-cm depth was monitored during the growing season and did not fall below -85 kPa. A variable, recommended nitrogen fertilizer treatment was included for corn on half the plot area, while nitrogen was withheld from the other half. Crop yields indicated that if chemical soil fertility and moisture are not limiting, yields are not likely to be depressed by the presence of a compacted zone along one side of the row. However, significant yield decreases were obtained for corn where fertilizer was limiting. This study suggests that any detrimental effects of wheel compaction on crop yields may be minimized substantially if wheel traffic is confined to the interrow area on one side of the row only, provided that additional crop stresses such as deficiencies of nutrients or water are not present.

Raghavan et al. (1979) compacted a Ste. Rosalie clay in Quebec with 1, 5, 10, and 15 passes over previously rototilled plots with three tractor tires, having ground contact pressures of 62, 41, and 31 kPa, respectively, to evaluate the effect of cumulative ground contact pressure on corn emergence, growth rate, biomass production and grain yield over a two-year period. Cumulative ground contact pressure was determined as the product of tire-ground contact pressure and the number of passes and ranged from 0-1.0 MPa. Corn yield decreases of 50 and 30% were observed for the two individual years with the higher cumulative contact pressures and multiple passes, but crop response to the variable ground contact pressure was different for the two years. In the first year, a wet season, corn emergence rate, growth rate, biomass production and grain yield decreased with increased cumulative ground contact pressure. In the second year, a dry

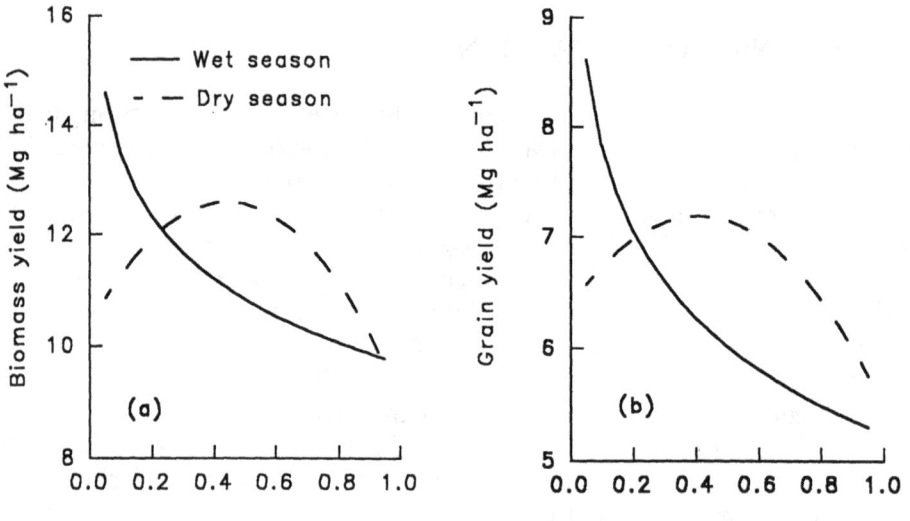

Fig. 5. Variation in corn biomass production (a) and corn grain yield (b) as related to cumulative ground contact pressure and seasonal precipitation (after Raghaven et al., 1979).

season, emergence rate, growth rate, biomass production and grain yield increased with increased ground contact pressure up to approximately 0.5 MPa and then decreased with additional ground contact pressure. This study demonstrates that compaction from wheel traffic can be detrimental to crop production, but that crop response will vary greatly in different years due to weather variation (Fig. 5).

Weather effects on crop response

Crop response to soil compaction can be highly sensitive to growing season weather conditions when the degree of compaction is significant but not extreme. Voorhees (1987) presented data on wheat yields from a clay loam soil where topsoil compaction reduced yields by about 20% when spring and early summer precipitation exceeded 200 mm. However, when early season precipitation was <200 mm, a yield increase due to compaction was noted (Fig. 6). The increased yield during the dry season was attributed to better seed-soil contact and improved soil water conditions. Voorhees (1987) presented this relationship also for soybeans based on nine crop-years of collected data (Fig. 6), providing more evidence for such an interaction between crop yield response to topsoil compaction and growing season precipitation. In the case of soybeans, May-August precipitation provided the best guide to crop response. At >400 mm precipitation, yields decreased with topsoil compaction, whereas at <300 mm

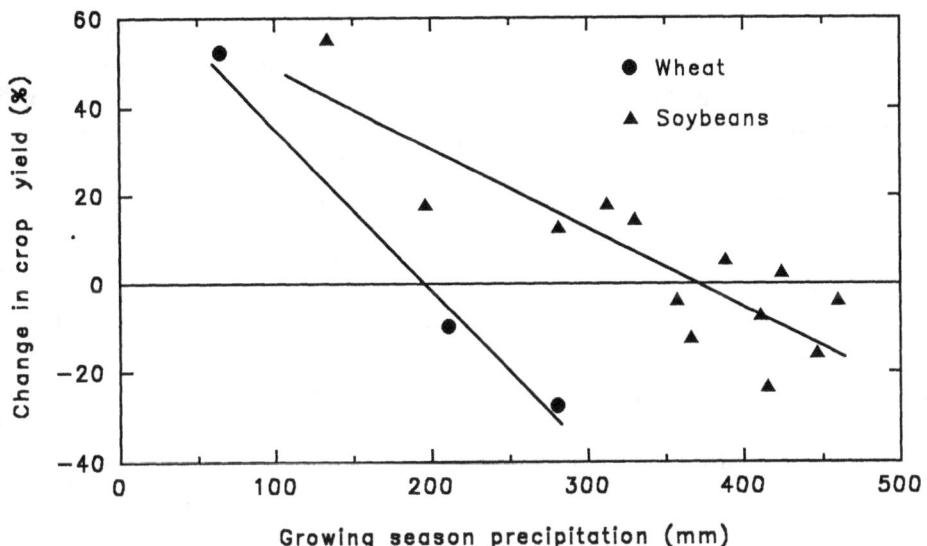

Fig. 6. Crop response to wheel track compaction on a clay loam soil, planted in tractor wheel track for wheat and three interrow passes for soybeans with an axle load of 3 Mg, in relation to growing season precipitation (after Voorhees, 1987).

precipitation yields increased. For both crops, precipitation during the active growing stage provided the closest correlation with yield. As with wheat, benefits of topsoil compaction during dry seasons appear to be related to less evaporative soil water loss. However, it should be noted that the increases in yields with compaction during periods of low precipitation, while showing a large relative increase, are in fact only small absolute increases over the non-compacted areas.

Bicki and Siemens (1990) presented data for corn and soybean yields over a four-year period on a silt loam soil. The effects of wheel traffic compaction (rear axle load 8.5 Mg) were compared for wheel traffic over the entire soil surface, wheel traffic in alternate interrows and no wheel traffic. Corn and soybean yields for the traffic treatments, averaged over the four years, were not significantly different. In individual years, corn yield varied according to soil water content and weather conditions. A positive corn response to compaction was observed during the dry years and a negative response was observed in the wetter seasons. In this experiment, soybeans showed little response to compaction.

Economic aspects of soil compaction

The total detrimental effect of soil compaction on crop production is difficult to assess in economic terms (see Chapter 23). While stand establishment and crop growth rate are commonly reduced due to compaction, there may also be a delay in crop maturity. This can increase production costs since harvesting corn at a higher grain moisture content results in increased drying costs. Compaction effects on crop production have also been hidden by a general increase in crop yields resulting from a gradual improvement in farming technology. Because of the increased use of heavy farm machinery, more attention must be given to soil compaction. This is particularly important in the temperate regions of North America because of the relatively short growing season and the need to utilize the complete season for maximum crop production.

An approach to assess soil susceptibility to compaction and to determine the economic aspects was proposed by Voorhees (1987). This assessment was based on basic soil mechanics and soil survey data to determine a soil's susceptibility to compaction. Then, based on long-term weather records, field research results and a general knowledge of prevailing cultural practices, a classification scheme could be established to assess the overall agro-economic consequences of soil compaction. This approach was designed to identify where changes need to be made in machinery management to avoid the detrimental consequences of compaction. Presently this classification scheme is in a skeletal form; however, efforts are in progress to model crop response to compaction, taking soil, crop and weather factors into consideration (Eradat Oskoui and Voorhees, 1990a).

SUBSOIL COMPACTION

Compaction in the subsoil, below normal tillage depth, has recently been identified as an additional crop production concern and is presenting special problems. The increase in farm machinery size, while reducing farm labor requirements and increasing efficiency of operation with regard to timeliness, has increased applied stress to the soil and the effective depth of penetration of stress (Söhne, 1958). Voorhees et al. (1978) reported increased penetrometer resistance to depths >30 cm, resulting from axle loads <4.5 Mg. Modern combine harvesters have axle loads of 10-20 Mg, depending on size and amount of grain in the storage bin. Manure spreaders, fertilizer and spray wagons and grain carts can have single axle loads >30 Mg. Topsoil compaction can be relatively easily corrected through tillage. However, subsoil compaction is not as easily corrected and is long lasting or may even be permanent (Håkansson et al., 1987).

A working group of scientists and engineers from Northern Europe and North America was organized to conduct field research to determine the effects of high axle loads on the extent and persistence of subsoil compaction over a range of soil and climatic conditions and to determine the effect of subsoil compaction on growth and yield of several crops. In North America, crops selected for investigation were limited to corn and soybeans. As part of this international working group, Voorhees et al. (1986) established two experimental sites in southern Minnesota. Axle loads of 9 and 18 Mg were uniformly applied over the entire surface of a Webster clay loam (38% clay) and a Ves clay loam (32% clay), and were compared with axle loads restricted to ≤4.5 Mg (check treatment). The Webster soil was formed in a wetter environment than the Ves soil and is classified as being poorly drained, while the Ves soil is classified as being moderately well drained. Soil water content at the time of axle loading was considered wet for the Webster soil, whereas axle load treatments were applied to the Ves soil in both the wet and dry states. After the initial axle load application, total plot areas were moldboard plowed to approximately 30 cm to alleviate topsoil compaction effects. Results from these trials show definite deleterious effects from heavy axle loads and also show an interaction between clay content and soil water content at the time of loading. Bulk density, penetrometer resistance and hydraulic conductivity below the 30-cm tilled zone were all influenced by the axle load treatments and were more pronounced with increased clay and soil water content. Four years after loading, compactive effects from the heavy axle loads were still observed down to 50 cm in spite of annual winter soil freezing to a depth of 90 cm.

Corn yields from these trials (Fig. 7) confirm that subsoil compaction will reduce crop yields (Voorhees et al., 1989). Yield reductions were most pronounced during the first year after axle load application and slowly deceased. First indications suggested that the subsoil compaction effect on crop yield would correct itself in about five years even though the soil physical parameters measured had not been restored to their original state. However, later measurements (data not published) taken during years of climatic stress, such as severe drought (year 7) or excess rain (year 9), again showed a 15% decrease in

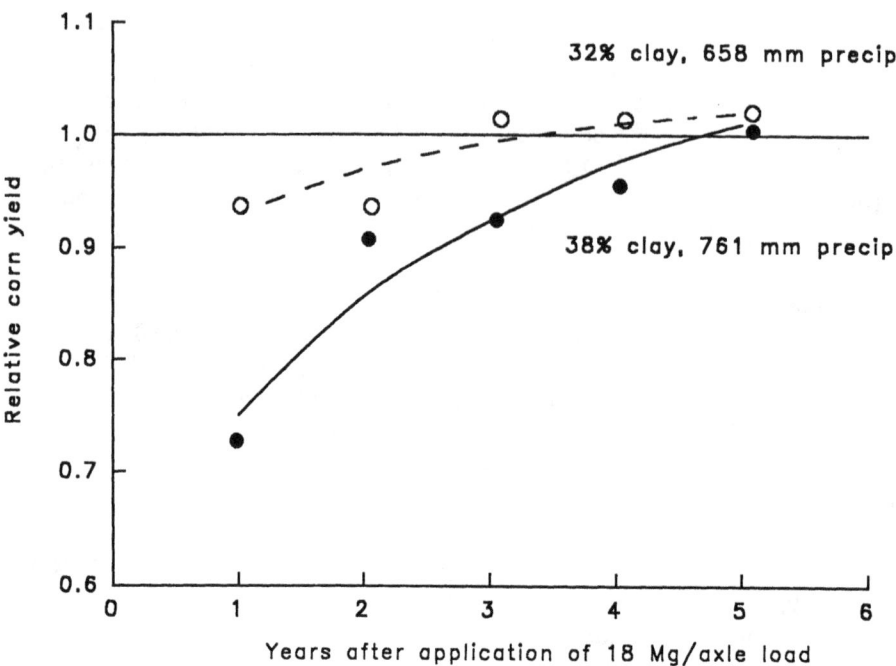

Fig. 7. Relative corn yield for five cropping seasons after high axle load (18 Mg) compaction as compared with the check treatment (axle loads ≤4.5 Mg) on two clay loam soils at two sites in Minnesota with different clay contents and long-term average precipitation (after Voorhees et al., 1989).

corn yield. Present ideas on the effect of subsoil compaction consider the possibility that these compacted areas may present lower crop production potential indefinitely as long as soil physical parameters remain affected. Soybean yield on these same plots was not reduced as consistently as corn yield, indicating different crop response to compaction (Johnson et al., 1990).

Similar results concerning subsoil compaction were presented by Gameda et al. (1985) for a clay soil (35% clay) in Ontario. Axle load treatments consisted of two levels, 10 and 20 Mg, plus a control, ≤6 Mg. Treatments were applied annually for three successive crop years. All compaction treatments were moldboard plowed after the application of axle load variables. Results from this study indicated that soil water content during compaction significantly affected subsoil bulk density distribution. Following compaction during dry soil conditions it appeared that only the 20-Mg axle load had significantly increased soil bulk density. In wet seasons, however, both the 10- and 20-Mg axle load treatments had significant effects on subsoil bulk density. The effect of the 10-Mg axle loading was more pronounced in the topsoil (Gameda et al., 1985), whereas in the subsoil the effect of the 20-Mg axle loading predominated. They concluded that the number of passes had a more predominant effect on topsoil bulk density (loading treatments with the 10-Mg axle load were applied twice), while total load effects prevailed in the subsoil. Moldboard plowing and overwinter freezing and thawing removed the effects of soil loading under low moisture conditions on topsoil bulk density, but were not as effective in reducing topsoil bulk density resulting from loading under wet soil conditions. For compaction under dry soil conditions, only the 20-Mg axle load treatment affected corn yields, whereas both loading treatments reduced corn yields under wet soil conditions.

Gameda et al. (1987) measured the recovery of a clay soil from a single incidence of heavy axle load compaction. Axle loads of 10 and 20 Mg were applied uniformly over the soil surface in the spring, and were followed by moldboard plowing to a depth of 20 cm and disc harrowing to a depth of 10 cm. Corn was then grown for three years and changes in bulk density to a depth of 60 cm and crop yield were monitored. Differences between the effects of loading treatments on soil bulk density decreased with time, but after three years the effect of both loading treatments was still apparent at depths between 30 and 40 cm. Crop growth and yields after three years were still significantly lower due to both axle load applications. The extent and trend of recovery in the subsoil suggested that wetting and drying and freezing and thawing were more effective in reducing bulk density increases which had resulted from the 10-Mg axle load than from the 20-Mg axle load. Comparison of relative yields suggested that the levels of compaction that existed in the subsoil may have been conducive to higher moisture availability in dry years, but may have led to aeration stresses, due to high soil water content, in years with above-average precipitation.

Lowery and Schuler (1991) conducted a study on two silt loam soils in Wisconsin and reported that corn grain yields were not affected after the third cropping season, even though some measured soil parameters were still adversely affected by high axle load compaction. The axle loads applied (8 and 12.5 Mg) were lower than in previously discussed trials. Again, plots were tilled by either chisel plowing or moldboard plowing after application of the loading treatments. Corn yield was significantly reduced in the first year for both soils and also in the second year following loading for one soil but not for the other soil. Soil water content at the time of load application was relatively low and probably limited crop response to loading treatments.

In the experiments cited above, which were associated with both subsoil compaction and, to a large extent, topsoil compaction, the respective compactive forces were applied over the entire soil surface area. In many cases, multiple passes have been made and the compactive forces were imposed at soil water conditions most conducive for compaction. These situations would rarely apply to production fields and they represent extreme situations. However, results from these trials do show the potential problems associated with soil compaction because heavy axle loads are applied annually to the soil in the spring during tillage and planting operations and also in the fall during harvest and primary tillage operations when soils may also be at a moisture content conducive to compaction. This is especially important with subsoil compaction in view of the demonstrated persistence over time.

TECHNIQUES ADOPTED TO REDUCE COMPACTION EFFECTS

The most effective way to reduce soil compaction is to reduce the compactive forces or load (see Chapter 20) and to control soil water content at the time of loading. Unfortunately, these techniques are not always feasible. Reducing the weight of farm machinery also means reducing machinery size and capacity, which in turn reduces efficiency for an individual operator. Farmers often do not have flexibility in controlling soil water content at the time of field operations. Timely planting is important and harvest is often a race against impending winter. Large machinery allows better timeliness of field operations (Fig. 8), but there is still the problem of prolonged wet periods which may extend through a planting or harvest season. This is especially critical during harvest because of the high axle loads involved and the urgency to harvest a crop. Short-term economic concerns (planting, harvesting, cultivation, etc.) will usually overrule possible long-term

Fig. 8. High-capacity combine harvester with an 8-row corn head (axle loads: front 15 Mg, rear 5 Mg).

costs associated with soil compaction. Models are being developed to assess these trade-offs (Eradat Oskoui and Voorhees, 1990b).

Controlled traffic, a concept in which all wheel traffic is restricted to permanent traffic lanes, could be a viable management system (see Chapter 22). Controlled traffic results in both a loose rooting zone and a firm traffic lane, thereby providing a good rooting environment for plant growth and good trafficability for timely field operations. This concept has not been widely accepted because many farmers perform tillage diagonally to previously planted rows which results in random wheel traffic. However, with no-till or ridge-till-planted row crops, most, if not all, wheel traffic is restricted to interrow areas. This is particularly true for ridge planting because any traffic directly over the rows destroys the ridges. Cultural practices for corn and soybeans requires many different field operations with different pieces of field equipment with different wheel spacings, which in effect results in most interrow areas being tracked during a cropping season, even in ridge-till-plant systems. Fig. 9, depicting three planter passes (13.7 m) in a 6-row ridge-till-plant system (left) and one planter pass (12.2 m) in a 16-row ridge-till-plant system (right), shows that nearly all corn interrows receive wheel traffic during the crop season. Production operations include: (1) planting; (2) fertilizer application, both a full width broadcast application of phosphorus

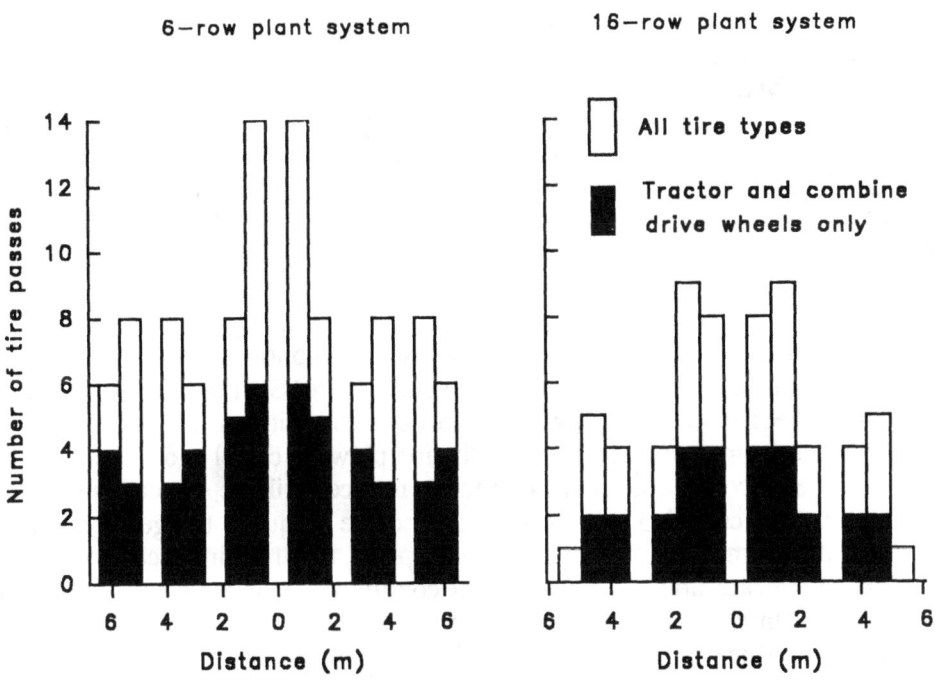

Fig. 9. Wheel track intensity and distribution in interrow zones for a 6-row (left) and a 16-row (right) ridge-till-plant system during one cropping season of corn (after Parsons et al., 1984).

and potassium and a 6- or 8-row injection of anhydrous ammonia; (3) full-width herbicide spray; (4) 6- or 8-row cultivation to re-establish ridges; (5) 6- or 8-row combine harvesting. The 16-row system includes a grain cart for unloading the combine in the field while the 6-row system assumes unloading at the field borders. Post-emergence spot spraying of weeds or a second cultivation was not included in Fig. 9 although both operations are common.

Harvest equipment often does not have wheel spacing that fits planting row spacing. The machinery industry is now responding to this need so that, ideally, planting, cultivating and harvesting can be done with all wheel traffic following the same paths each year. This would not necessarily reduce compaction, but would tend to concentrate it rather than subject most of the whole field to wheel traffic. Another feasible practice is to restrict all harvest transport traffic to the headlands. This results in more dead time for unloading the combine, but it would keep heavily loaded trucks and grain carts off most of the field area.

Increased awareness of the need for conservation of soil and water resources has caused a shift in attitudes about soil management practices. Conservation tillage or reduced tillage practices have gained considerable support by merit of their erosion control capabilities. Compaction problems associated with tillage, such as moldboard plowing with one tractor wheel in the plow furrow, the timeliness factor that forces tillage to be undertaken when the soil is too wet, and unnecessary secondary tillage for the sake of a fine smooth seedbed, should tend to decrease with conservation tillage. However, this has not always been the case because all field operations and associated wheel traffic must be considered, not just those involved with tilling the soil. Researchers and farmers have become concerned that continuous conservation tillage, especially no-till, may amplify soil compaction concerns (Voorhees and Lindstrom, 1983). There have been recommendations to plow or cultivate no-till fields every few years in order to alleviate surface compaction that may occur. Lindstrom et al. (1984) found higher bulk density and penetrometer resistance in surface layers with no-till compared to conventional moldboard plowing. Larney and Kladivko (1989) reported an improvement in soil structural conditions with reduced tillage or no-till on a soil with a low organic matter content and poor structural condition, but not on a soil with high organic matter content and stable structure. Significantly higher soil strength in wheel track interrows was also observed with no-till and ridge-till management systems compared to moldboard plow or chisel plow treatments. Hammel (1989) concluded that long-term reduced tillage management, in combination with cropping practices which require frequent tillage or cultural operations during moist soil conditions in the spring, may require occasional deep tillage to ameliorate adverse soil physical conditions. Mannering et al. (1987) stated that research is still necessary to determine how much tillage is really needed, considering that the natural physical structure and biological activity in the narrow seed zone could, at least to some extent, maintain adequate surface conditions for good germination and early growth.

As a last resort, subsoiling may be considered as an option (see Chapter 25). However, subsoiling is a high-energy operation and improvements obtained have been inconsistent and commonly of short duration. The reason for lack of consistent positive yield response may be insufficient depth of subsoiling or recompaction of loosened soil by subsequent wheel traffic or an actual deterioration of some soil properties by the subsoiling operation. For subsoiling to be effective, the soil must be relatively dry to allow shattering between the subsoiler shanks. Johnson et al. (1989) showed improved soil physical properties below the normal plowing depth following subsoiling in the fall when the soil was relatively dry. Results of this subsoiling persisted for only one year, despite efforts to control post-subsoiling traffic. Also, wheel traffic associated with normal spring tillage recompacted soil loosened by subsoiling.

Duval et al. (1989) observed long-term beneficial effects ten years after subsoiling on a Ste. Rosalie clay soil when no further traffic was imposed on the soil after subsoil operations. Compactive forces, from 15 passes over the soil surface with a tractor with a ground contact pressure of 62 kPa, were applied prior to tillage variables. Tillage variables applied after compaction were no-tillage, moldboard plowing to a depth of 20 cm followed by disc harrowing, chisel plowing to a depth of 30 cm with a narrow spear-point chisel on 30-cm centers, and subsoiling to 45-50 cm. Plot areas were then cropped to silage corn or allowed to remain fallow. All field work was done by hand. After ten years, the untilled plots showed mottling that was more abundant and closer to the surface than in the tilled treatments, while the subsoiled plots had a finer, less abundant and deeper mottling than the other treatments. Peds in the Ae horizon (20-33 cm) were finer in the subsoiled and chiseled treatments and coarser in the untilled and moldboard-plowed plots. The results obtained from these trials again show that compaction is very persistent in clay soils and that subsoiling can be beneficial, provided subsequent wheel traffic is controlled to permanent traffic lanes, otherwise random or intensive traffic patterns will rapidly recompact loosened soil.

While large, heavy machinery is often blamed for soil compaction problems, it also offers opportunities to minimize compaction. Larger capacity machinery means fewer wheel tracks per unit area of the field because of wider working width. If wheel track spacing can be standardized among different pieces of equipment, soil compaction problems can be minimized.

CONCLUSIONS

(1) Soil compaction from wheel traffic of modern agricultural machinery has been demonstrated to reduce yields of cultivated crops in the temperate region of North America. The influence of compaction on crop production is manifested in higher soil bulk density, increased soil strength and reduced pore volumes which, in turn, influence soil properties related to aeration, root exploration, and

water and nutrient uptake.

(2) The prime concern with soil compaction in the temperate regions of North America is the continuing increase in the size and weight of field equipment. Farm populations are decreasing and farm size is increasing; therefore, equipment size is increasing to allow the remaining farm population to handle the essentially constant area of cultivation efficiently. The change of cropping patterns to a predominately row crop culture also contributes to this problem. Another important factor in regard to soil compaction in the temperate regions of North America is the relatively short frost-free growing season. Early crop establishment and use of full season varieties is mandatory for maximum crop production. This often results in field work starting in the spring when field moisture conditions are conducive to soil compaction. Similarly, wet soils at harvest are also common.

(3) The soils in the region commonly have a relatively high clay content with expanding lattice clays and a relatively high organic matter content. Previously, it was generally thought that these soils were resistant to compactive forces due to the action of wetting and drying and freezing and thawing. Except for the few surface centimeters of soil, this has been found to be not true. In the case of freezing and thawing these soils undergo only one cycle per year below a depth of 10 cm.

(4) Surface compaction within the zone reached by tillage can be ameliorated with a combination of disruptive primary tillage and the action of freezing and thawing and wetting and drying. However, reduced tillage systems (particularly ridge-till and no-till-plant) are becoming more popular because of a combination of energy savings and soil erosion control. With this situation, surface compaction problems are becoming more prevalent. Many researchers are recommending an aggressive primary tillage every few years to overcome this problem.

(5) Subsurface compaction below the depth of the normal tillage zone has also been identified. Subsoil compaction is a function of axle load and has been increasing with the increased weight of equipment. Harvesting operations are mostly responsible for subsurface compaction because of the heavy combine harvesters and field transport equipment used. Compaction in the subsoil is resistant to amelioration from natural forces.

(6) Little progress has been made in reducing soil compaction problems. Reducing compactive forces on the soil during wet soil conditions is not always possible; presently the timeliness of field operation is considered more important than perceived compaction problems. Controlled traffic would appear to be a feasible approach and progress is being made on standardizing wheel spacing, but much work still needs to be done, particularly with combine harvesters and field transport equipment. Subsoiling has been demonstrated to be an ineffective cure for compaction. Therefore, in the temperate region of North America, soil compaction will remain a crop production problem until major changes have been made in field equipment design and cropping practices.

REFERENCES

Adams, E.P., Blake, G.R., Martin, W.P. and Boelter, D.H., 1960. Influences of soil compaction on crop growth and development. Trans. 7th Int. Cong. Soil Sci., Madison, WI, U.S.A., Vol. 1, pp. 607-615.

Akran, M. and Kemper, W.D., 1979. Infiltration of soils as affected by the pressure and water content at the time of compaction. Soil Sci. Soc. Am. J., 43: 1080-1086.

Angers, D.A., Kay, B.E. and Groenevelt, P.H., 1987. Compaction characteristics of a soil cropped to corn and bromegrass. Soil Sci. Soc. Am. J., 51: 779-783.

Bicki, T.J. and Siemens, J.C., 1990. Crop response to wheel compaction. Am. Soc. Agric. Eng., St. Joseph, MI, U.S.A., ASAE Pap. 90-1093, 12 pp.

Blake, G.R., Boelter, D.H., Adams, E.P. and Aase, J.K., 1960a. Soil compaction and potato growth. Am. Potato J., 37: 409-413.

Blake, G.R., Ogden, D.B., Adams, E.P. and Boelter, D.H., 1960b. Effect of soil compaction on development and yield of sugar beets. J. Am. Soc. Sugar Beet Technol., 11: 236-242.

Blake, G.R., Nelson, W.W. and Allmaras, R.R., 1976. Persistence of subsoil compaction in a Mollisol. Soil Sci. Soc. Am. J., 40: 943-947.

Campbell, R.B. and Moreau, R.A., 1979. Ethylene in a compacted field soil and its effect on growth, tuber quality, and yield of potatoes. Am. Potato J., 56: 199-210.

Carter, M.R., 1990. Relative measures of soil bulk density to characterize compaction in tillage studies on fine sandy loams. Can. J. Soil Sci., 70: 425-433.

Carter, M.R., Johnston, H.W. and Kimpinski, J., 1988. Direct drilling and soil loosening for spring cereals on a fine sandy loam in Atlantic Canada. Soil Tillage Res., 12: 365-384.

Duval, J., Raghavan, G.S.V., Mehuys, G.R. and Gameda, S., 1989. Residual effects of compaction and tillage on the soil profile characteristics of a clay-textured soil. Can. J. Soil Sci., 69: 417-423.

Eradat Oskoui, K. and Voorhees, W.B., 1990a. Economic consequences of soil compaction. Am. Soc. Agric. Eng., St. Joseph, MI, U.S.A., ASAE Pap. 90-1089, 22 pp.

Eradat Oskoui, K. and Voorhees, W.B., 1990b. Prediction of planting timeliness penalties for corn and soybeans in northern Corn Belt. Agron. Abstr., p. 153.

Fausey, N.R. and Dylla, A.S., 1984. Effects of wheel traffic along one side of corn and soybean rows. Soil Tillage Res., 4: 147-154.

Gameda, S., Raghavan, G.S.V., Theriault, R. and McKyes, E., 1985. High axle load compaction and corn yields. Trans. ASAE, 28: 1759-1765.

Gameda, S., Raghavan, G.S.V., McKyes, E. and Theriault, R., 1987. Subsoil compaction in a clay soil. II. Natural alleviation. Soil Tillage Res., 10: 123-130.

Garcia, F., Cruse, R.M. and Blackmer, A.M., 1988. Compaction and nitrogen placement effect on root growth, water depletion, and nitrogen uptake. Soil Sci. Soc. Am. J., 52: 792-798.

Håkansson, I., Voorhees, W.B., Elonen, P., Raghavan, G.S.V., Lowery, B., Van Wijk, A.L.M., Rasmussen, K. and Riley, H., 1987. Effect of high axle-load traffic on subsoil compaction and crop yield in humid regions with annual freezing. Soil Tillage Res., 10: 259-268.

Hammel, J.E., 1989. Long-term tillage and crop rotational effects on bulk density and soil impedance in northern Idaho. Soil Sci. Soc. Am. J., 53: 1515-1519.

Johnson, B.S., Erickson, A.E. and Voorhees, W.B., 1989. Physical conditions of a lake plain soil as affected by deep tillage and wheel traffic. Soil Sci. Soc. Am. J., 53: 1545-1551.

Johnson, J.F., Voorhees, W.B., Nelson, W.W. and Randall, G.W., 1990. Soybean growth and yield as affected by surface and subsurface compaction. Agron. J., 82: 973-979.

Kay, B.D., Grant, C.D. and Groenevelt, P.H., 1985. Significance of ground freezing on soil bulk density under zero tillage. Soil Sci. Soc. Am. J., 49: 973-978.

Larney, F.J. and Kladivko, E.J., 1989. Soil strength properties under four tillage systems at three long-term study sites in Indiana. Soil Sci. Soc. Am. J., 53: 1539-1545.

Lindstrom, M.J., Voorhees, W.B. and Onstad, C.A., 1984. Tillage system and residue cover effects on infiltration in northwestern Corn Belt soils. J. Soil Water Conserv., 39: 64-68.

Lowery, B. and Schuler, R.T., 1991. Temporal effects of subsoil compaction on soil strength and plant growth. Soil Sci. Soc. Am. J., 55: 216-223.

Mannering, J.V., Griffith, D.R. and Parsons, S.D., 1987. What will conservation tillage be like in twenty years? In: L.L. Boersma, D.E. Elrick, R.B. Corey, H.H. Cheng, T.C. Tucker, E.M. Rutledge, P.W. Unger, N.W. Foster, S. Kincheloe and P.H. Hsu (Editors), Future Developments in Soil Science Research. Soil Sci. Soc. Am., Madison, WI, U.S.A., pp. 351-360.

McKeague, J.A., Fox, C.A., Stone, J.A. and Protz, R., 1987. Effects of cropping system on structure of Brookston clay loam in long-term experimental plots at Woodslee, Ontario. Can. J. Soil Sci., 67: 571-584.

Parsons, S.D., Griffith, D.R. and Doster, D.H., 1984. Equipment, wheel spacing availability and adaptations for ridge-planted crops. Agric. Eng., 65: 10-14.

Phillips, R.E. and Kirkham, D., 1962. Soil compaction in the field and corn growth. Agron. J., 54: 29-34.

Raghavan, G.S.V., McKyes, E., Taylor, F., Richard, P. and Watson, A., 1979. The relationship between machinery traffic and corn yield reductions in successive years. Trans. ASAE, 22: 1256-1259.

Raghavan, G.S.V., Alvo, P. and McKyes, E., 1990. Soil compaction in agriculture: A view towards managing the problem. Adv. Soil Sci., 9: 1-36.

Soane, B.D., 1985. Traction and transport as related to cropping systems. Proc. Int. Conf. Soil Dynamics, Auburn, AL, U.S.A., Vol. 5, pp. 863-935.

Söhne, W., 1958. Fundamentals of pressure distribution and soil compaction under tractor tires. Agric. Eng., 39: 276-281, 290.

Stone, J.A., Allen, N.H.E. and Grant, C.D., 1985. Corn fertility treatments and surface structure of a poorly drained soil. Soil Sci. Soc. Am. J., 49: 1001-1004.

Taylor, J.H., 1985. Controlled traffic: A spin-off of soil dynamics research. Proc. Int. Conf. Soil Dynamics, Auburn, AL, U.S.A., Vol. 5, pp. 1101-1111.

Voorhees, W.B., 1983. Relative effectiveness of tillage and natural forces in alleviating wheel-induced soil compaction. Soil Sci. Soc. Am. J., 47: 129-133.

Voorhees, W.B., 1987. Assessment of soil susceptibility to compaction using soil and climatic data bases. Soil Tillage Res., 10: 29-38.

Voorhees, W.B. and Lindstrom, M.J., 1983. Soil compaction constraints on conservation tillage in the northern Corn Belt. J. Soil Water Conserv., 38: 307-311.

Voorhees, W.B., Carlson, V.A. and Senst, C.G., 1976. Soybean nodulation as affected by wheel traffic. Agron. J., 68: 976-979.

Voorhees, W.B., Senst, C.G. and Nelson, W.W., 1978. Compaction and soil structure modification by wheel traffic in the northern Corn Belt. Soil Sci. Soc. Am. J., 42: 344-349.

Voorhees, W.B., Nelson, W.W. and Randall, G.W., 1986. Extent and persistence of subsoil compaction caused by heavy axle loads. Soil Sci. Soc. Am. J., 50: 428-43.

Voorhees, W.B., Johnson, J.F., Randall, G.W. and Nelson, W.W., 1989. Corn growth and yield as affected by surface and subsoil compaction. Agron. J., 81: 294-303.

Soil Compaction in Crop Production
B.D. Soane and C. van Ouwerkerk (Eds.)
1994 Elsevier Science B.V. All rights reserved. .

CHAPTER 13

Responses of Tropical Crops to Soil Compaction

B. KAYOMBO[1] and R. LAL[2]

[1]Sokoine University of Agriculture, Department of Agricultural Engineering and Land Planning, Morogoro, Tanzania
[2]Ohio State University, Department of Agronomy, Columbus, U.S.A.

SUMMARY

Compaction of tropical soils, caused by mechanical land clearing followed by mechanized continuous cultivation and intensive cropping, is greatly accentuated by high-intensity tropical rains. In addition, there are also naturally occurring compacted soils and subsoil horizons. Compaction, caused by natural or anthropogenic factors, has detrimental effects on soil structure and on the growth and yield of a wide range of tropical crops. Furthermore, compaction can persist for a long time if adequate measures are not taken to minimize or ameliorate it.

Where feasible, land clearing should be done by the slash-and-burn method, so that most of the nutrients in the vegetation are returned to the soil without scraping it off or compacting it. Where mechanical land clearing is inevitable, due to economic or social circumstances, forest removal should be done by the use of the shear blade, whereby most of the roots and stumps are left in the ground intact, and the forest litter is not removed.

The no-tillage system with a crop residue mulch is a useful management practice to minimize soil compaction, provided it is adapted to the local soil and environment, and can be adopted as a package of all those cultural practices that make it work. Some experimental evidence indicates the ameliorating benefit of controlled traffic and controlled grazing in naturally compacted upland soils. The biological loosening of compact soils by cover crops is a potential substitute for mechanical loosening. In spite of high cost, the beneficial effects of mechanical loosening are often transient. However, mechanical loosening may be necessary for structurally inert soils of the arid and semi-arid tropics.

INTRODUCTION

The geographical limits of the tropics are to some extent arbitrary since the atmosphere is an entity and events in lower latitudes cannot be isolated from those of higher latitudes (Jackson, 1986). Nevertheless, this chapter limits the tropics to all regions of the earth's surface lying between 30° N and 30° S latitude, with the exception of those at very high altitude.

Soil compaction problems in the tropics arise from two distinct types of mechanised operations, firstly the use of heavy machinery in land clearing and secondly the use of machinery in subsequent crop production.

Mechanical land clearing in the tropics causes compaction of the surface soil horizons (Ahn, 1968; Seubert et al., 1977; Hulugalle et al., 1984; Dias and Nortcliff, 1985; Soane, 1986). Soil compaction in the tropics is further accentuated by intense tropical rains (Hudson, 1976; Kowal and Kassam, 1976), the presence of naturally compacted soil layers (Nicou et al., 1970; Macartney et al., 1971), and the widespread occurrence of soil horizons with a large percentage of gravel and plinthite (Smyth and Montgomery, 1962).

The magnitude of soil compaction problems resulting from increased use of large machinery and more intensive cultivation and their consequences on crop production, has not been as much researched in the tropics as in the temperate regions (Håkansson et al., 1988). However, the implementation of numerous large-scale mechanized land development programs is an inevitable consequence of attempts to increase food production rapidly. Yet the conclusions drawn from a critical appraisal of previously implemented large-scale agricultural development schemes in the tropics are far from encouraging (Baldwin, 1957; De Wilde, 1967). Failure of large-scale mechanized agricultural schemes in Africa, Latin America and tropical Asia has been attributed to many factors, including socio-economic conditions and poor infra-structure (Baldwin, 1957), insufficient baseline data to enable adequate planning for resource development and management (Bauer, 1978), and erratic rains. In addition to these exogenous factors, the ability of the soil to sustain economic yields under mechanized farming systems is obviously an important factor and needs to be adequately assessed. The importance of soil factors is basic to all planning. An important endogenous soil factor is the problem of soil compaction and its effects on soil and environmental degradation and on crop production.

EXPERIMENTAL EVIDENCE OF RESPONSES OF TROPICAL CROPS TO COMPACTION

Considerable evidence exists to show that soil compaction, originating either from anthropogenic or natural causes, exerts an enormous impact on the establishment, growth and yield of food and cash crops in tropical regions. The dominant crops vary in the different tropical regions. Rice (*Oryza sativa*) is a prominent staple food crop in Asia, whereas maize (*Zea mays*), wheat (*Triticum aestivum*), cowpea (*Vigna unguiculata*) and cassava (*Manihot esculenta*) are important food and often cash crops in tropical regions. Soybean (*Glycine max*) production is also expanding gradually in the tropics due to its vast potential as a source of edible oil and animal feed.

Growth and yield responses to soil compaction

Rice (Oryza sativa)

There are two types of rice. Paddy rice is grown under puddled irrigated conditions, and upland rice is grown in unpuddled rainfed conditions. Many studies on paddy rice grown in coarse-textured soils indicate that moderate soil compaction enhances better establishment of root growth, and produces higher grain yields due to a more favorable soil bulk density profile, a lower infiltration rate and higher surface retention of water (Ghildyal, 1978; Singh et al., 1980; Ogunremi et al., 1986a). Similar responses to compaction have been reported in upland rice for some soils (Kar et al., 1986), although in most instances soil compaction decreases seedling emergence, and causes stunted shoot and root growth (Ogunremi et al., 1986b).

Maize (Zea mays)

Experiments conducted in eastern Africa have shown that the germination and early growth of maize are severely restricted on untilled and shallow-tilled naturally compacted soils (Macartney et al., 1971). The poor crop performance is attributed to restricted root growth (Northwood and Macartney, 1971). Cultivation to 20-40 cm depth can significantly increase the total porosity of such soils, which consequently leads to deeper and stronger root systems and better ability to withstand drought.

The effects of gravel in the subsoil on root development of maize are widely recognized (Babalola and Lal, 1977a; Hulugalle et al., 1984). The presence of gravel increases mechanical resistance and reduces maize root growth by increasing the pore rigidity and diluting both water and nutrient contents of the soil (Babalola and Lal, 1977b; Maurya and Lal, 1980; Vine et al., 1981; Grewal et al., 1984). Loosening of gravel layers improves root development in naturally compacted horizons (Babalola and Lal, 1977b).

Mechanical land clearing and continuous mechanized farming causes significant deterioration of soil physical properties as compared to manual clearing or no-tillage, and consequently leads to reduced maize growth in tropical soils (Seubert et al., 1977; Lal, 1985a; Osuji, 1988; Ojeniyi, 1990). Soil compaction resulting from mechanical land clearing considerably reduces maize grain yield compared to land cleared by manual methods (Seubert et al., 1977; Hulugalle et al., 1984; Alegre et al., 1986). Kayombo and Lal (1986a) reported that a 4-pass treatment reduced emergence, plant height, leaf number, leaf area index and root growth of maize more on ploughed than on no-till plots in three consecutive growing seasons on an Alfisol in southwestern Nigeria (Table 1).

With manual seeding and harvesting operations simulating small landholder agriculture on Alfisols in western Nigeria, maize grain yields obtained with conventional tillage, based on soil inversion and mechanical loosening, can be lower than those for a no-tillage system, both at low and high levels of fertilizer

TABLE 1

Compaction effects on yield components of some crops (from Kayombo et al., 1986b)[1]

Roller treatment	Tillage	Maize	Cowpea	Soybean
		Grains/cob (g)	Grains/plant (g)	100-grain weight (g)
0	No tillage	117.9b	18.2a	36.1a
2	No tillage	75.3c	14.3b	25.5b
4	No tillage	60.4d	11.6c	21.7bc
0	Ploughed	156.3a	119.3a	32.4a
2	Ploughed	79.7c	11.3c	20.9c
4	Ploughed	58.8d	9.0d	14.9d

[1]Means followed by the same letters are not significantly different at $P_{0.05}$.

input (Lal, 1985a; Ike, 1986). Reduction of maize yield with conventional tillage is attributed to decline in soil structure, accelerated erosion and depletion of soil fertility (Lal, 1984, Kayombo and Lal, 1986a; Kayombo et al., 1986a; Onofiok, 1988). The problem of soil compaction in plow-based systems is greatly accentuated by motorized farm operations and vehicular traffic (Lal, 1985a). In semi-arid and arid tropical regions, deep mechanical cultivation (to 20-40 cm depth) of naturally compacted soils has often been shown to markedly increase grain yield (Northwood and Macartney, 1971; Nicou and Chopart, 1979; Nicou and Charreau, 1980; Willcocks, 1981).

Wheat (Triticum aestivum)

In large-scale irrigation projects, particularly in tropical Africa, intensive mechanized cultivation of the soil for wheat production leads to the development of a dense tillage pan (with soil bulk density as high as 1.70 Mg m⁻³) at about 15-20 cm depth. The pan restricts root penetration and water infiltration, resulting in reduced grain yields (Maurya, 1988; Gill and Aulakh, 1990). Loosening of these compacted pans by deep tillage may result in 3- to 8-fold increase in infiltration rate, reduced bulk density and improved grain yield (Bennie and Botha, 1986; Ahmed and Maurya, 1988; Daniel et al., 1988). Low wheat yields have also been reported from shallow, stony soils under reduced and no-tillage methods of seedbed preparation in semi-arid regions of Africa (Agenbag and Maree, 1988). Wheat yield can also be substantially reduced in headlands and wheel tracks (Fawcett et al., 1988). This is mainly due to the effect of high soil strength which inhibits root penetration. The continuing expansion of dryland wheat farming into semi-arid regions of tropical Africa and Australia has raised concern that the structural degradation of soils can seriously affect the sustainability of agricultural enterprises in these environments. Dalal (1982) and Dalal and Mayer (1986) observed that the major effect of continuous mechanized cultivation on several Vertisols is a decline of soil organic matter content, resulting in the reduction of chemical fertility and wheat yields. This is also associated with a decline in structural stability. Structural stability in these soils is strongly affected by their clay contents and the properties associated with the clay fraction (So et al., 1988).

Cowpea (Vigna unguiculata)
Studies on the effect of soil compaction on cowpea have been concerned with the response of its root system. Aina (1979) reported that, whilst tillage reduced soil bulk density at planting by 23%, adoption of a rotation of cowpea-cowpea after tillage was more effective in controlling recurrence of compaction of tilled soil than maize-maize and cassava-based rotations. Studies by Maurya and Lal (1979) and Willcocks (1981) have shown that the cowpea root system is not as seriously impeded by high soil density and strength below tillage depth as that of maize or soybean. While the mechanisms responsible for the resilience of cowpea roots are not well understood, this response has some positive agronomic significance. With naturally compacted soils, which form dense clods when ploughed, cereal crops can be preceded by cowpea in a rotation to take advantage of the beneficial soil loosening effects caused by the deep rooting cowpea. Cowpea root growth and subsequent grain yield are also not severely affected by mechanical land clearing, as was observed in southwestern Nigeria (Hulugalle et al., 1984).

There is, however, a limit to which a cowpea root system can penetrate high-density soil without adversely affecting yields. In a laboratory experiment, Onofiok (1989) found that severe soil compaction (1.70 Mg m^{-3}) significantly reduced leaf area, root dry matter accumulation and crop water-use efficiency. In field experiments, Kayombo et al. (1986b) observed a significant decrease in seedling emergence, height, leaf number and grain yield of cowpea as a result of four passes of a 2-Mg roller (452 kPa) (Table 1).

Soybean (Glycine max)
Vehicular traffic adversely affects seedling emergence, plant height, leaf number and shoot dry weight of soybean (Table 1) (Saini and Singh, 1980; Singh et al., 1980; Kayombo et al., 1986b). A high gravel concentration in the subsoil can also reduce the dry weight of soybean plants compared to those grown in gravel-free soil, as was observed in the southern Guinea savanna of Nigeria (Shannon, 1983). Soil compaction adversely affects root growth of soybean by reducing the length and weight of the roots (Maurya and Lal, 1979; Somapala and Willatt, 1979; Nogueira et al., 1983) and by decreasing nodulation (Singh et al., 1971; Katoch et al., 1983). Ortolani et al. (1982) measured up to 50% decrease in soybean grain yield as a result of machinery traffic causing compaction of a Latosol in Brazil. Similar effects have been reported from elsewhere (Katoch et al., 1983;

Kayombo et al., 1986b). Barbosa et al. (1989) measured up to 24% increase in yield as a result of deep tillage of a naturally compacted Alfisol in Bolivia. The benefits of deep tillage were attributed to deeper root penetration and hence access to subsoil water and nutrients. The nutrient status and water regime of the soil regulate the damaging effect of soil compaction on soybean grain yield. Nogueira et al. (1983) reported that soil compaction up to a bulk density of 1.75 Mg m^{-3} had no effect on soybean grain yield if nutrients and water were not limiting factors.

Cassava (Manihot esculenta)

Adequate development of cassava root tubers depends on a high total porosity and the ability of soil to accommodate voluminous tubers. Loose soil is obviously advantageous for cassava tuber development because roots first penetrate the soil and then enlarge at bulking (Onwueme, 1978). The adaptation of cassava to drought is partly explained by the ability of its feeder roots to penetrate deep into the soil and extract water (Ezumah, 1983). Compacted soils decrease the tuberous root/feeding root ratio and reduce yields. In addition, the adverse effects of soil water stress on plant growth and dry matter yield of cassava are accentuated if the cassava root system development is restricted by high bulk

Fig. 1. Cracks in the surface soil created by the growth of cassava tubers.

TABLE 2

Response of cassava tuber yield to soil compaction (from Kayombo and Lal, 1986b)

Roller treatment	Tillage	Cassava table yield (Mg ha⁻¹)	Tubers/plant (number)
0	No tillage	18.4	9.1
2	No tillage	16.3	9.5
4	No tillage	9.7	5.6
0	Ploughed	20.6	11.2
2	Ploughed	24.1	12.8
4	Ploughed	12.7	7.2
LSD ($P_{0.05}$)		3.1	6.5

density (Lal, 1981). Pre-plant field traffic consisting of 4 roller passes (452 kPa) has been shown to reduce plant height, leaf number, shoot dry weight and fresh tuber yield in both no-till and ploughed plots on an Alfisol in south-western Nigeria (Kayombo and Lal, 1986b; Table 2). Development of cassava tubers can also have some beneficial effects on soil structure (Fig. 1).

Other row crops

There are many other row crops of the tropics which are adversely affected by soil compaction. Two such crops are cotton (*Gossipium hirsutum*) and sorghum (*Sorghum bicolor*). There exists a considerable body of literature relevant to the effects of soil compaction on growth and yield of cotton (Barnes et al., 1971). Oni and Adeoti (1986) reported that increasing the number of tractor wheel passes (from 0 to 15) prior to tillage decreased the germination percentage and seed yield of cotton in northern Nigeria. Adverse effects on cotton were attributed to a significant decrease in root exploration as the number of tractor wheel passes increased.

Soil compaction also affects the production of sorghum. Ohu and Folorunzo (1989) reported that the total yield (plant + grain) of sorghum increased with increasing number of tractor wheel passes up to a certain level and then decreased with further increase in the number of passes. Slight compaction can be beneficial to some crops, particularly in a dry climate and with coarse-textured soils of low water holding capacity. In semi-arid and arid regions or if the weather conditions are abnormally dry, especially in loose soils, a certain amount of machinery traffic on the field can be beneficial to crop production. This is due to the combined advantages of better water storage, good germination and root proliferation and an enhanced uptake of water and nutrients by the crop (Kar et al., 1986; Onwualu and Anazodo, 1989).

FACTORS INFLUENCING RESPONSES OF TROPICAL CROPS TO SOIL COMPACTION

The consequences of widespread soil compaction problems in the tropics should be assessed in terms of soil degradation and agronomic production. The magnitude of adverse effects depends on soil type (including structural stability and subsoil properties), weather conditions and crop type (including various ways in which different plants respond to soil compaction hazard).

Soil type

Predominant soils of the humid tropics are, in order of importance, Oxisols, Ultisols and Alfisols. Low base status Oxisols and Ultisols cover about 50% of the land area and high base status Alfisols about 20%. In general, Oxisols have favourable structure, and some of them, for example, at Brasilia, Brazil, can be ploughed with heavy machinery even a day after a heavy rain with little aggregate disruption (Sanchez, 1976). The favorable structure of these soils is attributed to presence of very stable sand-sized granules or aggregates. The high aggregate stability is associated with a high clay content and the cementing or coating of aggregates with amorphous iron and aluminum oxides. Generally, Ultisols and Alfisols have less desirable structural properties than Oxisols because of lower clay contents in the A-horizon and the absence of well-developed clay films within peds. However, the clay films that coat these peds are anisotropic and generally increase aggregate stability, although this effect is seldom as strong as is observed in Oxisols. The coarse-textured topsoils of Ultisols and Alfisols are prone to serious erosion and compaction hazards, ultimately resulting in low crop productivity (Sanchez, 1976; Lal, 1985a).

Soils with a gravel horizon or stone line at shallow depth occur in abundance in West Africa (Segalen, 1969; Ahn, 1970), Latin America (Santamaria, 1965; Sanchez, 1981), South-East Asia (Panabokke, 1967; Grewal et al., 1984) and the tropics in general (Thomas, 1974). These gravel horizons have higher bulk densities than the surrounding soil and, therefore, have a profoundly negative effect on root penetration (Babalola and Lal, 1977a), often resulting in low crop yields (Agenbag and Maree, 1988).

In the sub-humid and semi-arid regions of Africa (i.e., precipitation $\geq$ potential evapotranspiration for 4-6 and 2-4 months of the year, respectively), uplands are dominated by "lateritic" and naturally compacted soils (Northwood and Macartney, 1971; Nicou and Charreau, 1980; Willcocks, 1981). Naturally compacted soils are also abundant in the Caribbeans (Martínez and Lugo-López, 1953) and elsewhere in the tropics. The high soil bulk density and the low infiltration rates exhibited by these soils predispose them to slaking under raindrop impact when exposed, leading to the development of an impermeable crust. As a result of this crust, runoff and soil compaction hazards are high (Kalms, 1977). Deep mechanical cultivation is necessary to produce any appreciable crop yield on such soils (Macartney et al., 1971; Willcocks, 1984).

Weather conditions

Tropical rains are characterized by high intensity and total amount (Morgan, 1974; Amezquita and Forsythe, 1975; Bols, 1978). Rainstorms with sustained intensity of 150-200 mm h^{-1} for 10-15 min are not uncommon (Lal et al., 1980). The median drop size (D_{50}) of tropical rains can be 2.5-3.5 mm (Aina et al., 1976; Kowal and Kassam, 1976; Lal, 1982). Tropical soils exposed to rainstorms of such high intensity and large drop size slake readily, develop an impermeable crust, and are prone to accelerated erosion and soil degradation.

Land clearing methods

Effects of methods of land development on soil and microclimate have been investigated in Nigeria (Lal and Cummings, 1979; Hulugalle et al., 1984), the

Fig. 2. Mechanical equipment for land clearing mounted on a crawler tractor (total mass about 30 Mg). The tree pusher and root rake shown here result in much more soil disturbance than does a shear blade which cuts trees at ground level with minimal soil disturbance.

TABLE 3

Average changes in soil bulk density and penetration resistance in the top 50 cm with duration of cultivation following land clearance by different methods (R. Lal, unpublished data)[1]

Method	Bulk density (Mg m⁻³)				Penetration resistance (kPa)			
	1978[2]	1979	1980	1981	1978[2]	1979	1980	1981
Traditional farming	0.64	1.06	1.07	1.27	21	96	52	132
Manual clearing	0.68	1.17	1.17	1.39	20	140	75	119
Shear blade	0.70	1.19	1.37	1.38	26	100	184	219
Tree pusher/ root rake	0.60	1.24	1.32	1.42	20	130	73	123

[1]Each figure is a mean of 25 separate analyses made in the dry season when soil moisture content was about 5% (w/w).
[2]Pre-clearance.

TABLE 4

Effects of clearing methods on between-row root length at the tasselling stage, and on yield of maize on a tropical Alfisol (after Hulugalle et al., 1984)

Clearing method	Root length (mm m⁻³) at depths (cm)		Grain yield (Mg ha⁻¹)
	0-10	10-20	
Manual	4400	700	4.44
Tree pusher	5300	800	4.09
Tree pusher/root rake	4300	1000	4.01
Shear blade	3900	500	4.17
LSD ($P_{0.05}$)	1000	400	0.22

Amazon basin (Seubert et al., 1977; Dias and Nortcliff, 1985), Ghana (Cunningham, 1963; Nye and Greenland, 1964), Malaysia (Daniel and Kulasingam, 1974) and Surinam (Van der Weert, 1974). Apart from manual methods, a wide range of mechanical equipment is used for land clearing (Fig. 2). The choice of equipment depends on several factors, including biophysical, socio-economic and logistic considerations. Most of these studies support the conclusion that mechanical land clearing, compared with manual clearing methods and forest control, is detrimental to soil quality and has adverse effects

TABLE 5

Maize grain yield (Mg ha⁻¹) as affected by land clearance methods (after Lal, 1981)

Clearing method	First season	Second season	Third season
Mechanical	4.67	1.44	2.88
Slash and burn	5.14	1.86	4.46
Slash	4.81	1.72	3.67
LSD (P$_{0.05}$)	0.69	0.71	0.98

on crop yield (Tables 3, 4, 5). The harmful effects of mechanical land clearing depend on various factors, including soil type, antecedent soil water content, tree density and root system, type and weight of the crawler tractor used, nature of the attachment (plane blade, tree pusher or shear blade), and above all, efficiency of the operator.

In the humid tropics, felling operations tend to be confined to the wet season owing to the ease of uprooting. However, clearance during the wet season can lead to severe problems due to several factors. Firstly, the available water-holding capacity of the surface layer of soils in the humid tropics is generally low (Lal, 1979). Therefore, the surface layer can reach saturation within a few hours after a heavy storm. Soil water content is the most dominant factor influencing compactability. Vehicular traffic on wet soils can lead to plastic displacement (deep rutting), puddling and smearing. Secondly, due to the low water-holding capacity of the surface layer of these soils, the surface layer can likewise reach permanent wilting point within a few days after a heavy rain. Ultra-desiccation of the surface layer can lead to crusting, resulting in a drastic decrease in the infiltration rate, in excessive runoff and accelerated soil erosion (Falayi and Lal, 1979). Thirdly, heavy rain on recently cleared soils is likely to cause additional compaction and downward movement of fine particles (Cunningham, 1963; Ahn, 1968).

Crop type

There are four types of crop responses to soil compaction: (1) yields remain unchanged; (2) drastic yield reduction; (3) yields increase; (4) yields follow an optimum curve. The first and second categories of response are most commonly observed (e.g., Kayombo and Lal, 1986b; Ohu and Folorunzo, 1989; Onwualu and Anazodo, 1989). When the range of porosities studied in a field experiment is large enough, an optimum bulk density can also be observed. The optimum bulk density, however, shifts to a higher total porosity when the water-holding capacity (clay and organic matter contents), the soil water content at compaction or the

TABLE 6

Optimal soil bulk densities for different soils for several tropical crops

Crop	Yield component	Optimal bulk density (Mg m⁻³)	Soil type	Reference
Maize	Grain	1.20	Oxisol	Onwualu and Anazodo (1989)
Upland rice	Root weight	1.60	Sandy loam	Kar and Varade (1972)
	Grain	1.60	Sandy loam	Ghildyal (1978)
	Grain	1.60	Alfisol	Ogunremi et al. (1986a)
	Grain	1.50	Ultisol	Ogunremi et al. (1986b)
Sorghum	Dry matter	1.58	Alfisol	Ohu and Folorunso (1989)
	Grain	1.58	Alfisol	Ohu and Folorunso (1989)
Cowpea	Grain	1.50	Alfisol	Ojeniyi (1989)
Cassava	Fresh tubers	1.35-1.50	Alfisol	Kayombo and Lal (1986b)

soil water content during crop growth increase. Crop requirements for nutrients and water are higher at high than at low yield levels. Therefore, the range of optimum porosities is smaller at high than at low yield levels. Generally, monocotyledonous crops are less vulnerable to low porosities than dicotyledonous crops. Optimum levels of bulk density for some tropical crops are shown in Table 6.

There are also large differences in the ability of different crops to tolerate mechanical stress in the soil. Crop responses to those soil properties which are indices of compaction differ widely among species (Tables 7 and 8). Cowpea, for example, is generally tolerant of moderate levels of soil compaction (Maurya and Lal, 1979; Willcocks, 1981). Lal et al. (1978, 1979) compared several possible cover crops for zero-tillage farming systems in the tropics and found significant differences among species to improve the structure of a compacted soil. Notable differences were observed also in the yield of succeeding food crops. Considering all aspects of crop and soil management, Lal et al. (1979) observed some favorable attributes of *Stylosanthes guianensis* in ameliorating compacted tropical Alfisols in Nigeria. This observation offers an attractive approach to alleviation of soil compaction: to select plants with sturdy roots able to penetrate semi-rigid soil. As yet, little systematic work has been done on these lines in the tropics.

CONTROL OF SOIL COMPACTION IN THE TROPICS

Control of soil compaction is a continual requirement in agriculture in order to maintain crop productivity. The conduct of field operations inevitably causes some compaction. Hence, a major task of soil management is, first, to minimize residual soil compaction to the extent possible and, second, to alleviate or rectify

TABLE 7

Regression equations and correlation coefficients relating soil physical properties to crop grain yield of maize, cowpea, soybean and cassava on a tropical Alfisol in western Nigeria (from Kayombo et al., 1986a; Kayombo and Lal, 1986a,b)[1,2]

Maize		
Dry bulk density (Mg m^{-3})	y = 17.34 - 8.84x	0.80***
Total porosity (ratio)	y = -6.33 + 23.74x	0.79***
Equilibrium infiltration rate (cm h^{-1})	y = 1.46 + 0.55x	0.82***
Soil water content (g g^{-1})	y = 2.75 + 11.98x	0.17 n.s.
Penetration resistance (MPa)	y = 7.80 - 1.68x	0.59*
Relative compaction (ratio)	y = 17.64 - 17.34x	0.80***
Specific volume (ratio)	y = 8.50 + 6.64x	0.75***
Cowpea		
Dry bulk density (Mg m^{-3})	y = 3.10 - 1.43x	0.73***
Total porosity (ratio)	y = 0.69 + 3.76x	0.72***
Equilibrium infiltration rate (cm h^{-1})	y = 0.57 + 0.02x	0.67**
Soil water content (g g^{-1})	y = 0.81 + 1.39x	0.12 n.s.
Penetration resistance (MPa)	y = 1.47 - 0.22x	0.47*
Relative compaction (ratio)	y = 3.12 - 2.77x	0.72***
Specific volume (ratio)	y = -1.12 + 1.09x	0.69**
Soybean		
Dry bulk density (Mg m^{-3})	y = 4.22 - 2.03x	0.74***
Total porosity (ratio)	y = 1.18 + 5.37x	0.72***
Equilibrium infiltration rate (cm h^{-1})	y = 0.47 + 0.04x	0.88***
Soil water content (g g^{-1})	y = 0.85 + 2.89x	0.13 n.s.
Penetration resistance (MPa)	y = 2.14 - 0.46x	0.58*
Relative compaction (ratio)	y = 4.29 - 3.98x	0.74***
Specific volume (ratio)	y = -1.66 + 1.50x	0.67**
Cassava		
Dry bulk density (Mg m^{-3})	y = 39.73 - 15.44x	0.62**
Total porosity (ratio)	y = 2.33 + 4.67x	0.60**
Equilibrium infiltration rate (cm h^{-1})	y = 11.75 + 9.30x	0.59*
Soil water content (g g^{-1})	y = 15.71 + 30.78x	0.24 n.s.
Penetration resistance (MPa)	y = 27.09 - 4.05x	0.42 n.s.
Relative compaction (ratio)	y = 40.45 - 30.67x	0.60**
Specific volume (ratio)	y = -1.53 + 9.98x	0.54*

[1]y = grain or tuber yield (Mg ha^{-1}); x = soil property, measured at 100 mm depth, except equilibrium infiltration rate.
[2]n.s., *, **, *** = not significant, significant at $P_{0.05}$, $P_{0.01}$ and $P_{0.001}$, respectively.

that unavoidable measure of compaction caused by traffic and tillage once it occurs.

TABLE 8

Multiple regression equations relating the yield of maize, cowpea, soybean and cassava to soil properties (from Kayombo et al., 1986b; Kayombo and Lal, 1986b)[1,2]

Maize	$y = -20.2 - 346.7v - 0.02f - 17.1g - 908.9d - 83.8c^2 + 0.4P^2$ $+ 654.7d^2 - 56.9w^2 + 0.6i$	0.97***
Cowpea	$y = 65.8 - 0.03i + 19.5w - 0.06P^2 - 73.4w^2 - 8.7c^2 + 24.5v$ $- 185.3d + 105.1d2$	0.81**
Soybean	$y = 1.6 + 0.001I^2 - 12.1w - 50.0w^2 - 0.2v^2 - 0.2g^2 + 0.2P$	0.96***
Cassava	$y = -1101.2 - 330.19^2 + 672.9w^2 + 1161.1f - 1.1f^2 + 42.8v^2$ $- 124.5w$	0.83***

[1]y = crop yield (Mg ha^{-1}); d = relative compaction (a ratio); v = specific volume (a ratio); i = equilibrium infiltration rate (cm h^{-1}); P = penetration resistance (MPa); w = gravimetric water content (g g^{-1}); g = dry bulk density (Mg m^{-3}); f = porosity (a ratio).
[2]** = significant at $P_{0.01}$; *** = significant at $P_{0.001}$.

Minimizing soil compaction hazards

The most obvious approach to minimizing soil compaction is the avoidance of all but essential traffic. This calls for reducing the number of operations involved in primary and secondary tillage, and using the most efficient implement at the most appropriate time so as to effect the desirable soil condition in a single pass rather than by a repeated sequence of passes. It has long been known that excessive soil manipulation, beyond that required to prepare a seedbed and control weeds, leads to yield loss and soil structure deterioration (Pereira and Jones, 1954; So et al., 1988). Overly intensive cultivation can cause soil degradation, including accelerated erosion by water and wind. Several attempts have been made to design appropriate field equipment for the soils and environments of the tropics (IITA, 1982; May, 1988). In the tropics, however, equipment design and operational efficiency have not advanced to a stage where field operations can be accomplished in one pass. Thus, other approaches must be explored to minimize soil compaction.

Crop residue mulch

Cultural practices that have proven effective in reducing compaction and erosion and in preserving soil organic matter content are based on the principle of maintaining a protective ground cover. A crop residue mulch, produced *in situ* or brought in, is an option for providing a protective cover, restoring soil properties, and preventing or reducing soil compaction (Lal, 1986). Benefits of mulching on soil structural improvement and crop yields have been documented for soils of West Africa (Okigbo, 1965, 1969; Lal, 1978; Mensah-Bonsu and Obeng, 1979; Maurya and Lal, 1980) and in the tropics in general (Lal, 1984, 1986).

An adequate amount of crop residue mulch enhances soil quality and maintains productivity. However, maintenance of a continuous ground cover with adequate mulch rate is easier advocated than achieved in practice. In the humid tropics, due to continuously high temperature and humidity throughout the year, residue mulches decompose rapidly (Lal et al., 1980). Termites also remove a considerable amount of crop residue mulch. It is therefore difficult to maintain a protective mulch layer in the tropical environment. Under these conditions, one of the practical means of obtaining mulch is through a no-tillage system where the residue from previous crop and dead weed growth may serve as *in-situ* mulch.

No-tillage farming

In no-tillage farming, seeds are planted in a narrow slit or hole, opened mechanically or by manually operated equipment and tools, in the sod suppressed by herbicides or in previous crop residue without primary or secondary tillage operations. The continuity of channels created by earthworms and other soil fauna, and macropores created by the decaying root system of previous crops are important factors in providing water transmission through the profile during high intensity tropical storms (Lal, 1985b; Derpsch et al., 1986). It is also the stability and continuity of these channels in the untilled soil that favour the development of deep root systems into the gravelly subsoil horizons (Maurya and Lal, 1980) which are otherwise difficult for roots of seasonal crops to penetrate (Babalola and Lal, 1977a,b; Vine et al., 1981; Grewal et al., 1984).

Although mechanical tillage involving ploughing and harrowing may improve water storage for the first few rains, it subsequently encourages compaction, runoff and erosion by splash and surface crusting (Collinet and Valentin, 1979). Moreover, mechanical tillage disrupts the continuity of pore space and creates an additional barrier by its smearing action in the plough sole layer. These negative attributes are, however, minimal in a no-tillage system. No-tillage systems can create a favourable soil temperature regime and improve soil structure by preventing slaking and raindrop impact (Lal, 1983). In Ghana, Baffoe-Bonnie and Quansah (1975) reported from their studies on Alfisols at Kumasi that no-tillage caused the least compaction, maintained high porosity, and had the lowest soil and water losses. Field experiments conducted at IITA, near Ibadan, Nigeria, on 4- to 5-ha agricultural catchments indicated that soil physical properties and chemical fertility declined substantially in the ploughed watershed after 6 years of continuous mechanized cultivation and 12 maize crops, while the decline was decidedly less in the no-tillage system (Lal, 1985a). A severe decline in the physical and chemical properties of the surface layer of the ploughed watershed was due to soil structural deterioration, which subsequently led to high erosion losses (Lal, 1984). Maize grain yields were higher following no-tillage than in the ploughed watershed throughout the 6 years (Lal, 1985a).

In Paraná, Brazil, Derpsch et al. (1986) and Roth et al. (1988) studied for 7 years the influence of tillage and cover crops on soil compaction, soil erosion and yield of several crops on an Oxisol. The tillage systems compared were no-tillage, minimum tillage and conventional tillage. Soil physical analysis showed higher bulk density in the top 20 cm under no-tillage, whereas conventional tillage, and to a lesser extent minimum tillage, led to the development of a plough pan at 20-30 cm depth. Infiltration rate was higher under no-tillage compared to conventional tillage and minimum tillage. The main factor influencing infiltrability was the formation of a surface seal, depending on the degree of soil cover, irrespective of the tillage system. The use of cover crops and crop rotation had a marked positive effect on the yield of maize, soybean and bean (*Phaseolus* spp.). Yields of wheat and soybean were higher under no-tillage and minimum tillage than conventional tillage systems. The conclusion drawn from these studies is that no-tillage in combination with adapted cover crops and crop rotations represents a production system which is efficient in reducing soil compaction, minimizing water runoff and consequently soil erosion, and in increasing crop yields (Derpsch et al., 1986; Roth et al., 1988).

The need to apply herbicides in no-tillage systems may result in negative ecological impacts and secondary effects on humans through consumption of contaminated products and damage to the environment. The effects of chemicals on soil and environment are not well understood, and neither are the pathways of herbicides and their degradative products through soil and water.

Controlled-traffic

Controlled traffic may involve zero traffic, reduced traffic, or guided traffic confined to specific positions in the field. The guided-traffic technique results in optimum seedbed conditions and a firmed traffic path, thereby providing satisfactory plant growth and trafficability for timely field operations. A controlled-traffic system that also combines the water conservation benefits of a judicious tillage method, is especially useful in a semi-arid tropical environment. This controlled-traffic/tillage system has variously been known as zonal tillage (Northwood and Macartney, 1971) or strip catchment tillage (Willcocks, 1984). In this system, mechanical cultivation by a narrow subsoil tine is restricted to a strip of about 30 cm width and at crop row intervals of about 75 cm (Fig. 3) (Northwood and Macartney, 1971; Willcocks, 1984; Tullberg, 1988). Although many areas of uncertainty remain, the results from limited field experiments demonstrate that controlled traffic can potentially: (1) reduce the fuel cost for crop establishment by at least 40%; (2) allow similar output and capacity from a tractor of at least 30% less power; (3) maintain yield without the necessity for

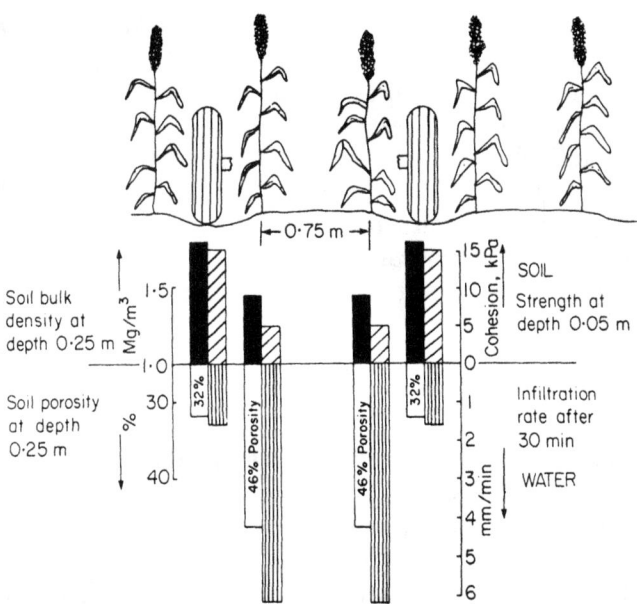

Fig. 3. Schematic of soil bulk density, soil strength, soil porosity and infiltration of water across a strip-tillage plot at sorghum harvest time in a Botswana Luvisol (from Willcocks, 1984).

deep tillage operations; (4) increase rainfall infiltration, and reduce runoff and erosion.

Controlled traffic can, thus, provide great economic benefits in extensive dryland grain production systems, but it does demand improved tractor/implement guidance. However, in relatively wetter areas and where rainfall is regular and uniformly distributed over the growing season, controlled traffic does not provide a crop yield advantage compared to conventional ploughing and ripping (Gill and Lungu, 1988).

Controlled grazing

Controlled grazing may involve no grazing (cut and stall feeding), reduced stocking rate or rotational grazing. On properly managed pastures with controlled grazing, soil compaction and erosion problems are generally less serious than on arable land planted to row crops. However, soils under heavily grazed pastures in Kenya (Pereira, 1973) and in northern Australia (Mott et al., 1979), can develop a surface crust and a low infiltration rate. Pereira (1973) compared infiltration and runoff characteristics of a heavily grazed catchment compared with that under controlled grazing. 40% of rainfall was lost as runoff from the heavily grazed watershed and the penetration of rainfall into the trampled soil was slight, with no wetting observed below 50 cm depth. On the other hand, the wetting front reached 50-125 cm deep on the watershed with controlled grazing and peak flows of runoff were accordingly reduced. A watershed management experiment at IITA, Ibadan, Nigeria, examined soil properties and runoff and erosion under grazed pastures (with a stocking rate of 3 cows ha^{-1}) in comparison with annual crops grown with and without *Leucaena* alleys (Lal, 1986). Results showed that soil bulk density and penetration resistance were lower while soil organic matter content was higher and soil structure superior in the surface soil of plots under pasture than those under maize. Consequently, there was less runoff and soil erosion under grazed pastures than under maize-cowpea rotation (Lal, 1986).

Harvest traffic

Mechanized harvesting is another major factor responsible for soil compaction, especially on headlands and turning points. Mechanized harvesting can result in serious compaction. In Puerto Rico, Shukla and Ravalo (1976, quoted by Howson, 1977) reported a severe problem of soil compaction due to transporting loaded carts in sugar cane (*Sachruum* spp) fields. In a field with a soil water content of 27.4% (w/w), the depth of significant compaction was 5 cm. When the soil water content in an adjacent field was 48.3% (w/w), the depth of significant compaction was 45 cm. Similar observations on the effects of harvest traffic on sugar cane-growing soils have been reported from Guyana (Howson, 1977). Soil compaction problems in sugar cane fields may also exist in Queensland (Australia), Paraná (Brazil), and elsewhere in the tropics. Lal (1985a) observed that crop stand, growth and yield of maize at the boundaries of the no-till watershed near the turning points of farm equipment were particularly poor. The soil appeared to be highly compacted for about 10-m wide strips on the headlands. The cumulative infiltration after 2 h was about 30% less near the turning points than in the rest of the field. Maize grain yield for the compacted region was 1.2 ± 0.4 Mg ha^{-1} in comparison with 1.9 Mg ha^{-1} for the uncompacted soil, i.e., a difference of 37%.

Avoiding compaction in land clearing

Clearly the risks of soil compaction in mechanical land clearing are so high that every possible means of avoiding the problem must be explored. Where feasible, forest removal should be done by the slash-and-burn method (Dias and Nortcliff, 1985) so that most of the nutrients in the vegetation are returned to the soil, without scraping off the topsoil or compacting it. Where clearing by mechanical means is the only practical alternative, a desirable method of forest removal is the use of the shear blade (Lal, 1986), whereby most of the roots and stumps are left in the ground intact, and the forest litter is not removed (Table 9). The soil water content at the time of land clearing and post-clearing management are also

TABLE 9

Effects of methods of deforestation on compaction and water characteristics of a tropical Alfisol (after Lal and Cummings, 1979)

Method of deforestation	Bulk density (0-10 cm) (Mg m^{-3})		Penetration resistance (0-4 cm) (kPa)		Hydraulic conductivity (0-10 cm) (cm min^{-1})		Equilibrium infiltration rate (cm h^{-1})	
	Before	After	Before	After	Before	After	Before	After
Mechanical	0.91	1.25	5	505	16.1	1.3	114	17
Slash and burn	0.86	1.12	66	284	15.2	5.0	44	5
Slash	0.89	1.13	59	183	9.8	4.6	141	62
LSD (P$_{0.05}$)	0.29		34.5		9.3		-	

TABLE 10

Effects of soil water content at the time of land clearance on bulk density (0-10 cm depth) and infiltration rate (after Ghuman and Lal, 1992)

Soil water content (m^3 m^{-3})	Bulk density (Mg m^{-3})	Infiltration rate (cm h^{-1})
0.11	1.36	62.0
0.16	1.44	2.2
0.20	1.49	3.3
LSD (P$_{0.05}$)	n.s.	43.6

important factors (Ghuman and Lal, 1992)(Table 10).

AMELIORATION OF COMPACTED SOILS

Due to the adverse effects of soil compaction on agricultural production in the tropics, recourse must be often made to both biological and mechanical measures for alleviating soil compaction.

Biological loosening of compact soils

Root systems of cowpea and pigeon pea (*Cajanus cajan*) are not seriously impeded by high soil densities and strength below tillage depth (Maurya and Lal, 1979; Willcocks, 1981; Hulugalle and Lal, 1985). A deep taproot system, given sufficient time, can alleviate moderate levels of soil compaction. Pereira and Beckley (1952) observed that regeneration of tropical grasslands in East Africa was more effective when protected by vegetative cover and less disturbed compared with deep ripping with heavy machinery. Full recovery, after various grass and leguminous crop covers, took 2 or more years of growing cover crops (Pereira and Thomas, 1961). Grazing capacity after full recovery was as good as or better than that obtained from ploughing and reseeding. At Muguga, Kenya, Pereira and Dagg (1967) reported that runoff and soil loss from Bermuda grass (*Cynodon dactylon*) pastures on a 10% slope in a 1000-mm rainfall region was negligible. At Ibadan, Nigeria, Lal et al. (1978, 1979) investigated the effects of a range of leguminous and grass fallows on restoration of a compacted and eroded Alfisol. Improvement in soil characteristics under *Bracharia, Paspalum, Cynodon* spp., *Pueraria, Stylosanthes, Stizolobium, Psophocarpus* and *Centrosema* were compared with that of a weed-fallow control. There were significant improvements in soil organic matter, total cation exchange capacity, infiltration rate, water retention at low suctions and soil bulk density under various grass and leguminous fallows compared with the control.

Considering all aspects of crop and soil management, Lal et al. (1979) recommended growing *Stylosanthes guianensis* on Alfisols in Nigeria. Also in south-western Nigeria, soil compaction caused by mechanical land clearing was alleviated by immediately seeding *Mucuna utilis* (Hulugalle et al., 1986). In Paraná, Brazil, Kemper and Derpsch (1981) and Derpsch et al. (1986) observed that growing cover crops increased water infiltration rates up to 416% on Oxisols and up to 629% on Alfisols when compared to wheat stubble. Cover crops also resulted in a friable soil tilth due to the biological loosening effect created by the root system.

Mechanical loosening of compact soils

Mechanical loosening of tropical soils has often become necessary to ameliorate soil compaction caused by: (1) land clearance by heavy machinery; (2) mechanized farming; (3) hard setting. A reclamation study conducted in 1980-1982 on an Ultisol in Peru that was severely compacted in 1972 (Seubert et al., 1977), showed that sub-soiling and chisel ploughing produced the highest corn and soybean yields (Alegre et al., 1986). In West Sumatra, Indonesia, Makarim et al. (1988) observed that deep tillage improved soil physical properties of a mechanically-cleared Oxisol. In Orange Free State, South Africa, Bennie and Botha (1986) reported that deep ripping of a continuously ploughed soil increased maize and wheat yields. Whilst compacted soil can be loosened for a short period by tillage operations, soil may resettle to a high density (Kemper and Derpsch, 1981; Daniel et al., 1988; Orellana et al., 1990). In some instances,

TABLE 11

Effects of no-tillage, shallow manual tillage and deep ploughing on yields (Mg ha^{-1}) of several crops grown on sandy Alfisols of West Africa (from Charreau, 1972)

Crop	No-tillage	Manual tillage (<5 cm)	Ploughing (15-25 cm)
Millet	1.31	1.74	1.60
Sorghum	1.52	2.42	1.88
Corn	1.86	3.49	3.21
Rice	1.62	2.36	2.80
Cotton	1.40	1.67	1.80
Peanuts	1.62	1.77	1.76

however, the residual effect of deep chiselling can persist up to the third consecutive growing season (Sharma and Mbogoni, 1988). Experiences with mechanized no-tillage systems on Alfisols in western Nigeria indicate the limitations towards the utilization of the so-called "kaolinitic" or "low-activity clay" soils for continuous crop production (Lal, 1985a). To overcome such soil limitations, periodic chiselling to a depth of about 50 cm in the row zone followed by sowing of a deep-rooted leguminous cover crop to avoid resettling of the loosened soil can improve infiltration and root development in the subsoil (Lal, 1985b).

In the sub-humid and semi-arid upland tropical regions dominated by naturally compacted soils, some mechanical cultivation is necessary for good crop establishment and yield (Hosegood, 1964; Northwood and Macartney, 1971; Charreau and Nicou, 1971; Charreau, 1972; Willcocks, 1984; Gill and Lungu, 1988; Barbosa et al., 1989). In Senegal, Charreau and Nicou (1971) and Charreau (1972) observed that shallow hoe cultivation decreased the bulk density of the first few centimeters from 1.6 to 1.4 Mg m^{-3}, whereas with tractor ploughing the same values of bulk density were obtained to a depth of 10-30 cm. Consequently, significant yield increases, as compared to the uncultivated plots, were obtained with both shallow hand cultivation and deep tractor tillage (Table 11). The data in Fig. 4 show that a decrease of 0.1 Mg m^{-3} in bulk density had a beneficial effect on root development and yield of sorghum and peanuts. Furthermore, there existed a close correlation between root density in the top 20 or 30 cm and grain yields. Beneficial effects of cultivation on root system development and yield of cereals have also been observed on naturally compacted Alfisols in Botswana (Willcocks, 1981, 1984) and Zambia (Gill and Lungu, 1988).

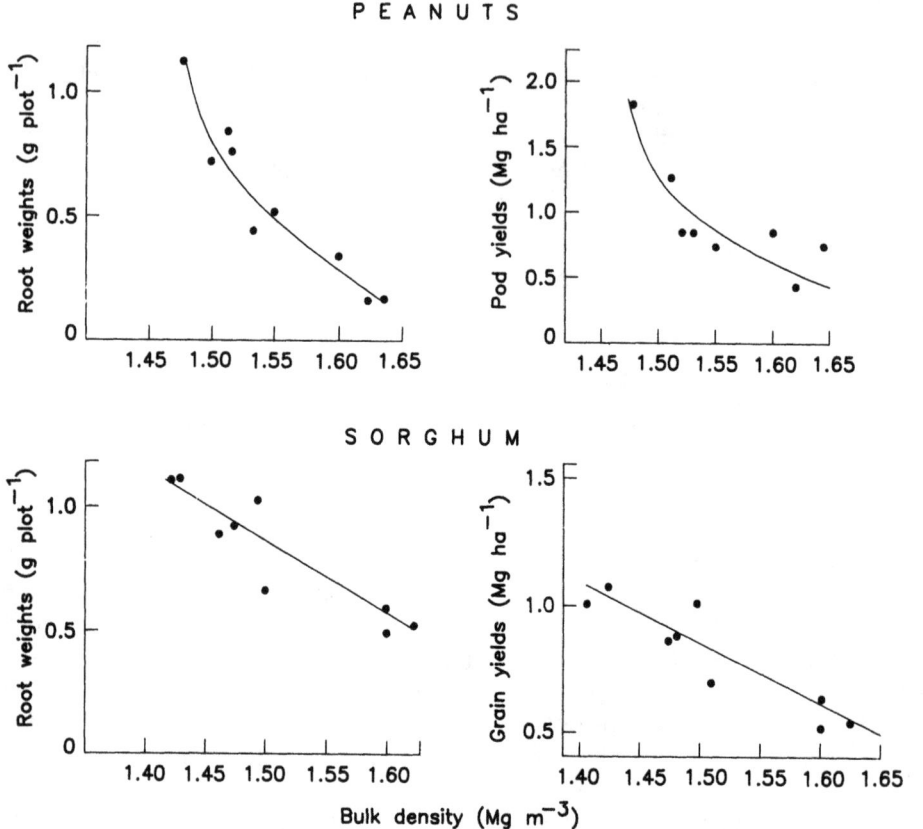

Fig. 4. Relationships between bulk density, root development and yields of peanuts (top) and sorghum (bottom) in sandy Alfisols at Bambey, Senegal (from Charreau and Nicou, 1971).

Examples of beneficial effects of soil compaction

Slight to medium levels of soil compactness have been found to be beneficial in some environments. In arid regions, compaction of highly permeable sandy soils may improve the efficiency of rainfall and irrigation utilization by reducing deep percolation losses and increasing water storage in the root zone. Compaction of coarse-textured soils assures some yield even though the yield level may be substantially lower than that expected from soils with favourable structure (Gulati et al., 1985a,b). In some arid regions of India, soil compaction is now adopted as an important cultural practice for paddy rice production because it minimizes the irrigation requirement by cutting down percolation losses, controls weeds and enhances efficient use of fertilizers by the rice crop (Ghildyal, 1978; Singh et al., 1980; Rami Reddy and Hukkeri, 1983; Kar et al., 1986). For soils of low water holding capacity, the Indian Agricultural Research Institute recommends the application of up to 12 passes with a heavy roller (450 kg, 45 x 90 cm) (Gupta, 1981).

CONCLUSIONS

(1) Mechanical land clearance in the tropics results in compaction problems, both with regard to increasing erosion and decreasing crop productivity. If mechanical land clearance is unavoidable, due to economic or social circumstances, it is likely that compaction problems could be materially reduced by a number of comparatively simple expedients which might be incorporated into contract specifications, even if they resulted in a slowing down of clearing rate and required greater management control. They include the use of the shear blade in clearing, the avoidance of wet season traffic and a reduction in topsoil movement.

(2) Soil compaction originating from other causes, e.g., continuous cultivation and intensive cropping, naturally compacted soils and subsoil gravel horizons, has equally detrimental effects on soil structure, crop growth and yield of a wide range of crops. Furthermore, soil compaction can persist for a long time if adequate measures are not taken to minimize or ameliorate it.

(3) In humid regions, the no-tillage system with a crop residue mulch is a useful technique to manage soil compaction, provided it can be adopted as a package of all cultural practices that make it work.

(4) There is some evidence for the ameliorative benefit of controlled traffic in naturally compacted upland soils, and also of controlled grazing.

(5) The biological loosening of compact soils by cover crops, woody shrubs and suitable tree species grown on plantations or in agro-forestry systems, is a viable option for improving soil structure.

(6) In addition to being expensive, the beneficial effects of deep tillage when undertaken to relieve compaction in the subsoil are often short-lived and transient.

(7) Considering the vastness of the tropical region, increased basic and applied research is urgently needed on the responses of crops to soil compaction and on developing promising compaction control measures.

REFERENCES

Agenbag, G.A. and Maree, P.C.J., 1988. The effect of tillage on root environment, plant development and yield of wheat (*Triticum aestivum*) in stony soil. Proc. 11th Conf. Int. Soil Tillage Res. Org. (ISTRO), Edinburgh, U.K., Vol. 2, pp. 531-536.

Ahmed, A. and Maurya, P.R., 1988. The effects of deep tillage on irrigated wheat production in a semi-arid zone of Nigeria. Proc. 11th Conf. Int. Soil Tillage Res. Org. (ISTRO), Edinburgh, U.K., Vol. 2, pp. 537-541.

Ahn, P.M., 1968. The effects of large scale mechanized agriculture on the physical properties of West African soils. Ghana J. Agric. Sci., 1: 35-40.

Ahn, P.M., 1970. West African Soils. Oxford University Press, London, U.K., 332 pp.

Aina, P.O., 1979. Tillage, seedbed configuration and mulching effects on soil physical properties and responses of cassava, cowpea and maize. Ife J. Agric., 1: 26-34.

Aina, P.O., Lal, R. and Taylor, G.S., 1976. Soil and crop management in relation to soil erosion in the rainforest of western Nigeria. In: C.R. Forster (Editor), Soil Erosion: Prediction and Control. Soil Conserv. Soc. Am., Ankeny, IA, U.S.A., Spec. Publ. 21, pp. 75-84.

Alegre, J.C., Cassel, D.K. and Bandy, D.E., 1986. Reclamation of an Ultisol damaged by mechanical land clearing. Soil Sci. Soc. Am. J., 50: 1026-1031.

Amezquita, E.C. and Forsythe, W.M., 1975. La probabilida diaria de las lluvias y la frecuencia horaria de su inicio como criterio para manejo de aguas y suelos en Turrialba, Costa Rica. (The daily probability of rainfall and the hourly frequency of its start as a criterion for soil and water management in Turrialba, Costa Rica). V. Congreso Latino Americano de la Ciencia del Suolo. IV Coloquio. Nacional Sobre Suelos Agosto, 1975. Medellin, Colombia (in Spanish).

Babalola, O. and Lal, R., 1977a. Sub-soil gravel horizon and maize root growth: I. Gravel concentration and bulk density effects. Plant Soil, 46: 337-346.

Babalola, O. and Lal, R., 1977b. Sub-soil gravel horizon and maize root growth: II. Effects of gravel size, inter-gravel texture and natural gravel horizon. Plant Soil, 46: 347-357.

Baffoe-Bonnie, E. and Quansah, C., 1975. The effect of tillage on soil and water loss. Ghana J. Agric. Sci., 8: 191-195.

Baldwin, K.D.S., 1957. The Nigerian Agricultural Project: An Experiment in African Development. Blackwell, Oxford, U.K., 221 pp.

Barbosa, L.R., Diaz, O. and Barber, R.G., 1989. Effects of deep tillage on soil properties, growth and yield of soya in a compacted Ustochrept in Santa Cruz, Bolivia. Soil Tillage Res., 15: 51-63.

Barnes, K.K., Carleton, W.M., Taylor, H.M., Throckmorton R.I. and Vanden Berg, G.E. (Editors), 1971. Compaction of Agricultural Soils. Am. Soc. Agric. Eng., St. Joseph, MI, U.S.A., 471 pp.

Bauer, F.H. (Editor), 1978. Cropping in North Australia: Anatomy of Success and Failure. Proc. 1st NARU Seminar, 24-27 August 1977, Darwin, N.T., Australia, 267 pp.

Bennie, A.T.P. and Botha, F.J.P., 1986. Effect of deep tillage and controlled traffic on root growth, water-use efficiency and yield of irrigated maize and wheat. Soil Tillage Res., 7: 85-95.

Bols, P.L., 1978. The Iso-erodent Map of Java and Madura. Soil Research Institute, Bogor, Indonesia, 38 pp.

Charreau, C., 1972. Problèmes posés par l'utilisation agricole des sols tropicaux par des cultures annuelles. (Problems caused by agricultural use of tropical soils under annual crops). Agron. Trop., 27: 905-929 (in French).

Charreau, C. and Nicou, R., 1971. L'amélioration du profil cultural dans les sols sableux et sablo-argileux de la zone tropicale seche Ouest-African et ses incidences agronomiques. (Amelioration of the soil profile in sandy and sandy loam soils of the dry tropical zone of West Africa and its agricultural effects). Agron. Trop., 20: 209-255, 903-978, 1183-1247 (in French).

Collinet, J. and Valentin, C., 1979. Analyse des différents facteurs intervenant sur l'hydrodynamique superficielle. Nouvelles perspectives. Applications agronomiques. (Analysis of different factors affecting the hydrodynamics of superficial soil layers. New perspectives. Agronomic applications). Cah. ORSTOM, Sér. Pedol., 7: 283-326 (in French).

Cunningham, R.K., 1963. The effect of clearing a tropical forest soil. J. Soil Sci., 14: 334-345.

Dalal, R.C., 1982. Organic matter content in relation to the period of cultivation and crop yields in some subtropical soils. Trans. 12th Int. Cong. Soil Sci., New Delhi, India, Vol. 6, p. 59.

Dalal, R.C. and Mayer, R.J., 1986. Long-term trends in fertility of soils under continuous cultivation and cereal cropping in southern Queensland. I. Overall changes in soil properties and trends in winter cereal yields. Aust. J. Soil Res., 24: 265-275.

Daniel, J.G. and Kulasingam, A., 1974. Problems arising from large-scale forest clearing for agricultural use: The Malaysian Experience. Planter (Malaysia), 51: 250-257.

Daniel, H., Jarvis, R.J. and Aylmore, L.A.G., 1988. Hardpan development in loamy sand and its effects upon soil conditions and crop growth. Proc. 11th Conf. Int. Soil Tillage Res. Org. (ISTRO), Edinburgh, U.K., Vol. 1, pp. 233-238.

De Wilde, J.C., 1967. Experiences with Agricultural Development in Tropical Africa, Vols. 1 and 2. John Hopkins, Baltimore, MD, U.S.A., 508 pp.

Derpsch, R., Sidiras, N. and Roth, C.H., 1986. Results of studies made from 1977 to 1984 to control erosion by cover crops and no-tillage techniques in Paraná, Brazil. Soil Tillage Res., 8: 253-263.

Dias, A.C.C.P. and Nortcliff, S., 1985. Effects of two clearing methods on the physical properties of an Oxisol in the Brazilian Amazon. Trop. Agric. (Trin.), 62: 207-212.

Ezumah, H.C., 1983. Agronomic considerations of no-tillage farming. In: I.O. Akobundu and A.E. Deutsch (Editors), No-tillage Crop Production in the Tropics. Int. Plant Protection Centre, Oregon State Univ., Corvallis, OR, U.S.A., pp. 102-110.

Falayi, O. and Lal, R., 1979. Effect of aggregate size and mulching on erodibility, crusting and crop emergence. In: R. Lal and D.J. Greenland (Editors), Soil Physical Properties and Crop Production in the Tropics. Wiley, Chichester, U.K., pp. 87-93.

Fawcett, R.G., Maynard, F.R., Pederson, N.R. and Hannay, J.N., 1988. The effect of traffic and other factors on wheat yields in south Australia. Proc. 11th Conf. Int. Soil Tillage Res. Org. (ISTRO), Edinburgh, U.K., Vol. 1, pp. 257-262.

Ghildyal, B.P., 1978. Effects of compaction and puddling on soil physical properties and rice growth. In: Rice and Soils. Int. Rice Res. Inst., Manila, Philippines, pp. 317-336.

Ghuman, B.S. and Lal, R., 1992. Effects of soil wetness at the time of land clearing on physical properties and crop response on an Ultisol in southern Nigeria. Soil Tillage Res., 22: 1-11.

Gill, K.S. and Lungu, O.I.M., 1988. Evaluation of tillage systems for *Zea mays* grown on a compacted Oxic Paleustalf. Proc. Int. Conf. Dryland Farming, Amarillo/Bushland, TX, U.S.A., pp. 559-561.

Gill, K.S. and Aulakh, B.S., 1990. Wheat yield and soil bulk density response to some tillage systems on an Oxisol. Soil Tillage Res., 18: 37-45.

Grewal, S.S., Singh, K. and Dyal, S., 1984. Soil profile gravel concentration and its effect on rainfed crop yields. Plant Soil, 81: 75-83.

Gulati, I.J., Laddha, K.C., Lal, R. and Gupta, R.P., 1985a. Effect of compacting sandy soil on soil physical properties and yield of guar (*Cyamopsis tetragonslec* L.). Trans. Indian Soc. Desert Technol. and Univ. Centre Desert Studies, 10: 19-23.

Gulati, I.J., Laddha, K.C., Lal, R. and Gupta, R.P., 1985b. Effect of compaction on nitrogen use efficiency and yield of bajra (*Pennisetum tyhoideum*) in a sandy soil of Rajasthan. Trans. Indian Soc. Desert Technol. and Univ. Centre Desert Studies, 10: 85-89.

Gupta, R.P., 1981. Management of highly permeable and low water retentive soils. Indian Agric. Res. Inst., S.P.C. Bull., 1: 1-4.

Håkansson, I., Voorhees, W.B. and Riley, H., 1988. Vehicle and wheel factors influencing soil compaction and crop response in different traffic regimes. Soil Tillage Res., 11: 239-282.

Hosegood, H.P., 1964. The effect of the Holt basin lister (mark X model) on soil structure. East Afr. Agric. For. J., 30: 26-30.

Howson, D.F., 1977. A recording cone penetrometer for measuring soil resistance. J. Agric. Eng. Res., 22: 209-212.

Hudson, N.W., 1976. Soil Erosion. Batsford, London, U.K., 320 pp.

Hulugalle, N.R. and Lal, R., 1985. Root growth of maize in a compacted gravelly tropical Alfisol as affected by rotation with a woody perennial. Field Crops Res., 13: 33-44.

Hulugalle, N.R., Lal, R. and Ter Kuile, C.H.H., 1984. Soil physical changes and crop root growth following different methods of land clearing in western Nigeria. Soil Sci., 138: 172-179.

Hulugalle, N.R., Lal, R. and Ter Kuile, C.H.H., 1986. Amelioration of soil physical properties by Mucuna after mechanized land clearing of a tropical rain forest. Soil Sci., 141: 219-224.

IITA, 1982. Farm inputs and equipment. Int. Inst. Trop. Agric., Ibadan, Nigeria, Ann. Rep., pp. 163-170.

Ike, I.F., 1986. Soil and crop responses to different tillage practices in a ferruginous soil in the Nigerian savanna. Soil Tillage Res., 6: 261-272.

Jackson, I.J., 1986. Climate, Water and Agriculture in the Tropics. Longman, London, U.K., 248 pp.

Kalms, J.M., 1977. Studies of cultivation techniques at Bouaké, Ivory Coast. In: D.J. Greenland and R. Lal (Editors), Soil Conservation and Management in the Humid Tropics. Wiley, Chichester, U.K., pp. 195-200.

Kar, S. and Varade, S.B., 1972. Influence of mechanical impedance on rice seedling root growth. Agron. J., 64: 80-81.

Kar, S., Samui, R.P., Prasad, J., Gupta, C.P. and Sabramanyam, T.K., 1986. Compaction and tillage depth combinations for water management and rice production in low-retentive permeable soils. Soil Tillage Res., 6: 211-222.

Katoch, K.K., Aggarwal, G.C. and Gary, F.C., 1983. Effect of nitrogen, soil compaction and moisture stress on nodulation and yield of soybean. J. Indian Soc. Soil Sci., 31: 215-219.

Kayombo, B. and Lal, R., 1986a. Effects of soil compaction by rolling on soil structure and development of maize in no-till and discing systems on a tropical Alfisol. Soil Tillage Res., 7: 117-134.

Kayombo, B. and Lal, R., 1986b. Influence of traffic-induced compaction on growth and yield of cassava (*Manihot esculenta,* Crantz). J. Root Crops, 12: 19-23.

Kayombo, B., Lal, R. and Mrema, G.C., 1986a. Traffic-induced compaction in maize, cowpea and soybean production on a tropical Alfisol after ploughing and no-tillage: Soil physical properties. J. Sci. Food Agric., 37: 969-978.

Kayombo, B., Lal, R. and Mrema, G.C., 1986b. Traffic-induced compaction in maize, cowpea and soybean production on a tropical Alfisol after ploughing and no-tillage: Crop growth. J. Sci. Food Agric., 37: 1139-1154.

Kemper, B. and Derpsch, R., 1981. Results of studies made in 1978 and 1979 to control erosion by cover crops and no-tillage techniques in Paraná, Brazil. Soil Tillage Res., 1: 253-267.

Kowal, J. and Kassam, A.H., 1976. Energy load and instantaneous intensity of rainstorm at Samaru, northern Nigeria. Trop. Agric. (Trin.), 53: 185-198.

Lal, R., 1978. Influence of within- and between-row mulching on soil temperature, soil moisture, root development and yield of maize in a tropical soil. Field Crops Res., 1: 127-139.

Lal, R., 1979. Physical characteristics of soils of the tropics. Determination and management. In: R. Lal and D.J. Greenland (Editors), Soil Physical Properties and Crop Production in the Tropics. Wiley, Chichester, U.K., pp. 7-44.

Lal, R., 1981. Effects of soil moisture and bulk density on growth and development of two cassava cultivars. In: E.R. Terry, K.A. Oduro and F. Caveness (Editors), Tropical Root

Crops: Research Strategies for the 1980's. Int. Devel. Res. Centre, Ottawa, Ont., Canada, pp. 104-110.

Lal, R., 1982. Effective conservation farming systems for the humid tropics. In: W. Kussow, S.A. El-Swaify and J. Mannering (Editors), Soil Erosion and Conservation in the Tropics. Am. Soc. Agron., Madison, WI, U.S.A., Spec. Publ. 43, pp. 57-76.

Lal, R., 1983. No-till Farming: Soil and Water Conservation and Management in the Humid and Sub-humid Tropics. Int. Inst. Trop. Agric., Ibadan, Nigeria, IITA Mono. 2, 64 pp.

Lal, R., 1984. Soil erosion from tropical arable lands and its control. Adv. Agron., 37: 183-248.

Lal, R., 1985a. Mechanized tillage systems effects on properties of a tropical Alfisol in watersheds cropped to maize. Soil Tillage Res., 6: 149-161.

Lal, R., 1985b. A soil suitability guide for different tillage systems in the tropics. Soil Tillage Res., 5: 179-196.

Lal, R., 1986. Conversion of tropical rainforest: Agronomic potential and ecological consequence. Adv. Agron., 39: 173-264.

Lal, R. and Cummings, D.J., 1979. Clearing a tropical forest. I. Effect on soil and microclimate. Field Crops Res., 2: 91-107.

Lal, R., Wilson, G.F. and Okigbo, B.N., 1978. No-till farming after various grasses and leguminous cover crops in a tropical Alfisol. I. Crop performance. Field Crops Res., 1: 71-84.

Lal, R., Wilson, G.F. and Okigbo, B.N., 1979. Changes in properties of an Alfisol produced by various crop covers. Soil Sci., 27: 377-382.

Lal, R., Lawson, T.J. and Anastase, A.H., 1980. Erosivity of tropical rains. In: M. De Boodt and D. Gabriels (Editors), Assessment of Erosion. Wiley, Chichester, U.K., pp. 143-151.

Macartney, J.C., Northwood, P.J., Dagg, M. and Dawson, R., 1971. The effect of different cultivation techniques on soil moisture conservation and the establishment and yield of maize at Kongwa, central Tanzania. Trop. Agric. (Trin.), 48: 9-23.

Makarim, A.K., Cassel, D.K. and Wade, M.K., 1988. Effects of land reclamation practices on physical properties of an acid infertile Oxisol. Soil Technol., 1: 195-207.

Martínez, M.B. and Lugo-López, M.A., 1953. Influence of subsoil shattering and fertilization on sugar cane production and soil infiltration capacity. Soil Sci., 75: 307-315.

Maurya, P.R., 1988. Performance of zero tillage in wheat and maize production under different soil and climatic conditions in Nigeria. Proc. 11th Conf. Int. Soil Tillage Res. Org. (ISTRO), Edinburgh, U.K., Vol. 2, pp. 769-774.

Maurya, P.R. and Lal, R., 1979. Effects of bulk density and soil moisture on radicle elongation of some tropical crops. In: R. Lal and D. J. Greenland (Editors), Soil Physical Properties and Crop Production in the Tropics. Wiley, Chichester, U.K., pp. 339-347.

Maurya, P.R. and Lal, R., 1980. Effects of no-tillage and ploughing on roots of maize and leguminous crops. Exp. Agric., 16: 185-193.

May, B.A., 1988. Agricultural engineering in Third World countries. The Agric. Eng., 43: 83-92, 112-126.

Mensah-Bonsu and Obeng, H.B., 1979. Effects of cultural practices on soil erosion and maize production in the semi-deciduous rainforest and forest-savanna transitional zones of Ghana. In: R. Lal and D.J. Greenland (Editors), Soil Physical Properties and Crop Production in the Tropics. Wiley, Chichester, U.K., pp. 509-519.

Morgan, R.P.C., 1974. Estimating regional variations in soil erosion hazard in Peninsular Malaysia. Malay. Nat. J., 28: 94-106.

Mott, J., Bridge, B.J. and Arndt, W., 1979. Soil seals in tropical tall grass pastures of northern Australia. Aust. J. Soil Res., 30: 483-494.

Nicou, R. and Chopart, J.L., 1979. Root growth and development in sandy and sandy clay soils of Senegal. In: R. Lal and D.J. Greenland (Editors), Soil Physical Properties and Crop

Production in the Tropics. Wiley, Chichester, U.K., pp. 375-384.

Nicou, R. and Charreau, C., 1980. Mechanical impedance related to land preparation as a constraint to food production in the tropics (with special reference to fine sandy soils in West Africa). In: Soil-related Constraints to Food Production in the Tropics. Int. Rice Res. Inst., Manila, Philippines, pp. 371-388.

Nicou, R., Seguy, L. and Hadad, G., 1970. Comparaison de l'enracinement de quatre variétés de riz pluvi en presence ou absence du travail du sol. (Comparison of the root development of four varieties of rainfed rice in presence or absence of soil tillage). Agron. Trop., 25: 639-659 (in French).

Nogueira, S., Dos, S.S. and Manfredini, S., 1983. Influencio do compactacáo do solo no desenvolvimento da soja. (Influence of soil compaction on the development of soybean). Pesq. Agrop. Bras., 18: 973-976 (in Portuguese).

Northwood, P.J. and Macartney, J.C., 1971. The effect of different amounts of cultivation on the growth of maize on some soil types in Tanzania. Trop. Agric. (Trin.), 48: 25-33.

Nye, P.H. and Greenland, D.J., 1964. Changes in soil after clearing tropical forest. Plant Soil, 21: 101-112.

Ogunremi, L.T., Lal, R. and Babalola, O., 1986a. Effects of tillage methods and water regimes on soil properties and yield of lowland rice from a sandy loam soil in southwest Nigeria. Soil Tillage Res., 6: 223-234.

Ogunremi, L.T., Lal, R. and Babalola, O., 1986b. Effects of tillage and seedling methods on soil physical properties and yield of upland rice for an Ultisol in southeast Nigeria. Soil Tillage Res., 6: 305-324.

Ohu, J.O. and Folorunzo, O.A., 1989. The effect of machinery traffic on the physical properties of a sandy loam soil and on the yield of sorghum in northeastern Nigeria. Soil Tillage Res., 13: 399-405.

Ojeniyi, S.O., 1989. Investigation of ploughing requirement of the establishment of cowpea (Vigna unguiculata). Soil Tillage Res., 14: 177-184.

Ojeniyi, S.O., 1990. Effect of bush clearing and tillage methods on soil physical and chemical properties of humid tropical Alfisols. Soil Tillage Res., 15: 269-277.

Okigbo, B.N., 1965. Effects of mulching and frequency of weeding on the performance and yield of maize. Nigerian Agric. J., 2: 7-9.

Okigbo, B.N., 1969. Maize experiments on the Nsukka Plains: III. Effects of different kinds of mulch on the yield of maize in the humid tropics. Agron. Trop., 28: 1036-1047.

Oni, K.C. and Adeoti, J.S., 1986. Tillage effects on differently compacted soil and on cotton yield in Nigeria. Soil Tillage Res., 8: 89-100.

Onofiok, O.E., 1988. Spatial and temporal variability of some soil physical properties following tillage of a Nigerian Paleustult. Soil Tillage Res., 12: 285-298.

Onofiok, O.E., 1989. Effect of soil compaction and irrigation interval on the growth and yield of cowpea on a Nigerian Ultisol. Soil Tillage Res., 13: 47-55.

Onwualu, A.P. and Anazodo, U.G.N., 1989. Soil compaction effects on maize production under various tillage methods in a derived savannah zone of Nigeria. Soil Tillage Res., 14: 99-114.

Onwueme, I.C., 1978. The Tropical Tuber Crops: Yams, Cassava, Sweet Potato and Cocoyams. Wiley, New York, U.S.A., 234 pp.

Orellana, M., Barber, R.G. and Diaz, O., 1990. Effects of deep tillage and fertilization on the population, growth and yield of soya during an exceptionally wet season on a compacted sandy loam, Santa Cruz, Bolivia. Soil Tillage Res., 17: 47-61.

Ortolani, A.F., Coan, O. and Salles, H.C., 1982. Influencia de compactacáo do solo no desenvolvimento da soja (Glycine max [L] Merril). (Influence of soil compaction on the development of soybean (Glycine max [L] Merril). Engenharia Agricola (Brazil), 6: 35-42 (in Portuguese).

Osuji, G.E., 1988. The effects of vehicular traffic on the growth pattern of maize on a tropical Alfisol. Proc. 11th Conf. Int. Soil Tillage Res. Organ. (ISTRO), Edinburgh, U.K., pp. 311-316.

Panabokke, C.R., 1967. Soils of Ceylon and Use of Fertilizer. Metro, Colombo, Sri Lanka, 151 pp.

Pereira, H.C., 1973. Landuse and Water Resources. Cambridge Univ. Press, Cambridge, U.K., 246 pp.

Pereira, H.C. and Beckley, V.R.S., 1952. Grass establishment on an eroded soil in a semi-arid African reserve. Emp. J. Exp. Agric., 21: 1-14.

Pereira, H.C. and Jones, P.A., 1954. A tillage study in Kenya coffee. II. The effect of tillage practices on the structure of soil. Emp. J. Exp. Agric., 22: 323-331.

Pereira, H.C. and Thomas, D.B., 1961. Productivity of tropical semi-arid thornshrub country under intensive management. Emp. J. Exp. Agric., 29: 269-286.

Pereira, H.C. and Dagg, M., 1967. Effect of tied ridges, terraces and grass leys on a lateritic soil in Kenya. Exp. Agric., 3: 89-98.

Rami Reddy, S. and Hukkeri, S.B., 1983. Effect of tillage practices on irrigation requirement, weed control and yield of lowland rice. Soil Tillage Res., 3: 147-158.

Roth, C.H., Meyer, B., Frede, H.G. and Derpsch, R., 1988. Effect of mulch rates and tillage systems on infiltrability and other soil physical properties of an Oxisol in Paraná, Brazil. Soil Tillage Res., 11: 81-91.

Saini, S.K. and Singh, J.N., 1980. Effect of soil compaction and planting depth on seedling emergence in soybean (*Glycine max* [L] Merril). Seed Res., 8: 127-131.

Sanchez, P.A., 1976. Properties and Management of Soils in the Tropics. Wiley, New York, NY, U.S.A., 618 pp.

Sanchez, P.A., 1981. Soils of the humid tropics. Studies in Third World Societies, 4: 347-410.

Santamaria, F., 1965. Geographical distribution of the "arrenife" (pavement) horizon in Venezuela. Bull. Soc. Venez. Cienc. Nat., 25: 350-354.

Segalen, P., 1969. Rearrangement of materials in soil and the formation of stone lines in Africa. Cah. ORSTOM, Ser. Pédol., 7: 113-131.

Seubert, C.E., Sanchez, P.A. and Valverde, C., 1977. Effect of land clearing methods on soil properties of an Ultisol and crop performance in the Amazon jungle of Peru. Trop. Agric. (Trin.), 54: 307-320.

Shannon, D.A., 1983. A study of factors responsible for variable growth of soybeans in the southern Guinea savannah of Nigeria. Ph.D. thesis, Cornell Univ., Ithaca, NY, U.S.A.,

Sharma, R.B. and Mbogoni, J.D.J., 1988. Residual effect of chiselling on growth and water status of rainfed soybeans. Proc. 11th Conf. Int. Soil Tillage Res. Org. (ISTRO), Edinburgh, U.K., Vol. 2, pp. 857-862.

Singh, A., Singh, J.N. and Tripathi, S.K., 1971. Effect of soil compaction on the growth of soybean (*Glycine max* [L] Merril). Indian J. Agric. Sci., 41: 422-426.

Singh, N.T., Patel, M.S., Singh, R. and Vig, A.C., 1980. Effect of soil compaction on yield and water use efficiency of rice in a highly permeable soil. Agron. J., 72: 499-502.

Smyth, A.J. and Montgomery, R.J., 1962. Soils and Land Use in Central Western Nigeria. Govt. Printer, Ibadan, Nigeria, 265 pp.

So, H.B., Cook, G.D. and Dalal, R.C., 1988. Structural degradation of Vertisols associated with continuous cultivation. Proc. 11th Conf. Int. Soil Tillage Res. Org. (ISTRO), Edinburgh, U.K., Vol. 1, pp. 123-128.

Soane, B.D., 1986. Process of soil compaction under vehicular traffic and means of alleviating it. In: R. Lal, P.A. Sanchez and R.W. Cummings (Editors), Land Clearing and Development in the Tropics. Balkema, Rotterdam, Netherlands, pp. 265-283.

Somapala, H. and Willatt, S.T., 1979. Effect of plant age on root penetration. In: R. Lal and D.J. Greenland (Editors), Soil Physical Properties and Crop Production in the Tropics. Wiley, Chichester, U.K., pp. 349-361.

Thomas, M.F., 1974. Tropical Geomorphology: A Study of Weathering and Land Form Development in Warm Climates. Wiley, New York, NY, U.S.A., 332 pp.

Tullberg, J.N., 1988. Controlled traffic in sub-tropical grain production. Proc. 11th Conf. Int. Soil Tillage Res. Org. (ISTRO), Edinburgh, U.K., Vol. 1, pp. 323-327.

Van der Weert, R., 1974. Influence of mechanical forest clearing on soil conditions and resulting effects on root growth. Trop. Agric. (Trin.), 51: 325-331.

Vine, P.N., Lal, R. and Payne, D., 1981. The influence of sands and gravels on root growth of maize seedlings. Soil Sci., 131: 124-129.

Willcocks, T.J., 1981. Tillage of clod-forming sandy loam soils in the semi-arid climate of Botswana. Soil Tillage Res., 1: 323-350.

Willcocks, T.J., 1984. Tillage requirements in relation to soil type in semi-arid rainfed agriculture. J. Agric. Eng. Res., 30: 327-336.

Soil Compaction in Crop Production
B.D. Soane and C. van Ouwerkerk (Eds.)
1994 Elsevier Science B.V. All rights reserved.   317

CHAPTER 14

Responses of Forest Crops to Soil Compaction

E.B. WRONSKI[1] and G. MURPHY[2]

[1]Dr. Ed Wronski and Associates, Canberra, Australia
[2]Forest Research Institute, Roturua, New Zealand

SUMMARY

Experiments conducted under controlled conditions in greenhouses and observations of tree growth in the field demonstrate that the compaction of soils during forestry operations reduces the rate of establishment of natural regeneration, reduces tree growth for periods spanning at least a decade and can have deleterious effects on tree form. The intensity of compaction is greatest during harvesting, but by judicious choice of harvesting systems, machines and running gear, based on considerations of slope and soil water conditions, the impact on the soil can be reduced. Other options available to reduce the intensity of compaction are the laying of slash beds on the main extraction trails, the grading of terrain into areas suitable for logging during the wet and dry periods of the year, based on soil type and drainage conditions, and, if necessary, the rehabilitation of soils by cultivation or deep ripping and the application of fertilizer.

INTRODUCTION

Forests provide man with edible fungi, honey, traditional and non-traditional medicinal plants, undergrowth for grazing domestic animals and, most importantly, wood products. It is now recognised, however, that the productivity of forests can be influenced by soil compaction resulting from vehicle traffic. Some research has been conducted on the effects of compaction after logging on the minor uses of forests, such as the effects on grasses and grazing (Garrison and Rummell, 1951), but most studies have focussed on the effect of soil disturbance and compaction on wood production. It is the intention in this chapter to do likewise.

The significance of compaction in forest management at a given site, depends on the maturity of the local forestry industry. The cutting down of wild or native forests occurs early in the development of the industry, generally in an environment of excess wood supply, and mistaken practices leading to site degradation may have little commercial impact on subsequent timber production. In time, potential limits to the wood supply become apparent, and an element of forest management is introduced. The forest becomes a regulated production forest. However, the cost of timber harvesting generally dominates the cost of wood production. It is difficult to justify even higher harvesting costs in order to reduce the risks of soil compaction.

In many countries, potential shortages of timber are now evident and as the value of forest products increases, productivity losses associated with crude harvesting practices can lead to significant monetary losses. Accordingly, the intensity of forest management increases, plantations are established and forest scientists take greater account of factors such as soil compaction in attempts to maintain or even increase site productivity.

There are differences between forest and most agricultural crops that make the effects of compaction in forest crops less obvious, more difficult to study and thus more difficult to overcome in management practices. In contrast to most agricultural systems, the area of the ground compacted by a forest vehicle's running gear is small in relation to the dimensions of the root system and canopy of the mature crop. Thus, if nutrient and water availability are not limiting factors to growth, as sometimes occurs on fertile sites in temperate regions (Turnbull et al., 1988), the productivity of the site may not be affected by compacted trails once canopy closure has occurred.

Assessing the impact of compaction during thinning on tree growth also poses special difficulties, in that it becomes necessary to separate the effects of the positive growth response due to reduced competition, from the possible negative effects of soil compaction. Moreover, compaction of soil during thinning operations occurs around an undisturbed root system generally extending well beyond the zone of compaction, which may mean that secondary effects on nutrient and water uptake are less. In addition, significant root grafting may occur in some stands (Wood and Bachelard, 1970), which is likely to mask localised compaction effects through the transfer of water and nutrients between rooting systems of felled and intact trees.

Given the above difficulties, the statistical problems associated with measuring growth differences between groups of individual trees and the presence of other factors which influence growth, forest scientists have often assessed the effects of soil compaction indirectly, through measurements of soil physical conditions in the field and observed effects of soil compaction on the growth of seedlings measured in greenhouses under controlled laboratory conditions. The control of the environment which is possible with this approach allows the investigator to monitor treatment effects, while holding all other variables constant. However, the applicability of such results to field conditions presents difficulties (Corns, 1988).

RESPONSES OF FOREST CROPS TO SOIL COMPACTION AND DISTURBANCE

Greenhouse pot trials

Studies of compaction effects on tree seedlings in the laboratory have shown that the germination and rate of seedling establishment on cores extracted from disturbed soils are generally suppressed. Typically, rates of establishment on cores of compacted field soils have been found to be only 60-80% of those on normal soils (Foil and Ralston, 1967; Hatchell et al., 1970). Such studies have also shown that some species are less affected by soil compaction than others. For example, Minore et al. (1969) found that the roots of lodgepole pine (*Pinus contorta*), Douglas fir (*Pseudotsuga menziessi*), and red alder (*Alnus rubra*) can grow in soil bulk densities that prevent the growth of Sitka spruce (*Picea sitchensis*), western red cedar (*Thuja plicata*) and western hemlock (*Tsuga heterophylla*).

In other studies, care has been taken to relate the growth response of seedlings to compaction under actual field conditions. Corns (1988) observed a significant reduction (up to 85% in total seedling weight) in the growth of white spruce (*Picea glauca*) and lodgepole pine seedlings with increasing bulk density of soil compacted to densities representative of various field conditions, viz., undisturbed control, clear-cuts 5-10 years after logging, and clear-cut roads and extraction trails immediately after logging. Growth reductions were more evident in lodgepole pine than in white spruce.

In a similar study, Sands et al. (1979) investigated the relationship in the field between root growth of radiata pine (*Pinus radiata*) and soil bulk density and soil strength. The interpretation of these data was assisted by greenhouse studies on the effect of soil compaction on root configuration and growth of seedlings (Sands and Bowen, 1978). They found significant reductions in the fresh and dry weights of roots and tops, root volumes and top height as bulk density increased from 1.35 to 1.60 Mg m^{-3}. Soil compaction also caused reductions in the length of the main root axis, first and second order laterals, and total root length. The number and diameter of first order laterals were increased by compaction. This response in root activity was similar to that observed in the field, which led to the conclusion that soil compaction during harvesting was a cause of reduced growth during the second and subsequent rotations.

In a variation of the normal pot trial, Wästerlund (1985) examined the effect of static pressure of 0-400 kPa applied for 10 min (equivalent to pressures applied by Swedish logging machines) to a range of soil cores prior to planting with Norway spruce (*Picea abies*) and Scots pine (*Pinus sylvestris*). After about 20 days, the lengths of primary roots, relative to the control, were found to show an almost linear decrease with increase of applied compression (Fig. 1). Norway spruce was, however, more impeded than Scots pine and secondary roots appeared to be more impeded than the primary roots.

Some care is needed when interpreting responses of seedlings grown in soil

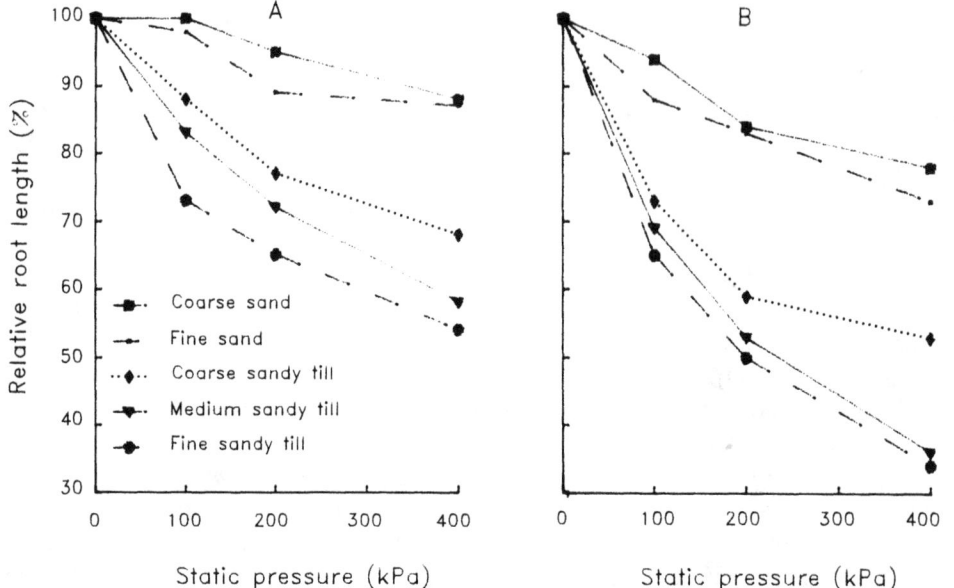

Fig. 1. Relative root length of pine (A) and spruce (B) seedlings about 20 days after sowing in five soils subjected to prior static pressure of 0-400 kPa (control = 100%) (from Wästerlund, 1985).

cores or reconstituted soils taken from the field, because associated with increases in bulk density during logging may be other changes in soil properties which effect growth. For example, Williamson (1990) examined seedling growth rates on mixed and compacted Xanthozem soils from Tasmania, Australia. He found that there was a complex interaction of many soil factors, both physical and chemical, which affected seedling growth. Growth rates on such soils were more dependent on organic matter and available nitrogen contents than on the level of compaction. Such soil attributes declined markedly with depth in the soil profile. This explained why even minor scuffing, involving the removal of surface organic matter during harvesting, rather than increase in bulk density, was causing difficulties for natural regeneration.

In summary, greenhouse pot trial experiments have demonstrated that compaction and relatively minor disturbance of the soil profile can cause serious reductions in seedling germination, establishment and growth. However, some species appear to be more sensitive to compaction than others.

Early growth of natural regeneration in clear-cuts

The poor rates of establishment and reduced growth observed on compacted soil in laboratory studies are also evident in naturally regenerated clear-cut. In the south-eastern United States, Hatchell et al. (1970) categorized compaction levels at forest sites into three classes: (1) undisturbed; (2) secondary extraction trails; (3) primary extraction trails (>10 passes). Only on secondary trails was the initial establishment of seedlings equal to or greater than that on undisturbed soils, and shoot growth was retarded on all degrees of disturbed soil during the first two years. By the third growing season, seedlings on primary trails were half to one third the dry weight of seedlings growing in undisturbed soil.

In naturally regenerated stands, the reduced growth observed in early years tends to persist for at least a decade. Terry and Campbell (1981) measured the growth of loblolly pine, six years after planting from seed, on four sites that had been disturbed and compacted by harvesting traffic but had received fertilization and bedding treatment. On all sites, height growth decreased when rut depth was >200 mm, with maximum growth reduction of 53% despite some site rehabilitation. Similarly, Smith and Wass (1979) observed growth reduction of 12-15% for Englemann spruce (*Picea englemanii*) and sub-alpine fir (*Abies lasiocarpa*) on and adjacent to contour extraction trails. Differences in total tree height between different areas reflected two major influences on growth: promptness of stocking (age) and site quality (growth rate). Trees on undisturbed areas were 1.3-8 years older than on extraction trails. However, the response to soil disturbance was not always negative, for on moderately coarse acid soils with cool, northerly aspects, disturbance led to 20% overall enhancement in height, indicating that, in certain circumstances, soil compaction can result in other changes to the physical and chemical environment that can enhance growth.

Froehlich et al. (1986) measured the growth response of 9-18-year old ponderosa pine (*Pinus ponderosa*) and 10-13-year old lodgepole pine to soil compaction in central Washington. For ponderosa pine at the mean increase in soil bulk density, the total height, diameter, and volume growth were reduced by 5, 8 and 20%, respectively. No significant correlation could be found between increasing intensity of compaction and growth of lodgepole pine.

Early growth of planted seedlings in clear-cuts

In the case of planted seedlings, the production losses associated with poor establishment are largely avoided and any growth losses observed due to compaction reflect changes in physical conditions that relate mainly to root growth. For example, Murphy (1983) observed that, though the survival rate of radiata pine seedlings planted on and off trails in New Zealand was similar after 4 years (75.5 and 78.0%, respectively), the height difference between on- and off-trail trees was 1.0 m (32%). However, there was a trend for the survival rate and height growth of trees planted on trails to decrease as the length of the trail and soil penetration resistance increased. Senyk (1990) reported similar findings for hemlock seedlings in Western Canada.

Similarly, Cochran and Brock (1985) found in Oregon that the height of ponderosa pine seedlings five years after planting was negatively correlated with increasing bulk density. Although bulk density accounted for less than half the variation in height growth, an 80% reduction in height of seedlings was observed between bulk densities of 0.92 and 1.12 Mg m^{-3}.

Williamson (1990) observed in pot trials that the effects of compaction on growth could be compounded by the removal of nutrients in organic matter during logging. Skinner et al. (1989), following field trials involving simulated logging damage (forest litter removal, topsoil removal and compaction), concluded that changes in physical conditions brought about by the removal of organic matter could compound growth losses by compaction. Relative to the undisturbed controls, tree volumes at age 4 years were less by about 25% where litter had been removed by hand, 65% where litter had been removed by machine, 70% where the topsoil had been removed and the subsoil compacted by two passes of a loader, and 80% where the topsoil was removed and the subsoil compacted by eight passes of a loader. Losses in productivity during the first 4 years resulted from combinations of nutrient loss through topsoil removal, changes in soil penetration resistance, and less favourable soil temperature regimes owing to the absence of forest litter.

Shishiuchi (1990) found that the height of Japanese cedar and Japanese larch seedlings planted on tractor skid trails was less than on undisturbed control areas 4-7 years after logging. Where the surface soil on the skid trail was absent, height growth was reduced by 38% for Japanese cedar and 26% for Japanese larch. However, if the surface soil was present, growth reductions for Japanese cedar were only 10%.

In certain situations, soil disturbance can have neutral or positive effects on the early growth of seedlings. Firth (1989) described the results from three extensive radiata pine growth trials established on clay loam, clay, and pumice soils in New Zealand. Growth on the undisturbed sites was almost the same as where disturbance and compaction were the most severe but, unexpectedly, the best growth in stem volume occurred where a moderate amount of disturbance and compaction had taken place (Fig. 2).

Growth responses after extraction thinning

The growth responses after thinning operations which cause compaction and deep ruts, will depend on the interaction between effects of soil compaction and root damage (Froehlich, 1972; Wronski, 1984a). Before considering such effects, it is helpful to appreciate the morphology and growth of the tree root system.

Numerous investigations indicate that the majority of fine absorbing roots of trees occur in the top 200 mm of soil (Bowen, 1964; Moir and Bachelard, 1969). On poorly draining soils, such as peat, up to 90% of the total root biomass occurs in the top 100 mm of soil. On fertile well-drained sites, roots tend to penetrate deeper than on infertile sites. However, on coarse infertile soils with poor water

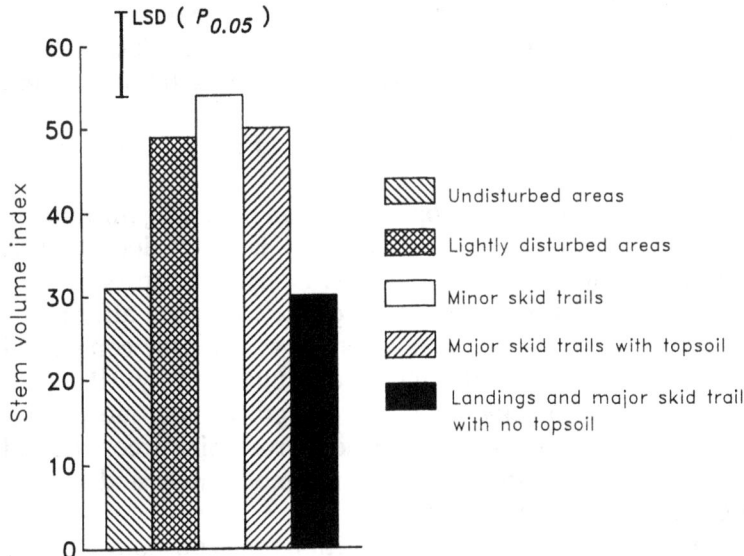

Fig. 2. Effect of soil disturbance and compaction on stem volume index of 4-year old radiata pine seedlings (from Firth, 1989).

retention properties, deeper sinker roots are also required to gain access to sufficient soil water in summer (Frazer and Gardiner, 1967; Sutton, 1969; Rytter, 1989). In addition, there appears to be some genetic influence on the depth of rooting. For example, the rooting system of spruce generally has a flatter plate structure than that of pines (Sutton, 1969).

Given the above relations between root morphology and soil conditions, the effect of compaction and soil disturbance is likely to be less on fertile than on infertile sites and greater in forests with genetic tendencies for a shallow root system. Forests planted on poorly draining soils are likely to suffer much greater damage to their root system during thinning because the root system is shallow and because such soils also have a lower bearing capacity.

Relationships have been developed to estimate the proportion of roots cut off by ruts as a function of tree density and the spacing of extraction tracks (Wronski, 1984a). However, direct evidence demonstrating the effect of such damage to the root system, as distinct from compaction effects, on tree growth is meagre. In a study of the effects of root pruning on seedlings, Nambiar (1983) observed it was only when more than half of the first order laterals of seedlings were removed that there was any significant ($>10\%$) reduction in growth.

Trees have a very high renewal rate for their fine root system (which makes up more than half the total root biomass), which for some species has been estimated to be 3-6 times a year (Agren et al., 1980; Van Praag et al., 1988; Van Veen et al., 1989). Thus, trees appear to have a root system larger than that required to maintain normal growth, presumably as insurance against occasional adverse conditions and to assist in competition with neighbours. This suggests that, while damage to the root system of individual trees during thinning might cause a temporary reduction in growth, the overall effect of root damage on growth may be small in view of the opportunistic nature of the root system and the tendency for root grafting to occur.

Evidence that compaction during thinning operations can have a significant and long-lasting effect on growth (Cochran and Brock, 1985), is better documented. Froehlich (1979) reported growth losses for residual Douglas fir trees of 14% where the soil was moderately compacted (11-40% of the root zone affected by ≥10% increase in soil bulk density) and 30% where it was heavily compacted (>40% of the root zone affected by ≥10% increase in bulk density). Residual western hemlock trees showed a 14% reduction in growth on moderately compacted soil. Froehlich (1979) also found that ponderosa pine trees showed a reduction in growth rate, over a 16-year period on moderately and heavily compacted soils, of 6% and 12%, respectively.

A 10-30% growth reduction for trees nearest extraction roads on stony infertile soils in Sweden has been observed by Wästerlund (1983). Growth reduction was greater for Norway spruce than Scots pine, but the growth reduction was less on good sites where roots were distributed to greater depths and were less damaged. In a similar study, Moehring and Rawls (1970) observed that it was only when skid trails passed immediately adjacent to three sides of trees that any reduction in growth of individual trees occurred.

Long-term effects on tree growth

The long-term effects of soil compaction and disturbance on tree growth depend on many factors: the intensity of the disturbance, the rate of recovery of the soil, the degree of damage to the root system, the tree species concerned, silvicultural practices, etc. The concept of "long-term", itself, will differ according to whether one is considering a fast-growing plantation in the tropics which is harvested on a ten-year cycle, or a natural temperate-zone forest managed on a 200-year rotation.

In relation to temperate-zone forests, long-term losses in timber volume of 6-15% (Froehlich, 1989) and 40-46% (Perry, 1964; Power, 1974) have been observed on skid trails compared to undisturbed sites. Douglas fir trees growing on the extraction trails in the Coastal Range of Oregon were found by Wert and Thomas (1981) to take 4 years longer to reach breast height than did trees growing on undisturbed areas. Height-age relationships for the extraction trails, "transition" zones (within a 3-m band either side of the extraction trail) and undisturbed areas, did not show significant differences between the transition zone and undisturbed areas. However, extraction trails produced significantly shorter trees (about 1 m). The difference in height remained consistent over a twenty-year period. On an equal area basis, they found that extraction trails produced only 26% of the volume and 59% of the number of trees compared with undisturbed areas 32 years after logging. Wert and Thomas also found 75% of the volume and 83% of the number of trees growing in the "transition" zone. These reductions resulted in an overall volume loss of 12% when applied over the entire area.

Not all disturbance, however, results in a growth loss. For example, Pfister (1969) reported that white pine trees below single-lane, out-sloped roads, grew faster than those within the stand. Growth may have been enhanced due to increased water-supplying capacity in the side-cast material.

Effects on tree form and wood quality

Although much of the research has concentrated on the effect of soil compaction and disturbance on forest crop productivity, some work has begun on the effect these have on the quality of the crop. Firth and Murphy (1989) found that the form of radiata pine trees growing "on-trail" on a clay loam soil in New Zealand was inferior to trees growing just "off-trail". An assessment of the trees when they were four years old showed that 47% of those "on-trail" were malformed to some extent, whereas 37% of those "off-trail" were malformed. By far the most prevalent type of malformation was severe butt sweep or toppling; 37% "on-trail" and 19% "off-trail".

Recent pilot studies of root rot in three newly thinned, 60-year-old Norway spruce stands on good sites in southern Sweden, showed a remarkable increase in the frequency of root rot as the distance to the extraction trail decreased (Berhardsson and Martinsson, 1986, quoted by Wästerlund, 1989a). Should such infections spread to the tree stem, reductions in wood quality may occur which could substantially detract from the value of the tree crop.

HARVESTING SYSTEMS IN FORESTS AND THEIR IMPACT ON THE SOIL

Machine and logging systems

A wide range of machine and logging systems has been developed to operate in the forest and each imposes a varying degree of intensity and distribution of compaction (Figs. 3 and 4). These systems can be divided into three main categories: (1) cable-based systems (logs suspended beneath cables); (2) ground-based skidding systems (logs dragged along the ground); (3) forwarder-based systems (log load carried by vehicle). Murphy (1984) has undertaken an extensive review of the relative impact of these systems and has concluded that tractor, skidder and forwarder operations typically expose the subsoil and compact 20-40% of the harvested area during clear-fell operations.

Fig. 3. A forwarder loading pine in an Australian plantation forest.

Fig. 4. A rubber-tyred skidder operating on soil of low strength where the fitting of wide tyres will reduce but not prevent compaction.

Disturbance tends to increase with the steepness of the country and wetness of the soil because once forward motion by the vehicle becomes difficult due to excessive disturbance, the disturbed area is bypassed.

Cable systems result in less compaction on the main extraction paths than skidder operations, but require larger landings and more access tracks. The overall area of disturbance of the harvested area is 5% less than for skidder operations (Miller and Sirois, 1986). The actual area of compacted soil may be significantly greater than the expressions of surface area given above as stresses in the soil during vertical compression spread laterally at depth (Garrison and Rummel, 1951; Wronski 1984a).

In partial or selection cutting, the amount of disturbance is at a maximum. Once removal intensities become sufficiently high for operators to pull the logs with a winchline rather than drive to them, disturbance declines. During thinning operations the area of deep disturbance is reduced to about 9-13% though somewhat higher levels of disturbance (23%) occur during clear felling operations (Wronski, 1984a).

The type of machine used can also influence the extent of compaction and disturbance. Larger machines tend to create a greater area of disturbance than smaller machines (Haupt, 1960). In wet conditions, the area of disturbance caused by tracked skidders is less than half the area disturbed by wheeled machines and flexible-tracked skidders cause less disturbance than do crawler tractor skidders (Froehlich et al., 1980; Murphy, 1982; Burger et al., 1985; Reisinger et al., 1988). However, in dry soil conditions, there appears to be little difference in effect between machines (Jakobsen and Moore, 1981).

During forest harvesting, the highest ground pressures are usually applied by the logging vehicle which transports the timber out of the forest. Ground contact pressures for these machines range from 30-200 kPa (Table 1). However, actual ground contact pressures of skidders can be at least twice those in Table 1, owing to the redistribution of forces when pulling logs up and down steep slopes (Lysne and Burditt, 1983). Moreover, the actual pressures transmitted to the

TABLE 1

Ground contact pressure for various logging systems (from Greacen and Sands, 1980)

System	Ground contact pressure (kPa)
Cable logging	0
Flexible-tracked skidder	30-40[1]
Crawler tractor	50-60[1]
Rubber tyre skidder	50-90[1]
Forwarder	90-200

[1]Also exerts high ground shear forces.

ground by tracked machines on flat ground may be up to three times greater than the average ground contact pressure, depending on the spacing of the road wheels supporting the track, the size of the track plates and the strength of the soil (Rowland, 1972).

In the case of skidders, or even of forwarders, when travelling upslope, large shear forces are applied to the ground. It has been demonstrated by Raghavan et al. (1977) that increases in soil bulk density at 25-45% wheel slip, were equivalent to those obtained by doubling the static loading. Such wheel slip is relatively common in skidder operations, implying that increases in soil bulk density and strength in upper soil horizons caused by skidders may be comparable to those caused by a forwarder, despite the lower ground contact pressure exerted by skidders.

The effect of such stresses on soil strength and root growth depends on the season during which operations are undertaken. In temperate climates, root growth tends to be minimal during winter, but in response to rising soil temperature and plant hormonal factors, increases to a maximum during spring and early summer (Sutton, 1969). Typically, soil compacted during winter operations in the absence of frost increases in strength by 200-300% during spring and summer due to a decline in soil moisture (Canarache, 1990). The strength of soil, in structural equilibrium with the forces applied by a vehicle's running gear, is about 10 times the applied ground contact pressure of the vehicle (Greacen and Sands, 1980). Thus, in spring and summer, i.e., the period of maximum root growth, the strength of a soil could be expected to be about 30 times the applied ground contact pressure of the vehicle's running gear.

Given that root growth begins to be inhibited at a soil strength of 2 MPa (Greacen and Sands, 1980), applied ground contact pressures during winter operations of as low as 60 kPa could be expected to cause reduced tree growth,

TABLE 2

Mechanized operations in Australian and New Zealand forests (from Greacen and Sands, 1980)

Site	Operation	Number of passes	Ground contact pressure (kPa)
Plantation	Ploughing and ripping	1	50-90
	Planting	1	50-90
	Thinning	6-300	0-200
	Clear felling	2-300	0-200
Native forest	Selection logging	2-50	50-80
	Clear felling	2-300	0-80

while for summer operations, ground contact pressures >200 kPa would be required. This is consistent with the observations of Wästerlund (1988, 1989a) that significant reductions in growth occurred in young trees whose root systems were compressed by vehicles with applied ground contact pressures of 60-90 kPa.

Both the intensity and the depth of soil compaction tend to increase with the number of vehicle passes, the greatest increase in compaction occurring during the first few passes. On primary skid trails (>10 passes), compaction can extend to depths >300 mm, but on secondary trails compaction does not seem to extend deeper than 250 mm on all but highly structured profiles (Moehring and Rawls, 1970; Wronski, 1984a; Gent and Morris, 1986). The intensity of compaction depends on the type of forestry operation, but in general the most intense compaction is likely to occur during the harvesting operation on the main extraction trails (Table 2).

Effect of forest traffic on soils

A soil is said to be compacted when it has experienced a reduction in pore space due to compression. It is considered to be puddled when a loss in pore space has occurred consequent to the application of both compressive and shear forces (Bodman and Rubin, 1948). The soil bulk density at which soil strength approaches 2 MPa and root growth is inhibited, varies between soil types. For most soils this occurs at a bulk density of about 1.5 Mg m^{-3}, but at clay contents >30%, root growth can become inhibited at bulk densities as low as 1.2 Mg m^{-3} (Canarache 1990; Williamson, 1990). On wet clay soils which develop ruts readily during trafficking, there may be no increase in bulk density following traffic. On some sub-plastic clay soils there may actually be increases in water content and porosity after trafficking, but structural changes occur which manifest themselves as a decrease in soil water potential and reduction in air permeability. On such soils, compaction may be greater during the drier period of the year (Williamson, 1990). Such soils also experience shrinkage during drying and a marked increase in soil strength (Dawidowski and Koolen, 1987).

It is well known that the forest soils which potentially have the highest degree of compactness and strength, are those which contain an extreme range of mineral partical size distribution, e.g., sandy clay loams and sandy clays with a high percentage of coarse sand (Sands and Bowen, 1978; Gent and Morris, 1986). Forest soils of low compactability are those with a very high clay content or a large percentage of gravel and rock fragments or organic matter, or fine soils with non-expanding clay minerals and low soil water content at the liquid limit (Froehlich et al., 1980; Howard et al., 1981). Soils with a high undisturbed bulk density in the field tend to have lower soil water content at the optimum level for compaction and also tend to be more compactable (Howard et al., 1981).

Laboratory studies have shown that sandy soils lacking in clay, puddle to greater bulk densities than silty clays (Bodman and Rubin, 1948), but clay soils, e.g., silty clay loams and loam, puddle more easily than coarse soils (Gent and Morris, 1986). Such findings are consistent with the field observations of Moehring (1970), that soil disturbance tends to be more pronounced on clay soils than on light sandy soils and that the effects of such disturbance tend to be less permanent on clay soils.

In areas having well developed mature soil profiles, e.g., much of Australia, coarse material often overlies a clay subsoil which tends to inhibit drainage. Soils which have a water content at field capacity above their liquid limits and above their optimum water content for compaction, are more likely to be compacted and puddled than those that are below this water content. Soils falling into this category are essentially coarse soils with liquid limits <25% (w/w) resting above a duripan or other drainage impediment (Howard et al., 1981). Such soils also suffer from the deleterious potential mixing of the clay layer with the coarse upper layer (Miller and Sirois, 1986).

Most soils have an optimum water content for compaction at a matric potential >-33 kPa. At soil water contents just above the optimum, the soil weakens and the potential for puddling increases. For coarse loamy soils, the optimum level for compaction occurs at a matric potential as low as -100 kPa (Howard et al., 1981). Matric potentials just above clay subsoils generally do not fall below -35 kPa during the rainy season of the year (Wronski, 1984b). Taking into account the lowering of matric potentials in the soil profile above the drainage impediment due to internal drainage, this implies that all moderately deep soils finer than a sandy loam, resting on a clay subsoil, have a very high potential for puddling and compaction during wet weather and explains why loam soils with clay subsoils are generally the most severely disturbed after logging in wet conditions (Miller and Sirois, 1986).

Rates of recovery of compacted soil

The natural recovery of compacted forest soils takes place mainly by volume changes associated with frost heave, freezing-thawing and the wetting and drying of soil containing a moderate amount of clay. To a lesser extent the natural biological activity of worms, grubs and beetles can also play a role in movement of soil particles. Other than biological activity, there appears to be no process that can alleviate the compaction of soils low in clay such as sands (Sands et al., 1979). Significant compaction has been observed in skid trails not used for decades (Vanderheyden, 1981, quoted by Froehlich and McNabb, 1984; Wert and Thomas, 1981; Jakobsen, 1983; Froehlich et al., 1985). The concentration of roots in these compacted zones is always less than in undisturbed soil and those roots present tend to be thicker and closer to the surface. It is generally observed that more rapid rates of recovery occur in the surface soil, possibly because near the surface the extremes in temperature and water content are greater, causing more shrinkage and swelling, greater amounts of organic matter are incorporated into the soil and there is more aeration of the soil (Mace, 1971, quoted by Froehlich and McNabb, 1984; Dickerson, 1976; Thorud and Frissell, 1976; Shishiuchi, 1990). Deeper in the profile (200-500 mm depth), soil in primary skid trails has been found to remain compacted for 30-40 years (Power, 1974; Vanderheyden, 1981, quoted by Froehlich and McNabb, 1984; Wert and Thomas, 1981).

MANAGEMENT PRACTICES AIMED AT REDUCING SOIL COMPACTION

Effects of organic matter

In forestry operations, there is considerable potential to utilise the large amounts of organic matter available during harvesting operations to alleviate the impact of such operations on the soil. Johnson et al. (1979) observed that in areas where surface organic matter (leaf litter) had not been removed prior to skidding operations, soil strength within the skidder wheel rut was 20% less after the operation than in tracks where the litter had been removed. Organic matter has a high elasticity under compression forces and reduces soil compactability by increasing the resistance to deformation and/or by increasing rebound effects (Soane, 1990).

The structure of organic matter in the soil has an important influence on the level of puddling which may occur during harvesting operations. On poorly draining soils, a highly structured root mat generally develops in the surface 150-200 mm, which can lead to substantial (50-70%) increases in soil strength under relatively dry conditions (Björkhem et al., 1975, quoted by Mellgren, 1982; Wingate-Hill and Jacobsen, 1982; Hassan and Sirois, 1984; Wästerlund, 1989b). However, during heavy rain the contribution of roots to the strength of the surface appears to be negligible (Wronski et al., 1990) and, if the strength of the surface layer is to be retained, operations need to stop for a few hours until the surface layer has drained. On such soils, cultural practices, such as ploughing prior to forest establishment, or applying hot burns to remove debris after harvesting (Raison, 1980), are also to be avoided as these destroy the root mat, leading to a reduced bearing capacity of the soil and, in the case of hot burns, higher compactability of the soil through the removal of organic matter in the surface horizons.

In many harvesting operations, there is opportunity to place slash on the extraction tracks, thereby increasing the organic matter content of potentially disturbed soils on the skid trails. In both forwarder and skidder extraction operations, the slash can result in a substantial increase (>20%) in the bearing capacity of the soil (Wronski et al., 1990).

When tracks are fitted over the wheels of forwarders, the effect of a thick elastic slash bed is to distribute the applied loads more evenly over the soil surface, thereby reducing the maximum ground contact pressure. However, even when tracks are not fitted, there are beneficial effects, presumably because slash reduces the stress gradients near the edges of extraction trails and after a few passes acts as a strong fabric introduced into the surface soil.

Planning of operations and operational controls

Restricting machine movement to designated tracks can substantially reduce the area of compacted trails to about 10% of the harvest area, which amounts to a reduction in the area of compacted soil of about two-thirds and represents the best option for reducing the area of disturbed trails (Froehlich et al., 1981; Murphy, 1982). Studies have shown that the loss in skidder productivity caused by such practices is <10% (Bradshaw, 1979; Froehlich et al., 1981; Tesch and Lysne, 1983). However, the costs of planning the harvesting operation are increased.

Myhrman (1990) found that rut depths, formed by 6-wheeled harvesters and 8-wheeled forwarders, more than doubled when turning around corners compared to driving straight ahead. This implies that planning skid trails with few turns will reduce the severity of soil disturbance.

On sensitive catchments with duplex or other poorly draining soils, lateral water flow occurs and the detailed planning of skid tracks requires a knowledge of the relative soil wetness across catchments. O'Loughlin et al. (1990) have developed a computer mapping technique that determines the lateral flow regime and thence the relative degree of soil wetness over a catchment. Input parameters for the model are estimated evapotranspiration rates and a detailed topographic map. However, the high management costs associated with application of the technique limit its use to the most valuable and intensely managed forests.

Grading terrain into areas suitable for logging in the wet and dry seasons of the year is another technique that aims at minimizing compaction and soil disturbance (Greene and Stuart, 1985; Karr et al., 1987). In colder regions of the northern hemisphere, soils prone to compaction problems when wet are usually logged during winter when the ground is frozen. In more temperate or tropical regions, such soils are logged in summer when drier conditions prevail. One problem with this approach is that there is no adequate compaction test for predicting vehicle compaction during various seasons. The standard Proctor test generally over-estimates maximum bulk densities and can under-estimate the optimum water content, while compressibility tests do not take into account the effect of multiple passes and wheel slip on compaction. Other problems are that soil water content is usually highly variable and difficult to estimate. Limiting operations on the basis of soil water content is difficult to administer and is disruptive to harvest scheduling and thus very costly (Froehlich and McNabb, 1984).

Despite these difficulties, forest researchers in British Columbia, Canada, (Lewis and Carr, 1989) have developed a rating system for the compaction hazard based on soil type. Skeletal soils with >70% coarse fragments or sands and loamy sands, are considered to have a low compaction hazard rating, loams a medium hazard rating, while silty and clay soils are given a high hazard rating. On soils of moderate or low hazard rating, it is proposed that up to 20% of the ground surface may be disturbed or compacted, falling to <10% on sites with a high hazard rating.

On wet clay soils in the radiata pine plantations of Australia, a very effective control on compaction and puddling has been the introduction of a 100-mm limit on allowable rut depths. The application of this regulation has resulted in the detailed evaluation of soil bearing capacity, the selection of processing equipment that can most effectively lay slash on extraction tracks and a move to reduce the ground contact pressures of forwarders by fitting dual and wide tyres, in order to avoid the costs and inconvenience of interruption of operations (Wronski et al., 1990).

Machine and running gear selection

Soil disturbance and compaction can be reduced by using cable systems. In such operations, compaction occurs mainly on the landings close to the roads and thus is more easily ameliorated than in other operations. On very moist flat sites the use of torsion suspension skidders is generally recommended. Significant advantages of these machines are their greater traction and load capacity, which result in fewer passes being required to move the same amount of timber relative to conventional skidders. However, high operation costs associated mainly with track maintenance tends to discourage their use (Burger et al., 1985; Reisinger et al., 1988).

An alternative to the torsion suspension skidder in difficult conditions is the fitting of wide tyres or dual wheels to conventional skidders, which can reduce ground contact pressures to <30 kPa. Several studies have demonstrated that wide tyres not only reduce soil disturbance, but can increase productivity on wet and steep terrain (Hassan and Sirois, 1984; Mellgren and Heidersdorf, 1984), with disturbance to the soil occurring mainly where the vehicle turns. However, the fitting of such tyres does not always reduce the intensity of compaction, because with the greater traction developed by wide tyres, contractors tend to increase the skidder loading. This and the high cost of fitting wide tyres can, to some extent, be overcome by downsizing the skidder when fitting wide tyres (Greene and Stuart, 1985; Novak, 1988).

In general, the area of disturbed and compacted soils declines when wide tyres are fitted (Mellgren and Heidersdorf, 1984) and resulting bulk densities do not seem to be significantly greater than those caused by a cable yarding logging system (Sauder and Myles, 1985). Wider tyres are also currently being fitted to forwarders in north-eastern Canada, Australia and Scandinavia to reduce soil damage. In a Swedish test comparing different machines on the basis of rut formation, Myhrman (1990) found that an 8-wheeled forwarder, fitted with

TABLE 3

Equipment type based on the limitations of soil, water and slope conditions assuming frozen ground conditions in winter (from Anonymous, 1987)

Soils and water conditions[1]	Summer logging		Winter logging	
	Max. slope (%)	Equipment type	Max. slope (%)	Equipment type
Coarse-textured on dry moist and wet sites	30	Rubber-tyred skidder	30	Rubber-tyred skidder
	40	Crawler tractor	40	Crawler tractor
	50	LGP track type	50	Cable systems
Medium-textured on dry and moist sites	30	LGP rubber-tyred	30	Rubber-tyred skidder
	40	LGP track type	40	Crawler tractor
Medium-textured on wet sites	20	LGP rubber-tyred	50	LGP track type
	30	LGP track type + cable systems	51	Cable systems
Fine-textured on wet sites; organic soils on dry, moist and wet sites	-	Cable systems only	-	Cable systems only

[1]Coarse textured = coarse fragments (>2 mm) >70% (v/v) (fragmental); medium textured = coarse fragments (skeletal) 35-70% (v/v); fine textured = sandy, loamy, silty, clayey with coarse fragments <35% (v/v).

600-mm wide tyres, formed ruts twice as deep as the same forwarder fitted with 800-mm tyres.

Those factors which influence the effect on soils of available machines and systems, are usually compiled by the Forest Services in different countries in order to develop guidelines for the choice of equipment in various terrains. An example of such guidelines for Canada (frozen ground in winter) is given in Table 3.

Another innovative approach to logging very wet sites is the development by Canadian and American loggers of the "shovel logging" system. This involves a tracked excavator/loader traversing back and forth along trails parallel to the road verge, lifting logs furthest from the roadside across the trail, and stacking them closer to the roadside. The process is then repeated on an adjacent track closer to the roadside. This system results in the minimum of trafficking along well defined tracks by a very low ground contact pressure machine, carrying negligible payload.

The selection of the most appropriate logging equipment for a site depends on: (1) an evaluation of the day-to-day variation in soil strength; (2) the extent to which the soil loses strength during trafficking; (3) the number of machine passes required on the main extraction trails; (4) the ground contact pressure of the machines to be used for timber processing and extraction. Until recently such information has been combined on the basis of past experience in a subjective way. However, the need to protect certain sensitive water catchments, the introduction of legally binding codes of logging practice and rapid changes in forest technology, require the development of quantitative methods for evaluating and predicting the interaction between logging machines and the soil (Wronski et al., 1990).

Amelioration of compacted forest soils

The effects of compaction on tree growth may be ameliorated by tillage and ripping, the incorporation of litter and slash into the soil and the application of mulches and fertilizer. The choice of ameliorative measures applied depends on the intensity of compaction, soil type and the value of the forest.

Reduced growth of planted radiata pine seedlings on heavily damaged extraction trails and landings, was noticed on pumice soils and clay soils in New Zealand and on sandy soils in South Australia during the late 1960's and early 1970's. As the value of the forest was high, deep ripping and fertilizer application were recommended to overcome this problem (Berg, 1975; Ballard, 1978; Sands et al., 1979).

The main benefit that arises from ripping is the development of low-resistance root pathways to soil water at depth, which reduces drought conditions during summer. The addition of fertilizer assists growth and survival by supporting rapid root growth to deeper soil layers. Various winged rippers have been developed to rip compacted soils to depths of at least 700 mm on forest landings and major skid trials. Such ameliorative measures tend to be more successful for coarse soils than fine soils which form very large clods (Moehring, 1970; Croton, 1986).

In a detailed study of various tillage implements for forestry operations, including brush blades, rock rippers, disc harrows and winged subsoilers, Andrus and Froehlich (1983) concluded that there is a critical depth, which varies with soil conditions and tine geometry, below which the tine of a brush blade or rock ripper compresses the soil about it and does not shatter the soil. This critical depth decreases with increases in soil water content and clay content. The problem is overcome by adding a wing, which can vary in width from 300-1500 mm, to the base of the ripper tine (Spoor and Godwin, 1978), which then lifts and shatters the soil above. Such winged rippers have the added advantage that they allow greater spacing between tines and are less likely to accumulate slash. In areas where there is a deep structured soil and compaction is confined to the

TABLE 4

Tillage implements recommended for various site conditions (from Andrus and Froehlich, 1983)

Tillage implement	Site conditions allowing effective tillage (all soils)				
	Soil water status	Condition	Logging debris on trail (%)	Max. slope of trail	
				Down (%)	Across (%)
Disk harrow	Dry to moist	No rocks	None	20	10
Brush blade	Dry	Not on clay with cobbles and rocks	Minimal amounts	30	10
Winged subsoilers and rock rippers	Dry	Not on clay with rocks and boulders	Minimal to large amounts	40	15

surface, the use of disc harrows and brush blades suffices to restore productivity (Table 4).

Other data on the benefits of tillage on seedling growth have been tabulated by Froehlich and McNabb (1984, quoting Andrus, 1982). These data demonstrate growth responses to deep ripping and tillage of 17-70% and increases in survival of 10-40%. Quite often the equipment required to ameliorate compaction is as heavy as the equipment that did the damage. Hence, ameliorative measures should be undertaken only when soils are dry.

Shishiuchi (1990) showed that the application of 0.1 kg of granular NPK fertilizer to individual Japanese cedar seedlings planted on bare skid trails, more than doubled their height compared with unfertilized trees on the same trails. Despite the large improvement in growth, fertilized trees were still 30% less in height than unfertilized trees planted on undisturbed areas.

For forests having a lesser value, the application of natural mulches, wood ash or the spreading of the seed of nitrogen-fixing plants on compacted soils are generally recommended (Carr, 1987). The addition of mulch increases the amount of water available to seedlings while simultaneously reducing soil strength by maintaining a higher soil water content. In a comparative study of the rehabilitation of compacted podzolic soils in native forest of Western Australia, Schuster (1979) found that increased growth and survival occurred only when compacted soil was ripped or fertilizer was applied. Mulching with bark debris was beneficial only if combined with ripping and fertilizer application. However, the benefit of sowing a nitrogen-fixing ground cover crop, in the hope that the natural mulching and increase in soil nitrogen would be beneficial, was negligible. Schuster also concluded that the application of nutrients by heaping and burning of logging debris, while increasing initial survival rates, had little effect on long-term growth unless the soil was also ripped.

CONCLUSIONS

(1) Compaction of soils can reduce the establishment and growth of forest tree seedlings but the effects differ between species. The negative effects of compaction can persist for decades.

(2) During thinning operations, both soil compaction and damage to the root system of the remaining stand can occur but the effects of compaction are likely to be greater and more persistent.

(3) Forest operations which result in the application of highest ground contact pressures and traffic intensity are harvesting and thinning. If operations are conducted during wet conditions, machines with ground contact pressures as low as 60 kPa can still produce negative effects.

(4) Forest operations on sandy soils will result in more intense and prolonged compaction than those on clay soils, although the disturbance on clay and loam soils tends to be more pronounced, especially if there is a drainage impediment.

(5) There is considerable potential to reduce the impact of traffic on forest soils by adding organic matter through the laying of slash on the main extraction tracks prior to trafficking and avoiding cultural practices such as burning and ploughing on poorly draining soils which reduce their bearing capacity.

(6) Intensive management of forest operations, involving restricting machine movement to designated tracks chosen to avoid wet and/or steep areas on a catchment, and scheduling operations on soils suitable for logging in summer and winter, can also substantially reduce soil compaction.

(7) Consideration of the potential impact of machinery on soils well before operations commence, can also be helpful as this allows the most appropriate logging system for the terrain to be selected.

(8) If necessary, the ground contact pressure of machines can be reduced by fitting wider tyres. Alternatively, crawlers or flexible-tracked machines can be selected, and, if the terrain is too steep or too wet, cable systems can be employed.

(9) If significant soil compaction occurs, despite the measures undertaken to reduce it, ameliorative measures, such as deep ripping with winged tines, can be undertaken subsequently.

REFERENCES

Agren, G.I., Axelsson, B., Flower-Ellis, J.G.K., Linder, S., Persson, H., Staaf, H. and Troeng, E., 1980. Annual carbon budget for a young Scots pine. In: Structure and Function of Northern Coniferous Forests - An Ecosystem Study. Swedish Natural Sci. Res. Council, Stockholm, Sweden, Ecol. Bull. (Stockholm) 32, pp. 307-313.

Andrus, C.W., 1982. Tilling compacted soils following ground-based logging in Oregon. M.Sc. thesis, Oregon State Univ., Corvallis, OR, U.S.A., 170 pp.

Andrus, C.W. and Froehlich, H.A., 1983. An evaluation of four implements used to till compacted forest soils in the Pacific North-West. Oregon State Univ., Corvallis, OR, U.S.A., For. Res. Lab., Res. Bull. 45, 12 pp.

Anonymous, 1987. Ground skidding guidelines with emphasis on minimizing site disturbance. B.C. Min. For. Lands, Engineering Branch and Silvicultural Branch, Victoria, B.C., Canada, 58 pp.

Ballard, R., 1978. Use of fertilizers at establishment of exotic forest plantations in New Zealand. N.Z. J. For. Sci., 8: 70-104.

Berg, P.J., 1975. Developments in the establishment of second rotation radiata pine at Riverhead Forest. N.Z. For., 20: 276-282.

Bernhardsson, A. and Martinsson, L. 1986. Rötfrevkens i högproductiva granskogar. (Frequency of wood rot in highly productive pine forests). Student thesis, Swedish Univ. Agric. Sci., School For. Eng., Skinnskatteberg, Sweden, 17 pp. (in Swedish).

Björkhem, U., Lundeberg, G. and Scholander, J., 1975. Rotförekomst och tryckhållfasthet i skogsmark. Rotkartoring och plattbelastningsförsök i gallringsbestånd au gran. (Root distribution and compressive strength in forest soils. Root mapping and plate loading tests in thinning-stage stands of Norway spruce). Royal College Forestry, Stockholm, Sweden, Res. Note 22, 60 pp. (in Swedish).

Bodman, G.B. and Rubin, J., 1948. Soil puddling. Soil Sci. Soc. Am. Proc., 13: 27-36.

Bowen, G.D., 1964. Root distribution of *Pinus radiata*. CSIRO Div. Soils, Adelaide, Australia, Div. Rep., 14 pp. (cited from For. Abstr., 26: 4878).

Bradshaw, G., 1979. Preplanned skid trails and winching versus conventional harvesting on a partial cut. Oregon State Univ., For. Res. Lab., Corvallis, OR, U.S.A., Res. Note 62, 4 pp.

Burger, J.A., Perumpral, J.V., Kreh, R.E., Torbet, J.L. and Minaei, S., 1985. Impact of tracked and rubber-tyred tractors on a forest soil. Trans. ASAE, 28: 369-373.

Canarache, A., 1990. PENETR - A generalized semi-empirical model estimating soil resistance to penetration. Soil Tillage Res., 16: 51-70.

Carr, W.W., 1987. Restoring productivity on degraded forest soils: Two case studies. Canadian Forest Service and British Columbia Ministry of Forests and Lands, FRDA Rep. 002, 21 pp.

Cochran, P.H. and Brock, T., 1985. Soil compaction and initial height growth of planted ponderosa pine. USDA Forest Service, Res. Note PNW-434, 4 pp.

Corns, I.G.W., 1988. Compaction by forestry equipment and effects on coniferous seedling growth on four soils in the Alberta foothills. Can. J. For. Res., 18: 75-84.

Croton, J.T., 1986. Effect of soil moisture and density on rehabilitation ripping. Alcoa (Western Australia), Environmental Res. Note 12, 19 pp.

Dawidowski, J.B. and Koolen, A.J., 1987. Changes of soil water suction, conductivity and dry strength during deformation of wet undisturbed samples. Soil Tillage Res., 9: 169-180.

Dickerson, B.P., 1976. Soil compaction after tree length skidding in Northern Mississippi. Soil Sci. Soc. Am. J., 40: 965-966.

Firth, J.G., 1989. How soil disturbance impacts tree growth. N.Z. For. Indust., July 1989, pp.

38-39.

Firth, J. and Murphy, G., 1989. Skidtrails and their effect on the growth and management of young *Pinus radiata*. N.Z. J. For. Sci., 19: 22-28.

Foil, R.R. and Ralston, C.W., 1967. The establishment and growth of loblolly pine seedlings on compacted soils. Soil Sci. Soc. Am. Proc., 31: 565-568.

Frazer, A.I. and Gardiner, J.B.H., 1967. Rooting and stability in Sitka Spruce. Forestry Commission, London, U.K., Bull. 40, 28 pp.

Froehlich, H.A., 1972. Soil compaction: Implication for young-growth management. In: Managing Young Forests in the Douglas-fir Region. Oregon State Univ., School of Forestry, Corvallis, OR, U.S.A. Vol. 4, pp. 49-63.

Froehlich, H.A., 1979. Soil compaction from logging equipment. Effects on growth of young ponderosa pine. J. Soil Water Conserv., 34: 276-278.

Froehlich, H.A., 1989. Soil damage, tree growth, and mechanization of forest operations. In: Seminar on the Impact of Mechanization of Forest Operations on the Soil. Louvain-la Neuve, Belgium, 11-15 September 1989, pp. 76-86.

Froehlich, H.A., Azevedo, J., Cafferata, P. and Lysne, D., 1980. Predicting soil compaction on forested land. USDA Forest Service, Equip. Dev. Centre, Missoula, MT, U.S.A., 120 pp.

Froehlich, H.A., Aulerich D.E. and Curtis, R., 1981. Designing skidtrail systems to reduce soil impacts from tractive logging machines. Oregon State Univ., For. Res. Lab., Corvallis, OR, U.S.A., Res. Pap. 44, 13 pp.

Froehlich, H.A. and McNabb, D.S., 1984. Minimizing soil compaction in Pacific Northwest forests. In: E.L. Stone (Editor), Forest Soils and Treatment Impacts. Proc. 6th Conf. North American Forest Soils, June 1983, Soc. Am. Foresters/Dept. Forestry, Wildlife and Fisheries, Univ. Tennessee, Knoxville, TN, U.S.A., pp. 159-192.

Froehlich, H.A., Miles, D.W.R. and Robbins, R.W., 1985. Soil bulk density recovery on compacted skid trails in central Idaho. Soil Sci. Soc. Am. J., 49: 1015-1017.

Froehlich, H.A., Miles, D.W.R. and Robbins, R.W., 1986. Growth of young *Pinus ponderosa* and *Pinus contorta* on compacted soil in central Washington. For. Ecol. Manage, 15: 285-294.

Garrison, G.A. and Rummell, R.S., 1951. First-year effects of logging on Ponderosa pine forest range lands of Oregon and Washington. J. For., 49: 708-713.

Gent, J.A. and Morris, L.A., 1986. Soil compaction from harvesting and site preparation in the Upper Gulf coastal plain. Soil Sci. Soc. Am. J., 50: 443-446.

Greacen, E.L. and Sands, R., 1980. Compaction of forest soil: A review. Aust. J. Soil Res., 18: 163-189.

Greene, W.D. and Stuart, W.B., 1985. Skidder and tire size effects on soil compaction. Southern J. Appl. For., 9: 154-157.

Hassan, A.E. and Sirois, D.L., 1984. Performance of a skidder with dual tyres on wetland. Am. Soc. Agric. Eng., St. Joseph, MI, U.S.A., ASAE Pap. 84-1552, 18 pp.

Hatchell, G.E., Ralston, C.W. and Foil, R.R., 1970. Soil disturbances in logging. J. For., 68: 772-775.

Haupt, H.F., 1960. Variation in aerial disturbance produced by harvesting methods in Ponderosa pine. J. For., 58: 634-639.

Howard, R.F., Singer, M.J. and Frautz, M.A., 1981. Effects of soil properties, water content and compactive effort on the compactibility of selected Californian forest and range soils. Soil Sci. Soc. Am. J., 45: 231-236.

Jakobsen, B.F., 1983. Persistence of compaction effects in a forest Krazhozem. Aust. For. Res., 13: 305-308.

Jakobsen, B.F. and Moore, G.A., 1981. Effects of two types of skidders and of slash cover on soil compaction by logging of Mountain Ash. Aust. For. Res., 11: 247-255.

Johnson, J.A., Hillstrom, W.A., Miyata, E.S. and Shetron S.G., 1979. Strip selection method of mechanised thinning in northern hardwood pole size stands. Michigan Technological Univ., Ford Forestry Centre, Res. Note 27, 13 pp.

Karr, B.L., Hodges, J.D. and Nebeker, T.E., 1987. The effect of thinning methods on soil physical properties in North-Central Mississippi. Southern J. Appl. For., 11: 110-112.

Lewis, T. and Carr, W.W., 1989. Developing Timber Harvesting Prescriptions to Minimise Site Degradation. B.C. Ministry of Forests, Timber Harvesting Subcommittee, March 1989, 224 pp.

Lysne, D.H. and Burditt, A.L., 1983. Theoretical ground pressure distribution of log skidders. Trans. ASAE, 26: 1327-1331.

Mace, Jr., A.C., 1971. Recovery of forest soils from compaction by rubber tyred skidders. Univ. Minn., St Paul, MN, U.S.A., For. Res. Note 226, 4 pp.

Mellgren, P.G., 1982. High flotation tyres for logging machines. Pulp Paper Can., 83: 27-32.

Mellgren, P.G. and Heidersdorf, E., 1984. The use of high flotation tyres for skidding on wet and/or steep terrain. FERIC, Canada, Tech. Rep. TR-57, 48 pp.

Miller, J.H. and Sirois, D.L., 1986. Soil disturbance by skyline yarding vs. skidding in a loamy hill forest. Soil Sci. Soc. Am. J., 50: 1579-1583.

Minore, D., Smith, C.E. and Woolard, R.F., 1969. Effects of high soil density on seedling root growth of seven northwestern tree species. USDA Forest Service, Res. Note PNW-112, 6 pp.

Moehring, D.M., 1970. Forest soil improvement through cultivation. J. For., 68: 328-331.

Moehring, D.M. and Rawls, I.K., 1970. Detrimental effects of wet weather logging. J. For., 68: 166-167.

Moir, W.H. and Bachelard, E.P., 1969. Distribution of fine roots in three *Pinus radiata* plantations near Canberra, Australia. Ecology, 50: 658-662.

Murphy, G., 1982. Soil damage associated with production thinning. N.Z. J. For. Sci., 12: 281-292.

Murphy, G., 1983. *Pinus radiata* survival, growth and form four years after planting off and on skidtrails. N.Z. J. For., 28: 184-193.

Murphy, G., 1984. A survey of soil disturbance caused by harvesting machinery in New Zealand plantation forests. N.Z. Forest Service, Forest Res. Inst., FRI Bull. 69, 9 pp.

Myhrman, D., 1990. Factors influencing rut formation from forestry machines. Proc. 10th Int. Conf. ISTVS, Kobe, Japan, Vol. 2, pp. 467-475.

Nambiar, E.K.S., 1983. Interplay between nutrients, water, root growth and productivity in young plantations. For. Ecol. Manage, 30: 213-232.

Novak, W., 1988. Downsizing skidders with high flotation tires. Can. For. Indust., 108: 41-46.

O'Loughlin, E.M., Vertessy, R.A., Dawes, W.R. and Short, D.L., 1990. The use of predictive hydrologic modelling for managing forest ecosystems subject to disturbance. IUFRO Meeting, XIX World Congress, Montreal, Q., Canada, Vol. B, pp. 252-266.

Perry, T.O., 1964. Soil compaction and Loblolly pine growth. Tree Planters' Note 67, 9 pp.

Pfister, R.D., 1969. Effect of roads on growth of western white pine plantations in northern Idaho. USDA Forest Service, Res. Pap. INT-65, 8 pp.

Power, W.E., 1974. Effects and observations of soil compaction in the Salem District. USDI-BLM, Salem, OR, U.S.A., Tech. Note.

Raghavan, G.S.V., McKyes E. and Chasse, M., 1977. Effect of wheel slip on soil compaction. J. Agric. Eng. Res., 22: 79-83.

Raison, R.J., 1980. Possible site deterioration associated with slash-burning. Search Sci. Tech. Aust. N.Z., 11: 68-72.

Reisinger, T.W., Simmons, G.L. and Pope, P.E., 1988. The impact of timber harvesting on soil properties and seedling growth in the south. Southern J. Appl. For., 12: 58-67.

Rowland, D., 1972. Tracked vehicle ground pressure and its effects on soft ground performance. Proc. 4th Int. Conf. ISTVS, Stockholm, Sweden, Vol. 1, pp. 353-383.

Rytter, L., 1989. Distribution of roots and root nodules and biomass allocation in young intensively managed grey alder stands on a peat bog. Plant Soil, 119: 71-79.

Sands, R. and Bowen, G.D., 1978. Compaction of sandy soils in Radiata pine forests. II. Effects of compaction on root configuration and growth of Radiata pine seedlings. Aust. For. Res., 8: 163-170.

Sands, R., Greacen, E.L. and Gerard, C.J., 1979. Compaction of sandy soils in Radiata pine forests. I. A penetrometer study. Aust. J. Soil Res., 17: 101-113.

Sauder, B.J. and Myles, D.V., 1985. Low ground pressure tyres for skidders. Canadian Forestry Service, Research and Technical Services, Ottawa, Ont., Canada, Rep. DPC-X-20, 24 pp.

Schuster, C.J., 1979. Rehabilitation of soils damaged by logging in south-west Western Australia. Western Australia For. Dept., Res. Pap. 54, 7 pp.

Senyk, J.P., 1990. Effects of ground-based forest harvesting operations on soils and tree productivity. Proc. 10th Int. Conf. ISTVS, Kobe, Japan, Vol. 2, pp. 567-575.

Shishiuchi, M., 1990. The effect of tractor logging on physical properties of forest soil and the growth of planted seedlings. Proc. 10th Int. Conf. ISTVS, Kobe, Japan, Vol. 2, pp. 477-485.

Skinner, M.F., Murphy, G., Robertson, E.D. and Firth, J.G., 1989. Deleterious effects of soil disturbance on soil properties and the subsequent early growth of second-rotation radiata pine. In: W.J. Dyck and C.A. Mees (Editors), Research Strategies for Long-term Site Productivity. Proc. IEA/BE A3 Workshop, Seattle, WA, U.S.A., August 1988. IEA/BE A3 Rep. 8. For. Res. Inst., New Zealand, Bull. 152, pp. 201-212.

Smith, R.B. and Wass, E.F., 1979. Tree growth on and adjacent to contour skidroads in the subalpine zone, south-eastern British Columbia. Environment Canada, Canadian Forestry Service, Victoria, B.C., Canada, 28 pp.

Soane, B.D., 1990. The role of organic matter in soil compactibility: A review of some practical aspects. Soil Tillage Res., 16: 179-201.

Spoor, G. and R.J. Godwin., 1978. An experimental investigation into the deep loosening of soil by rigid tines. J. Agric. Eng. Res., 23: 243-258.

Sutton, R.F., 1969. Form and development of conifer root systems. Commonwealth Agricultural Bureaux, Farnham Royal, U.K., Tech. Comm. 7, 131 pp.

Terry, T.A. and Campbell, R.G., 1981. Soil management considerations in intensive forest management. In: Forest Regeneration - Proceedings. Am. Soc. Agric. Eng., St. Joseph, MI, U.S.A., Publ. 10-81, pp. 98-105.

Tesch, S.D. and Lysne, D.H., 1983. Skidding treetops attached to merchantable logs: Effects on ground based logging production. Oregon State Univ., For. Res. Lab., Corvallis, OR, U.S.A., Res. Note 73, 6 pp.

Thorud, D.B. and Frissell, S.S., 1976. Time changes in soil density following compaction under an oak forest. Univ. Minnesota, St. Paul, MN, U.S.A., For. Res. Note 257, 4 pp.

Turnbull, C.R.A., Beadle, C.L., Bird, T. and McLeod D., 1988. Volume production in intensively managed eucalypt plantations. APPITA, 41: 447-450.

Vanderheyden, J., 1981. Chronological variation in soil density and vegetative cover of compacted skid trails of the Western Oregon Cascades. M.Sc. thesis, Oregon State Univ., Corvallis, OR, U.S.A., 139 pp.

Van Praag, H.J., Sougnez-Remy, S., Weissen, F. and Carletti, G., 1988. Root turnover in a beech and a spruce stand in the Belgian Ardennes. Plant Soil, 105: 87-103.

Van Veen, J.A., Merckx, R. and Van de Geijn, S.C., 1989. Plant and soil related controls of the flow of carbon from roots through the soil microbial biomass. Plant Soil, 115: 179-188.

Wästerlund, I., 1983. Kanttrådens tillväxtförluster vid gallring p.g.a. jordpackning och rotskador i stickväg. (Growth reduction of trees near strip roads resulting from soil compaction and damaged roots - a literature survey). Sver. Skogsvardsforb. Tidskr., 81: 97-109 (in Swedish with English abstract).

Wästerlund, I., 1985. Compaction of till soils and growth tests with Norway spruce and Scots pine. For. Ecol. Manage, 11: 171-189.

Wästerlund, I., 1988. Damages and growth effects after selective mechanical clearing. Scand. J. For. Res., 3: 259-272.

Wästerlund, I., 1989a. Effects of damages on the newly thinned stand due to mechanized forest operations. In: Seminar on the Impact of Mechanization of Forest Operations on the Soil. Louvain-la-Neuve, Belgium, 11-15 September 1989, pp. 165-175.

Wästerlund, I., 1989b. Strength components in the forest floor restricting maximum tolerable machine forces. J. Terramech., 26: 177-182.

Wert, S. and Thomas, B.R., 1981. Effects of skid roads on diameter, height, and volume growth in Douglas-fir. Soil Sci. Soc. Am. J., 45: 629-632.

Williamson, J.R., 1990. The Effects of Mechanized Harvesting Operations on Soil Properties and Site Productivity. Forestry Commission, Tasmania, Australia, 193 pp.

Wingate-Hill, R. and Jakobson, B.F., 1982. Increased mechanisation and soil damage in forests: A review. N.Z. J. For. Sci., 12: 380-393.

Wood, J.P. and Bachelard, E.P., 1970. Root grafting in radiata pine stands in the Australian Capital Territory. Aust. J. Bot., 18: 251-259.

Wronski, E.B., 1984a. Impact of tractor thinning operations on soils and tree roots in a Karri forest, Western Australia. Aust. For. Res., 14: 319-332.

Wronski, E.B., 1984b. Prediction of soil trafficability in a Karri forest, Western Australia. Aust. For. Res. 15: 367-380.

Wronski, E.B., Stodart, D.M. and Humphries, N., 1990. Trafficability assessment as an aid to planning logging operations. APPITA, 43: 18-22.

Soil Compaction in Crop Production
B.D. Soane and C. van Ouwerkerk (Eds.)
343

CHAPTER 15

Responses of Perennial Forage Crops to Soil Compaction

J.T. DOUGLAS

Scottish Centre of Agricultural Engineering, SAC, Penicuik, U.K.

SUMMARY

An increasingly large amount of wheel traffic, from a variety of machines and transporters, is necessary for management and harvesting of grass and legume forage species. The soil compaction problem in perennial forage crop production has been recognised in both farming practice and in a number of research activities. Research has shown that any significant deterioration in soil conditions brought about by wheel traffic leads to impaired crop performance and yield, particularly if the traffic is imposed on wet soil or if the soil remains relatively wet during the main period of crop growth. The influence of traffic on land in forage cropping and the subsequent effects on yield, can be moderated by use of alternative machinery or traffic management systems. These alternatives are discussed, together with techniques for improvement of over-compacted grassland. Research has indicated that measures taken to avoid compaction are more likely to be beneficial than most ameliorative operations.

INTRODUCTION

Soil compaction problems may arise in perennial forage crops as a result of frequent and often heavy vehicle traffic over a number of years. Attention is given here primarily to research in north-western Europe, but the problems are likely to be found widely, particularly where such crops are grown in areas of high summer rainfall or under irrigation. The presence of a high organic matter content, a root mat and stable structure in the surface layers of soil under perennial forage crops may mitigate the adverse effects of vehicle traffic. However, there is increasing evidence that crop productivity may be seriously reduced as a result of traffic-induced compaction in addition to any adverse effects attributable to direct crop damage from the passage of wheels.

Type of crop

Perennial grass and legume species are the most important crops managed for conservation as silage (by anaerobic fermentation), pellets or hay (by de-hydration), or for purposes of zero or mechanical grazing (cutting and transport of fresh material to livestock).

Fields of grass are cut commonly between one and five times per year for silage or drying, or usually only once for hay. A large number of perennial grasses are used for conservation, either alone or in mixtures, including ryegrass (*Lolium perenne* L.), timothy (*Phleum pratense* L.), smooth bromegrass (*Bromus inermis* Leyss.), cocksfoot or orchardgrass (*Dactylis glomerata* L.), reed canarygrass (*Phalaris arundinacea* L.) and fescues (*Festuca* spp.). The types of grass or grass mixtures used by growers, and the use to which the herbage is put in particular farming systems, differ between climatic regions.

The most important forage legumes are lucerne (alfalfa) (*Medicago sativa* L.), red clover (*Trifolium pratense* L.) and white clover (*Trifolium repens* L.); other perennial species used to a lesser extent, include birdsfoot trefoil (*Lotus corniculatus* L.) and sainfoin (*Onobrychis viciifolia* Scop.).

Lucerne is normally gathered for hay after one or two cuts, though high yields for silage over three or four cuts each year can be obtained on well-suited soils. Lucerne is best adapted to well-drained, highly fertile soils and is relatively intolerant of conditions associated with poor drainage (Miller, 1984). In low rainfall areas of Africa and North America, lucerne is often grown on irrigated land with several cuts for hay being taken each year. Although in optimum soil conditions the extensive taproot system of lucerne renders the plants tolerant of drought, taproot development may fail in compacted soil; in such circumstances, crop production may be impaired or may cease. Pure stands of red clover will provide sufficient material for conservation generally only in the first year or two after establishment; red clover is most productive on well-drained land but is more tolerant of soil wetness than lucerne. White clover is only suited for conservation if cultivated in a mixture with forage grasses; this species, which is shallower rooting than either lucerne or red clover, requires relatively moist soil conditions.

Crop management and machinery

Each stage of forage crop establishment, management and harvest, necessitates operations involving vehicle traffic. The sequence of operations for seedbed preparation is generally similar to that for annual cereal crops, and includes primary and secondary tillage, and rolling the seedbed; for small-seeded crops, the latter may be necessary both before and after planting. On established swards in temperate regions, rolling is often necessary in the spring months to push stones into the soil surface and to level wheel ruts and mole-hills. Fertilizer application by tractor-mounted or trailed spreaders is required on up to four occasions per year depending on the number of cuts and on the adopted fertilizer policy. In livestock enterprises, the requirement for disposal of animal waste can lead to additional traffic from tractors and trailed slurry carriers. The necessity for frequent field spreading of animal slurry is usually greatest in intensive dairy or beef cattle systems in which the stock is over-wintered indoors and where slurry storage capacity is limited; in those circumstances, traffic at 6-15 m spacing from tractors and tankers, ranging in capacity from 3 to 15 Mg, occurs at the end of winter or in early spring when the land is relatively wet and highly compactable.

The area of a field covered by wheel traffic during forage crop management is dependent on the track width and working width of machines, tyre width, type of crop, and frequency of harvests and other field operations. Invariably, the total area covered annually by traffic on forage crops is considerably larger than that during cultivation of annual crops (Schuler, 1989). For example, in grassland for silage, the entire area of a field will be tracked each year up to nine times by tractor wheels alone (Frost, 1984). A single lucerne harvest may cover up to 70% of the field surface (Grimes et al., 1978) or 48% if some degree of traffic control is employed (Meek et al., 1988).

The greatest intensity (load per wheel per area per time) of vehicular activity occurs during crop harvesting operations (Douglas and Campbell, 1987). The working width of mowers is usually in the range 1.6-2.8 m and, therefore, the traffic associated with both cutting and lifting the crop recurs at that spacing across a field. Lifting the cut material typically involves a forage harvester working in combination with a high-sided trailer, or the use of a round-baling machine followed by collection and transportation by another tractor. Schuler (1989) reported a range of annual traffic intensities for forage crops (110-186 Mg km ha^{-1}) which was approximately double that for arable row crops (32-88 Mg km ha^{-1}).

As with all agricultural machinery, the sizes of power units, powered equipment and transportation units vary considerably between farms and types of farming system. Typical vehicle masses for forage production in Europe are as follows: tractor 3-4 Mg, mower 1 Mg, harvester or baler 1.5 Mg, trailer (loaded) 4 Mg, slurry tanker (loaded) 5 Mg. Farmers with relatively large areas of conservable crop, and contracting companies, frequently use equipment which has a mass that is 50-100% larger than average.

Perception of the soil compaction problem

Identification by farmers of wheel-induced soil compaction effects on the productivity of their forage conservation fields can be a difficult task because so many other factors can influence the performance of perennial crops, e.g., botanical composition, sward and stock management, direct wheel damage to the plants, pests and diseases, weather patterns and soil fertility. The difficulty is exacerbated by the general absence of a direct economic value for forage and on-farm consumption of forage, which leads to a lack of yield monitoring by growers. Furthermore, the long-established maxim that agricultural soil benefits from a period under grass (Batey, 1988) may confuse judgement on the likely condition of land under perennial species. In their review of soil limitations to grassland production in England and Wales, Thomas and Evans (1975) concluded that low aeration of the surface layer of the soil, due to compaction by vehicles and livestock, was a major obstacle to better grassland production.

A number of on-farm observations of impaired regrowth in wheel tracks in red clover swards in the United Kingdom linked the crop response to direct damage to the plants (Castle and Watson, 1974; Copeman and Younie, 1981). However, in the absence of soil information, the indirect effect of altered soil properties could not be implicated.

A survey of the compaction problem in Scotland (Soane, 1987) revealed that slurry and dung spreading were thought to be damaging to the soil by 41% of farmers; on dairy farms, located primarily on wetter soils in western Scotland, the proportion was 55%. The survey also showed that there was a greater perception of compaction damage with increasing size of farm, which Soane attributed to the use of heavier machines and less opportunity to avoid compaction risks by keeping off the land directly after wet weather. Twenty-five percent of grassland farmers believed that compaction led to herbage yield losses and a similar

Fig. 1. Grass silage bale collection and evidence of soil and ryegrass sward damage by wheel traffic in Scotland.

percentage declared that they did not know if there was a penalty from over-compaction.

As with annual crops, soil compaction of a degree which is limiting to productivity, can occur in fields of perennial forage crops, both in specific wheel track positions and, relatively more uniformly, over entire fields. In land under perennial crops, however, there is an increased probability of the latter situation because of year-to-year accumulations of wheel traffic patterns. Due to farmers' routines or preferences, the direction of traffic in a particular field may be generally consistent from year to year; however, the position of wheeltracks within that direction is likely to be totally random (Fig. 1.).

The causes and incidence of soil compaction by vehicles have been reviewed thoroughly by Soane (1983) and Håkansson et al. (1988). In fields where forage crops are grown for conservation, soil damage by wheels can be manifest, as distinct ruts or depressions, intensive uniform compaction, tyre tread imprints, or as surface smearing. In each case a deterioration in soil properties may arise which is potentially detrimental to crop productivity.

EXPERIMENTAL EVIDENCE OF COMPACTION

Studies of soil compaction under perennial forage crops have tended to concentrate more on crop performance aspects than on the state of the soil on which the crops were grown. By-and-large, changes in soil properties (e.g., bulk density and strength) after traffic were reported, rather than effects on those conditions in the soil which directly influence crop responses (e.g., aeration, water status and temperature).

Soil aspects

Under grass
The major proportion of research into the response of grassland soils to compactive forces of traffic has taken place in north-western Europe (particularly in Scandinavia and also in the Netherlands and the United Kingdom) where grass for conservation is a crop of major importance, grown on soils that are prone to extended periods of wetness.

Tveitnes and Njøs (1974) reported that compaction by a tractor on three soils in western Norway led to a small increase in the proportion of large (>6 mm) soil aggregates and reduced the proportion of air-filled pore space from 19 to 12% (v/v). Also in Norway, Gaheen and Njøs (1977) identified impaired soil physical properties after tractor traffic, imposed at two levels of wetness, on poorly drained loam soil under timothy grass (Table 1). Timothy root length was positively correlated with infiltration rate and saturated hydraulic conductivity, and negatively correlated with soil vane shear strength.

A review of a number of soil compaction studies in Sweden (Eriksson et al.,

TABLE 1

Soil physical properties corresponding to different compaction treatments in a timothy grass crop (from Gaheen and Njøs, 1977)

Treatment[1]	Infiltration rate (mm min[-1])	K_s (mm min[-1])	Bulk density (Mg m[-3])	Air volume at -10 kPa (%, v/v)	Vane shear strength (kPa)
A0B0	6.0	2.00	1.31	10	60
A0B1	0.8	0.33	1.44	2	60
A1B0	5.7	5.70	1.12	21	30
A1B1	0.9	0.66	1.22	15	40

[1]A0 = wet soil; A1 = moist soil; B0 = no compaction other than at harvest operations; B1 = one pass of tractor.

1974) contains reference to soil bulk density increases in hay fields generated by single- and double-pass applications of harvest traffic in dry conditions by a small (1.8 Mg) tractor and a loaded forage wagon (3.6 Mg). In Norway, Myhr and Njøs (1983) compared a number of tractor traffic activities on "mineral" and "organic" soils. They found that elimination of traffic resulted in decreased soil strength and increased air-filled pore volume (Table 2). Neither halving the intensity of normal traffic, nor using dual rear tyres, significantly reduced compaction effects on the soil pore space relative to those from normal traffic; the moderated traffic treatments were associated with slightly decreased soil shear strength.

Douglas and Crawford (1989, 1993) reported large increases in soil bulk density and vane shear strength, and reductions in soil porosity and permeability, following wheel traffic on a clay loam in Scotland; the effect of wheels was largest in the first season after reseeding, and increased as tyre/soil contact pressure was increased. Changes in bulk density and vane shear strength were closely related to the amount of compactive effort applied over two years (calculated as the

TABLE 2

Air-filled porosity (%,v/v) at 0.10-0.15 m depth after different traffic treatments on grassland on different soil types (after Myhr and Njøs, 1983)

Soil type	Traffic treatment					
	Zero	1/2 Normal	Normal	Dual wheels	Normal + 1/2	LSD
Mineral soil	15.5	13.7	13.1	13.2	12.8	0.9
Organic soil	14.3	11.9	11.7	12.1	11.5	0.9

product of ground contact pressure and number of passes). In a concurrent traffic management experiment, comparing conventional, reduced ground contact pressure and zero-traffic systems for silage, Douglas et al. (1992a) observed soil compaction arising from field operations in the conventional and reduced ground contact pressure systems during the first year after sowing. In particular, wheel ruts of 32 mm depth were formed in the conventional system in the spring by a tractor used for herbicide application; in contrast, no ruts were observed in the reduced ground contact pressure system where the same size of tractor, fitted with wide, low-inflation pressure tyres was used. Traffic during the three mowing and harvesting operations resulted in soil compaction in both the conventional and the reduced ground contact pressure systems but it was considerably greater in the former than in the latter (Fig. 2). In the zero-traffic system, in the absence of wheel activity, the soil was considerably less dense and soil bulk density did not change significantly in the course of the experiment. The differences in bulk density profiles were created largely during traffic activity in the first year. An important consequence of the greater soil bulk density after conventional traffic was that the air-filled pore volume was consistently lower during winter and spring than that of the soil in the two alternative systems. This caused decreased aeration and lower soil temperatures, such that they were below optimum for

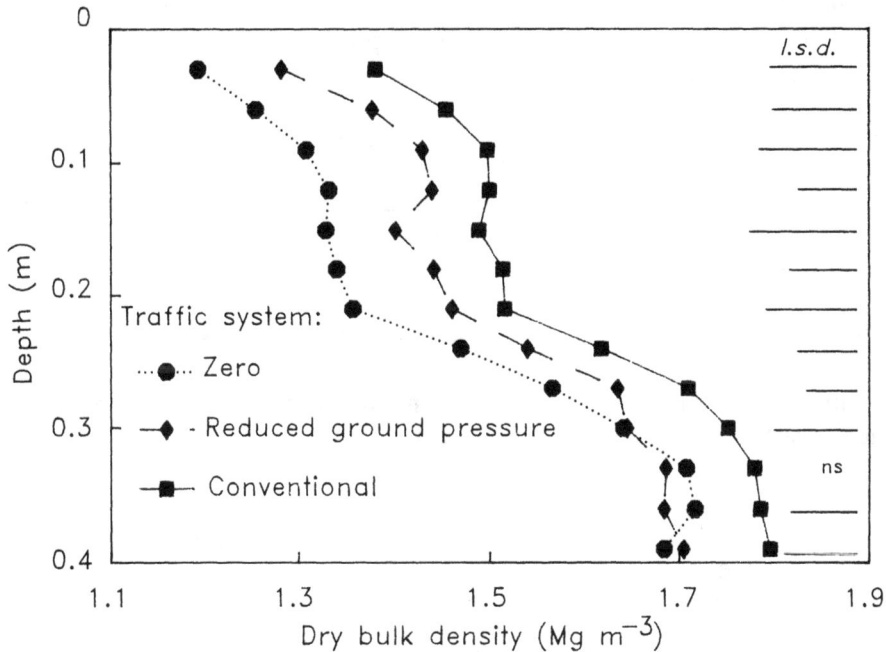

Fig. 2. Variation with depth in bulk density at the end of the fourth year of a traffic systems experiment on grassland for silage (from Douglas et al., 1992a).

TABLE 3

Soil bulk density and drainable porosity after post-harvest traffic on red clover and clover/ryegrass swards (from Frame, 1985)

Treatment	Bulk density (Mg m^{-3})	Air-filled porosity (%, v/v)
Wheeling		
Control (nil)	1.24	9.1
Two tractor passes	1.36	2.6
Four tractor passes	1.40	1.8
SED	0.017	0.52
Sward		
Red clover	1.35	3.6
Clover/ryegrass	1.34	3.6
SED	0.008	0.37

unimpaired root growth and function, and for nutrient utilisation.

Under clover and clover/grass mixtures

Relatively little work on the presence or development of soil compaction under pure clover (*Trifolium* spp.) stands has been reported. In Denmark, Rasmussen and Møller (1981) found that for harvest traffic, both increased ground contact pressure (from 0 to 216 kPa) and number of passes (from 1 to 3) resulted in increased soil shear strength in the 0-10-cm layer. Air-filled porosity at 2-6 and 12-16 cm depth decreased with increased ground contact pressure. By wheel tracking after harvests with a 3.38-Mg tractor, Frame (1985) produced significant increases in bulk density, and associated reductions in air-filled porosity (Table 3).

There is no evidence that the degree of soil damage by wheels is greater under clover swards than under grass, though the stronger cut stems and crownal area and the more dense root mat of the latter might be expected to offer larger mechanical resistance to traffic-induced stresses.

Under lucerne (alfalfa)

Sheesley et al. (1974) reported increased strength in a sandy loam due to compaction by wheel traffic, particularly in the 15-30-cm zone, to a depth of almost 60 cm. In Germany, Haass and Simon (1975) measured soil bulk density and porosity changes under a lucerne stand after application of 200 kPa ground contact pressure at a range of wheel slip; they reported significant reductions in total and air-filled pore volumes at 0-10 cm depth resulting from traffic at zero wheel slip, and further reductions as slip was increased from 10-15% to 45-50%. In a comparison of zone production systems on a sandy loam soil in California,

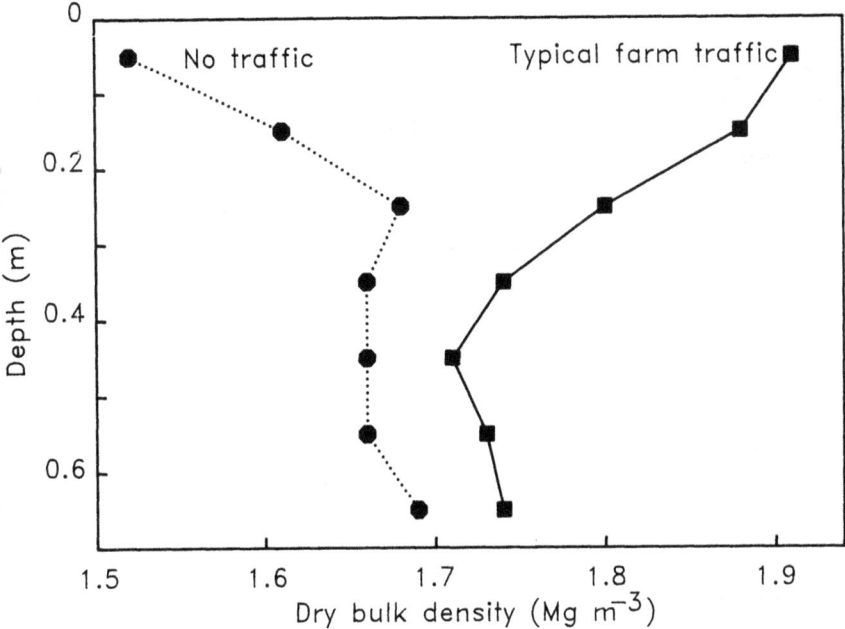

Fig. 3. Bulk density profiles in 1984 after no traffic and simulated typical farm traffic in lucerne (alfalfa) cropping (after Meek et al., 1988).

Meek et al. (1988) obtained high bulk density values (1.9 Mg m⁻³) in the 0-15-cm layer, following a traffic regime designed to simulate that employed by local farmers, even when it was applied when the soil was relatively dry. In contrast, a significantly lower bulk density of 1.6 Mg m⁻³ could be maintained in the total absence of wheel traffic (Fig. 3). Harvest traffic reduced hydraulic conductivity at the 2-11-cm depth by 60% compared to that of no harvest traffic treatments (Meek et al., 1989).

Effects on the crop

The vast majority of research findings has indicated a strong negative relationship between soil compaction and herbage dry matter yield. Furthermore, there have been indications of impaired crop quality (protein content) from crops grown on over-compacted soil.

Grass

In Scandinavia, yield penalties from wheel-induced soil compaction have been well documented. A loss of 15% dry matter occurred in western Norway (Tveitnes and Njøs, 1974). Eriksson et al. (1974) reported a 5-t ha⁻¹ (20%) loss of hay dry matter over 4 years of their experiment in Sweden; the effects of

TABLE 4

Grass dry matter yields for separate cuts in Norway (from Myhr and Njøs, 1983)

Region	Cuts per year	Cut number	Compaction treatment		
			None (Mg ha⁻¹)	Normal (%)[2]	High[1] (%)[2]
Southern	2	1	6.91	98	98
		2	4.05	97	95
	3	1	4.96	97	94
		2	3.18	93	92
		3	1.65	86	80
Northern	1	1	6.94	102	98
	2	1	5.58	93	90
		2	2.46	86	81

[1]"High" treatment = normal tractor traffic plus 1/2 normal traffic.
[2]Relative figures ("None" = 100%).

compaction were particularly evident in the third and fourth years of the trial. At seven locations in southern Norway, Myhr and Njøs (1983) found a mean loss of 9% dry matter from three cuts in the most intense of three traffic treatments; for two sites in northern Norway the mean loss after compaction was 13% dry matter from two cuts (Table 4).

In Iceland, traffic by a light tractor decreased hay yield by 6-25% in the year following its application (Óskarsson, 1975). Luten et al. (1983) compared the effects of wheel traffic from a 6.6-Mg tractor/forage wagon combination on a Dutch polder clay soil on a range of grass species and recorded yield penalties of 9-12% for perennial ryegrass and 18-19% for timothy and cocksfoot. They did not comment on the relative resistance of species to direct plant damage or to conditions for growth after soil compaction. In Scotland, Frame and Merrilees (1990) reported no difference in response to wheel traffic between diploid and tetraploid cultivars of perennial ryegrass. Declining herbage yield with increased number of machinery passes and tyre/soil contact pressure has been demonstrated at various locations in Europe. For example, Rasmussen and Møller (1981) showed that, in first-year grass, post first-cut compaction could lead to a yield (total of second and fourth cuts) as low as 46% of that from a no-traffic control (Fig. 4a). The same study indicated that for one to three wheel passes, the greatest yield decrease occurred between 0 and 69 kPa ground contact pressure. Further increments in ground contact pressure to 216 kPa tended to result in smaller yield penalties than occurred when the number of wheel passes was increased at each ground contact pressure. Effects of traffic on yield were smaller in the second year of the same crop, when a yield depression of 29% was measured after three passes at 216 kPa ground contact pressure.

Relatively heavy traffic by a tractor/slurry tanker combination (total load 11 Mg), imposed in early spring on a clay loam soil in Northern Ireland, significantly reduced yield at first harvest (by 13%); after three passes of that equipment, the loss of dry matter increased to 33% (Frost, 1988a). No adverse effects of traffic were evident at two subsequent cuts, despite additional tractor-plus-trailer passes at first and second harvests. In another experiment, Frost (1988a) observed that yield compensation at the second cut after primary growth had been markedly impaired by severe wheel track damage. The dearth of soil characterisation in Frost's comprehensive comparison of timing and number of passes of wheel traffic, renders it difficult to attribute yield reductions either to soil compaction or to plant damage.

In Scotland, traffic during winter or spring, or at first harvest, significantly diminished plant growth and productivity in the first and second years after reseeding, the extent of which was related to the amount of applied compactive effort (Douglas and Crawford, 1991). The impact of soil compaction on yield was largest during the primary growth phase in spring, when the soil tended to remain relatively wet, irrespective of whether the soil was compacted in the previous year or in the current season. It has been observed frequently that compaction effects are larger in the early life of swards than when the crop is more mature

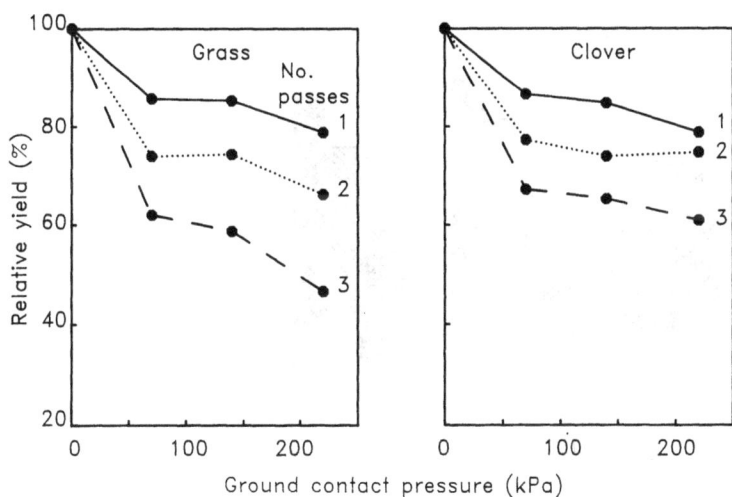

Fig. 4. Relationship between ground contact pressure applied to the soil by tractor traffic and relative yield from second to fourth cuts of first-year grass (left) and clover (right) (from Rasmussen and Møller, 1981).

(Rasmussen and Møller, 1981; Zhezmer et al., 1990; Douglas and Crawford, 1991), probably as a consequence of poorly-developed root mats offering insufficient protection to both plants and soil.

During the drier summer months, regrowth was less susceptible to the adverse conditions in relatively dense soil (Douglas et al., 1992a). The compensation of yield losses at the first cut by enhanced regrowth reported by Frost (1988a), and considered to be the result of plant and/or soil recovery from damage, was detected by Douglas et al. (1992a) only in summer periods of atypically low rainfall; such conditions tended to disadvantage the swards which yielded most at the first cut.

Clover and clover/ grass mixtures

Research on the effects of soil compaction on clover performance has indicated responses similar to those observed with grass crops (Rasmussen and Møller, 1981; Frame, 1985) (Fig. 4b). However, Edwards (1989) reported a larger negative yield response with red clover than with Italian ryegrass to increasing soil strength. In mixed swards, reduced populations of clover plants (Håkansson, 1973) and reductions in clover yield (Davies and Hughes, 1980) have been observed. Frame (1987) confirmed his earlier observations that, on average, increasing the number of wheel passes had a larger detrimental effect on yield than varying the timing of post-harvest traffic between up to 3 and up to 6 days after harvesting, and that companion grasses failed to protect clover from the damaging impact of wheels (Fig. 5). Specific susceptibilty of clover species to conditions in over-dense soil has not been elucidated clearly, due to confounding

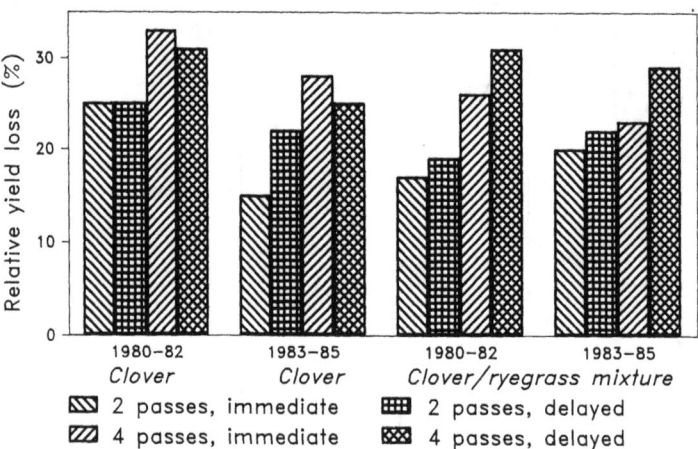

Fig. 5. Red clover yield losses, relative to a no-traffic control, after traffic on clover and clover/ryegrass swards up to 3 days (immediate) and 6 days (delayed) after harvesting (after Frame, 1985, 1987).

TABLE 5

Lucerne (alfalfa) dry matter yields (Mg ha[-1])[1] after no traffic and pre-plant, repeated and simulated farm traffic treatments (from Meek et al., 1988)

Traffic treatment	1983	1984	1985	Total
None	19.7[a]	26.2[a]	25.2[a]	71.1
Pre-plant	19.5[a]	26.0[a]	24.5[b]	70.0
Repeat (pre-plant + harvests)	15.9[b]	21.5[c]	21.8[c]	59.2
Farm pattern	18.8[a]	22.6[b]	22.5[c]	63.9

[1]Column values with the same superscript are not significantly different (Duncan's multiple range test, 5% level).

effects of direct plant damage, lack of winter hardiness, disease, and the relatively short life-span of clovers.

Lucerne (alfalfa)

Impaired productivity of lucerne crops following wheel traffic has been reported from the U.S.A. and Europe. On a loess loam in Germany the consequence of both shoot injury and soil compaction was a yield decline of 16% (Haass and Märtin, 1975a). By comparing the effects of wheel tracking between rows of lucerne plants with in-row traffic, Haass and Märtin (1975b) showed that soil compaction influences were secondary to the contribution from direct damage to the plant stand. Regrowth vigour on relatively compact soil was restricted only in periods of exceptional soil wetness. In the U.S.A., Meek et al. (1988) reported a yield loss of 17% over three years, arising from repeated (pre-plant plus harvest) wheel traffic covering 100% of the plot area, and a loss of 10% after simulated conventional farm traffic that covered 48% of the plot area (Table 5). In that experiment, significantly fewer fine roots were detected in over-compacted soil than in traffic-free soil in the second year of production (Rechel et al., 1990). The detrimental effect of compaction on root growth resulted in poorer growth rates and biomass production (Rechel et al., 1987). Šantrůček (1989) reported reductions in yield and nutritive value (content of crude protein, potassium, phosphorus and calcium) when a clay loam/loamy soil in former Czechoslovakia was compacted to a bulk density >1.28 Mg m[-3].

Relationships betweeen soil properties and forage crop yield

The dominant conclusion from research at a variety of locations, on different soil types and with a number of important forage species, is that soil compaction does limit perennial forage crop productivity. Generally, reductions in yield after traffic have been attributed to deterioration in soil physical conditions such as

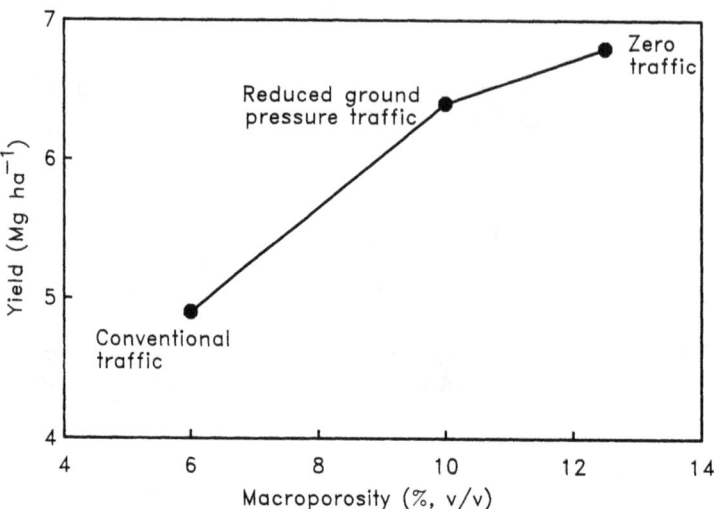

Fig. 6. Relationship between first-cut grass yield from three traffic systems, averaged over 4 years (after Douglas et al., 1992a), and soil macroporosity at 50-100 mm depth (after Koppi et al., 1992).

increased bulk density or decreased air-filled porosity. For example, in a comparison of traffic systems, Douglas et al. (1992a) found that first-cut yields of ryegrass were heavier when the volume of soil macropores was larger (Fig. 6).
Douglas and Crawford (1989) demonstrated that, in second-year ryegrass, both

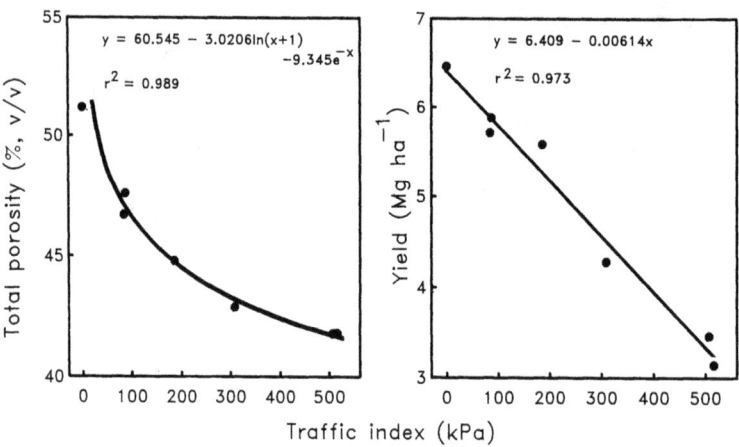

Fig. 7. The effect of increasing traffic index (tyre/soil contact pressure x number of passes) on soil porosity (left) and second-year ryegrass grass yield (right) (after Douglas and Crawford, 1989).

soil porosity and herbage yield decreased with increased amount of wheel traffic (Fig. 7). In both of the foregoing experiments, compaction effects were greatest on springtime (primary) grass growth, in a period when the soil was relatively wet and, when compact, exhibited a relatively high degree of water saturation.

Overall, the causative factors of compaction and the changes induced in the soil strongly suggest that measures can be taken, or strategies adopted, to eliminate or minimise the impact of traffic on forage crops and production efficiency.

AVOIDANCE AND AMELIORATION OF COMPACTION

Practical strategies for avoiding compaction on land managed for conservable perennial forage crops, which are essentially similar to those described elsewhere for annual crops, have been summarised for grassland by Douglas (1989). The high frequency of traffic activity required on forage crops, dictates that some degree of soil damage by wheels is practically unavoidable when conventional machinery is employed. However, the research reviewed above has aided identification of alternative approaches to traffic management capable of minimising, or even eliminating, soil compaction. Foremost among these alternatives are reduced ground contact pressure systems, in which vehicles, machines and implements are fitted with larger-than-standard tyres with low inflation pressures, in order to reduce tyre/soil contact pressure and, thereby, the potential for compaction.

Fig. 8. Tractor, forage harvester and silage trailer used in a reduced ground contact pressure system (30-200 kPa) (from Douglas and Campbell, 1987).

TABLE 6

Annual herbage yields from 2.4-m wide zero-traffic beds and from reduced ground contact pressure and conventional traffic systems (from Douglas et al., 1992a)

Traffic system	Dry matter (Mg ha $^{-1}$)				
	1986	1987	1988	1989	Average
Zero (bed area)	11.2	14.9	14.7	14.6	13.9
Reduced ground contact pressure	11.7	13.5	15.3	14.5	13.7
Conventional	10.9	11.3	12.7	12.7	11.9
LSD	ns	1.03	1.33	0.92	

Low ground contact pressure tyres have been proved to be effective in enhancing grass yields, relative to those obtained after using standard tyres, when fitted to a tractor/slurry tanker combination (Frost, 1988b), or to the complete set of equipment necessary for production of grass for silage (Fig. 8). Douglas et al. (1992a) compared a reduced ground contact pressure system, applying commercially available tyres suitable for operation at inflation pressures ranging from 30 to 200 kPa (depending on machine), with a conventional traffic system applying standard tyres with inflation pressures ranging from 160 to 480 kPa. A consistent benefit in yield accrued from the former system (Table 6).

On wet soils of low bearing capacity in Ireland, a low ground contact pressure silage harvesting system was successful in moderating soil damage and in enabling harvesting to be completed where conventional systems failed due to sinkage

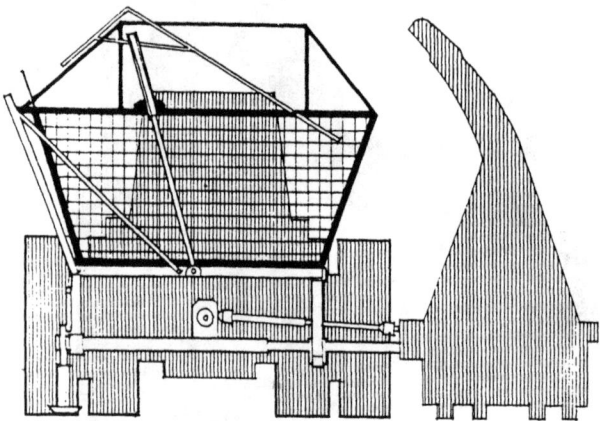

Fig. 9. Diagrammatic rear view of low ground contact pressure forage harvesting equipment (from Forristal, 1989).

(Forristal, 1989). The key components of Forristal's machinery were wide or dual tyres on the tractor and a tipping hopper mounted on the rear of the tractor (Fig. 9).

Simulations of reduced ground contact pressure traffic promoted good yields in Dutch (Gabriels et al., 1990) and Russian (Zhezmer et al., 1990) trials; the former authors reported yield reductions at ground contact pressures >80 kPa, and the latter at >150 kPa on wet peat soil and >200 kPa on dryland meadows.

A zone production system for alfalfa (lucerne), comprising traffic-free areas between permanent traffic lanes, has been advocated by Meek et al. (1988). A perennial ryegrass sward on 2.4-m wide zero-traffic beds consistently out-yielded a sward in a conventional traffic system over four years (Douglas et al., 1992a) (Table 6). However, there was no overall advantage when the uncropped wheel track area (approximately 17%) in the zero-traffic system was taken into account.

Another approach with potential for reducing wheel-induced soil damage involves reducing the area of the field receiving traffic and/or reducing the intensity (mass x number of passes per area) of the traffic. In forage cutting, both the area trafficked and the intensity of the traffic can be significantly diminished by using wide mowers, raking together two adjacent swaths, and thereby reducing by half the area covered by relatively heavy harvesting equipment (Schuler, 1989). Unfortunately, no benefit will accrue from larger-than-average machines if, as is usually the case, compactable land receives complete traffic coverage over a succesion of cuts and seasons.

Methods for ameliorating over-compacted grassland soil

In annual crop management in the U.K., the technique of occasionally loosening the subsoil, in which the soil is displaced and disturbed but is not inverted or brought to the surface, is in common usage. For perennial crops, such as grass, non-inverting cultivation is possible for both the topsoil and the subsoil. At a number of locations in England, worthwhile increases in grass yield were obtained only when subsoil loosening was carried out under suitably dry soil conditions without damaging the sward; a shallower loosening (mini-subsoiler) was ineffective (Farrar and Marks, 1978).

More recently, a number of researchers have evaluated a slant-legged loosener (Paraplow). In a two-year (1987-1988) study in southern England (D. Scholefield, personal communication, 1990), loosening by Paraplow after the first silage harvest resulted in an annual yield loss of approximately 25%, which was attributed to damage to the grass root system. Smith et al. (1990) compared untreated clay loam soil (compacted by stock) with paraplowed soil. Loosening generated a 13% increase in dry matter yield at the first cut over three years, but no significant enhancement of total yield in each of the years. Frost (1988a) used a Paraplow to 30-35 cm depth and yields were reduced at the following cut and, within one year of loosening, the smaller soil strength values obtained, reverted to the larger values recorded in unloosened areas. In New Zealand, Chapman and Allbrook (1987) reported a yield benefit from loosening by a grassland subsoiler to 45 cm depth, attributable to increased root penetration. However, when land was treated in summer, root pruning caused problems of crop desiccation. It is in the drier summer months that loosening by whatever implement is most effective in terms of altering the soil structure by shattering large units. A widely recognised problem associated with grassland loosening is the creation of irregularities in ground surface level that can severely limit the subsequent use of mowers.

Shallow slitting (to a depth of 15 cm) by a spiked "aerator", proved to be very effective in alleviating the detrimental effects of cattle treading on a clay loam soil in Wales (Davies et al., 1989). This relatively inexpensive and easily implemented treatment may also have a useful application in wheel-damaged grassland.

The effects of adverse soil conditions on yield can be overcome by increasing rates of application of fertilizer-nitrogen (Gooderham, 1977). However, it has also been observed that on compacted land the response of grass to nitrogen can be impaired to such an extent that additional nitrogen cannot compensate fully for soil damage in neither hay (Tveitnes and Njøs, 1974) nor first-cut silage crops (Douglas and Crawford, 1993) (Fig. 10). In such circumstances, the extra nitrogen which is not utilised by the crop may be lost through leaching, runoff, or volatilization.

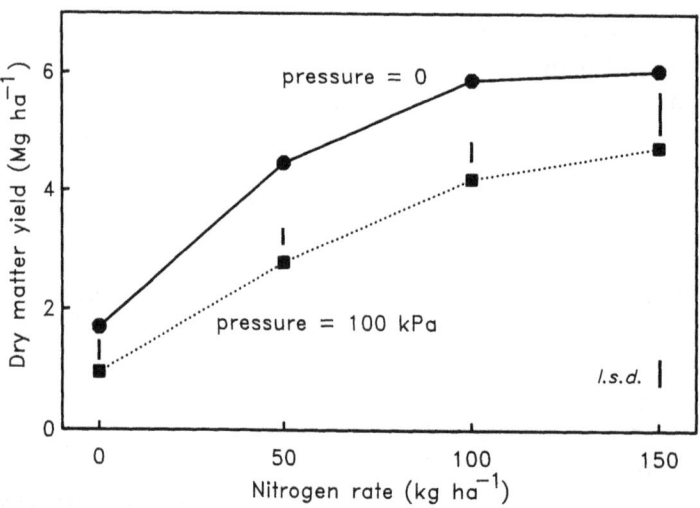

Fig. 10. Ryegrass yield response, at first cut, to rate of fertilizer-nitrogen after application of zero and 100 kPa tyre/soil contact pressure; averaged over 3 years (after Douglas and Crawford, 1993).

Sward replacement by reseeding, usually preceded by ploughing to loosen compacted topsoil, is another course of action appropriate to over-compacted grassland. However, this approach will be discouraged for the forseeable future in temperate agriculture because ploughing of grassland promotes the mineralization and mobilization of nitrogen, which can lead to potential contamination of water supplies.

Soil compaction in grassland can be alleviated to some extent by natural processes such as freeze/thaw cycles (Frame, 1985; Myhr and Njøs, 1983), earthworm activity (Douglas et al., 1986) and root proliferation (Tisdall and Oades, 1980; Douglas et al., 1992b). Angers et al. (1987) reported that long-term cropping with bromegrass had the beneficial effect of generating an increase in inter-aggregate porosity, which could persist at stress levels greater than those normally encountered in the field. There is no evidence, however, of the afore-mentioned processes promoting complete recovery of soil structure from a relatively over-compacted state to a condition closer to optimum.

CONCLUSIONS

(1) Evidence from research to date has demonstrated that: (1) most, if not all, perennial forage crops are sensitive to the growth-limiting conditions which can often prevail in over-compacted soil; (2) practical methods for minimising compaction in soil under forage crops have been devised; (3) measures taken to avoid over-compaction are more likely to be successful than ameliorative operations.

(2) There are still a number of areas in which further research and development could lead to significant advances in the reduction of the impact of soil compaction on perennial forage crop production. These areas include: (1) the acceptance of slurry by the soil and the utilisation of its supplied nutrients by the crop; (2) amelioration of compacted soil; (3) alternative machinery systems enabling widespread adoption of controlled-traffic systems; (4) consequences and benefits of reduced nitrogen inputs; (5) the compaction problem on irrigated forage crops.

REFERENCES

Angers, D.A., Kay, B.D. and Groenevelt, P.H., 1987. Compaction characteristics of a soil cropped to corn and bromegrass. Soil Sci. Soc. Am. J., 51: 779-783.
Batey, T., 1988. Soil Husbandry. Soil and Land Use Consultants, Aberdeen, U.K., 157 pp.
Castle, M.E. and Watson, J., 1974. Red clover silage for milk production. J. Brit. Grassland Soc., 29: 101-108.
Chapman, R. and Allbrook, R.F., 1987. The effects of subsoiling compacted soils under grass - a progress report. Proc. Agron. Soc. New Zealand, 17: 55-58.
Copeman, G.J.F. and Younie, D., 1981. Feed quality and utilisation of red clover swards. In: R.B. Murray (Editor), Legumes in Grassland. Proc. 5th Study Conf. Scott. Agric. Colls., Peebles, U.K., pp. 53-58.

Davies, A., Adams, W.A. and Wilman, D., 1989. Soil compaction in permanent pasture and its amelioration by slitting. J. Agric. Sci., Camb., 113: 189-197.

Davies, W.E. and Hughes, L., 1980. The effect of wheeled vehicles on the yield and persistency of red clover/timothy stands. Welsh Plant Breeding Sta., Aberystwyth, U.K., Ann. Rep. 1979, pp. 50-52.

Douglas, J.T., 1989. Avoiding soil damage from wheel traffic in silage fields. Scott. Agric. Coll., Perth, U.K., Tech. Note 198, 2 pp.

Douglas, J.T. and Campbell, D.J., 1987. Conventional, low ground pressure and zero traffic systems in ryegrass grown for silage, 1985-86. Scott. Inst. Agric. Eng., Penicuik, U.K., Dep. Note SIN/489, 27 pp.

Douglas, J.T. and Crawford, C.E., 1989. Effect of wheel-induced compaction on grass yield and nitrogen uptake, 1988. Scott. Centre Agric. Eng., Penicuik, U.K., Dep. Note 19, 16 pp.

Douglas, J.T. and Crawford, C.E., 1991. Wheel-induced soil compaction effects on ryegrass production and nitrogen uptake. Grass Forage Sci., 46: 405-416.

Douglas, J.T. and Crawford, C.E., 1993. The response of a ryegrass sward to wheel traffic and applied nitrogen. Grass Forage Sci., 48: 91-100.

Douglas, J.T., Jarvis, M.G., Howse, K.R. and Goss, M.J., 1986. Structure of a silty soil in relation to management. J. Soil Sci., 37: 137-151.

Douglas, J.T., Campbell, D.J. and Crawford C.E., 1992a. Soil and crop responses to conventional, reduced ground pressure and zero traffic systems for grass silage production. Soil Tillage Res., 24: 421-439.

Douglas, J.T., Koppi, A.J. and Moran C.J., 1992b. Changes in soil structure induced by wheel traffic and growth of perennial grass. Soil Tillage Res., 23: 61-72.

Edwards, L. M., 1989. Dry matter growth performance of red clover and Italian ryegrass as cover crops spring-seeded into fall-seeded winter rye in relation to soil physical characteristics. J. Soil Water Conserv., 44: 243-247.

Eriksson, J., Håkansson, I. and Danfors, B., 1974. The effect of soil compaction on soil structure and crop yields. Swed. Inst. Agric. Eng., Uppsala, Sweden, Bull. 354, 101 pp.

Farrar, K. and Marks, M.J., 1978. Subsoiling experiments on grassland in the West Midland Region. Exp. Husb., 34: 26-39.

Forristal, D., 1989. A low ground pressure silage harvesting system for small farms on low trafficability soils. Proc. 11th Int. Cong. Agric. Eng., Dublin, Ireland, pp. 223-230.

Frame, J., 1985. The effect of tractor wheeling on red clover swards. Res. Dev. Agric., 2: 77-85.

Frame, J., 1987. The effect of tractor wheeling on the productivity of red clover and red clover/ryegrass swards. Res. Dev. Agric., 4: 55-60.

Frame, J. and Merrilees, D.W., 1990. The effect of tractor wheeling on the productivity of perennial ryegrass (*Lolium perenne* L.) swards. Proc. 13th Gen. Meeting Eur. Grassland Fed., Banská Bystrica, Czechoslovakia, Vol. 1, pp. 170-174.

Frost, J.P., 1984. Some effects of machinery traffic on grass yield. In: J.K. Nelson and E.R. Dinnis (Editors), Machinery for Silage. British Grassland Soc. Occ. Symp., 17: 18-25.

Frost, J.P., 1988a. Effects on crop yields of machinery traffic and soil loosening. Part 1. Effects on grass yield of traffic frequency and date of loosening. J. Agric. Eng. Res., 39: 301-312.

Frost, J.P., 1988b. Effects on crop yields of machinery traffic and soil loosening. Part 2. Effects on grass yield of soil compaction, low ground pressure tyres and date of loosening. J. Agric. Eng. Res., 40: 57-69.

Gabriels, P.C.J., Arts, W.B.M. and Bosma, A.H., 1990. Grassland production related to traffic-induced soil compaction. Proc. 13th Gen. Meeting Eur. Grassland Fed., Banská Bystrica, Czechoslavakia, Vol. 1, pp. 195-199.

Gaheen, S.A. and Njøs, A., 1977. Effect of tractor traffic on timothy (*Phleum pratense* L.) root system in an experiment on a loam soil. Agric. Univ. Norway, Dept. Soil Fert. Manage.,

Rep. 93, 12 pp.

Gooderham, P.T., 1977. Some aspects of soil compaction, root growth and crop yield. Agric. Progress, 52: 33-44.

Grimes, D.W., Sheesley, W.R. and Wiley, P.L., 1978. Alfalfa root development and shoot regrowth in compact soil of wheel traffic patterns. Agron. J., 70: 955-958.

Haass, J. and Märtin, B., 1975a. Der Einfluss eines spezifischen Raddruckes von 1.2 und 2.0 kp/cm² sowie von Pflegemassnahmen auf Wachstum und Ertrag der Luzerne (*Medicago media* Pers.). (The effect of a specific wheel pressure of 1.2 and 2.0 kp/cm² as well as of after-cultivation on growth and crop yields of alfalfa (*Medicago media* Pers.)). Arch. Acker-Pflanzenb. Bodenkd., 19: 711-722 (in German).

Haass, J. and Märtin, B., 1975b. Der Einfluss von Bodenverdichtung (Ap-Horizont) auf Wachstum und Ertrag von Luzerne (*Medicago media* Pers.) ohne Befahren der Pflanzen. (The effect of soil compaction (Ap horizon) on the growth and crop yield of alfalfa (*Medicago media* Pers.) without running over the plants). Arch. Acker-Pflanzenb. Bodenk., 19: 723-733 (in German).

Haass, J. and Simon, W., 1975. Einfluss von Raddruck und Triebradschlupf bei differenzierter Bodenfeuchte auf Luzerne und Luzernegras. (Alfalfa and alfalfa grass as influenced by wheel pressure and driving wheel slip at varied soil moisture levels). Arch. Acker-Pflanzenb. Bodenkd., 19: 905-915 (in German).

Håkansson, I., 1973. Tung körning vid skörd av slåttervall. Tre försök på Röbäcksdalen 1969-1972. (Effect of heavy machinery when harvesting ley crops. Three field experiments in northern Sweden 1969-1972). Lantbrukshögskolan, Uppsala, Sweden, Rapp. Jord. Avdel., 33, 22 pp. (in Swedish).

Håkansson, I., Voorhees, W.B. and Riley. H., 1988. Vehicle and wheel factors influencing soil compaction and crop response in different traffic regimes. Soil Tillage Res., 11: 239-282.

Koppi, A.J., Douglas, J.T. and Moran, C.J., 1992. An image analysis evaluation of compaction in grassland. J. Soil. Sci., 43: 15-25.

Luten, W., Roozeboom, L. and Remmelink, G.J., 1983. Invloed van berijden op produktie en persistentie van grassoorten. (Effect of driving on yield and persistence of grass species). Proefsta. Rundv., Schapen en Paarden, Lelystad, Netherlands, Rapp. 90, 18 pp. (in Dutch).

Meek, B.D., Rechel, E.A., Carter, L. and DeTar, W.R., 1988. Soil compaction and its effects on alfalfa in zone production systems. Soil Sci. Soc. Am. J., 52: 232-236.

Meek, B.D., Rechel, E.A., Carter, L. and DeTar, W.R., 1989. Changes in infiltration under alfalfa as influenced by time and wheel traffic. Soil Sci. Soc. Am. J., 53: 238-241.

Miller, D.A., 1984. Forage Crops. McGraw-Hill, New York, NY, U.S.A., 530 pp.

Myhr, K. and Njøs, A., 1983. Verknad av traktorkjøring, fliere slåttar og kalking på avling og fysiske jordeigenskapar i eng. (Effects of tractor traffic, number of cuts and liming on yields and soil physical properties in Norwegian grasslands). Norges Landbrukshøgskole, Inst. Jordkultur, Melding 126, 14 pp. (in Norwegian).

Óskarsson, M., 1975. Faktorer som påvirker graesmarkens varighed og avling. (Factors influencing persistence and yield of grassland). Nord. Jordbruksforsk., 57: 189-190 (in Danish).

Rasmussen, K.J. and Møller, E., 1981. Genvaekst efter fortørring af graesmarksafgrø der. II. Jordpakning i forbindelse med høst og transport. (Regrowth after pre-wilting of grassland crops. II. Soil compaction in connection with harvest and transport). Tidsskr. Planteavl., 85: 59-71 (in Danish).

Rechel, E.A., Carter, L.M. and DeTar, W.R., 1987. Alfalfa growth response to a zone-production system. 1. Forage production characteristics. Crop Sci., 27: 1029-1034.

Rechel, E.A., Meek, B.D., DeTar, W.R. and Carter, L.M., 1990. Fine root development of alfalfa as affected by wheel traffic. Agron. J., 82: 618-622.

Šantrůček, J., 1989. Kvalita píce vojtěšky v závislosti na zhutnění a kultivaci půdy. (Quality of lucerne forage in relation to soil compaction and cultivation). Rostl. Výr., 35: 1101-1107 (in Czech).

Schuler, R.T., 1989. Compaction implication in machinery management. Am. Soc. Agric. Eng., St. Joseph, MI, U.S.A., ASAE Pap. 89-1017, 7 pp.

Sheesley, R., Grimes, D.W., McClellan, W.D., Summers, C.G. and Marble, V., 1974. Influence of wheel traffic on yield and stand longevity of alfalfa. Calif. Agric., 28: 6-8.

Smith, C.A., Frame, F., Merrilees, D.W. and Whytock, G.P., 1990. Alleviation of soil compaction due to animal treading (poaching) in long-term grassland. Proc. 13th Gen. Meeting Eur. Grassland Fed., Banská Bystrica, Czechoslavakia, Vol. 1, pp. 211-214.

Soane, B.D. (Editor), 1983. Compaction by agricultural vehicles: A review. Scott. Inst. Agric. Eng, Penicuik, U.K., Tech. Rep. 5, 95 pp.

Soane, B.D., 1987. Over-compaction of soils on Scottish farms: A survey. Scott. Inst. Agric. Eng., Penicuik, U.K., Res. Summary 3, 18 pp.

Thomas, R. and Evans, C., 1975. Field experience on grassland soils. In: Soil Physical Conditions and Crop Production. HMSO, London, U.K., MAFF, Tech. Bull. 29, pp. 112-124.

Tisdall, J.M. and Oades, J.M., 1980. The management of ryegrass to stabilise aggregation of red-brown earth. Aust. J. Soil Res., 18: 415-422.

Tveitnes, S. and Njøs, A., 1974. Køyreskadeforsøk på eng under Vestlandstilhøve. (Soil compaction problems on grassland in West Norway). Forskning Försök Landbruket., 25: 271-283 (in Norwegian).

Zhezmer, N.V., Zotov, A.A., Dedaev, G.A., Shevtzov, A.V. and Kozlov, V.V., 1990. Effect of agricultural machinery operations on the grassland soil productivity and soil conditions. Proc. 13th Gen. Meeting Eur. Grassland Fed., Banská Bystrica, Czechoslavakia, Vol. 1, pp. 77-78.

Soil Compaction in Crop Production
B.D. Soane and C. van Ouwerkerk (Eds.)
©1994 Elsevier Science B.V. All rights reserved.

CHAPTER 16

Role of Soil and Climate Factors in Influencing Crop Responses to Soil Compaction in Central and Eastern Europe

J. LIPIEC[1] and C. SIMOTA[2]

[1]Polish Academy of Sciences, Institute of Agrophysics, Lublin, Poland
[2]Research Institute of Soil Science and Agrochemistry, Bucharest, Romania

SUMMARY

In Central and Eastern Europe, due to the recent increase in field traffic, soil compaction has become one of the major factors affecting crop production. The variations in crop response to soil compaction associated with crop type, soil characteristics and weather conditions, are reviewed. The importance of soil water in influencing soil compactability and crop response to soil compaction is illustrated. The influence of fertilization on crop response to soil compaction and the response of plant roots are briefly discussed.

INTRODUCTION

Recent increases in the mechanization of agronomic practices is one of the major factors causing soil compaction in Central and Eastern Europe (Stoinev, 1975; Butorac and Turšić, 1978; Haman, 1979; Lhotský, 1984a; Canarache et al., 1984a,b; Kovda, 1987; Rusanov, 1988; Petelkau, 1989a; Várallyay and Lesztak, 1990; Domżał et al., 1991; Lhotský et al., 1991; Lipiec et al., 1991). This effect is enhanced by the increasing mass of farm vehicles. For example, in the former U.S.S.R., since 1922 the mass and power of tractors used have increased 15 and 20 times, respectively (Kravchenko, 1986), while in former Czechoslovakia since 1955 the mass of tractors and tillage implements have increased by 68 and 200%, respectively (Valigurská and Lhotský, 1985). In the region under discussion, the mass of presently available wheel tractors varies from 3 to 13 Mg and the mass of combine harvesters from 7 to 18 Mg. In the former U.S.S.R., relatively heavy vehicles are used on large farms, where commonly used crawler tractors exert ground contact pressures from 150 to 200 kPa, wheel tractors from 200 to 300 kPa, and the ground contact pressure of combine harvesters exceeds 300 kPa (Rusanov, 1988).

Another important factor influencing soil compaction is vibration of agricultural vehicles (Popov et al., 1977; Timofeyev, 1978; Haman, 1989). Popov et al. (1977), using a seismograph, showed that the vibration in the soil under a wheel tractor is of low frequency and sinusoidal, while under a crawler tractor the vibration is of high frequency and non-sinusoidal. In a sandy loam soil, at a depth of 80 cm, the frequency and transversal amplitude of the vibration were 1.4-1.6 Hz and 0.5 mm, respectively, under a wheel tractor (mass 7.5 Mg) and 25 Hz and 1.0 mm, respectively, under a crawler tractor (mass 6.5 Mg). The compactive effect of the vibration forces is likely to be smaller in fine- than in coarse-textured soils, owing to the higher damping capacity of fine-textured soils.

The cumulative percentage of soil covered by wheelings varies in a wide range, depending on season and crop type. For example, in the former U.S.S.R., from 30 to 80% of the field area was covered by the wheelings of tractors at seedbed preparation for agricultural crops alone (Kovda, 1987), while some parts of the field received 3-9 wheelings. In Poland, for the whole growing season, up to 250% (cereals) and 500% (root crops) of the field area is covered by wheelings (Domżał et al., 1991). With root crops, almost half of the compaction due to field traffic occurs at harvesting.

SOIL RESPONSE

The compactive effect of vehicular pressure is closely influenced by the soil water status (Petelkau, 1984; Lipiec and Tarkiewicz, 1986; Medvedev and Tzybulko, 1987; Bondarev et al., 1988; Okhitin et al., 1991). To avoid excessive compaction for potatoes (*Solanum tuberosum*), Ermich and Hofmann (1984) established the limits of soil water content for various ground contact pressures in relation to the content of fine particles <6 μm (Fig. 1).

These data agree with the findings of Petelkau (1984), who suggested that the ground contact pressure of farm vehicles should be limited in spring to 50 kPa on sandy soils and 80 kPa on loamy and clay soils when the soil water content is 85-95% of the field capacity, while during summer and autumn, when the soil water content is below 70% of the field capacity, the admissible values of ground contact pressure for sandy, loamy and clay soils are 80, 150 and 200 kPa, respectively. However, at a soil water content corresponding to the field capacity or higher, the admissible ground contact pressure for soils of various texture is only 40-50 kPa (Bondarev et al., 1988).

Gaponenko (1987) indicated that in spring another important factor conducive to soil compaction is a fine structure induced by the seedbed preparation. Soil structure was one of the main criteria in the classification of the susceptibility of soils to compaction in Hungary (Várallyay and Lesztak, 1990). In this classification, the assessment of soil structure and its stability was based mainly on data concerning soil texture, cementing compounds (carbonates, sesquioxides), organic matter content and salinity-alkalinity properties. On sandy soil, compaction can be frequently enhanced by a high soil water content in the

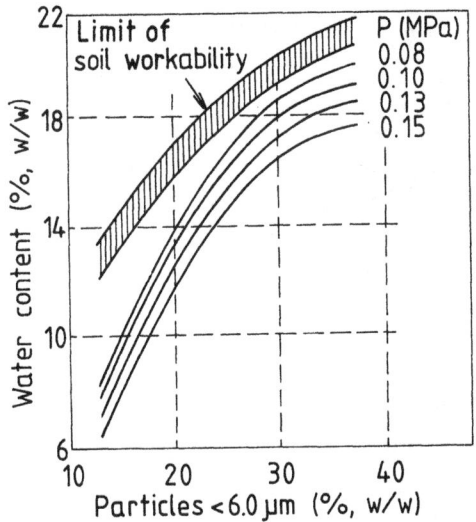

Fig. 1. Critical soil water content for the avoidance of harmful soil compaction in relation to the content of fine particles <6 μm and ground contact pressure (from Ermich and Hofmann, 1984).

subsurface layer while the surface layer is workable (Petelkau et al., 1983).

Okhitin et al. (1991) showed that repeated passes of a tractor weighing 5.3 Mg caused significant increases in soil deformation, mostly in the plough layer. However, subsoil compaction may occur when vehicles with axle loads >4 Mg are used (Petelkau, 1984).

Soil deformation influences several aspects of the soil environment, such as soil air (Gliński and Stępniewski, 1985), water (Walczak, 1977), strength (Lipiec et al., 1991), heat (Lipiec et al., 1991) and biological activity (Haman, 1979; Stępniewska et al., 1990) which, in turn, affect crop growth and production. An example of the response of penetration resistance and air-filled porosity to the bulk density in relation to the matric water potential is shown in Fig. 2.

Through reduced infiltration and increased runoff, wheel traffic may induce water erosion and result in ponded water on the surface in spring and after heavy rain (Petelkau, 1983; Lhotský, 1984a). On some soils, irrigation is also an important factor. The effect can be indirect through a higher soil water content at traffic, as well as direct by the compacting action of the water itself. A high compactive effect to a depth of 80-90 cm was caused by flood irrigation on chestnut soils (Bondarev, 1987). On salt-affected and grassland soils this effect was even found to 200 cm depth.

CROP RESPONSE

Since the relationship between soil compaction and crop growth is a complex

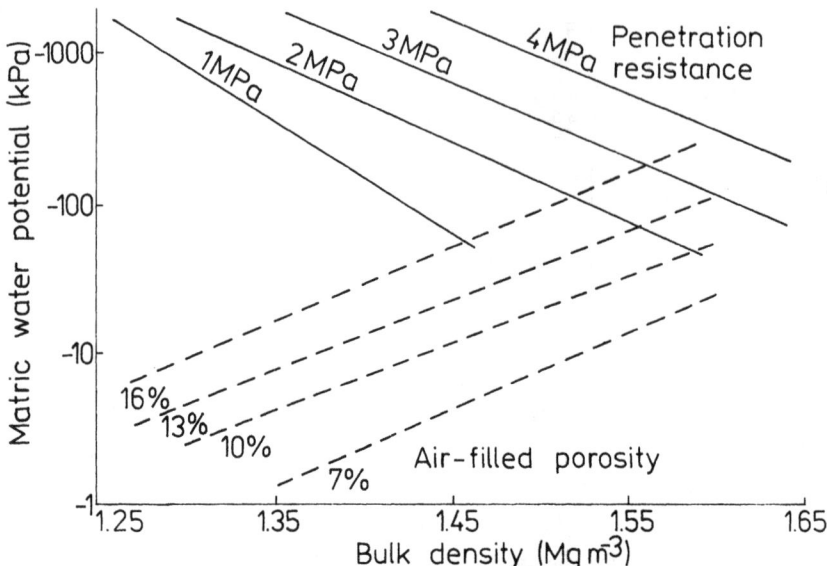

Fig. 2. Levels of penetration resistance (MPa) and air-filled porosity (%, v/v) of silty loam soil in relation to bulk density and matric water potential (after Lipiec et al., 1991).

system of interacting functional relationships, there is no universal index describing the crop response. The yield decrease in overcompacted soil is frequently attributed to excessive mechanical impedance (Shipilov, 1982; Pabin and Sienkiewicz, 1984; Pabin and Włodek, 1986; Lipiec et al., 1991) via restriction of the root penetration and, consequently, the reduction of water and nutrient uptake or insufficient aeration (Gliński and Stępniewski, 1985; Pfleger and Unger, 1985). The interrelations are strongly dependent on soil characteristics, crop type and weather conditions and their importance is frequently limited to specific growth conditions (Gliński and Lipiec, 1990).

Many researchers indicate that under most conditions the soil water content or the soil water potential appear to be of particular importance because they directly affect crop growth and indirectly affect other significant factors, such as aeration, mechanical impedance and soil temperature (Pittelkow and Unger, 1988; Lipiec et al., 1991).

Crop emergence and establishment

In the former U.S.S.R., wheel traffic at seedbed preparation was found to decrease the emergence of various field crops by 20-40% (Kovda, 1987). Pabin and Sienkiewicz (1984) found that the negative effect of soil compaction on the emergence of sugar beet *(Beta vulgaris)* in moist soil was greater at greater sowing depth (Fig. 3). However, in dry soil the emergence of the same crop was earlier

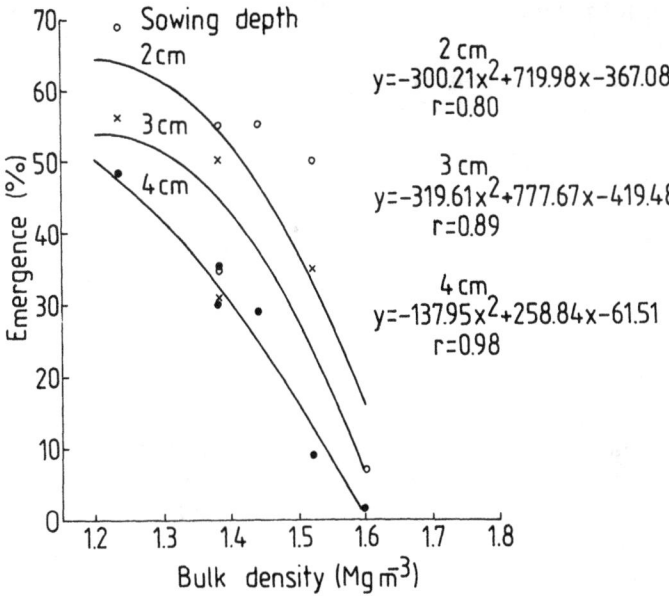

Fig. 3. Emergence of sugar beet in relation to soil bulk density in the 0-25-cm layer and sowing depth (after Pabin and Sienkiewicz, 1984).

and more uniform in the compacted than in the loose soil, presumably due to better water availability (Tindzhyulis and Brazauskas, 1987).

Reduced emergence of rye (*Secale cereale*) due to soil compaction, resulted in a 0.05-0.8 Mg ha[-1] lower yield (Slesarev, 1987). Droese et al. (1983) indicated that the emergence of oats (*Avena sativa*) was relatively less affected by soil compaction than the number of ears and the yield. Restricted crop emergence was simultaneously conducive to weed development (Shipilov, 1982).

Soil compaction may also affect the plant population at harvest and the resistance to lodging. It was shown that on silty loam, at bulk densities of 1.35 and 1.58 Mg m^{-3}, the plant population of spring barley *(Hordeum vulgare)* at harvest was lower than the sowing density by 17.6 and 20.8%, respectively, while the resistance to lodging (on a scale from 1 to 9; 9 being the maximum resistance to lodging) was 4.4 and 8.6, respectively (Lipiec et al., 1991).

Crop yield

Cereals
Excessive pre-sowing compaction resulted in reduced growth, both of shoots and roots of spring barley. A further response to increasing soil compaction above optimum consisted in a reduced abundance of spikes (Beran, 1988) and number of tillers per m^2 (Nugis, 1987) and, subsequently, in a reduction of the

leaf area index and dry matter formation in the aerial parts (Beran, 1988; Lipiec et al., 1991). Further compaction caused the grain weight per spike to decline and, at the highest compaction level, compensatory growth failed to occur.

In most experiments the final crop yield was the main indicator of crop response to soil compaction. The yield response to soil compaction of small-grained cereals such as barley, wheat (*Triticum aestivum*), rye and oats, was frequently parabolic (Krüger, 1970; Dechnik et al., 1982; Droese et al., 1982; Śmierzchalski et al., 1984; Petelkau et al., 1985; Pabin and Włodek, 1986; Petelkau and Seidel, 1986; Afanasyev et al., 1987; Nugis, 1987; Slesarev, 1987; Beran, 1988; Petelkau et al., 1988a; Lipiec et al., 1991).

Figs. 4 and 5 show relationships between grain yield of cereal crops and soil bulk density in the plough layer. The data indicate that very pronounced optima occurred in those experiments where a wide range of bulk densities was compared. This was confirmed in other experiments with a relatively narrow range of traffic intensities (Trzecki, 1974; Kollar et al., 1976; Špaldon and Karabinová, 1978; Sienkiewicz et al., 1988; Śmierzchalski et al., 1988a,c), in which no response or slightly positive effects on the yield were obtained. Figs. 4 and 5 also show that the absolute values of the optimum bulk density are higher for coarse soils, which may result partly from higher textural density. Rendzina soils in Estonia gave the highest yield of cereals at bulk densities of 1.22-1.32 Mg m^{-3} (Nugis, 1987). It is interesting to note that on humic chernozem (fine loam), bulk density values were low and yield response of spring barley was only small (Fig. 4), even though the treatments included passes of heavy tractors weighing from 7.5 to 12.5 Mg (Shipilov, 1982).

Petelkau et al. (1988b) and Petelkau (1989b) developed regression models for describing the response of grain yield of winter rye to bulk density in upper, middle and lower parts of the plough layer of loamy sand and in the upper subsoil layer. For the plough layer this response is illustrated in Fig. 6. In the former U.S.S.R., many researchers related crop response to soil compaction to the ground contact pressure exerted by vehicular equipment (Afanasyev et al., 1987; Bondarev et al., 1988; Dedaev et al., 1988; Rusanov, 1988). The mean decrease in the yields of spring and winter cereals and forage crops in relation to the ground contact pressure exerted by various tractors from numerous experiments in Belorussia, Estonia, Lithuania, Ukraine, Moscow and Voronezh Regions, Krasnodar Territory and Tula Region is shown in Table 1. The relatively large coefficients of variation indicate the influence of other factors affecting crop response to soil compaction.

Rusanov (1991) reported the difference in yield response due to either crawler or wheel tractors. The average results from experiments in various regions of the former U.S.S.R. indicated that crawler tractors DT-74 and DT-75M (mass 6.5 Mg, ground contact pressure 50 kPa) caused a reduction in crop yield by 11%, whereas wheel tractors T-150K and K-700 (mass 7.5 and 11.8 Mg, ground contact pressure 80 and 180 kPa, respectively) caused a yield reduction of 20-25%. The

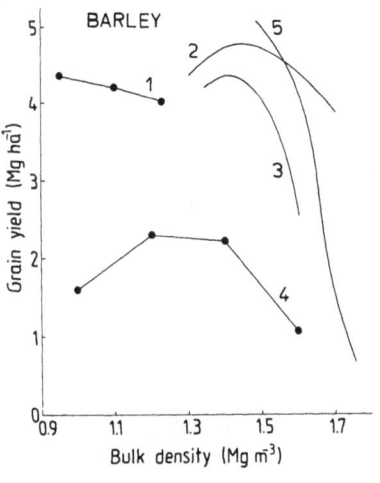
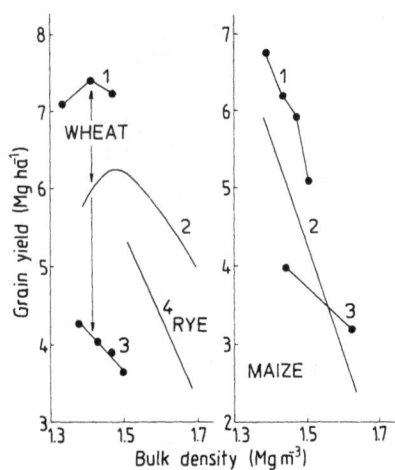

Fig. 4. (left). Relationships between the grain yield of spring barley and soil bulk density in the plough layer. 1. fine loam, Voronezh Region, Russia (after Shipilov, 1982); 2. loamy sand, Germany (after Petelkau and Seidel, 1986); 3. silty loamy sand, Poland (after Pabin and Włodek, 1986); 4. light loam, Lithuania (after Tindzhyulis and Brazauskas, 1987); 5. loam, former Czechoslovakia (after Lhotský et al., 1991). Curves without dots were derived from regression models of long-term experiments.

Fig. 5. (right). Relationships between the grain yield of winter wheat, winter rye and maize, and soil bulk density in the plough layer. Left: 1. clay loam, former Czechoslovakia (after Kollar et al., 1976); 2. loamy sand, Germany (after Petelkau and Seidel, 1986); 3. chernozem, Romania (after Sin et al., 1989); 4. loamy sand, Germany (after Blank and Petelkau, 1988). Right: 1. chernozem, Romania (after Sin et al., 1989); 2. heavy loam, Romania (after Canarache et al., 1984b); 3. loamy sand, Ukraine (after Maliyenko, 1984). Curves without dots were derived from regression models of long-term experiments.

greater reductions in yield caused by wheel tractors were found not only in the wheel tracks but also in the space between wheel tracks (Puponin and Matyuk, 1987).

Under Romanian conditions, the yield of maize *(Zea mays)* and wheat (Canarache et al., 1984a,b; Sin, 1984; Sin et al., 1989) grown on fine-textured soils, decreased linearly with increasing soil bulk density (Fig. 5). This is attributable not only to specific crop responses but also to high initial soil compactness due to the farming system, as well as to droughty climatic conditions (A. Canarache, personal communication, 1990). The simulation model developed by Simota and Canarache (1988) predicted grain maize yields with differences between the observed experimental and simulated yield of < 15%. Research work in Ukraine (Maliyenko, 1984) showed that whole-plot compaction of sandy soil by tractor wheels after planting reduced maize grain yield by 20%, whereas when

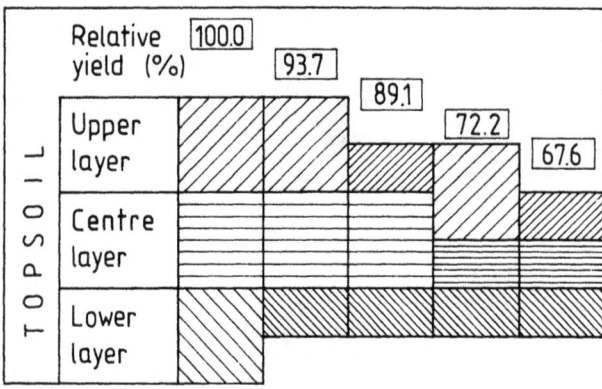

Fig. 6. Influence of bulk density differences in upper, centre and lower topsoil layers (0-10, 10-20, 20-30 cm depth) on the grain yield of winter rye. Wide hatched: optimum bulk density; narrow hatched: 0.15 Mg m⁻³ overcompaction (from Petelkau et al., 1988b).

compaction was confined to the interrow zone, the yield reduction was only 10%. In another study, moderate interrow compaction resulting in bulk densities of 1.23-1.30 Mg m^{-3}, did not affect the yield of maize cobs (Kuznetzov, 1978).

TABLE 1

Mean yield depression of spring and winter cereals and forage crops due to tractor passes in 30 experiments (from Rusanov, 1991)

Tractor mass (Mg)	Tractor type	Ground contact pressure (kPa)	Number of passes	Yield depression (%, w/w)	Coefficient of variation (%)
3.2	Wheel	150	1	7.8	80.8
			2-3	14.1	66.5
			4-5	19.8	80.5
6.1	Crawler	150	1	7.5	40.0
			2-3	13.3	62.3
			4-5	19.2	53.1
7.5	Wheel	180	1	12.3	58.5
			2-3	23.3	35.4
			4-5	30.3	39.2
12-13	Wheel	200	1	16.3	49.9
			2-3	26.8	31.7
			4-5	35.1	28.6

A significant negative yield response of cereal crops to soil compaction was also found in numerous model experiments in Yugoslavia (Turšić, 1982; Butorac, 1982a,b; Racz and Butorac, 1983). Under the same site conditions, maize was more sensitive to soil compaction than spring barley (Shevlagin, 1966; Krüger, 1970; Slesarev, 1987) or wheat (Sin et al., 1989). Śmierzchalski et al. (1984) reported that the optimum bulk density was higher for winter wheat than for winter rye, spring barley and spring oats.

Small-grained cereals were less sensitive to soil compaction than wide-row crops such as soybean *(Glycine hispida)* (Butorac and Turšić, 1978), legume field peas *(Pisum sativum)* (Pabin and Włodek, 1986; Gaponenko, 1987) and sugar beet (Sin et al., 1989). Pabin and Włodek (1986) reported that a statistically significant decrease in the yield of spring barley and peas grown on silty loamy sand, occurred at bulk densities >1.55 Mg m^{-3} and >1.49 Mg m^{-3}, respectively.

The differences among crops largely depend on the compaction level. Puponin and Matyuk (1987) reported that for spring barley and some grasses the response to moderate soil compaction was similar, while in heavily compacted soil the yield depression was 33% for barley and 15% for grasses.

The response of cereals to soil compaction varied markedly among varieties (Nugis, 1987). Due to varietal differences (2-4 varieties), the yield depression on compacted loamy sand soil ranged from 6 to 15% for spring barley, from 13 to 23% for spring wheat and from 10 to 23% for spring oats. On sandy loam soil, the corresponding ranges were 12-18%, 13-16% and 23-37%, respectively. The response of cereals to subsoil compaction was similar to that of potatoes (Puponin and Matyuk, 1987).

Root crops
Root crops are traditionally regarded as particularly sensitive to compaction. The importance of soil compaction on growth of these crops is enhanced because their harvestable products grow underground.

Sugar beet was shown to be one of the crops most sensitive to overcompaction in Romania (Sin et al., 1989), former Czechoslovakia (Lhotský, 1984a) and Poland (Pabin and Sienkiewicz, 1984). Fig. 7 illustrates the relationship between sugar beet yield and bulk density. The yield decrease of the crop was much greater when a compact layer in clay soils occurred at a depth of <22 cm than at a depth of 30 cm (Lhotský, 1984a). On a fertile chernozem, compaction in the upper part of the plough layer was considered as a main reason for a drop in crop yield (Kuznetzova and Danilova, 1987). The reduction in root yield of the crop was accompanied by a decrease in sugar content (Sin et al., 1989). A significant reduction in the crop yield when compacted after sowing was attributed to decreased emergence (Pabin and Sienkiewicz, 1984).

From analysis of some experiments with potatoes on sandy and loamy soils (Ermich and Hofmann, 1984; Maliyenko, 1984; Starczewski et al., 1984; Nugis, 1987), it follows that increments in bulk density in the plough layer of 0.05, 0.1,

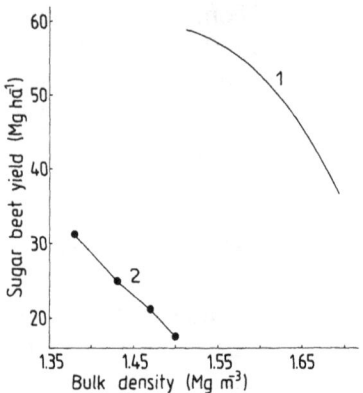

Fig. 7. Relationship between sugar beet yield and soil bulk density in the plough layer. 1: sandy loam, Poland (after Pabin et al., 1991); 2: chernozem, Romania (after Sin et al., 1989). Curve 1 was derived from a regression model of long-term experiments.

0.15 and 0.2-0.25 Mg m^{-3} resulted in yield reductions of 0.4-4.4, 4.2-4.8, 9.7-21.9 and 14.5-25.5%, respectively. However, on clay soil the tuber yield reduction following the same increase in bulk density was higher (Ermich and Hofmann, 1984). Yield response was greater with post- than with pre-planting compaction and with whole-plot compaction than with only interrow or row compaction (Maliyenko, 1984). Starczewski et al. (1984) reported that, compared to gross yield, marketable yield of potatoes (tubers >35 mm) was reduced by 36.9-80.9%, depending on the compaction level (rolling before planting). Studies by Maliyenko (1984) in Ukraine indicated significantly greater incidence of potato scab in soil subjected to compaction. Interrow cultivation in strongly compacted plots led to increases in tuber yield of >2 Mg ha^{-1}, while in a very loose soil yield decreases of similar magnitude were found (Starczewski et al., 1984). Ermich and Hofmann (1984) reported that, to avoid yield reductions of >5%, the ground contact pressure at early planting of potatoes should be as low as ≤60 kPa. To achieve this, these authors suggested the use of dual wheels and low tyre inflation pressure.

The yield of carrot *(Daucus carota)* was depressed by up to 17% by moderate compaction (1 pass of a light, 1.9-Mg tractor) on the whole area on silty loam soil at seedbed preparation (Kęsik, 1985). In addition, the proportion of small roots (shorter than 100 mm and smaller than 25 mm in diameter) and forked roots, as well as dry matter content, increased (Kęsik, 1985; 1990; 1991). The deformed roots were unsuitable for processing (Racz, 1973).

Oil seed rape, radish and peas

These species are also sensitive to compaction. In field experiments, seed yield of winter oil seed rape *(Brassica napus)* (Śmierzchalski et al., 1988b) and whole plant yield of oil radish (*Raphanus sativus var. oleiferus*) (Blank and Petelkau, 1988) decreased linearly with increasing soil bulk density in the plough layer. An adverse whole-plant yield response of peas to soil compaction was mainly associated with low air-filled porosity in the growing season (Pabin and Włodek, 1986).

Clover, alfalfa and grassland

An increase in bulk density of light loam induced by wheel traffic at seedbed preparation and at tillering of barley with undersown clover, resulted in higher yield of clover in the following year compared to that from untrafficked soil (Trzecki, 1974). This effect was ascribed to less lodging of barley (nurse crop) and better emergence and rooting of clover. In a pot experiment, the growth and nutrient uptake of red clover from compacted soil were better than those of alfalfa (Butorac, 1982a).

On a clay loam soil in former Czechoslovakia, the yield response of alfalfa to increasing bulk density from 0.97 to 1.70 Mg m^{-3}, was closely related to variety (Šantrůček, 1989a,b). Var. Palava gave the highest forage yields in the bulk density range 1.20-1.24 Mg m^{-3}, while with var. Rambler 1 a pronounced optimum was not found. The latter variety was higher in nutrient content and lower in fibre content than was the var. Palava.

The hay yield in grasslands was only slightly affected by a single pass of a tractor exerting a ground contact pressure of 100 kPa, while with ground contact pressures of 200 and 300 kPa the yield was reduced by 19-20% (Dedaev et al., 1988). This reduction was largely attributed to a decreasing density of the canopy through direct damage to the plants and smaller plant height. The magnitude of the damage was modified by rainfall and air temperature in the periods between cuttings. Under controlled conditions, the presence of a compacted layer diminished the dry matter yield of ryegrass (*Lolium multiflorum*) by 5.1% when the water supply was adequate and by 15.3% when drought was imposed (Unger et al., 1988).

Fruit shrubs

Interrow compaction (5 and 10 passes of a 1.9-Mg tractor) resulted in a slightly lower raspberry (*Rubus idaeus*) yield and lower specific weight of fruit, chlorophyll content and intensity of photosynthesis (Wieniarska et al., 1987). The negative yield response and the reduction in root growth with increasing compaction of silty loam were greater in raspberry than in northern red berry (*Ribes schlechtendalli*) and black currant (*Ribes nigrum*) (Kęsik and Stanek, 1983).

Effect of soil type

Numerous studies performed under the same weather conditions revealed that crop responses to compaction are considerably affected by soil type and the associated differences in soil texture and structure (Droese et al., 1975, 1983, 1986; Butorac and Turšić, 1978; Turšić, 1982; Lhotský, 1984b; Śmierzchalski et al., 1984; Afanasyev et al., 1987; Nugis, 1987; Lehfeldt, 1988). This effect also depends on the type of crop.

The relationship between the yield of winter oil seed rape and bulk density was parabolic for sandy and loamy soils and negative linear for loess soil (Droese et al., 1986), while in the case of oats the response was parabolic for all three soils (Droese et al., 1983). An increased compaction of gley-podzolic and brown soil of sandy loam texture led to a significant yield decrease of winter wheat and soybean when bulk density was >1.6 Mg m^{-3}, while the compaction of loamy red and loamy sand alluvial soils did not cause any significant changes in the yield (Butorac and Turšić, 1978). However, the negative yield response of barley to compaction was much greater in the red, gley-podzolic and brown soils than that on the alluvial soil (Turšić, 1982). Śmierzchalski et al. (1984) reported that the wheat yield response to increasing bulk density was parabolic in sandy soil and, in the low compaction range, linearly positive in loamy and silty soils.

In eastern Germany, Lehfeldt (1988) showed that the negative effect of soil compaction on the root growth of winter wheat, winter barley, winter rye, spring barley, oats, oil seed radish *(Raphanus chinensis)*, alfalfa, white lupin *(Lupinus albus)* and potatoes, was greater on sandy than on loamy moraine soils. In Poland, the yield response of spring barley to heavy compaction was more negative on loamy sand than on silty loam in dry periods (Lipiec et al., 1991). However, in Belorussia, crop growth was more depressed by vehicular traffic on loamy than on sandy soils (Afanasyev et al., 1987). According to these authors, this may be due to the fact that Belorussian sandy soils are rather resistant to compaction because of their favourable granulometric distribution and surface roughness of the grains. Ermich and Hofmann (1984) showed that the reduction of potato tuber yield due to compaction was greater on highly productive, fine-textured soils than on sandy loam soil (Fig. 8).

A greater depression of the tuber yield due to compaction on more productive soils was confirmed by Nugis (1987). However, in chernozems (Central Chernozem Zone, Russia), which have a high natural fertility (organic matter content 5.5-6.0%, w/w, to a depth of approximately 120 cm) and a high resistance to deformation, the yield of maize cobs was not significantly affected by compaction (Kuznetzova, 1987). Greater compactive effects occurred in chernozems with a lower humus content, where vehicular traffic may increase bulk density not only in the plough layer but also in the subsoil (Gaponenko, 1987). Effects of compaction were less permanent in chernozems containing hydromica and montmorillonite, owing to swelling and shrinkage forces (Gorbunov, 1974). Four swelling-shrinkage cycles decreased bulk density in the chernozem by 0.33 Mg m^{-3} but in poorly structured sod-podzolic soil only by 0.06 Mg m^{-3} (Kuznetzova and Danilova, 1987).

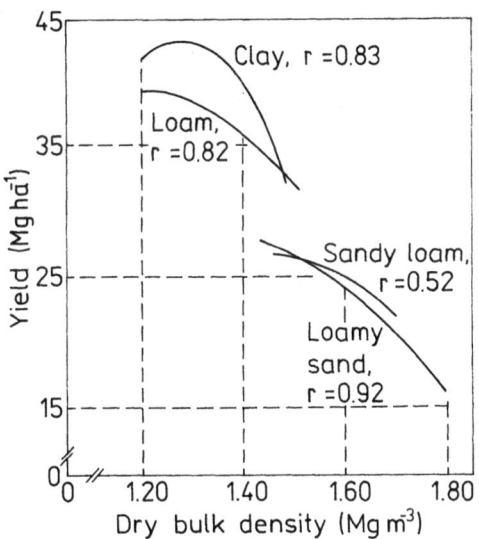

Fig. 8. Effect of compaction on the tuber yield of potatoes on soils of various texture (after Ermich and Hofmann, 1984).

The above relationships between the potato tuber yield and soil bulk density, and those for other crops presented in the preceding section of this chapter, show high variability of "optimum" bulk density for the same crop, depending on soil texture and soil structure. This suggests that when relating compaction effects among soils to bulk density, their composition should be taken into account. For winter wheat, Petelkau and Kunze (1980) established a quantitative relationship between the optimum bulk density in the Ap horizon, y, and the content of fine particles <6 μm (%, w/w), x,:

$$y = 1.56 - 0.006 \; x \; \pm \; 0.05 \tag{1}$$

Recent studies in Poland (Lipiec et al., 1991) showed that the "degree of compactness", being a percentage of a certain reference bulk density for the same soil (Håkansson, 1990), can be useful for characterizing the state of compactness from the point of view of soil physical conditions for crop production. In an experiment with pre-sowing compaction of two soils of different texture, the maximum crop yield of spring barley was obtained at the same degree of compactness (Fig. 9). Such response of the crop yield to the degree of compactness has proved to be similar to that obtained in Scandinavian field experiments (Håkansson, 1983; Riley, 1988), as well as in pot experiments in Yugoslavia (Racz and Butorac, 1983).

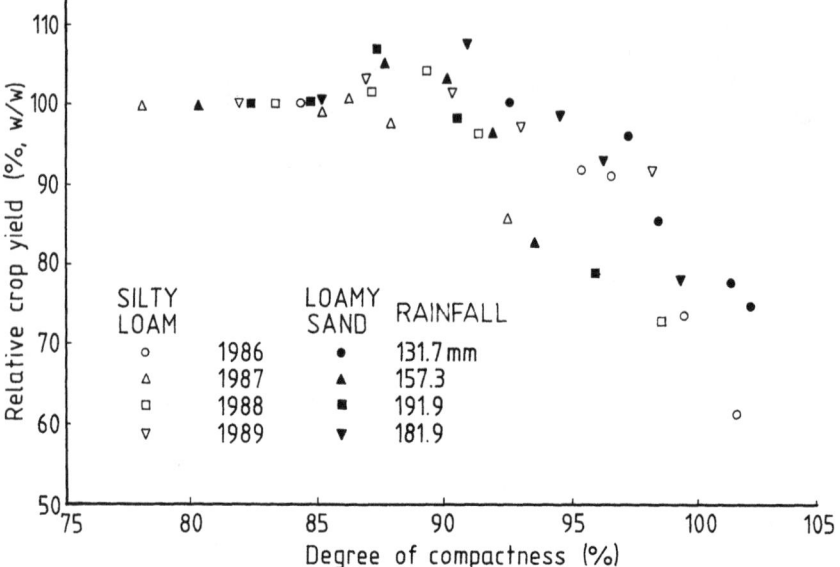

Fig. 9. Relative grain yield of spring barley (100% = control) in different seasons, in relation to the degree of compactness. The rainfall for the whole growth period is indicated (after Lipiec et al., 1991).

Effects of climatic conditions

The main climatic factors which can modify the crop response to vehicular traffic are rainfall through changes in soil water content and air temperature through frost action.

In the northern part of the region under discussion, high soil water content in spring resulting from winter rain or snow, enhances soil compactability (Puponin, 1984; Rusanov, 1991). Therefore, according to the GOST State Standards of the former U.S.S.R., the permissible ground contact pressure exerted by machinery in field work at a water content of >90% (w/w) of field capacity, is as low as 80 kPa (Rusanov, 1991). In addition, a drainage system for removing the excess water and obtaining workable conditions, is a very important factor (Puponin, 1984). The soil water status is of great importance at the time of harvest of vegetable crops, sugar beet and potatoes, which frequently requires the use of heavy transport equipment in wet conditions (Bondarev et al., 1987).

In the Moscow Region, where during the growing season a relatively high soil water content prevails, the effect of long-term (8 years) compaction of sandy loam soils by one or two passes of various tractors (with ground contact pressures of 150-250 kPa) every year, was cumulative and resulted in increasing reduction of crop yield in successive years (Puponin and Matyuk, 1987). The mean yield decrease of some cereals, forage and root crops was 5.1% after one year and 12.1

and 15% after 3 and 8 years, respectively.

Crops show wide variations in response to compaction between seasons. Compaction of heavy loam soil under wet conditions and harrowing prior to sowing did not sufficiently loosen the soil and, consequently, drastically reduced crop emergence (Rusanov, 1991). However, under relatively dry soil conditions, moderate compaction after sowing increased upward water movement and improved growth conditions for winter wheat (Špaldon and Karabinová, 1978; Rusanov, 1991), maize (Stoinev, 1975) and sugar beet (Tindzhyulis and Brazauskas, 1987).

Yield reductions due to intensive field traffic were less in wet than in dry seasons for winter oil seed rape in Poland (Śmierzchalski et al., 1988b) and for spring barley in the Voronezh Region in Russia (Shipilov, 1982; Bondarev et al., 1987) and in Poland (Kossowski et al., 1991). A pronounced effect of a dry season was found for spring barley grown on chernozem in the Voronezh Region (Fig. 10).

Another factor is rainfall distribution during the growing season. In Poland, the negative effect of excessive soil compaction on yield of spring barley was greater in seasons with low rainfall in the period between sowing and stem shooting (Kossowski et al., 1991) and, in the case of peas, in seasons with irregular rainfall distribution (Pabin and Włodek, 1986).

In clay soils, rainfall and resulting soil water content changes may decrease bulk density of compacted soil through swelling and shrinking processes. High self-loosening capability of fertile chernozems was reported by Kuznetzova and Danilova (1987). In some areas of the former U.S.S.R. with a continental climate, compaction effects can be alleviated by frost action (Sudakov et al., 1984;

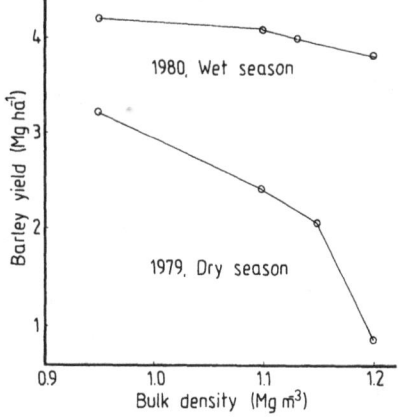

Fig. 10. Relationship between grain yield of spring barley and bulk density of fine loam soil in wet and dry seasons (above and below long-term average precipitation, respectively) (after Shipilov, 1982).

Medvedev and Tzybulko, 1987). This effect depends on the compaction level. In chernozems, soil compaction was significantly alleviated during winter when bulk density was <1.30 Mg m^{-3}. However, at bulk density 1.35-1.40 Mg m^{-3}, the loosening effect was of minor importance (Medvedev and Tzybulko, 1987).

Effect of fertilizer application

Attempts were undertaken to counteract the negative effects of soil compaction by increasing the level of fertilization (Butorac and Turšić, 1978; Turšić, 1982; Tindzhyulis and Brazauskas, 1987; Kęsik, 1991). In a severely compacted soil, higher levels of NPK fertilization usually did not prevent a yield depression of cereals (Butorac and Turšić, 1978; Turšić, 1982), soybean (Butorac and Turšić, 1978) and carrots (Kęsik, 1991). This can be explained partly by uneven distribution of roots in the compacted soil (Lipiec et al., 1991) and the resulting nutrient and water uptake. However, the combination of moderate compaction and higher fertilization had a positive effect on the yields of winter wheat and soybean (Butorac and Turšić, 1978). Organic farmyard manure at high rates (80-120 Mg ha^{-1}) was recommended by Bondarev et al. (1987) to increase the soil resistance to compaction.

Plant roots

Growing roots cause an increase in bulk density in the surrounding soil (Gliński and Lipiec, 1990). However, roots of some plants can perforate compact layers and create easily accessible pathways for the roots of a subsequent crop. A similar effect caused by lupin and lucerne as catch crops on sandy and loamy soils was reported by Lehfeldt (1988), in contrast to wheat roots which could not grow through the dense layer in direct-drilled (no-tillage) soil (Werner and Graul, 1989).

A common response of root systems to increasing bulk density is a decreased root mass (Petelkau et al., 1987, 1988b; Lehfeldt, 1988; Pittelkow et al., 1988; Cieśliński et al., 1990), retarded root penetration (Dannowski, 1983) and shallower rooting depth (Unger and Steinert, 1983; Petelkau et al., 1985; Droese et al., 1988; Lehfeldt, 1988; Unger et al., 1988; Lipiec et al., 1990, 1991). The ability of roots to grow in compacted soil varies with species (Petelkau et al., 1985).

In loose loamy sand (bulk density 1.35-1.38 Mg m^{-3}), about 90% of the roots of winter wheat and spring barley were found in the 0-50-cm layer, while in the same soil compacted to a bulk density of 1.66 Mg m^{-3}, 90% of the roots were present in the 0-15-cm layer (Petelkau et al., 1985).

The roots of spring barley grown in severely compacted soil were characterized by a higher degree of flattening, tortuous growth, distorted epidermal cells and radially enlarged cortex cells (Lipiec et al., 1991). These changes in the root system affect the uptake of nutrients and water. A restricted root system resulted in a lower phosphorus uptake (Słowik and Soczek, 1971) and reduced water uptake from deeper soil layers (Unger et al., 1988). However, when water supply was sufficient, soil compaction could provide a better opportunity for a restricted root system to absorb more water and to increase the water use efficiency, owing to a higher unsaturated hydraulic conductivity and greater water movement towards the roots (Lipiec et al., 1988). The root response to soil compaction depends to a high degree on the pore size distribution, especially of pores having a diameter greater than that of the roots, and on pore continuity (Pfleger and Werner, 1986; Lehfeldt and Koszinski, 1989; Gliński and Lipiec, 1990). In direct-drilled (no-tillage) soil, despite the greater bulk density, the root penetration into the subsoil was greater than in the tilled soil, owing to a greater proportion of continuous channels (Lehfeldt and Koszinski, 1989). Pfleger and Werner (1986) indicated that an increase in volume of continuous pores in the compacted layer by as little as 0.84% (v/v), resulted in a 25% increase in root length of spring barley. This effect was mainly attributed to increased gas diffusion and improved penetrating power of the roots.

Pfleger and Unger (1985) have shown that for loess soil a bulk density of 1.60 Mg m^{-3} can be regarded as the limit for root penetration of spring wheat, annual ryegrass and red clover, and 1.70 Mg m^{-3} for melilot *(Melilotus officinalis)*. Gätke (1984) found that root growth of barley is largely affected by mechanical impedance and poor aeration, while shoot growth is mainly dependent on the soil water content.

CONCLUSIONS

(1) The variable crop responses to compaction by agricultural vehicles depend on interactions between crop type, soil type, weather conditions and the degree of compactness. In these interactions the soil water status plays a very important role.

(2) In Central and Eastern Europe, root crops, such as sugar beet, as well as maize, seem to be the most sensitive to soil compaction. Under the same site conditions, small-grained cereals, such as barley and wheat, are less sensitive to compaction than soybeans and peas. On compacted soil, the yield depression of small-grained cereals may differ for different varieties.

(3) Excessive soil compaction resulted in lower quality of harvested produce, for example, decreased marketable yield of potatoes (tubers >35 mm), lower specific weight of strawberry fruit, lower sugar content in sugar beet, and a higher proportion of shorter and deformed roots of carrots.

(4) The effect of soil type on crop response to soil compaction is related to its composition, chemical fertility and resulting compactability. Field studies showed that the "degree of compactness", being independent of soil texture, can be useful to compare compaction effects on crop production in different soils.

(5) Under the climatic conditions of Central and Eastern Europe, compaction of dry soil at planting time increased upward water movement and improved growth conditions. The final crop yield from compacted soil tended to be less reduced in wet seasons and in seasons with favourable rainfall distribution than in dry seasons.

(6) The wide variation in crop responses to soil compaction implies that greater attention should be paid to define under which site conditions various levels of soil compactness will have a negative or favourable effect on crop production. Such information will provide a suitable basis for the development of models that would be helpful in understanding and describing the complex interactions.

ACKNOWLEDGEMENT

The help of Mrs. Ewa Jaroszyńska in the computer preparation of the manuscript is gratefully acknowledged.

REFERENCES

Afanasyev, N.I., Podobedov, I.I. and Orda, A.I., 1987. Dyernovo-podzolistyye-pochvy Byelorussii. (Sod-podzolic soils in Belorussia). In: V.A. Kovda (Editor), Pereuplotneniye Pakhotnykh Pochv. (Compaction of Arable Soils). Akad. Nauk U.S.S.R., "Nauka", Moscow, U.S.S.R., pp. 46-59 (in Russian).

Beran, P., 1988. Vliv stupňovaného utuženi půdy na tvorbu výnosu ječmene jarniho. (The influence of soil compaction on the yield of spring barley). In: Sbornik Přednášek Mezinarodni ČSSR-NDR, VUZZP Praha, Czechoslovakia, pp. 65-71 (in Czech).

Blank, P. and Petelkau, H., 1988. Einfluss differenzierter Bodenbearbeitung zu Winterroggen und Ölrettich-Zwischenfrucht in einer Getreidefruchtfolge auf die Lagerungsdichte eines anlehmigen Sandbodens sowie deren Wirkung auf pflanzenbauliche Parameter und den Ertrag. (Influence of tillage under winter rye and oil radish as intercrop in crop rotation on bulk density of loamy sand and plant parameters and yield). Arch. Acker- Pflanzenbau Bodenkd., 32: 601-609 (in German).

Bondarev, A.G., 1987. Oroshayemyie pochvy. (Irrigated soils). In: V.A. Kovda (Editor), Pereuplotneniye Pakhotnykh Pochv. (Compaction of Arable Soils). Akad. Nauk U.S.S.R., Moscow, pp. 139-143 (in Rusian).

Bondarev, A.G., Rusanov, V.A. and Medvedev, V.W., 1987. Zaklucheniye (Summary). In: A. Kovda (Editor), Pereuplotneniye Pakhotnykh Pochv. (Compaction of Arable Soils). Akad. Nauk U.S.S.R., "Nauka", Moscow, U.S.S.R., pp. 205-209 (in Russian).

Bondarev, A.G., Sapozhnikov, P.M., Utkaieva, V.F. and Shepotiev, V.N., 1988. Izmeneniye fizicheskykh svoystv i plodorodya pochv pri ikh uplotnieny dvizhytelamy sielskokhazyaystvennoy tyekhniki. (Changes in soil physical properties and productivity as related to compaction by agricultural vehicles). Sbornik Nauchnykh Trudov, VIM, Moscow, U.S.S.R., 118: 46-57 (in Russian).

Butorac, A., 1982a. Proučavanje reakcije crvene djeteljine (Trifolium pratense) i lucerne (Medicago sativa) u model pokusima na zbijenost tla u kombinaciji s gnojidbom. (Studies on red clover and lucerne response in model experiments on soil compaction in combination with fertilization). Poljoprivredna Znanstvena Smotra - Agriculturae Conspectus Scientificus, 58: 19-37 (in Croatian).

Butorac, A., 1982b. Reakcija jare zobi *(Avena sativa)* i ozime raži *(Secale cereale)* u model pokusima na zbijenost tla i gnojidbu. (Response of summer oats and winter rye in model experiments on soil compaction and fertilization). Poljoprivedna Znanstvena Smotra - Agriculturae Conspectus Scientificus, 59: 225-241 (in Croatian).

Butorac, A. and Turšić, I., 1978. Uzajamni odnosi zbijenosti tla i gnojidbe u uzgoju ozime psenice i soje. (Interaction of soil compaction and fertility in growing winter wheat and soybean). Agronomski Glasnik, 4: 749-763 (in Croatian).

Canarache, A., Cioroianu, F., Colibaş, I., Colibaş, M., Dumitru, R., Florescu, C.I., Horobeanu, I., Pătru, V., Simota, H., Stanescu, S. and Trandafirescu, T., 1984a. Einfluss der Bodenverdichtung auf die Produktivität der Ackerböden Rumäniens. (Effect of soil compaction on the productivity of arable soils in Romania). Tag.-Ber., Akad. Landwirtsch.-Wiss. D.D.R., Berlin, Germany, 227: 41-48 (in German).

Canarache, A., Colibaş, I., Colibaş, M., Horobeanu, I., Pătru, V., Simota, H. and Trandafirescu, T., 1984b. Effect of induced compaction by wheel traffic on soil physical properties and yield of maize in Romania. Soil Tillage Res., 4: 199-213.

Cieśliński, Z., Miatkowski, Z. and Sołtysik, A., 1990. Influence of agromeliorative ploughing on some physical properties of soil and the distribution of the roots of wheat. Zesz. Probl. Post. Nauk Roln., 385: 73-80.

Dannowski, M., 1983. Methode zur Ermittlung der Durchwurzelbarkeit unterschiedlich verdichteten Bodens. (A method for determining the suitability of differently compacted soils for root penetration). Tag.-Ber., Akad. Landwirtsch.-Wiss. D.D.R., Berlin, Germany, 215: 165-172 (in German).

Dechnik, I., Lipiec, J. and Tarkiewicz, S., 1982. Influence of wheel traffic on physical properties of sandy soil and crop yields. Proc. 9th Int. Conf. Int. Soil Tillage Res. Org. (ISTRO), Osijek, Yugoslavia, pp. 183-188.

Dedaev, G.A., Shevtzov, A.V., Kozlov V.V. and Kobylchenko, E.S., 1988. Vlijaniye khodovykh sistem selskokhozyaystvennykh mashin na pochvy i urozhaynost lugovykh travostoyev. (Influence of tractive systems of agricultural machines on soil and productivity of meadow grasses). Sbornik Nauchnykh Trudov, VIM, Moscow, U.S.S.R., 118: 98-103 (in Russian).

Domżał, H., Gliński, J. and Lipiec, J., 1991. Soil compaction research in Poland. Soil Tillage Res., 19: 99-109.

Droese, H., Radecki, A. and Śmierzchalski, L., 1975. Reakcja roślin uprawnych na stopień zagęszczenia gleby. I. Pszenica ozima. (Response of particular crops to soil compactness level. I. Winter wheat). Rocz. Nauk Roln., A, 101, 2: 107-121 (in Polish).

Droese, H., Radecki, A. and Śmierzchalski, L., 1982. Reakcja roślin uprawnych na stopień zagęszczenia gleby. II. Zyto ozime. (Response of particular crops to soil compactness level. II. Winter rye). Rocz. Nauk Roln., A, 105, 3: 43-55 (in Polish).

Droese, H., Radecki, A. and Śmierzchalski, L., 1983. Reakcja roślin uprawnych na stopień zagęszczenia gleby. III. Owies. (Response of particular crops to soil compactness level. III. Oats). Rocz. Nauk Roln., A, 105, 2: 97-109 (in Polish).

Droese, H., Radecki, A. and Śmierzchalski, L., 1986. Reakcja roślin uprawnych na stopień zagęszczenia gleby. IV. Rzepak ozimy. (Response of particular crops to soil compactness level. IV. Winter rape). Rocz. Nauk Roln., A, 106, 2: 91-102 (in Polish).

Droese, H., Radecki, A. and Witkowski, F., 1988. Wpływ zagęszczenia gleby na system korzeniowy niektórych roślin uprawnych. (Soil density effect on the root system of some crops). Zesz. Probl. Post. Nauk Roln., 356: 63-68 (in Polish).

Ermich, D. and Hofmann, B., 1984. Grenzwerte der Druckbelastung des Ackerbodens zur Verhinderung von Schadverdichtungen bei der Pflanzbettbereitung zu Kartoffeln. (Limits to the pressure on arable soils to prevent harmful compaction during seedbed preparation for potatoes). Tag.-Ber., Akad. Landwirtsch.-Wiss. D.D.R., Berlin, Germany, 227: 157-163 (in German).

Gaponenko, V.S., 1987. Tzentralnaya i pravobereznaya lesostep. (Central and right-bank zone of forest steppe). In: V.A. Kovda (Editor), Pereuplotneniye Pakhot-nykh Pochv. (Compaction of Arable Soils). Akad. Nauk U.S.S.R., "Nauka", Moscow, U.S.S.R., pp. 105-114 (in Russian).

Gätke, C.R., 1984. Zur Wirkung unterschiedlicher bodenphysikalischer Einflussgrössen auf die Wurzelausbreitung und die pflanzliche Stoffproduktion. (Effect of different soil physical factors on root penetration and crop biomass production). Tag.-Ber., Akad. Landwirtsch.-Wiss. D.D.R., Berlin, Germany, 227: 177-183 (in German).

Gliński, J. and Lipiec, J., 1990. Soil Physical Conditions and Plant Roots. CRC Press, Boca Raton, FL, U.S.A., 250 pp.

Gliński, J. and Stępniewski, W., 1985. Soil Aeration and Its Role for Plants. CRC Press, Boca Raton, FL, U.S.A., 230 pp.

Gorbunov, N.I., 1974. Mineralogiya i Koloidnaya Khimiya Pochv. (Mineral and Colloidal Chemistry of Soils). Akad. Nauk U.S.S.R., "Nauka", Moscow, U.S.S.R., 314 pp. (in Russian).

Håkansson, I., 1983. On the reasons for influences of heavy machinery on crop yield. Mezinarodni Vedecke Symposium "Změny Půdniho Prostředi ve Vztahu k Intenzifikačnim Faktorům", Brno, Czechoslovakia, pp. 57-66.

Håkansson, I., 1990. A method for characterizing the state of compactness of the plough layer. Soil Tillage Res., 16: 105-120.

Haman, J., 1979. Biological aspects of soil compaction. Zesz. Probl. Post. Nauk Roln., 220: 151-166.

Haman, J., 1989. Soil compaction by working elements of tillage implements. Abstracts Int. Conf. "Soil Compaction as a Factor Determining Plant Productivity", 5-9 June 1989, Lublin, Poland, p. 76.

Kęsik, T., 1985. The influence of soil compaction caused by tractor passes on the yield and some geometrical features of carrot roots. Proc. 3rd Int. Conf. "Physical Properties of Agricultural Materials and their Influence on the Design and Performance of Agricultural Machines and Technologies", Prague, Czechoslovakia, August 19-23, pp. 451-456.

Kęsik, T., 1990. Influence of loess soil compaction on the crop and some morphological and physical features of the carrot roots. Zesz. Probl. Post. Nauk Roln., 385: 81-96.

Kęsik, T., 1991. Response of carrot to soil compaction and increased fertilization. I. Root yield and structure). Zesz. Probl. Post. Nauk Roln., 389: 139-145.

Kęsik, T. and Stanek, R., 1983. Reakcja krzewów porzeczek i malin na zagęszczenie gleby. (Response of berries and raspberries to soil compaction). Mat. Konf. Kom. K.N.O. PAN "Agrotechnika sadów ze szczególnym uwzglednieniem terenów erodowanych", Lublin, Poland, 1983: 41-44 (in Polish).

Kollar, B., Kováč, K., Pilat, A. and Zigo, J., 1976. Vplyv obrabania půdy na jej objemovu hmotnost, vlhkost a urodu zrna ozimnej psenice. (Effect of soil cultivation on soil bulk density, soil water and grain yield of winter wheat). Rostl. Vyr., 22: 929-942 (in Czech).

Kossowski, J., Lipiec, J. and Tarkiewicz, S., 1991. The response of spring barley yield to the degree of soil compactness related to meteorological conditions. Zesz. Probl. Nauk Roln., 396: 81-87.

Kovda, V.A., 1987. Predisloviye (Foreword). In: V.A. Kovda (Editor), Pereuplotneniye Pakhotnykh Pochv. (Compaction of Arable Soils). Akad. Nauk U.S.S.R., "Nauka", Moscow, U.S.S.R., pp. 3-4 (in Russian).

Kravchenko, V.I., 1986. Uplotneniye pochv mashinami. (Soil compaction by machinery). Akad. Nauk Kazachskoy S.S.R., Alma Ata, U.S.S.R., 96 pp. (in Russian).

Krüger, W., 1970. Über den Einfluss unterschiedlicher Bodenverdichtung auf einige bodenphysikalische Eigenschaften und das Pflanzenwachstum. (The effect of different degrees of soil compaction on some soil physical properties and plant growth). Albrecht-Thaer-Archiv., 14: 613-623 (in German).

Kuznetzov, N.G., 1978. Sokhraneniye plodorodiya pochvy pri vozdeystviy na neye khodovykh sistem traktorov i rabochykh organov mashin. (Effects of tractive systems of tractors and working elements of machines on soil and maintenance of its productivity). Vestnik Selskokh. Nauki, 7: 115-118 (in Russian).

Kuznetzova, I.V., 1987. Chernozemnyye pochvy C.CH.O. (Chernozem soils in the Central Chernozem Zone). In: V.A. Kovda (Editor), Pereuplotneniye Pakhotnykh Pochv. (Compaction of Arable Soils). Akad. Nauk U.S.S.R., "Nauka", Moscow, U.S.S.R., pp. 86-97 (in Russian).

Kuznetzova, I.V. and Danilova, V.I., 1987. Samorazuplotneniye raznykh tipov pochv pod vliyaniyem protzesov nabukhaniya-usadki. (Effect of plough pan compaction on root penetration of various crops grown on sandy and loamy soils). In: V.A. Kovda (Editor), Pereuplotneniye Pakhotnykh Pochv. (Compaction of Arable Soils). Akad. Nauk U.S.S.R., "Nauka", Moscow, U.S.S.R., pp. 189-193 (in Russian).

Lehfeldt, J., 1988. Auswirkungen von Krumenbasisverdichtungen auf die Durchwurzelbarkeit sandiger und lehmiger Bodensubstrate bei Anbau verschiedener Kulturpflanzen. (Effects of upper subsoil compaction on root penetration of various crops grown on sandy and loamy soils). Arch. Acker-Pflanzenb. Bodenkd., 32: 533-539 (in German).

Lehfeldt, J. and Koszinski, S., 1989. Auswirkungen von Verdichtungsstrukturen in der Krumenbasis von Moränenstandorten auf die Durchwurzelungsintensität des Bodens. (Effects of compacted structures in the upper subsoil layer of sandy-loamy diluvial soils on the intensity of root growth). In: Vorträge wissensch. Tagung FZB, 26-29 Juni 1989, Müncheberg, Germany, pp. 199-207 (in German).

Lhotský, J., 1984a. Problema uplotneniya pakhotnykh zemel i yeye resheniye v ChSSR. (Problems of compaction of arable soils and its alleviation in Czechoslovakia). Scientia Agriculturae Bohemslovaca, 16: 155-164 (in Russian).

Lhotský, J., 1984b. Stand der Entwicklung von Methoden zur Ermittlung und Beseitigung von Schadverdichtungen in den Böden der CSSR. (Present state of developing methods to determine and eliminate harmful soil compaction in Czechoslovakia). Tag.-Ber., Akad. Landwirtsch.-Wiss. D.D.R., Berlin, Germany, 227: 215-220 (in German).

Lhotský, J., Beran, P., Paris, P. and Valigurská, L., 1991. Degradation of soil by increasing compression. Soil Tillage Res., 19: 287-295.

Lipiec, J. and Tarkiewicz, S., 1986. The effect of moisture on the crushing strength of aggregates of loamy soil at various density levels. Pol. J. Soil Sci., 19: 27-31.

Lipiec, J., Kubota, T., Iwama, H. and Hirose, J., 1988. Measurement of plant water use under controlled soil moisture conditions by the negative pressure water circulation technique. Soil Sci. Plant Nutr., 34: 417-428.

Lipiec, J., Kania, W. and Tarkiewicz, S., 1990. The effect of wheeling on physical characteristics of soils and rooting of some cereals. Zesz. Probl. Post. Nauk Roln., 385: 97-105.

Lipiec, J., Tarkiewicz, S., Kossowski, J. and Håkansson, I., 1991. Soil physical properties and growth of spring barley related to the degree of compactness of two soils. Soil Tillage Res., 19: 307-317.

Maliyenko, A.M., 1984. Reaktziya kartofelya i kukuryzy na uplotneniye dernovo-podzolistoy supeschanoy pochvy propashnym kolesnym traktorom tyagovovo klassa 1,4. (Response of potatoes and maize to compaction of sod-podzolic sandy soil by row-crop wheeled tractor of haulage class 1,4). Sbornik Nauchnykh Trudov, VIM, Moscow, 102: 118-121 (in Russian).

Medvedev, V.W. and Tzybulko, V.G., 1987. Obosnovaniye dopustimykh urovneynagruzki

MTA na pochvu na primere chernozemnykh pochv USSR. (Justification of admissible levels of MTA loads on chernozem soils of the USSR as an example). In: V.A. Kovda (Editor), Pereuplotneniye Pakhotnykh Pochv. (Compaction of Arable Soils). Akad. Nauk U.S.S.R., "Nauka", Moscow, U.S.S.R., pp. 173-181 (in Russian).

Nugis, E., 1987. Dernovo-podzolistyye pochvy Estonii. (Sod-podzolic soils in Estonia). In: V.A. Kovda (Editor), Pereuplotneniye Pakhotnykh Pochv. (Compaction of Arable Soils). Akad. Nauk U.S.S.R., "Nauka", Moscow, U.S.S.R., pp. 35-45 (in Russian).

Okhitin, A.A., Sudakov, A.V., Lipiec, J. and Tarkiewicz, S., 1991. Deformation of silty loam soil under the tractor tyre. Soil Tillage Res., 19: 187-195.

Pabin, J. and Sienkiewicz, J., 1984. Wpływ zagęszczenia gleby i głębokości siewu nasion na wschody i plonowanie buraków cukrowych. (Effect of soil bulk density and sowing depth on emergence and yields of sugar beets). Roczn. Gleb. XXXV, No. 3-4: 75-86 (in Polish).

Pabin, J. and Włodek, S., 1986. Wpływ zagęszczenia gleby lekkiej na niektóre laściwosci fizyczne oraz na plonowanie peluszki i jeczmienia jarego. II. Gęstość, zwięzłość i porowatość powietrzna gleby a plony roślin. (Effect of compaction of a light soil on its physical properties and on yields of field pea and spring barley. II. Bulk density, penetration resistance and air-filled porosity of the soil and crop yields). Pam. Puł., 88: 87-99 (in Polish).

Pabin, J., Sienkiewicz, J. and Włodek, S., 1991. Effect of loosening and compacting on soil physical properties and sugar beet yield. Soil Tillage Res., 19: 345-350.

Petelkau, H., 1983. Ursachen, Enstehung und Prinzipien zur Einschränkung von Bodenstrukturschäden. (Causes and development of soil structure damage and principles of its reduction). Tag.-Ber., Akad. Landwirtsch.-Wiss. D.D.R., Berlin, Germany, 215: 39-48 (in German).

Petelkau, H., 1984. Auswirkungen von Schadverdichtungen auf Bodeneigenschaften und Pflanzenertrag sowie Massnahmen zu ihrer Minderung. (Effects of harmful compaction on soil properties and crop yields and measures to reduce compaction). Tag.-Ber., Akad. Landwirtsch.-Wiss. D.D.R., Berlin, Germany, 227: 25-34 (in German).

Petelkau, H., 1989a. Aufgaben und Lösungswege zur Minderung von schädlichen Bodenverdichtungen durch Mechanisierungsmittel der Feldwirtschaft. (Objectives and possibilities for alleviation of harmful soil compaction by mechanization of agronomic practices). Arb. Mech. Pflanzen-Tierprod. Schlieben, 43, pp. 93-114 (in German).

Petelkau, H., 1989b. Die Beurteilung des Verdichtungszustands der Krume abgeernteter Ackerflächen als wichtigste Steuergrösse für die Grundbodenbearbeitung. (Evaluation of the state of compactness in the topsoil of harvested fields as the most important parameter in controlling primary tillage). In: Vörtrage anlässlich der wissensch. Tagung des FZB, 26-29 Juni 1989, Müncheberg, Germany, pp. 172-180 (in German).

Petelkau, H. and Kunze, A., 1980. Die Lagerungsdichte des Bodens als wesentliche Steuerungsgrösse für die Bodenbearbeitung. (Bulk density as a relevant factor in the control of soil tillage). In: Bodenbearbeitungs-Forschung: Druckbelastung der Ackerböden, Teil 1, Univ. Halle-Wittenberg, Germany, Wissenschaftliche Beiträge 14: 35-65 (in German).

Petelkau, H. and Seidel, K., 1986. Bearbeitbarkeit und Befahrbarkeit von Ackerböden in Abhängigkeit von der Bodenfeuchte. (Workability and trafficability of arable soils as influenced by soil water). Tag.-Ber., Akad. Landwirtsch. Wiss. D.D.R., Berlin, Germany, 246: 46-54 (in German).

Petelkau, H., Dannowski, M., Gätke, C.R. and Seidel, K., 1983. Bodenbearbeitungssteuerung. (Control of soil tillage). Wiss. Jahresbericht Forsch. Zentrum Bodenfruchtbarkeit, Müncheberg, Germany, FZB-Report 1983, pp. 21-28 (in German).

Petelkau, H., Gätke, C.R., Dannowski, M. and Seidel, K., 1985. Zusammenhänge zwischen Bodenlagerungsdichte und pflanzlicher Stoffproduktion. (Relationship between bulk density of soil and crop production). Wiss. Jahresbericht Forsch. Zentrum Bodenfruchtbarkeit, Müncheberg, Germany, FZB-Report 1985, pp. 24-29 (in German).

Petelkau, H., Kalk, W.D., Bosse, O., Gätke, C.R., Dannowski, M. and Seidel, K., 1987. Strukturschonende Bodenbearbeitung und Einschränkung schädlicher Bodenverdichtungen. (Conservation tillage and the reduction of harmful compaction). Wiss. Jahresbericht Forsch. Zentrum Bodenfruchtbarkeit, Müncheberg, Germany, FZB-Report 1987, pp. 24-30 (in German).

Petelkau, H., Gätke, C.R., Dannowski, M., Seidel, K. and Augustin, J., 1988a. Bodenphysikalische Grundlagen für die Steuerung der Grundbodenbearbeitung. (Soil physical fundamentals for control of primary tillage). In: Erhöhung der Bodenfruchtbarkeit und der Erträge durch wissenschaftlichen Fortschritt. Müncheberg, Germany, pp. 362-379 (in German).

Petelkau, H., Seidel, K., Gätke, C.R. and Dannowski, M., 1988b. Prinziplösung für die Steuerung der Grundbodenbearbeitung. (Principles of controlling primary tillage). Wiss. Jahresbericht Forsch. Zentrum Bodenfruchtbarkeit, Müncheberg, Germany, FZB-Report 1988, pp. 86-93 (in German).

Pfleger, I. and Unger, H., 1985. Die Abhängigkeit des Wurzelwachstums von der Bodendichte und Feuchteversorgung. (Dependence of root growth on soil density and water supply). Tag.-Ber., Akad. Landwirtsch.-Wiss. D.D.R., Berlin, Germany, 231: 209-220 (in German).

Pfleger, I. and Werner, D., 1986. Einfluss röhrenförmiger Hohlräume in Bodenpasten unterschiedlicher Dichte und Feuchte auf das Wurzeldurchdringungsvermögen von Sommergerste. (Effects of tube-shaped hollow spaces in soil pastes of different density and water content on the penetrating power of spring barley). Arch. Acker-Pflanzenbau Bodenkd., Berlin, 30: 259-268 (in German).

Pittelkow, U. and Unger, H., 1988. Über die Auswirkungen von Brückenzonen in Krumenbasisverdichtungen auf die Erträge bei Unterschieden im Bodenwasserangebot und der Bodendichte (Ergebnisse aus der Jenaer Bodenmodellanlage). (The effects of bridging zones in compacted upper subsoil on crop yield at different levels of water supply and soil compaction - results from the Jena soil model experiment). Arch. Acker-Pflanzenb. Bodenkd., Berlin, 32: 389-395 (in German).

Pittelkow, U., John, K. and Körbs, P., 1988. Über die Auswirkungen von Brückenzonen in Krumenbasisverdichtungen auf Durchwurzelung und Wasserentzug bei differenzierter Bodenfeuchte und -dichte (Ergebnisse aus der Jenaer Bodenmodellanlage). (The effects of bridging zones in compacted upper subsoil on root penetration and water uptake at different soil water contents and bulk densities - results from the Jena soil model experiment). Arch. Acker- Pflanzenb. Bodenkd., Berlin, 6: 379-387 (in German).

Popov, A.I., Nugis, E. and Makhlak-Suyts, A. Kh., 1977. Vozdeystviye koles mashin na pochvu. (The effect of machine wheels on soil). Zemledeliye, 2: 77-79 (in Russian).

Puponin, A.I., 1984. Obrabotka pochvy v intensivnom zemlediliy nechernozemnoy zony. (Soil tillage in the intensive agriculture of the non-chernozem zone). Kolos, Moscow, U.S.S.R., pp. 3-184 (in Russian).

Puponin, A.I. and Matyuk, N.S., 1987. Dernovo-podzolistyye pochvy Podmoskovya. (Sod-podzolic soils in the Moscow Region). In: V.A. Kovda (Editor), Pereuplotneniye Pakhotnykh Pochv. (Compaction of Arable Soils). Akad. Nauk U.S.S.R., "Nauka", Moscow, pp. 27-34 (in Russian).

Racz, I., 1973. Utjecaj fizikalnih svojstava tla na rast i razvoj korijena mrkve. (The influence of soil physical properties on growth and development of carrot roots). Poljoprivredna Znanstvena Smotra - Conspectus Agriculturae Scientificus, 30: 263-276 (in Croatian).

Racz, I. and Butorac, A., 1983. Utjecaj zbijenosti tla na rast, razvoj i prinos nekih kultura. (The effect of soil compaction on growth, development and yield of some agricultural crops). Poljoprivredna Znanstvena Smotra, 62: 491-500 (in Croatian).

Riley, H., 1988. Cereal yields and soil physical properties in relation to the degree of compactness of some Norwegian soils. Proc. 11th Conf. Int. Soil Tillage Res. Org. (ISTRO), Edinburgh, U.K., Vol. 1, pp. 109-114.

Rusanov, V.A., 1988. Osnovnye polozhenya ispolzovannyye pri razrabotke GOSTov po normam i metram i metodam otzenki vozdeystviya dwizhiteley na pochvy (GOST 26955-86, 26953-86 i 26954-86). (Main prerequisites used in the elaboration of quantitative norms and methods for the evaluation of the effect of running gear on soil - GOST 26955-86, 26953-86 and 26954-86). Sbornik Nauchnykh Trudov, VIM, Moscow, 118: 6-45 (in Russian).

Rusanov, V.A., 1991. Effects of wheel and track traffic on the soil and on crop growth and yield. Soil Tillage Res., 19: 141-143.

Šantrůček, J., 1989a. Kvalita píce vojtěšky v závislosti na zhutnění a kultivaci půdi. (Quality of lucerne forage in relation to soil compaction and cultivation). Rostl. Výr., 35: 1101-1107 (in Czech).

Šantrůček, J., 1989b. Vliv kypření a zhutňování půdy vojtěškovych porostů na tvorbu výnosu píce. (Effect of soil loosening and compaction of lucerne lands on the formation of forage yield). Rostl. Výr., 35: 1151-1160 (in Czech).

Shevlagin, A.I., 1966. Reaktziya selskokhozyaystvennykh kultur na razlichnuyu plotnost slezheniya pochvy. (Response of agricultural crops to various soil compaction). Sbornik Ref. Mezhdunar. Nauch. Simpoz., Brno, Czechoslovakia, pp. 93-162 (in Russian).

Shipilov. M.A., 1982. Vliyaniye uplotneniya pochv na urozhay. (The effect of soil compaction on crop yields). Zemledeliye, 11: 17-19 (in Russian).

Sienkiewicz, J., Jabłonski, W. and Włodek, S., 1988. Działanie wałowania na niektóre fizyczne własciwosci gleby lekkiej i plonowanie żyta. (Effect of rolling on some physical properties of light soil and yield of rye). Zesz. Probl. Post. Nauk Roln., 356: 215-222 (in Polish).

Simota, C. and Canarache, A., 1988. Effects of induced compaction on soil water balance and crop yields estimated with a deterministic simulation model. Proc. 11th Conf. Int. Soil Tillage Res. Org. (ISTRO), Edinburgh, U.K., Vol. 1, pp. 391-396.

Sin, G., 1984. Rationalisierung der Grundbodenbearbeitung und Auswirkungen der Bodenverdichtung. (Rationalization of primary tillage and soil compaction effects). Tag.-Ber., Akad. Landwirtsch.-Wiss. D.D.R., Berlin, Germany, 227: 97-101 (in German).

Sin, G., Ionita, St., Terbea, M. and Boruga, I., 1989. Einfluss der Bodenverdichtung auf Wachstum und Ertrag von Feldkulturen und auf einige Parameter beim Pflügen. (Influence of soil compaction on growth and yield of field crops and on some working parameters at ploughing). In: Vorträge wissensch. Tagung FZB, 26-29 Juni 1989, Müncheberg, Germany, pp. 181-188 (in German).

Slesarev, V.N., 1987. Chernozemy Zapadnoy Sibiri. (Chernozems in West Siberia). In: V.A. Kovda (Editor), Pereuplotneniye Pakhotnykh Pochv. (Compaction of Arable Soils). Akad. Nauk U.S.S.R., "Nauka", Moscow, U.S.S.R., pp. 127-138 (in Russian).

Słowik, K. and Soczek, Z., 1971. Wpływ różnego stopnia zagęszczenia gliny średniej na szybkość pobierania ^{32}P przez truskawki odmiany Regina. (Influence of various compaction of medium loam on uptake rate of ^{32}P by strawberry var. Regina). Zesz. Probl. Post. Nauk Roln., 112: 313-316 (in Polish).

Śmierzchalski, L., Droese, H. and Radecki, A., 1984. Reakcja roślin zbożowych na zagęszczenie gleby. (Responses of cereal plants to the degree of soil compaction). Zesz. Probl. Post. Nauk Roln., 305: 183-187 (in Polish).

Śmierzchalski, L., Droese, H. and Radecki, A., 1988a. Wpływ ukladu gleby na plonowanie zbóż jarych. (Effect of soil compaction on the yield of spring cereals). Zesz. Probl. Post. Nauk Roln., 356: 237-246 (in Polish).

Śmierzchalski, L., Droese, H. and Radecki, A., 1988b. Wpływ układu gleby na plonowanie rzepaku ozimego. (Effect of soil compaction on yield of winter oil seed rape). Zesz. Probl. Post. Nauk Roln., 356: 259-270 (in Polish).

Śmierzchalski, L., Droese, H. and Szujecka, W., 1988c. Wpływ ukladu gleby na plonowanie zbóż ozimych. (Effect of soil compaction on the yield of winter cereals). Zesz. Probl. Post. Nauk Roln., 356: 247-257 (in Polish).

Špaldon, E. and Karabinova, M., 1978. Vliv utláčania půdy po zasiati pšenice na polnú vzchadzavost a prezimovanie. (Effect of soil compaction after sowing of wheat on crop establishment and wintering). Acta Fytotechnika, 33: 55-68 (in Czech).

Starczewski, J., Droese, H. and Śmierzchalski, L., 1984. Wpływ uprawy roli i zagęszczenia gleby na plony ziemniaków. (Effect of tillage and soil compaction on potato yields). Rocz. Nauk Roln., 106, A, 1: 67-81 (in Polish).

Stępniewska, Z., Lipiec, J., Dąbek-Szreniawska, M., Bennicelli, R. and Stępniewski, W., 1990. The influence of anaerobic conditions on enzymatic activity in brown-grey (Lessivé) soil (horizon Ap). Folia Soc. Sci. Lublinensis, s. Geogr., 1: 53-59.

Stępniewski, W., Bennicelli, R., Stępniewska, Z. and Przywara, G., 1990. The content and the uptake of Mn and Fe by winter rye in different soil oxygen conditions. Folia Soc. Sci. Lublinensis, s. Geogr., 1: 61-67.

Stoinev, K., 1975. Vliyaniye na obrabotkata toreneto i priseytbenoto upltnyavanye na usluzhen chernozem stalnitza vrkhu dobira ot tzarevitzata i pshenitzata. (Effect of tillage, fertilizer application and pre-sowing compaction of leached smolnitza on maize and winter wheat). Soil Sci. Agrochem., 10: 119-120 (in Bulgarian).

Sudakov, A.V., Okhitin, A.A., Kuznetzova, E.P. and Bankin, M.P., 1984. K metodike izucheniya uplotniayushchego deistviya khodovykhi sistem selskokhozyaistvennoy tekhniki. (Method for studying the compactive effect of tractive systems of agricultural vehicles). Sbornik Nauchnykh Trudov, VIM, Moscow, U.S.S.R., 102: 121-128 (in Russian).

Timofeyev, A.I., 1978. Perspektivy razvitiya zemledelcheskikh mashin. Voprosy mekhanizatzi selskokhazyaystvennogo proizvodstva. (Prospects of the development of agricultural machinery. The problems in mechanization of agricultural production). Sbornik Nauchnykh Trudov. T. KHY, 15: 5-7 (in Russian).

Tindzhyulis, A.P. and Brazauskas, R.B., 1987. Dyernovo-podzolistyie suglinistyie pochvy Litvy. (Sod-podzolic soils of silty loam texture in Lithuania). In: V.A. Kovda (Editor), Pereuplotneniye Pakhotnykh Pochv. (Compaction of Arable Soils). Akad. Nauk U.S.S.R. "Nauka", Moscow, U.S.S.R., pp. 60-66 (in Russian).

Trzecki, S., 1974. Wpływ ugniatającego działania kół ciągnika na właściwości gleby i plonowanie jęczmienia z wsiewką koniczyny. (Effect of soil pressure by tractor wheels in spring on soil properties and yields of barley with undersown clover). Rocz. Nauk Roln., 100: 111-120 (in Polish).

Turšić, I., 1982. Utjecaj zbijenosti tla i mineralne gnojidbe na prinos jarogo jecma. (The effect of soil compaction and mineral fertilization on the yield of spring barley). Poljoprivredna Znanstvena Smotra-Agriculturae Conspectus, 58: 39-48 (in Croatian).

Unger, H. and Steinert, P., 1983. Erste Ergebnisse aus Modelluntersuchungen zu einer neuen Prinziplösung für die Überbrückung von Schadverdichtungen an der Krumenbasis schwerer Böden. (Preliminary results from model experiments on a new solution for overcoming detrimental compaction in the lower part of the topsoil of heavy soils). Tag.-

Ber., Akad. Landwirtsch.-Wiss. D.D.R., Berlin, Germany, 215: 221-232 (in German).

Unger, H., Pittelkow, U. and Schulze, R., 1988. Auswirkungen von Verdichtungen und N-Tiefapplikation auf Bodenparameter, Wurzelwachstum und Erträg. Ergebnisse aus der Jenaer Bodenmodellanlage. (Effects of compaction and deep placement of nitrogen on soil parameters, root growth and yield. Results of the Jena soil model experiment). Arch. Acker-Pflanzenb. Bodenkd., Berlin, Germany, 32: 293-301 (in German).

Valigurská, L. and Lhotský, J., 1985. Vliv różné intenzity zhutnĕní na zmĕny fyzikálnich vlastnosti pŭdy. (The effect of various compaction intensities on changes in soil physical properties). Rostl. Výr., 31: 603-612 (in Czech).

Várallyay, G. and Lesztak, M., 1990. Susceptibility of soils to physical degradation in Hungary. Soil Technol., 3: 289-298.

Walczak, R., 1977. Model investigations of water binding energy in soils of different compaction. Zesz. Probl. Post. Nauk Roln., 197: 11-44.

Werner, D. and Graul, W., 1989. Einfluss reduzierter Bodenbearbeitung auf den Strukturzustand eines Produktionsschlages mit Lössböden. (Effect of reduced soil tillage on soil structure of a production area on loess soil). In: Vorträge wissensch. Tagung FZB, 26-29 Juni 1989, Müncheberg, Germany, pp. 189-198 (in German).

Wieniarska, J., Lipecki, J., Stanek, R. and Kęsik, T., 1987. The effects of soil compaction due to machinery operation on a raspberry plantation. Fruit Science Reports, XVI, 2: 71-78.

PART E
===

VEHICLE AND TRAFFIC SYSTEMS IN CROP PRODUCTION

Soil Compaction in Crop Production
B.D. Soane and C. van Ouwerkerk (Eds.)
©1994 Elsevier Science B.V. All rights reserved.

391

CHAPTER 17

Quantification of Vehicle Running Gear

F.G.J. TIJINK

Institute of Agricultural Engineering (IMAG-DLO), Wageningen, Netherlands

SUMMARY

To assess the capability of off-road vehicles to compact the soil, the interaction between the running gear and the soil on which it operates must be understood and quantified. The distribution of stresses and the size and shape of the contact area are crucial. They depend on the structural properties of the running gear and of the soil. Consequently, the interaction between soil and vehicle is very complex.

This chapter outlines the major structural aspects of tyres and tracks and focusses on the distribution of radial (normal) and tangential (shear) stresses in the contact area, including the relevant factors of influence. Finally, techniques for characterizing running gear in relation to compaction problems that are suitable for designing, researching and advising on machinery, are discussed.

INTRODUCTION

Running gear represents the link between vehicle and soil. To assess the compaction capability of a tyre or track, the temporal and spatial distribution of radial (normal), tangential (shear) and vibration stresses and the factors that influence these stresses must be quantified (Soane, 1985). To be able to model soil compaction, it is also important to know the stress propagation and the soil's response to applied stresses.

In view of the extremely complex structure of the pneumatic tyre and its interaction with the soil, it is unlikely that a complete and satisfactory theory of tyre behaviour will be available in the near future (Pacejka, 1979, 1981; Wong, 1989). However, as a result of the extensive amount of empirical data now available, some guidance can be given in characterizing running gear in relation to soil compaction.

This chapter will be restricted to forces applied on a horizontal soil surface by running gear in steady-state, linear movement. Before dealing with the interaction between running gear and soil and the techniques for characterizing this interaction, the major structural aspects of tyres and tracks will be discussed.

TYRES

Tyre construction

Fig. 1 shows the most important components of a pneumatic tyre. The carcass consists of cord layers (plies) laid in specific directions. It is flexible and carries the inflation stresses and makes the tyre strong. Carcass strength depends on the properties of the cords used, the angle between the cord layers, and the number of plies in the carcass. For detailed information on materials and their properties, reference is made to Clark (1981).

Conventional tyres have a cross-ply construction. In this type of tyre, the successive cord layers cross each other at an angle of 60-80°. However, radial-ply tyres are now increasingly used for tractor drive wheels. In these tyres, the cords run radially from bead to bead. To improve stability, a stiff tangential belt has been included. This construction ensures relatively flexible sidewalls, combined with a well-braced tread.

The bead must keep the inflated tyre on the rim seat and presses the tyre against the rim flanges. Therefore, the bead usually contains a coil of steel wires. The sidewall protects the carcass. The rubber of the sidewall must have excellent resistance against fatigue and ageing.

The tread is bound to the carcass and the pattern is moulded in the tread. Tyre operating performance and driving qualities also depend on the type of tread pattern.

Tyres used in agriculture

A wide variety of tyre types and sizes is available. Tyre standards contain about 250 different sizes for agricultural use. Moreover, there are very many tread patterns, constructions and qualities. Consequently, there are more than 2000 different agricultural tyres.

Each tyre type has its own special fields of application. The tread pattern is based on the functional requirements of the tyre and is therefore a good visual indication of the expected use of the tyre.

A uniform code-marking system has been adopted (Table 1), which makes it easier to identify the various types of agricultural tyres. This system was set up in the U.S.A. by the Tyre and Rim Association (TRA) and has been adopted by the American Society of Agricultural Engineers (ASAE). The intention was to stamp the code on every tyre, but in practice most manufacturers mark their tyres with trade names. Sometimes the code marking is given in manufacturers' catalogues, below the trade identifications.

A simpler identification code is used in Germany. Tractor rear tyres carry the

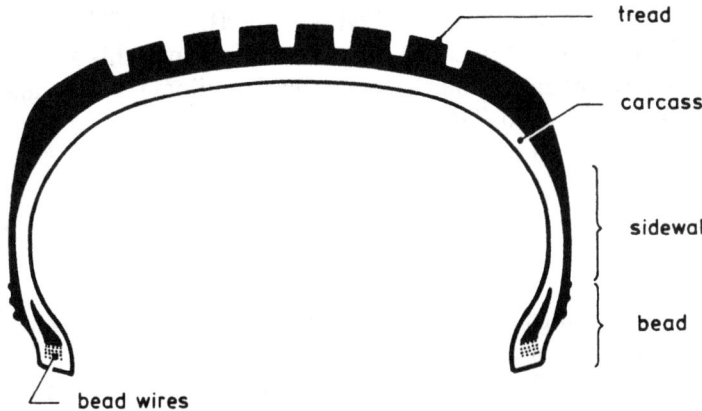

Fig. 1. Main components of a pneumatic tyre.

code AS (Ackerschlepper = agricultural tractor) and agricultural trailer tyres the code AW (Ackerwagen = agricultural trailer). The code AM (Ackermaschine = agricultural machine) is used for implement tyres. Tractor front tyres have the code AS-Front.

Another important group are flotation tyres. There is no clear definition of these tyres, but in general they are wider than one-third of their diameter. "Terra-

TABLE 1

International tyre code

	Tyre type	Code marking
Rear Tractor	Regular Agriculture	R1
	Rice and Cane	R2
	Industrial and Sand	R3
	Industrial-Lug Type	R4
Front Tractor	Single Rib	F1
	Regular Agriculture	F2
	Industrial Rib	F3
Garden Tractor	Regular Garden	G1
	Intermediate Tread	G2
	Rib Tread	G3
Implement	Rib Tread	I1
	Utility	I2
	Traction Implement	I3
	Plough	I4
	Plough	I5
	Smooth Tread	I6

Tires" (Goodyear), "Low Ground-Pressure" tyres (LIM), "TWIN" tyres (Trelleborg) and "Extra Wide" tyres (Michelin), fit more or less into this group. Only some of these tyres have been standardized by the Scandinavian Tire and Rim Organisation (STRO, 1990) and TRA (1991). The biggest flotation tyre (73x44.00-32) has a width of 1.10 m and an overall diameter of 1.86 m.

Tyre specifications

Tyre markings
A tyre can have up to 15 markings but every tyre carries at least two markings to indicate size designation and construction.

The size of early tyres was specified by quoting the section width, B, and the nominal rim diameter, d, both specified in inches (Fig. 2). These tyres had an aspect ratio, H/B, of about 1.0. At present, because of the steady stream of new types of agricultural tyres coming on the market and the fact that tyres from other fields of work are also used in agriculture, there are many size specifications. The most important size code systems are given in Table 2.

Originally, the tensile strength of the carcass was specified by the number of plies of cotton. Currently, other materials with a greater tensile strength than cotton are used in tyre construction, and the strength is expressed in ply-rating (PR). This index indicates the ratio of tyre strength to cotton strength; it does not

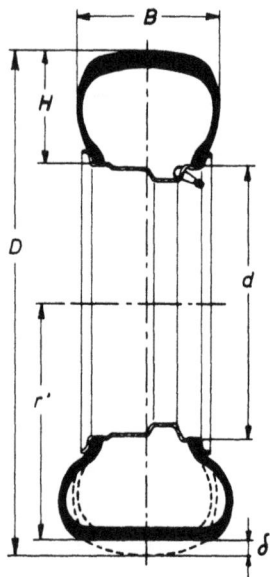

Fig. 2. Tyre and rim description. B = tyre section width; δ = tyre deflection; D = overall tyre diameter; H = tyre section height; d = rim diameter; r' = static loaded radius.

TABLE 2

Systems for designating tyre size

Size system	Tyre use	Examples of size markings
B-d (B, d in inches)	Tractor - steering wheel (non driven)	11.00-16
	- drive wheel (H/B=0.8)	16.9-38 20.8R38
	- drive wheel cultivation work	9.5-44
	Combine harvester drive wheel	24.5-32
	Implement tyre	6.50-16
B-d (B in mm; d in inches)	Passenger car	155R13
B/ H/B -d (B, d in inches; H/B in %)	Tractor drive wheel Implement tyre	18.4/70R38 16.0/70-20 19.0/55-17
B/ H/B -d (B in mm; d in inches; H/B in %)	Tractor drive wheel Implement tyre Passenger car	520/70-38 650/60-38 600/55-26.5 185/70-R13
DxB-d (D, B, d in inches)	Flotation tyre	44x41.00-20 73x44.00-32

necessarily state the number of plies. PR values indicate the maximum tyre inflation pressure allowed[1]. On the latest generation of radial-ply tractor drive wheel tyres, the PR code has been replaced by a service description and a symbol for inflation pressure. Service descriptions (Load Index and Speed Symbol) have been standardized (STRO, 1990; ETRTO, 1991; TRA, 1991; WdK, 1991).

Load Index (LI) indicates tyre load capacity in a number code. Tyre catalogues generally contain tables showing LI and load capacity. LI can have values between 0 and 279; the corresponding load capacities range from 45 kg to 136 Mg.

---

[1] In tyre specifications, tyre inflation pressure is usually expressed in bar (kg cm^{-2}) or psi (lb in^{-2}), while scientists are adopting kPa (kN m^{-2}) as the unit of inflation pressure. It should be noted that 1 bar $\approx$ 14 psi $\approx$ 100 kPa.

Fig. 3. Markings on a modern tractor rear tyre.

Speed symbols up to 40 km h^{-1} are important for agricultural purposes. The speed symbol for this field of application consists of the letter A and a number between 1 and 8. This number multipied by 5 shows the permitted driving speed. For example, A6 means a permitted driving speed of 30 km h^{-1}. Recently, agricultural tyres suitable for a speed of 50 km h^{-1} have been introduced. They have the letter B as speed symbol.

Ever more manufacturers are marking their tyres also with a star symbol for inflation pressure. In Europe (STRO, 1990; WdK, 1991), one star means that the load index is based on an inflation pressure of 160 kPa, two stars mean an inflation pressures of 240 kPa and three stars mean an inflation pressure of 320 kPa. In the U.S.A., these stars correspond with inflation pressures of 120 kPa, 160 kPa and 210 kPa, respectively (TRA, 1991).

Fig. 3 shows an example of a European symbol-marked tractor rear tyre:

18.4R38 146 A8 *

where 18.4 = tyre section width of 18.4 inches (= 467 mm); R = radial tyre (cross-ply tyres are marked with a dash (-) instead of the letter R); 38 = rim diameter of 38 inches (=965 mm); 146 A8 = service description (146 = Load Index: load capacity of 3,000 kg; A = Agriculture; 8 = Speed Symbol for 40 km h^{-1}); * = service description, indicating that the load index is based on an inflation pressure of 160 kPa.

Tyre load capacity
Load capacity depends on tyre volume (size), inflation pressure, carcass strength and operating conditions (driving speed, driven, steered, etc.). Generally, load capacities are given at a driving speed of 30 or 40 km h^{-1}. Load capacities

at 30 km h^{-1} are the basis for all field operations (ETRTO, 1991). Detailed information can be found in tyre standards and manufacturers' specifications. Tyre width increases load capacity much more than overall diameter (Söhne, 1969).

Tyre inflation pressure strongly influences load capacity. Most agricultural tyres, with an aspect ratio H/B of about 0.8, show the following relationship between wheel load W and inflation pressure p_i:

$$W = cp_i^{0.585} \tag{1}$$

where W = wheel load (kN); c = a constant, dependent on tyre type; p_i = tyre inflation pressure (kPa). When used in dual formation, the load capacity is 1.76 W. In triple fitment, the multiplication factor is 2.46.

Empirical relationships between load capacities and tyre parameters (e.g., width, diameter, aspect ratio, inflation pressure and volume) have been presented by Bekker (1960), Sitkei (1969), Sitkei and Söhne (1969), Steiner and Söhne (1979) and Perdok and Arts (1987).

It should be noted that in most countries the axle load for road travel is limited to 10 Mg. Consequently, wheel load is limited to 5 Mg. To carry such a load, a small tyre needs a higher inflation pressure than a big tyre. A very common implement tyre (16.0/70-20; with B = 0.42 m and D = 1.08 m) needs an inflation pressure of 620 kPa to carry 5 Mg at 30 km h^{-1}, while the biggest tyre (73x44.00-32; with B = 1.10 m and D = 1.86 m) can carry the same load at 70 kPa inflation pressure.

Tyre deformations

Under a static load, a tyre shows radial deformation and sidewall bulging, while a moving tyre can also show tangential carcass and lug deformations.

Radial deformations

Measurements by Knight and Green (1962) on surfaces ranging from firm to soft, show that deflection is not uniform laterally. As a rule, a cross-ply tyre shows inward deformation only, whereas a radial tyre also shows outward deformation. The carcass of a cross-ply tyre will be shortened as a result of inward deformation. The radial tyre has a stiff belt in the tangential direction, which distributes the deformations over the whole circumference, thus causing the outward carcass deformation. Steiner (1979) found outward deformations at about three-quarters of the rotation angle of a towed 13.6R28 tractor rear tyre.

Tyre deflection. Radial deformation is usually measured in a static situation and is called deflection, δ. Static tyre deflection depends on inflation pressure, load

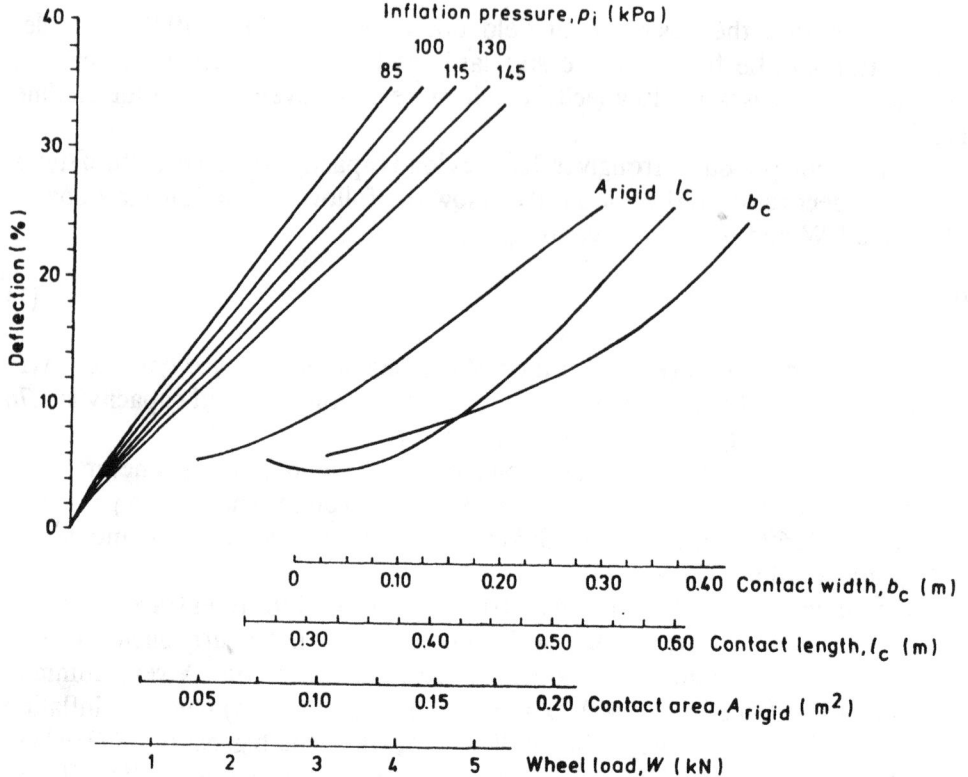

Fig. 4. Generalized tyre deflection chart of a 18.4-38 tractor rear tyre on a rigid surface.

and tyre stiffness, and on the character of the supporting surface. Some manufacturers provide "generalized deflection charts". These charts give the deflection behaviour of a specific tyre for a range of inflation pressures and wheel loads (Fig. 4).

In order to prevent carcass damage, manufacturers set limits to permissible tyre deflection. Normal agricultural tyres, with an aspect ratio H/B of about 0.8, typically have a deflection limit δ_{max} = 0.20H. Tyres with a smaller aspect ratio have a greater deflection limit (e.g., δ_{max} = 0.25H for a 800/65R32 tractor rear tyre). According to Ageikin (1987), variable inflation tyres may be operated for short durations at an extreme deflection of δ_{max} = 0.35H.

Many authors have measured tyre deflection behaviour. They include Matthews and Talamo (1965), Sitkei (1969), Sonnen (1970), Sharon (1975), Abeels (1976), Laib (1979), Plackett (1984) and Kising (1988). Sonnen (1970) measured remarkable differences in the deflection curves for loading and unloading. Sitkei (1969) and Bolling (1987) also measured maximum radial deflection on deformable soil. They found that deflection decreases with increasing sinkage.

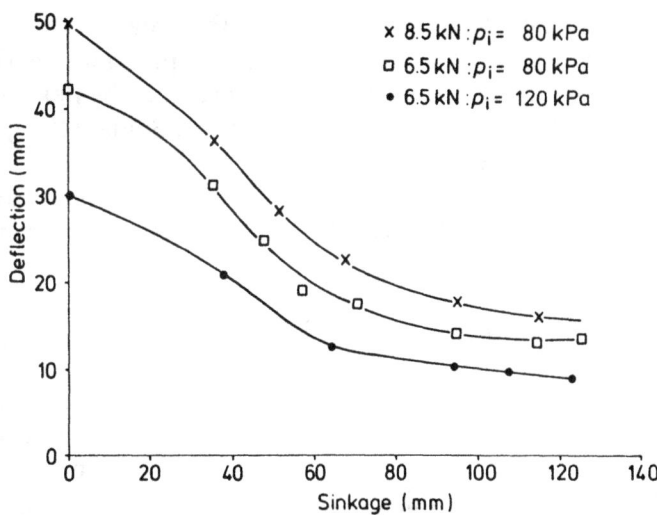

Fig. 5. Influence of sinkage on tyre deflection, for three combinations of wheel load and inflation pressure (after Sitkei, 1969).

Fig. 5 reveals that tyre-soil interaction depends on soil strength parameters as well as on tyre parameters. Tyre stiffness is crucial.

Tyre stiffness. Tyre stiffness, C, is defined as the ratio of tyre load, W, and deflection on a rigid surface, δ:

$$C = W\delta^{-1} \tag{2}$$

The extensive research and analyses performed by Sharon (1975), Kising (1988) and Lines and Murphy (1991) show that inflation pressure and tyre volume (size) greatly affect the stiffness of agricultural tyres. Stiffness increases almost linearly with inflation pressure. Typical values are $C = 250$ kN m^{-1} and $C = 450$ kN m^{-1} at inflation pressures of 80 kPa and 200 kPa, respectively. Within the normal range of tyre loads, driving speeds, ply ratings, rim widths and lug heights, these characteristics influence tyre stiffness only slightly. Despite the great dependence on inflation pressure, tyre stiffness can vary widely, because many other parameters are influential. At an inflation pressure of 100 kPa, the stiffness values of tractor rear tyres can range between 200 and 300 kN m^{-1}. These differences are mainly the result of tyre volume (size).

Tyre contact areas on hard and soft surfaces. The tyre contact area is the portion of the tyre in contact with the supporting surface. Interruptions in the contact area that result from tread patterns are considered to be part of the contact area

(ISTVS, 1968). There is no single unified theory to describe the area of contact between a tyre and a rigid roadway (Browne et al., 1981). Approximate and empirical methods to calculate tyre contact areas have been developed by Bekker (1956), Krick (1969), Sitkei (1969), Komandi (1976), Steiner and Söhne (1979), Painter (1981) and by Upadhyaya and Wulfsohn (1988).

In a simple approach, a tyre can be seen as a torus. This implies that the contact area on a firm surface (A_{rigid}) is elliptical. From the geometry we can write:

$$A_{rigid} = (\pi/4)(\text{contact length})(\text{contact width}) \approx \pi\delta(DB)^{0.5} \tag{3}$$

This simple model explains the following findings of Browne et al. (1981): (1) tyre deflection is the most important variable governing the contact area of a tyre; (2) at fixed deflection (inflation pressure and tyre load vary) the contact area will remain effectively constant. From this model we can also see that tyre contact area increases as the aspect ratio H/B decreases.

The dynamic contact area of a tyre is somewhat larger than the static contact area (Vanden Berg and Gill, 1962; Seitz, 1968). However, according to Van Eldik Thieme et al. (1981) this effect may be ignored.

The contact area of tractor tyres with a closed centre tread is about 30% smaller than that of tyres with an open centre tread (Söhne, 1952). This is mainly because open-centred tyres are stiffer. When B >0.35D, the contact width will be larger than the contact length.

McKyes (1985) proposed the following rule of thumb to estimate tyre contact area:

$$A_{rigid} = BD/4 \tag{4}$$

For a wide range of tyre sizes, a good correlation was found between the contact areas calculated with this equation and the contact areas provided by a tyre manufacturer (Fig. 6). Note that the biggest tyre, 1.10 m wide and 1.86 m in diameter, has a contact area $A_{rigid} = 0.5$ m².

Measurements by Tijink and co-workers on implement tyres revealed that if the wheel loads are in agreement with tyre specifications, the contact area is fairly constant in an inflation pressure range between 80 and 300 kPa. This can be explained by eqn. (1) (at constant deflection).

On a deformable surface, the tyre contact area (A_{soft}) is always larger than on a rigid surface (A_{rigid}). Methods to measure the size and shape of the tyre contact area on soft terrain are mainly conducted under static or quasi-static conditions (Söhne, 1957; Smith and Dickson, 1990). Models for the contact area on soft soils have been published by Bekker (1956), Wong (1989) and Schwanghart (1990). For rut depths >0.05D, McKyes (1985) gave a simple rule of thumb: $A_{soft} = BD/2$.

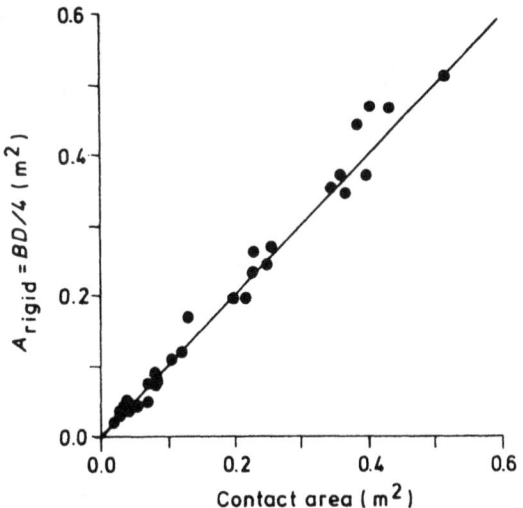

Fig. 6. Calculated contact area, using eqn. (4), and contact areas provided by a tyre manufacturer for a wide range of tyre sizes. The straight line is the 1:1 line.

Tangential carcass and lug deformations

Tangential carcass deformations are not only deformations in the sense of strains, but are also relative displacements in relation to a fixed system of coordinates. Displacement depends on tyre construction, load, inflation pressure and torque. Steiner (1979) found that for a towed tyre tangential carcass deformation is almost symmetrical. A radial tyre shows displacement all around the circumference, whereas a cross-ply tyre has half a circumference free from relative displacement.

True measurements of tangential lug deformations are unknown and difficult to perform. Relative displacement of lugs in the mutual contact area consists of lug bending and carcass deformation. Lug movements in the mutual contact area have been measured by Cegnar and Fausti (1961), Wann and Reed (1962) and Steiner (1979). Comparison of measured and theoretical curves shows that when the lugs enter the contact area, they lag behind the carcass to a certain extent as a result of friction between lug and surface. When the lugs leave the contact area, their relative displacement is greater because of their elastic reaction. This behaviour agrees with visual observations of the direction of lug movement.

Tyre deformations in the mutual contact area can influence wheel slip. On a firm surface, slip is influenced by carcass and lug deformation. Analyses by Steiner (1979) show that the influence of tyre deformations on slip should not be ignored and that radial tyres have smaller slip percentages because these tyres deform less than a cross-ply tyre (Table 3).

TABLE 3

Measured slip components (%) on a rigid surface and estimated slip components on a deformable soil (from Steiner, 1979)

Surface	Tyre	Slip caused by				
		Carcass deformation	Lug deformation	Soil deformation	Slip in the contact area	Total
Rigid	13.6-28	1.9	5.8	-	-	7.5
	13.6R28	0.6	4.4	-	-	5.2
Soil	13.6-28	1.2	4.3	8.0	7.5	21.0
	13.6R28	0.4	2.6	7.5	7.5	18.0

TRACKS

The track was first conceived of as a portable railway that is laid down in the front of the vehicle, travelled over and picked up again. The application of the concept of movable roadways to off-road vehicles led to the development of various track-laying vehicles. Tracked vehicles are used on a large scale in earth moving, resource industries and military operations. They can carry heavy loads and have excellent mobility and traction. In recent decades, the number of crawler tractors used in crop production in western Europe and America has fallen sharply, mainly because of their limited driving speed and prohibited use on metalled roads. However, crawler tractors are still important in agriculture in eastern Europe (see Chapter 16).

The running gear of a tracked vehicle transmits loads to the ground through a system of roadwheels running over a mobile road which is the track. Two groups of tracks can be distinguished: (1) rigid tracks; (2) flexible tracks.

Conventional crawler tractors have rigid steel tracks which cannot flex upwards when in contact with a deformable terrain. These rigid tracks have restricted speeds and no suspension system. Military tracked vehicles, designed for high-speed operations, have flexible tracks and a track suspension system.

Traditionally, the nominal ground pressure (NGP), which is defined as the weight of the vehicle divided by the contact area, is used as an important characteristic of tracked vehicles. However, in this chapter, the term "average ground pressure" is being used with a view to promoting consistency with usage for tyres. It should be noted that, just like "NGP", "average ground pressure" denotes the *calculated* average value of the vertical component of the stress in the soil-wheel or soil-track contact area. It should be be kept in mind that the distribution of the stresses in the soil-wheel or soil-track contact area is far from uniform.

Crawler tractors in the 75-300-kW engine power range, generally have an average ground pressure of 40-60 kPa. Special tracks, with lower average ground pressure values, can be delivered as an option. Smaller tractors may have an average ground pressure of 20-40 kPa.

In the mid-1980s a new track-type tractor was introduced in agriculture (see Chapter 21). Its running gear consists of a "rubber" belt as track, one rigid sprocket at the rear, one tyre as idler at the front, and 4 roadwheels in between, attached to a suspension system. The track tension must be very high ($\approx$track load) because of the friction-based transmission between sprocket and belt. The suspension system can be adjusted such that all 6 roadwheels have the same static load. This rubber-belted tractor, which weighs 15 Mg, is permitted to travel on metalled roads, even within the European Community (EC) where the total weight of wheeled tractors is limited to 14 Mg. In recent years, more makes of rubber-belted tractors have come on the market.

At present, rubber-belt tracks for crawler tractors with even lower average ground pressures and rubber wheel tracks are being developed. Rubber wheel tracks, i.e., a half-track system that contains an endless rubber belt, may be used instead of the standard tractor drive wheel tyres in special applications. For example, they may be an interesting option for use on traffic lanes in potato and vegetable cropping. Fitting such a system on the tractor is no more difficult that replacing a standard drive wheel (Tijink and Arts, 1992).

STRESSES IN THE WHEEL-SOIL INTERFACE

The performance of a wheel depends entirely on the soil-wheel interaction. The geometry of the contact area is only part of this interaction: the entire stress situation at the interface, i.e., magnitude, orientation and distribution of stresses is also important.

In the contact area between the soil and a towed or driven device, radial (normal) stresses as well as tangential (shear) stresses occur.

Stresses in the contact area between soil and rigid wheel or roller

Hegedus (1965) measured normal stress distribution in the wheel-soil interface and the influence of wheel load and slip on this distribution. The distribution of both normal and shear stresses in the wheel-soil interface has been measured by Onafeko and Reece (1967) and Krick (1969, 1971).

Radial stress distribution
Maximum radial stress does not occur directly under the wheel axle as suggested by the plate sinkage analogy of Bekker (1956). It actually occurs in

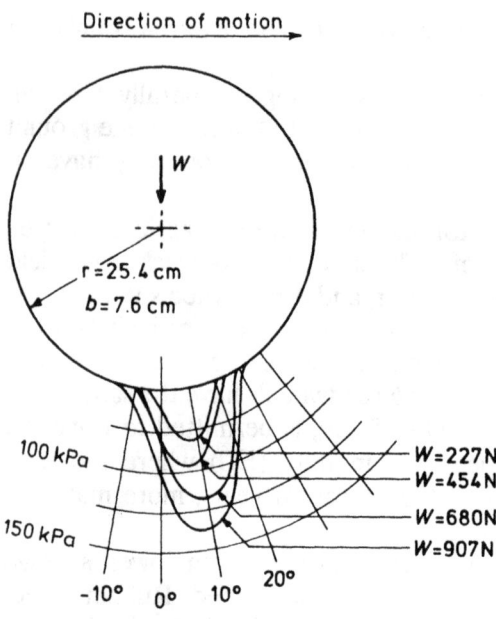

Fig. 7. Radial stress distributions under a wheel on dry sand for different wheel loads at zero slip (after Hegedus, 1965).

front of the axle. Experimental data show that on deformable soil the position of the maximum radial stress is very stable over a wide range of slip values (Hegedus, 1965; Krick, 1971). On dry, compacted sand and on unyielding sand, the maximum radial stress shifts forwards at increasing wheel slip (Hegedus, 1965, Onafeko and Reece, 1967). This effect may result from a "soil transport" phenomenon underneath the wheel as part of wheel slip. Hegedus (1965) and Krick (1971) found an increase in the maximum radial stress at increasing wheel load (Fig. 7).

The stress distribution is not uniform laterally. At the edge of the wheel rim there is a concentration of stress which decreases at increasing slip (Table 4).

TABLE 4

Ratio of the maximum radial stress (σ_r) at the edge of the wheel rim to the maximum radial stress (σ_m) in the middle of the wheel rim (Krick, 1969)

	Towed	Slip = 12 %	Slip = 25 %	Slip = 40 %
σ_r/σ_m	1.8	1.6	1.5	1.4

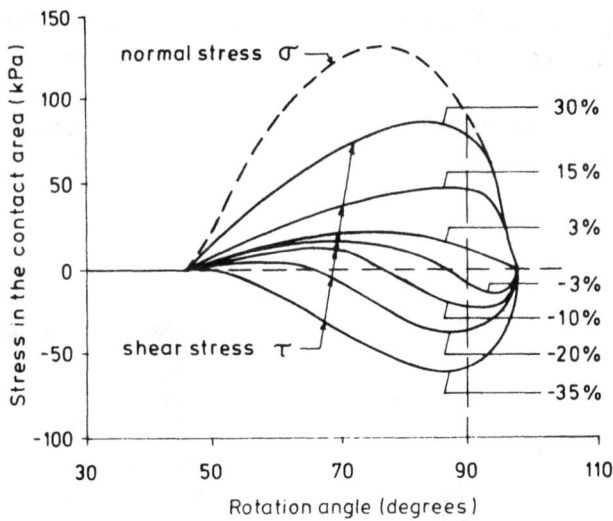

Fig. 8. Influence of wheel slip on stress distribution in the longitudinal centre plane of a rigid wheel (W = 7 kN; B = 0.20 m; D = 0.88 m) on soft clay loam (after Krick, 1971).

Tangential stress distribution

In the contact area of a towed wheel, the shear stress changes direction: having occurred in the direction opposite to the wheel rotation (positive), it now occurs in the same direction in which the wheel rotates (negative). A driven wheel has only positive shear stress.

The magnitude of shear stress strongly depends on slip (Fig. 8). The ratio of shear stress to normal stress increases at increasing wheel slip (Wiendieck, 1968; Krick, 1971).

Stresses in the tyre contact area

The essential difference between a rigid wheel and a pneumatic tyre is the flexibility of the tyre. When the tyre moves in steady-state on soft soil, there will be an equilibrium between the tyre-deflecting forces and the rut-forming forces.

Average ground pressure
Average ground pressure is defined as:

$$p_g = W/A \tag{5}$$

where p_g = average ground pressure; W = vertical wheel load; A = area of the tyre-road interface.

Average ground pressure on a rigid surface. This parameter is extremely useful when designing and selecting tyres and machinery. It is very convenient for quickly assessing the compaction capability of a complete fleet of machinery (Tijink, 1991; see also Chapter 19).

In terramechanics, this average ground pressure is often given as the sum of tyre inflation pressure p_i and a pressure p_c for carcass stiffness:

$$p_{rigid} = p_i + p_c \qquad (6)$$

Eqn. (6) is less accurate than eqn. (5) because the carcass stiffness is not a constant. An increase in inflation pressure (at a constant wheel load) results in a decrease in carcass stiffness pressure, p_c. At high inflation pressures, p_c can even have negative values. An increase in wheel load (at a constant p_i) results in a small increase in p_c (Söhne, 1952). Eqn. (6) is useful as a rule of thumb in situations where the deflection behaviour of the tyre is not known. German advisory services assume that $p_{rigid} = 100$ kPa is achieved at inflation pressures of 70 kPa for cross-ply tyres and 80 kPa for radial tyres. In the Netherlands, similar rules of thumb are used for quickly estimating the average ground pressure, for example, $p_{rigid} = 1.25p_i$. Eqn. (5) is preferred when the deflection characteristics are known.

Karafiath and Nowatzki (1978) presented an equation which includes the effect of inflation pressure on tyre stiffness:

$$p_{rigid} = c_1 p_i + p_c \qquad (7)$$

where p_{rigid} = average ground pressure; p_i = inflation pressure; p_c = average pressure transmitted by the carcass at $p_i = 0$; c_1 = constant expressing the effect of the carcass stiffness of the tyre.

At $c_1 = 1$, eqns. (7) and (6) are equal. Tyre deflection characteristics must be known before calculations can be made with eqn. (7). Here too, eqn. (5) is preferred.

Average ground pressure on deformable soil. Owing to the larger contact area, the average ground pressure is less on deformable soil than on a rigid surface. Extensive research on multipass behaviour (Holm, 1972; Tijink and Koolen, 1985; Tijink, 1988) revealed that when a tyre passes in a track for the second time, wheel sinkage is less than at the first pass. Consequently, the contact area is smaller and the average ground pressure is greater than when the tyre passes for the first time. On very soft soil, the soil deforms, not the tyre. Therefore, the average ground pressure can be considerably lower than the inflation pressure (Söhne, 1952, 1963; Baganz and Kunath, 1963).

The average ground pressure on deformable soil is not helpful when selecting tyres. It is only useful if both tyre and soil deformation properties are included. However, general characteristics on soil deformation are very scarce.

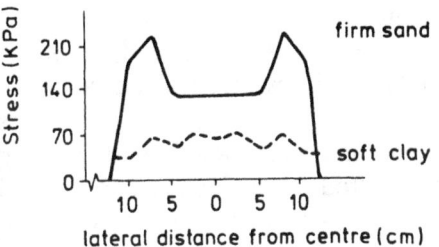
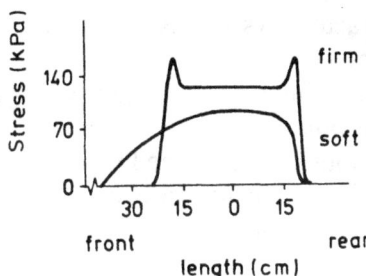

Fig. 9. Lateral and longitudinal stress distribution under an 11-38 towed smooth tyre at an inflation pressure of 100 kPa for firm and soft soils (after Vanden Berg and Gill, 1962).

Radial stress distribution in the tyre contact area

Radial stress distribution is far from uniform and deviates appreciably from the average ground pressure. According to Vanden Berg and Gill (1962), the maximum stresses are at least twice the average stresses.

Smooth tyre on a rigid surface. Normal stress distributions along the longitudinal and lateral axes have been given by Vanden Berg and Gill (1962), Seitz (1968), Van Eldik Thieme and Pacejka (1971) and Plackett (1984). Stress peaks occur at the edges (Fig. 9). Stress increases abruptly at the front of the tyre, but also decreases sharply at the rear of the contact area.

In the centre plane, the stress distribution is fairly uniform and depends on inflation pressure. Along the lateral axis, high peak stresses occur at the edges as a result of bending stresses in the tyre shoulders. These peak stresses hardly decrease with decreasing inflation pressure. At low inflation pressure, the normal stresses under the sidewalls can be more than 5 times the inflation pressure.

Measurements by Seitz (1968) and Plackett (1984) indicate that the maximum stresses for radial tyres are 10-20% below those for cross-ply tyres. This agrees with measurements made by Tijink and co-workers on 20.0/70-20 implement tyres.

Smooth tyre on deformable soil. Radial stresses underneath deflecting smooth tyres are lower than under rigid wheels. On soft soil, normal stress distribution is more uniform than on firm soil along both the longitudinal and lateral axes (Fig. 9). Maximum radial stress decreases with increasing soil deformability. A smaller wheel load (at constant inflation pressure) and a lower inflation pressure (at constant wheel load) result in a decrease in maximum normal stresses, while the shape of the stress distribution is fairly constant (Krick, 1971). Similar to rigid wheels, the normal stress underneath smooth tyres on deformable soil is hardly influenced by wheel slip.

Lugged tyre on deformable soil. Tyres with tread patterns exhibit extremely complicated stress distributions. The normal stress at any point depends on soil characteristics and tyre parameters (inflation pressure, dynamic load, structural characteristics, tyre driving or braking torque, tyre side forces, speed, etc.). No general functional relationship is available; only qualitative trends can be given.

Trabbic et al. (1959) and Liang and Yung (1966) measured normal stress distribution across the lug face, on the undertread and on the leading and trailing lug faces. An increase in inflation pressure resulted in an increase in normal stresses in the middle of the contact area and a decrease in normal stresses at the edges of the circumference of the contact area. Normal stress on the leading lug face and the undertread increased at higher pull, while normal stress on the lug face and the trailing side decreased.

Extensive research at Auburn in the U.S.A., reviewed by Burt et al. (1987, 1990), shows the effects of soil firmness and wheel load on the normal stress on tyre lugs of an 18.4-38 and an 18.4R38 tractor rear tyre. Fig. 10 shows the influence of wheel load on normal stress on a lug. Dynamic load ranges from 25 to 100% of the rated load capacity. On firm soils, an increase in wheel load decreases normal stress on a lug at the tyre centre (Fig. 10 left), while normal stress increases on a lug at the tyre edge (Fig. 10 right). This agrees with the change in normal stress in lateral direction underneath a smooth tyre on a rigid road. Increasing the wheel load (at constant inflation pressure) raises the bending moment at the tyre shoulders, resulting in greater stresses underneath the sidewalls and in smaller stresses in the tyre centre plane. On loose soils, normal stresses on the lugs at the tyre centre and at the tyre edge increase with increasing wheel load. This agrees with the effect of load on normal stress observed underneath rigid wheels and smooth tyres on soft soils.

On firm soils, the ratio of normal stress to inflation pressure is typically 2-2.5

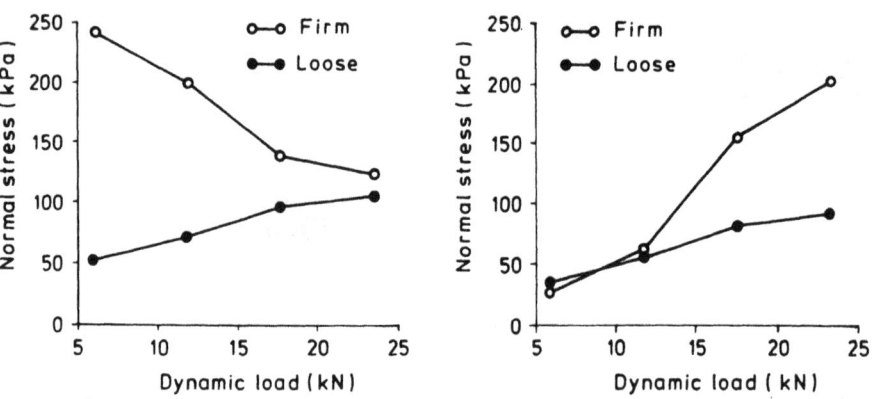

Fig. 10. Effects of dynamic load on normal stress on a lug when averaged across the contact angle of 260-270°, for firm and soft soils. Tyre: 18.4R38 tractor rear tyre at 110 kPa inflation pressure, 18.5% slip and 0.15 m s^{-1} forward speed. Left: at the tyre centre. Right: at the tyre edge (from Burt et al., 1990).

for a radial-ply tyre and 5-7 for a cross-ply tyre. These differences must be caused by the differences in sidewall flexibility between radial and cross-ply constructions. Increasing the wheel load (at constant inflation pressure) increases the normal stress at the undertread.

Tangential stress distribution in the tyre contact area
Various authors have measured tangential stress in the tyre contact area: Liang and Yung (1966), Seitz (1968), Krick (1969, 1971), Van Eldik Thieme and Pacejka (1971), Burt et al. (1987, 1990) and Oida et al. (1990).

Smooth tyre on a rigid surface. Shear stresses generally show peaks on the edges of the longitudal contact. An increase in inflation pressure increases the maximum shear stress. Shear stress becomes positive through the action of a driving torque. With increasing driving torque, the maximum shear stress moves backwards. The stress peak becomes negative as a result of the action of a breaking torque.

Smooth tyre on deformable soil. Along both the longitudinal and the lateral axes, the stress distribution for tyres is more uniform than for rigid wheels (Krick, 1971). At small slip values, the stress still increases towards the edge of the tyre but at high slip this effect disappears. Krick (1971) found almost equal maximum tangential stress at slip values of 5.5, 12.8 and 33.1%.

Lugged tyre on deformable soil. Tangential stress on the lug face, the undertread, the leading lug side and the trailing lug side, have been measured by Liang and Yung (1966) for an 11-28 tractor rear tyre used on sand. Stresses were measured at two pulls. An increase in pull resulted in a decrease in tangential stress on the lug face and the lug sides, while tangential stress on the undertread increased.
   Burt et al. (1990) found that the ratio of tangential to normal stress on the lugs was hardly influenced by dynamic load. The value of this ratio was not constant along the lateral axis. Tangential stress on the lug surface was frequently found to exceed normal stress.

STRESSES IN THE TRACK-SOIL INTERFACE

The performance of a tracked vehicle is directly related to the normal and shear stress distributions in the track-terrain interface. Experimental evidence has shown that the actual stress distribution under tracks is usually far from uniform, especially in the case of flexible tracks. This can be explained by the basic concept of a track system: a number of rigid roadwheels running over a track. Consequently, the average ground pressure is not suitable for characterizing the performance of flexible tracks on soft ground.

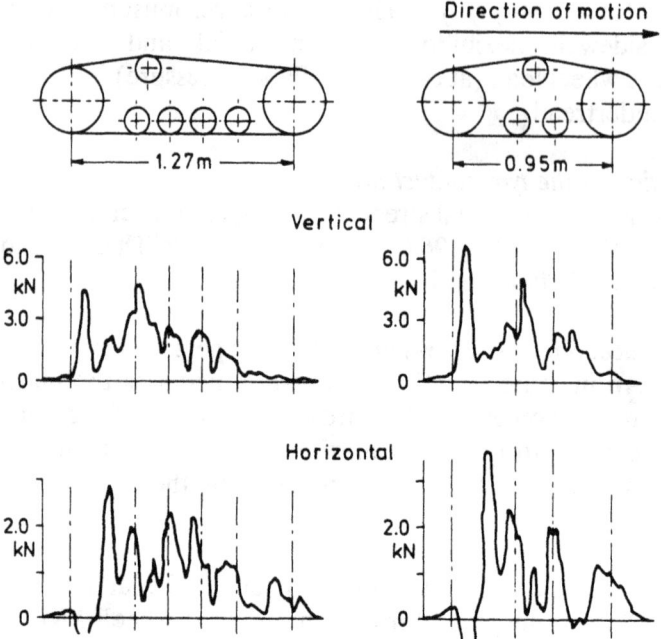

Fig. 11. Vertical and horizontal force distribution underneath two types of rigid tracks on loose sand. Left: track load = 13.14 kN; drawbar pull = 7.66 kN; slip = 15.4%. Right: track load = 13.14 kN; drawbar pull = 7.81 kN; slip = 20.0% (after Wills, 1963).

Stresses underneath rigid tracks

According to Söhne (1970) and Chancellor (1976), maximum normal stress is two to three times the average ground pressure. Measurements by Wills (1963) give an impression of the magnitude and distribution of stresses underneath rigid tracks on loose sand. Fig. 11 shows that a lower average ground pressure results in: (1) a lower maximum normal stress; (2) a lower maximum shear stress; (3) lower slip at the same tractive effort. Fig. 11 also illustrates that the drawbar pull has a great influence on the distribution of the normal stress. Maximum normal stress will increase and move backwards by the action of drawbar pull. Guskov (1968) found that a high drawbar pull resulted in approximately doubling the normal stress at the end of the track. Unfortunately, in many research reports neither the site where the measurements were undertaken nor the methods used are given.

Stresses underneath flexible tracks

The inadequacy of the average ground pressure as a general track characteristic has been pointed out by various authors, including Bekker (1956) and Rowland (1972). Rowland suggested replacing the average ground pressure by the mean maximum pressure (MMP), which is defined as the mean value of the maxima occurring under all the roadwheels. Rowland also proposed a set of empirical equations to predict MMP for various types of tracks. These equations include important vehicle and track parameters: vehicle weight, number of roadwheels, roadwheel diameter and track width. Although Rowland's approach includes important design aspects, it does not include soil conditions. Consequently, normal stresses cannot be calculated.

The stress distribution in the track-soil contact area depends on soil properties and track parameters such as track load, track dimensions, drawbar pull, track type (rigid/flexible), grouser shape, number of roadwheels, track tension, roadwheel diameter, track suspension system, and position of the driving sprocket. For detailed reviews on soil-flexible track interaction, reference should be made to Karafiath and Nowatzki (1978), Yong et al. (1984) and Wong (1989).

Maximum normal stress decreases with decreasing vehicle weight, increasing number of roadwheels, increasing roadwheel diameter and increasing track tension. Wong and Preston-Thomas (1986) found a significant effect of suspension characteristics on track performance on soft terrains. This may be explained by the influence of the suspension system on the load distribution. Analyses by Wong et al. (1984), Wong (1986) and Wong and Preston-Thomas (1986), showed that the effect of the number of roadwheels and track tension on track performance predominated on soft terrains. On rigid surfaces these effects were less pronounced.

Measurements on the stress distribution in the track-soil interface of recently introduced rubber-belted tractors are very scarce. Our measurements on a prototype rubber-belted wheel track showed normal stress peaks that are more than 3 times the average ground pressure.

CHARACTERISTICS FOR SELECTION OF RUNNING GEAR

To characterize the compaction capability of running gear it would be ideal to have information about the complete tyre-soil interaction: the dynamic distribution of normal and shear stresses over time. At present, tyre mechanics and soil mechanics are far away from this ideal. Further research on this subject could improve vehicle and traffic systems. It could also result in better empirical characteristics for the interaction between running gear and soil.

For quick examinations it is sufficient to concentrate on a more or less empirical approach. From the above, it may be concluded that the following factors help to reduce the compaction capability of tyres: (1) low inflation pressure; (2) low average ground pressure on a rigid surface; (3) small loads; (4) low tyre stiffness; (5) radial tyres; (6) low wheel slip; (7) low lugs. Among these factors, inflation pressure and average ground pressure on a rigid surface are the most important.

The compaction capability of tracks decreases with: (1) lower average ground pressure; (2) smaller load; (3) a greater number of roadwheels; (4) greater track tension; (5) larger roadwheel diameter; (6) the presence of an adequate suspension system.

For quick estimations of rigid track performance it is sufficient to focus on the average ground pressure. To examine flexible track systems on soft terrain, more track characteristics have to be included. Special attention should be paid to the number of roadwheels and track tension.

CONCLUSIONS

(1) The interaction between running gear and soil is determined by the structural properties of both the running gear and the soil.

(2) The major characteristics for selecting tyres are the inflation pressure and the average ground pressure on a rigid surface.

(3) For quickly estimating rigid track performance, the average ground pressure is a suitable parameter. However, when examining flexible track systems on soft terrain, the number of roadwheels and track tension should also be taken into account.

REFERENCES

Abeels, P.F.J., 1976. Tire deflection and contact studies. J. Terramech., 13: 183-196.
Ageikin, I.S., 1987. Off-the-road Wheeled and Combined Traction Devices. Amerid Publishing Co., New Delhi, India, 202 pp.
Baganz, K. and Kunath, L., 1963. Einige Spannungs- und Verdichtungsmessungen unter Schlepperlaufwerken. (Some measurements of stresses and compaction beneath tractor running gear). Agrartech., 13: 180-182 (in German).
Bekker, M.G., 1956. Theory of Land Locomotion. The University of Michigan Press, Ann Arbor, MI, U.S.A., 520 pp.
Bekker, M.G., 1960. Off-the-Road Locomotion. The University of Michigan Press, Ann Arbor, MI, U.S.A., 220 pp.
Bolling, I., 1987. Bodenverdichtung und Triebkraftverhalten bei Reifen - Neue Mess- und Rechenmethoden. (Soil compaction and tractive performance of tyres - New methods of measurement and calculation). Ph.D. thesis, Tech. Univ. München, Germany, 274 pp. (in German).
Browne, A., Ludema, K.C. and Clark, S.K., 1981. Contact between the tire and roadway. In: S.K. Clark (Editor), Mechanics of Pneumatic Tires. US Government Printing Office, Washington, DC, U.S.A., pp. 249-363.
Burt, E.C., Bailey, A.C. and Wood, R.K., 1987. Effects of soil and operational parameters on soil-tire interface stress vectors. J. Terramech., 24: 235-246.
Burt, E.C., Bailey, A.C. and Wood, R.K., 1990. Tire dynamic load effects on soil-tire interface and soil profile stresses. Proc. 10th Int. Conf. ISTVS, Kobe, Japan, Vol. 1, pp. 159-169.
Cegnar, A. and Fausti, F., 1961. Movements under the contact area of radial and conventional tires. Trans. ASAE, 4: 224-225.

Chancellor, W.J., 1976. Compaction of soil by agricultural equipment. Univ. California, Davis, CA, U.S.A., Bull. 1881, 53 pp.

Clark, S.K. (Editor), 1981. Mechanics of Pneumatic Tires. US Government Printing Office, Washington, DC, U.S.A., 931 pp.

ETRTO, 1991. Standards Manual. The European Tyre and Rim Technical Organisation, Brussels, Belgium, Section A (Agricultural Tractor and Implement Tyres), 32 pp.

Guskov, V., 1968. The effect of drawbar pull on the rolling resistance of track-laying tractors. J. Terramech., 5: 27-32.

Hegedus, E., 1965. Pressure distribution under rigid wheels. Trans. ASAE, 8: 305-308, 311.

Holm, C., 1972. Das Verhalten von Reifen beim mehrmaligen Überfahren einer Spur auf nachgiebigen Boden und der Einfluss auf die Konzeption mehrachsiger Fahrzeuge. (The multi-pass behaviour of tyres on soft soil and its influence on the design of multi-axle vehicles). VDI-Verlag, Düsseldorf, Germany, Fortschritt-Berichte VDI, Reihe 14, No. 17, 132 pp. (in German).

ISTVS, 1968. Glossary of Terrain-Vehicle Terms. J. Terramech., 5: 53-69.

Karafiath, L.L. and Nowatzki, E.A., 1978. Soil mechanics for off-road vehicle engineering. Trans Tech Publications, Clausthal, Germany, 515 pp.

Kising, A., 1988. Dynamische Eigenschaften von Traktor-Reifen. (Dynamic properties of tractor tyres). VDI-Verlag, Düsseldorf, Germany, Fortschritt-Berichte VDI, Reihe 14, No. 40, 212 pp. (in German).

Knight, S.J. and Green, A.J., 1962. Deflection of a moving tire on firm to soft surfaces. Trans. ASAE, 5: 116-120.

Komandi, G., 1976. The determination of the deflection, contact area, dimensions, and load carrying capacity for driven pneumatic tires operating on concrete pavement. J. Terramech., 13: 15-20.

Krick, G., 1969. Radial and shear stress distribution under rigid wheels and pneumatic tires operating on yielding soils with consideration of tire deformation. J. Terramech., 6: 73-98.

Krick, G., 1971. Die Wechselbeziehungen zwischen starrem Rad, Luftreifen und nachgiebigen Boden. (The interrelations between rigid wheels, pneumatic tyres and soft soil). Ph.D. thesis, Tech. Univ. München, Germany, 158 pp. (in German).

Laib, L., 1979. On the dynamic behaviour of agricultural tires. J. Terramech., 16: 77-85.

Liang, T. and Yung, C., 1966. A microscopic study of tractive performance of a lugged tire operating on sand. Trans. ASAE, 9: 513-515.

Lines, J.A. and Murphy, K., 1991. The stiffness of agricultural tyres. J. Terramech., 28: 49-64.

Matthews, J. and Talamo, J.D.C., 1965. Ride comfort for tractor operators, III. Investigation of tractor dynamics by analogue computer simulation. J. Agric. Eng. Res., 10: 93-108.

McKyes, E., 1985. Soil Cutting and Tillage. Elsevier, Amsterdam, Netherlands, Developments in Agricultural Science 7, 217 pp.

Oida, A., Satoh, A., Itoh, H. and Triratanasirichai, K., 1990. Three-dimensional stress distributions on tire-sand contact surface. Proc. 10th Int. Conf. ISTVS, Kobe, Japan, Vol. 1, pp. 229-240.

Onafeko, O. and Reece, A.R., 1967. Soil stresses and deformations beneath rigid wheels. J. Terramech., 4: 59-80.

Pacejka, H.B., 1979. Tyre tractors and vehicle handling. Int. J. Vehicle Design, 1: 1-23.

Pacejka, H.B., 1981. Analysis of tire properties. In: S.K. Clark (Editor), Mechanics of Pneumatic Tires. US Government Printing Office, Washington, DC, U.S.A., pp. 721-870.

Painter, D.J., 1981. A simple deflection model for agricultural tyres. J. Agric. Eng. Res., 26: 9-20.

Perdok, U.D. and Arts, W.B.M., 1987. The performance of agricultural tyres in soft soil conditions. Soil Tillage Res., 10: 319-330.

Plackett, C.W., 1984. The ground pressure of some agricultural tyres at low load and with zero sinkage. J. Agric. Eng. Res., 29: 159-166.

Rowland, D., 1972. Tracked vehicle ground pressure and its effect on soft ground performance. Proc. 4th Int. Conf. ISTVS, Stockholm, Sweden, Vol. 1, pp. 353-384.

Schwanghart, H., 1990. Measurement of contact area, contact pressure and compaction under tires in soft soil. Proc. 10th Int. Conf. ISTVS, Kobe, Japan, Vol. 1, pp. 193-204.

Seitz, N., 1968. Die Kräfte in der Bodenberührungsfläche schnell rollender Reifen. (The forces in the contact area of fast moving tyres). VDI-Verlag, Düsseldorf, Germany, Fortschritt-Berichte VDI, Reihe 12, No. 19, 60 pp. (in German).

Sharon. I., 1975. Untersuchungen über die Schwingungseigenschaften grossvolumiger Niederdruckreifen. (Research on the vibration characteristics of high-volume, low-pressure tyres). Ph.D. thesis, Tech. Univ. Berlin, Germany, 141 pp. (in German).

Sitkei, G., 1969. Die Kennzahlen von AS-Reifen und die Probleme der Bereifung. (Characteristics of agricultural tractor tyres and the problems of tyre selection). Proc. Int. Conf. ISTVS, Essen, Germany, Vol. 3, pp. 23-43 (in German).

Sitkei, G. and Söhne, W., 1969. Beziehungen zwischen den Kenngrössen von Acker-schlepperreifen auf fester Fahrbahn. (Relationships between the characteristics of agricultural tractor tyres and a rigid roadway). Grundl. Landtech., 19: 29-32 (in German).

Smith, D.L.O. and Dickson, J.W., 1990. Contributions of vehicle weight and ground pressure to soil compaction. J. Agric. Eng. Res., 46: 13-29.

Soane, B.D., 1985. Traction and transport systems as related to cropping systems. Proc. Int. Conf. Soil Dynamics, Auburn, AL, U.S.A., Vol. 5, pp. 863-935.

Söhne, W., 1952. Die Kraftübertragung zwischen Schlepperreifen und Ackerboden. (The transmission of forces between tractor tyres and field soil). Grundl. Landtech., 3: 75-87 (in German).

Söhne, W., 1957. Der Reifen auf dem Acker. (The tyre on the field). Forschungsanstalt für Landwirtschaft, Braunschweig, Germany, 70 pp. (in German).

Söhne, W., 1963. Beitrag zur Mechanik des Systems Fahrzeug-Boden unter besonderer Berücksichtigung der Ackerschlepper. (Contribution to the vehicle-soil system, with special reference to the agricultural tractor). Grundl. Landtech., 17: 5-16 (in German).

Söhne, W., 1969. Agricultural engineering and terramechanics. J. Terramech., 6: 9-30.

Söhne, W., 1970. Fahrwerksart, Triebkraft und Rollwiderstand geländegängiger Fahrzeuge bei unterschiedlig tragfähigen Böden. (Running gear, tractive force and rolling resistance of terrain vehicles on soils of different bearing capacity). Tech. Univ. München, Germany, Inst. Landmaschinen, Institutsveröffentlichungen, Vol. 1, pp. 141-169 (in German).

Sonnen, F.J., 1970. Über den Einfluss von Form und Länge der Aufstandsfläche auf die Zugfähigkeit und den Rollwiderstand von AS-Reifen. (On the influence of shape and length of the contact area on the tractive performance and the rolling resistance of agricultural tractor tyres). Ph.D. thesis, Tech. Univ. Braunschweig, Germany, 177 pp. (in German).

Steiner, M., 1979. Analyse, Synthese und Berechnungsmethoden der Triebkraft-Schlupf-Kurve von Luftreifen auf nachgiebigen Boden. (Analysis, synthesis and calculation methods of the tractive force-slip curve of pneumatic tyres on soft soil). Ph.D. thesis, Tech. Univ. München, Germany, 190 pp. (in German).

Steiner, M. and Söhne, W., 1979. Berechnung der Tragfähigkeit von Ackerschlepperreifen sowie des Kontaktflächenmitteldruckes und des Rollwiderstandes auf starrer Fahrbahn. (Calculation of the load capacity of agricultural tractor tyres, the average ground contact pressure and the rolling resistance on a rigid roadway). Grundl. Landtech., 29: 145-152 (in German).

STRO, 1990. Databok. (Data Book). The Scandinavian Tire and Rim Organisation, Malmö,

Sweden, Section T (Lantbrucksdäck), 40 pp. (in Swedish).

Tijink, F.G.J., 1988. Load-bearing processes in agricultural wheel-soil systems. Ph.D. thesis, Agric. Univ. Wageningen, Netherlands, 173 pp.

Tijink, F.G.J., 1991. The influence of wheel configuration on ground pressure of agricultural vehicles. Proc. 5th Europ. Conf. ISTVS, Budapest, Hungary, Vol. 1, pp. 89-96.

Tijink, F.G.J. and Arts, W.B.M., 1992. The field performance of a new track system. Proc. Int. Conf. Agric. Eng, AGENG, Uppsala, Sweden, pp. 107-108.

Tijink, F.G.J. and Koolen, A.J., 1985. Prediction of tire rolling resistance and soil compaction, using cone, shear vane, and a falling weight. Proc. Int. Conf. Soil Dynamics, Auburn, AL, U.S.A., Vol. 4, pp. 800-813.

TRA, 1991. Yearbook. Tire and Rim Association, Akron, OH, U.S.A., Section 4, 24 pp.

Trabbic, G.W., Lask, K.V. and Buchele, W.F., 1959. Measurements of soil-tire interface pressures. Agric. Eng., 40: 678-681.

Upadhyaya, S.K. and Wulfsohn, D., 1988. Relationship between tire deflection characteristics and soil-tire contact area. Am. Soc. Agric. Eng., St. Joseph, MI, U.S.A., ASAE Pap. 88-1005, 23 pp.

Vanden Berg, G.E. and Gill, W.R., 1962. Pressure distribution between a smooth tire and the soil. Trans. ASAE, 5: 105-107.

Van Eldik Thieme, H.C.A. and Pacejka, H.B., 1971. The tire as a vehicle component. In: S.K. Clark (Editor), Mechanics of Pneumatic Tires. US Government Printing Office, Washington, DC, U.S.A., pp. 545-839.

Van Eldik Thieme, H.C.A., Dijks, A.J. and Bobo, S., 1981. Measurements of tire properties. In: S.K. Clark (Editor), Mechanics of Pneumatic Tires. US Government Printing Office, Washington, DC, U.S.A., pp. 541-720.

Wann, R.L. and Reed, I.F., 1962. Studies of tractor tire-tread movement. Trans. ASAE, 5: 130-132.

WdK, 1991. AS-Treibradreifen mit Betriebskennung. (Drive wheels of agricultural tractors with service description). Wirtschafsverband der deutschen Kautschukindustrie e.v. (WdK), Frankfurt, Germany, Leitlinie 157, Blatt 1-4, 19 pp. (in German).

Wiendieck, K.W., 1968. A theoretical evaluation of the shear-to-normal stress ratio at the soil-wheel interface. J. Terramech., 5: 9-25.

Wills, B.M.D., 1963. The measurement of soil shear strength and deformation moduli and a comparison of the actual and theoretical performance of a family of rigid tracks. J. Agric. Eng. Res., 8: 115-131.

Wong, J.Y., 1986. Computer aided analysis of the effects of design parameters on the performance of tracked vehicles. J. Terramech., 23: 95-124.

Wong, J.Y., 1989. Terramechanics and Off-Road Vehicles. Elsevier, Amsterdam, Netherlands, 251 pp.

Wong, J.Y. and Preston-Thomas, J., 1986. Parametric analysis of tracked vehicle performance using an advanced computer simulation model. Proc. Inst. Mech. Eng., 200: 101-114.

Wong, J.Y., Garber, M. and Preston-Thomas, J., 1984. Theoretical prediction and experimental substantiation of the ground pressure distribution and tractive performance of tracked vehicles. Proc. Inst. Mech. Eng., 198: 265-285.

Yong, R.N., Fattah, E.A. and Skiadas, N., 1984. Vehicle Traction Mechanics. Elsevier, Amsterdam, Netherlands, Developments in Agricultural Engineering 3, 307 pp.

Soil Compaction in Crop Production
B.D. Soane and C. van Ouwerkerk (Eds.)
417

CHAPTER 18

Quantification of Traffic Systems in Crop Production

H. KUIPERS[1] and J.C. van de ZANDE[2]

[1]Bennekom, Netherlands
[2]Research Station for Arable Farming and Field Production of Vegetables (PAGV), Lelystad,
Netherlands

SUMMARY

Numerous criteria for quantifying the intensity of field traffic are reviewed and consideration
is given to their theoretical and practical relevance. Their selection and utilization depend
largely upon the particular application. For general purposes, the product of load and loading
time per unit area, the Field Load Index (FLI), is considered to be the most appropriate.
Loading 1 ha with 1 bar during 1 s requires roughly 30 t h. From the comparison of the results
of different methods of quantifying field traffic on the basis of the same dataset, it appeared
that there is good agreement between FLI and other parameters, such as Traffic Intensity (TI)
and Compaction Risk Factor (CRF). Higher mechanization levels are attended by lower
average values of TI, CRF and FLI.

Except for transport of harvest products and slurry, the weight of the power source is the
first and often the main determinant for the total load. In turn, the weight of the power source
is mainly dependent on the available power. The weight/power ratio for animal traction is 10-
20 times higher than for tractors. For some important crops in highly mechanized farming, the
time spent in the field by wheeled machinery presently ranges from 5 to 10 h ha^{-1} year^{-1}.

In arable cropping the loading events can be categorized in four groups: (1) seedbed
preparation and sowing (seedbed loading), (2) between sowing and harvest (rootbed loading),
(3) during harvest (harvest loading), (4) tillage between harvest and the next seedbed
preparation (post-harvest loading). Based on data from the U.S.A. and the Netherlands, it has
been estimated that for an annual load of 40 t h ha^{-1}, 10% was exerted on the seedbed, 10%
on the rootbed, 55% at harvest and 25% in post-harvest tillage. On grassland, the yearly
loading by field traffic is usually somewhat lower, but here there is normally no loosening effect
of tillage operations.

Calculations about the theoretical position of wheelmarks made in the Netherlands, indicate
that up to 50% of the field area may be loaded with >5 t h ha^{-1} for seedbed and rootbed
loading together. Usually, the headlands are more frequently loaded than the field itself. For
ploughing operations this may result in 3 times higher figures than for the field itself.

The few data available suggest that the load distribution in the field may well be an
important clue to understand crop responses to soil compaction.

INTRODUCTION

In any study of the incidence of compaction problems in commercial crop production, the distribution of vehicle traffic must be quantified. While in many field experiments traffic treatments are applied uniformly over the experimental plots, under commercial crop production systems traffic is usually markedly non-uniform. For many field operations, traffic follows a regular and repetitive pattern, depending on the size of the tractors and the working width of the implements involved. However, in some operations, such as the transport of harvested products (e.g., grain, straw), the traffic of the heavily loaded vehicles involved may be on a near-random distribution.

Estimating either the overall or the distributional effects of field traffic on crop production requires knowledge of both the compaction capability of the vehicle and the soil compactability at the time of each traffic event. In the field, loadings will occur spotwise under a hoof or stripwise under a vehicle tyre. All the spotwise and stripwise loadings of one field trip, e.g., occurring during one harrowing, can be integrated into a "loading event" for that specific field. As a next step, all loading events occurring at the same stage of crop production, e.g., at the harvest of a crop, can be integrated into a specific group and, finally, all groups together (seedbed, rootbed, harvest and post-harvest loadings) can be integrated for a whole year, to estimate the annual traffic load.

The single loading event can be considered as the lowest, the annual traffic as the highest level of integration. Each level of integration requires a specific approach, but parameters should be selected carefully to facilitate the integration to the next higher level, or the differentiation into the next lower one. This chapter deals mainly with the two higher levels, the annual traffic load and the four groups of loading events.

PARAMETERS FOR QUANTIFYING FIELD TRAFFIC

Many types of parameters to quantify field traffic have been proposed, varying from very simple to very complex, and each type has certain advantages and disadvantages. They may be classified roughly into six categories as defined below.

The parameters in categories (1)-(4) provide, with varying degrees of accuracy, estimates of the compactive capability of field traffic but are unable to predict the likely actual soil or crop responses because they include no information as to the compactability of the soil at the time of the traffic. Such information is included in the parameters in categories (5) and (6). The equations given are reproduced as originally specified although, in several cases, the terminology and units employed do not conform to current usage.

(1) Overall coverage

These parameters (e.g., % coverage, rut length per hectare) indicate the total coverage of wheels for a single traffic event, or a series of traffic events for a whole year. They provide no information on the compaction risk of such traffic.

Rut length (Frese, 1969)

The total length of the two ruts made by the rear wheels of a tractor pulling an implement is simply calculated according to the equation:

$$RL = 20/WW \tag{1}$$

where RL = total rut length (km ha^{-1}); WW = working width of the implement (m).

Traffic Intensity (Lumkes, 1984)

This parameter is the proportion of the area covered with wheel tracks during a sequence of single field operations. The width of the tracks is actually measured in the field. Traffic Intensity is calculated according to the equation:

$$TI = 100 \ (\Sigma t)/w \tag{2}$$

where TI = Traffic Intensity (%); t = width of individual wheel tracks (m); w = width of the field (m).

The parameters RL and TI show the proportion of the field area which is trafficked, but not to what extent (ground contact pressure or wheel load) or where in the field. Differences between farming systems can be shown using these parameters, but they do not indicate the risk of soil compaction unless the farming systems compared have a similar technology level.

(2) Distribution of coverage

Traffic Effect (Arndt and Rose, 1966)

Overall coverage gives no indication of the distribution of wheel traffic over the field. Maps and histograms provide opportunities to show the number of passes within a specified time interval for a whole field or a modular unit of a field (see section "Distribution of loading across the field", Fig. 7). Numerical equations can indicate coverage in specified traffic bands. The following equations show the accumulation of traffic intensity in different traffic bands, T_1, T_2, T_3, for conventional tractors with single front (f) and back wheels (b), and for rowcrop tractors with dual (d) front and rear wheels (r), for one planting (p) and three cultivation actions (c_1, c_2, c_3):

$$T_1 = r_{c1} + r_{c2} + r_{c3} \tag{3}$$

$$T_2 = (2f + 2b)_p + d_{c1} + d_{c2} + d_{c3} \tag{4}$$

$$T_3 = (2f + 2b)_p \tag{5}$$

Thus, it is possible to quantify the stripwise distribution of different types of wheels. The traffic history of the field as a whole can be summarised by assembling information for adjacent, 46-cm wide strips, as shown in the following relationship:

$$\text{crop}/T_1/\text{crop}/T_2/\text{blank}/T_3 \tag{6}$$

In this approach, only the different types of wheels are quantified and no indication is given of any compaction risk parameter, such as wheel load or ground contact pressure.

(3) Individual wheel hazard

The compaction hazard for single wheels can be calculated by a number of parameters but these do not usually take into account the cumulative effect where wheels on a given vehicle run in the same wheel track.

Load Index (Freitag, 1979)
A representative of this category of traffic parameters is the Load Index, which is defined as:

$$LI = WL/(WD * TW) \tag{7}$$

where LI = Load Index (lb in^2); WL = wheel load (lb); WD = wheel diameter (in); TW = tyre or wheel width (in).

With this formula, comparative information can be given on the potential effect of wheeling with different vehicles, e.g., tractors, trucks, military vehicles, cars, etc. It has been shown that LI is related to wheel sinkage, rut formation and cone index.

(4) Overall traffic hazard
A number of parameters have been proposed which estimate the overall compaction hazard during single or multiple traffic events by including information on the weight of the vehicle, combined with an estimate of the amount and distribution of the traffic over the field.

Mobility Index (Gill and Vanden Berg, 1968)
This parameter is defined by the particulars of the equipment concerned. Since

the results can be calculated for individual wheels, it is also possible to quantify the effects for wheels mounted in tandem. The Mobility Index is calculated as follows:

$$MI = c_1 [\{(c_2*EW/TW * WD * NW) + WL - c_3\} (c_4 * EP) - c_5] \tag{8}$$

where MI = Mobility Index; EW = equipment weight (lb); TW = tyre or wheel width (in); WD = wheel diameter (in); NW = number of wheels; WL = wheel load (1000 lb); EP = engine power (hp); $c_1...c_5$ = constants.

Traffic Intensity (Eriksson et al., 1974; Raghavan et al., 1979)
This parameter is generally used in the Scandinavian countries. Eriksson et al. (1974) defined it as:

$$TI = (10/WW) WL \tag{9}$$

where TI = Traffic Intensity (t km ha^{-1}); WW = working width of the implement (m); WL = wheel load (t).
In Canada, an alternative definition for Traffic Intensity was given by Raghavan et al. (1979):

$$TI = NP * IP \tag{10}$$

where TI = Traffic Intensity (kPa); NP = number of passes; IP = inflation pressure (kPa).

Mechanization Degree (Perdok and Van de Werken, 1983)
This parameter defines the amount of energy applied in the field as:

$$MD = EP * FT \tag{11}$$

where MD = Mechanization Degree (kWh ha^{-1}); EP = engine power (kW); FT = field time (h ha^{-1}).

Compaction Risk Factor (Van de Zande, 1991)
This parameter has been proposed to express the effect of several factors of relevance to the compaction risk of field operations:

$$CRF = (c_1*WL * IP) / (c_2*WW * TW) \tag{12}$$

where CRF = Compaction Risk Factor (kN^2 ha^{-1}); WL = wheel load (kg); IP = tyre inflation pressure (kg cm^{-2}); WW = working width of the implement (cm); TW = tyre or wheel width (cm); c_1, c_2 = constants.

Field Load Index (Kuipers, 1986)
This parameter characterizes the overall loading on a field as:

$$FLI = W * LT \tag{13}$$

where FLI = Field Load Index (t h ha^{-1}); W = weight of vehicle plus implement
(t); T = field time of the vehicle (h ha^{-1}).

(5) Estimated soil response

These parameters include an assessment of both the compactive capability of
the field traffic and the compactability of the soil at the time of the traffic.

Bulk Density (Raghavan and McKyes, 1978)
The bulk density to be expected at a certain date after planting may be derived
from the following model equation:

$$BD = c_1 + c_2 * \ln (NP * IP) - c_3 * \ln (MC) + c_4 * SD + c_5 * DAP \tag{14}$$

where BD = dry bulk density (g cm^{-3}); NP = number of passes; IP = tyre
inflation pressure (kg cm^{-2}); MC = soil water content (g cm^{-3}); SD = soil depth
(cm); DAP = days after planting; $c_1...c_5$ = constants.

(6) Estimated crop response
These parameters estimate the effect on crop yield of the predicted soil
responses to the traffic.

Crop Yield Loss (Arvidsson and Håkansson, 1991)
An estimate of the crop yield loss due to field traffic may be calculated as a
result of the following 4 components: (1) recompaction of the topsoil after
ploughing; (2) damage to the structure of the topsoil persisting after ploughing,
being a function of traffic intensity and clay content; (3) subsoil compaction,
being a function of traffic intensity; (4) traffic in a growing crop. Changes in the
degree of compactness, D, were computed from moisture content, MC, wheel
track length, AL, tyre inflation pressure, IP, total vehicle weight, TW, and number
of passes, NP, according to the equation:

$$D = f(MC,AL,IP,TW,NP) \tag{15}$$

resulting in an effect on crop yield (YL) as a function of the degree of
compaction (D):

$$YL = f(D) \tag{16}$$

For a specific case this results in:

$$YL = 0.00154 * TI * CF * CC \tag{17}$$

where YL = yield loss (%); TI = Traffic Intensity (t km ha^{-1}); CF = correction factor, ranging from 0 to 1.5, depending on soil water content and tyre inflation pressure; CC = clay content (%, w/w).

OBSERVATIONS OF FIELD TRAFFIC INTENSITY

Ever since mechanization was introduced into agricultural practice, there has been interest and concern about the distribution of the traffic load exerted on the soil. This interest commenced at the time of animal power. Von Nitzsch and Holldack, quoted by Scheibe (1935), found that penetrometer resistance in the footprints of horses was 50% higher than in tractor ruts at 15 cm depth and 17% higher at 20 cm depth. However, footprints had a less dominant effect on soil compaction than tractor tyres because of their scattered distribution. In mechanized agriculture the footprints have disappeared and all the loading is in continuous wheel marks.

Frese (1969) mentioned two major aspects of the changes in field loading caused by the evolution of farm mechanization: the reduction of total rut length, or the area covered by wheel tracks, and the decreasing weight/power ratio with newer and bigger tractors. In sugar beet growing in Germany, the total area covered with tracks during the year decreased from 580 to 310% and total rut length decreased from 278 to 86 km ha^{-1} as tractor power increased from 12 to 67 kW.

In a thorough review of soil compaction by agricultural vehicles, Soane et al. (1982) presented data on the upward trend in tractor power and mass (up to 250 kW and 30-35 t) and the high mass of vehicles for transportation and spreading of lime and slurry (sometimes >20 t). The review also contains data on the area covered with wheel tracks and the number of passes in these wheel tracks. In a single cereal growing season in Byelorussia, >80% of the field was covered by tractor wheelings, with up to 9 passes. For sugar beet, even 11 passes were made on part of the area. Seedbed preparation for spring barley in Scotland resulted in a coverage of 91%, including overlap. In California, a single harvesting operation of lucerne covered 75% of the field with wheel tracks. Successive harvestings resulted in up to 20 wheel passes a year for certain parts of the field.

For small grains in Sweden, Håkansson (1984) mentioned an annual area of wheel tracks ranging from 300 to 500% of the field area. Traffic Intensity (Eriksson et al., 1974) ranged from 120 to 220 t km ha^{-1}. Frese (1969) reported that in Germany, by the end of the 1960s, the product of wheel track length and tractor weight amounted to about 200 t km ha^{-1}. As a general figure for annual field loading in arable cropping, Håkansson and Danfors (1981) mentioned 350 t km ha^{-1}. Domżał et al. (1991) indicated that in Poland the coverage by wheel tracks for cereals and root crops was up to 250% and 500% of the field area, respectively. Harvest operations were responsible for about one third of the total area covered by wheel tracks in cereals and for half of the total area in root crops.

For farms in the southwestern part of the Netherlands, Van de Zande (1991) found that the annual coverage with wheel tracks decreased slightly as farm size increased. For a rotation with three crops it decreased from 365 to 333% of the field area when farm size increased from <35 ha to 1200 ha. For potatoes, sugar beet and winter wheat, the areas covered with wheel tracks were 473, 394 and 252% of the field area, respectively. In southern Australia, where small grains are grown on large farms (2000 ha or more) and big tractors (average 120 kW) are used, Halloway and Dexter (1990) reported lower traffic intensities than are common in Europe, with an average of 62.7 t km ha^{-1} and a total coverage by tracks of 165%. When overlaps are taken into account, 52.5% of the area was not touched by wheels in a small-grain cropping season.

ANALYSIS OF QUANTIFICATION OF TRAFFIC HAZARD

The data usually included in the quantification of field traffic systems refer to: (1) the weight of vehicles and implements, W; (2) the distance travelled per unit area, L/A; (3) the ground contact pressure, p; (4) the area covered with wheel tracks, a; (5) the number of passes, n.

The relationships between these different aspects become clearer if we introduce three additional characteristics: (1) the contact area between the tyre and the soil, c; (2) the loading time of a surface element, t; (3) the time spent by the vehicle in the field, T.

From the above definitions it follows that:

$$W = cp \tag{18}$$

and

$$c/a = t/T \tag{19}$$

The combination of eqns. (18) and (19) gives:

$$WT = atp \tag{20}$$

or

$$WT/A = pta/A \tag{21}$$

This relationship indicates that the product of weight and time spent per unit area equals the product of contact pressure, loading time and the fraction of the area covered with tracks.

In eqns. (18)-(21), the distance travelled per unit area, L/A, and the number of passes, n, do not appear directly. However, if we divide L/A by the travelling speed, we obtain T/A. The number of passes is directly related to the loading time, t.

For a set of identical wheels, the description given above is more or less complete. Generally, however, many different wheel passes have to be added up. This is no problem numerically, but factors which do not appear in the addition may influence the overall soil response. For example, it is obvious that the effect of a light loading is different if it is preceded by either no loading or by a heavy loading. Nevertheless, it may well be that the effect of a combination of a heavy and a light loading is almost independent of the order in which they occur.

In the compaction risk factor, CRF (Van de Zande, 1991), three factors which are positively related to compaction, W, p and a/A, are multiplied. Correlative research will be the best tool to identify whether this or another combination of the factors involved is most suitable for the prediction of field compaction. However, in this chapter the effects of loading on the soil will not be discussed.

In eqns. (20) and (21), the left-hand parts describe the loading of the field as a whole, and the right hand parts express the distribution of the loading across the field. In a field situation where loading times are short and ground pressures not too widely different for the various vehicles involved, the product of load and loading time per unit area, which is the left hand side of eqn. (21), seems to be a reasonable overall characterization of the "traffic load". Introduced by Kuipers (Kuipers, 1986) this product is called Field Load Index . A Field Load Index of 1 bar s is roughly equivalent to 30 t h ha^{-1}. From an energy point of view this is a reasonable approach. For a certain field operation, a specific amount of energy is required. Since energy output is related to engine or body weight, the required energy can be expressed as the product of weight and loading time. This product is easily calculated and is also a sound physical parameter, the impulse.

Summarizing, it may be predicted from theoretical considerations that for the calculation of practical values of different traffic intensities on the scale of a farmer's field, it is sufficient to know (1) the total weight of the power source, tools and crew involved in the different field operations; (2) the speed; (3) the working width of the implements or the time spent per unit area; (4) the tyre dimensions; (5) the inflation pressure of the tyres.

COMPARISON OF FOUR TRAFFIC PARAMETERS

To demonstrate and compare the different results obtained with four traffic parameters, calculations were undertaken using a dataset of traffic information for farms of different size, collected by Van de Zande (1983) for potatoes, sugar beet and winter wheat crops. The four traffic parameters included a criterion of area coverage, i.e., Traffic Intensity (Lumkes, 1984), and three criteria of compaction risk: (1) Compaction Risk Factor (Van de Zande, 1991); (2) Traffic

TABLE 1

Results for four traffic parameters for potatoes, sugar beet and winter wheat crops on farms of different size in the Netherlands (after Van de Zande, 1983)[1]

	Farm size (ha)					
	<35	35-55	55-80	80-200	1200[2]	Mean

I. Traffic Intensity (ha ha^{-1} year^{-1}), after Lumkes (1984)

Potatoes	5.4^a	4.7^b	4.4^b	4.9^{ab}	4.1^b	4.7^x
Sugar beet	3.9^{fhi}	3.8^{fh}	4.5^{bh}	3.4^{fh}	2.6^{fi}	3.9^y
Winter wheat	2.9^{ki}	2.5^{km}	2.2^m	2.2^m	2.9^{ki}	2.5^z
Mean 3 crops	3.7^p	3.6^p	3.4^p	3.4^p	3.3^p	3.5

II. Compaction Risk Factor (kN^2 ha^{-1} year^{-1}), after Van de Zande (1991)

Potatoes	2.1^a	2.1^a	2.3^a	2.4^a	2.4^a	2.3^x
Sugar beet	1.4^f	1.7^{fh}	1.9^{ah}	1.6^{fh}	1.5^{fh}	1.7^y
Winter wheat	0.8^k	1.1^l	0.8^k	0.9^{kl}	1.1^l	0.9^z
Mean 3 crops	1.2^p	1.6^t	1.5^{pt}	1.6^{pt}	1.7^t	1.5

III. Traffic Intensity (t km ha^{-1}), after Eriksson et al. (1974)

Potatoes	268.2^a	256.4^a	262.0^a	251.1^a	269.3^a	260.7^x
Sugar beet	172.8^f	199.7^{fh}	236.1^{ah}	174.1^{fh}	163.5^{fhj}	197.8^y
Winter wheat	111.5^k	125.4^{ko}	104.2^k	111.0^k	162.3^{oj}	115.1^z
Mean 3 crops	159.5^p	190.7^p	182.0^p	172.0^p	205.3^p	178.1

IV. Field Load Index (t h ha^{-1}), after Kuipers (1986)

Potatoes	66.2^a	61.0^a	64.8^a	60.3^a	68.8^a	63.7^x
Sugar beet	40.8^f	45.8^{fh}	53.8^h	38.2^f	38.8^{fhj}	45.5^y
Winter wheat	26.5^{kl}	31.1^{ol}	24.2^k	25.9^{kl}	39.3^{oj}	27.5^z
Mean 3 crops	38.3^p	45.2^p	42.8^p	39.9^p	51.0^p	42.3

[1] Superscripts a-o denote pairs of groups significantly different at $P_{0.10}$; p-t denote pairs of groups significantly different within the mean 3 crops row only at $P_{0.10}$; x-z denote pairs of groups significantly different within the mean column at $P_{0.10}$.
[2] One farm only.

Intensity (Eriksson et al., 1974); (3) Field Load Index (Kuipers, 1986). The results are shown in Table 1.

All criteria showed much higher values for potatoes than for winter wheat, with sugar beet intermediate. Apart from the results for the single 1200-ha farm, the area covered was highest for the smallest farms and decreased as farm size increased. In contrast, the three criteria of compaction risk gave lowest values for the smallest farm size and highest values for 35-55-ha farms, tending to then decrease as farm size increased further. The data from the single large 1200-ha farm indicate that, although the area covered may be low, the compaction risk according to all three criteria was even higher than on the 35-55-ha farms. These results illustrate the importance of farm size and crop type in influencing compaction risk and indicate the need for further research to compare the validity of different traffic parameters.

The similarity of the four traffic parameters is demonstrated by the high correlation coefficients in the relationship between FLI and Traffic Intensity (Lumkes, 1984), CRF (Van de Zande, 1991) and TI (Eriksson et al., 1974), which were found to be 0.88, 0.95 and 0.97, respectively (n = 15).

INFLUENCE OF MECHANIZATION ON FIELD LOADING

Curiously, when the level of mechanization increases, the changes in field loading are more related to the distribution of the load than to ground pressures. The average pressure under the shoes of a standing man of 75 kg will be about 0.15 bar, a walking man may well exert a pressure of 1 bar (Van Wijk, 1980). For a standing horse of 750 kg, the average contact pressure on the soil will be about 0.75 bar if we assume that a hoofprint has about the same contact area as a footprint (e.g. 250 cm^2). For a walking horse the pressure will be two times higher and for a pulling animal still more. Generally, contact pressures of tractor tyres range from 1 to 2 bar; for transport vehicles they may be a few times higher.

Scheibe (1935) quotes figures of Von Nitzsch and Holldack published in 1930, which indicate that the hoofprints of a walking horse cover 10 m^2 per 100 m of travel. If we assume that the steps of the ploughman and the horse have the same length and that their footprints on the soil have about the same size, then our estimate of the total area of footprints and hoofprints together is 15 m^2 per 100 m of travel. This total area corresponds to 60 and 50% of the area of the field when using one-furrow ploughs of 25 and 30 cm working width, respectively.

Ploughing with a tractor-drawn 4-furrow plough with a working width of about 1.5 m, results in two, 6.7-km long ruts per hectare. If the tyre width is 45 cm, then the ruts will cover 60% of the area. For tyres of 35 cm, just fitting in the open furrow, the figure would be 47%. An obvious difference, however, is that making the footprints with horse ploughing required about 10 h, whereas the ruts will be made in 1 h. The combined weight of the horse, the plough and the man could have been 500 or 600 kg and the tractor with plough and driver perhaps ten times as much. Obviously the weight of the operational unit and the operation time, both essential characteristics of the loading event on the scale of a farmer's field, changed drastically.

In field traffic, loading times tend to be less than 1 s, which is short enough for further compaction to occur when the loading is repeated (Aboaba, 1969; Tijink, 1988; Horn, 1989). This means that, in field traffic, loading time is sufficiently short to be included in a characterization of field loading. If, for example, a vehicle weighing 5 t works for 1 h on 1 ha, while exerting a ground contact pressure of 100 kPa, and the tracks cover the whole area once, the loading time is 0.18 s. If only 10% or even 1% of the area were loaded, the loading times would increase to 1.8 and 18 s, respectively. We would then have reached the situation where loading times are long enough to reach near-equilibrium and are no longer relevant.

WEIGHT OF POWER SOURCES

If hand labour or animal traction is used, the weight per power unit is restricted and will range from 50 kg for a light person to 1000 kg for a heavy horse or a pair of bullocks. The available power would be approximately 0.05 and 1 kW, respectively.

Tractors, powered by internal combustion engines, range from small two-wheeled implements steered by a walking person, to big machines of >250 kW. Frese (1969) mentions that at that time the weight/power ratio for 66-kW

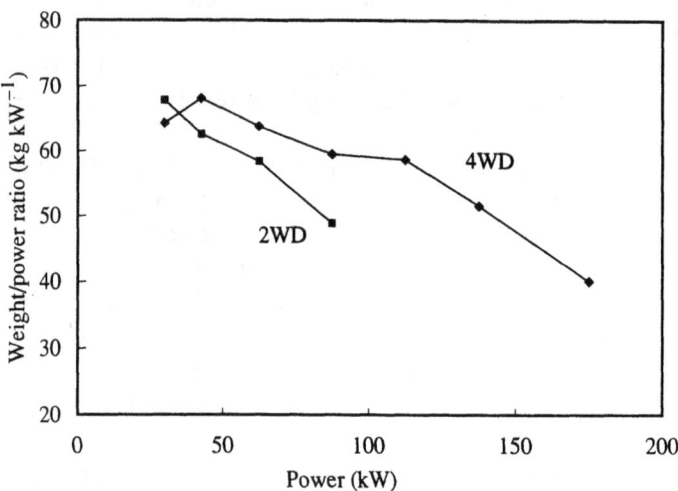

Fig. 1. Average weight/power ratio for different makes of 2- and 4-wheel drive tractors available in the Netherlands (1990).

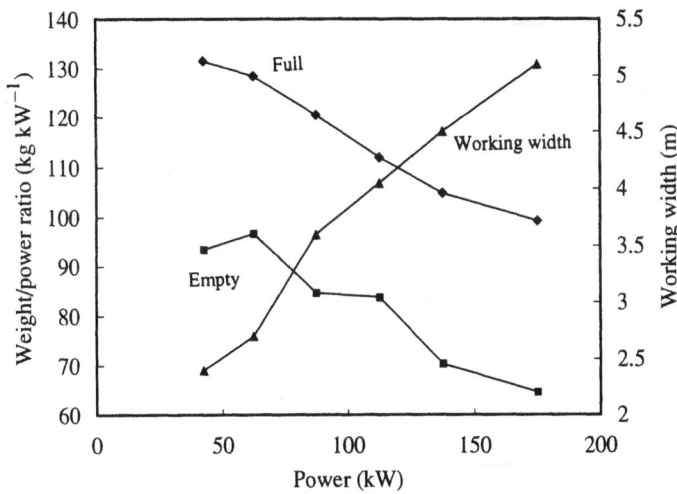

Fig. 2. Weight/power ratio and working width for combine harvesters (with empty and full grain tank) available in the Netherlands (1990).

tractors was 68 kg kW^{-1}. Göhlich (1984) found that the weight/power ratios of small 2-wheel drive and large 4-wheel drive tractors were 90 and 50 kg kW^{-1}, respectively. Fig. 1 shows the weight/power ratio values for 2- and 4-wheel drive tractors presently available in the Netherlands, as calculated from manufacturers' leaflets and retailers' data. With increasing engine power the weight/power ratio of 4-wheel drive tractors decreases from 68 kg kW^{-1} (50-kW tractors) to 44 kg kW^{-1} (175-kW tractors). For similar engine power, 2-wheel drive tractors have a more favourable weight/power ratio than 4-wheel drive tractors.

The same tendency was found for large self-propelled equipment, such as combine harvesters (Fig. 2). With increasing working width the weight/power ratio decreases from 95 kg kW^{-1} for 3-m wide to 65 kg kW^{-1} for 6-m wide combine harvesters with empty grain tanks. A full grain tank will increase these ratios by about 40 kg kW^{-1}.

WEIGHT OF IMPLEMENTS

The weight of drawn implements is limited by the weight of the power source. The pulling force that animals can produce more or less continuously is 10 to 15% of their weight. For 2-wheel drive tractors under field conditions a reasonable figure for the pulling force is 50% of the rear axle load, or 1/3 of the tractor weight. This means, if the rolling resistance coefficient for the pulled load is put at 0.1 (Tijink, 1988), that animals can pull about their own weight and tractors at least 3 times their weight. Most implements will be considerably lighter

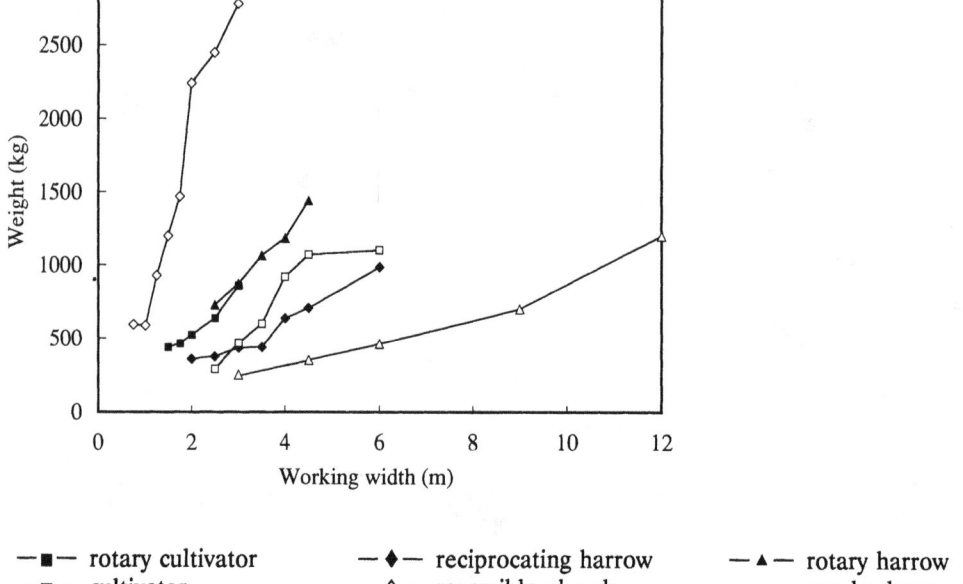

—■— rotary cultivator   —♦— reciprocating harrow   —▲— rotary harrow
—□— cultivator   —◇— reversible plough   —△— weeder harrow

Fig. 3. Weight-working width relationship for different tillage implements available in the Netherlands (1990).

than these maximum values.

Modern tractor ploughs weigh 400-500 kg per m working width for one-way ploughs and up to twice as much for reversible ploughs (see Fig. 3). The relationship between the weight of tillage implements and tractor weight is shown in Table 2. The weight of sowing and planting equipment amounts to about 250 kg per meter working width.

TABLE 2

Implement weight as a proportion of the weight of the tractor (after IMAG, 1984)

Implement	Number of models	Weight (% of tractor weight)
One-way plough	16	15-39
Reversible plough	25	19-59
Field cultivator	10	16-27
Rotary harrow	14	22-35
Rotary cultivator	2	30-35
Rotary mower	25	15-28
Mower/conditoner	3	19-49

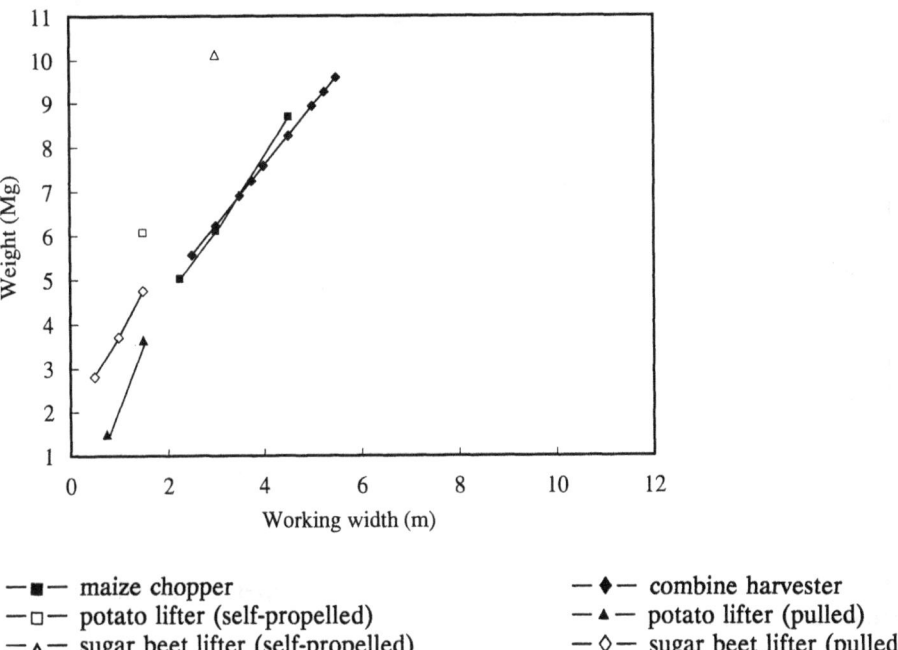

Fig. 4. Weight-working width relationship for different harvest equipment available in the Netherlands (1990).

Harvest operations are primarily transport processes, in which the compaction risk is clearly different if 6 t ha^{-1} of small grains is to be harvested or 60 t ha^{-1} of sugar beet. In harvest operations, different power sources may be used for separate steps of the process, or special, self-propelled machines may be used which combine some or all steps of the process. Such combinations are likely to be heavier than a tractor with the same power, but lighter than the total weight of a set of tractor-implement combinations which would be required for the same job. Fig. 4 illustrates the relationships between working width and weight for harvest equipment.

FIELD TIME

Standard task times for specific field operations as reported in labour studies, are considerably higher then net field time because they include preparation time, time to travel to the field, repair time, turning time, and time to recharge or unload machines. For the loading of field soils only the total time spent in the field is relevant. For the major part of the field the values calculated from forward speed and working width will apply, but on the headlands more time will be spent.

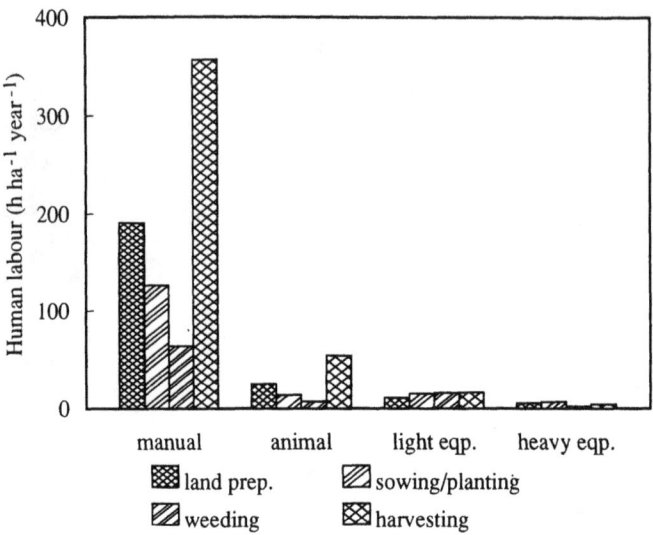

Fig. 5. Human labour requirement for different technology levels, averaged for all crops (after Van Heemst et al., 1981).

Total time spent in the field depends strongly on the type of crop and the level of mechanization. In Fig. 5, which is based on a detailed study of the literature (87 references) by Van Heemst et al. (1981), data on field labour requirement for different technology levels is presented. Van de Zande (1988) presented data on field labour for different technology levels in Thailand, specified per region and

TABLE 3

Number of passes and field time (h ha⁻¹) required for different crops and groups of loading events in the Netherlands, based on IMAG-data service computer programs (PAGV, 1990)[1]

Crop	Seedbed		Rootbed		Harvest		Post-harvest		Total	
	n	t	n	t	n	t	n	t	n	t
Winter wheat	2	1.9	9	4.8	4	6.0	3	6.4	18	19.1
Spring wheat	2	1.9	5	3.7	4	6.0	2	4.6	13	16.2
Sprong barley	2	1.9	3	1.6	4	5.5	4	5.5	13	14.5
Silage corn	2	2.3	3	1.7	2	6.0	2	3.8	9	13.8
Sugar beet	3	3.3	9	6.3	2	7.8	2	3.8	16	21.2
Potatoes	3	5.3	20	11.3	2	12.8	2	3.7	27	33.1
Onions	5	3.8	13	6.6	3	20.6	2	3.7	23	34.7

[1] n = number of passes; t = field time.

cropping system. These data also show that in developing countries the required amount of field labour is several orders of magnitude larger than in industrialised countries.

For the Netherlands, Table 3 shows average field time requirements and number of equipment passes for different crops as grown on an average (50 ha) Dutch farm with a common level of mechanization.

Dumas and Renoll (1983) reported field times in cotton production of 8.0 h ha^{-1} for a subsoil-bedding system and 9.2 h ha^{-1} for a mouldboard plough system, with 20 and 21 machine trips, respectively. For the operations from ploughing through sowing, Michel et al. (1985) reported for dry beans 3.2-5.2 h ha^{-1}, for silage corn 3.0-5.4 h ha^{-1} and for sugar beet 2.9-5.4 h ha^{-1}. The number of operations was 5-8 for dry beans and sugar beet and 5-10 for silage corn.

From data for the southwestern part of the Netherlands (Van de Zande, 1983), it has been calculated that the total annual field time for the major part of a farmer's field varied from 5.1 to 9.3 h ha^{-1} for winter wheat, from 7.3 to 15.1 h ha^{-1} for sugar beet and from 12.5 to 20.8 h ha^{-1} for potatoes (Table 4). The average number of field trips was 13, 14 and 17, respectively.

Summarizing, these data suggest that for arable crops, at the present level of mechanization, total annual field time for 15-20 machine passes ranges from 5 to 20 h ha^{-1}, which is 1% or less of the time reported by Curfs (1976) for vegetable production in the tropics at a low level of mechanization. For grassland in the Netherlands, where mineral fertilizer and slurry are applied and where the grass is mown two times, about 10-12 passes are required, with a total annual field time of 5-6 h ha^{-1}.

TABLE 4

Field times (h ha^{-1} year^{-1}) for potatoes, sugar beet and winter wheat crops on farms of different farm sizes[1]

	Farm size (ha)					
	< 35	35-55	55-80	80-200	1200[2]	Mean
Potatoes	20.8[a]	17.3[b]	16.3[bc]	16.3[b]	12.5[e]	17.1[x]
Sugar beet	13.8[f]	12.8[f]	15.1[cf]	9.8[f]	7.3[f]	13.2[y]
Winter wheat	9.3[kl]	8.3[kl]	7.3[lm]	6.1[lm]	5.2[m]	7.9[z]
Mean 3 crops	12.8[p]	12.6[pt]	11.8[pt]	10.3[pt]	8.5[t]	11.9

[1]Superscripts a-o denote pairs of groups significantly different at $P_{0.10}$; p-t denote pairs of groups significantly different within the mean 3 crops row only at $P_{0.10}$; x-z denote pairs of groups significantly different within the mean column at $P_{0.01}$.
[2]One farm only.

GROUPS OF LOADING EVENTS

For the soil, each separate field operation is a loading event. The sequence in which these events normally occur, makes it possible to classify them in distinct groups (Soane, 1985). On arable land, a first group of loading events includes all traffic after the main tillage operation through sowing or planting. The result of these loadings is the condition of the arable layer at the start of crop growth. These loadings will be called "seedbed loadings". The second group of operations includes all traffic after planting or sowing and before harvest. These loadings may change the soil structure in the arable layer during the growth of the crop and they will therefore be called "rootbed loadings".

The third group consists of the "harvest loadings". Harvest operations do not influence the growth of the present crop, but usually have such a negative influence on the condition of the arable layer, that tillage operations are required to recondition the soil. These tillage operations constitute the fourth group of loading events, the "post-harvest loadings", which result in a new starting condition for the next crop, and in a renewed, protective, shock-absorbing layer for the prevention of subsoil compaction by field traffic.

On grassland, there are many rootbed loadings and several harvest loadings each year. However, there are normally no loosening operations and, therefore, the effects of loading events will accumulate from year to year. These mechanical loadings interact with the influence of the footprints of grazing cattle and the pressures exerted on the soil by resting animals. For this type of loading the loading times are so much higher than for field traffic, that another approach is required. However, this is beyond the scope of this chapter.

Seedbed loadings

In a study carried out in Wyoming by Michel et al. (1985), 4-7 operations were required for seedbed preparation and sowing of sugar beet and dry beans and 4-9 operations for silage corn. The time required was 2.2-3.4 h ha⁻¹ for sugar beet, 2.5-3.2 h ha⁻¹ for dry beans and 2.7-3.4 h ha⁻¹ for silage corn. In this study a 52-kW tractor was used, the weight of which can be estimated at 3 t. Assuming an additional weight of 20% of the tractor weight for the implements, the total weight would be around 3.6 t. Therefore the Field Load Index for these seedbed loadings most likely ranged from 8 to 12 t h ha⁻¹.

For bed preparation, bed conditioning and planting, the time requirement reported by Dumas and Renoll (1983) for cotton in Alabama was 1.1 h with a 45-kW tractor. Under the same assumptions as above this leads to a Field Load Index of only 3.6 t h ha⁻¹.

For the Netherlands, the data presented by Van de Zande (1983) have been recalculated to distinguish between the different groups of loading events for the

TABLE 5

Field Load Index (t h ha⁻¹) for groups of loading events for different crops and farm sizes in the Netherlands (after Van de Zande, 1983)

Crop	Loading event	Farm size (ha)				
		<35	35-55	55-80	80-200	1200
Potatoes	Seedbed	8.2	6.4	8.7	7.5	10.5
	Rootbed	8.0	9.1	6.7	7.3	6.7
	Harvest	40.1	33.8	43.8	36.8	35.8
	Post-harvest	17.4	14.9	10.0	9.8	15.7
Sugar beet	Seedbed	3.2	3.6	5.1	6.4	6.0
	Rootbed	4.9	6.4	5.3	5.7	3.4
	Harvest	26.0	30.6	35.2	21.2	19.3
	Post-harvest	9.7	9.3	8.5	5.7	10.1
Winter wheat	Seedbed	4.5	3.2	2.5	3.4	8.8
	Rootbed	2.2	2.5	2.7	3.1	2.0
	Harvest	15.7	17.7	12.6	13.4	5.6
	Post-harvest	6.9	8.2	4.9	9.0	25.7

winter wheat, sugar beet and potato crops, for different farm size categories, (Table 5). For seedbed loadings the FLI ranges from 6.4 to 10.5 t h ha⁻¹ for potatoes, from 3.2 to 6.0 t h ha⁻¹ for sugar beet and from 2.5 to 8.8 t h ha⁻¹ for winter wheat.

Rootbed loadings

Michel et al. (1985) reported time requirements for sugar beet, dry beans and silage corn of 3.1 , 2.5 and 2.0 h ha⁻¹, respectively. The corresponding field load indexes were 11.7, 9.5 and 7.6 t h ha⁻¹, respectively. The figures found by Dumas and Renoll (1983) add up to 2.2 h ha⁻¹ for rootbed loading events. For spraying of insecticides a heavier tractor was used than for the other operations. Taking this into account, the estimated Field Load Index is 10.0 t h ha⁻¹. For the Netherlands, average figures for potatoes, sugar beet and winter wheat are 7.6, 5.1 and 2.5 t h ha⁻¹, respectively (Table 4).

Harvest loadings

According to Dumas and Renoll (1983), the harvest of cotton in two pickings required about 2 h ha⁻¹ with a 100-kW tractor. This means that the Field Load

Index was at least 14 t h ha⁻¹. The data from the survey by Van de Zande (1983) for the harvest of potatoes, winter wheat and sugar beet can be translated into Field Load Indices of, on average, 13.0, 26.5 and 38.0 t h ha⁻¹, respectively (Table 5).

Post-harvest loadings

Michel et al. (1985) reported average times for ploughing of 1.95 and for chiselling of 0.78 h ha⁻¹, which is equivalent to FLI values of about 7.3 and 2.9 t h ha⁻¹, respectively. The cotton field studied by Dumas and Renoll (1983) was prepared by stalk cutting, disking and ploughing. Together, these three operations required 2.17 h ha⁻¹ and the FLI was about 7.0 t h ha⁻¹. The ploughing operation alone required 1.05 h ha⁻¹ or 3.4 t h ha⁻¹. The survey by Van de Zande (1983) shows average post-harvest loadings of 5-26 t h ha⁻¹, depending on crop and farm size (Table 5).

For ploughing with a two-furrow plough, pulled by a 22-kW tractor, an FLI of 8.6 t h ha⁻¹ can be calculated (Scheibe, 1935). For a one-furrow plough drawn by

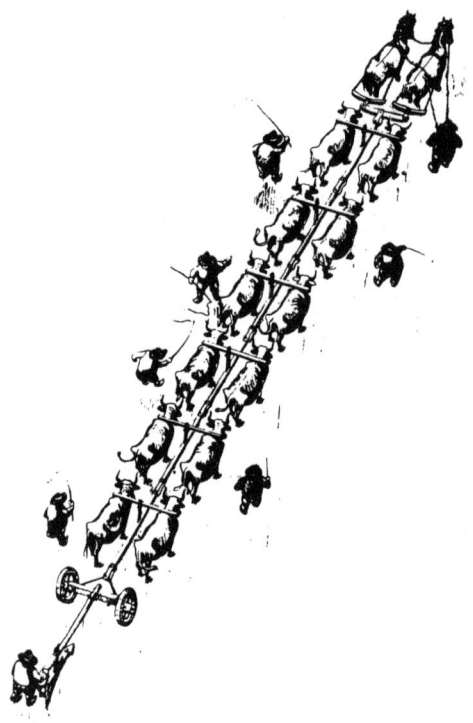

Fig. 6. Early example of excessive soil loading (after Jirlow, 1958)

TABLE 6

Estimated data on field loading at ploughing for different mechanization levels[1]

	One horse (1)	Two horses (1)	14 animals (2)	22-kW tractor (1)	52-kW tractor (3)	90-kW tractor (4)
Weight (t)	0.6	1.2	5.0	2.2	3.6	6.7
Power (kW)	0.6	1.2	5.0	22	52	90
Coverage (%)	40	80	560	50	50	60
Ground pressure (bar)	1-4	1-4	1-4	1.5	1.5	1.2
Loading time (h ha⁻¹)	11	12	12	3.8	2.0	1.0
FLI (t h ha⁻¹)	6.6	14.7	60	8.6	7.3	6.7

[1](1) Scheibe (1935); (2) Jirlow (1958); (3) Michel et al. (1985); (4) Van de Zande (1983).

two horses, the estimated FLI amounts to 14.7 t h ha⁻¹. The ploughing operation described by Jirlow (1958), where a one-furrow plough in South Sweden was pulled by 2 horses and 12 oxen and accompanied by 8 men (Fig. 6), may be regarded as an extreme example of high loading at ploughing. Especially in some developing countries, operations of this kind still take place (Hall, 1992).

In Table 6 some estimated relevant data for field loading at ploughing are summarized. They indicate that, for animal traction, the FLI is proportional to the weight of the animals because field time is constant. The higher weight of powered traction is compensated for by the shorter field time. However, heavier tractors reduce the FLI only slightly or not at all, as is demonstrated in Table 5 by the FLI figures for the post-harvest loadings on the 1200-ha farm.

COMPARISON OF THE GROUPS OF LOADING EVENTS

According to Van de Zande (1983), harvest loading ranged from 33 to 43 t h ha⁻¹ for potatoes, from 19 to 35 t h ha⁻¹ for sugar beet and from 6 to 18 t h ha⁻¹ for winter wheat, which is 70-80%, 50-60% and 15-50%, respectively, of the yearly loading. For cotton in Alabama (Dumas and Renoll, 1983) this figure was 40%. Generally, the levels of rootbed and seedbed loading were similar.

When comparing the highest and the lowest loadings between the different groups of loading events, great differences occur. Differences between highest and lowest loading are as much as 5 times the lowest value, e.g., post-harvest loading in winter wheat where intensive post-harvest cultivation occurs on the large 1200-ha farm. The best and the worse case options can be demonstrated by calculating the extremes of possible values of loading events per crop. Differences between these values range from 55 to 81 t h ha⁻¹ for potatoes, from 31 to 90 t h ha⁻¹ for sugar beet and from 15 to 55 t h ha⁻¹ for winter wheat, which are differences of 38, 128 and 145% of the total mean, respectively (Table 1).

Summarizing, it may be stated that, for Dutch conditions and for the crop rotation studied, a reasonable overall estimate is an annual loading of 40 t h ha^{-1}, of which 10% is on the seedbed, 10% on the rootbed, 55% at harvest and 25% in post-harvest tillage. However, for other crops and other mechanization levels these figures may be quite different.

TRAFFIC LOADINGS ON GRASSLAND

For broadcasting fertilizer, mowing, tedding and windrowing, relatively light implements are used; they weigh no more than 10-20% of the tractor weight. Moreover, for these operations field time is rather short. For the Netherlands, common figures are: 0.2 h ha^{-1} for broadcasting fertilizer, 1.0 h ha^{-1} for mowing and 1.0, 0.7 and 0.5 h ha^{-1}, respectively, for three times tedding. If a 35-kW tractor is used for these operations, the total weight of tractor and implements will be about 2.5 t h ha^{-1} and these five trips together will then be equivalent to about 8.5 t h ha^{-1}. For collecting and transportation of the partly dried grass 1.5 h ha^{-1} is required, which may add another 5 t h ha^{-1} to the total loading.

For slurry application, heavy machines are common. For a reasonable capacity of 6 t of slurry, the average load during the operation will be about 6 t (tractor + half-filled tank). The time requirement is low, e.g., 0.25 h ha^{-1}, and, therefore, the FLI will be about 1.5 t h ha^{-1}.

The total annual traffic load depends on the number of mowings and slurry applications. With one fertilizer application, two mowings (each followed by 3 times tedding, collecting and transportation) and one slurry application, the estimate for the whole year would be 28 t h ha^{-1}. This is only about 70% of the yearly traffic load of arable land. However, since grassland normally is not loosened by specific operations, the figures suggest that there is perhaps more reason for concern about compaction by field traffic on grassland, than in the ploughed layer of arable land (see Chapter 15).

DISTRIBUTION OF LOADING ACROSS THE FIELD

In order to gain more insight into the way that field traffic covers the soil surface with wheel tracks, several labour-consuming drawing techniques have been used to visualize the spatial distribution of traffic patterns (Arndt and Rose, 1966; Soane et al., 1982; Lumkes, 1984), of which Fig. 7 shows three examples.

Use of these techniques has been introduced in computerized bookkeeping programmes with some graphical possibilities to present the data in "rut maps" (Fig. 8) which give the distribution of wheel load and inflation pressure over a unit width of field, subdivided into 10-cm wide strips (e.g., Lumkes, 1989; Van de Zande, 1991). Arvidsson and Håkansson (1991) also use this kind of calculated wheel load distribution in combination with the degree of compactness in an economic context. The codes used in Fig. 8 are given in Table 7.

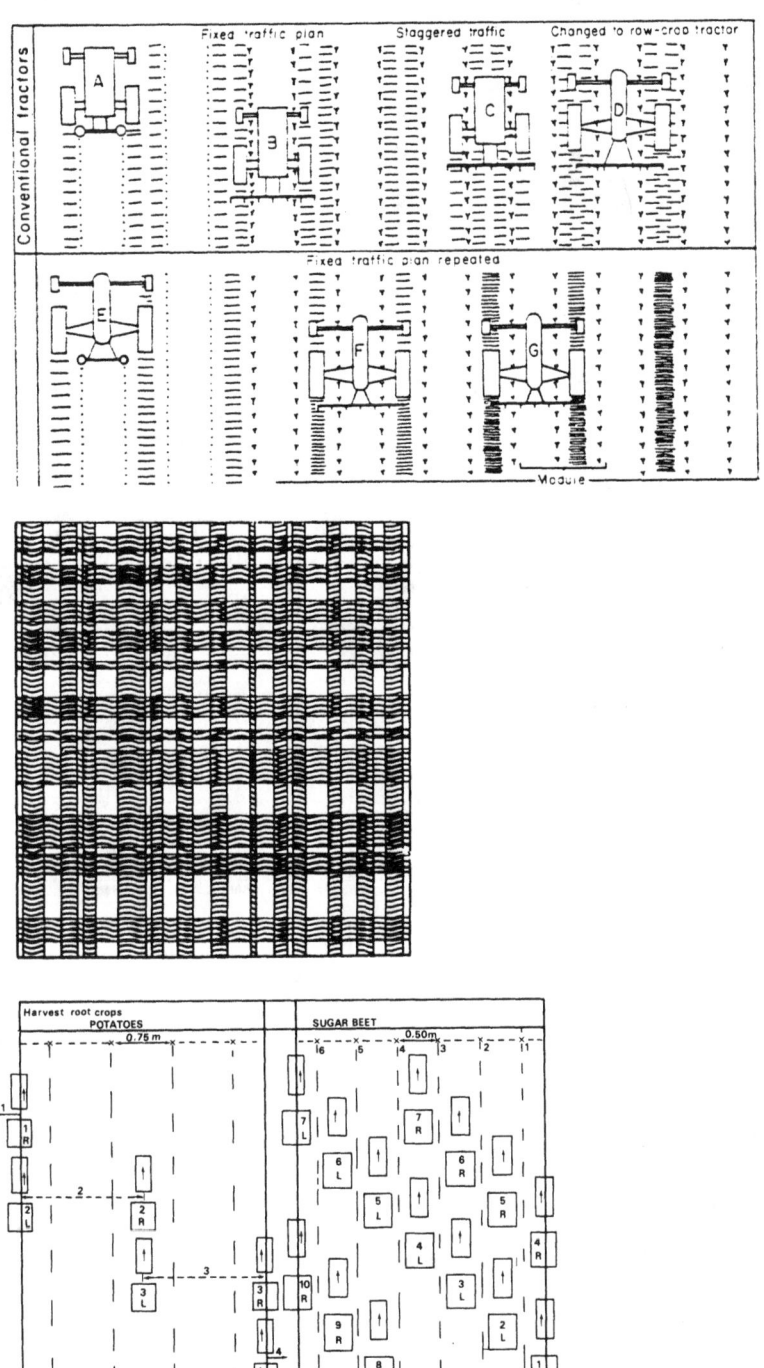

Fig. 7. Different ways in which field traffic patterns have been visualized in the past. Top: Arndt and Rose (1966); centre: Soane et al. (1982); bottom: Lumkes (1984).

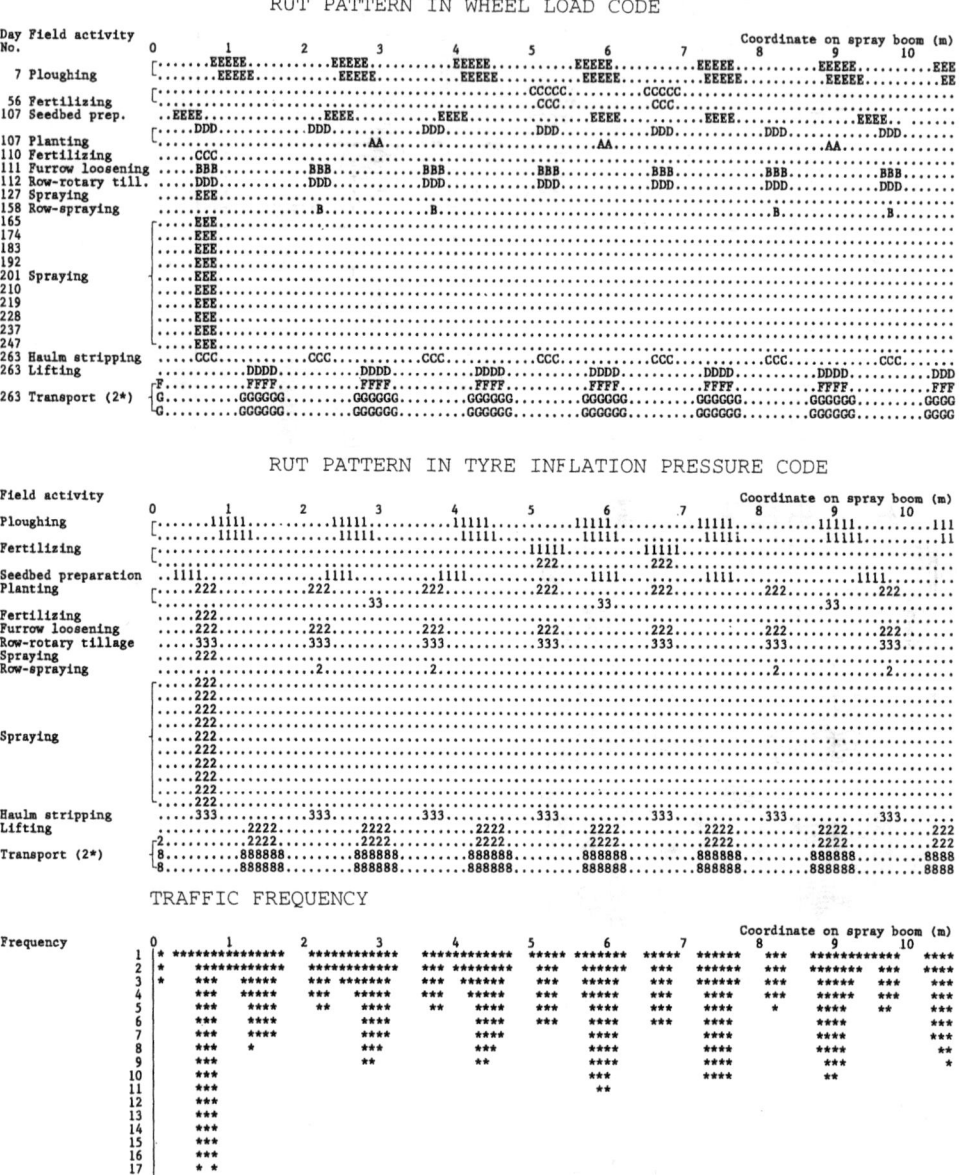

Fig. 8. Computed map indicating traffic intensity in terms of wheel load (top), tyre inflation pressure (centre) and traffic frequency (bottom) of field operations on a specific date (Julian day No.) on a 10.5-m wide strip (i.e., spray boom width), divided into 10-cm wide strips, needed to grow potatoes on a 150-ha farm in 1982 (from Van de Zande, 1991).

TABLE 7

Codes used for tyre inflation pressures and wheel loads (see Fig. 8)

Tyre inflation pressure (bar)	Code	Wheel load (kg)	Code
0.5-1.0	0	0- 500	A
1.0-1.5	1	500-1000	B
1.5-2.0	2	1000-1500	C
2.0-2.5	3	1500-2000	D
2.5-3.0	4	2000-2500	E
3.0-3.5	5	2500-3000	F
3.5-4.0	6	3000-3500	G
4.0-4.5	7	3500-4000	H
4.5-5.0	8		

Nowadays, more advanced new technologies, such as Geographical Information Systems (GIS) can be used to present graphically place-dependent data, such as wheel load, inflation pressure, soil water content, soil type, etc. The concept of "farming by soil" requires place-dependent information to apply fertilizer and pesticides, as well as a position registration attribute. In these systems, the location of wheel tracks, tramlines and the exact position of the equipment during the field operation, can be monitored (Robert, 1989). In connection with farm registration and management computer programmes, these data can be incorporated into field-specific databases. These new technical opportunities open perspectives of gaining more exact information on field traffic patterns.

On the main part of the field (i.e., the field minus the headlands), the position of the wheel tracks is related to the working width of the machines used. Van de Zande (1983, 1991) calculated the field loading for different crops for adjacent 10-cm wide strips and concluded that the total number of loadings per year varied from 0 to 8 for winter wheat, from 0 to 13 for sugar beet and from 0 to 17 for potatoes. His data allow for an estimate of the build-up of this loading in the course of the year, expressed in t h ha^{-1} (Table 8).

The highest loading, between 215 and 225 t h ha^{-1}, occurs where the ruts of the spraying machine for potatoes coincide with earlier loadings, but this maximum covers only 1.7% of the field area. Values of FLI >145 t h ha^{-1} were calculated for only 2.5% of the field area. Before harvest about half of the total field area is covered by wheel tracks and about one quarter has been subjected to loadings >15 t h ha^{-1}. This is only one example, taken from an 80-ha farm in the southwestern part of the Netherlands in one year with only three crops. However, it indicates that, despite intensive loading of the loosened soil, in the period of crop growth the major part of the arable layer may well be left in a favourable

TABLE 8

Proportion (%) of the surface with a Field Load Index (FLI) higher than the indicated value, for winter wheat, sugar beet and potatoes[1] (after Van de Zande, 1983, 1991)

FLI (t h ha⁻¹)	A			B			C			D		
	W	S	P	W	S	P	W	S	P	W	S	P
0	29	54	48	52	58	48	69	78	76	85	86	79
5	22	25	45	49	30	45	58	60	72	81	78	76
15	0	0	11	28	14	22	40	33	58	61	64	67
25			0	8	7	20	36	33	58	39	52	64
35				5	7	16	31	21	46	35	24	56
45				2	7	7	18	20	41	28	21	46
55				2	7	3	10	19	32	16	19	32
65				2	6	2	8	9	13	10	9	22
75				2	3	2	4	7	9	7	8	20
85				2	3	2	3	7	4	5	7	17
95				2	2	2	2	7	2	3	7	10
105				0	0	2	2	7	2	2	7	4
115						2	0	5	2	0	5	2
125						2		4	2		4	2
135						2		4	2		4	2
145						2		2	2		2	2
155						2		0	2		0	2
195						2			2			2
205						1			2			2
215						0			1			2
225									0			0

[1]W = winter wheat; S = sugar beet; P = potatoes; A = seedbed preparation and sowing; B = A + crop protection; C = B + harvesting; D = C + ploughing.

condition. This view is supported by the high crop yields obtained on this farm.

For the ploughing operation, detailed figures for the time spent on the headlands were given by Scheibe (1935). His calculations, based on extensive field measurements, show that for both one-furrow horse-drawn ploughs and two-furrow tractor ploughs, total loading of the headlands was about 3 times the loading of the field.

In harvest operations the crop is transported to the headlands and then along the headlands towards the entry of the field. This implies an increasing loading in the direction of that entry and if there is only one entry, the whole harvest has to pass here, which may lead to very high loading figures and loading times long enough to induce a near-equilibrium situation in the soil.

The distribution of the loadings in the course of the year and across the field may well be an essential clue for understanding crop reactions on a field scale.

primarily was aimed at indicating the different aspects involved and the way in which they may be quantified.

CONCLUSIONS

(1) To quantify field traffic, numerous criteria have been proposed. Their selection and application depend largely on the type of problem under investigation. Theoretical considerations suggest that for loadings of short duration (<1 s) imposed by field traffic, the Field Load Index (FLI), i.e., the product of weight and loading time per unit area (t h ha-1), is likely to be an effective criterion for quantifying the compaction risk from field traffic on the scale of a farmers' field.

(2) The Field Load Index is in good agreement with other parameters for quantifying field traffic, such as Traffic Intensity, TI (Eriksson et al., 1974) and Compaction Risk Factor, CRF (Van de Zande, 1991).

(3) Higher mechanization levels are attended by lower average values of TI, CRF and FLI, owing to increased working widths and shorter loading times per unit area.

(4) In mechanized arable farming, average figures for the annual field traffic load are: 42 t h ha^{-1} (FLI), 178 t km ha^{-1} (TI) and 1.48 kN^2 ha^{-1} (CRF). The average distribution of the loading over the year is: seedbed loading 10%, rootbed loading 10%, harvest loading 55% and post-harvest loading 25%.

(5) On grassland the annual traffic intensity is about 70% of that on arable land, but the effects may last longer as on grassland there is normally no soil loosening by tillage. Consequently, increases in compactness may accumulate over several years.

(6) Near field entries loading may well be so frequent, that an equilibrium (as in static loading) is reached.

(7) The distribution of wheel tracks across the field during traffic results in a loading pattern with high compaction risks on a small part of the area, and low risks on a relatively large part.

(8) The irregular loading pattern which occurs in the field is likely to provide a key for understanding the reactions of individual plants to spatial differences in the severity of soil compaction.

REFERENCES

Aboaba, F.O., 1969. Effects of time on compaction of soils by rollers. Trans. ASAE, 12: 302-304.

Arndt, W. and Rose, C.W., 1966. Traffic compaction of soil and tillage requirements. J. Agric. Eng. Res., 11: 170-187.

Arvidsson, J. and Håkansson, I., 1991. A model for estimating crop yield losses caused by soil compaction. Soil Tillage Res., 20: 319-332.

Curfs, H.P.F., 1976. Systems development in agricultural mechanization with special reference to soil tillage and weed control. Wageningen Agric. Univ., Wageningen, Netherlands, Communication 76-5, 180 pp.

Domżał, H., Gliński, J. and Lipiec, J., 1991. Soil compaction research in Poland. Soil Tillage Res., 19: 99-109.

Dumas, W.T. and Renoll, R., 1983. Fuel and time data for Alabama cotton production practices. Trans. ASAE, 26: 399-400.

Eriksson, J., Håkansson, I. and Danfors, B., 1974. The effect of soil compaction on soil structure and crop yields. Swedish Inst. Agric. Eng., Uppsala, Sweden, Bull. 354, 101 pp.

Freitag, 1979. History of wheels for off-road transport. J. Terramech., 16: 49-68.

Frese, H., 1969. Aktuelle Probleme der Bodenbearbeitung. (Present-day problems in soil tillage). Archiv Deutsche Landwirtschafts Gesellschaft (D.L.G.), 44: 53-73 (in German).

Gill, W.R. and Vanden Berg, G.E., 1968. Soil Dynamics in Tillage and Traction. USDA, Washington, DC, U.S.A., Agric. Handbook 316, 511 pp.

Göhlich, H., 1984. The development of tractors and other agricultural vehicles. J. Agric. Eng. Res., 29: 3-16.

Håkansson, I. and Danfors, B., 1981. Effect of heavy traffic on soil conditions and crop growth. Proc. 7th Int. Conf. ISTVS, Calgary, Canada, Vol. 1, pp. 239-253.

Håkansson, I., 1984. Jordpackning. (Soil compaction). Statens Lantbruksinf. 4, 11 pp. (in Swedish).

Hall, A., 1992. Ploughman's Progress. Farming Press Books, Ipswich, UK, 147 pp.

Halloway, R.E. and Dexter, A.R., 1990. Traffic intensity on arable land on the Eyre Peninsula of South Australia. J. Terramech., 27: 247-259.

Horn, R., 1989. Strength of structured soils due to loading - A review of processes on macro- and microscale; European aspects. In: W.E. Larson, G.R. Blake, R.R. Allmaras, W.B. Voorhees, and S.C. Gupta (Editors), Mechanics and Related Processes in Structured Agricultural Soils. Kluwer, Dordrecht, Netherlands, pp. 9-22.

IMAG, 1984. Inst. Agric. Eng. (IMAG), Wageningen, Netherlands, Bull. Nos. 235-1036, August 1963-May 1984 (in Dutch).

Jirlow, R., 1958. Den Svenska plogens historia. (The history of the Swedish plough). Kungl. Skogs- Lantbruksakad. Tidskrift, 97: 121-151 (in Swedish).

Kuipers, H., 1986. Soil compaction in arable farming. Trans. 13th Congr. Int. Soc. Soil Sci., Hamburg, Germany, pp. 310-327.

Lumkes, L.M., 1984. Traffic intensity. In: F.R. Boone (Editor), Experiences with Three Tillage Systems on a Marine Loam Soil, II: 1976-179. PUDOC, Wageningen, Netherlands, Agric. Res. Rep. 925, pp. 12-23.

Lumkes, L.M., 1989. Het field traffic model. (The field traffic model). PAGV Jaarboek 1987/'88. (PAGV Yearbook 1987/'88). Res. Stat. Arable Farming Field Prod. Vegetables (PAGV), Lelystad, Netherlands, Publ. 43, pp. 333-337 (in Dutch with English summary).

Michel, J.A., Jr., Fornstrom, K.J. and Borrelli, J., 1985. Energy requirements of two tillage systems for irrigated sugar beets, dry beans and corn. Trans. ASAE, 28: 1731-1735.

PAGV, 1990. Kwantitatieve Informatie 1990-1991. (Quantitative Information 1990-1991). Res. Stat. Arable Farming Field Prod. Vegetables (PAGV), Lelystad, Netherlands, Publ. 53, 186 pp. (in Dutch).

Perdok, U.D. and Van de Werken, G., 1983. Power and labour requirements in soil tillage - a theoretical approach. Soil Tillage Res., 3: 3-25.

Raghavan, G.S.V. and McKyes, E., 1978. Statistical models for predicting compaction generated by off-road vehicular traffic in different soil types. J. Terramech., 15: 1-14.

Raghavan, G.S.V., McKyes, E., Taylor, F., Richard, P. and Watson, A., 1979. Vehicular traffic effects on development and yield of corn (maize). J. Terramech., 16: 69-76.

Robert, P.C., 1989. Land evaluation at farm level using soil survey information systems. In: J. Bouma and A.K. Bregt (Editors), Land Qualities in Space and Time. PUDOC, Wageningen, Netherlands, pp. 299-311.

Scheibe, J., 1935. Untersuchungen über die Vorzüge und Nachteile des Beet- und

Kehrpflügens. (Investigations into the advantages and disadvantages of one-way and reversible ploughing). Inaug. Diss. Friedrich-Wilhelms-Univ. Berlin, Germany, 80 pp. (in German).

Soane, B.D., 1985. Traction and transport as related to cropping systems. Proc. Int. Conf. Soil Dynamics, Auburn, AL, U.S.A., Vol. 5, pp. 863-935.

Soane, B.D., Dickson, J.W. and Campbell, D.J., 1982. Compaction by agricultural vehicles: a review. III. Incidence and control of compaction in crop production. Soil Tillage Res., 2: 3-36.

Tijink, F.G.J., 1988. Load-bearing processes in agricultural wheel-soil systems. Ph.D. thesis, Wageningen Agricultural University, Wageningen, Netherlands, 173 pp.

Van de Zande, J.C., 1983. Berijding en bodemverdichting van bouwland op West Zuid-Beveland. (Field traffic and soil compaction of arable land in West Zuid-Beveland). Inst. Cultuurtech. Waterhuish. (ICW), Wageningen, Netherlands, Nota 1462, 62 pp. (unpublished) (in Dutch).

Van de Zande, J.C., 1988. Labour requirement data base for different crops and regions in Thailand. Centre World Food Studies (SOW), Staff Working Paper SOW-87-14, Wageningen/Amsterdam, Netherlands, 86 pp.

Van de Zande, J.C., 1991. Computed reconstruction of field traffic patterns. Soil Tillage Res., 19: 1-15.

Van Heemst, H.D.J., Merkelijn, J.J. and Van Keulen, H., 1981. Labour requirements in various agricultural systems. Quart. J. Int. Agric., 20: 178-201.

Van Wijk, A.L.M., 1980. Playing conditions of grass sports fields. Ph.D. thesis, Wageningen Agricultural University, Wageningen, Netherlands, Agric. Res. Rep. 903, 124 pp.

Soil Compaction in Crop Production
B.D. Soane and C. van Ouwerkerk (Eds.)

CHAPTER 19

Benefits of Low Ground Pressure Tyre Equipment

G.D. VERMEULEN[1] **and U.D. PERDOK**[2]

[1]Institute of Agricultural Engineering (IMAG-DLO), Wageningen, Netherlands
[2]Wageningen Agricultural University, Department of Soil Tillage, Wageningen, Netherlands

SUMMARY

The concept of the application of low ground pressure to reduce over-compaction of soil is discussed at the level of a uniaxial compression test, a single wheel, a vehicle and a traffic system. Generally, reduction of the pressure on the soil leads to lower levels of soil compactness. This principle is easily demonstrated in the laboratory. However, in a low ground pressure farming system, many factors other than the average ground pressure of tyre equipment determine the compaction levels obtained throughout the year. Nevertheless, traffic system experiments on a field scale generally show positive responses of the topsoil condition and the yield of most crops to substituting a low ground pressure traffic system for a conventional traffic system. Other benefits of low ground pressure traffic systems and opportunities for their application are briefly discussed.

INTRODUCTION

The basic concept of using low ground pressure on field soils is that over-compaction and plastic deformation of the soil by agricultural field traffic can be avoided or reduced by the application of running gear that spreads the vertical load of the machinery used evenly over a relatively large ground contact area. Thus, stress levels within the soil can be kept low and the bulk density can be kept unchanged or be limited to values below those at which soil properties clearly limit crop growth.

In this chapter, the theoretical aspects and experimental evidence of the effects of low ground pressure on soil structure are discussed first, using results of uniaxial compression tests, investigations with single wheels and with agricultural vehicles. Next, investigations on soil and crop responses to low ground pressure traffic systems are discussed. Some attention is given to other effects of low ground pressure and to the practicality of its application.

EFFECTS OF UNIAXIAL PRESSURE ON SOIL

Uniaxial, confined compression is often used as a simple means of characterizing the behaviour of soil under increasing average principal stress (see Chapters 2 and 3). A typical example of soil behaviour is shown schematically in Fig. 1. The average porosity of loose soil in a sample generally decreases linearly with the logarithm of the pressure, showing parallel lines for different water contents at compression. For undrained samples, the compaction stops when most of the air has been pressed out and the soil has become near-saturated. Under wheels, where the soil is partly confined, plastic deformation of the soil occurs easily in this wet condition. When, in the laboratory, initially loose soil is pre-compacted, uniaxial compression results in little further compaction until the pre-compaction pressure is reached; thereafter the resulting porosity again follows the virgin compression curve. Similar behaviour is found in undisturbed, structured field samples, but here the resistance to mechanical pressure is a result of both the level of pre-compaction and additional structural strength developed in time by chemical and other bonds. In this chapter, the initial resistance of soil to mechanical pressure is designated as structural strength (kPa).

The results of uniaxial compression tests show that the structural strength of the soil determines its response to mechanical pressure and that structural strength decreases with increasing soil water content. Freshly loosened soils always exhibit very low structural strength, whereas soils receiving no tillage may exhibit high structural strength.

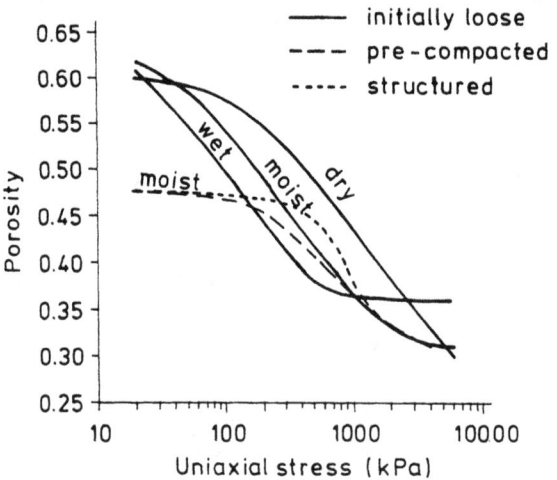

Fig. 1. Compression curves for initially loose, pre-compacted and undisturbed, structured soil at different soil water status (schematized).

In order to establish allowable normal stress levels in the field, knowledge of the relationships between soil properties and plant growth is required (see Chapter 11), but also information on the structural strength of the soil at the time of traffic events is essential. According to Lebert (1989), high structural strength in sandy soils (clay content <15%, w/w) usually coincides with high bulk densities, causing adverse conditions for plant growth. In structured soils (clay content >15%, w/w), high structural strength may also occur at low bulk densities when the soil has a high degree of aggregation (dense, stable aggregates).

Translation of laboratory results to principal stress situations in the field is complicated by the fact that stress duration, stress path and degree of soil confinement appear to have a substantial effect on the level of compaction obtained.

THE AVERAGE GROUND PRESSURE OF A SINGLE WHEEL AND ITS EFFECTS ON SOIL

Average ground pressure

Average ground pressure is defined as the average vertically downwards contact stress which a single, rubber-tyred wheel exerts on the soil surface. Average ground pressure is calculated from the ground contact area, i.e., the contact area between the soil and the tyre, and the wheel load. Usually, the distribution of the vertically downward contact stress (= normal contact stress) over the contact area is not uniform, but is a result of the local equilibrium of forces resulting from soil and tyre deformations. For simplicity of calculations, the peak pressures under lugs are not taken into account. Söhne (1952) reconstructed the normal contact stress distribution under a 170-20 implement tyre with a load of 0.39 Mg for rigid and deformable surfaces from measured ground contact areas and normal stress at a depth of 8 cm, for the first and second wheel pass (Fig. 2). For soft soil conditions and a rigid tyre (high inflation pressure, stiff), the normal contact stress distribution is non-uniform and mainly determined by the rigid, round shape of the tyre and the reaction forces of the deformed soil. For a hard surface and a soft tyre (low inflation pressure, flexible), the distribution is approximately uniform and mainly determined by the rigid, flat surface and the reaction forces of the flexible tyre. For soft soil conditions and a soft tyre, the normal contact stress distribution is moderately uniform for a first pass and more uniform for a second pass over the same soil. Similar results were obtained by direct measurement of normal contact stresses by Vanden Berg and Gill (1962), Krick (1969) and Burt and Wood (1987). Recently, Koolen et al. (1992) estimated normal contact stress distributions of the currently largest tyres (66x43-25) on a deformable surface from measurements of normal stress at 15 cm depth. Also

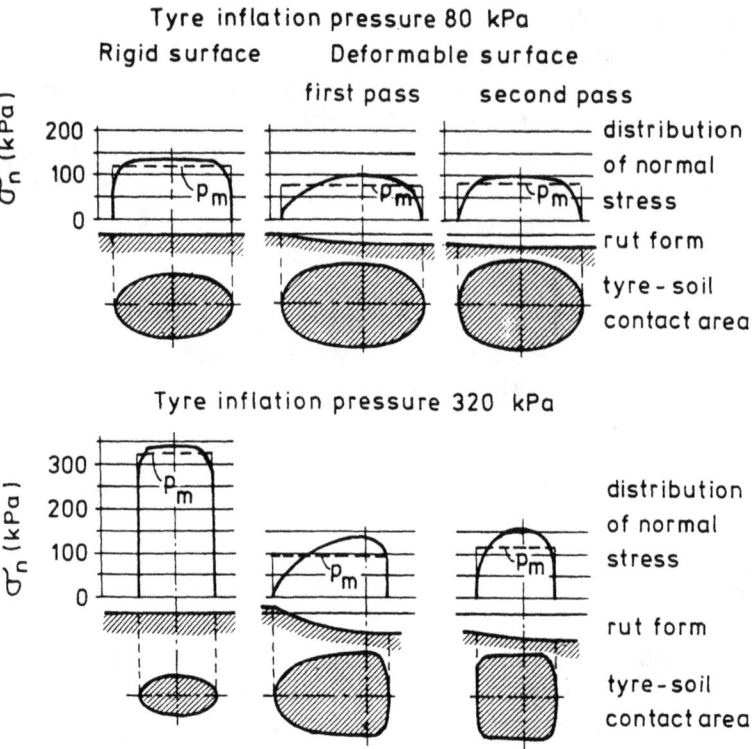

Fig. 2. Ground contact areas and associated average ground pressure (p_m), distribution of normal stress (σ_n) and rut shape along the longitudinal centerline of a moving 170-20 tyre for two tyre inflation pressures, two wheel passes, and for rigid and deformable surfaces (after Söhne, 1952).

for this large tyre size, reduction of the tyre inflation pressure at constant wheel load (increasing deflection), resulted in more uniformly distributed normal contact stresses and a lower average ground pressure.

Decline of normal stress with depth

The effect of normal contact stresses at the soil surface on normal stress in deeper soil layers was mathematically described by Söhne (1953) for three normal contact stress distributions. Although the model is only valid for an isotropic medium and small deformations and does not account for shear stresses, it explains that high normal stresses in the soil profile can be avoided by applying a sufficiently low average ground pressure, preferably combined with a uniform normal contact stress distribution. Fig. 3 shows the decline of maximum normal stress with depth under the centre of a tyre operating on

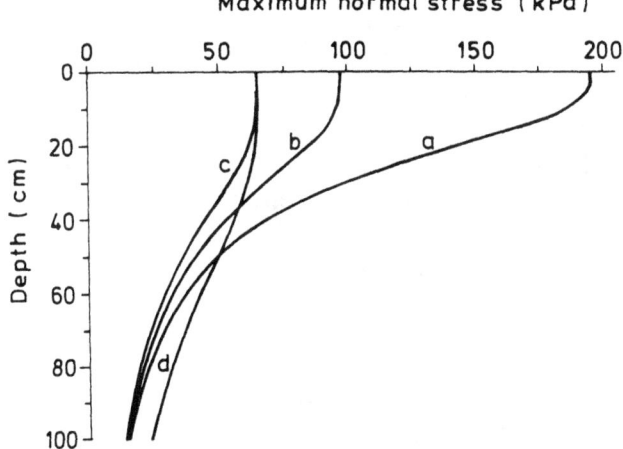

Fig. 3. Reduction of maximum normal stress with depth under four theoretical tyres with different wheel load (P), tyre inflation pressure (p_i), contact area (c.a.) and normal contact stress distribution (s.d.), calculated according to Söhne (1953), using a concentration factor of 5. (a) Small tyre: P = 2.0 Mg, c.a. = 0.15 m², non-uniform s.d. (high p_i); (b) large, stiff tyre: P = 2.0 Mg, c.a. = 0.30 m², non-uniform s.d. (low p_i); (c) large, flexible tyre: P = 2.0 Mg, c.a. = 0.30 m², uniform s.d. (low p_i); (d) largest flexible tyre: P = 4.0 Mg, c.a. = 0.60 m², uniform s.d. (low p_i).

deformable soil, for four simplified theoretical cases. The curve for a small tyre with non-uniform normal contact stress distribution (a) shows high maximum normal stress values at shallow depth and a rapid decline with depth. The same load on a wider, stiff tyre (b) shows a much lower level of maximum normal stress at shallow depth, but a slow decline with depth. Additional spreading of the load by using a wider, flexible tyre (c), again reduces the maximum normal stress at shallow depth. A further increase of the load at constant average ground pressure (d) results in equal normal stress levels at shallow depth, but higher normal stress levels in the deeper layers. Ultimately, the wheel load determines the normal stress level in the deeper layers, but this will never exceed the maximum normal contact stress level. This example shows that, in theory, the normal stress levels could be kept lower than the soil's strength by limiting the wheel load at fixed average ground pressure, by lowering the average ground pressure at fixed load or a combination of both measures.

Standard average ground pressure

Uniaxial compression tests showed that the compaction behaviour of a specific soil is determined mainly by the maximum principal stress, the stress path, the stress duration and the degree of confinement it is subjected to. According to the theory for decline of stresses with depth for a moving tyre on a soil profile with specific structural strength, these factors are determined at all depths by the load, the stress distribution under the tyre and the forward speed.

The forward speed can easily be kept constant during tyre comparisons. However, evaluating the compaction effect of different average ground pressures and wheel loads, will most likely give inconsistent results when all possible combinations of load and stress distributions under tyres are compared. Therefore, a standard average ground pressure, that can be reliably measured and that features comparable stress distributions (i.e., the ratio between the maximum stress and average ground pressure) for different tyres of the same family of tyres is introduced here. The standard average ground pressure is defined as the average ground pressure of a tyre at its maximum allowable deflection (at its loading capacity for 30 km h^{-1}) on a rigid surface, with the restriction that the tyre has the appropriate inflation pressure, in agreement with its flexibility (ply rating). Thus defined, the standard average ground pressure characterizes the maximum compactive potential of the tyre at a specific tyre inflation pressure because: (1) a lower load than the loading capacity (partial loading) reduces the contact stresses; (2) application of the tyre on a deformable surface results in a more uniform stress distribution and increased contact area which lead to lower maximum principal stress.

A practical advantage of this definition of standard average ground pressure is that for tyres within a family of tyres, good relationships exist between tyre size (width), inflation pressure and loading capacity (Perdok and Arts, 1987) and between inflation pressure and standard average ground pressure (Vermeulen et al., 1988). Therefore, within a family of tyres with similar design features, the inflation pressure may be considered to be a good indication for the compactive potential of the tyre.

Effects of the average ground pressure of a single wheel on soil

Compaction under wheels, as a result of the stresses imposed on the soil, must be taken into account when considering the maintenance of soils at an optimum level of compactness. Acceptable increases in compactness may be slight or nil for soils which receive no tillage (e.g., grassland, arable subsoils and topsoils under zero tillage). However, for soils which are initially in a loose condition, some compaction may be desirable to achieve the optimum level of compactness. In an arable situation, temporary overcompaction of the topsoil may be acceptable after harvest because the effect can be repaired by subsequent tillage operations.

Soil behaviour during uniaxial compression confirms that when the structural strength exceeds the applied stress, no increase in compactness will occur. For example, when a wheel runs on a dry clay soil under grass, no rut will be formed, even at a high wheel load and a high average ground pressure. Recent single-wheel experiments with varying standard average ground pressure on grassland (Vermeulen et al., 1993) clearly show that, even on soils with a modest structural strength, a reduction of standard average ground pressure can eliminate significant rut formation. To predict the possible occurence of undesired increases in compactness in the subsoil, the stress levels at the interface of topsoil and subsoil should be considered, and compared with the structural strength of the subsoil. A practical method to estimate these stress levels was developed by Koolen et al. (1992), who fitted the Fröhlich equation to a dataset from normal stress measurements in the field with pressure cells located at the interface of the topsoil and the subsoil (at 30 cm depth for Dutch field conditions). For a variety of wheel loads, tyre inflation pressures and tyre sizes in the larger size range, the field data fitted well when the tyre reaction forces were approximated by those of a circular plate with a load equal to the wheel load and of such a diameter that its average ground pressure was equal to 2 times the tyre inflation pressure. Koolen's empirical equation was used to calculate possible combinations of wheel load and tyre inflation pressure for five levels of maximum normal stress at 30 cm depth. The results show (Fig. 4) that the combination of low maximum normal stress (<100 kPa) and wheel loads >1.35 Mg, is only possible when using tyres with low inflation pressures (<100 kPa).

On recently loosened soil, wheel traffic will always increase the bulk density because such a soil has a very low structural strength. This was confirmed by Soane et al. (1986), who concluded that it was extremely difficult to traffic

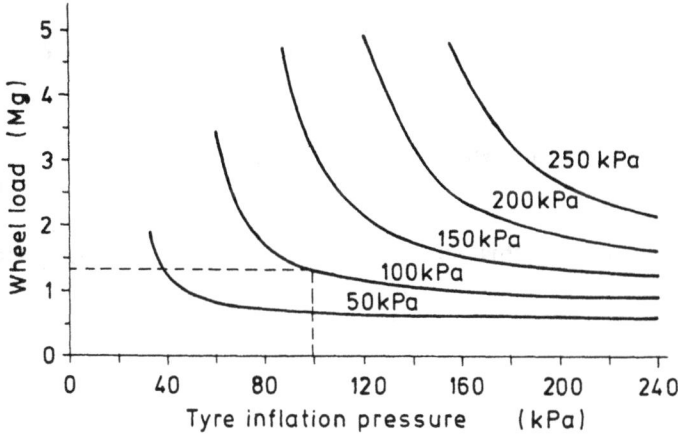

Fig. 4. Predicted normal stress at the interface of topsoil and subsoil (30 cm depth) in relation to wheel load and tyre inflation pressure for normal, moist soil conditions (after Koolen et al., 1992).

loamy sand which was freshly loosened to a depth of 45 cm, without considerable re-compaction, even with a light tractor (mass 3.5 Mg) having wide tyres and extremely low tyre inflation pressure (30 kPa). For the specific condition tested, bulk densities after traffic were higher than before the soil was loosened in all cases. Smith and Dickson (1990) reported on a traffic experiment with a variety of single wheels on sandy loam which was freshly loosened to a depth of 50 cm. In this experiment, average ground pressures were measured in the field without accounting for different contact stress distributions, which might explain the inconsistency of the dry bulk density responses to the average ground pressure applied. However, when comparing results obtained with tyres at approximately similar contact stress distributions, the experiment did show that higher bulk densities are obtained with increasing average ground pressure.

In an arable situation with a loose topsoil in spring, considerable increases in compactness in the topsoil may be accompanied by a lack of any changes in compactness in the subsoil. Steinkampf and Sommer (1989) reported on field experiments where soil stress was measured under 23.1R26, 30.5LR32 and 73x44.00-32 tyres at inflation pressures of 240, 130 and 120 kPa, respectively, all with a high wheel load of 6.8 Mg. Their measurements show that when larger tyres are used to carry the same load, the soil stress clearly decreases, both at 20 and 40 cm depth, on ploughed as well as on unploughed loamy silt soil. However, for the specific condition tested, the porosity of the soil was affected by the applied stress only in the ploughed topsoil. After a wheel pass with the smallest tyre at 240 kPa inflation pressure, the porosity was clearly lower than after a pass with one of the larger tyres at 120 or 130 kPa inflation pressure.

EFFECTS OF THE AVERAGE GROUND PRESSURE OF A VEHICLE ON SOIL

Vehicles often have more than one wheel which follows the same track. This means multiple wheelpasses with differently loaded and differently tyred wheels at short time intervals. Multi-wheelpass effects have been studied mostly on initially loose soils. In this situation, every additional wheelpass causes an incremental increase in soil compactness. However, the effect decreases with the number of passes, until after a certain number of passes an equilibrium degree of compactness is reached.

Raghavan et al. (1976) used a single empirical equation to calculate the bulk density from applied average ground pressure, pre-compaction pressure and water content of the topsoil. From the results of their experiments, they concluded that the average ground pressure, p, could be replaced by n x p (n = number of passes) to express the multipass effect for up to 10-15 passes, depending on soil type. This implies that, at the same number of wheel passes, low ground pressure (LGP) vehicles (with wider tyres or more tyres on the same axle), can be expected to produce lower levels of compactness than high ground pressure (HGP) vehicles. When a vehicle has been converted to low ground pressure by increasing the number of wheels that follow the same track (i.e., more axles), average ground pressure is lower, but the number of wheel passes is higher. Because of the multipass effect, this wheel arrangement would be less efficient in avoiding high levels of compactness in the topsoil than wide tyres and dual wheel arrangements.

Rüdiger (1990) characterized the level of soil compactness by the penetration resistance below the rut. He modelled the effects of passes of 6 different vehicles with different wheel loads (1.6-3.7 Mg), number of axles and tyre inflation pressures (50-350 kPa) on an initially loose, sandy soil profile. The number of wheel passes ranged from 2 to 14. Fig. 5 shows the results of model calculations for three alternative wheel configurations with a total load of 6.4 Mg, the penetration resistance being expressed as the relative compactness (%). The two LGP alternatives (b and c) both result in a lower relative compactness than that following two passes of the conventional tyre inflated to 300 kPa. Supporting the total load on more axles (c) clearly lowers the level of soil compactness in deeper layers, but is less efficient just below the soil surface.

In addition to these studies on initially loose soil, there is a need to study

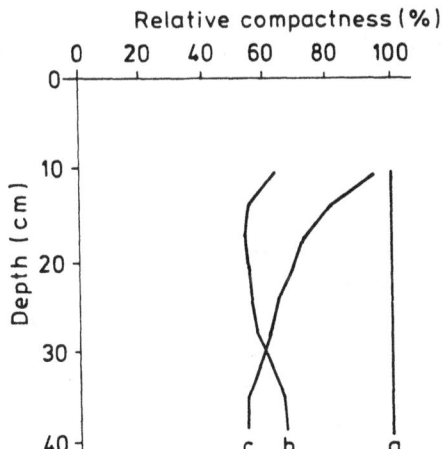

Fig. 5. Predicted relative compactness (based on penetration resistance a = 100) for three alternative wheel configurations to transport a load of 6.4 Mg. (a) 2 passes with a conventional tyre (tyre load 3.2 Mg, p_i = 300 kPa); (b) 2 passes with a wider tyre (tyre load 3.2 Mg, p_i = 75 kPa); (c) 8 passes with a conventional tyre (tyre load 0.8 Mg, p_i = 75 kPa) (after Rüdiger, 1990).

the role of the structural strength of soil in influencing the response to multiple wheel passes (Soane et al., 1981). Such studies would be particularly useful to aid the prediction of compaction effects of vehicles with different LGP wheel arrangements on permanent grassland or other land with crops grown under a zero-tillage regime, and within the subsoil of arable land.

SOIL AND CROP RESPONSES TO LOW GROUND PRESSURE TRAFFIC SYSTEMS

In the context of a traffic system integrated within a farming system, one has to consider the summation of compaction effects on a field scale and the resulting crop effects. The following factors are important: (1) different vehicles operate in succession on the same field; (2) traffic events occur at different times of the year and the crop cycle, at which both the soil conditions and the desired level of compactness may be different; (3) soil conditions during a traffic event are spatially variable, both horizontally and vertically; (4) tillage operations may be used to alleviate the compaction effects; (5) natural processes occuring in the soil influence the soil condition; (6) different crops may require different soil conditions; (7) the soil condition at the start of the system will influence the results of subsequent traffic.

The effects of practical traffic systems on soil compactness are not easily investigated in a systematic way due to the complexity of the whole system. As a result, the few field-scale experiments which have been conducted were focussed on the application of low ground pressure systems in strategic farming systems and on the interacting effects of some of the factors mentioned above.

Systems experiments

Håkansson et al. (1985) reported on field experiments on heavy clay soils in Sweden. The treatments included traffic by tractors in spring with either single rear wheels with p_i = 100-120 kPa or dual rear wheels with p_i = 50-60 kPa. Annual tillage included primary tillage to a depth of 20-25 cm, followed by 2-5 harrowing operations. Spring-sown crops yielded 6% higher under low ground pressure traffic than under high ground pressure traffic. At zero ground pressure (no traffic), yields were 26% higher than under high ground pressure traffic.

In the Netherlands, a full-size low ground pressure (LGP) traffic system for arable farming was compared with the currently used, high ground pressure (HGP) traffic system and a zero-traffic (ZGP) system (Vermeulen and Klooster, 1992). The effects of the three traffic systems were studied from 1986 to 1989 in a field experiment on a marine clay loam soil, in a crop rotation

Fig. 6. Fertilizer application in spring in a low ground pressure traffic system, using an 82-kW tractor with a standard average ground pressure of 50 kPa.

consisting of winter wheat, sugar beet, onions and ware potatoes. Soil tillage included annual primary tillage to a depth of 25 cm.

Tyres were selected on the basis of tyre inflation pressures, such that the average ground pressure had maximum values (= standard average ground pressure) as shown in Table 1 (Vermeulen et al., 1988). Fig. 6 shows 650/60-38

TABLE 1

Standard average ground pressures (p_s) and tyre inflation pressures (p_i) applied in an experiment comparing high ground pressure (HGP) and low ground pressure (LGP) traffic systems (kPa) (from Vermeulen and Klooster, 1992)

Tyres	HGP		LGP	
	p_s	p_i	p_s	p_i
Seedbed preparation equipment	100	80	50	40
Trailers and implements	300	240	100	80
Tractors and other equipment	200	160	100	80

TABLE 2

Average air-filled porosity (m^3 m^{-3}) at -10 kPa matric water potential of the topsoil in the growing season (1986-1989) in relation to the use of high ground pressure (HGP), low ground pressure (LGP) and zero ground pressure (ZGP) traffic systems (from Vermeulen and Klooster, 1992)

Crop	HGP	LGP	ZGP[1]
Winter wheat	0.130	0.136	0.178
Sugar beet	0.108	0.113	0.142
Onions	0.104	0.106	0.143
Potatoes[2]	0.102	0.104	0.140

[1]Traffic lanes not included.
[2]Below the ridges.

tyres fitted on the rear axle of an 82-kW tractor to allow fertilizer to be applied in spring with an average ground pressure <50 kPa, i.e., with a tyre inflation pressure of 40 kPa. The zero ground pressure system was practiced on 2-m wide strips, which were situated in between permanent traffic lanes. Essentially the same machinery was used as in the HGP and LGP systems, but the track width was adapted.

The soil response to the three traffic systems was measured throughout the year. In summer, compared to the HGP system, the LGP system showed only small improvements in average air-filled porosity of the topsoil (Table 2). Larger differences in air-filled porosity between the LGP and the HGP systems were found below wheel ruts caused by seedbed preparation equipment (Lerink, 1990). Average air-filled porosities in the ZGP situation were clearly higher than those in the HGP or the LGP system.

From porosity and oven-dry tensile strength measurements made by Lerink (1990) during root crop harvest operations in the autumn, it was observed that plastic soil deformation in wheel ruts was less in the LGP system than in the HGP system. During the winter, after primary tillage, the soil roughness in the HGP system was found to be greater than in the LGP system. The smoothest surface was found in the ZGP situation. This indicates that the soil structure effects persisted during the winter. Residual effects were also clearly noticeable when making potato ridges in spring. With increasing average ground pressure exerted by the traffic system, the proportion of fine soil in the ridges decreased and the mean weight diameter of the soil aggregates increased from 4.7 mm for ZGP and 5.4 mm for LGP to 6.0 mm for HGP. The experiment did not yield conclusive evidence about the effects of the traffic systems on the subsoil.

With the exception of wheat, crop yields increased with lower average

TABLE 3

Summary of average relative total crop yields (%, w/w) for high ground pressure (HGP), low ground pressure (LGP) and zero ground pressure (ZGP) traffic systems (from Vermeulen and Klooster, 1992)

Crop	HGP	LGP	ZGP[1]
Wheat (grain)	100	101	97
Sugar beet (total, fresh)	100	104	108
Onions (total, fresh)	100	106	109
Potatoes (total, fresh)	100	103	110
Root crops combined	100	104	109

[1]Traffic lanes not included.

ground pressure levels (Table 3). Despite the relatively small improvements in average air-filled porosity in the LGP system, half of the potential yield increase, i.e., the difference between yields of the ZGP system and the HGP system, was realized by the LGP system for the root crops. For all crops, yields were highest when the air-filled porosity at -10 kPa matric water potential of the topsoil was around 0.14 m^3 m^{-3}. This suggests that the level of soil compactness after the ZGP treatment may have been sub-optimal for wheat.

The number of work-days suitable for seedbed preparation in spring was evaluated for the three traffic systems. A work-day was defined as a day on which the surface layer of the soil is dry enough to make a seedbed and the soil at 15 cm depth is dry enough to prevent the wheels of the tractor from inducing an air-filled porosity of either <0.08 or <0.10 m^3 m^{-3} (Table 4). The results show that the average ground pressure level of traffic systems can strongly affect the number of days suitable for seedbed preparation where the lower part of the topsoil dries relatively slowly, as was the case on the experimental site. Therefore, adapting the average ground pressure level to the prevailing conditions for the combined actions of tillage and traffic is more appropriate than waiting for drier soil conditions.

Chamen et al. (1990) conducted a field experiment with normal (N), low ground pressure (L) and zero (Z) traffic systems when growing winter wheat on a clay soil (59.5% clay). Initially, the soil was loosened to a depth of 40 cm. In the experimental years, 1983-1986, two tillage regimes were applied: shallow cultivation (SC) and no cultivation/direct drilling (DD). In this experiment, the average ground pressure level was also controlled by the tyre inflation pressure. In the N system, tyres were used that required 100-250 kPa inflation pressure to carry the applied loads. For the L system, tyres were selected such that tyre inflation pressures <50 kPa were possible at all times. The Z system was practiced on untrafficked beds in between permanent traffic lanes. Cone

TABLE 4

Number of suitable work-days in a 60-day period in the spring of 1986 and 1987, for the combined actions of tillage and traffic (from Vermeulen and Klooster, 1992)

Traffic system	Aeration target[1]	Seedbed preparation (depth 2 cm)		Potato plantbed preparation (depth 10 cm)	
		1986	1987	1986	1987
HGP	0.08	0	8	0	8
LGP	0.08	16	35	12	30
ZGP	0.08	17	35	12	30
HGP	0.10	0	2	0	2
LGP	0.10	0	11	0	11
ZGP	0.10	17	35	12	30

[1]Minimum air-filled porosity (m^3 m^{-3}) at 15 cm depth under wheel ruts.

penetration resistance and soil bulk density generally increased with increasing average ground pressure in the layer between 10 and 40 cm depth. This experiment showed that the random bulk density and mechanical resistance of an initially loosened soil tend to increase more rapidly with time under a LGP system than under a conventional system because of the larger area of soil stressed during each operation. However, in a LGP system, these properties ultimately stabilize at a lower level than in a conventional system. Fig. 7 shows the stabilized bulk density profiles for the three traffic systems at the end of the experiment. The average yield of winter wheat over the experimental years (Table 5) was not significantly different for the N and L traffic systems and the tillage regimes. The yield for the Z system tended to be lower because of manganese deficiency in the crop associated with the low bulk density found in this treatment.

Douglas et al. (1992a) compared a reduced ground pressure traffic system (R) in perennial ryegrass for silage with a conventional traffic system (C) and a zero ground pressure system (Z) in Scotland. The effects of the traffic systems were studied from 1986 to 1989 on a clay loam soil. The experiment started after 20 cm deep ploughing, harrowing and grass-sowing. Each year three cuts of grass were harvested.

In the R system, the conventional tyres used in the C system were replaced by commercially available larger-than-standard tyres for which recommended inflation pressures at the maximum occuring wheel load were relatively low. For the C and R systems, the average ground pressure was calculated from typical wheel loads and corresponding ground contact areas measured in the field, with the tyres inflated to the recommended pressures at maximum occurring

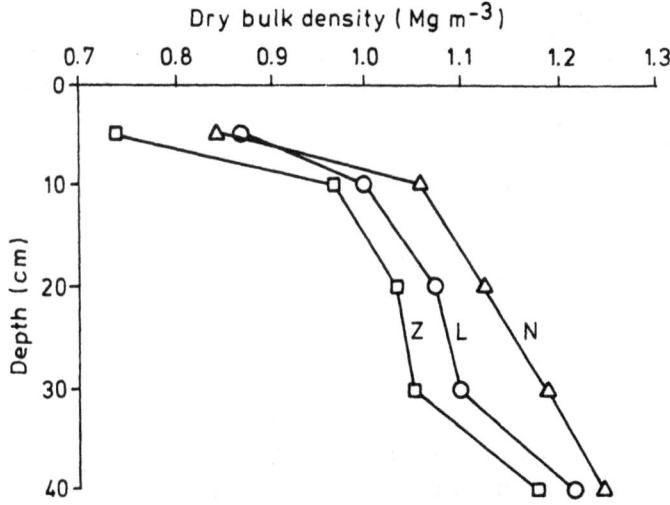

Fig. 7. Bulk density profiles for normal (N), low ground pressure (L) and zero (Z) traffic systems in winter wheat (after Chamen et al., 1990).

load. The inflation pressures ranged from 30-200 kPa in the R system and from 160-480 kPa in the C system. The Z system was practiced on 2.4-m wide strips in between permanent traffic lanes. Similar equipment as in the C and R systems was modified to operate from the permanent traffic lanes.

After the first year of the experiment, the bulk density in the topsoil (0-25 cm) consistently increased with increasing average ground pressure level of the

TABLE 5

Annual crop yields (Mg ha^{-1}) of winter wheat for normal (N), low ground pressure (L) and zero (Z) traffic systems and shallow cultivation (SC) and direct drilling (DD) treatments (after Chamen et al., 1990)

Year	N		L		Z[1]	
	SC	DD	SC	DD	SC	DD
1983	8.9	8.8	8.7	8.7	7.3	7.5
1984	8.0	8.0	8.6	8.4	7.7	7.7
1985	5.6	5.5	5.5	5.4	4.9	5.6
1986	6.6	6.5	7.5	6.9	7.2	6.6
Total	29.1	28.8	30.3	29.4	27.1	27.4

[1]Traffic lanes not included.

traffic system (Fig. 8). The bulk density for the C system and in particular the R system tended to decrease with time in the 0-6 cm layer, possibly because of faunal activity (Koppi et al., 1992). Vane shear strength at 4 cm depth, measured after each harvest, was consistently and significantly ($P < 0.05$) larger in the C than in the R system. In contrast to bulk density, the vane shear strength tended to increase with time.

Total dry matter yields obtained in each year of the experiment are presented in Table 6. Over the 4-year period, dry matter production was 16% higher in the Z system and 15% higher in the R system than in the C system, which was due mainly to impaired primary growth in the C system during wet soil conditions in spring. Compared with the C system, nitrogen use-efficiency in the Z and R systems was enhanced substantially. Therefore, losses of nitrogen to the environment could be markedly less in the Z and R systems.

In another 4-year perennial grass experiment with traffic regimes, consisting of the application of zero pressure, or pressures of 40 kPa and 100 kPa on the entire field once a year, Douglas et al. (1992b) found that in the 40-kPa and 100-kPa treatments the bulk density near the soil surface decreased with time. They attributed this phenomenon to the formation and subsequent persistence of pore space associated with the growth of grass roots.

In eastern Germany, Döll (personal communication, 1991) found that systematic application of LGP vehicles had a positive effect on the yield of wheat and the efficiency of nitrogen uptake by the crop. Petelkau (1986) showed that, on coarse-textured soil, traffic can have a considerable residual

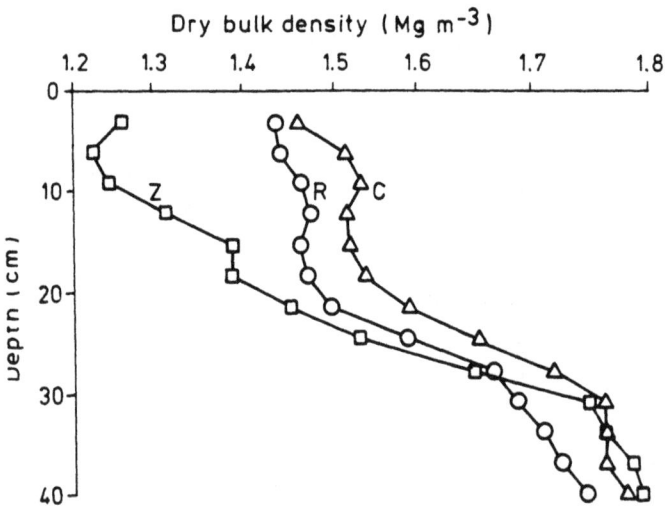

Fig. 8. Bulk density profiles for conventional (C), reduced ground pressure (R) and zero ground pressure (Z) traffic systems in perennial grass for silage (after Douglas et al., 1992a).

TABLE 6

Annual dry matter yield (Mg ha⁻¹) of grass for silage for conventional (C), reduced ground pressure (R) and zero ground pressure (Z) traffic systems (from Douglas et al., 1992a)

Year	C	R	Z[1]
1986	11.0	11.7	11.2
1987	11.3	13.5	14.9
1988	12.8	15.3	14.7
1989	12.7	14.4	14.6
Total	47.8	54.9	55.4

[1]Traffic lanes not included.

effect on yields. He made measurements in traffic lanes where vehicles had passed 23 times in a timespan of two years, with average ground pressure levels of 100, 300 and 500 kPa, respectively. Residual soil structural deterioration after soil loosening to a depth of 18 cm on former traffic lane locations, caused yield reductions of oats of 4, 16 and 18% (w/w), respectively.

Discussion

In current traffic systems with relatively narrow tyres, compaction effects in arable farming may be avoided by using tramlines or other existing traffic methods where wheels do not compact the crop zones, particularly in arable farming with an intensive tillage regime. In zero traffic systems this damage may be restricted to permanent traffic lanes. The latter possibility is discussed in Chapter 22. The application of wide low ground pressure tyres is a response to a positive decision to avoid over-compaction under wheel ruts (Soane, 1985). After all passes involved in field operations up to the time of sowing, soil conditions should be suitable for plant growth. Although a logical step would be to bring the whole field to this condition, currently the policy of avoiding severely overcompacted areas and establishment of "average good" conditions in the field is practised. At harvest time, compaction of the topsoil could be allowed to the extent that it can be removed by tillage. Several experiments have shown that residual soil compaction effects, observed in winter and spring, increase with increasing average ground pressure of the traffic system. The relative importance of the direct and residual compaction effects for crop establishment and growth was not investigated in relation to the application of low average ground pressure.

The system experiments with perennial grass (Scotland) and the zero tillage regime for winter wheat (England) showed that, as compared to a normal ground pressure system, a low ground pressure system results in lower random ground pressure system, a low ground pressure system results in lower random bulk densities in uncultivated soil after a period of transition from the loose initial soil condition to a condition which is in equilibrium with the average ground pressure of the traffic system. In Scotland this happened as a result of wheel passes within the first year. In England, due to differences in tyre width and associated field coverage, equilibrium was reached after the first year for low ground pressure and after two years for high ground pressure. The development of structural strength with time in the absence of increased bulk density was very evident in Scotland, where the bulk density in the top 12 cm of the soil even decreased with time, owing to natural causes (grassroots, fauna).

For topsoils which are tilled annually (Netherlands), the wider ruts made by low ground pressure (LGP) traffic before sowing or planting covered a larger area of the field than the narrower ruts resulting from the current high ground pressure (HGP) traffic, which might explain the small differences in bulk density found between LGP and HGP. Probably the equilibrium of soil strength and compactive potential of the traffic systems was never reached in the period from primary tillage to sowing. None of the experiments gave evidence for changes in the condition of the subsoil, despite the application of axle loads up to 8 Mg.

In simple quantitative expressions of traffic incidence (see Chapter 18), no accurate prediction can be made as to the actual increases of compactness which may occur. In some situations no compaction may result. The system experiments described above, which utilize low ground pressure tyres, show that this may occur on soils which exhibit appreciable structural strength. In such situations, methods to quantify traffic incidence may give misleading indications of compaction potential. A better indication of compaction potential is gained from the highest standard average ground pressure and the highest wheel load used in the system in relation to the strength profile of the soil at the time of traffic.

The experiments showed mostly positive but relatively small responses of the structure of the topsoil and, for all crops, an increase in yield by substituting a LGP system for a conventional traffic system. At all locations the soils were clayey, soil conditions were often wet during traffic events and macroporosities after wheel passage over tilled soil were very low, both for LGP and HGP. This indicates that the soil was compacted to a point close to saturation where plastic soil deformation occurs easily. In the Netherlands, the amount of plastic deformation was found to be less for LGP than for HGP traffic (Lerink, 1990). The occurrence of clear residual effects of the traffic systems, found after primary and secondary tillage in the Netherlands, supports this view. Similar observations on the shallow tilled winter wheat plots in England indicate that the same feature occurred there in the top 10 cm of the soil. This effect might be responsible for the significant yield differences between LGP and HGP found for root crops and the same tendency for winter wheat, despite the fact that the soil porosities were very similar. The differences in yield between low ground pressure systems and zero traffic were dependent on the crop. Root crops benefit consistently from the notably looser and more friable soil conditions found under the Z system with a regular tillage regime, but winter cereal crops tend to yield lower under Z than under conventional systems. It is of importance that the lower winter wheat yields under the Z system were accompanied by symptoms of unbalanced nutrition, such as manganese deficiency in England and oversupply of nitrogen in the Netherlands, since these effects are recognized as being associated with sub-optimum levels of soil compactness. Perennial grass in Scotland yielded almost equally under the low ground pressure and zero traffic systems. The observation that yield differences between systems were mainly due to the harvested quantities during the first cut (wet growing conditions) suggests that the macroporosity was too low for the conventional traffic system, but sufficient to ensure aeration under wet conditions for both the reduced ground pressure and the zero traffic systems.

The systems experiments were all conducted on clayey soil, for which development of structural strength with time may be expected. The applied "low" inflation pressures, and associated standard average ground pressures, were more or less arbitrarily chosen. In general, on most soil types a maximum standard average ground pressure of <100 kPa is expected to prevent compaction problems by vehicles, provided working conditons are reasonable. Therefore, low ground pressure tyre equipment may be best defined as tyre equipment having a standard average ground pressure <100 kPa. However, there is substantial evidence that structural strength may vary considerably between sites and soil profile layers and with soil water content (Lebert, 1989). Moreover, structural strength also depends on the local tillage regime applied (Larson et al., 1988; Hammel, 1991). Therefore, it may be expected that allowable average ground pressure levels, such as published in general terms by Söhne (1953), Fekete (1977), Perdok and Terpstra (1983), Petelkau (1986) and Grecenko (1989), could best be determined for small regions with comparable soil profiles and soil management systems, rather than for groups of soils distributed over a large region.

OTHER EFFECTS OF LOW GROUND PRESSURE

The application of LGP on soft soil generally leads to less rut formation, less rolling resistance of towed tyres and improved traction of driven tyres (McLeod et al., 1966; Danfors, 1977; Perdok, 1978; Perdok and Arts, 1987).

Perdok and Tijink (1990) stressed that rolling resistance and tractive performance primarily depend on soil condition (Fig. 9). On firm, dry soil, a good estimate of the rolling resistance coefficient, ρ (rolling resistance/load

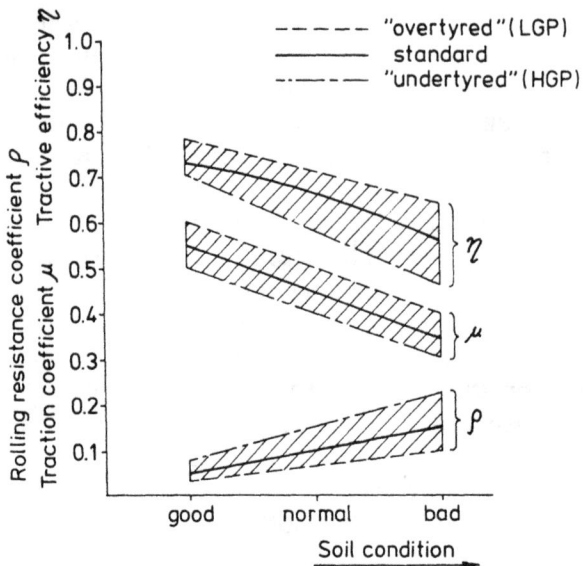

Fig. 9. Relationship between tractive performance and soil condition, and the effect of tyre size selection (from Perdok and Tijink, 1990).

ratio), is 0.05; on loose and moist soil, ρ may increase to 0.15. Low ground pressure ("overtyred") may lower ρ by a factor of two thirds. At a wheel slip of 0.15-0.20, the traction coefficient, μ (pull/load ratio), ranges from about 0.55 for good soil conditions to about 0.35 for soft soils. LGP may increase μ by 0.05. Rolling resistance and wheel slip represent considerable energy losses in converting axle power into drawbar pull. For standard tyres, the tractive efficiency, η (power available at the drawbar/power available at the axle), ranges from 0.56 to 0.73. LGP may improve the tractive efficiency by about 0.10   for soft soils. LGP tyres also tend to exhibit the maximum tractive efficiency at lower wheel slip than standard tyres (McLeod et al., 1966). Therefore, the use of LGP may further improve the tractive efficiency where low slip is required.

The lower levels of soil compaction and plastic deformation associated with LGP as compared to normal traffic, also cause the energy consumption for tillage operations to be somewhat lower. Vermeulen and Klooster (1992) found an energy saving of 8% for 25-cm deep tillage when applying LGP the year round. However, for shallow tillage, Chamen et al. (1990) found no differences in energy consumption between LGP and conventional, high ground pressure traffic systems.

Siefkes and Göhlich (1991) found that the dynamic behaviour of wide tyres, when used at an inflation pressure <80 kPa, is always superior to those of conventional tyres and dual tyres. This improves the driving comfort of the operator, especially on rough terrain.

PRACTICALITY OF LOW GROUND PRESSURE TYRE EQUIPMENT

To evaluate the practicality of equipping a traffic system with low ground pressure tyre equipment, it is necessary to define the target level of maximum average ground pressure of the system. It was shown earlier that the average ground pressure of a vehicle in the field is equal to or lower than its standard average ground pressure, which is directly related to the inflation pressure used. Therefore, the standard average ground pressure or the tyre inflation pressure may be used as a design criterion for a low ground pressure traffic system. We define a low ground pressure traffic system as a system using vehicles having a standard average ground pressure of 100 kPa, which corresponds to a tyre inflation pressure of about 80 kPa. It is also assumed that the vehicles will have the same loaded weights and have the same capacities as current vehicles. For special situations, when the soil typically exhibits low strength (loose soils and low strength grasslands), we assume that a standard average ground pressure of 50 kPa is required, which corresponds to an inflation pressure of 40 kPa. It is common practice that in these situations lower-than-normal loads and capacities are applied. Therefore, we assume that this principle is also acceptable in a low ground pressure system.

Common standard average ground pressures of current agricultural vehicles are in the range of 100-600 kPa and axle loads may be up to 20 Mg. To a certain extent, the standard average ground pressure (and tyre inflation pressure) can be reduced by using bigger tyres. Examples of the relationship between tyre size, tyre inflation pressure and load capacity for traction tyres are presented in Fig. 10. The highest load that can be accommodated on a traction tyre with an inflation pressure of 80 kPa is 3.8 Mg, by using the largest available tyre sizes. Insufficient space provided for tyre mounting or tyre size restrictions for agricultural reasons, obviously imposes limits on the tyre load that can be applied. The development of low section height (low aspect ratio) tyres has helped to realize low average ground pressure at restricted tyre diameter.

An additional possibility to achieve a low average ground pressure is the utilization of the manufacturer's allowances for lower inflation pressure or higher tyre load at low speed. For low speed in the field, depending on type of use, tyre construction and manufacturer, greater tyre deflection is allowed than at 30 km h^{-1}. For general usage at lower speeds (<20 km h^{-1}) and for most types of tyres, an increase in loading capacity of 20% is allowed. This speed allowance can be used when road traffic is usually restricted to occasional travelling over short distances, when road transport is performed with empty hoppers or bins only and when the maximum load occurs occasionally in the field only, for instance during lifting implements at the headlands. Tyre load

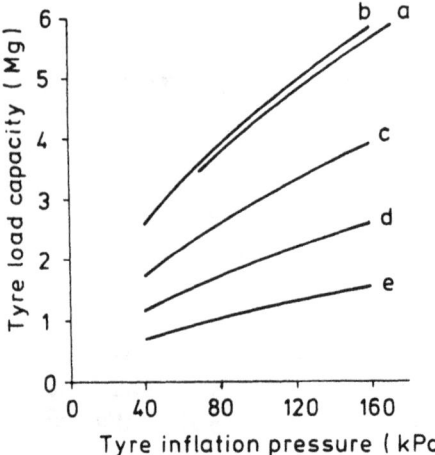

Fig. 10. Relationship between tyre load capacity and tyre inflation pressure recommended for 30 km h[-1] of 73x44.00-25 (a), 32L R 32 A8 (b), 20.8 R 38 A8 (c), 16.9 R 34 A8 (d) and 12.4 R 28 A8 (e) tyres, according to TRA (1991) and ETRTO (1991) standards.

capacities are sometimes allowed to be up to 50% higher (<8 km h[-1]) for field work (ETRTO, 1991; TRA, 1991). Central tyre inflation systems (CTIS) give additional flexibility to apply fully these specific speed allowances at low inflation pressure in the field and to use regular inflation pressures, recommended for 30 or 40 km h[-1] on the road (Tijink, 1991). Generally, using low speeds will make it possible to accommodate up to 5 Mg wheel load (10 Mg axle load) on the largest tyres currently available (1.10 m wide) at an inflation pressure of 80 kPa. It should be noted that the pressure distribution in the tyre-soil contact area in this "overloaded" state may be different from the pressure distribution in the standard average ground pressure situation.

When the use of bigger tyres alone does not give sufficient possibilities to achieve a low standard average ground pressure, further reduction is possible by applying more tyres (duals or more axles). Obviously, because of the unavailability of tyres wider than 1.10 m, this needs to be done when converting vehicles having axle loads >10 Mg to a low ground pressure configuration (see also Chapter 20).

Tractors

The wheel and axle loads of tractors may vary considerably, depending on: (1) the unladen weight and weight distribution; (2) the possible application of ballasting weights; (3) the implement's weight and work force requirements; (4) the method of implement attachment; (5) the possible use of support wheels of implements; (6) the implement's centre of gravity relative to the

TABLE 7

Typical current minimum inflation pressure, $p_{i(min)}$, and average ground pressure, p_g, for the fully loaded rear axle of four-wheel drive tractors, at two forward speeds

Engine power (kW)	Unladen tractor weight (Mg)	Maximum rear axle load[2] (Mg)	Maximum implement weight (Mg)	Typical standard tyre sizes	30[1] km h^{-1}		20 km h^{-1}	
					$p_{i(min)}$ (kPa)	p_g (kPa)	$p_{i(min)}$ (kPa)	p_g (kPa)
40	2.6	4.0	1.5	12.4R36	230	290	170	210
				13.6R36	200	250	150	190
60	3.9	5.4	2.0	14.9R38	250	310	190	240
				16.9R34	190	240	140	180
80	5.1	7.0	2.5	16.9R38	270	340	200	250
				18.4R38	210	260	150	190
100	5.8	8.3	3.0	18.4R38	280	350	200	250
				20.8R38	200	250	150	190
120	6.4	9.9	3.5	20.8R38	270	340	200	250

[1]For A6 tyres; 40 km h^{-1} for A8 tyres.
[2]Standard front ballast weights mounted.

tractor axles; (7) the mode of operation (working position vs. transport position). To ensure a low average ground pressure for the tractor, it is sufficient to consider the tractor-implement combinations or operating situations that exhibit the highest axle load for the rear and the front axle, respectively. The actually occurring highest axle loads depend on the fleet of machinery associated with the tractor. The TRACTYRE computer programme (Van Raay, 1992) was developed to determine correct "low ground pressure" tyres for a particular tractor and the associated machinery.

In the context of this chapter, a study was made of 248 four-wheel drive tractors to derive the typical current average ground pressure situation for the rear axle (Table 7), using standard tyres on 40-, 60-, 80-, 100- and 120-kW tractors. Maximum rear axle load was assumed to occur when ballasting the three-point hitch at 600 mm distance behind the lower links (centre of gravity of the implement) up to a point where the load on the front axle was reduced to 20% of the tractor weight (the minimum to maintain steerability). Next, it was calculated what tyre sizes, with diameters similar to the standard tyres, should be mounted on the tractor rear axle to obtain a standard average ground pressure of 100 kPa (80 kPa tyre inflation pressure), without compromising the maximum rear axle load (Table 8). The restrictions for speed and road travel given for the different tyres are valid when the axle bears the maximum load. In practice, most of the time the tractor rear axle will be much less heavily loaded and the limitations may not be relevant.

For work on loose soils before sowing or planting and for work on grassland with a very low bearing capacity, the inflation pressure should be reduced to 40 kPa. Often, the same tyres as for a standard average ground pressure of 100 kPa can be used because in this situation high torque (pulling force) is not required and the axle loads (including implement weight) are usually much lower than maximum. For example, we consider the 80-kW tractor with 30.5LR32 rear tyres (Table 8). At an inflation pressure of 40 kPa, these tyres have a load capacity of 2.42 Mg (30 km h^{-1}). Under these circumstances, for speeds up to 20 km h^{-1}, the load capacity is 2.90 Mg (20% higher). Thus the maximum rear axle load is 5.8 Mg, which corresponds with an applicable implement weight of 1.7 Mg (without front weights), which is sufficient for secondary tillage, fertilizer spreading and sowing with a reasonable capacity.

Ploughing with a low ground pressure tractor presents the problem that tyres

TABLE 8

Some tyre sizes for the rear axle of four-wheel drive tractors suitable to carry the maximum axle load at an inflation pressure of 80 kPa, at specified speed limits

Engine power (kW)	Max. axle load (Mg)	Standard tyre diameter (mm)	General usage <30 km h^{-1}		Little road travel <20 km h^{-1}		No road travel or CTIS, <8 km h^{-1}	
			Size	Dia.	Size (mm)	Dia.	Size (mm)	Dia. (mm)
40	4.0	1450-1500	600/60-30.5	1465	18.4R28	1490	16.9R30	1475
			520/70R30	1490	480/70R34	1536	480/70R30	1434
60	5.4	1575-1600	700/55-34	1634	23.1R26	1605	520/70R34	1592
			800/50-34	1650	600/65-34	1644	480/79R38	1637
			16.9R34[1]	1575	14.9R38[1]	1600		
80	7.0	1675-1750	30.5LR32	1820	24.5R32	1800	23.1R30	1700
			18.4R38[1]	1750	800/50-34	1650	650/60-38	1745
					16.9R38[1]	1675	67x34.00-30	1720
100	8.3	1750-1835	20.8R38[1]	1835	30.5LR32	1820	24.5R32	1800
			580/70R38[1]	1777	800/55-30.5	1650	66x44.00-25	1689
					18.4R38[1]	1750		
120	9.9	1835	23.1R34[1]	1833	20.8R38[1]	1835	30.5LR32	1820
							73x44.00-32	1864

[1]Dual wheel arrangement.

wider than 450 mm will not fit in the furrow, even when wide-furrow ploughs are used. However, Vermeulen et al. (1987) found that the uniformity of the work and the degree of soil inversion were not adversely affected with tyres of up to 650 mm in width, despite compaction of some of the freshly ploughed soil. Tyres wider than 650 mm can still be used if the last plough share is fitted with an attachment that slices a strip of soil of the required extra width from the landside at about half the ploughing depth and deposits it in the furrow, so that a wide, flat surface is prepared for the tyre (Lischka, 1985).

The passage of wheels in between rows of a standing crop presents a definite restriction on the tyre width to be applied. For cultivation work and spraying, even the standard tyres of tractors are usually too wide and often a light tractor with special narrow width, high inflation pressure tyres is used for this purpose. When using these tyres, low average ground pressure can best be achieved by applying the largest possible diameters combined with minimising the load per tyre through the use of light equipment and dual or triple tyre arrangements with the tyres aligned according to the row spacing. Particularly useful are special cultivation tractors (tool carrier type) with four large diameter wheels. To achieve an average ground pressure of 100 kPa, it is estimated that the load per tyre (260 mm wide) should not exceed 0.5 Mg. This corresponds with a maximum implement weight of 0.27 Mg for a 40-kW standard tractor with dual rear tyres and 0.90 Mg for a triple wheel arrangement. The option of having compacted spraying lanes is often considered acceptable because of the large working width of sprayers. Also in the case of a tractor pulling harvesting machinery for root crops, rear axle wheels pass in between the plant rows. The rear axle of a tractor pulling a two-row potato harvester can be converted to low ground pressure by applying dual wheels aligned according to the plant row spacing. Achieving low ground pressure with common 3- and 6-row sugar beet harvesters is difficult because of the heavy implement weight, the larger lifting and pulling forces involved and, consequently, the heavier tractor required. However, by harvesting the beets before the tyres pass the location, wide tyres could be applied to achieve low ground pressure. Possibilities for 6-row sugar beet harvesting are: (1) mounting the machinery in the rear hitch while driving the tractor backwards; (2) using self-propelled harvesting machinery.

Transport vehicles

The wheel load of drawn transport equipment depends on the empty vehicle weight, the payload capacity and the vertical load on the trailer hitch. The current average ground pressure situation of transport equipment with 16/70-20 tyres (418 mm wide and 1075 mm in diameter) was described by Tijink (1991), based on data of 35 tipping trailers and 40 slurry tankers with single, steered tandem and steered triple axles (Table 9). With the exception of some

TABLE 9

Typical current average ground pressure situation for trailers and slurry tankers with 16/70-20 tyres (after Tijink, 1991)

Payload capacity (Mg)	Maximum load on trailer wheel (Mg)			Average ground pressure (kPa)		
	1 axle	2 axles	3 axles	1 axle	2 axles	3 axles
4	2.4	-	-	210	-	-
6	3.3	-	-	300	-	-
8	4.3	2.3	-	400	210	-
10	5.2	2.8	-	490	270	-
12	-	3.4	2.5	-	320	220
16	-	4.5	3.2	-	440	290
20	-	-	4.0	-	-	360
25	-	-	4.9	-	-	450
30	-	-	5.8	-	-	530

single-axle slurry tankers, it is not yet common to construct drawn transport equipment such that the largest available tyre size can be used. Usually a tyre of approximately 0.6 m width and 1.35 m in diameter is the largest possible size for these vehicles. For normal agricultural use, at speeds up to 30 km h^{-1}, these tyres can accomodate a load of about 3.1 Mg at 80 kPa inflation pressure.

By using a CTIS, a loading capacity of about 4.2 Mg is possible in the field at speeds <10 km h^{-1}. Another possibility to increase the payload of drawn transport equipment is to use a weight transfer facility (WTF) in combination with the trailer-CTIS. With a WTF system, nearly the full payload will be supported by the trailer on the road. In the field the WTF shifts as much of the payload as possible to the tractor, using the additional loading allowance of the tractor wheels when operated in the field at low speed. The maximum possible loads for single-, tandem- and triple-axled trailers using a tyre inflation pressure of 80 kPa in the field are summarized in Table 10. Fig. 11 shows a combination of a tractor and a slurry tank, where the application of three axles, a CTIS and a WFT system, makes it possible to transport 22 m^3 slurry (= 22 Mg) over the field using a tyre inflation pressure of 80 kPa. Self-propelled transport vehicles typically suited for low ground pressure are the "swamp" vehicles, available with 3, 4 or 5 wheels (having a single front wheel for 3- and 5-wheel vehicles). Usually the largest high flotation tyres (66x43.00-25 or 73x44.00-32) are fitted on these vehicles. Recently, these vehicles are being used with slurry tanks and slurry injection equipment in Dutch agriculture. For field use only (filling at the edge of the field), the maximum permissible slurry volumes are about 3.5, 7 and 9 m^3, respectively, for the 3-, 4-

TABLE 10

Maximum permitted vehicle loads when using tyres of 0.60 m width and 1.35 m diameter, at an inflation pressure of 80 kPa, corresponding to an average ground pressure of about 100 kPa (after Tijink, 1991)

Tyre system and speed limit	Maximum trailer axle load[1] (Mg)	Number of axles	Maximum gross vehicle weight (Mg)	Maximum payload (Mg)
Conventional				
30 km h[-1] for general use	6.2	1	6.9	5.5
		2	13.6	10.9
		3	19.1	15.3
CTIS, no weight transfer				
8 km h[-1] for fieldwork	8.4	1	8.4	6.7
		2	16.8	13.4
		3	25.2	20.1
CTIS + WTF				
8 km h[-1] for fieldwork	8.4	1	10.0	8.0
		2	20.0	15.0
		3	30.0	22.0
73x44.00-32 tyres 8 km h[-1] for fieldwork[2]	10.0	1	10.0	8.0

[1]In the field.
[2]Tyre width 1.10 m, tyre diameter 1.86 m; use at high payload restricted to fieldwork.

and 5-wheeled vehicles, used with a tyre inflation pressure of 80 kPa (Tijink, 1990). Lischka (1985) described a two-axle truck with high flotation tyres, for transport of harvested sugar beet. The loading capacity of the truck was 9 Mg when using an inflation pressure of 100 kPa.

Self-propelled harvesters

Currently, the largest sugar beet harvesters, combine harvesters and field choppers for grass and silage maize are fitted with large tyre sizes. However, because of the high axle loads, inflation pressures of about 150-200 kPa are used. The axle load for the largest sugar beet harvesters may be 15-20 Mg. In general, the maximum axle load that can be accommodated with a tyre inflation pressure of 80 kPa is 10 Mg. Therefore, the number of axles and the distribution of the load over the axles need to be adapted to convert these largest machines to low average ground pressure without compromising the

Fig. 11. The use of three axles on a trailer (front- and rear-axles steered) and systems for central inflation control and weight transfer allows field transport of 22 m³ (22 Mg) slurry at a tyre inflation pressure of 80 kPa.

capacity and the functions of the machine. Lighter harvesters can often be converted directly to low average ground pressure. Removal of bulk hoppers from sugar beet harvesters and collection of the beet in a tipping trailer is a practical way to convert even the largest harvesters to low average ground pressure. A problem with combine harvesters and field choppers for grass and silage maize that have a maximum axle load of <10 Mg, is the restricted tyre width that can be utilised in connection with maximum permitted width during road transport. A solution for this problem can be the application of dual wheels with a quick connection system. Transport of these wheels on the road could be done on a small trailer, as is done for the header.

ECONOMICS

The attractiveness of LGP for farmers in arable farming and horticulture is ultimately determined by the profitability of its application (see Chapter 23). A case study on the economics of a Dutch 60-ha arable farm (Janssens, 1991), using data from a field experiment (Vermeulen and Klooster, 1992), showed that for an intensive cropping system with a large proportion of sugar beet, potatoes and onions, the LGP traffic system showed a marginal improvement of farm profits as compared to HGP. It may be expected that in vegetable production on a farm scale, the profitability of LGP will be more satisfactory, owing to the high value of the produce. Expected improvements in timeliness by the application of LGP have not been evaluated, partly because taking advantage of this aspect may trade off with soil quality and yield effects. However, improvement in timeliness might be a factor of great economic importance, for instance during the harvesting of sugar beet. Experience shows that LGP has the potential to improve the efficiency of fertilizer usage and to lower the required input of direct energy in the farming system. These factors are currently of minor economic importance, but are expected to become increasingly important. When the subsoil is loosened on a regular basis to alleviate compaction, LGP may be a very effective way to reduce the cost of the farming system by making subsoiling superfluous. LGP could possibly be used in soil amelioration to re-compact subsoils of agricultural land to an optimal physical condition, directly after subsoiling, whereafter the soil is allowed to gain additional structural strength naturally. This technique has received little research attention, but it has the potential to yield a reasonable bearing capacity and a sound physical condition of the subsoil, on which a low ground pressure farming system can be applied.

CONCLUSIONS

(1) The principle that the level of soil compactness increases with the mechanical pressure exerted on the soil is easily demonstrated in a uniaxial compression test. When applied to the soil condition in a traffic system, factors such as the non-uniform normal contact stress and soil water content distribution in horizontal and vertical directions under a wheel, structural strength of the soil, plastic soil deformation, multipass effects, soil tillage and climatic impact on the soil, interfere with the principle.

(2) The benefits of lowering the average ground pressure of tyre equipment as compared with current average ground pressures, include less topsoil compaction effects and, thus, smaller yield reductions for most crops, without compromising the number of work days available. The availability of cost-effective engineering solutions for axle loads up to 10 Mg, makes low ground pressure (standard average ground pressure < 100 kPa) a realistic option to avoid over-compaction of the soil.

(3) Theoretically, but not yet proven experimentally, low ground pressure tyre equipment, in combination with relatively high axle loads, should also avoid over-compaction effects below the topsoil, provided that this part of the soil profile exhibits a reasonable strength.

REFERENCES

Burt, E.C. and Wood, R.K., 1987. Three-dimensional tire deformation on deformable surfaces. Trans. ASAE, 30: 601-604.

Chamen, W.C.T., Chittey, E.T., Leede, P.R., Goss, M.J. and Howse, K.R., 1990. The effect of tyre/soil contact pressure and zero traffic on soil and crop responses when growing winter wheat. J. Agric. Eng. Res., 47: 1-21.

Danfors, B., 1977. Jordpackning - hjülutrüstning. (Soil compaction - wheel equipment). Jordbrukstekniska Inst., Meddelande 368, 53 pp. (in Swedish, with English summary).

Douglas, J.T., Campbell, D.J. and Crawford, C.E., 1992a. Soil and crop responses to conventional, reduced ground pressure and zero traffic systems for grass silage production. Soil Tillage Res., 24: 421-439.

Douglas, J.T., Koppi, A.J. and Moran, C.J., 1992b. Alteration of the structural attributes of a compact clay loam soil by growth of a perennial grass crop. Plant Soil, 139: 195-202.

ETRTO, 1991. Standards manual. The European Tyre and Rim Technical Organisation, Brussels, Belgium, Section A (Agricultural Tractor and Implement Tyres), 32 pp.

Fekete, A., 1977. Some observations on the contact pressure of tyres. Zesz. Post. Nauk Roln., 183: 125-130.

Grecenko, A., 1989. Some engineering aspects of preventing excessive soil compaction. Proc. 4th Eur. Conf. ISTVS, Wageningen, Netherlands, Vol. 1, pp. 62-68.

Håkansson, I., Henriksson, L. and Gustafsson, L., 1985. Experiments on reduced compaction of heavy clay soils and sandy soils in Sweden. Proc. Int. Conf. Soil Dynamics, Auburn, AL, U.S.A., Vol. 5, pp. 995-1009.

Hammel, K., 1991. Mit richtiger Bereifung Strukturschäden vermeiden. (Prevent soil structure damage with proper tyres). Bio-land 18: 28-31 (in German).

Janssens, S.R.M., 1991. Rendabiliteit van een verminderde bodembelasting; Bedrijfseconomische evaluatie van een lagedruk-berijdingssysteem. (Profitability of applying lower loads to the soil; An economic evaluation of a low ground pressure system at farm level). Res. Sta. Arable Farming Field Prod. Vegetables (PAGV), Lelystad, Netherlands, Rep. 127, 57 pp. (in Dutch with English summary).

Koolen, A.J., Lerink, P., Kurstjens, D.A.G, Van den Akker, J.J.H. and Arts, W.B.M., 1992. Prediction of aspects of soil-wheel systems. Soil Tillage Res., 24: 381-396.

Koppi, A.J., Douglas, J.T. and Moran, C.J., 1992. An image analysis evaluation of soil compaction in grassland. J. Soil Sci., 43: 15-25.

Krick, G., 1969. Druck und Schubverteilung unter Rädern und Reifen auf nachgiebigem Boden unter Berücksichtigung der Reifendeformation. (Pressure and shear stress distribution under wheels and tyres on deformable soil with reference to tyre deformation). Proc. 3rd Int. Conf. ISTVS, Essen, Germany, Vol. II, pp. 50-75 (in German).

Larson, W.E., Gupta, S.C. and Culley, J.L.B., 1988. Changes in bulk density and pore water pressure during soil compression. In: J. Drescher, R. Horn and M. de Boodt (Editors), Impact of Water and External Forces on Soil Structure. Catena, Cremlingen, Germany, Catena Suppl. 11, pp. 123-128.

Lebert, M., 1989. Beurteilung und Vorhersage der mechanischen Belastbarkeit von Ackerböden. (Judgement and prediction of the resistance of arable land to mechanical loading). Bayreuther Bodenkd. Ber., Band 12, 131 pp. (in German).

Lerink, P., 1990. Prediction of the immediate effects of traffic on field soil qualities. Soil Tillage Res., 16: 153-166.

Lischka, A., 1985. Versuche und praktische Erfahrungen mit Terrareifen. (Experiments and practical experiences with Terratyres). In: Lischka, A. and Isensee, E., 1985. Terrareifen. (Terra tyres). Rationalisierungs-Kuratorium Landwirtsch., Kartei 2.1.2.1., pp. 117-124 (in German).

McLeod, H.E., Reed, I.F., Johnson, W.H. and Gill, W.R., 1966. Draft, power efficiency, and soil-compaction characteristics of single, dual and low-pressure tires. Trans. ASAE, 9: 41-44.

Perdok, U.D., 1978. A prediction model for the selection of tyres for towed vehicles on tilled soil. J. Agric. Eng. Res., 23: 369-383.

Perdok, U.D. and Terpstra, J., 1983. Berijdbaarheid van landbouwgrond: Bandspanning en bodemverdichting. (Trafficability of agricultural land: Tyre inflation pressure and soil compaction). Landbouwmech., 34: 363-366 (in Dutch).

Perdok, U.D. and Arts, W.B.M., 1987. The performance of agricultural tyres in soft soil conditions. Soil Tillage Res., 10: 319-330.

Perdok, U.D. and Tijink, F.G.J., 1990. Developments in IMAG research on mechanization in soil tillage and field traffic. Soil Tillage Res., 16: 121-141.

Petelkau, H., 1986. Grenzparameter für die Bodenbelastung beim Einsatz von Traktoren und Landmaschinen aus der Sicht der Bodenfruchtbarkeit. (Criteria for soil loading by tractors and machinery in relation to soil fertility). Tag.-Ber., Akad. Landwirtsch.-Wiss. DDR, Berlin, Germany, 250: 25-36 (in German).

Raghavan, G.S.V., McKyes, E., Amir, I., Chasse, M. and Broughton, R.S., 1976. Prediction of soil compaction due to off-road vehicle traffic. Trans. ASAE, 19: 610-613.

Rüdiger, A., 1990. Quantitative Bewertung der Bodenbelastung durch Radfahrwerke. (Quantifying soil loading by running gear). Agrartech., 40: 10-12 (in German).

Siefkes, T. and Göhlich, H., 1991. Dynamische Eigenschaften von Breit- und Zwillingsreifen. (Dynamic properties of wide and dual tyres). Landtech., 46: 340-343 (in German).

Smith, D.L.O. and Dickson, J.W., 1990. Contributions of vehicle weight and ground pressure to soil compaction. J. Agric. Eng. Res., 46: 13-29.

Soane, B.D., Blackwell, P.S., Dickson, J.W. and Painter, D.J., 1981. Compaction by agricultural vehicles: A review. II. Compaction under tyres and other running gear. Soil Tillage Res., 1: 373-400.

Soane, B.D., 1985. Traction and transport systems as related to cropping systems. Proc. Int. Conf. Soil Dynamics, Auburn, AL, U.S.A., Vol. 5, pp. 863-935.

Soane, G.C., Godwin, R.J. and Spoor, G., 1986. Influence of deep loosening techniques and subsequent wheel traffic on soil structure. Soil Tillage Res., 8: 31-237.

Söhne, W., 1952. Die Kraftübertragung zwischen Schlepperreifen und Ackerboden. (Power transmission between tractor tyres and soil). Grundl. Landtech., 3: 75-87 (in German).

Söhne, W., 1953. Druckverteilung im Boden und Bodenverformung unter Schlepperreifen. (Pressure distribution in the soil and soil deformation under tractor tyres). Grundl. Landtech., 5: 49-63 (in German).

Steinkampf, H. and Sommer, C., 1989. Druck- und Verdichtungsmessungen im Feld unter gross-volumigen Reifen. (Field measurements of pressure and compaction under large-volume tyres). In: H. Schwanghart (Editor), 1989. Reifen landwirtschaftlicher Fahrzeuge. (Tyres of Agricultural Vehicles). VDI/MEG Kolloquium Landtech., 7, pp. 156-169 (in German).

Tijink, F.G.J., 1990. Rijden over land. 2. Ontwikkelingen op voertuiggebied. (Field traffic. 2. Developments in vehicles). Landbouwmech. 41/11: 7-9 (in Dutch).

Tijink, F.G.J., 1991. The influence of wheel configuration on ground pressure of agricultural vehicles. Proc. 5th Eur. Conf. ISTVS, Budapest, Hungary, Vol. 1, pp. 89-96.

TRA, 1991. Tyre and Rim Association, Akron, OH, U.S.A., Yearbook, Section 4, 24 pp.

Vanden Berg, G.E. and Gill, W.R., 1962. Pressure distribution between a smooth tire and the soil. Trans. ASAE, 5: 105-107.

Van Raay, B, 1992. Het Tractyre programma. (The Tractyre computer programme). Inst. Agric. Eng. (IMAG), Wageningen, Netherlands, Int. Rep., 10 pp. (in Dutch).

Vermeulen, G.D. and Klooster, J.J., 1992. The potential of a low ground pressure traffic system to reduce soil compaction on a clayey loam soil. Soil Tillage Res., 24: 337-358.

Vermeulen, G.D., Arts, W.B.M. and Fluit, J., 1987. Tillage and compaction effects of ploughing with wide tyres. Inst. Agric. Eng. (IMAG), Wageningen, Netherlands, Int. Note 294, 10 pp.

Vermeulen, G.D., Arts, W.B.M. and Klooster, J.J., 1988. Perspective of reducing soil compaction by using a low ground pressure farming system; selection of wheel equipment. Proc. 11th Conf. Int. Soil Tillage Res. Org. (ISTRO), Edinburgh, U.K., Vol. 1, pp. 329-334.

Vermeulen, G.D., Arts, W.B.M., Verwijs, B.R. and Van Maanen, J., 1993. Berijdingsmogelijkheden veengrasland. II. Afstemming van bandspanning en draagkracht. (Possibilities of traffic on peat grassland. II. Adjustment of tyre inflation pressure to bearing capacity). In: H. Snoek (Editor), Proc. Themadag Grasland en Berijding (Theme day Grassland and Traffic), 17 June, 1993, Res. Sta. Dairy Farming (PR), Lelystad, Netherlands, pp. 27-33 (in Dutch).

Soil Compaction in Crop Production
B.D. Soane and C. van Ouwerkerk (Eds.)
479

CHAPTER 20

Benefits of Limited Axle Load

I. HÅKANSSON[1] and H. PETELKAU[2]

[1]Swedish University of Agricultural Sciences, Department of Soil Sciences, Uppsala, Sweden
[2]Fürstenwalde, Germany

SUMMARY

Many farm vehicles have high weights and axle loads. Unless the ground contact pressure is extremely low, this leads to high stresses and detrimental compaction in deep subsoil layers. Repeated passes cause cumulative effects. Subsoil compaction is very persistent, maybe permanent, and loosening can only partly alleviate the effects. In field experiments, vehicles with high axle loads have reduced crop yield for decades and caused detrimental environmental effects. Therefore, subsoil compaction is an urgent soil conservation problem, and the axle load of off-road vehicles must be limited. This means that heavy machines must have many axles and wheels.

In the former U.S.S.R., a State Standard was adopted for agricultural vehicles concerning maximum ground contact pressures and stresses exerted in the subsoil. In some other countries, recommendations concerning upper limits on the axle or wheel load have been developed. Subsoil compaction can also be reduced by the use of suitable tyres and light-weight vehicles, the combination of field operations, a good organization of the field traffic, many entrances to the fields, a suitable field geometry and special transport lanes. Most of these measures would also considerably reduce topsoil compaction.

INTRODUCTION

During recent decades, parallel to a decrease in manpower and an increase in farm size, the weight of machines in agriculture and forestry has continuously increased. The largest farms and the heaviest machines are found in semi-arid areas, such as the drier parts of the American prairies and similar regions in the former U.S.S.R. and Australia. However, even in such areas, the soils are sometimes moist and sensitive to compaction when heavy machines are used.

In the U.S.A., a common agricultural tractor will have an engine power of 250 kW and a total weight of 20 Mg. Although not common, there are some units with an engine power of >500 kW and a total weight of >40 Mg. Grain trailers

Fig. 1. When fully loaded and equipped, this heavy grain trailer used in the American corn belt has an axle load of 25 Mg. A still larger model with an axle load of 40 Mg is available. (Photo: Ward B. Voorhees).

with up to 40 Mg on one axle are used (Fig. 1). In the former U.S.S.R., tractors with an engine power of 200 kW and a weight of 13 Mg are common.

Fig. 2. Sugar beet harvester weighing 32 Mg when loaded, and equipped with only four wheels. (Photo: Peter Schulz).

Even in humid areas, very heavy machines are sometimes used, although the average machine weight is somewhat lower than in semi-arid areas. Harvesting and transport vehicles in particular are frequently very heavy. West-European combine harvesters for cereals may weigh >20 Mg and sugar beet harvesters >30 Mg when loaded (Fig. 2), and these machines are used even in regions with a wet harvest period. Slurry tankers may have a volume of >20 m³ and a loaded weight of 25 Mg. Heavy trucks are also sometimes used in fields for transport purposes.

Occasionally, farmers' fields are trafficked by construction vehicles or military vehicles weighing >50 Mg. Most modern forestry vehicles are also very heavy. In northern Europe they usually weigh 6-14 Mg when empty and 15-30 Mg when loaded. In other parts of the world the vehicles may differ greatly from those in Europe, but the weights are similar.

With respect to subsoil compaction, the total weight of a machine is generally not the most influential parameter. The axle or wheel load is usually more decisive.

EVIDENCE OF WEIGHT EFFECTS ON SUBSOIL COMPACTION

Prediction of stress distribution and soil compaction under wheels and tracks

Several authors have predicted the stress distribution under running gear of various machines and in some cases, on this basis, have also predicted the machinery-induced soil compaction. The classical works by Söhne (1953, 1958) may be mentioned as examples, as well as more recent works by Perumpral et al. (1971), Carpenter et al. (1985), Smith (1985), Bolling (1986), Jakobsen and Dexter (1989) and Olsen (1994).

The predictions show that the stresses in the soil under loaded wheels decrease with depth. If the ground contact pressure is the same, the decrease is faster under wheels with small loads (= small contact areas) than under wheels with high loads (= large contact areas). Depending on the soil properties, the stresses may be more or less concentrated downwards or spread sideways. This is accounted for by a concentration factor that generally increases with the soil water content; the wetter the soil, the more the stress is concentrated and the deeper it penetrates.

Olsen (1994) divided the soil under the ground contact area into three zones, an upper zone where the vertical stress is nearly the same as the ground contact pressure, an intermediate zone where it decreases at a relatively high rate and depends on both ground contact pressure and wheel load, and a deep zone where the stress decreases very slowly with depth and depends almost exclusively on the wheel load (Fig. 3). When using normal farm vehicles, the incidence of compaction in the plough layer is mainly determined by the ground contact pressure and in deep subsoil layers by the axle load. In the upper part of the subsoil, both ground contact pressure and axle or wheel load are of importance.

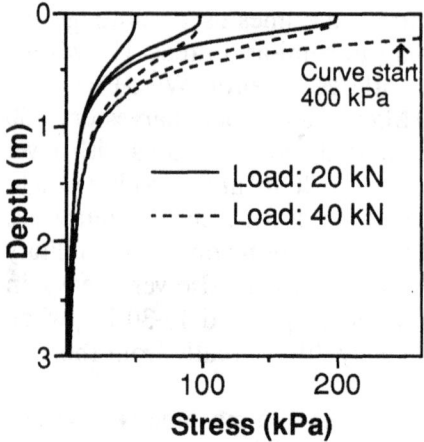

Fig. 3. Predicted vertical normal stress under the centre of a circular loaded area. Curves are drawn for two loads and three areas, viz., from left to right, 0.4, 0.2 and 0.1 m² (from Olsen, 1994).

Two loaded wheels mounted close together (e.g. dual wheels) interact, and although they efficiently reduce topsoil compaction, they are less efficient in reducing subsoil compaction. However, the more widely the wheels are spaced, the less is the interaction. Thus, to avoid compaction in deep subsoil layers, heavy vehicles should have many wheels spaced widely apart, preferably on separate axles. Unfortunately, this often makes the design of the machines more complicated and expensive. It also increases the number of wheelings and, thus, it may not be effective in decreasing topsoil compaction, unless the ground contact pressure is also reduced.

Measurements of stresses and compaction in the subsoil as a result of traffic with heavy vehicles

In many investigations, the stress in the ground contact area of wheels or tracks and the stress at various depths in the soil have been measured (e.g., Bolling, 1986; Horn et al., 1987; Taylor and Burt, 1987; Van den Akker et al., 1994). Such measurements, however, are difficult and available information is still limited. Generally, the stress in the ground contact area is not uniformly distributed. However, the variations level out in the subsoil, and here the measured stresses are usually in reasonable agreement with predicted stresses. The incidence of machinery-induced subsoil compaction has been established in many studies by measurements of various parameters, such as soil bulk density, total porosity, macro-porosity or penetration resistance (e.g., Danfors, 1974; Voorhees et al., 1986; Dannowski, 1987). Unfortunately, investigations of this kind are very time-

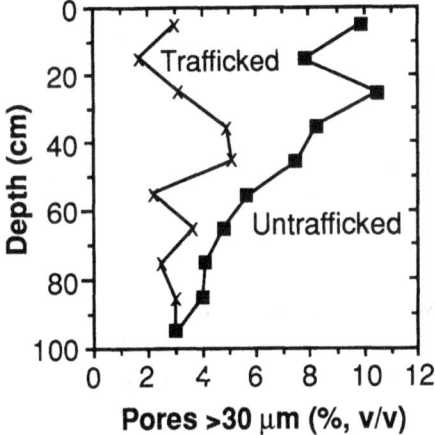

Fig. 4. Effect of intensive traffic with military vehicles weighing 50 Mg on the amount of coarse pores (>30 μm) in a clay soil (from Eriksson, 1976).

consuming and individual studies are often limited. However, most results are fairly consistent and quite clear conclusions can still be drawn.

Available information demonstrates that the axle load is a crucial factor influencing the depth of subsoil compaction. After applying traffic on the soil surface, significant compaction has usually been observed to a depth of about 30 cm at an axle load of 4 Mg, 40 cm at 6 Mg and 50 cm at 10 Mg. Still higher axle loads or very heavy tracked vehicles have caused compaction to 1 m depth (Gliemeroth, 1948; Dumbeck, 1984; Fig. 4). Danfors (1990) observed that compaction of the 30-50-cm layer slightly decreased when the tyre inflation pressure in wheels loaded by 4-6 Mg was reduced from 150 to 100 or 50 kPa.

If deep ruts are formed in the soil, even lighter vehicles may cause subsoil compaction, since in such cases the plough layer provides less protection. The same applies to tractor wheels running in the open furrow during mouldboard ploughing, which is often the most serious cause of subsoil compaction.

Tracked vehicles usually have a low ground contact pressure, but the stress is concentrated under the roadwheels, where it can be several times higher than the ground contact pressure (Corcoran and Gove, 1985). Therefore, under normal agricultural conditions, tracks seem to be less efficient in reducing soil compaction than in improving the possibilities of trafficking wet soils. Thus, Danfors (1974) found that an 18-Mg tracked vehicle caused at least as great compaction at 50-120 cm depth as a wheeled vehicle loaded with 16 Mg on a tandem axle unit. However, studies in the former U.S.S.R. indicated that subsoil compaction from tracked vehicles with modern running gear and relatively uniform ground contact pressure, is less severe than from wheeled vehicles (Rusanov, 1987).

The results of the investigations mentioned here generally agree with the stress predictions previously described. This shows that the stress predictions provide a useful basis for conclusions concerning the influence of vehicle weight, axle load, wheel arrangement or ground contact pressure on subsoil compaction. When predicting soil compaction, a certain error in estimated stress may, nonetheless, be less important than a comparable error in soil water content.

Effects of repeated wheel passes on soil compaction

In mechanized crop production, the total wheel track area throughout a year is often several times the field area (Table 1). In row crops, the wheelings are often concentrated in temporary traffic lanes, which are sometimes exposed to >40 wheel passes in a single season. Even with moderate wheel loads and ground contact pressures, repeated traffic may lead to deep compaction and persistent negative crop responses (Petelkau and Dannowski, 1990).

Some investigations indicate that repeated traffic may cause an increasingly deep soil layer to lose its elasticity and to function like a solid punch compacting a fresh layer. Therefore, with each wheel pass the soil compaction penetrates ever deeper. To restrict this effect to acceptable areas, permanent traffic lanes (controlled traffic) is a possible solution (see Chapter 22).

Persistence of subsoil compaction

Soil compaction becomes more persistent the deeper it penetrates. Factors generally considered to alleviate compaction are wetting/drying, freezing/thawing, biological activity and tillage. Both frequency and intensity of the influences of all these factors rapidly decrease with depth.

Recent experience indicates that the efficiency of freezing in alleviating subsoil compaction was previously over-estimated. Presumably, parallels were drawn with

TABLE 1

Tracked area, relative to the field area (%), associated with techniques commonly used in crop production (Petelkau, 1987)

	Primary tillage	Pota-toes[1]	Sugar beet[1]	Winter wheat[1]	Silage maize[1]
Area with wheel tracks (%)	81	98	89	61	66
Cumulative tracked area (%)					
Total	211	552	981	481	344
With ground contact pressure >200 kPa	18	311	457	225	228
With wheel load >2 Mg	74	213	171	95	127

[1]Primary tillage not included.

the drastic effects of freezing in the topsoil. However, in laboratory studies, Hedberg (1976) found that many intensive freezing/thawing cycles were required to bring a compacted soil into a new, loose equilibrium state. In field studies, Van Ouwerkerk (1968) found no loosening of a compacted sandy loam subsoil in the Netherlands during a 6-year period. In a clay loam in Minnesota, Blake et al. (1976) found no alleviation of subsoil compaction over a 9-year period, in spite of annual freezing to about 1 m depth. In a forest area with sandy loam in Minnesota, Thorud and Frissell (1976) observed that artificial compaction was alleviated after 9 years in the 0-7-cm layer but persisted in the 15-22-cm layer. Froelich et al. (1985) obtained similar results in loamy soils in Idaho 23 years after compaction.

In order to establish the effects on soils and crops from traffic by vehicles with high axle load, an international series of field experiments was carried out at 25 locations in humid areas of Europe and North-America (Håkansson et al., 1987). At most locations the normal frost depth was >0.5 m. The whole surface of the experimental plots was covered once or four times by wheel tracks from a heavy vehicle on one occasion, when the soil water content was near field capacity. The axle load usually was 10 Mg and the tyre inflation pressure 300 kPa. Subsequently, the axle load was limited to 5 Mg and all experimental plots were treated uniformly, including annual ploughing to a depth of 20-25 cm. No subsoiling was done. In most cases, significant compaction was established to 50 cm depth or more (Fig. 5). An increase in the penetration resistance in the subsoil remained nearly unaltered throughout an 11-year period in both sandy soils and clay soils (Etana and Håkansson, 1994).

All observations mentioned above originate from freeze/thaw areas and demonstrate that in such areas, even in clay soils, subsoil compaction persists for decades. Presumably it is more persistent in non-swelling soils and in regions without freezing. Thus, subsoil compaction is always very persistent, and in sandy soils or tropical areas it is probably permanent.

Crop response to subsoil compaction

In the international series of experiments, the high axle-load traffic also caused persistent negative crop responses. The mean response on plots covered four times by wheel tracks is shown in Fig. 6. However, the response varied between sites and years. At individual sites, the effects sometimes disappeared for one or two years, and then re-appeared. This may have been caused partly by experimental errors and partly by weather variations.

Large negative crop yield responses were obtained during the first two years, probably since the heavy experimental traffic initially also led to a coarser structure of the plough layer. Under comparable climatic conditions, a series of similar experiments was carried out in Sweden with much lighter vehicles, so that the compaction was mainly restricted to the plough layer. Even in this series the

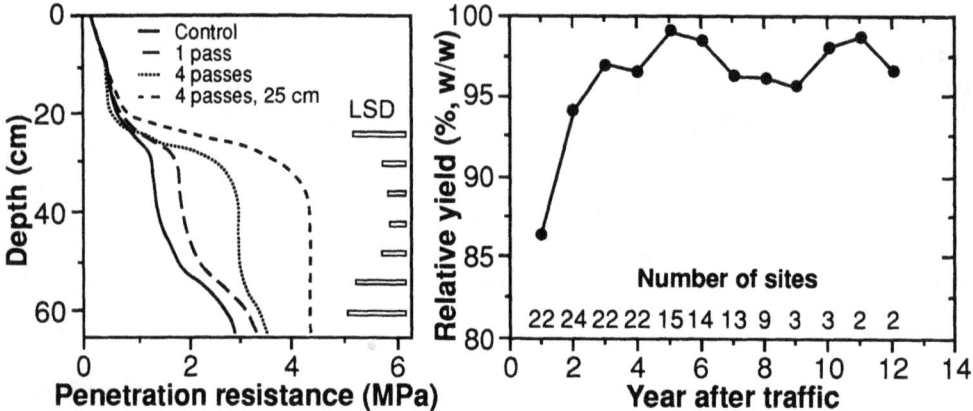

Fig. 5 (left). Typical results of measurements in experiments where traffic with an axle load of 10 Mg was applied. The diagram shows the penetration resistance at various depths in a sandy soil in Denmark 6 years after traffic with a loaded dump-truck, applied either on the soil surface or after removal of the 25 cm deep plough layer (from Schjønning and Rasmussen, 1994).

Fig. 6 (right). Residual effects on relative crop yield (yield without heavy traffic = 100%) of four passes on one occasion by vehicles having a load of 10 Mg on a single axle or 16 Mg on a tandem axle unit. Mean values from experiments in seven countries in freeze/thaw areas of Europe and North America (data from Håkansson et al., 1987, with the inclusion of some recent data).

direct effect of the wheeling on soil bulk density was eliminated by ploughing immediately afterwards but, in spite of that, soil structure remained coarser and crop yield was reduced in subsequent years. This effect, however, disappeared within five years (Håkansson et al., 1988).

ı It may be assumed that in the high-axle load experiments the plough layer compaction also disappeared within five years and that the negative crop yield response after the fourth year was due to subsoil compaction alone. If we assume that plough layer compaction caused the major part of the crop response in the first year and a minor part in years 2-4, the remaining part, attributable to subsoil compaction, was nearly constant throughout the experimental period. This indicates that crop response to subsoil compaction may persist for decades, and part of it may be permanent.

The experiments with high axle loads also included plots which received one pass by the heavy vehicle, and here the mean crop response was about one-fourth of that for four passes. This demonstrates that repeated traffic causes cumulative effects, and this probably applies up to much higher traffic intensities than those used in the experiments.

When a farmer uses vehicles with high axle loads for most field operations, it means that, on average, one wheeling is applied over his entire field within one or two years, and four wheelings within six or eight years. After 20 years, the total number of wheelings may be three times as high as on the most intensively trafficked plots in the experiments. Since the effects are very persistent and seem to cumulate up to a very high traffic intensity, the yield reduction also may be expected to be three times as large.

From the results obtained, no simple rules can be derived concerning the sensitivity of different soils to subsoil compaction. Large negative crop responses were obtained on sites with heavy clay soils as well as on sites with sandy soils. So far, all soils must be regarded as sensitive. Of course differences occur, but they are likely to be as large within groups of similar soils as between major soil groups. Since most farms have several soil types, nearly all of them will probably have areas where traffic with heavy machines under moist conditions is detrimental. However, the drier the climate, the longer the periods when the soils are not damaged.

In some experiments in the international series, traffic with even higher axle loads was applied (Voorhees et al., 1986; Gameda et al., 1987). This increased subsoil compaction and reduced crop yield even more than the axle load of 10 Mg applied on most sites.

It is well established that the topsoil is usually overloosened by ploughing, and that a moderate recompaction improves crop growth. It cannot be excluded that even subsoil compaction may have positive effects. By means of mathematical modelling, Jakobsen et al. (1989) predicted a positive crop response under certain conditions, although in general the mean effect was negative. This indicates that it is necessary to collect more detailed information on the effects of subsoil compaction for various combinations of soils, crops and climates.

Subsoil loosening

It is a general opinion that large areas of agricultural land have been damaged by subsoil compaction, to a great extent caused by pressure and slip by tractor wheels running in the open furrow when ploughing. In order to alleviate these effects, numerous techniques for loosening compacted subsoil layers have been developed.

Subsoil loosening is always expensive, and it can seldom completely ameliorate a compacted subsoil. Usually it is only possible to reduce crop yield losses caused by subsoil compaction by one third to one half (Werner, 1989). When most successful, the mechanical loosening starts a prolonged regenerating process. Generally, a prerequisite for a persistent effect is that subsequent loading of the soil is largely reduced (Soane et al., 1986). Otherwise a rapid re-compaction may occur and the state of compactness existing before loosening may be reached or even exceeded in a short time (Table 2).

TABLE 2

Change in soil bulk density caused by different wheel pressures during a series of six tillage operations for one crop after subsoil loosening of dense sandy loam soils (J. Lehfeldt, personal communication, 1985)

Depth (cm)	Ground contact pressure (kPa)	Max. wheel load (Mg)	Bulk density (Mg m³)		
			Before loosening	After loosening	After trafficking
25-30	100	1.7	1.71	1.59	1.72
	300	3.4	-	-	1.77
	500	3.4	-	-	1.88
35-40	100	1.7	1.73	1.60	1.70
	300	3.4	-	-	1.73
	500	3.4	-	-	1.78
45-50	100	1.7	1.66	1.64	1.68
	300	3.4	-	-	1.69
	500	3.4	-	-	1.72

In former East Germany, a "segment plough" for partial loosening of the subsoil has been developed (Gätke, 1983; Böttcher et al., 1988; Unger et al., 1988). From the compacted layer in the bottom of each furrow a 10-cm wide slit

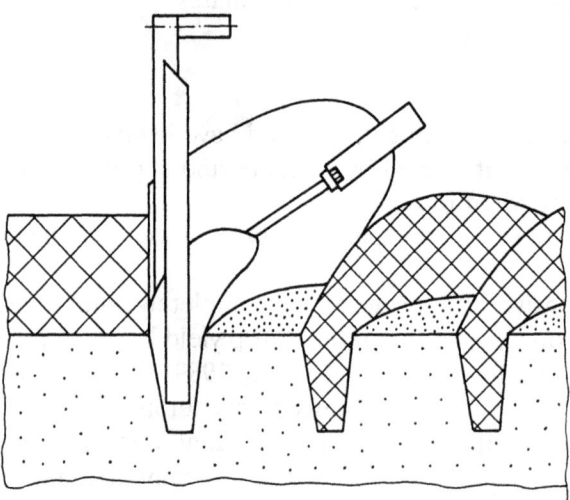

Fig. 7. The function of the "segment plough".

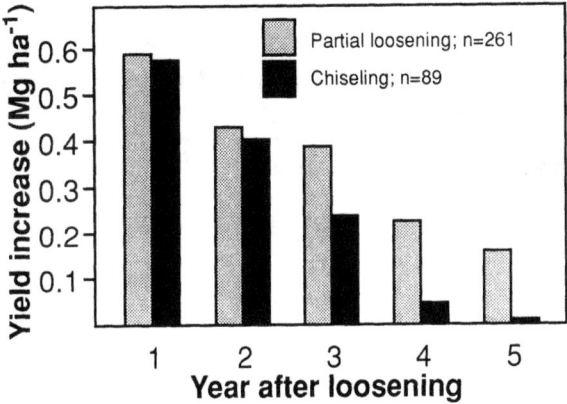

Fig. 8. Yield increase in cereal crops after partial subsoil loosening with a segment plough and after traditional subsoiling by chiseling. Mean yield in unloosened control plots ≈ 4 Mg ha⁻¹. (B. Böttcher and J. Lehfeldt, personal communication, 1990).

is cut out by a chisel attached to the plough and is filled with topsoil (Fig. 7). The slits form pathways for the roots to deeper layers, whereas the unloosened parts of the subsoil support the loads of the machines. Yield results are shown in Fig. 8. Similar results were reported by Kooistra and Boersma (1994).

The experience concerning subsoil loosening can be summarized as follows: (1) it is better to avoid over-compaction of the subsoil by protecting it against mechanical overloading than to loosen it periodically; (2) loosening should not be undertaken without diagnostic evidence of detrimental compaction; (3) after loosening, the subsoil must be protected against mechanical overloading until the structure has stabilized.

BENEFITS OF REDUCED VEHICLE WEIGHTS

Ecological benefits

Besides reduced crop yield, subsoil compaction may have many other ecological consequences. Root growth is impaired, which obstructs the utilization of plant nutrients and increases the risk of leaching (Table 3). The function of the soil as a filter for various substances may deteriorate and the microbial decomposition of pesticides may be hampered.

Compaction reduces the saturated hydraulic conductivity of the soil. This may lead to the development of wet areas in the fields, impaired soil workability and increased erosion. When a freshly tilled loose soil layer lies on top of a compacted, impermeable layer, intensive rainfall may cause a downhill slide of the whole loose layer.

TABLE 3

Ten-year accumulated values of water and nitrogen balance in a field trial with suction lysimeters at Müncheberg (former East Germany) on Albic Luvisol (after Petelkau et al., 1988)

	Loose-optimal soil density	Moderately compacted	Relative[1]
Water balance (mm)			
Rainfall	5427	5427	
Change of soil water	+155	+135	
Percolation	1390	1734	125
Evapotranspiration	3882	3558	92
Nitrogen balance (kg ha^{-1})			
Fertilization	1690	1690	
Uptake by plants	1452	1254	86
Leaching	311	469	151
Crop yield (Mg ha^{-1})			
Shoot dry matter[2]	87.5	74.7	85.0

[1]Loose-optimal = 100%.
[2]Winter cereals + cruciferous catch crop alternating with spring-sown cereals.

Thus, subsoil compaction is a serious soil conservation issue and a long-term threat to soil productivity. Since it may impair the productivity of soils for generations to come, it is not only an economic problem. Therefore, agricultural machinery should be designed and used in such a way that unacceptable subsoil compaction is avoided. This means, foremost, that the axle load must be kept within acceptable limits.

Even if a machine does not cause unacceptable subsoil compaction, a low weight may still be advantageous with regard to topsoil compaction. The benefits of this, however, vary considerably. For tillage operations, the working width usually increases in direct proportion to the tractor weight and, if traffic intensity is expressed in Mg km ha^{-1}, the difference between tractors of various weights is small. For harvesting or transport operations, on the other hand, the traffic intensity may sometimes increase proportionally to the weight of the machines.

The use of machines with low weights and ground contact stresses reduces the sensitivity to high soil water content. This may result in great timeliness benefits for critical field operations and in higher crop yields, better utilization of plant nutrients and reduced leaching. It may also reduce the need for loosening of the soil and, thus, increase the feasibility of reduced tillage and the possibilities for double-cropping or growing of catch crops.

Economic benefits

The machine plus labour costs generally decrease with increasing machine size up to a rather large size. However, calculations for heavy machines in Sweden show that the "soil compaction costs" may sometimes be as high as the machine and labour costs together. Thus, a machinery system can only be economically optimized if also the soil compaction costs are considered.

Håkansson (1985) and Håkansson and Danfors (1988) estimated that on Swedish farms growing barley, oats and wheat, the soil compaction costs due to slurry spreading in early spring with 5-10-m^3 tankers, could amount to Skr 230-1880 ha^1 ($\approx$ US $ 40-330 ha^{-1}), depending on soil type, soil water content, vehicle weight, spreading width, field size and value of subsequent crops. Nilsson (1990) estimated the total soil compaction costs for different harvesters, when harvesting canning peas in a district with an annual pea area of 6,800 ha, at Skr 0.8-1.7 million ($\approx$ US $ 130,000-290,000), depending on the weight of the harvester and on the working width, wheel arrangement, tyre size and inflation pressure.

ADVANTAGES AND APPLICATION OF THE RESTRICTED AXLE-LOAD CONCEPT

Axle load restrictions have been established for public highways, even though these are constructed to resist heavy traffic. However, the thickness of the pavement is limited, and excessively high axle loads may deform underlying, softer layers. The load limit is usually 6-12 Mg per axle, depending on the quality of the road. For tandem axles, the total load permitted is usually less than twice the load permitted on a single axle, since at some depth the compacting effects of the two axles start interacting. This indicates that the axle load is of great importance for the deformation of deep subsoil layers and therefore it should be limited also for arable fields and other off-road areas.

The highest acceptable axle load for arable land varies with soil and climatic conditions. However, for practical reasons load limits must be common for large regions and chosen in such a way that subsoil compaction does not become unacceptable when the vehicles are used to their full capacity under the wettest conditions. Once a farmer has invested in a machine, he will use it to its full capacity as long as soil trafficability permits, irrespective of compaction hazards.

Arable soils are essential for human life. Maintenance of their productivity should be a responsibility for the whole society and must not be left entirely to individuals. Society has adopted rules for the protection of soils from chemical hazards, and should also consider rules to protect them from mechanical overloading.

Until now, such rules have been developed only in the former U.S.S.R. They were given the status of a State Standard that is now adopted at least in Russia.

TABLE 4

Maximum permissible ground contact pressure (GP) and vertical normal stress (VS) at a depth of 0.5 m exerted by running gear of vehicles on loamy soils and clay soils, according to the State Standard of the former U.S.S.R. (Bondarev et al., 1987)

Soil water (0-30 cm) (% of field capacity)	GP (kPa)		VS at 0.5 m depth (kPa)	
	Spring	Summer	Spring	Summer
>90	80	100	25	30
70-90	100	120	25	30
60-70	120	140	30	35
50-60	150	180	35	45
<50	180	210	35	50

Specifications are laid down for the maximum permissible ground contact pressure of the running gear of vehicles and the estimated maximum vertical normal stress at a depth of 50 cm (Bondarev et al., 1987; Nugis and Lehtveer, 1987; Rusanov, 1988). The limits depend on soil texture and soil water content (Table 4). For wheeled vehicles, the maximum ground contact pressure is calculated by multiplying the ground contact pressure, as estimated from the wheel load and the ground contact area, with a non-uniformity factor of 1.5. The

Fig. 9. A 15-m³ tanker for spreading of slurry manure, equipped with triple axles to reduce subsoil stresses and a 12-m wide spreading boom to reduce the number of tracks.

ground contact area is measured on a rigid surface and adjusted for the increase obtained when a 50-mm deep rut is formed. The normal vertical stress at a depth of 50 cm under the centre of the ground contact area, depends on the magnitude and shape of the ground contact area and on the ground contact pressure; it is estimated with methods developed in civil engineering to calculate sinkage of foundations. When applying this standard, wheel loads are limited indirectly by the admissible normal stresses at 50 cm depth; by the method used, the ground contact pressure distribution and the characteristics of the running gear are also considered.

In Sweden, a load limit of 6 Mg per axle for farm vehicles was recommended by soil scientists several years ago and it has been adopted by many farmers and machinery manufacturers (Fig. 9). In former East Germany, officially acknowledged recommendations have been worked out (Petelkau, 1984, 1986). Recommended ground contact pressure limits range from 50 kPa on moist sandy soils to 200 kPa on dry clay soils and wheel load limits from 1.5 to 2 Mg for machines with standard tyres. For other wheel equipment the load limits are adjusted. When applying the State Standard of the former U.S.S.R. for vertical normal stress in the subsoil to vehicles with standard tyres, load limits similar to those mentioned in this paragraph are obtained.

There is a general need for guidelines for maximum subsoil stresses and axle or wheel loads for off-road vehicles. However, although the principles may be universal, the limits must be set with regard to regional soil and climatic conditions. Preferably, they should be developed in an international joint effort, since this probably is a prerequisite to interest important machinery manufacturers in the issue. This should be done as soon as possible, before the world-wide impairment of soils has gone too far.

OTHER OPPORTUNITIES FOR REDUCING SUBSOIL COMPACTION IN PRACTICE

As indicated above, when estimating machinery costs, the soil compaction costs should also be considered. Especially for field operations carried out under moist conditions, this would lead to the use of lighter machines. Tractors and other machines are often much heavier than necessary, simply because the farmers use the same machines for many types of work. If compaction costs were considered, more specialized equipment would often be used. Farmers could achieve this by making use of contractors or by cooperating with neighbours. For some operations much lighter machines than those commonly used have, in fact, been developed. Fig. 10 shows an example from forestry. Very light machines are also available for spraying and fertilizing in agriculture and are widely used in some countries, e.g., in the U.K. (Smith and Dickson, 1990).

In heavy harvesting and transport operations, e.g., when harvesting forage or sugar beet or spreading fertilizer or manure, the intensity of machinery traffic is

Fig. 10. A hand-steered, light-weight forestry vehicle. The inset shows tracks formed by an ordinary, heavy forwarder under wet conditions. (Photo: J. Palm).

TABLE 5

Effects of soil conservation measures[1] relative to conventional management in a 5-year crop rotation (potatoes - winter rye + catch crop - silage maize - winter rye - winter barley + catch crop) on a 50-ha field with a length/width ratio of 2.0 (H. Petelkau and U. Barkusky, unpublished results, 1990)

	Relative values[2]
Yield	118
Production costs	107
Financial result	176
Total trafficked area	89
Parts of total with ground contact pressure >200 kPa	18
Parts of total with wheel load >2 Mg	25

[1]*Machinery for conventional management:* 160-kW tractors for primary tillage, 66-kW tractors for other operations, cereal harvesters with 5.1 m cutting width, maize harvesters with 2.4 m cutting width, two-row potato harvesters, 8-Mg trailers for field transport and manuring.
Soil conservation measures: Use of the same machines equipped with dual wheels, low ground pressure tyres or tracks, wheel loads limited to 2 Mg (whenever possible), transverse transport lanes (bare) at 300-m intervals, combination of operations and reduced ploughing depth.
[2]Conventional management = 100.

often strongly influenced by field length and shape. Fields that are too long may lead to the use of excessively large vehicles. Alternatively, when using normal-size vehicles, there will be excessive traffic with loaded or empty vehicles over the fields.

According to investigations in former East Germany (Barkusky, 1990), a field length of about 300 m is usually optimal. Longer fields should be divided or broken by transport lanes, where products can be transferred between field and road vehicles. The costs of these soil conservation measures will be compensated for by improved crop yields (Table 5). A suitable organization of the traffic and a large working width may considerably reduce the traffic intensity. Much can be gained by matching the working width, the vehicle size and the field length, and by arranging many entrances to the fields.

The use of optimum wheel equipment is an important option to reduce soil compaction. This is discussed in greater detail in Chapter 17. Increasing the number of axles and wheels on heavy machines reduces the subsoil compaction. However, heavy machines with many axles and wide, low ground pressure tyres must have steerable axles (Fig. 11). Tracks of new design for agricultural tractors may offer opportunities to reduce compaction.

Fig. 11. Heavy vehicles with more than one axle and wide tyres with low ground contact pressures must have steerable axles, as employed in this bulk trailer with a load carrying capacity of 10 Mg.

Tillage operations make up a large part of the annual field traffic. It may be necessary to plough a compacted soil, but at the same time more traffic is applied (Table 1). If, during all other field operations, compaction is sufficiently reduced, ploughing may be excluded, and this may further reduce compaction. A technique to reduce subsoil compaction during mouldboard ploughing is to use half the normal depth for the last plough body, so that the tractor wheels run at a shallower depth (Petelkau et al., 1978). During seedbed preparation, fertilizing and sowing, the most obvious way to reduce soil compaction is to combine operations whenever possible.

CONCLUSIONS

(1) Traffic with high axle-load vehicles may cause serious subsoil compaction and this can only be slightly reduced by the use of low ground pressure tyres. The deeper the compaction occurs, the longer it persists. Even on clay soils in areas with annual freezing, subsoil compaction persists for decades, and in sandy soils or warmer climates it may be permanent. In field experiments, crop growth was still reduced more than ten years after high axle-load traffic and repeated passes caused cumulative effects. Subsoil compaction can only be partly alleviated by deep tillage.

(2) For soil conservation reasons, axle or wheel loads should be limited, but permissible limits depend on soil water conditions. In the former U.S.S.R., a State Standard for maximum admissible stresses in the subsoil has been adopted, and in some other countries axle or wheel load limits have been recommended. Guidelines for load restrictions should preferably be developed in a joint international effort.

REFERENCES

Barkusky, U., 1990. Kalkulative Ermittlung des Einflusses von Schlaggrösse, -form und Besatz mit Bewirtschaftungshindernissen unter besonderer Berücksichtigung der schlagspezifischen Verdichtungsdisposition der Bodensubstrate auf materielle und finanzielle technologische Aufwendungen. (Calculative determination of technical and economic consequences of field size and shape, and of obstacles to management, with special regard to the sensitivity of the soils to compaction). Ph.D. thesis, Humboldt-Univ., Berlin, Germany, Sektion Pflanzenproduktion, 103 pp. (in German).

Blake, G.R., Nelson, W.W. and Allmaras, R.R., 1976. Persistence of subsoil compaction in a Mollisol. Soil Sci. Soc. Am. J., 40: 943-948.

Bolling, I., 1986. How to predict soil compaction from agricultural tires. J. Terramech., 22: 205-223.

Bondarev, A.G., Medvedev, V.W., Rusanov, A.V. and Sudakov, A.V., 1987. Kommentarii k GOST 26955-86, Technika selskochozjaistvennaja mobilnaja. Normy dopustimogo vozdeistvija dvizjitelei na pochvu. (Comments on GOST 26955-86, Mobile farm machines - standards for admissible effects of running gear on the soil). Zemledelie, 9: 29-30 (in Russian).

Böttcher, B., Baur, A., Herzog, R. and Lehfeldt, J., 1988. Untersuchungen zum Arbeitseffekt von Werkzeugen zur partiellen Vertiefung der Ackerkrume. (Studies on the performance of tools for partial deepening of the topsoil). Arch. Acker- Pflanzenb. Bodenkd., 32: 279-291 (in German).

Carpenter, T.G., Fausey, N.R. and Reeder, R.C., 1985. Theoretical effect of wheel loads on subsoil stresses. Soil Tillage Res., 6: 179-192.

Corcoran, P.T. and Gove, D.S., 1985. Understanding the mechanics of track traction. Proc. Int. Conf. Soil Dynamics, Auburn, AL, U.S.A., Vol. 4, pp. 664-678.

Danfors, B., 1974. Packning i alven. (Compaction in the subsoil). Swedish Inst. Agric. Eng., Uppsala, Sweden, Rep. S 24, 91 pp. (in Swedish, with English summary).

Danfors, B., 1990. Soil compaction in the subsoil. Proc. Int. Conf. Agric. Eng., Berlin, 24-26 October. VDI-Gesellschaft Agrartechnik, Verein Deutscher Ingenieure, Düsseldorf, Germany, pp. 27-29.

Dannowski, M., 1987. Die Auswirkungen wiederholter Belastung von Spuhrbahnen durch Fahrwerke auf Trockenrohdichte und Durchwurzelbarkeit eines stark lehmigen Sandbodens. (Effects of repeated loading of traffic lanes by running gear on bulk density and rootability of a loamy sand). Arch. Acker- Pflanzenb. Bodenkd., 31: 573-581 (in German).

Dumbeck, G., 1984. Einfluss aussergewöhnlicher Druckbelastung auf das Bodengefüge und die Durchwurzelung. (Effect of extraordinary ground pressure on soil structure and root development). Mitt. Deutsch. Bodenkd. Gesellsch., 40: 61-62 (in German).

Eriksson, J., 1976. Influence of extremely heavy traffic on clay soil. Grundförbättring, 27: 33-51.

Etana, A. and Håkansson, I., 1994. Persistence of subsoil compaction and crop response after traffic with high axle-load vehicles in Sweden. Soil Tillage Res., 29: 167-172.

Froelich, H.A., Miles, D.W.R. and Robbins, R.W., 1985. Soil bulk density recovery on compacted skid trails in central Idaho. Soil Sci. Soc. Am. J., 49: 1015-1017.

Gameda, S., Raghavan, G.S.V., McKyes, E. and Theriault, R., 1987. Subsoil compaction in a clay soil. I-II. Soil Tillage Res., 10: 113-122 and 10: 123-130.

Gätke, C.R., 1983. Das Prinzip des Segmentpflügens als wirksame Methode zur nachhaltigen Strukturverbesserung des krumennahen Unterbodens auf D-Standorten. (The principle of segment ploughing as an efficient method for durable structure improvement in the upper part of the subsoil on diluvial sites). Tag.-Ber., Akad. Landwirtsch.- Wiss. DDR, Berlin, Germany, 215: 189-196 (in German, with English summary).

Gliemeroth, G., 1948. Selbstverschuldete Strukturstörungen des Bodens unter besonderer Berücksichtigung des Schlepperraddrucks. (Man-induced soil structure deterioration, with special consideration of tractor wheel ground pressure). Berichte Landtech. 1948: II, pp. 19-54 (in German).

Håkansson, I., 1985. Jordpackning - nu vet vi kostnaden. (Soil compaction costs can be estimated). Lantmannen, 5: 22-24 (in Swedish).

Håkansson, I. and Danfors, B., 1988. The economic consequences of soil compaction by heavy vehicles when spreading manure and municipal waste. Proc. CIGR-Seminar on Storing, Handling and Spreading of Manure and Municipal Waste. Swedish Inst. Agric. Eng., Uppsala, Sweden, pp. 13:1-13:10.

Håkansson, I., Voorhees, W.B., Elonen, P., Raghavan, G.S.V., Lowery, B., Van Wijk, A.L.M., Rasmussen, K. and Riley, H., 1987. Effect of high axle-load traffic on subsoil compaction and crop yield in humid regions with annual freezing. Soil Tillage Res., 10: 259-268.

Håkansson, I., Voorhees, W.B. and Riley, H., 1988. Vehicle and wheel factors influencing soil compaction and crop response in different traffic regimes. Soil Tillage Res., 11: 239-282.

Hedberg, D., 1976. Volymförändringar hos jordarter efter upprepad frysning och tining. (Changes of the volume of soils after repeated freezing and thawing). Ph.D. thesis, Univ. Stockholm, Sweden, 109 pp. (in Swedish, with English summary).

Horn, R., Burger, M., Lebert, M. and Badewitz, G., 1987. Druckfortpflanzung in Böden unter langsam fahrenden Traktoren. (Stress propagation in the soil under slowly moving tractors). Z. Kulturtech. Flurber., 28: 94-102 (in German, with English summary).

Jakobsen, B.F. and Dexter, A.R., 1989. Prediction of soil compaction under pneumatic tyres. J. Terramech., 26: 107-119.

Jakobsen, B.F., Dexter, A.R. and Håkansson, I., 1989. Simulation of the response of cereal crops to soil compaction. Swedish J. Agric. Res., 19: 203-212.

Kooistra, M.J. and Boersma, O.H., 1994. Subsoil compaction in Dutch marine sandy loams: loosening practices and effects. Soil Tillage Res., 29: 237-247.

Nilsson, K., 1990. Packningsskador vid konservärtskörd - ekonomiska konsekvenser och åtgärder för att minska packningen. (Estimation of the economic consequences of soil compaction when harvesting canning peas). Swedish Univ. Agric. Sci., Uppsala, Sweden, Div. Soil Manage., Rep. 79, 16 pp. (in Swedish, with English summary).

Nugis, E. and Lehtveer, R.V., 1987. Predelnie pokazateli fizicheskogo sostojanija pochv. (Limiting indexes of soil physical properties). Zemledelie, 9: 18-20 (in Russian).

Olsen, H.J., 1994. Calculation of subsoil stresses. Soil Tillage Res., 29: 111-123.

Perumpral J.V., Liljedahl, J.B. and Perloff, W.H., 1971. A numerical method for predicting the stress distribution and soil deformation under a tractor wheel. J. Terramech., 8: 9-22.

Petelkau, H., 1984. Auswirkungen von Schadverdichtungen auf Bodeneigenschaften und Pflanzenertrag sowie Massnahmen zu ihrer Minderung. (Effects of harmful compaction on soil properties and crop yield and measures for its reduction). Tag.-Ber., Akad. Landwirtsch.-Wiss. DDR, Berlin, Germany, 227: 25-34 (in German, with English summary).

Petelkau, H., 1986. Grenzparameter für die Bodenbelastung beim Einsatz von Traktoren und Landmaschinen aus der Sicht der Bodenfruchtbarkeit. (Limits to the ground pressure of tractors and farm machinery from the point of view of soil fertility). Tag.-Ber., Akad. Landwirtsch.-Wiss. DDR, Berlin, Germany, 250: 25-36 (in German, with English summary).

Petelkau, H., 1987. Durch Fahrwerke landwirtschaftlicher Mechanisierungsmittel verursachte Schadwirkungen und Vorschläge zu ihrer Verminderung. (Soil deterioration caused by running gear of farm machinery and proposals for its reduction). Kongr.-Tag. Ber., Martin-Luther-Univ., Halle-Wittenberg, Germany, Wissensch. Beiträge 1987/11 (S 59), pp. 106-119 (in German).

Petelkau, H. and Dannowski, M., 1990. Effect of repeated vehicle traffic in traffic lanes on soil physical properties, nutrient uptake and yield of oats. Soil Tillage Res., 15: 217-225.

Petelkau, H., Bosse, O. and Marschler, R., 1978. Einige Ergebnisse der ackerbaulichen Erprobung des Aufsattel-Beetpfluges B 550 und des B 550 in Kombination mit dem Saatbettbereitungsgerät B 601. (Some results of field tests of the mounted plough B 550, and of the B 550 in combination with the seedbed preparation implement B 601). Agrartech., 28: 246-248 (in German).

Petelkau, H., Gätke, C.R., Dannowski, M., Seidel, K. and Augustin, J., 1988. Bodenphysikalische Grundlagen für die Steuerung der Grundbodenbearbeitung. (Soil physical fundamentals of primary tillage control). Tag.-Ber. "Erhöhung der Bodenfruchtbarkeit und der Erträge durch wissenschaftlichen Fortschritt", Forschungszentrum für Bodenfruchtbarkeit, Müncheberg, Germany, pp. 362-378 (in German).

Rusanov, V.A., 1987. Trebovanija k technikie. (Requirements on the agricultural machinery). Zemledelie, 9: 20-23 (in Russian).

Rusanov, V.A., 1988. Osnovnyje polozjenija, ispolzovannyje pri razrabotkie GOST ov po normam i metodam otsenki vozdiejstvija dvizjitelej na pochvu. (Basic studies for the development of "Standards and methods for the evaluation of influences of running gear on soil" [GOST 26955-86, 26953-86 and 26954-86]). Sbornik nauchnykh trudov VIM, 118: 6-45 (in Russian).

Schjønning, P. and Rasmussen, K.J., 1994. Danish experiments on subsoil compaction by vehicles with high axle load. Soil Tillage Res., 29: 215-227.

Smith, D.L.O., 1985. Compaction by wheels: a numerical model for agricultural soils. J. Soil Sci., 36: 621-632.

Smith, D.L.O. and Dickson, J.W., 1990. Contribution of vehicle weight and ground pressure to soil compaction. J. Agric. Eng. Res., 46: 13-29.

Soane, G.C., Godwin, R.J. and Spoor, G., 1986. Influence of deep loosening techniques and subsequent wheel traffic on soil structure. Soil Tillage Res., 8: 231-237.

Söhne, W., 1953. Druckverteilung im Boden und Bodenverformung unter Schlepperreifen. (Pressure distribution in the soil and soil deformation under tractor tyres). Grundl. Landtech., 5: 49-63 (in German).

Söhne, W., 1958. Fundamentals of pressure distribution and soil compaction under tractor tires. Agric. Eng., 39: 276-281, 290.

Taylor, J.H. and Burt, E.C., 1987. Total axle load effects on soil compaction. J. Terramech., 24: 179-186.

Thorud, D.B. and Frissell, S.S., 1976. Time changes in soil density following compaction under an oak forest. Univ. Minnesota, St. Paul, MN, U.S.A., Minnesota For. Res. Note 257, 4 pp.

Unger, H., Werner, D., Pittelkow, U. and Reich, J., 1988. Bodenkundlich-pflanzenbauliche Grundlagen einer Verfahrenslösung zur Melioration von Strukturschäden in der Krumenbasis von Lö- und V-Standorten. (Soil science and agronomic fundamentals of structure amelioration in the upper part of the subsoil of loess and degraded profiles). Tag.-Ber. "Erhöhung der Bodenfruchtbarkeit und der Erträge durch wissenschaftlichen Fortschritt", Forschungszentrum für Bodenfruchtbarkeit, Müncheberg, Germany, pp. 305-314 (in German).

Van den Akker, J.J.H., Arts, W.B.M., Koolen, A.J. and Stuiver, H.J., 1994. Comparison of stresses, compactions and increase of penetration resistances caused by a low ground pressure tyre and a normal tyre. Soil Tillage Res., 29: 125-134.

Van Ouwerkerk, C., 1968. Two model experiments on the durability of subsoil compaction. Neth. J. Agric. Sci., 16: 204-210.

Voorhees, W.B., Nelson, W.W. and Randall, G.W., 1986. Extent and persistence of subsoil compaction caused by heavy axle loads. Soil Sci. Soc. Am. J., 50: 428-433.

Werner, D., 1989. Zielstellung und Ergebnisse mechanischer Eingriffe in die Bodenstruktur. In: Bodenstrukturforschung, Krumenbasisbearbeitung auf D-, Lö- und V-Standorten. (Goals and results of mechanical soil structure amelioration In: Soil Structure Research, Tillage of the Upper Part of the Subsoil of Diluvial, Loess and Degraded Profiles). Forschungszentrum für Bodenfruchtbarkeit, Müncheberg, Germany, pp. 14-20 (in German).

Soil Compaction in Crop Production
B.D. Soane and C. van Ouwerkerk (Eds.)
501

CHAPTER 21

Benefits of Tracked Vehicles in Crop Production

D.C. ERBACH

USDA-ARS, National Soil Tilth Laboratory, Ames, IA, U.S.A.

SUMMARY

There is concern about compaction of soils by wheeled agricultural equipment, and especially by equipment with high axle loads. Tracked vehicles have been and continue to be used to some extent in crop production. Development of rubber-belt tracks has removed many of the disadvantages of steel tracks relative to rubber tires. This has caused renewed interest in tracked equipment. Loading of the soil by trafficking with wheeled or tracked equipment is a complex dynamic process that is not well understood. Evaluation of the stresses applied to the soil by traffic with wheels and tracks and of the resulting soil compaction, shows that tracked vehicles can be used to reduce compaction, which tends to improve crop emergence, growth and yield. Effective management of field traffic with equipment used in crop production requires an improved understanding of the mechanics of soil loading by wheels and tracks.

INTRODUCTION

During the early 1950s the average mass of a wheeled vehicle was around 3 Mg, but by the 1980s the mass of 4-wheel drive tractors ranged from 10 to 20 Mg. Since the 1960s, mass and power of tractors have increased by 60-80%, while the area of the footprint of the running gear has increased by only 20%. Studies indicate that the heavier the load on the soil, the more detrimental the effect of traffic on crop yields. To minimize negative effects of traffic-induced soil compaction, the former U.S.S.R. established a State Standard, which dictated that, when the soil water content is >90% of the water content at field capacity, ground contact pressure exerted by wheel- or track-type field equipment should not exceed 80 kPa (Rusanov, 1991).

Söhne (1958) used a modified form of the Boussinesq equation to predict pressure transmission into soil beneath tires. He found that, for a constant ground contact pressure on a given soil, an increase in tire load increases the depth to which a given stress occurs. He also predicted that, for a constant tire load, an increase in inflation pressure increases stress at a given depth. From these predictions, he concluded that stress near the soil surface is a function mainly of ground contact pressure and that subsoil stress is a function primarily of total load. The degree to which soil compacts is influenced by soil properties and applied load. Relevant factors include soil water content, vehicle parameters and mode of vehicle movement, i.e., whether travelling in a straight line or turning (Braunack, 1986b).

The ultimate objective in designing tractive devices for agricultural equipment is to provide the required tractive effort with minimum detrimental effects on soil structure (Reaves and Cooper, 1960). Tracked vehicles have been investigated as a means of reducing soil compaction. They also have been developed to improve trafficability, decrease chances of bogging down and increase tractive efficiency (Gray, 1975; Baldwin, 1978; Quick, 1990). Tracked equipment operates more effectively than wheeled vehicles over rough terrain. Trafficability over wet and sandy soils is improved by distribution of the mass over a greater area.

During the past 70 years, there have been considerable changes in the proportion of tracked tractors in use in the U.S.A. For the period 1925-1966, annual tracked tractor sales were normally between 6 and 10% of all tractors produced. The only period in which tracked tractor sales accounted for a larger percentage was 1942-1945 (during World War II) and 1956. The maximum, 25%, occurred in 1943. The number of track-laying tractors shipped by manufacturers peaked at 43,000 in 1944. However, by 1967, according to U.S.A. census statistics, tracked tractors accounted for only 3.8% of all agricultural tractors in the U.S.A. In the U.K., the use of track-type tractors is even lower. Soane et al. (1981) reported that track-type tractors in Britain account for only approximately 1% of agricultural tractors sold.

In some other countries, particularly in southern and eastern Europe, the proportion of track-type tractors in use has been much greater. In the 1960s, in Italy and the former U.S.S.R., 24 and 43%, respectively, of the agricultural tractors were equipped with tracks (Soane et al., 1981). Although the number of tracked tractors is generally declining, the numbers in use in eastern Europe remain higher than in western Europe. In Hungary, about 7% of all tractors are 52-kW tracked vehicles, while in Germany, France, Spain, the U.K. and Italy the overall proportion of tracked vehicles in total tractor sales amounts to about 4%. In former Yugoslavia, steel-tracked tractors account for about 1.5% of total tractors. Increased mobility and draft power are reported to be the primary reasons, rather than reduced compaction, for using tracked tractors.

While the world-wide use of tracked tractors in crop production is likely to remain of limited importance, there is continued interest in their potential benefits, as well as ongoing research into their development. Current use of tracked tractors is generally confined to certain areas and special purposes, for instance, in sugar cane production, rice harvesting or seedbed preparation.

Tracked vehicles have obvious disadvantages, but they also have many advantages. Traditionally, tracked equipment has used steel tracks. Based upon equipment designed primarily for industrial work at low speeds, agricultural steel-tracked tractors are characterized by low speed, but also by high drawbar pull (Brixius and Zoz, 1976). Operating speed of steel-tracked vehicles is limited by power loss, noise, vibration, and wear. Many of the disadvantages of steel tracks have been overcome by recent developments in rubber tracks.

TRACKS USED IN CROP PRODUCTION

A track drive system generally includes a track, a drive wheel, an idler wheel and several midwheels. The track is either a chain of rigid links, or a flexible, solid or pneumatic belt. The track is driven by the drive wheel, usually at the rear of the track, passes around an idler wheel, usually at the front of the track, and then passes under the midwheels located between the driver and idler wheels. All wheels transmitting force to the soil through the track are called roadwheels. In most agricultural applications, the driver, the idler and the midwheels are roadwheels.

Track-type drive systems are used on tractors and other self-propelled agricultural equipment, such as applicators of chemicals and harvesting machines. Mobility is the major reason that tracks are used on harvesters. The need to harvest crops despite wet conditions has prompted development and use of track-type traction and flotation aids such as half-track attachments for tractors and combine harvesters (Fig. 1) and track-type running gear for transport equipment. These modifications are regularly used for rice harvest and for harvest of small grains and row crops when weather conditions require. Track-type devices may reduce rutting of soil, and thus less tillage is needed to prepare the soil for the next crop. Fig. 2 shows the relative difference in surface deformation caused by

Fig. 1. Eight-row maize harvester, with a mass of 14.0 Mg when the grain tank is empty, fitted with steel half-tracks for mobility in wet soil.

Fig. 2. Relative surface deformation caused by a large, track-type tractor with a mass of 12.9 Mg and a ground contact pressure of 20 kPa (horizontal path), and a much lighter four-wheeled vehicle with a mass of 1.9 Mg and a ground contact pressure of 220 kPa (diagonal path).

vehicles with different ground contact pressures. The path of a track-type tractor with a mass of 12.9 Mg and a ground contact pressure of 20 kPa, crosses the picture horizontally. This path is intersected by that of a wheel-type vehicle with a mass of 2.3 Mg and a ground contact pressure of 240 kPa. The surface deformation, or rut depth, was much greater for the vehicle with the greater ground contact pressure even though it was much lighter. Presumably, less surface deformation relates to less soil compaction.

Fig. 3. Tractor with a mass of 12.9 Mg and extended steel tracks (1.02 m wide and 3.5 m long) with a ground contact pressure of 20 kPa.

Steel tracks

Traditionally, tracks have been made of steel (Figs. 3 and 4). To improve tractive efficiency and flotation, many grouser designs have been developed. Sometimes, track pads are used to improve on-road mobility by reducing damage to road surfaces and to damp vibration. Tracks may be classified as rigid or flexible. Rigid tracks have links constructed so as to prevent upward flexing of the track between roadwheels. Flexible tracks allow some upward flexing, the magnitude of which depends upon track tension. Rudiger and Kohler (1987) reached no definitive conclusions regarding the proportion of total load sup-

Fig. 4. Steel-tracked tractors used in the People's Republic of China generally have a power of 45-60 kW and a mass of 5-7 Mg. Top: track design; bottom: tractor pulling a moldboard plow.

ported by sections of track between wheels of track-laying vehicles. However, they found that a tensioned track tended to carry more of the load on loose soils than on medium compact soil. Rigid tracks apply a more uniform load to the soil surface than flexible tracks. Peak stresses under the roadwheels of flexible tracks may be several times the stress between wheels. The condition of the surface on which the track is operating greatly influences the surface stress distribution. Peak stresses tend to increase with soil firmness.

Belted tracks

Evans and Gove (1986) reported the development of a rubber-belt track. The belt is made of wire-reinforced rubber, which is constructed much like a tire. To improve traction, grousers are molded to the outer surface of the belt, and lugs are molded to the inner surface to serve as guides and to prevent lateral movement of the track relative to driver and idler. The roadwheels for the rubber-belt track may be solid or pneumatic. This track allows on-road mobility similar to that of rubber tires. However, bituminous road surfaces may be damaged when a rubber-belt tracked machine is steered. Tractors with rubber-belt tracks (Fig. 5) for agricultural use are being marketed commercially in the U.S.A., the U.K. and elsewhere; to date, sales worldwide amount to about 2000 machines. There is considerable interest in the use of rubber-belt tracked tractors on the heavy soils in Europe. Rubber-tracked systems are under development for other agricultural equipment. Combine harvesters (Fig. 6), grain wagons (Fig. 7) and other types of tracked machinery fitted with rubber-belt equipment, are commercially available in limited quantities.

Fig. 5. Tractor with a mass of 13.6 Mg and rubber-belt tracks (0.6 m wide and 2.6 m long) with a ground contact pressure of 40 kPa.

Fig. 6. Combine harvester for maize, with a mass of 12.6 Mg and rubber-belt tracks with a ground contact pressure of 50 kPa.

Pneumatic tracks

During the late 1960s, a pneumatic track, i.e., a cross between the pneumatic tire and the steel track, was developed in Italy. The pneumatic track is long and oval like a conventional track but is made of rubber and has an inner circumference which is reinforced with a belt of nylon. Taylor and Burt (1975)

Fig. 7. Grain wagon equipped with rubber-belt tracks. When filled with grain, the wagon has a mass of 23 Mg and a ground contact pressure of 65 kPa.

reported that pneumatic tracks had the potential to increase productivity by allowing increased speed and on-road mobility. There has, however, been little commercial use of this concept.

BENEFITS OF USING TRACKS

Tractors with tracks have most commonly been used under conditions requiring high drawbar pull over extended periods of time (Brixius and Zoz, 1976). Advantages of tracks include: (1) higher tractive coefficient, which requires lower vehicle mass; (2) higher tractive efficiency, which requires smaller engines and drivelines; (3) a long, narrow track contact patch of a larger area, which results in lower ground contact pressure (Janzen et al., 1985).

Tractive efficiency

Pull/mass ratios are greater for tracked tractors than for wheeled tractors, mainly because the ground contact area for tracks is greater than for wheels. Taylor and Burt (1975) compared performances of a steel track, a pneumatic track and a pneumatic tire under controlled conditions, and concluded that steel and pneumatic tracks exceeded the tire in traction performance as measured by the pull/mass ratio. Tracked tractors have a somewhat greater mass/axle power ratio at their design travel speed than wheeled tractors (Brixius and Zoz, 1976).

Osborne (1971) and Domier et al. (1971) found that a tracked vehicle had a higher tractive efficiency than either two-wheel drive or four-wheel drive tractors. Culshaw and Dawson (1987) compared the performance of a rubber track with that of a radial tire. The coefficient of traction at maximum efficiency was always greater for the track than for the wheel. A significant advantage in effective power was shown for a tractor with a rubber-belt track, compared with a four-wheel drive tractor, when both were operated on a concrete surface (Furleigh and Leviticus, 1990). Measured drawbar power under both untilled and tilled soil conditions was also greater for the track-type tractor.

Bekker (1958) proposed that, because of differences in soil shear pattern for closely and widely spaced grousers, the pull/mass ratio of a track-type tractor could be improved by increasing the space between track grousers. With closely spaced grousers, a shallow volume of soil was sheared and with widely spaced grousers the volume of disturbed soil extended deeper and farther from the grouser. No discussion of the impact on soil compaction was given.

Comparing a two-wheel drive 67-kW tractor with a 52-kW tracked tractor, Osborne (1971) found that when each was appropriately loaded, the tracked tractor had a better rate of work than the wheeled tractor. Because the tracked tractor operated at lower speeds, a wider implement was necessary, especially in coarser textured soils, to provide an adequate load to enable high power to be developed. Ground conditions had less effect on performance of tracked tractors than of wheeled tractors. Osborne (1971) reported that the rolling resistance of a tracked tractor was negligible. Brixius and Zoz (1976) found that, for tractors of similar mass, a four-wheel drive tractor had a 22% greater productivity than a tracked tractor. The pull/mass ratio and the tractive efficiency of wheeled tractors are improved by all-wheel drive. Osborne (1971) concluded that the greater coefficient of traction for four-wheel drive tractors was due to compression of soil by the front wheels, which created better surface conditions for traction for the rear wheels.

From a theoretical analysis, Guskov (1969) concluded that: (1) an increase in forward speed of a tracked tractor leads to a higher tractive effort at constant slip, or to less slip at the same tractive effort; (2) the optimal range of working speeds for tractors with drawbar pulls of 25-35 kN is 7-8 km h^{-1} on mineral soils and 6-7 km h^{-1} on peat soils; (3) the optimal range of working speeds could be increased by improving the ground drive of the tractor and reducing the mechanical losses in the transmission and track assembly. A friction-drive rubber track produced about 25% more pull than a radial tractor tire (Culshaw, 1988).

Trafficability

Garber and Wong (1981a) used a terrain stiffness factor, a track suspension stiffness factor, and a track tension-spring stiffness factor to estimate differences of pressure distribution under tracks. This system provided useful information about the potential mobility of new vehicle designs, related to sinkage and motion resistance.

Yong et al. (1976) designed a passive grouser to reduce environmental damage to the soil surface. The grouser initially contacts the soil surface in a straight bearing posture. At low levels of slip, drawbar pull with the passive grouser was only slightly less than that with more aggressive grousers. As slip increased, the drawbar pull of the aggressive grousers increased considerably relative to that of the passive grouser, but the sinkage and damage to the soil surface also increased considerably.

Culshaw and Dawson (1987) reported that the rolling resistance of a rubber track was greater than that for a tire, for all but very soft surfaces. They also found that on hard surfaces the contact pressure under the track had a very uneven distribution. For machines of similar mass, Culshaw (1988) found that a vehicle with rubber tracks produced twice the pull of a wheeled tractor. Tractive efficiencies of the two machines were similar but the tracked vehicle caused less rutting of a soft soil. In an evaluation of traction aids, Southwell (1964) concluded that half-tracks were an effective means of increasing traction. However, they have the disadvantage of being difficult to remove and, if not removed, they cause a significant increase in rolling resistance when the tractor is lightly loaded and operated at high speed.

Reece and Adams (1966) concluded that for vehicles with a ground contact pressure which approaches the bearing capacity of the soil, a small amount of slip can increase sinkage at the rear of the contact area. They stated that this increase can be minimized by increasing the contact area and extending the contact over a long narrow strip.

An advantage of reduced soil compaction is reduced draft of tools used in subsequent tillage operations. Rusanov (1991) reported increased draft for moldboard plowing of up to 65% when soil was compacted by tractors, and up to 90% when soil was severely compacted by traffic with trucks and harvesting machines.

Soil compaction

Many studies indicate that tracked vehicles compact soil less than wheeled vehicles (Reaves and Cooper, 1960; Soane, 1973; Taylor and Burt, 1975; Janzen et al., 1985; Bashford et al., 1988; Erbach et al., 1988; Rusanov, 1991). However, other studies report little difference between tracked and wheeled vehicles with respect to their compaction effects (Dilts and Clark, 1975; Brixius and Zoz, 1976; Burger et al., 1985). Burger et al. (1985) found that, despite a three-fold difference in ground contact pressure, changes in soil density and porosity caused by an unloaded, rubber-tired log skidder were not greater than those caused by a steel-tracked crawler.

Taylor and Burt (1975) observed that the long, narrow footprint of a track (compared to the large, oval footprint of a tire) disturbs and compacts a smaller proportion of the soil surface. Soil bulk density after passage of the tractive devices was greater for the tire than for either a steel or pneumatic track.

Chancellor (1976) stated that, for tire and crawler tractors of equal mass, the ground contact pressure exerted on the soil surface by the tire will be more than twice as great as that exerted by the track. Not only do wheels compact the soil deeper, but the pressure exerted is not entirely vertical. Some of the pressure is absorbed laterally and tangentially as it is transmitted deeper into the soil.

In a study by Brown et al. (1992), wheeled tractors with ground contact pressures of about 125 kPa created a more compacted condition in the topsoil than tracked vehicles with ground contact pressures of 30-40 kPa. However, below a depth of 125 mm, differences in soil compaction were minimal. Soil trafficked by both tractive devices had a lower air-filled pore space and hydraulic conductivity than non-trafficked soil. Braunack (1986a) reported that the passage of tracks resulted in increased soil bulk density and cone penetration resistance and in decreased saturated hydraulic conductivity. He concluded that soil structure damaged by trafficking with tracked vehicles could be partially regenerated by wetting and drying cycles.

The ground contact pressure exerted by a vehicle is one of the most important performance parameters affecting soil compaction. The rolling resistance, the tractive properties of a tractive device, and the soil deformation beneath the device, are all related to the distribution of the pressure over the supporting area of a wheel or track. Because actual ground contact pressures are difficult to measure or calculate, the average ground contact pressure is usually used. P_{av}, the average pressure under a track, is commonly calculated with the following equation (Sofiyan and Maximenko, 1965):

$$P_{av} = m/2bl \qquad (1)$$

where m = tractor mass; b = track width; l = contact length of the track, commonly measured as the distance from the axis of the frontmost roadwheel to the axis of the rearmost roadwheel.

An indication that this method of determining ground contact pressure may be inadequate, is shown by work of Reed (1958) and of Reaves and Cooper (1960). They measured stress levels in soil below tracks that were approximately twice the average ground contact pressure. Also, Baganz and Kunath (1963) found the changes in bulk density after the passage of a tracked vehicle to be approximately twice as great as was to be expected on the basis of the estimated ground contact pressures. Vibrations transmitted through the rigid track were thought to account for the differences. Bulk density of soil trafficked with a rubber-belt track tractor was less than when trafficked with a rubber tire of a four-wheel drive tractor (Bashford et al., 1988). This was especially true for the subsoil.

The intensity and depth of compaction caused by equipment traffic are affected by soil conditions. Gassman et al. (1989) used a field experiment and finite element analysis to examine the depth of compaction. For a theoretical soil profile, consisting of layers with densities similar to those measured in the field, differences in compactness were not significant among soils subjected to simulated pressures of four tractors (both tracked and wheeled). The model used, demonstrated that strain from an applied load is distributed more evenly and extends to a greater depth for a uniform soil profile than for a layered soil. They concluded that more accurate simulation of vehicle loading is required to accurately predict compaction due to vehicle loading.

Reed (1940) concluded that rubber tires tend to compact soil more than either wheels with steel lugs or tracks. Reaves and Cooper (1960) found that, for both tires and tracks, maximum stresses occurred at a depth of 76 mm beneath the center of the tractive device. Stress levels decreased from this point both laterally and vertically. It was reported that for the conditions evaluated, a 300-mm wide track compacted the soil less than a 330-mm wide tire. The rubber tire caused stresses in the soil that were generally twice those caused by the track. Taylor and Burt (1975) found that, for a given normal load, soil stress and soil bulk density at a given depth are less for loads applied with a track than with a tire. Similarly, Erbach et al. (1988) reported that track-type tractors tended to affect soil bulk density, penetration resistance, aggregate size and also maize growth, less than wheel-type tractors. By using a soil strain gauge (Erbach et al., 1991), Kinney et al. (1992) measured significantly greater soil strain beneath a tractor with single rear wheels than beneath a tractor of equal mass but equipped with steel tracks. The effects of individual ground wheels on soil strain were apparent for depths of <200 mm. However, Culshaw and Dawson (1987) evaluated traffic effects for four soil conditions and they reported that soil bulk densities before and after passage of a rubber track and a radial tire were not significantly different.

The action of a tractive device can cause changes in soil properties other than compaction. Braunack (1986b) noted that soil subjected to tracked-vehicle traffic, when compared with non-trafficked soil, showed a decrease in topsoil strength even though bulk density increased and saturated hydraulic conductivity decreased.

Compaction under conventional pneumatic tires is related to load, ground contact pressure, wheel slip, tire dimensions, carcass construction, inflation pressure, forward speed and number of passes (Soane et al., 1981). Many of these factors also affect soil compaction by pneumatic track-layers. To reduce the incidence of compaction, Soane et al. (1981) suggested that it was desirable to lower the ground contact pressure of tires on field soils to <200 kPa and preferably to <100 kPa. However, with load-carrying vehicles, this can result in a considerable increase in the cost of the running gear. Vehicles with conventional wheeled systems with a mass >12 Mg, are likely to cause appreciable compaction below the depth of normal cultivation, regardless of the ground contact pressure.

Fekete (1972) reported that maximum compaction caused by a track, occurred at a depth equal to approximately half the width of the track. For common agricultural track-layers this would be at a depth of about 200 mm. Soane (1973) investigated changes of strength and packing state occurring under a 505-mm wide track with a ground contact pressure of 35 kPa. Changes in packing state and cone resistance were detectable to a depth of about 200 mm below the original soil surface although the maximum intensity of compaction was much less than those under conventional rubber front and rear tires.

Evans and Gove (1986) concluded that a rubber belt track can reduce soil compaction because less machine weight is required for a given drawbar pull, less area is disturbed because of the long, narrow contact patch, and lower vertical pressures are possible because of the large ground contact area.

Traffic with track-type and wheel-type tractors significantly reduced porosity in the surface 200 mm of soil (Brown et al., 1992). The greatest effect was found for pores >60 μm in diameter. Reduction in volume of large pores was greater for wheel-type than for track-type tractors, but the difference was not statistically significant.

Crop response

Evidence for crop responses to compaction caused by wheeled tractors and other agricultural equipment is reviewed in Chapters 11-16. Complex relationships between crop responses and both high and low levels of compaction often give conflicting results, which are related to weather and other variables. Because crops do not respond directly to soil density, there is no clear relationship between the level of soil compaction and crop growth. However, crops do respond to the extent to which their needs for heat, water, nutrients, oxygen and light are satisfied. Therefore, crop response to compaction will occur only when the compaction restricts the supply of an input needed by the plant. Reductions in crop emergence and growth are often associated with vehicle traffic for tillage, fertilizer and pesticide application and harvest. However, when soil water content is limiting and seed/soil contact is poor, more rapid and better emergence of crops (and weeds) is sometimes observed in the wheel tracks. This is associated with the concept of an optimum level of compaction, which is related to soil and weather conditions (see Chapter 12). Where track-type tractors are used, over-compaction is likely to occur less frequently but, under conditions of high soil water content following planting, reduced crop emergence is occasionally observed in track positions (Fig. 8).

Erbach et al. (1988) found that track-type tractors, when used to pull a field cultivator for seedbed preparation, had less effect on soil conditions (Table 1) and on maize yield (Table 2) than wheel-type tractors. In a field experiment conducted on a poorly drained and moderately slowly permeable silt-loam soil that responded to soil compaction, maize emerged more slowly, grew more slowly, and yielded less in tractor paths than in non-trafficked areas. The decrease

Fig. 8. Evidence of lower soil permeability to water and reduced emergence of barley which had been drilled using a 48-kW track-type tractor with a mass of 5.6 Mg and a ground contact pressure of 31 kPa.

TABLE 1

Effects of tractor traffic and type of tractive device on soil condition in the top 0.3-m layer of soil after maize emergence (after Erbach et al., 1988)

	Bulk density (Mg m^{-3})	Cone index (kPa)	Soil water content (m^3 m^{-3})
Location			
Tractor path	1.42	756	0.299
Non-trafficked area	1.34	600	0.281
Tractive device			
Track	1.40	754	0.297
Wheel	1.43	762	0.302
LSD (P$_{0.05}$)	0.02	35	0.003

in growth and yield tended to be greater as ground contact pressure increased. Bulk density to a depth of 30 cm was greater in the tractor path than in the non-trafficked soil and tended to be greater for the wheel-type tractors than for the track-type tractors. In the upper 7.5 cm of the soil profile, larger aggregate sizes were found in the tractor path than in non-trafficked areas. Increased numbers of large soil aggregates in the tractor path were also reported by Rusanov (1991).

TABLE 2

Effects of tractor traffic for seedbed preparation tillage and of type of tractive device on maize growth (after Erbach et al., 1988)

	Emergence rate index (% day^{-1})	Plant population (plants ha^{-1})		Plant height (m)	Yield (Mg ha^{-1})	Grain water content (%, w/w)[1]
		Emerged	Harvest			
Location						
Tractor path	9.5	56000	57400	2.02	9.0	20.9
Non-trafficked	10.5	58100	58000	2.14	10.4	20.6
Tractive device						
Track	9.8	58200	57600	2.07	9.3	20.8
Wheel	9.2	54800	57100	1.98	8.7	21.0
LSD (P$_{0.05}$)	0.3	1200	n.s.	0.03	0.3	0.2

[1]Wet basis.

He found that the weight of soil aggregates >50 mm in diameter increased by 4% in the path of a track-type tractor and by 31% in the path of a wheel-type tractor. Large soil aggregate size may have delayed plant emergence and contributed to slow growth and low yields. This experiment, however, cannot differentiate soil bulk density effects deeper in the soil profile from the effects of aggregate size near the soil surface.

From the results of numerous experiments in the former U.S.S.R., Rusanov (1991) concluded that crop yield response depends mainly on the magnitude of pressure exerted on the soil by wheel-type or track-type equipment and on the number of passes of the wheel or track along the same path. Crop yield response to soil compaction treatments differed among years. In a year when wheat was sown into relatively dry soil, yield was greatest from the treatment which induced the greatest soil bulk density. In all other years, yields, compared with that of the control plot, decreased with increased soil compaction. The benefit of using track-type equipment was associated with the lower ground contact pressure applied by that equipment. When soil was compacted with track-type equipment with 150 kPa ground contact pressure, the yield reduction was approximately half the reduction that occurred when soil was compacted with 200 kPa ground contact pressure. Yield reductions of grain and forage crops were as large as 25% when wheeled tractors with 200 kPa ground contact pressure were used and 11% with track-type tractors exerting 160 kPa, as compared with yields when wheeled tractors with 80 kPa ground contact pressure were used.

PROBLEMS WITH THE USE OF TRACKS

Because of certain limitations, tracked vehicles virtually disappeared from use in U.S.A. agriculture between 1960 and 1990. The two major limiting factors were maintenance costs and restricted movement on hard-surfaced roads.

Operator comfort

Slow speed and rough ride limit the agricultural use of tracked tractors. Traditionally, tracked tractors are noisy and less comfortable to drive than wheeled tractors and therefore operator fatigue is greater. This results in reduced field efficiency and lowered productivity. However, there is no inherent reason that a tracked tractor cannot be designed to provide as much operator comfort as a wheeled tractor.

Use on roads

Under present conditions, many U.S.A. farmers run operations that are several kilometers apart and depend upon the same tractors and equipment for work in all fields. Steel-tracked tractors are not adapted for, and often are not allowed to, travel on hard-surfaced roads. Therefore, in the U.S.A., wheeled tractors, with higher speeds and good on-road mobility, have replaced steel-tracked tractors. Pneumatic and rubber-belt tracks do not have the limitations of steel tracks with respect to on-road travel.

Turning

Slower moving tracked tractors may require more time to turn than do wheeled tractors and therefore field efficiency is reduced. When tracked vehicles are turned in a short radius on soft soil they cause berming, or mounding, of the soil. This is usually undesirable for agricultural purposes. Use of articulation to improve steering characteristics of tracked vehicles has been tried but has not received commercial acceptance. Recently, a "concept" model has been built in which each wheel of an articulated four-wheel drive tractor was replaced by a rubber-belt track assembly. This tractor is expected to be suitable for row crop use as well as for tillage.

Use in row crops

Wheel-type tractors have greater ground clearance and the spacing between the wheels can usually be adjusted on the axle, a feature not presently available on commercial track-type tractors. These factors limit the suitability of tracked tractors for row-crop use. In particular, damage caused to crops when turning track-type tractors at the end of fields may be excessive.

Proper ballasting

As drawbar pull on a tracked vehicle increases, weight is shifted toward the rear of the tractor. This weight-shift causes an increased ground contact pressure (Brixius and Zoz, 1976) relative to that calculated from the track contact area based upon the track width and length. Therefore, if the benefits of reduced ground contact pressure of a tracked tractor are to be obtained, it is important that tractors be properly ballasted and the load properly hitched.

Expense

In the U.S.A., between 1971 and 1975, the sale of tracked vehicles declined and the sale of four-wheel drive tractors tripled, even though track-layers were shown to give the highest tractive efficiency (Brixius and Zoz, 1976). Disadvantages, such as operator discomfort, low road mobility and high cost were main reasons for the sales decline.

Tracked tractors tend to have a high initial cost per unit of drawbar-pull, and the cost to maintain steel tracks is high. For instance, in a study examining tires and tracks in agriculture, Brixius and Zoz (1976) found that the owning and operating costs for a four-wheel drive tractor were 33% less than for a track-type tractor. They concluded that results of compaction studies and cost/benefit analyses support the use of four-wheel drive tractors rather than steel track-layers.

ADVANTAGES AND DISADVANTAGES OF TRACKED VEHICLES

Do tracked vehicles compact soil less than do wheeled vehicles? As yet, not all variables affecting soil compaction in current agricultural practices are fully understood. When comparing a tracked vehicle with a wheeled vehicle, the different designs, ground contact pressures and mechanisms influence the way in which the soil is affected.

The actual stress applied to the soil by tires and tracks is not well documented. Commonly, inflation pressure is used as the estimated ground contact pressure of pneumatic tires. The ground contact pressure of tracks is usually assumed to be the tractor mass divided by the area of the tracks in contact with the ground. The contact area is assumed to be the width of the tracks times the length between the axles of the drive wheel and the idler wheel. Garber and Wong (1981a,b) developed a method of terrain-vehicle analysis to improve estimates of pressure distribution under tracks as affected by soil condition and by the design of the track-suspension system. The track, suspension and track tensioning device were all considered. The roadwheel arrangement has a considerable effect on ground contact pressure distribution. Increasing the number of road wheels reduces maximum ground contact pressure and improves uniformity of pressure distribution. Increasing the stiffness of the suspension and increasing the track tension, reduces the levels of maximum pressure. The predicted pressure distribution beneath the track was most uniform in soft soil and very non-uniform in very firm terrain. Garber and Shwartzman (1984) analyzed performance factors, such as sinkage, of a tracked vehicle under uneven ground contact pressure conditions. They concluded that uneven ground contact pressure could affect slip-sinkage quite noticeably.

If tracked machines exert lower pressure on the soil surface than do wheeled machines, then tracked machines should compact soil less than do wheeled machines. But loading of the soil as a result of traffic with either a track or a wheel, is not well understood. Although the duration of loading with a track is greater than that with a wheel, with both devices loading time is quite short and applied stress is dynamic.

To solve the problem of soil compaction, experiments must be conducted on tracked vehicles and their mechanics, track and grouser types, and on new structural designs. For large farming operations in certain conditions, it may be economically beneficial to operate rubber-belt and pneumatic tracklayers. In the long run, their better on-road mobility and decreased compaction effects may offset their greater operating costs. However, more studies need to be undertaken on the influence of their turning characteristics on soil, and on their operating weight, inflation pressure of pneumatic tracks, and track size.

Although their disadvantages are obvious, scientific studies show numerous advantages of tracked vehicles over wheeled vehicles, including: (1) less compaction of the soil; (2) greater rate of plant emergence; (3) increased plant populations; (4) increased plant growth; (5) greater crop yield.

Pneumatic rubber tracks were developed to overcome problems with steel tracks. These tracks give on-road mobility while exceeding the rubber tire in tractive performance. However, there has been little commercial adoption of pneumatic tracks. Rubber-belt tracks also overcome many of the problems of steel tracks and are being commercially accepted. With the beneficial compaction qualities of the tracked vehicle, it seems that tracked vehicles can increase productivity and might become an important agricultural tool.

Taylor and Gill (1984), in describing management practices for controlling soil compaction, stated that "attempts to relate wheel traffic directly to crop yields often end in frustration". Yield is the result of a complex chain of events and, therefore, can only be indirectly correlated with traffic.

As yet, there is an incomplete understanding of the effects of track-type equipment on soil compaction and of the responses of crops to soil compaction. However, there is much research in progress and, because of the concern about the use of large equipment in crop production, there is considerable innovative work underway on the development of equipment for tracked tractors. This exciting research and development will lead to options and recommendations that farmers can use to be more productive while improving and sustaining their soil resource.

CONCLUSIONS

(1) Tracked vehicles apply lower ground contact pressure to the soil surface than conventional tires and, therefore, cause less soil compaction, resulting in improved crop growth. They also have benefits in tractive efficiency and trafficability.

(2) Traditionally, tracked vehicles have disadvantages associated with operator comfort, use on roads, turning in the field, use in row crops and cost. They have been used primarily for low-speed, high-draft applications.

(3) The disadvantages of tracked vehicles have been considered so important that currently only a small proportion of agricultural vehicles are equipped with tracks.

(4) The development of rubber tracks, along with improvements in operator comfort, has removed many of the disadvantages of steel tracks relative to rubber tires and has led to renewed research and development on tracked vehicles for crop production.

REFERENCES

Baganz, K. and Kunath, L., 1963. Einige Spannungs- und Verdichtungsmessungen unter Schlepperlaufwerken. (Some measurements of stress and compaction under tractor wheels and tracks). Dtsch. Agrartech., 13: 180-182 (in German).

Baldwin, N., 1978. An Old Motor Kaleidoscope of Farm Tractors. Old Motor Magazine, London, 96 pp.

Bashford, L.L., Jones, A.J. and Mielke, L.N., 1988. Comparison of bulk density beneath a belt track and tire. Appl. Eng. Agric., 4: 122-125.

Bekker, M.G., 1958. Performance improvement in track-type tractors. Agric. Eng., 39: 630-632.

Braunack, M.V., 1986a. The residual effects of tracked vehicles on soil surface properties. J. Terramech., 23: 37-50.

Braunack, M.V., 1986b. Changes in physical properties of two dry soils during tracked vehicle passage. J. Terramech., 23: 141-152.

Brixius, W.W. and Zoz, F.M., 1976. Tires and tracks in agriculture. Trans. ASAE, 85: 2034-2044.

Brown, H.J., Cruse, R.M., Erbach, D.C. and Melvin, S.W., 1992. Tractive device effects on soil physical properties. Soil Tillage Res., 22: 41-52.

Burger, J.A., Perumpral, J.V., Kreh, J.L., Torbert, J.L. and Minaei, S., 1985. Impact of tracked and rubber-tired tractors on a forest soil. Trans. ASAE, 28: 369-373.

Chancellor, W.J., 1976. Compaction of soil by agricultural equipment. Univ. California, Davis, CA, U.S.A., Div. Agric. Sci., Bull. 1881, 53 pp.

Culshaw, D., 1988. Rubber tracks for traction. J. Terramech., 25: 69-80.

Culshaw, D. and Dawson, J.R., 1987. The performance of a simple rubber track for an agricultural vehicle. Inst. Eng. Res., Silsoe, U.K., Div. Note DN 1382, 21 pp.

Dilts, R.A. and Clark, S.J., 1976. Depth effect of soil compaction from wide tires. Am. Soc. Agric. Eng., St. Joseph, MI, U.S.A., ASAE Pap. MC-76-103.

Domier, K.W., Friesen, O.H. and Townsend, J.S., 1971. Traction characteristics of two-wheel drive, four-wheel drive and crawler tractors. Trans. ASAE, 14: 520-522.

Erbach, D.C., Melvin, S.W. and Cruse, R.M., 1988. Effects of tractor tracks during secondary tillage on corn production. Am. Soc. Agric. Eng., St. Joseph, MI, U.S.A., ASAE Pap. 88-1614, 15 pp.

Erbach, D.C., Kinney, G.R., Wilcox, A.P. and Abo-Abda, A.E., 1991. Strain gage to measure soil compaction. Trans. ASAE, 34: 2345-2348.

Evans, W.C. and Gove, D.S., 1986. Rubber belt track in agriculture. Am. Soc. Agric. Eng., St. Joseph, MI, U.S.A., ASAE Pap. 86-1061, 14 pp.

Fekete, A., 1972. Some observations on the contact pressure of tyres. Proc. Int. Conf. Perspectives of Agricultural Development, Warsaw, Poland, Vol. 1, pp. 163-173.

Furleigh, D.D. and Leviticus, L.I., 1990. Relating rubber belt track and 4WD field performance using Nebraska test data. Am. Soc. Agric. Eng., St. Joseph, MI, U.S.A., ASAE Pap. 90-1591, 21 pp.

Garber, M. and Shwartzman, M., 1984. Agriculture tracked vehicle-soil interaction under uneven contact pressure conditions. J. Terramech., 21: 261-271.

Garber, M. and Wong, J.Y., 1981a. Prediction of ground pressure distribution under tracked vehicles. I: An analytical method for predicting ground pressure distribution. J. Terramech., 18: 1-23.

Garber, M. and Wong, J.Y., 1981b. Prediction of ground pressure distribution under tracked vehicles. II: Effects of design parameters of the track-suspension system on ground pressure distribution. J. Terramech., 18: 71-79.

Gassman, P.W., Erbach, D.C. and Melvin, S.W., 1989. Analysis of track and wheel soil compaction. Trans. ASAE, 32: 23-29.

Gray, R.B., 1975. The Agricultural Tractor: 1855-1950. Am. Soc. Agric. Eng., St. Joseph, MI, U.S.A., 154 pp.

Guskov, V.V., 1969. Effect of forward speed on the drawbar performance of a track-laying tractor. J. Agric. Eng. Res., 13: 203-209.

Janzen, D.C., Hefner, R.E. and Erbach, D.C., 1985. Soil and corn response to track and wheel compaction. Proc. Int. Conf. Soil Dynamics, Auburn, AL, U.S.A., Vol. 5, pp. 1023-1038.

Kinney, G.R., Erbach, D.C. and Bern, C.J., 1992. Soil strain under three tractor configurations. Trans. ASAE, 35: 1135-1139.

Osborne, L.E., 1971. A field comparison of the performance of two- and four-wheel drive and tracklaying tractors. J. Agric. Eng. Res., 16: 46-61.

Quick, G.R., 1990. Australian Tractors. Am. Soc. Agric. Eng., St. Joseph, MI, U.S.A., 167 pp.

Reaves, C.A. and Cooper, A.W., 1960. Stress distribution in soils under tractor loads. Agric. Eng., 41: 20-21, 31.

Reece, A.R. and Adams, J., 1966. One aspect of tracklayer performance. Trans. ASAE, 9: 6-9, 13.

Reed, I.F., 1940. A method of studying soil packing by tractors. Agric. Eng., 21: 281-282, 285.

Reed, I.F., 1958. Measurement of forces on track-type tractor shoes. Trans. ASAE, 1: 15-18.

Rudiger, A. and Kohler, U., 1987. Abschätzung des mittleren Bodendrucks unter Gleisbandfahrwerken. (Estimation of the average soil pressure below tracked vehicles). Agrartech., 37: 76-78 (in German).

Rusanov, V.A., 1991. Effects of wheel and track traffic on the soil and on crop growth and yield. Soil Tillage Res., 19: 131-143.

Soane, B.D., 1973. Techniques for measuring changes in the packing state and cone resistance of soil after the passage of wheels and tracks. J. Soil Sci., 24: 311-321.

Soane, B.D., Blackwell, P.S., Dickson, J.W. and Painter, D.J., 1981. Compaction by agricultural vehicles: A review. II. Compaction under tyres and other running gear. Soil Tillage Res., 1: 373-400.

Söhne, W., 1958. Fundamentals of pressure distribution and soil compaction under tractor tires. Agric. Eng., 39: 276-281, 290.

Sofiyan, A.P. and Maximenko, Y.I., 1965. The distribution of pressure under a tracklaying vehicle. J. Terramech., 2: 11-16.

Southwell, P.H., 1964. An investigation of traction and traction aids. Trans. ASAE, 7: 190-193.

Taylor, J.H. and Burt, E.C., 1975. Track and tire performance in agricultural soils. Trans. ASAE, 18: 3-6.

Taylor, J.H. and Gill, W.R., 1984. Soil compaction: State-of-the-art report. J. Terramech., 21: 195-213.

Yong, R.N., Fattah, E.A. and Youssef, A., 1976. Performance of a passive grouser-track system. Trans. ASAE, 85: 2045-2052.

Soil Compaction in Crop Production
B.D. Soane and C. van Ouwerkerk (Eds.)
©1994 Elsevier Science B.V. All rights reserved.

CHAPTER 22

Development and Benefits of Vehicle Gantries and Controlled-Traffic Systems

J.H. TAYLOR

USDA-ARS, National Soil Dynamics Laboratory, Auburn, AL, U.S.A.

SUMMARY

Controlled traffic is a concept which makes it possible to optimize soil physical conditions for both plants and machines by permanently zoning a field into traffic lanes and crop zones. The controlled-traffic concept has evolved from the original idea of reducing compaction in the crop zone into a complete soil compaction management system. While most of the early research was done with modified tractors, gantry vehicles are ideal for deriving the full benefits of the controlled-traffic concept. The development of gantry vehicles is traced from the 1850's to today's modern versions. An overview of international research is presented which confirms the wide scope of the problem and indicates a rather general agreement on methods of solution.

INTRODUCTION

Traffic-induced soil compaction is a problem that has developed rapidly since World War II wherever agriculture has become highly mechanized. A cycle of tillage and traffic developed that has caused serious problems in both energy consumption and soil condition. Subsoiling is an energy-intensive operation and, if a heavily loaded wheel follows it, soil compaction may be more severe than before subsoiling. Even when done in a properly designed cultural system, subsoiling is treating the symptoms and intensifying the disease. Most of the deep tillage undertaken is needed because of compaction caused by traffic from heavy machinery. It is imperative that more be done to prevent soil compaction and thereby reduce the need to reclaim highly compacted soil.

Soil conditions required for good crop growth are not generally conducive to good flotation or tractive efficiency of machinery, and soil conditions suited to good flotation and tractive efficiency are not conducive to good crop growth. A crop production system consisting of untilled traffic lanes and untrafficked crop

Fig. 1. Modified conventional equipment used for the early controlled-traffic research in Alabama, U.S.A.

zones has great potential for improving both the crop zone and the traffic lanes rather than continuing the frustrating compromise required by conventional crop production systems.

Controlled traffic is a concept; the gantry is a vehicle. The unique concept and purpose of controlled traffic probably originated in the 1950's. The development of wide-frame (gantry) vehicles started a century earlier. Halkett's (1858) description of permanent tracks for steam traction engines was probably an extension of the idea of railroads for steam locomotives. It certainly was intended to improve the flotation and maneuverability of those massive machines. Controlled traffic was initiated to stop wheel traffic compaction in the crop zone and it eventually expanded to a complete soil compaction management concept.

In this chapter, the controlled-traffic concept and the gantry vehicle will be discussed as separate entities. It is important to understand that gantry farming is not necessarily controlled-traffic farming. While the gantry would seem to be the ideal vehicle for implementing the controlled-traffic concept, it is not the only way. Modified tractors (Fig. 1) were used for many years before the first gantries were built for controlled traffic. Also, most gantries are probably still used as harvesting aids, and very few gantry farming systems would qualify as controlled traffic according to the definition of that concept.

CONTROLLED TRAFFIC

Taylor (1983) defined controlled traffic as a crop production system in which the crop zone and the traffic lanes are distinctly and permanently separated. This system establishes traffic lanes that are not deep tilled and are used for wheel paths year after year. The lanes become compacted, improving tractive efficiency, flotation and timeliness of operations, while the untrafficked crop zone, if initially well prepared, tends to stay that way without annual deep tillage. The crop zone of a controlled-traffic crop production system may be managed by conventional tillage or by conservation tillage methods. The machinery may be wide-frame (gantry) or conventional (cantilevered). Crops need moist soil with a low level of compaction; tires need highly compacted, dry soil. Controlled traffic is a concept that makes it possible to optimize soil conditions for each of these directly opposed requirements in the same field.

Soil compaction is inherently neither good nor bad; it is just one more factor that must be under management's control. More recent evolution of the controlled-traffic concept (Taylor, 1989) indicates that the crop zone can be further subdivided into desired levels of soil compaction, depending upon final use in the cropping scheme. As Fig. 2 illustrates, the controlled-traffic concept combined with a gantry machine has the potential for achieving the desired "management" of soil compaction. In the 1980's, several gantry vehicles were built to achieve this purpose. One of these is shown in Fig. 3.

Taylor (1986) established a rationale for controlled traffic as a soil compaction management system. The rationale was based primarily on research results from three soil-tire relationships: (1) effect of shape of the traction device; (2) effect

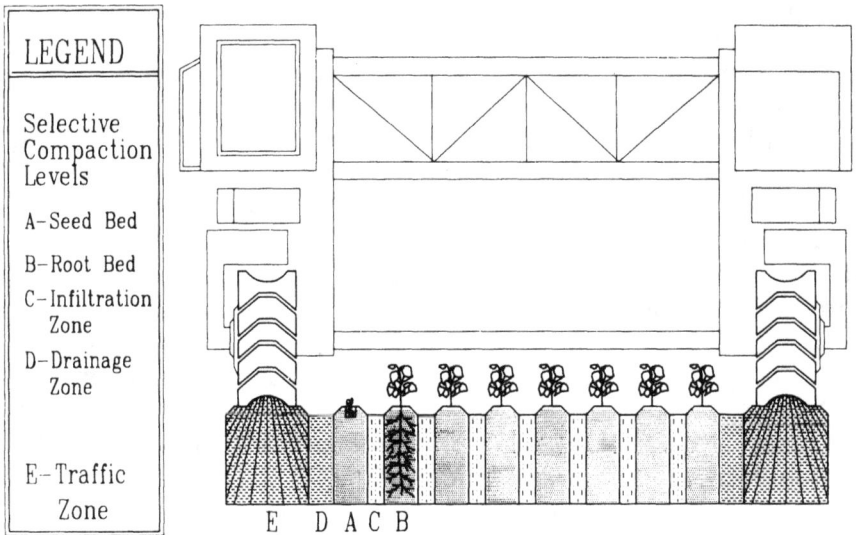

Fig. 2. The gantry vehicle combined with the controlled-traffic concept makes "management" of compaction possible.

Fig. 3. This 8-m wide gantry used by the Agricultural Research Service (ARS) of the United States Department of Agriculture (USDA), was designed for research, not as a prototype tractor.

of total load; (3) effect of multi-passes. Traction-device footprints which are long and narrow in the direction of travel improve traction and tractive efficiency, while reducing both the volume and the intensity of soil compaction. This shape can be obtained from tracks or multi-axle vehicles (Taylor and Burt, 1975). Larger total loads on an axle compact soil deeper, even when surface pressures and inflation pressures are held constant (Taylor et al., 1980). In a tilled soil, a tire was run four times in the same rut. The first pass was the critical one, producing 75% of the bulk density change and 90% of the sinkage. Therefore, reducing the number of passes is of limited benefit (Taylor et al., 1982).

Controlled traffic was initiated to eliminate compaction caused by wheel traffic in the crop zone. Cooper et al. (1969) discussed the early field work in Alabama, U.S.A., on effects of traffic compaction on cotton. They found reduced air and water infiltration, increased runoff and reduced plant root activity, any of which can contribute to reduced yields. Deep tillage performed in 1959 gave yield increases over conventional tillage of 450 kg ha^{-1} in 1960, 205 kg ha^{-1} in 1961 and 80 kg ha^{-1} in 1962. Three years of wheel traffic had essentially eliminated the effect of the deep-tillage treatment. Experiments were then designed to determine the feasibility of controlled traffic.

Dumas et al. (1975) reported average yield increases of 15% due to controlled traffic in cotton over a 4-year period, regardless of the type of tillage imposed. However, deep-tilled plots with no traffic yielded 56% more than conventionally tilled plots with tractor and sprayer traffic. Williford (1980) discussed the wide-bed cotton production system developed at Stoneville, MS, U.S.A., during the 1970's. It has proven to be the most economical system for the sandy loam soils of the Delta region. The system was designed to restrict wheel traffic to specific traffic zones throughout the season. When used as a controlled-traffic system, annual deep tillage was not required to maintain yield levels.

In the former U.S.S.R., Dvortsov and Polyak (1979) reported to the FAO that they found many cases where 30-45% of the field surface was compacted by the tractor running gear. Some of the surface was passed over several times. They reported soil compaction to depths of 100 cm and a gradual, irreversible compaction of the substrata. In the Netherlands, controlled-traffic research was conducted for several years using modified conventional tractors with a wheel span of 3.0 m (Lamers et al., 1986). Compared to conventional random traffic systems, the controlled traffic gave consistent yield increases of up to 10%. However, they concluded that the yield increases were insufficient to compensate for the loss of production on the traffic lanes. They calculated that a minimum span of 10 m would be necessary to justify controlled traffic economically. Beginning in 1984, they concentrated their research on low ground pressure (LGP) tires for management of soil compaction.

At the Federal Agricultural Research Center (FAL), Braunschweig, Germany, Sommer et al. (1988) used controlled traffic as part of their conservation tillage research. The cropping system consisted of rotating corn, sugar beet, barley, and flax. The wheel spacing was 2.5 m and the firm traffic lanes provided mobility for field equipment, such as sprayers, when fields with conventional traffic patterns were much too soft and wet. Their results included increased yields, reduced erosion and improved trafficability.

In England, Chamen et al. (1988, 1992) undertook research with a 12-m wide gantry. In earlier research, using 3-m wide equipment, they found that energy requirements for crop establishment could be reduced by up to 70% by eliminating traffic from the cropped area. They also found that the loose seedbed created by zero traffic triggered a manganese deficiency, causing reduced wheat yields. While this deficiency was corrected by applying manganese sulphate, they stated that using a roller to slightly firm the seedbed would do as well. An increase in bulk density and cone index was found in the topsoil (10 cm) under their LGP tire system, producing a higher tillage draft than with conventional high pressure tires. This is a clear indication that account must be taken of the area wheeled per pass with low pressure tires, not just tire inflation pressure.

In Scotland, Dickson and Campbell (1990) compared a 2.8-m wide, modified tractor on permanent traffic lanes with a conventional traffic system (both with ploughing and direct-drill tillage systems) for winter barley. The zero-traffic system gave increased yields and more uniform ripeness. The emergence of barley was depressed by transient waterlogging under direct-drilling, especially in the conventional traffic system. Similar traffic treatments were also compared for potatoes (Dickson et al., 1992), with yield increases of 14% being found for the zero-traffic system. In the above two studies, measurements of draft forces for primary cultivation following zero-traffic, showed reductions of 14 and 40% for barley and potato land, respectively, compared to following conventional traffic systems. Douglas et al. (1992) reported that the nitrogen uptake was enhanced by the less compacted soil in zero-traffic plots, resulting in yield increases for ryegrass grown for silage. Enhanced nitrogen uptake with less loss to atmosphere and ground water by denitrification and leaching is an important factor in today's environment.

In Israel, Hadas (1987) found controlled traffic especially effective in soils of the semi-arid, subtropical region, where organic matter content and structural stability are low, and compaction occurs under rather small traffic loads. He reported that soil compaction reduced air porosity, water retention, hydraulic conductivity, root proliferation and root activity. He found potential yield increases of 10-15% for controlled traffic in field crops (cotton, alfalfa, wheat) and he expected similar results for high-cost vegetable crops. However, root or bulb crops were expected to have even higher marketable yield increases under Israeli conditions.

In Australia, reports of several controlled-traffic and gantry-farming projects have been published. Arndt and Rose (1966) contrasted the patterns of soil compaction caused by draft animals and the new mechanized systems in Australia. The units of soil compaction by animals are small, randomly distributed disks, which require uniform tillage treatments over the entire field for correction. In the mechanized systems, the units of soil compaction are long, continuous bands, clearly defined for preferential treatment. However, "because of tradition the whole field is ploughed each year." Arndt and Rose concluded that "the overall efforts can now be logically reduced by reducing the area treated each year; and since the extent of this area depends on the traffic system, further advances in minimum tillage are expected from improved designs in tillage-traffic systems."

Tullberg and Murray (1987), at Queensland Agricultural College, reported on research to assess the effects of controlled traffic on tractor energy losses and tillage requirements as well as soil and crop effects. For subtropical, dryland grain production they concluded that controlled traffic can: (1) reduce the fuel cost of crop establishment by at least 40%; (2) allow similar output and capacity from a tractor of at least 30% less power; (3) maintain yields without the necessity for deep tillage operations; (4) increase rainfall infiltration, which also reduces runoff and erosion in some circumstances.

At the Institute for Irrigation and Salinity Research, Tatura, V., Australia, Adem and Tisdall (1984) have developed a successful soil management system. When removed from pasture and used for commercial tomato production, the productivity of irrigated red-brown soils in northern Victoria lasts only about 3 years because of loss of water-stable aggregates and organic carbon. The "Tatura System" used minimum tillage on permanent, raised beds, with traffic confined to the furrows. Sunflowers were double-cropped with cereals. The winter cereal not only provided a crop but also roots to improve soil structure and mulch for the summer crop to conserve water, reduce soil temperature, protect surface soil from impact of raindrops, ensure emergence and encourage earthworm activity. The beds were moistened by capillary action from shallow water in the furrow.

Lyasko (1982) reported that increased soil compaction by vehicles in the former U.S.S.R., reduced the yields of barley, winter wheat, winter barley, potatoes, peas and oats. A threshold index for vehicle compaction was suggested which could be used for the design and use of agricultural machines. Given additional research in this area, management of compaction by vehicles will be improved.

Carter et al. (1988) confirmed that removal of traffic from crop zones doubled the infiltration rate of California soils and permitted elimination of primary tillage. However, they also found that "tillage" reduced water infiltration rates in traffic-free crop zones, possibly due to disruption of macro-pore water channels. The ability of natural forces to recover soil structure and tilth in some soils, is a sufficient incentive to seek production systems that eliminate as much traffic and tillage as possible.

Compaction problems can also be serious in forest soils (see Chapter 14). Miles (1978), after studying the damage done to forest soil by logging operations, found that the major compaction occurred where primary and secondary skid trails were located. He concluded that compaction reduction by restricting traffic should be a part of the forest management system. Froehlich (1979) observed soil compaction in Oregon forests. Sixteen years after logging operations, he found soil compaction levels in the old skid trails at approximately the same level as in new skid trails in nearby current logging operations. He also suggested traffic control as the remedy. Froehlich concluded that "the growth reduction caused by compaction is great enough to indicate that existing skid trails should be used to the extent possible in future harvests. Repeated entries into a stand markedly increase the area covered by skid trails, and the impact is cumulative."

Timeliness may well be the major economic benefit of permanent traffic lanes. Spoor et al. (1988) and Spoor and Miller (1989) studied controlled traffic for sugar beet and discussed some of the timeliness effects, leading to improved plant establishment, development of quality beets, ease of harvesting in wet seasons and resulting soil conditions for the following crop. Sugar beet yields decline 4-5% for each week's delay in planting past the optimum date. Controlled traffic reduced planting delays, due to rains, by 4 days for direct-drilled beets. However, rain on the 5th and subsequent days prevented further drilling for a period of 20 days. Traffic lanes reduced harvester and trailer rolling resistance, and traffic-free crop zones improved beet quality and enabled more efficient beet lifting at harvest time.

Schumacher and Froehlich (1989) described a spatially-variable system for the application of chemicals, which was successfully designed, built and tested at South Dakota State University. The system monitored the location of field position in a controlled-traffic grid system and released chemicals according to a predetermined rate. The authors stated that the system "requires" the use of a controlled-traffic field pattern. This marks the beginning of the predicted use of controlled traffic for "mapping" fields for various objectives (Taylor, 1986).

GANTRY VEHICLES

Webster's Dictionary defines the gantry as a "framework supported at each end so that it spans a distance." Various other generic terms (spanner, bridge, wide-frame) have been used by investigators and are also defined by this brief definition of a gantry. Halkett (1858) described his system of guideway agriculture, which enabled all the farm operations to be performed by steam power.

The system consisted of "laying down at intervals of fifty feet or more, permanent and parallel guideways or rails, by which a locomotive cultivator, carrying the motive power, is supported and guided, and to the underside of which are attached the various implements to be used". Henry Grafton's 1860 "System of Steam Culture" (Spence, 1960) proposed to move over the field, without rails, in "well worn paths where the soil was packed firm" (Fig. 4). This project, like many others, may have never moved beyond the planning stage.

Dozens of gantry vehicles were proposed, and many were built during the early part of the steam power era. Evidently, this was followed by a long period of benign neglect of gantry vehicle design. By the end of World War I, internal combustion engines had replaced steam engines as the prime mover for agriculture. Many, perhaps most, gantry vehicles in the world today were designed

Fig. 4. Grafton's gantry vehicle, 1860 (from Spence, 1960).

and are used for harvesting aids. No attempt will be made to review the history and development of these vehicles.

Numerous articles on "gantry" agriculture have been published in the former U.S.S.R., where traffic-induced soil compaction problems may be greater than in any other country. Large farms and large equipment combined with susceptible soils and weather conditions have created significant compaction problems. Molosnov (1985) reported on two major scientific conferences held in 1983 and 1984. There were reports from some institutes of spanners actually built and used in research, while others reported on schemes planned for the future. Electrification of the spanners was a central theme, and the objective of the conferences was to develop a unified, scientific concept for implementing the program.

Lazovski (1984) undertook a historical, strategic review of gantries in agriculture. He credits the idea of agro-gantries to M. A. Pravotorov in the early 1950's but states that, while other countries have gantries in production testing stages, the former U.S.S.R. has left the idea to a few amateurs who were enthusiastic about the concept. After analyzing the future needs of agriculture in the former U.S.S.R., Lazovski decided they could only be met by use of "gantry" agriculture.

In the former U.S.S.R., Levchuk et al. (1979) and Rabochev et al. (1979) found the area of the field packed by traction devices to be 81% for winter rye and 91% for sugar beets. This caused soil compaction, disrupted soil structure and aggravated erosion. In addition, they found yield reductions of 45% in corn, 50% in sugar beet and 35% in sunflowers. In spite of these good reasons, it appears that it was the anticipation of a future labor shortage that encouraged the Soviet scientists to develop plans for controlling traffic and increasing productivity by the use of "bridge" systems. These systems would permit field operations to be undertaken at any time, thus ensuring undamaged soil structure, non-injury of plants, a minimum of hand labor, use of electrical energy, and simultaneous operations of tillage, fertilizing, harvesting, etc. There is no report of a "bridge" actually having been built in the former U.S.S.R., but it was considered in the planning for combating labor and energy shortages of the future.

Hutchinson (1968), Soane et al. (1972) and Soane (1975), at the Scottish Institute of Agricultural Engineering, reported on their controlled-traffic research program. Following the trend during the late 1950's and early 1960's for potato yields to increase when chemical weed-control was substituted for inter-row cultivations, they developed a self-propelled implement carrier to study the effects of traffic on potato production. Studies of traditional practices in the area revealed 91% coverage of the land by tractor tires. A 24% increase in potato yield was reported for zero traffic over normal traffic.

In England, one form of gantry development is represented by the "Monotrail" (Dowler, 1977) and described by Rutherford (1979). Invented by British farmer David Dowler, the toolbar gantry was 12 m long, and initially the basic power

Fig. 5. The Dowler gantry was developed on a small-grain farm in England. (Photo: Dowler Gantry Systems, Ltd.)

unit (hydraulic pump and motors), were from a commercially available self-propelled windrower. The machine had two hydraulically driven wheels and two caster wheels and moved endwise down roads and along the end of fields. In 1989, a modern version of this gantry (Fig. 5) became commercially available from Dowler Gantry Systems, Ltd.[1] This gantry system was developed primarily for small-grain farming.

Tillett and Holt (1987) of the Horticultural Engineering Group at Silsoe Research Institute, Silsoe, U.K., developed a 9-m wide, tracked vehicle. The vehicle was initially designed as a harvesting aid but has since been used for planting and for several other tasks in vegetable production. The quality of the produce sent to the market is of primary importance. Gantries straddling a wide bed can significantly improve the quality of crops, such as cauliflower, where repeated selective harvest is necessary.

In Israel, gantry development has progressed to the stage of commercial availability. The early model was called the MERHAV tractor and developed primarily for cotton production by Granot, a cooperative of kibbutzim. The commercial unit is called the F.P.U. (Field Power Unit). It is a nominal 6-m wide machine, designed for 6 rows of cotton. The hydrostatic 4-wheel drive gantry has sufficient power for general row-crop production systems. Ashot Ashkelon Industries, Ltd.[1] has produced more than a dozen of these vehicles. Most are used on kibbutzim in Israel, but three units are undergoing trials in Georgia,

---

[1]Mention of a trademark or vendor does not constitute a guarantee or warranty of the product by the U.S. Department of Agriculture and does not imply its approval to the exclusion of other products or vendors that may also be suitable.

Fig. 6. The Field Power Unit (F.P.U.) was initially developed for cotton production in Israel. (Photo: Ashot Ashkelon Industries, Ltd.).

U.S.A. The University of Georgia has initiated an extensive controlled-traffic research program utilizing the F.P.U. as an implement carrier. Fig. 6 shows a recent model of the F.P.U. produced in Israel.

Japan has worked with gantries for many years. The Japanese Automatic Field Work Apparatus or "Country Crane" was described by Kisu (1976). The gantry was 20 m wide and operated on concrete rails. This was a second-generation machine based on several years experience with an earlier model. This machine was completely automatic. It could be programmed to leave the storage area, go into the field, undertake tilling, puddling of paddy rice fields, or dusting and return to the storage area without manual assistance.

The National Agricultural Research Center in Tsukuba, Japan (Miyazawa et al., 1987), has the most sophisticated gantry system devised to date. The gantry is electrically powered and operates on concrete rails on 12-m spacings. It is completely automated and under computer control from the storage area to a pre-selected soil plot. All operations needed for rice production are carried out by this gantry. Special equipment, including robotic manipulators, are in various stages of development for use on this gantry. The scientists are planning the development of a practical field system on rubber tires with an isolated energy system and a spot location detection system.

Hilton and Bowler (1986) discussed Australian gantry activity and commercial potential. As elsewhere, most past activity was concerned with gantry vehicles developed as harvesting aids. However, these investigators identified irrigated cotton and tomatoes as two crops where gantry-based production systems would be advantageous and profitable. Their review covered development of gantry vehicles internationally and discussed advantages and disadvantages of gantry systems.

Spratt (personal communication, 1982) of P.S.I. Engineering[1] in central Tennessee, U.S.A., constructed a gantry with 33.5-m wheel spacing and 4-wheel drive. The wheel motors only served to "index" the machine from one set of rows to the next, where the tool carrier operated laterally along the gantry length. This gantry was intended for both crop production and in-field processing of vegetables and other crops needing some processing. The tool carrier platform rotated the implements at the end of the pass, preparing them for a return pass.

The Agricultural Research Service of the United States Department of Agriculture developed performance specifications and contracted for construction of two gantries for controlled-traffic research in Alabama and California. The California gantry was approximately 10 m wide with provisions for width adjustment; the Alabama gantry (Fig. 3) had a width of 8 m. Both machines were designed for research, not as prototype gantries for future farmer use. Because of possible research requirements, these vehicles were built with extremely strong frames and powerful engines. They have 4-wheel drive with hydrostatic motors in each wheel. Because of the high vehicle gross weight, special traffic lanes were prepared at each location. The lanes were on elevated soil, which was pre-compacted to a high level of bulk density.

A special class of gantries that should be mentioned are those designed for greenhouse use. Greenhouse gantries are usually designed for the same functions as field gantries but operate under covered and more restricted space. Taylor (1986), in discussing controlled traffic and gantry farming, stated "... the change from random field culture to fixed patterns lays the foundation for the future industrialization of agriculture". The greenhouse gantry is at the forefront of this effort. Greenhouse gantries are usually on rails, which solves one dimension of the spatial location problem. Computer control of all operations eliminates the need for human operators in unfavourable environments caused by temperature, humidity, chemicals, etc. The Japanese gantry discussed earlier (Miyazawa et al., 1987), was built for research and directed toward the development of an agricultural robot for farming in the future. A similar gantry has already been constructed for the greenhouse at Shimane University in Japan. An extensive discussion of greenhouse gantry development in the United Kingdom was published by Sharp (1983).

BENEFITS OF CONTROLLED-TRAFFIC AND GANTRY SYSTEMS

The most commonly expected benefit of any change in cultural practices is increased yields. In this respect, controlled traffic has been disappointing to the casual observer. While there are specific cases of spectacular yield increases, the

---

[1]Mention of a trademark or vendor does not constitute a guarantee or warranty of the product by the U.S. Department of Agriculture and does not imply its approval to the exclusion of other products or vendors that may also be suitable.

average worldwide increase is probably less than 10%. Obviously, the primary reason for this is that annual deep tillage is widely undertaken to correct temporarily the compaction problems arising under conventional traffic systems. Uncontrolled traffic following deep tillage promptly starts the recompaction process.

The most commonly realized benefit of controlled traffic is reduced crop production and machinery costs (see Chapter 24). Under a controlled-traffic regime a lower tillage requirement can be expected to reduce the operating and capital costs of implements and tractors. Such a reduction in machinery costs is economically more reliable than would be a corresponding increase in crop yields. Research has shown that, for most conditions, deep tillage can be stopped when traffic is controlled. Draft for topsoil tillage is reduced and tractive efficiency and flotation are improved, so that less powerful machinery may be purchased.

Removal of wheel traffic from crop zones has several beneficial results: (1) less tillage is needed; (2) infiltration rates may be significantly increased; (3) water storage capacity is increased; (4) runoff and erosion may be eliminated; (5) drainage is improved; (6) crop residue and stubble can be more easily kept on the surface; (7) soil tilth improves with time rather than deteriorating, as it does under conventional culture.

Timeliness of operations, such as planting, spraying and harvesting, from firm traffic lanes, can significantly improve both the quantity and the quality of the produce. This timeliness factor is so critical for high-value crops that it may be the greatest economic benefit of controlled traffic.

A gantry vehicle can fully exploit the controlled-traffic concept and such a vehicle, operating on firm traffic lanes, provides a stable platform. This stable platform makes it possible to control the many variables involved in the management of the soil physical conditions. In addition, wide spray booms, planting and tillage depths, and harvesting machines, can all be accurately controlled. Plant spacing and row width become a management decision and are not dictated by wheel width.

Improved timeliness and accuracy also make multi-cropping possible, and this can stretch growing seasons by weeks or months. If a winter crop is seeded into a summer crop 30 days before harvest, a farmer can enjoy a 13-month growing season each year. Surface soil is often at optimum moisture content for planting long before the field will support conventional equipment. A gantry on permanent traffic lanes can incorporate seed in what might be called "wet tillage" (Monroe and Taylor, 1989).

Substitution of the fixed traffic patterns of gantries for the random traffic of conventional cropping systems, provides several opportunities. The enhanced compatability with automated and non-random operations makes it practical to bury trickle irrigation lines and erect stakes or wires for crop support. "Mapping" of fields for yields, weeds and pests, soil moisture, fertilizer needs, etc., will permit a higher level of management and "mapping" will be possible with the non-random patterns of the gantry system. Automatic steering controls would allow the gantry operator to relax or to turn his attention to important non-steering activities. Completely automatic steering developments are more feasible with fixed traffic patterns. This development will allow continuous operation and permit gantries to operate in environments unsuitable for human employment.

There may be some perceived disadvantages in changing from conventional tractors to gantry farming systems. As a universal tool carrier, the gantry must have more sophisticated hitches and controls, which are expensive. The management level for gantry farming would probably be higher, requiring skilled operators with extensive crop expertise. Transporting gantry vehicles requires more skill and attention than conventional tractors and equipment. Also, the necessity for mechanisms to provide a lengthwise transport mode increases complexity and cost. In-field operations, while conducive to automation, would also be less tolerant of operator error. Field efficiency for gantry vehicles may be a problem on small or irregular-shaped fields, and turn-rows or headlands will need careful management to prevent excessive loss of fertile area. Control of surface water which collects in the compacted traffic lanes, could be more difficult.

Harvesting operations with gantries used primarily as harvesting aids are well developed. Ironically, harvesting and handling of the produce are probably the least developed segments of a controlled-traffic/gantry farming system. Development and mounting of practical harvesting systems must be augmented by a system to move the harvested products economically from the harvester to suitable collection points while controlling traffic-induced soil compaction.

Because of the above factors, most of the initial investment and application of gantry systems will probably come from producers of high-value crops, such as vegetables, nursery produce, orchards-fruits and flowers, especially when irrigation-dependent. These producers are accustomed to gantry-type vehicles and to bed-farming and the management level is already high.

Gantry use in greenhouse operations, usually on some type of rails, may well be the path of development of the vehicle systems, which then could be modified for outdoor field operations.

CONCLUSIONS

(1) Traffic-induced soil compaction can be a problem throughout the world wherever agriculture is mechanized. The first pass of a tire is the critical one, and the greater the total load on each axle, the deeper is the resulting soil compaction.

(2) High levels of soil compaction are generally best for traction and transport operations, while low levels are more desirable for crop growth. Controlled traffic is a concept which makes it possible to optimize soil conditions for both tires and crops.

(3) Within the crop zone, different levels of soil compaction may be desired to create proper micro-environments for seeds, roots, water infiltration and drainage. Management, not elimination, of compaction is the objective.

(4) Controlled traffic is a concept; the gantry is a vehicle. Gantry farming is not necessarily controlled traffic, but the gantry would seem to be the ideal vehicle for realization of the full benefits of the controlled-traffic concept.

(5) The use of a gantry vehicle for controlled-traffic farming provides a stable platform for management of all soil-crop operations. It also provides enhanced compatibility with automated and non-random operations such as "mapping" fields, burial of trickle irrigation lines and automatic steering or robotic operations.

(6) The improved timeliness of critical operations, such as planting, spraying and harvesting, which can be achieved when operating on permanent traffic lanes, may well be the greatest economic advantage of the controlled-traffic concept. In future research, evaluation of timeliness benefits should receive a high priority.

REFERENCES

Adem, H.H. and Tisdall, J.M., 1984. Management of tillage and crop residues for double cropping in fragile soils of southeastern Australia. Soil Tillage Res., 4: 577-587.

Arndt, W. and Rose, C.W., 1966. Traffic compaction of soil and tillage requirements. J. Agric. Eng. Res., 11: 170-187.

Carter, L., Meek, B. and Rechel, E., 1988. Zone production research with wide tractive research vehicle. Proc. 11th Conf. Int. Soil Tillage Res. Org. (ISTRO), Edinburgh, U.K., Vol. 1, pp. 221-225.

Chamen, W.C.T., Vermeulen, G.D., Campbell, D.J., Sommer, C. and Perdok, U.D., 1988. Reduction of traffic-induced soil compaction by using low ground pressure vehicles, conservation tillage and zero traffic systems. Proc. 11th Conf. Int. Soil Tillage Res. Org. (ISTRO), Edinburgh, U.K., Vol. 1, pp. 227-232.

Chamen, W.C.T., Watts, C.W., Leede, P.R. and Longstaff, D.J., 1992. Assessment of a wide span vehicle (gantry) and soil and cereal crop responsses to its use in a zero traffic system. Soil Tillage Res., 24: 359-380.

Cooper, A.W., Trouse, A.C., Jr. and Dumas, W.T., 1969. Controlled traffic in row crop production. Proc. 7th Int. Congr. Agric. Eng. (CIGR), Baden-Baden, Germany, Section III, Theme 1, pp. 1-6.

Dickson, J.W. and Campbell, D.J., 1990. Soil and crop responses to zero- and conventional-traffic systems for winter barley in Scotland, 1982-1986. Soil Tillage Res., 18: 1-26.

Dickson, J.W., Campbell, D.J. and Ritchie, R.M., 1992. Zero and conventional traffic systems for potatoes in Scotland, 1987-1989. Soil Tillage Res., 24: 397-419.

Douglas, J.T., Campbell, D.J. and Crawford, C.E., 1992. Soil and crop responses to conventional, reduced ground pressure and zero traffic systems for grass silage production. Soil Tillage Res., 24: 421-439.

Dowler, D., 1977. Improvements in or relating to agricultural implements (The Monotrail). British Patent Specification GB 1578857.

Dumas, W.T., Trouse, A.C., Jr., Smith, L.A., Kummer, F.A. and Gill, W.R., 1975. Traffic control as a means of increasing cotton yields by reducing soil compaction. Am. Soc. Agric. Eng., St. Joseph, MI, U.S.A., ASAE Pap. 75-1050.

Dvortsov, E.F. and Polyak, A.Y., 1979. High-powered tractors and their implements, including aspects of their impact on the soil. Food and Agriculture Organization of the United Nations, Economic Commission for Europe, Agriculture/Mechanization Rep. 80, (FAO/ECE/AGRI/WP.2/27), New York, NY, U.S.A.

Froehlich, H.A., 1979. Soil compaction from logging equipment: Effects on growth of young ponderosa pine. J. Soil Water Conserv., 34: 276-278.

Hadas, A., 1987. Controlled traffic research in Israel. Acta Hort., 210: 43-47.

Halkett, P.A., 1858. On guideway agriculture: Being a system enabling all the operations of the farm to be performed by steam power. J. Soc. Arts, London, 7: 41-53.

Hilton, D.J. and Bowler, L.J., 1986. Prospects for the use of gantry systems in Australian agriculture. Proc. Conf. Agric. Eng., Adelaide, SA, Australia, pp. 156-162.

Hutchison, P.S., 1968. The development of a self-propelled tool frame. Scot. Inst. Agric. Eng., Penicuik, U.K., Dept. Note SSN/21, 6 pp.

Kisu, M., 1976. Automatic Field Work Apparatus. Inst. Agric. Machinery, Saitama, Japan, Special Rep., March 30.

Lamers, J.G., Perdok, U.D., Lumkes, L.H. and Klooster, J.J., 1986. Controlled traffic farming systems in The Netherlands. Soil Tillage Res., 8: 65-76.

Lazovski, V.V., 1984. Primenenie mobil'nykh mostovykh sistem. (The use of mobile gantry systems). Mekhanizatsiya i Elektrifikatsiya Sel'skogo Khozyaistva, 2: 3-5 (in Russian).

Levchuk, N.S., Godunov, I.M. and Nechitailo, N.G., 1979. Obosnovanie mostovykh skhem sel'skokhozyaistvennykh agregatov. (The basis of bridge systems for agricultural units). Mekhanizatsiya i Elektrifikatsiya Sotsialisticheskog Sel'skogo Khozyaistva, 7: 6-8 (in Russian).

Lyasko, M.I., 1982. Uplotnyayushchee vozdeistvie s.-kh. traktorov i mashin na pochvu i metody ego otsenki. (The compaction action of agricultural tractors and machines on the soil and methods of its evaluation). Traktor i Sel'khoz-mashiny, 10: 7-11 (in Russian).

Miles, J.A., 1978. Soil compaction produced by logging and residual treatment. Trans. ASAE, 21: 60-62.

Miyazawa, F., Yoshida, T., Sawamura, A. and Taniwaki, K., 1987. Gantry system. Proc. Symp. Agric. Mech. Int. Coop. High Tech. Era, Tokyo, Japan, pp. 109-114.

Molosnov, M.F., 1985. Elektrifitsirovannyie mostovye agregaty dyla rastenie-vodstva. (An electrified gantry unit for plant production). Mekhanizatsiya i Elektrifikatsiya Sel'skogo Khozyaistva, 4: 7-10 (in Russian).

Monroe, G.E. and Taylor, J.H., 1989. Traffic lanes for controlled traffic cropping systems. J. Agric. Eng. Res., 44: 23-31.

Rabochev, I.S., Bakhtin, P.U., Gavalov, I.V. and Aksenenko, V.D., 1979. Umen'shenia otritsatel'nogo vozdeistviya mobil'nykh agrigatov na pochvu. (Reducing the adverse effect of vehicles on the soil). Vestnik Sel'skokho-zyaistvennoi Nauki, Pochvennyi Institut Imeni V. V. Dokuchaeva, 4: 90-94 (in Russian).

Rutherford, I., 1979. The 'Monotrail' - A new system for mechanised arable farming in Great Britain. Proc. 9th Int. Cong. Agric. Eng. (CIGR), East Lansing, MI, U.S.A., Pap. III-3-1.

Schumacher, J.A. and Froehlich, D.P., 1989. Computer controlled chemical application in controlled traffic fields. Am. Soc. Agric. Eng., St. Joseph, MI, U.S.A., ASAE Pap. 89-1606.

Sharp, J.R., 1983. The design of gantry systems for protected crops. Nat. Inst. Agric. Eng., Silsoe, U.K., Rep. 41, 63 pp.

Soane, B.D., 1975. Studies on some soil physical properties in relation to cultivations and traffic. In: Soil Physical Conditions and Crop Production. Min. Agric., Food and Fisheries, HMSO, London, U.K., Tech. Bull. 29, pp. 160-183.

Soane, B.D., Hutchison, P.S. and Campbell, D.J., 1972. The effect of traffic and weed control methods on soil conditions and potato yields. Scot. Inst. Agric. Eng., Penicuik, U.K., Dept.

Note SSN/102, 18 pp.

Sommer, C., Dambroth, M. and Zach, M., 1988. The mulch-seed concept as a part of conservation tillage and integrated crop production. Proc. 11th Conf. Int. Soil Tillage Res. Org. (ISTRO), Edinburg, U.K., Vol. 2, pp. 875-879.

Spence, C.C., 1960. God Speed the Plow. Univ. Illinois Press, Urbana, IL, U.S.A., 183 pp.

Spoor, G., Miller, S.M. and Breay, H.T., 1988. Timeliness and machine performance benefits from controlled traffic systems in sugar beet. Proc. 11th Conf. Int. Soil Tillage Res. Org. (ISTRO), Edinburgh, U.K., Vol. 1, pp. 317-322.

Spoor, G. and Miller, S., 1989. Soil management opportunities for improving productivity. Brit. Sugar Beet Rev., 57: 52-55.

Taylor, J.H., 1983. Benefits of permanent traffic lanes in a controlled traffic crop production system. Soil Tillage Res., 3: 385-395.

Taylor, J.H., 1986. Controlled traffic: A soil compaction management concept. Trans. ASAE, 95: 1090-1098.

Taylor, J.H., 1989. Controlled traffic research: An international report. Proc. 11th Int. Cong. Agric. Eng. (CIGR), Dublin, Ireland, Vol. 3, pp. 1787-1794.

Taylor, J.H. and Burt, E.C., 1975. Track and tire performance in agricultural soils. Trans. ASAE, 18: 3-6.

Taylor, J.H., Burt, E.C. and Bailey, A.C., 1980. Effect of total load on subsurface soil compaction. Trans. ASAE, 23: 568-570.

Taylor, J.H., Trouse, A.C., Jr., Burt, E.C. and Bailey, A.C., 1982. Multipass behavior of a pneumatic tire in tilled soils. Trans. ASAE, 25: 1229-1231, 1236.

Tillett, N.D. and Holt, J.B., 1987. The use of wide span gantries in agriculture. Outlook Agric., 16: 63-67.

Tullberg, J.N. and Murray, S.T., 1987. Controlled traffic tillage and planting. Farm Mechanization Centre, Qld. Agric. College, Lawes, Qld, Australia, Report to the National Energy Research Development and Demonstration Council, Canberra, Australia, 54 pp.

Williford, J.R., 1980. A controlled-traffic system for cotton production. Trans. ASAE, 23: 65-70.

PART F
=======================================

ECONOMIC ASPECTS OF SOIL COMPACTION AND ITS CONTROL

Soil Compaction in Crop Production
B.D. Soane and C. van Ouwerkerk (Eds.)

CHAPTER 23

Economics of Modifying Conventional Vehicles and Running Gear to Minimize Soil Compaction

K. ERADAT OSKOUI[1], D.J. CAMPBELL[2], B.D. SOANE[2] and M.J. McGREGOR[3]

[1]USDA-ARS North Central Soil Conservation Centre, Morris, MN, U.S.A.
[2]Scottish Centre of Agricultural Engineering, SAC, Penicuik, U.K.
[3]Rural Resource Management Department, SAC, Edinburgh, U.K.

SUMMARY

Accurate predictions of the net economic benefits to be gained from the adoption of modifications to vehicles and their running gear are required if compaction problems in commercial crop production are to be overcome. The modifications most readily available to the farmer are: (1) reduction of load on running gear; (2) reduction of ground contact pressure by using wider tires or additional wheels and axles; (3) reduction in the area of fields subjected to traffic, particularly when soils are wet. The likely economic success of these options will depend closely on farm and vehicle size. Apart from the need to quantify, in economic terms, the benefits in crop yield and quality from such changes in management, it is essential to identify and evaluate all primary and secondary costs. Site factors, such as soil type, cultivation system, rainfall, cropping system and farm size, are likely to have a profound influence on the economics of the system. The extrapolation of experimental data obtained on small plots to a farm-scale basis requires great caution, due to the dominant influence of vehicle traffic distribution on overall changes in soil compactness.

A number of computer-based models, capable of analysing the whole-farm economic responses to changes in machinery management, have been developed in Canada, the U.S.A. and Sweden. Such models are powerful tools for exploring the multi-factorial interactions between the numerous machinery options and site factors, but there is a shortage of accurate data to verify predictions.

INTRODUCTION

Soil compaction is recognised as a fundamental problem in the commercial production of many agricultural and forest crops throughout the world. While there is much to be learnt about the interactions between vehicle running gear and the soil and also between the resulting soil conditions and crop growth, such studies do not provide the full information required by farmers to assess the influence of compaction control on the profitability of their enterprises. The selection by farmers of the most appropriate size and type of currently available agricultural vehicles and implements, with their associated running gear, must depend on economic principles. Clearly, to guide compaction management in commercial agriculture, analyses in terms of benefit-cost and benefit/cost ratio[1] are necessary, if research results are to have any meaning in the commercial context. In particular we need to know: (1) the costs of implementing possible modifications to vehicles and their running gear; (2) the effect of such modifications on farm revenue; (3) the effect of soil, crop, farm size and weather factors on overall farm profitability.

The economy of scale and the reduction in labour costs associated with the widespread use of very large, heavy machinery, may result in management systems in which the penalties, in terms of crop value losses due to compaction, exert an appreciable influence on overall profitability. This chapter will review the rather limited amount of work published and summarize some unpublished work.

APPROACHES TO ECONOMIC ANALYSIS

The commercial importance of a quantitative economic approach to compaction problems was recognised by Gill (1971). He summarised, for a wide range of crops, the numerous ways in which compaction problems relate, directly or indirectly, to both decreases in farm income and increases in production costs. He forecast that "It would be enlightening to analyse the entire system in greater detail so that the economic value of each component could be established". Although not yet achieved in full, such analyses can be attempted by determining the change in profitability due to the adoption of specified soil compaction management measures, using the generalized equation:

$$\Delta P = \Delta O - \Delta C \tag{1}$$

where ΔP = change in enterprise profitability; ΔO = change in economic output attributed to a specified change in the level of compaction; ΔC = extra cost of obtaining the specified change in level of compaction.

The interactions between machinery management systems, soil conditions and crop responses are so complex that the solution of eqn. (1) requires a large body of reliable data concerning many compaction-related factors which need to be

---

[1]The term "benefit-cost" (i.e., benefit minus cost) is used synonymously with "net benefit" and is expressed in currency units. The term "benefit/cost ratio" is used to indicate the relative magnitude of benefit and cost items involved in a particular comparison. Both terms are relevant to economic studies and to decision taking by farmers. For comparative purposes, all currency values quoted in the literature have been converted to US $ at the exchange rates prevailing in September 1992, without adjustment for inflationary changes since the time of the study concerned.

TABLE 1

Compaction-related factors affecting the economic output and costs associated with changes in the use of machinery and associated running gear

Related to output	Related to costs
Crop yield	PRIMARY COSTS
Crop quality	
	Purchase and operating costs of machinery
	Purchase and operating costs of running gear
	SECONDARY COSTS
	Tractive efficiency and trafficability
	Power and fuel requirements
	Produce drying, handling and selection requirements
	Timeliness of field operations
	Fertilizer and pesticide requirements
	Tillage draft force
	Tillage tool wear
	Tillage requirement (primary, secondary, subsoiling)
	Erosion hazard

quantified in economic terms, including both primary and secondary costs (Table 1). In this context, primary costs relate to the purchase and standard operating costs of machinery and running gear, whereas secondary costs include all the components of crop production costs which are influenced by changes in the level of soil compaction. In many respects, the types of data tabulated in Table 1, especially those concerning the complex interaction of cost-related factors, is still largely lacking for whole-farm situations. Many field experiments concerning compaction management have not been designed for easy interpretation in economic terms and the results obtained on small plots may have little relevance to whole-farm situations without careful adjustment to take account of the different distribution of wheel traffic (Gunjal and Raghavan, 1986).

The results of experiments have to be interpreted with great caution and the direct summation of the profit penalties associated with compaction cannot yet be made with any degree of confidence. This explains why the benefit-cost studies which have been reported in the literature have been few and have tended to concentrate on limited aspects, e.g., optimum tractor size (Gunjal et al., 1987) or a particular farming operation, such as harvesting (Håkansson et al., 1988), or have dealt with the matter at a general level (Stephens, 1990; Eradat Oskoui and Voorhees, 1991a).

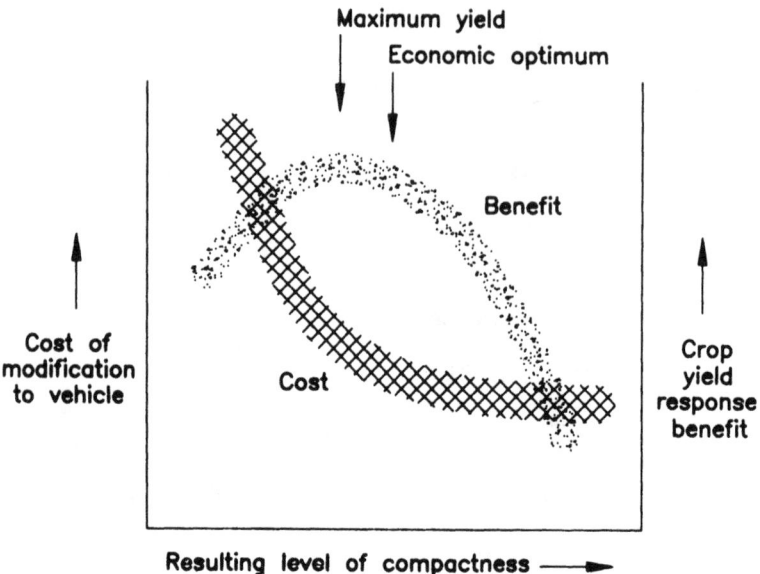

Fig. 1. Conceptual diagram of the relationship between the cost of vehicle modification, crop yield response benefit and resulting level of soil compactness.

The objective of any economic analysis of compaction is to assess the net benefit derived from modifications to the type and management of vehicle running gear. A conceptual diagram (Fig. 1) suggests that costs associated with vehicle modification, tend to increase as the level of soil compactness decreases. However, as the level of compactness decreases, the crop yield benefit rises to a maximum at an optimum level of compactness (see Chapter 12). Further decreases in compactness will result in sub-optimal conditions and progressive loss of yield. The level of compactness corresponding to the economic optimum (maximum net benefit), may be above the level at which maximum yields are obtained. Unfortunately, the practical evidence in economic terms for these relationships is, as yet, largely lacking.

Direct effects of soil compaction on crop output

The shape of the response curve of crop yield to soil compaction (shown in simplified form in Fig. 1) has been found to depend on soil type, crop type, and nutrient, oxygen and water supply during the growing period (see Chapters 8, 11 and 12). Any economic analysis must therefore take into account site factors, such as soil type (Fig. 2) and weather during the growing season (Voorhees, 1977; Raghavan et al., 1978; Bicki and Siemens, 1990; W.B. Voorhees, 1990, personal communication).

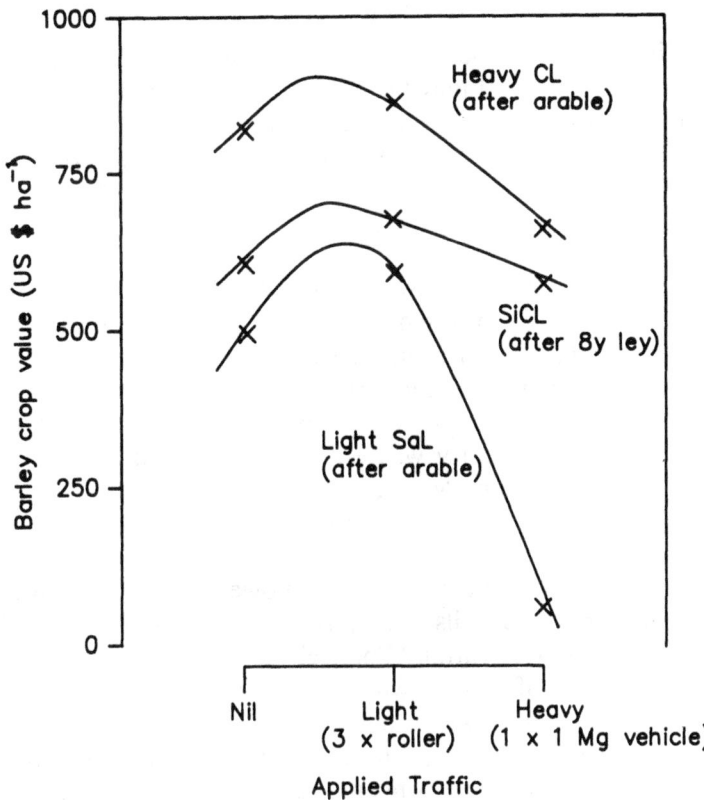

Fig. 2. Response of barley crop value to 3 levels of applied traffic on 3 soil types in the U.K. (after Kubota and Williams, 1967).

Attention must also be paid to crop quality, since compaction has been found to lower the quality and uniformity of many crops and may reduce market value because the produce is misshapen (Voorhees et al., 1977; Campbell, 1982) or dirty (Subotic and Stanacev, 1982). In cereals, delayed or irregular ripening due to compaction can seriously reduce the market value of the grain and increase drying costs (Campbell and McGregor, 1989). Increases in grain water content attributed to compaction amount to 8% in barley (W.S. Reid, personal communication, 1983) and 4.5% in corn (Gaultney et al., 1982).

Indirect effects of soil compaction on crop production costs

In addition to direct effects on crop output, the evaluation of a number of secondary costs (Table 1) is necessary if a rigorous analysis is to be attempted.

Diseases, pests and weeds

The costs of controlling diseases, pests and weeds, are an appreciable proportion of total production costs and may be influenced by compaction. For example, compaction is known to be associated with many root diseases (Raghavan et al., 1982), particularly in soybeans (Moots et al., 1988) and peas (Vigier and Raghavan, 1980).

Fertilizer requirements

Crops growing on compacted soils may show a greater fertilizer requirement, especially for nitrogen, than do those growing on loose soils (Bakermans and De Wit, 1970; Jaggard, 1977; Swaine, 1981. The uptake of nitrogen on compacted soils after no-till was less than after plowing for spring barley (Smith et al., 1984) and winter wheat (Cannell et al., 1980). In grass silage production, Douglas et al. (1992) showed that nitrogen use efficiency was enhanced by the adoption of either a reduced ground contact pressure or a zero traffic system.

Power requirements

Compaction resulting from wheel traffic can raise the power requirement, and hence the costs, of cultivating both topsoils and subsoils (see Chapter 25) and may increase the required depth of cultivation (Chancellor, 1976). Voorhees and Hendrick (1977) found that the draft for plowing increased by 92% following heavy compaction, while after even one wheel pass, fuel consumption was 19% higher than without traffic. The draft for deep tine cultivation after potatoes was about 40% less after zero-traffic than after conventional traffic (Dickson et al., 1992). Wheel traffic necessitates additional cultivation to eliminate wheel ruts to achieve a level seedbed (Elliot, 1979) and also increases the draft requirement for subsoiling (Carter, 1980).

Tool and tire wear

The wear on all soil engaging tools, and the costs of replacements, will increase in compacted soils, particularly those cultivator tines and drill coulters which run in line with a tractor wheel. Tires subject to additional wheelslip on compacted soils suffer additional wear (Voorhees and Hendrick, 1977).

Timing of operations

Any variation in either the time at which an operation can start or its duration, will influence profitability, especially any delays in planting (Eradat Oskoui, 1981; Knowles and Audsley, 1983). A practical methodology for calculating timeliness penalties has been prepared by Witney and Elbanna (1985).

Erosion hazard

In the U.S.A., considerable effort is expended to assess the economic consequences of erosion, part of which is attributed to compaction, causing a marked reduction in water intake (see Chapter 7). In the U.K., Harrison Reed (1983) also attributed erosion problems to wheel traffic, soil losses sometimes exceeding 100 Mg ha^{-1}, with considerable economic cost.

Importance of regional differences

Evidence from field experiments and observations on farms has confirmed the importance of regional differences in soil, climate and cropping factors in any economic analysis. For example, in the responses of crops to subsoil compaction in Minnesota (Voorhees et al. 1989), Wisconsin, U.S.A. (Lowery and Schuller, 1991) and Quebec, Canada (Gameda et al. 1994), pronounced differences were observed at different sites. Miller (1986) reported that soil compaction was the most important economic component of soil degradation affecting crop production in Quebec, in contrast to Ontario where the dominant component was soil erosion. In Europe, compaction problems have been most evident in northern countries (see Chapter 1), where a high soil water content during seedbed preparations and harvesting operations can lead to severe soil damage.

POSSIBLE MODIFICATIONS TO VEHICLES, RUNNING GEAR AND MACHINERY MANAGEMENT SYSTEMS TO MAXIMISE PROFIT

A knowledge of the mechanical processes involved in soil compaction (see Chapters 2, 3 and 4), confirms that the most important options under the control of the farmer are likely to be: (1) reduction of wheel or track load (see Chapter 20); (2) reduction of ground contact pressure (see Chapters 19 and 21); (3) use of extended axles and gantries for zero-traffic cropping (see Chapters 22 and 24); (4) reduction of the number of passes and, hence, of the area receiving traffic (see Chapter 25); (5) avoidance of field traffic at periods of high soil water content (see Chapter 25).

Within each option there are a number of alternative approaches. Wheel or track load can be reduced by: (1) using a vehicle of lower power or capacity; (2) improved vehicle design; (3) use of lightweight material in construction, such as aluminum, plastics (Dann, 1983) and ceramics (Anon., 1984); (4) use of new technology, e.g., low-volume spraying and stripper-header harvesting of cereals (Price, 1989).

Reduction in ground contact pressure can be achieved by: (1) addition of dual, triple or cage wheels (Gee-Clough et al., 1981; Dickson et al., 1983); (2) use of wider tires; (3) use of tracks or half-tracks (see Chapter 21). Reduction in the number of passes can be achieved by both improved control of field traffic and the use of combined implements, e.g., cultivator-drill combinations. The area receiving traffic may be reduced by the use of temporary tramlines, especially in cereal crops.

All of these options need to be fully quantified in economic terms. In this chapter, emphasis will be placed on the economics of those modifications to vehicles and running gear which lead to reductions in load and ground contact pressure, with mention of the economic aspects of some other options in machinery management.

ECONOMIC ANALYSIS TECHNIQUES

A wide range of techniques is available to undertake economic analyses with varying degrees of precision and complexity. Simple budgeting techniques, such as gross margins and partial budgets, can be used for whole-farm analyses. Partial budgets are a suitable technique where a simple pair-wise comparison (e.g., make the modification or not) is required. The object of the analysis is to calculate the expected change in profit associated with proposed changes in management. The analysis contains only those income and expense items which will change if the proposed modification in management (e.g., a change in tractor running gear) is implemented. Such studies provide useful guidelines on likely economic responses. However, because of lack of data, it is rarely, if ever, possible to include all the relevant compaction-related factors listed in Table 1 and, consequently, the results must represent only a partial estimate of the full economic implications of the proposed change. A more detailed analysis might indicate either greater or smaller advantages of the modification. Because most analyses are usually based on data from only a limited number of sites and seasons and do not take into account the residual effects of compaction, they lack widespread applicability. However, they may provide valuable information for the area in which they were developed.

Models

The development of computer-based modeling procedures (Arvidsson and Håkansson, 1991; Lavoie et al., 1991; Eradat Oskoui and Voorhees, 1992), makes it possible to provide predictive capabilities for a range of multi-factor comparisons (e.g., farm size, soil type, seasonal rainfall, etc.). Such models may be primarily concerned with estimating crop economic output or with costs of making modifications or, in a few cases, with a full analysis of benefit-cost relations. However, the practical application of an economic model will depend on the validity of the input data and the predictive relationships. Because this complex subject is still at an early stage, there have been few attempts to validate the predictions of economic models of compaction by comparison with independent data obtained from realistic, whole-farm situations. Computer-based predictions should therefore be regarded as provisional estimates.

Types of models
A whole-farm model can be a simple farm budget, a simulation model, an

expert system or involve a mathematical programming approach. However, because mathematical programming techniques are optimising, it is possible to miss near-perfect systems where benefit-cost relations are not at optimum level.

Fully mechanistic models to predict soil and crop responses to traffic do not appear to be possible with present knowledge. As a result, present models tend to rely on statistical relationships which are found to give an appropriate level of reliability within the defined limits of the model.

Lavoie et al. (1991) model

A number of linear programming submodels were developed to maximise net whole-farm income in response to the yield loss implications of different tractor sizes, farm sizes and weather conditions for two crops (grain corn and corn silage), grown in Quebec with three cultural practices (conventional tillage, reduced tillage, crop rotation). Additional submodels were developed for wheat, barley, oats and tame hay[1] when cultivated conventionally with only one tractor size. The data for yield responses of grain corn and silage corn to different traffic regimes were obtained from Raghavan et al. (1978), with adjustment from small-plot to whole-farm basis, according to procedures developed by Gunjal and Raghavan (1986) and Gunjal et al. (1987). Machinery operation costs and product sale prices were based on five-year averages. A statistical relationship was used to estimate the effect of different weather conditions on the number of working days available for field operations. Estimated costs included insurance, fuel, lubricants, repair and maintenance components, with interest on borrowing set at 14.8%.

Arvidsson and Håkansson (1991) model

This computer-based model uses empirical data from a large number of compaction experiments throughout Sweden. It is designed for advisory and educational use, at individual farm level, and predicts the economic consequences of variation in the machinery used for all types of field operations for a wide range of arable and ley crops. The total yield loss attributable to compaction can be computed for a number of specific field situations, such as: (1) recompaction after plowing; (2) persistence of compaction after subsequent plowing; (3) subsoil compaction; (4) compaction due to traffic in the growing crop, e.g., ley crops. From basic site and machinery input data, the model calculates the distribution of traffic separately for the headlands and for the main area of the field. Changes in the degree of compactness in designated areas of the field are estimated by regression equations involving soil water status expressed on an arbitrary scale from 1 (very dry) to 5 (very moist), axle load, inflation pressure, tire width and the number of passes. The overall crop output from the field is then calculated.

---

[1]Hay from perennial forage species of grass and/or legumes which have been intentionally seeded by farmers.

Eradat Oskoui and Voorhees (1992) model

A mathematical programming model to predict crop output from changes in machinery management was developed by Eradat Oskoui and Voorhees (1992). The budgeting technique (eqn. 1), along with simulation, was used to produce various soil, weather, cultivation system and wheel management scenarios and an expert system was employed to evaluate and rank the outcome of each scenario.

The model was developed to quantify the differences in benefit/cost and benefit-cost relations between the modified and a standard reference management system. Initially the benefit/cost difference was restricted to changes in yield (both positive and negative), but the importance of changes in fuel consumption, environmental aspects (e.g., erosion), maintenance costs (implements, tires, etc.) was recognised, although these factors were not included in the first version of the model. Crop yield was estimated from air-filled pore space at planting using a skewed parabolic (logarithmic) relationship ($r^2 = 0.96$) (Eradat Oskoui and Voorhees, 1991b). Air-filled pore space at planting was calculated from the value of bulk density predicted by the model developed by Gupta and Larson (1982), using values for the stress concentration factor calculated from the cone index. The cone index was calculated from soil parameters by the method of Witney et al. (1984). Tests on the validity of the Eradat Oskoui and Voorhees (1992) model are in progress.

For a given set of soil, climate, crop and management factors, and a vehicle having four equal-sized tires, the variation in the yield component of output, O_y, was defined by:

$$O_y = f(W, Ac, A, N) \tag{2}$$

where W = machine weight; Ac = tire contact area; A = overall field area affected by compaction; N = number of tires. This relationship includes most of the factors concerning the load, configuration and management of running gear, which are under the control of the farmer.

As a standard reference machinery management system, the model used the planting of corn under typical conditions for the Northern Corn Belt in the U.S.A. on 455 ha of medium-textured soil in a season of medium rainfall (400 mm annual). The 222-kW tractor, of 16.3 Mg total mass (including ballast), was fitted with four equal-sized (20.8-38 10PR), low ground contact pressure tires and it pulled a planter of 12 m width. Using predictive relationships for wheel/soil and soil/plant interactions, the program calculated for the area of the wheel tracks: (1) bulk density; (2) air-filled pore space at planting; (3) crop yield. For the standard system, the overall loss of crop output due to compaction in the tractor wheel tracks at planting was estimated to be US $ 1.97 ha^{-1} or US $ 894 for 455 ha. The model was then used to estimate the influence of possible variation in site factors within the Northern Corn Belt, such as soil texture (5, 20, 35% clay), cultivation system (plowed, no-till) and seasonal rainfall (<280, 400 and >520 mm annually). All 18 combinations of the site variables were then used to evaluate the effects on income of three types of tractor wheel modification: (1) wheel number (singles, duals, triples); (2) drive type (2-wheel, 4-wheel); (3) type of tire (conventional, low ground contact pressure).

Estimation of costs of machinery modifications

While it usually is simple to establish the purchase price of alternative vehicles or running gear, consideration must also be given to any differences in lifetime costs (depreciation), repairs and maintenance, fuel and labour requirements, compared with the standard equipment. Such costs should be discounted to give the total discounted cost, which is then amortised for the expected lifetime area capability or, alternatively, to cost per unit time worked. Campbell and McGregor (1989) used hourly cost as the unit for comparison. However, since obtaining an accurate hourly cost can be difficult for whole-farm comparisons, Eradat Oskoui and Voorhees (1992) preferred "hectare years" and they expressed costs and benefits in US $ ha^{-1}. Godwin et al. (1992) also expressed machinery costs on an area basis.

In comparisons of machinery costs, it is important to include interest charges on capital used for purchase of equipment and costs of tax, insurance, labour, fuel and oil, in order to obtain realistic costs for machinery operation (Campbell and McGregor, 1989; Lavoie et al., 1991; Godwin et al., 1992).

EXAMPLES OF ECONOMIC ANALYSES

Examples of different types of economic analysis will be reviewed according to the types of changes in machinery management which were studied. In some studies, observational or experimental data have been analysed for a relatively simple range of conditions. In contrast, studies undertaken with computer-based models consider a large number of hypothetical situations with interacting relationships, involving machinery, soil and weather factors. Both types of study have value in indicating the extent and application to compaction problems of current economic information.

Effect of changing tractor power

The choice of tractor power (with its close relationship to size, total weight and axle load) is a fundamental decision concerning virtually all mechanization systems. The economics of tractor power selection have in the past been based on purely mechanization efficiency principles, with the crop output penalty due to high axle loads being largely ignored. More recently, this situation has changed, with considerable interest now being attached to the need to optimise all factors, including compaction, which are related to tractor power.

The weight of tractors used for a specific operation can be reduced by selecting a lower power model or attempting to use a machine with a lower weight/power ratio. Any reduction in weight of tractors and other self-propelled vehicles must, however, be related to the need for acceptable tractive efficiency. According to data supplied by manufacturers, the average weight/engine power ratios for crawler, 4-wheel drive and 2-wheel drive tractors available in the U.S.A., are approximately 127, 71 and 66 kg kW^{-1}, respectively. The most likely way in which tractor weight can be reduced, is to choose a lower power model than would otherwise be considered appropriate, in the belief that overall profitability will increase.

Effect on crop value of tractor and implement size for plowing and seedbed operations in Sweden

The effects on crop output of using three sizes of tractor (with a mass of 4.5, 6.0 or 7.6 Mg) for moldboard plowing on a soil having 20% (w/w) clay, with 3, 4 and 5-furrow plows weighing 0.7, 1.3 and 1.4 Mg, respectively, were estimated by Danfors et al. (1992), using the Arvidsson and Håkansson (1991) model. When the tractors were running on conventional tires, the total yield losses associated with the use of small, medium and large tractors, were equivalent to US $ 14, 13 and 17 ha^{-1}, respectively. A corresponding study on the effects of using small, medium or large combined implements and tractors running on conventional tires for seedbed operations, showed that estimated total yield losses were US $ 12, 13 and 14 ha^{-1}, respectively. These results indicate that, for the conditions compared, the loss of yield associated with heavier vehicles was not compensated for by wider working widths.

Effect of tractor power on economics of green pea production in Quebec

The influence of tractor power on the economic output of green pea production was studied in Quebec by Gunjal and Raghavan (1986), using data from small-plot experiments with 2-wheel drive tractors of 32 and 97 kW power in 1980 (wet season) and 1982 (dry season), respectively. By varying the number of passes from 0 to 15, the calculated cumulative ground contact pressures varied from 34 to 684 kPa. The results obtained were converted to a whole-farm situation and predictive relations were used to estimate the effect on economic yield loss of variation in tractor power from 20 to 140 kW (Fig. 3). The validity of the extrapolation of tractor power from 97 to 140 kW must clearly be regarded as tentative until confirmed experimentally.

As tractor power increased from 20 to 100 kW for the whole-farm situation, the yield loss decreased. This contrasted with the results for the small plots (i.e., 100% wheel coverage), where the opposite effect was obtained. The optimum tractor size, in terms of minimizing both yield losses and total economic costs, was found to be about 105 kW. However, the authors stressed that the predicted whole-farm results depended strongly on the actual whole-farm wheel traffic

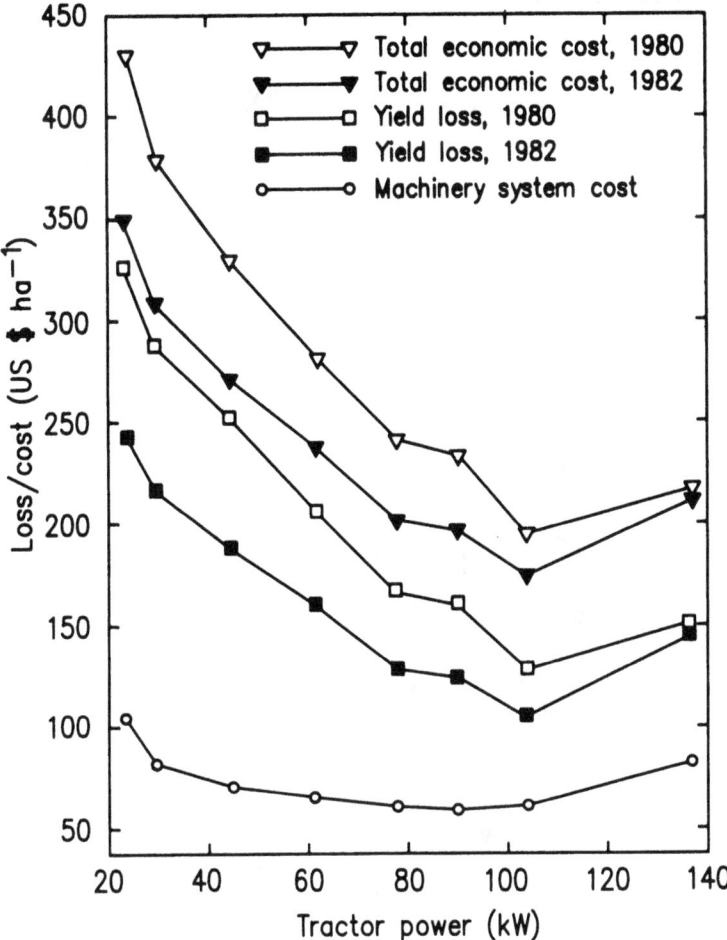

Fig. 3. Relationship (whole-farm basis) between machinery system size and green pea yield losses, machinery costs and total economic costs (after Gunjal and Raghavan, 1986).

distribution, details of which are not widely available. For a typical commercial cultivation system involving 5 field operations, the estimated yield losses attributable to machinery traffic were between 1.8 and 3.8 times the cost of the machinery, which stresses the dominant role of yield losses due to compaction in the overall economics of green pea production.

Effect of tractor power on economics of corn production in Quebec

Gunjal et al. (1987) summarise a number of field experiments on the effects of whole-plot, pre-emergence tractor traffic on the yield of grain corn and corn silage in both wet and dry years. They adjusted the yield response data obtained

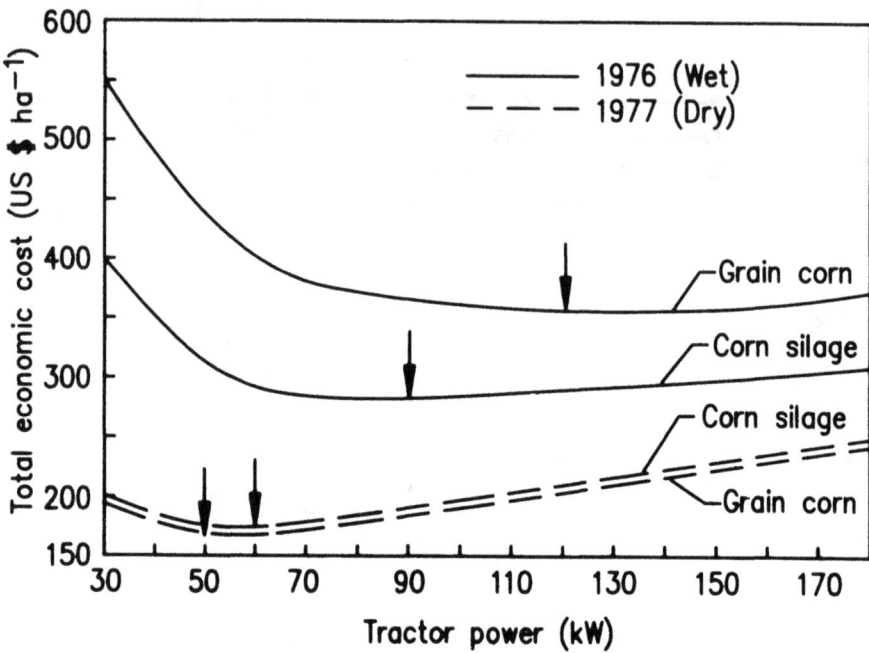

Fig. 4. Relationship between total economic costs and tractor power for grain corn and silage corn in two seasons (↓ indicates optimum tractor power) (after Gunjal et al., 1987).

from small plots to a whole-field basis by applying an equation to predict the actual compacted area from tractor power. Estimated farm-level yields for corn silage and grain corn declined as tractor power increased. The authors attributed this to the linear increase in contact pressure with increasing tractor size and the concurrent logarithmic decrease in the actual compacted area, although the evidence for these relationships was not discussed. They estimated the costs of operating tractors of different size, taking into account fixed costs, operating costs and timeliness costs. Total economic costs were obtained by adding machinery costs for cultivation and crop establishment to the losses in yield attributable to the compaction during three pre-emergence operations (cultivating, chiseling and seeding). The relationships between total economic costs and tractor power for grain corn and corn silage in two seasons are shown in Fig. 4, together with the calculated optimum tractor power for each condition.

Lavoie et al. (1991) found that the optimum tractor size for a conventional cultivation system was 60 kW for a dry year, 100 kW for an average year and 140 kW for a wet year. On a 200-ha farm in an average season, the use of a 60- or 140-kW tractor instead of a 100-kW tractor, would result in reductions in net income of US $ 3.5 ha⁻¹ and US $ 5.4 ha⁻¹, respectively. At lower than optimum tractor size, overall soil compaction on a field basis tends to be increased because of narrow working width and close wheel tracks, whereas at higher than optimum tractor size, overall soil compaction is increased by the effect of high loads carried by wheels. In a wet season, yield losses for both grain corn and silage corn were much greater with small tractors than large tractors (Fig. 5). Lavoie et al. (1991) stressed that in a wet season a high-power tractor should be equipped with the widest implements it could handle, if full advantage is to be obtained from the reduced wheel traffic effect with a high-power tractor. In a dry season, however, the extra compaction resulting from a small (60-kW) tractor gave a greater yield increase than was the case for larger tractors (Fig. 5), which was

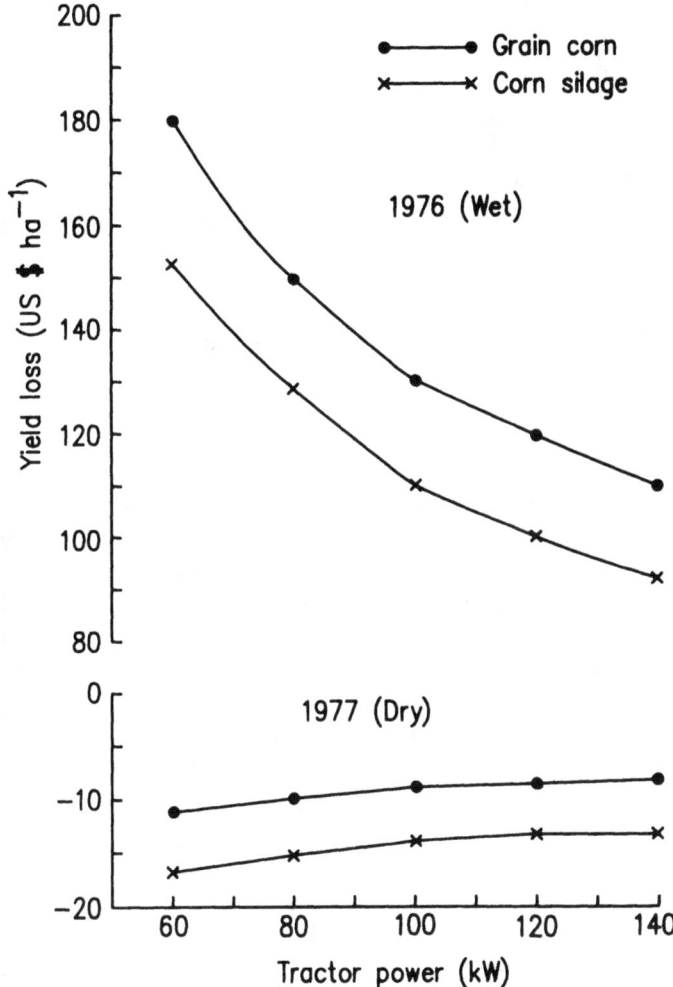

Fig. 5. The effect of tractor power in conventionally cultivated grain corn production on crop yield losses attributable to compaction in both wet and dry seasons (after Lavoie et al., 1991).

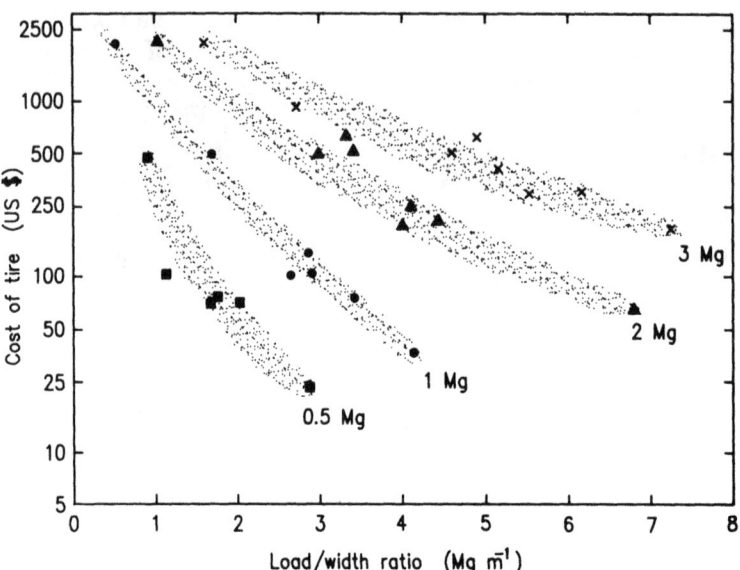

Fig. 6. Relationship between cost of tires of different load/width ratio for loads of 0.5, 1, 2, and 3 Mg (U.K. data, 1985).

attributed to the advantage of a more compact soil in a dry season.

Effect of changing ground contact pressure

Purchase costs of wide, low ground contact pressure tires (U.K. data)

A study undertaken in the U.K. showed that the purchase cost of wide tires of 0.5, 1, 2 and 3 Mg nominal carrying capacity, was inversely and logarithmically related to the load/width ratio (Fig. 6). As the width increases, load/width ratio decreases and cost increases very rapidly, so that the use of very wide tires would have to attract a very large increase in crop output to be justified. Any additional wear on wide, low ground pressure tires leading to more frequent replacement than for conventional tires would incur still higher costs. Apart from cost, there may be restrictions to the use of very wide tires such as their unsuitability for plowing if the plow-side wheels have to travel in the furrow, although recently plows have been developed which allow the use of tractor wheels up to 0.7 m wide (Anon., 1990).

Purchase costs of wide, low ground contact pressure tires (U.S.A. data)

Eradat Oskoui (1993) conducted a survey of over 1200 common agricultural tires used in the U.S.A. Tire cost was exponentially related to tire width (Fig.7) for conventional ($r^2 = 0.81$), low ground pressure ($r^2 = 0.79$ and $r^2 = 0.90$ for manufacturers A and B, respectively) and specialised tires ($r^2 = 0.93$), e.g., "heavy

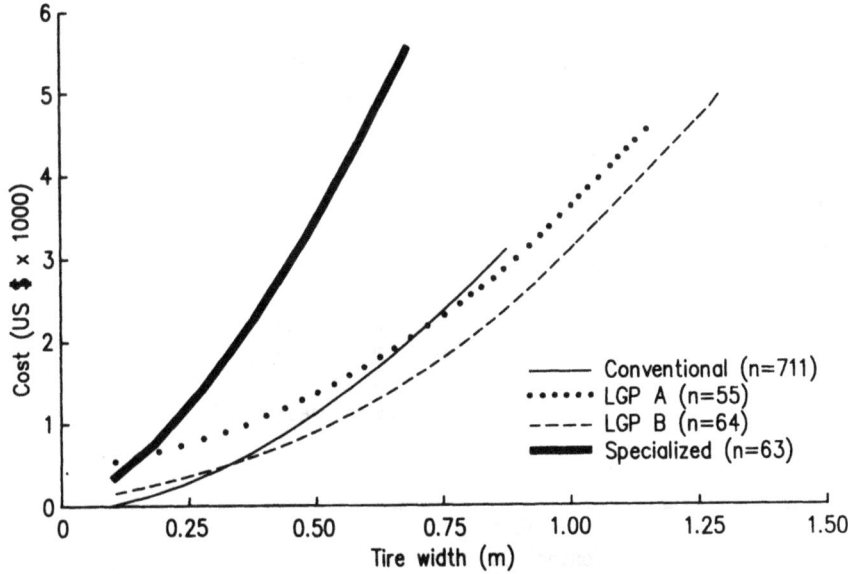

Fig. 7. Influence of tire width on the cost of tires of all makes and types available in the U.S.A. at 1991 prices. "LGP A" and "LGP B" are low ground contact pressure tires from different manufacturers (from Eradat Oskoui, 1993).

duty" and "long life". The limit for tire width, and hence contact area, was found to be about 0.9 m for conventional tires and 1.3 m for low ground contact pressure tires.

Since the actual tire parameters which are most relevant to compaction are tire-soil contact area and load carrying capacity, a new parameter called the "contact coefficient" was defined by Eradat Oskoui (1993) as the product of tire-soil contact area at 75 mm sinkage (at the manufacturer's recommended inflation pressure and ply rating). Tire cost was also found to be exponentially related to the contact coefficient (Fig. 8). The inclusion of ply rating considerably increased the precision in predicting tire cost from the contact area for low ground contact pressure tires ($r^2 = 0.92$ and $r^2 = 0.96$ for manufacturers A and B, respectively), conventional tires ($r^2 = 0.83$) and specialised tires ($r^2 = 0.95$).

No significant difference in cost of conventional tires of similar contact coefficient was found for different manufacturers. However, low ground contact pressure tires from one manufacturer (labelled "LGP A" in Fig. 8) were appreciably more expensive than the products of other manufacturers (labelled "LGP B"). The prices of specialized tires were about 35% greater than of conventional tires (Figs. 7 and 8).

Tires with high contact coefficients are very expensive (Fig. 8) since, as the tire width increases, the ply rating also may have to be increased to sustain the

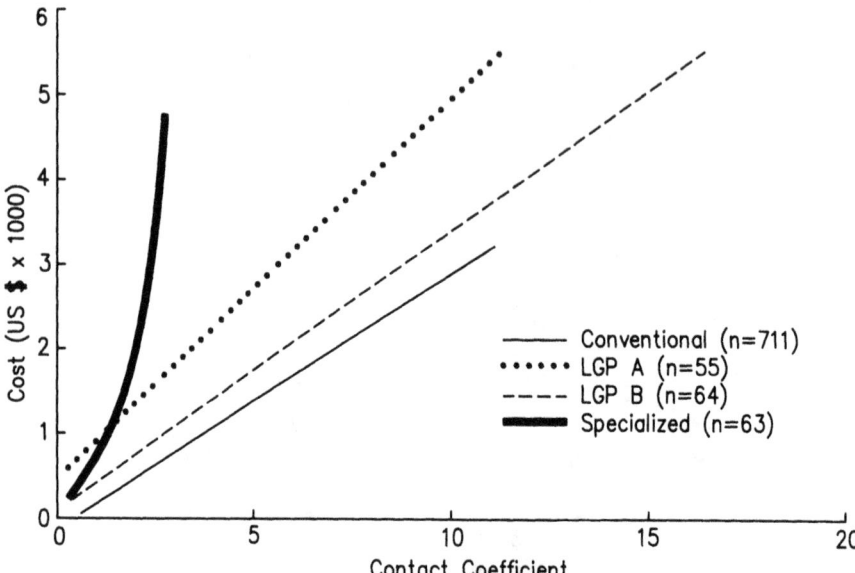

Fig. 8. Influence of the contact coefficient on the cost of tires of all makes and types available in the U.S.A. at 1991 prices. "LGP A" and "LGP B" are low ground contact pressure tires from different manufacturers (from Eradat Oskoui, 1993).

recommended load. The average predicted contact coefficient of standard tires fitted to tractors, increases rapidly with tractor engine power from about 1 at 15 kW up to about 4 at 150 kW, but only reaches about 4.5 at 250 kW. If these tractors had been fitted with low ground contact pressure tires, very much higher values of contact coefficient would have been involved (Fig. 8).

Benefit-cost relations of low ground contact pressure tires for grass silage production in Scotland

In a long-term field study in Scotland, Douglas (1989) compared the yield of grass silage following the use of a tractor, forage harvester and silage trailer fitted with either conventional or low ground contact pressure tires. The use of low ground contact pressure tires showed a yield value advantage of US $ 252 ha^{-1} at the first cut, thereby allowing the purchase cost of the alternative running gear (US $ 9,000) to be recovered at the first cut from an area of only 36 ha.

Benefit-cost relations for low ground contact pressure tires for a combine harvester and transport vehicles in cereal harvesting

On cereal farms of 100, 200 and 300 ha, the economic benefits of substituting low ground contact pressure tires for conventional tires on a combine harvester (5.3 m cut), tractor (90 kW) and trailer (10 Mg capacity, single or tandem axle), were estimated in a simulation study by Godwin et al. (1992), using the Arvidsson

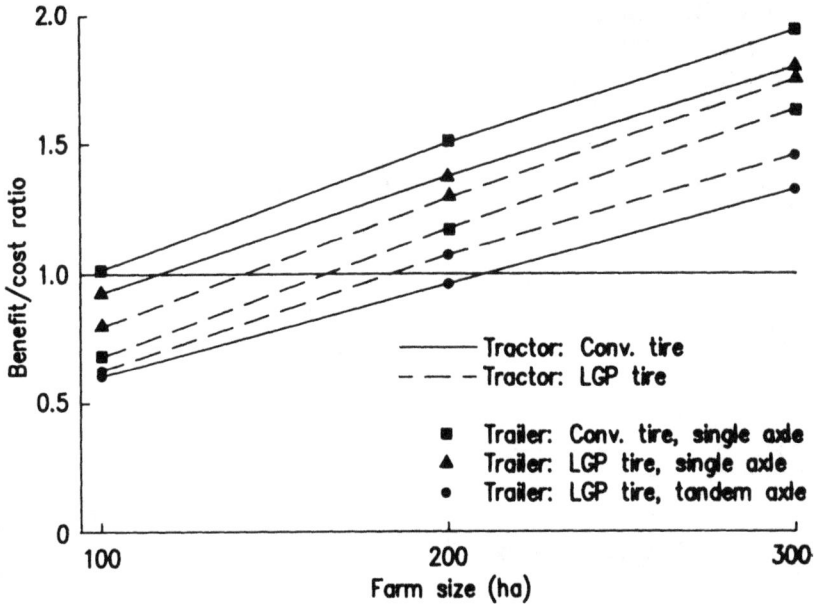

Fig. 9. The influence of farm size on the benefit/cost ratio for conventional and low ground contact pressure tires and single and tandem axles fitted to tractor/trailer combinations used for transport after a cereal combine harvester with low ground contact pressure tires (after Godwin et al., 1992).

and Håkansson (1991) model. The additional costs for low ground contact pressure tires on the combine harvester and transport vehicles were US $ 44 ha^{-1} for a 100-ha farm, but only US $ 16 ha^{-1} for a 300-ha farm.

The study included a comparison of the costs of different post-harvest remedial cultivation systems (minimum tillage, moldboard plowing, subsoiling). The depths of these cultivations were adjusted according to the estimated depth of ruts resulting from the combine harvester (52 mm for conventional tires; 43 mm for low ground contact pressure tires).

The economic yield of the following crop was calculated using the model to predict the effects of the type of tires on the combine harvester and tractors and the number of axles on the trailers. The estimated advantage of using low ground contact pressure tires was calculated from the estimated yield benefit plus reduced tillage benefit minus the additional costs. Results were reported in terms of both net benefit and benefit/cost ratio. Net benefit values were strongly dependent on farm size and were usually negligible on farms <100 ha. On 200- and 300-ha farms, changing to low ground contact pressure tires on the combine alone, gave benefit-cost values of US $ 3.7 ha^{-1} and US $ 5.0 ha^{-1}, respectively. The overall benefit/cost ratio values of fitting low ground contact pressure tires or a tandem axle to the tractor/trailer combination, were again strongly related to farm size (Fig. 9). If, in practice, the use of low ground contact pressure tires could avoid the need for remedial subsoiling, the net benefit of using these tires on all harvesting vehicles would increase to about US $ 30 ha⁻¹ for all farm sizes.

Use of multiple rear wheels on tractors

The costs of multiple wheels for tractors or trailers can be much reduced if standard size tires can be used rather than wide tires. Eradat Oskoui (1993) showed that, to achieve a contact coefficient of 14, it is cheaper to fit dual conventional tires, each costing US $ 2000 and having a contact coefficient of 7, than to fit single low ground contact pressure tires, at US $ 6,800 each, even when the cost of the rims is also taken into account. This illustrates a possible

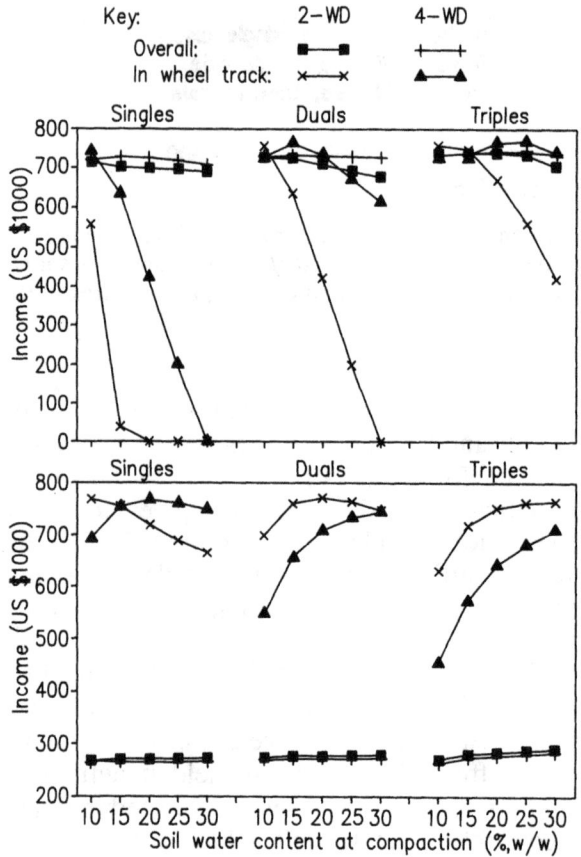

Fig. 10. The effect of soil water content at compaction on the income from corn yield on 455 ha, either in the wheel tracks or overall, for both 2-wheel and 4-wheel drive tractors, fitted with either single or dual or triple wheels. Top: no-till, medium soil in a medium rainfall growing season; bottom: plowed medium soil in a dry season (after Eradat Oskoui and Voorhees, 1992).

application of the concept of contact coefficient in the selection of tires.

Similarly, the addition of steel cage wheels (Gee-Clough et al., 1981; Dickson et al., 1983) is also comparatively cheap. However, any fitting or removal of conventional or other wheels that may be necessary for operational reasons (e.g., for legal reasons on public roads or when changing to other duties, such as plowing), must incur costs in view of the time taken.

Response in output of corn to the use of multiple rear wheels on a tractor used for planting in the U.S.A.

The computer-based model developed by Eradat Oskoui and Voorhees (1992) was used to estimate the effect on income from corn (both in the wheel tracks and overall) of using single, dual and triple rear wheels on a 222-kW tractor (2-wheel and 4-wheel drive) used for planting. The multi-factor capability of the model enabled the effects of two cultivation systems (no-till, plowing), three growing seasons (wet, medium, dry), five values of soil water content at the time of compaction (10-30%, w/w) and three soil textures (heavy, medium, light) to be included in the analysis. A selection of the data (Fig. 10) illustrates that the increase in tire/soil contact area and associated reduction in ground contact pressure with dual and triple wheels, had a pronounced effect on yield in the wheel tracks, depending on soil and weather conditions. Where aeration tended to be deficient (no-till, medium rainfall), the use of duals and triples was very effective in increasing income from corn growing in the wheel tracks. Where soil water was deficient (plowed, dry season), the greater compaction from single wheels resulted in higher yields in the wheel tracks, but this had virtually no influence on overall income.

Use of multiple front wheels

For work on weak soils, e.g., seedbeds, there may be economic justification for using wide tires or multiple tires on the front axle of a tractor (Campbell and Dickson, 1984). A further possibility is to mount four additional conventional front wheels and tires on a frame between and ahead of the standard front wheels, thus greatly reducing the ground contact pressure of these wheels. In the U.K., such a system is commercially available at a cost of US $ 2,400.

Use of additional axles

Certain vehicles, such as large transport trailers and tankers, can be readily fitted with an extra axle so that additional wheels are mounted in tandem. This will increase the ground contact area, reduce the total volume of ruts (Campbell et al., 1986) and cause less rolling resistance than if the additional wheels were mounted on the same axle. The cost of purchasing and fitting additional axles to trailers and tankers must be added to the cost of the extra tires, but twin axles are now commonly employed where large loads have to be carried in towed vehicles.

Use of tracks and half-tracks

Tracks and half-tracks, whether of steel or rubber construction, offer opportunities of reducing ground contact pressure (see Chapter 21) but, since they are generally much more expensive than wheels of corresponding load capacity, their use tends to be restricted to applications where high traction and wet-ground trafficability are important.

At higher power levels, the price per kW for a crawler tractor is appreciably higher than for wheeled tractors. Rubber-tracked crawler tractors are about 50% more expensive than equivalent 4-wheel drive tractors and about 25% more than equivalent steel-tracked crawlers.

Some farmers consider tracks to be economically justified for high draft operations, such as plowing in certain regions having clay soils. Adverse crop responses to the use of tracked vehicles for seedbed preparation have sometimes been reported for peas (Dawkins et al., 1981) and barley (Soane, 1970). Obviously, the economics of using tracks must be based on a complex analysis. However, this appears not to have been attempted.

Changes in tractor mechanisms and operating conditions

Gill (1971) pointed out that a number of comparatively cheap modifications to tractor mechanisms, such as a differential lock, might have appreciable benefit in reducing the compaction resulting from tractor use. A feature of 2- and 4-wheel drive mechanisms is the difference in the ratio of front to rear axle load, which is usually about 0.6 and 0.4 for very small and large 2-wheel drive tractors, respectively, and greater than 1 for 4-wheel drive tractors. The average ground contact pressure under a 4-wheel drive tractor will therefore tend to be lower than under a comparable 2-wheel drive tractor of the same weight. The extra cost of a 4-wheel drive tractor over a 2-wheel drive tractor is about 10% in the U.S.A. and about 20% in the U.K. In the U.S.A., 2-wheel drive tractors fitted with "front-wheel-assist" (i.e., driven standard-size front wheels) generally cost about 7% more than those without that option.

The influence of the wheel-drive type used on tractors at planting on the income from corn in the Northern Corn Belt in the U.S.A., was estimated by Eradat Oskoui and Voorhees (1992) (Fig. 10). Where soils were suffering from deficient aeration (i.e., after no-till in a medium rainfall growing season), income was appreciably higher from the crop in the wheel tracks for 4-wheel drive than 2-wheel drive tractors. However, when the soil water status was low (i.e., after plowing in a dry season), the extra compaction under 2-wheel drive tractors resulted in increased income. This demonstrates the complexity of the soil/crop/machine/weather interactions which must be taken into account in economic analyses.

The speed of vehicles has an important influence on the economic efficiency of field operations and, in addition, may affect the intensity of the resulting compaction. In general, tractor speeds are increasing and studies have tended to show that compaction is less at higher vehicle speeds (see Chapter 25).

An ability to adjust tire inflation pressure from the cab to suit working conditions could have important effects on compaction, which might be cost effective. Tests on radial tires operated at inflation pressures less than the minimum recommended, showed small advantages in terms of reduced compaction and increased yield of barley (Campbell et al., 1984). Although this practice cannot be recommended because of the danger of and the cost associated with damage arising from the additional flexing of the tire carcass, there are opportunities to explore the application of inflation pressure control to suit soil or road conditions.

Effect of reducing the number of vehicle passes and area covered

Common-sense restriction of vehicle traffic in fields, especially at planting and harvest, can often be achieved at little or no cost. In cereal crops in Europe, there is widespread use of temporary tramlines in which all vehicle wheels for sowing, fertilizing and spraying operations run in the same tracks. Rance (1982) showed that, in 351 fields in England, the adoption of tramlines resulted in a mean increase in crop output of US $ 130 ha^{-1}. While the losses in the tramline area (3-5% of the total field) were about 100%, this was more than compensated for by increased yields (10-20%) over the rest of the field. Since that time, the most frequently used tramline spacing has increased from 12 to 18 m and some farmers are using 24 m spacing.

The elimination or combination of soil working operations may be cost effective. Gunjal and Raghavan (1986) estimated that the reduction of machinery passes from 5 to 4 in a pea crop, could increase net benefit by US $ 28 ha^{-1} (from savings of US $ 15 ha^{-1} in machinery and profits of US $ 13 ha^{-1} by increased yield). Changing the cultivation system from conventional to minimum or no-till systems will reduce traffic appreciably. Machinery size has a dominant effect on the number of passes per unit area and any farm-scale analysis related to the effects of different size (power) of machinery must utilise realistic data concerning the field distribution of wheel tracks, including headlands.

Effect of extended axles and gantries

The availability of techniques for growing crops in traffic-free conditions (see Chapter 22), have given rise to concern as to the economic implications (see Chapter 24). The cost of extending the length of tractor axles can be negligible, compared with other production costs. Axles on tractors up to 250 kW can usually be extended up to about 3.5 m, compared to a maximum of about 1.5 m for 120-kW tractors. However, commercially-available gantry-type tractors cost about US $ 250,000 and are generally considered too expensive for conventional farm crops, although some specialized vegetable or fruit farms may be able to justify such investment.

Economic analysis by Campbell and McGregor (1989) of data obtained with machines with their axles extended to a 2.8-m track, showed that the gross margin advantage of a field-scale zero-traffic system for winter barley, using a full-scale 12-m wide gantry, would be US $ 92 ha^{-1} under plowing and US $ 46 ha^{-1} under no-till.

Effect of soil water status at the time of compaction

The effect of soil water status on the incidence of compaction in the field is well documented (see Chapter 25). Clearly, this effect must be included in any model study of the economics of machinery management because of the variation in weather conditions. Also, farmers have some choice as to when field operations are undertaken and can, to some extent, refrain from allowing vehicle traffic on the soil when the water content is high enough to cause deep rutting, smearing and excessive compaction. However, at both planting and harvest, delays may impose a timeliness penalty and decisions have to be taken regarding the economics of either carrying out the operation on schedule or delaying it in the hope that drier conditions will prevail. There is need for a more exact method of assessing the benefit-cost relations of timeliness decisions of this type.

In a simulation study by Danfors et al. (1992), the effects on yield losses of conventional and low ground contact pressure tires used for plowing, were related to the soil water status, expressed on a arbitrary scale from 1 (very dry) to 5 (very moist) (Fig. 11). The yield losses attributable to changes in soil water status within the range 3.0-4.0, were as large or larger than the yield losses attributed to the two types of tires. This emphasizes the economic importance of the soil water content at the time traffic is applied.

Effect of wet and dry growing seasons

Although seasonal weather effects, particularly rainfall, are generally out of the control of the farmer, there are many important relationships between seasonal weather and the economic responses to modifications to vehicle management which must be taken into account when using predictive models.

Eradat Oskoui and Voorhees (1992) used their model to predict the influence of seasonal rainfall in the Northern Corn Belt of the U.S.A. on overall income from corn, in relation to type of cultivation and soil texture (Table 2). Under no-till, the highest income was obtained after a medium growing season rainfall. Wet and dry seasons resulted in decreases in income, especially on the heavy soil, for which income was nil. On plowed soil, however, the effect of increases in seasonal rainfall were always accompanied by large increases in income, with income also tending to increase as the clay content increased. These results

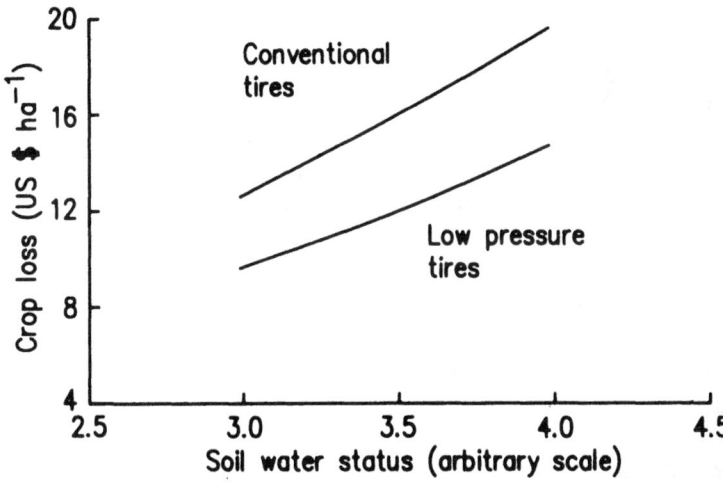

Fig. 11. The effect of soil water status at plowing on crop yield loss for tractors fitted with either conventional or low ground contact pressure tires. Soil water status ranged from 1 (very dry) to 5 (very moist), with 3 representing the optimum for plowing (after Danfors et al., 1992).

highlight the importance of including seasonal weather as a variable in any study on the economics of vehicle modification; they show that periods of deficiency or excess in either aeration or soil water, will give rise to entirely different income responses.

TABLE 2

Effect of soil texture, cultivation system and growing season rainfall on estimated overall income from corn, following use of a standard reference machinery management system (222-kW, 2-wheel drive tractor) for planting (after Eradat Oskoui and Voorhees, 1992)

Cultivation system	Growing season weather	Annual rainfall (mm)	Overall income (US $ ha⁻¹)		
			Light soil (5% clay)	Medium soil (20% clay)	Heavy soil (35% clay)
No-till	Wet	>520	1450	1297	0
	Medium	400	1582	1538	1429
	Dry	<280	1055	1319	1297
Plowed	Wet	>590	1121	1319	1407
	Medium	400	681	857	1033
	Dry	<280	220	571	681

CONCLUSIONS

(1) Reduction in the value of crops attributable to soil compaction can, in certain circumstances, amount to several hundred US dollars per hectare.

(2) Factors, such as farm size, soil type, seasonal weather and cultivation systems, may have a dominant influence on the economic responses to changes in machinery management and must be included in any model study.

(3) The costs of avoiding or minimising compaction through the adoption of changes in the running gear and management of vehicles, may be economically acceptable but more work is needed to evaluate these changes in greater detail in full-scale studies of farming systems.

(4) Extreme caution is needed when extrapolating, in economic terms, the results of small-scale traffic experiments to commercial whole-farm crop production.

(5) Changes in the management of vehicles and their running gear will have both direct and indirect effects on the overall profitability of whole-farm enterprises. However, many of these effects have not been quantified and thus the full economic implications of such changes have not been evaluated.

(6) There is a need for more holistic systems-based analyses which would allow net farm income to be determined more accurately by including the costs of both vehicle modification and amelioration of compaction through cultivation.

(7) Computer-based models are powerful tools for predicting economic responses to changes in machinery management under multi-factor situations and their use should be extended to include long-term, whole-farm analyses of all operations involving field machinery throughout the crop production cycle. However, such models need validation and should be mutually compared rather than used in isolation.

REFERENCES

Anon., 1984. Diesels lose their cool. New Scientist, 101: 18.

Anon., 1990. New plough lets you use wide tyres. Farmers Weekly, 113: 41.

Arvidsson, J. and Håkansson, I., 1991. A model for estimating crop yield losses caused by soil compaction. Soil Tillage Res., 20: 319-332.

Bakermans, W.A.P. and De Wit, C.T., 1970. Crop husbandry on naturally compacted soils. Neth. J. Agric. Sci., 18: 225-246.

Bicki, J.J. and Siemens, J.C., 1990. Crop response to wheel compaction. Am. Soc. Agric. Eng., St. Joseph, MI, U.S.A., ASAE Pap. 90-1093, 12 pp.

Campbell, D.J., 1982. A review of the clod problem in potato production. J. Agric. Eng. Res., 27: 373-395.

Campbell, D.J. and Dickson, J.W., 1984. Effect of four alternative front tyres on seedbed compaction by a tractor fitted with a rear wheel designed to minimise compaction. J. Agric Eng. Res., 29: 83-91.

Campbell, D.J. and McGregor, M.J., 1989. The economics of zero traffic systems for winter barley in Scotland. Proc. 11th Int. Cong. Agric. Eng., Dublin, Vol. 3, pp. 1743-1747.

Campbell, D.J., Dickson, J.W. and Ball, B.C., 1984. Effect of under-inflation of tractor tyres on seedbed compaction and winter barley establishment and yield. J. Agric. Eng. Res., 29: 151-158.

Campbell, D.J., Dickson, J.W., Ball, B.C. and Hunter, R., 1986. Controlled seedbed traffic after ploughing or direct drilling under winter barley in Scotland, 1980-1984. Soil Tillage Res., 8: 3-28.

Cannell, R.Q., Belford, R.K., Gales, K., Dennis, C.W. and Prew, R.D., 1980. Effects of water logging at different stages of development on the growth and yield of winter wheat. J. Sci. Food Agric., 31: 117-132.

Carter, L.M., 1980. Wheels are costly. Am. Soc. Agric. Eng., St Joseph, MI, U.S.A., ASAE Pap. PR/80.

Chancellor, W.J., 1976. Compaction of soil by agricultural equipment. Univ. California, Div. Agric. Sci., Davis, CA, U.S.A., Bull. 1881, 53 pp.

Danfors, B., Ilskog, E., Håkansson, I. and Arvidsson, J. 1992. Analysis of agricultural field equipment systems with respect to soil compaction effects. Proc. AGENG 92, Uppsala, Sweden, Pap. 920909, 7 pp.

Dann, R.T., 1983. Agricultural equipment: Cash squeeze slows development. Machine Design, (Sept. 8): 43-51.

Dawkins, T.C.K., Hebblethwaite, P.D., McGowan, M. and King, J., 1981. Soil physical conditions and the pea crop. Soil Water, 9: 19-21.

Dickson, J.W., Campbell, D.J. and Henshall, J.K., 1983. An assessment of seedbed compaction by open, flat-lugged, steel tractor wheels. J. Agric. Eng. Res., 28: 45-60.

Dickson, J.W., Campbell, D.J. and Ritchie, R.M., 1992. Zero and conventional traffic systems for potatoes in Scotland, 1987-1989. Soil Tillage Res., 24: 397-419.

Douglas, J.T., 1989. Avoiding soil damage from wheel traffic in silage fields. Scott. Agric. College, Edinburgh, U.K., Tech. Note T198, 2 pp.

Douglas, J.T., Campbell, D.J. and Crawford, C.E., 1992. Soil and crop responses to conventional, reduced ground pressure and zero traffic systems for grass silage production. Soil Tillage Res., 24: 421-439.

Elliott, J.G., 1979. The price of loaded wheels. Soil Water, 7: 9-11.

Eradat Oskoui, K., 1981. Agricultural mechanisation systems analysis - Tractor power selection for tillage operations. Ph.D thesis, Univ. Edinburgh, Edinburgh, U.K., 425 pp.

Eradat Oskoui K., 1993. Economic optimization of tire selection criteria. Am. Soc. Agric. Eng., St Joseph MI, U.S.A., ASAE Pap. 93-1032.

Eradat Oskoui, K. and Voorhees, W.B., 1991a. Economic consequences of soil compaction. Trans. ASAE, 34: 2317-2323.

Eradat Oskoui K. and Voorhees, W.B., 1991b. Modelling crop response to soil compaction. Am. Soc. Agric. Eng., St Joseph MI, U.S.A., ASAE Pap. 91-1556, 9 pp.

Eradat Oskoui K. and Voorhees, W.B., 1992. Development of COMPAC: Compaction optimization and management for agricultural crops. Am. Soc. Agric. Eng., St Joseph MI, U.S.A., ASAE Pap. 92-1528, 18 pp.

Gameda, S., Raghavan, G.S.V., McKyes, E., Watson, A. K. and Mehuyes, G., 1994. Long term effects of a single incidence of high axle load compaction on a clay soil in Quebec. Soil Tillage Res., 29: 173-177.

Gaultney, L., Krutz, G.W., Steinhardt, G.C. and Liljedahl, J.B., 1982. Effects of subsoil compaction on corn yields. Trans. ASAE, 25: 563-569, 575.

Gee-Clough, D., Aggrawal, S., Jayasundera, M.L., Singh, A., Tiangco, V.M. and Shah, N.G., 1981. Recent research into vehicle performance in wetland conditions. Proc. 7th Int. Conf. ISTVS, Calgary, Canada, Vol. 1, pp. 205-237.

Gill, W.R., 1971. Economic assessment of soil compaction. In: K.K. Barnes, W.M. Carleton,

H.M. Taylor, R.I. Throckmorton and G.E. Vanden Berg (Editors), Compaction of Agricultural Soils. Am. Soc. Agric. Eng., St. Joseph, MI, U.S.A., pp. 431-458.

Godwin, R.J., Kerr, D.McM., Kuthan, E. and Håkansson, I., 1992. An economic evaluation of wheel/tyre systems for cereal harvesting. Proc. AGENG 92, Uppsala, Sweden, Pap. 920101, 18 pp.

Gunjal, K.R. and Raghavan, G.S.V., 1986. Economic analysis of soil compaction due to machinery traffic. Appl. Eng. Agric., 2: 85-88.

Gunjal, K.R., Lavoie, G. and Raghavan, G.S.V., 1987. Economics of soil compaction due to machinery traffic and implications for machinery selection. Can. J. Agric. Econ., 35: 591-603.

Gupta, S.C. and Larson, W.E., 1982. Modeling soil mechanical behaviour during tillage. In: D.M. Kral, S. Hawkins, P.W. Unger and D.M. Van Doren (Editors), Predicting Tillage Effects on Soil Physical Properties and Processes. Am. Soc. Agron./Soil Sci. Soc. Am., Madison, WI, U.S.A., Spec. Publ. 44, pp. 151-178.

Håkansson, I., Voorhees, W.B. and Riley, H., 1988. Vehicle and wheel factors influencing soil compaction and crop response in different traffic regimes. Soil Tillage Res., 11: 239-282.

Harrison Reed, A., 1983. The erosion risk of compaction. Soil Water, 11: 29, 31, 33.

Jaggard, K.W., 1977. Effects of soil density on yield and fertiliser requirement of sugar beet. Ann. Appl. Biol., 86: 301-312.

Knowles, D. and Audsley, E., 1983. The effect of timeliness of sowing on crop yields. Nat. Inst. Agric. Eng., Silsoe, U.K., Div. Note DN 1194, 19 pp.

Kubota, T. and Williams, R.J.B., 1967. The effects of changes in soil compaction and porosity on germination, establishment and yield of barley and globe beet. J. Agric. Sci., Camb., 68: 227-233.

Lavoie, G., Gunjal, K. and Raghavan, G.S.V., 1991. Soil compaction, machinery selection, and optimum crop planning. Trans. ASAE, 34: 2-8.

Lowery, B. and Schuller, R.T., 1991. Temporal effects of subsoil compaction on soil strength and plant growth. Soil Sci. Soc. Am. J. 55: 216-223.

Miller, M.H., 1986. Soil degradation in Eastern Canada: Its extent and impact. Can. J. Agric. Econ., 33: 7-18.

Moots, C.K., Nickell, C.D. and Gray, L.E., 1988. Effects of soil compaction on the incidence of *Phytophthora megasperma* f.sp. *glycinea* in soybean. Plant Disease, 72: 896-900.

Price, J.S., 1989. Future developments in stripper harvesting. Proc. 11th Int. Cong. Agri. Eng., C.I.G.R., Dublin, Ireland, Vol. 3, pp. 2023-2029.

Raghavan, G.S.V., McKyes, E., Gendron, G., Borglum, B.K. and Lee, H.H., 1978. Effects of tire contact pressure on corn yield. Can. Agric. Eng., 20: 34-37.

Raghavan, G.S.V., Taylor, F., Vigier, B., Gauthier, L. and McKyes, E., 1982. Effect of compaction and root rot disease on development and yield of peas. Can. Agric. Eng., 24: 31-34.

Rance, D., 1982. Pointers to profitable winter barley. ICI, Agric. Div. Publ., 31 pp.

Smith, K.A., Elmes, A.E., Howard, R.S. and Franklin, M.F., 1984. The uptake of soil and fertiliser nitrogen by barley growing under Scottish conditions. Plant Soil, 76: 49-57.

Soane, B.D., 1970. The effects of traffic and implements on soil compaction. J. Proc. Inst. Agric. Eng., 25: 115-126.

Stephens, L.E., 1990. Soil compaction: Is it worth the cost? Am. Soc. Agric. Eng., St. Joseph, MI, U.S.A., ASAE Pap. 90-1075, 6 pp.

Subotic, B. and Stanacev, S., 1982. Effect of growth of basic tillage and additional soil compaction after harvest on the yield, deformations and diseases of sugar beet roots. Proc. 9th Conf. Int. Soil Tillage Res. Organ. (ISTRO), Osijek, Yugoslavia, pp. 243-248.

Swaine, R.W., 1981. The field problem - compaction and yield. Proc. SAWMA Conf. Soil

Compaction, Causes and Cures, Silsoe, U.K., Pap. 1, 4 pp.

Vigier, B. and Raghavan, G.S.V., 1980. Soil compaction effects in clay soils on common root rot of canning peas. Can. Plant Disease Survey, 60: 43-45.

Voorhees, W.B. 1977. Soil compaction: How it influences moisture, temperature, yield, root growth. Crops Soils Mag., 29: 7-10.

Voorhees, W.B. and Hendrick, J.G., 1977. Compaction: Good and bad effects on energy needs. Crops Soils Mag., 29: 11-13.

Voorhees, W.B., Holt, R.F. and Orr, P., 1977. Basic and applied implications of wheel traffic and soil compaction in the potato industry. Am. Potato J. Abstr., pp. 3-4.

Voorhees W.B., Johnson J.F., Randall G.W. and Nelson, W.W., 1989. Corn growth and yield as affected by surface and subsoil compaction. Agron. J., 81: 298-303.

Witney, B.D. and Elbanna, E.G., 1985. Simulation of crop yield losses from untimely establishment. Res. Dev. Agric., 2: 105-117.

Witney, B.D., Elbanna E.B. and Eradat Oskoui, K., 1984. Tractor power selection with compaction constraints. Proc. 8th Int. Conf. ISTVS, Cambridge, U.K., Vol. 2, pp. 761-769.

Soil Compaction in Crop Production
B.D. Soane and C. van Ouwerkerk (Eds.)
©1994 Elsevier Science B.V. All rights reserved.   569

CHAPTER 24

Economics of Gantry- and Tractor-based Zero-Traffic Systems

W.C.T. CHAMEN[1], E. AUDSLEY[1] and J.B. HOLT[2]

[1]Silsoe Research Institute, Silsoe, U.K.
[2]Bedford, U.K.

SUMMARY

The economics of zero-traffic systems are discussed with particular reference to the operating characteristics of gantries. Differences between gantry- and tractor-based systems and systems which use a mixture of these machines, are examined. The effect of machine designs, speed, mobility, draught requirements, payloads and traffic lane yield losses are discussed. These considerations are included in a case study on a hypothetical farm in East Anglia growing combinable crops. Five mechanization systems of contrasting traffic intensity were compared: (1) conventional practice, using tractors of 56, 75 or 112 kW; (2) a 6-m tractor-based zero traffic system, using 75- or 112-kW tractors; (3) a partial gantry system, using both a low-powered, wide-working gantry (12 m, 70 kW, 5.8 kW m^{-1}) and conventional 56- or 75-kW tractors; (4) a two-width gantry system, using a low-powered, wide-working gantry (12 m), a high-powered narrow-working gantry (6 m, 125 kW, 21 kW m^{-1}) and a similar width 163-kW gantry-based harvester; (5) a 6-m gantry system, using only the high-powered, narrow working gantries. On medium soil there was a 9% difference in profitability between systems, with a 6-m tractor-based zero-traffic system being the least profitable and a system using two widths of gantry providing the greatest return. The conventional system was intermediate on both medium and heavy soil, but on the latter, the range of differences approached 30%. The greatest profit was generated by a 6-m gantry system. All the zero-traffic systems relied on an increase in yield to maintain their position relative to conventional practice. The profitability of gantry systems would be increased if the capital cost of the equipment were reduced or the engine power increased. Farms of 400-500 ha are needed to justify use of the zero-traffic systems considered in the case study.

INTRODUCTION

Both in the U.K. and worldwide, farms have become more mechanised, and the complexity, cost and weight of field machinery have increased. Increases in the weight of such machinery have created the need for larger, more powerful tractors to ameliorate the soil damage which these machines have caused. Considerable

Fig. 1. A 12-m span gantry shown operating with a power-driven cultivator.

research has been undertaken to investigate the extent of this damage, and a number of solutions to the problem have been proposed. The zero-traffic approach is one which aims to separate permanently the cropped and wheeled areas of a field (Taylor, 1989) (see Chapter 22). This can be achieved either with gantries, or with modified tractor-based equipment. Gantries, unlike tractors, are machines whose implements normally work within the span of their widely spaced pairs of tandem wheels (Fig. 1). They generally operate by working parallel to the long side of a field and after each pass, move sideways by various means to the adjacent strip of ground. One of the traffic lanes is then used again during the return pass across the field. In this way gantries create single uncropped traffic lanes at a centre distance equal to their span. Tractor-based zero-traffic systems, on the other hand, create double tracks, neither of which are used on adjacent passes across the field. These systems also require the implement width to match the centre distance of the double tracks, or to be in multiples of it. Systems which use tractors for certain operations and gantries for others are also common, particularly for crops susceptible to seedbed compaction or direct mechanical damage.

The objective of this chapter is to consider the cost of gantries and other systems in relation to their working dimensions, the operations performed and the types of crops grown. The operating costs[1] and work rates compared with

---

[1]Where costs are given in currency units they have been converted from UK £ to US $ at the exchange rate applicable to early September 1992. Such values should be used for making relative comparisons between machinery systems rather than for the precise evaluation of actual costs.

conventional equipment are also discussed and the whole-farm economics of a hypothetical farm growing combinable crops in the East Anglia region of the U.K are investigated.

OPERATIONAL FACTORS AFFECTING MACHINE DESIGN FOR ZERO TRAFFIC SYSTEMS

Table 1 shows how the specifications of vehicles intended for field crop production vary considerably according to their use. The greater the range of tasks the vehicle needs to perform, the greater its complexity and cost. For almost all operations, excessive vehicle weight is a disadvantage as it, in turn, increases the size, cost and weight of other components, such as the engine and running gear. Any resulting increase in the width of running gear will cause more land to be lost to traffic lanes. Similarly, increases in vehicle weight will also be accompanied by reduction in the load-carrying capacity, and therefore in the work rate. Additional torque would also be necessary to negotiate slopes, because even on level land, gradients of perhaps 15% are often encountered in stepping up from a field to a farm track. The weight of the main spanning beam of a gantry generally increases as the square of the span. Its width is normally determined by factors such as the row spacing of the crops to be grown, the width capable of being harvested efficiently, the size of the labour team required, the handling and transport of produce and the manoeuvrability in and between fields.

The implement width of tractor-based systems generally needs to be in excess of 3 m, to ensure that the area of land lost to traffic lanes is within economic limits. Unfortunately, this generally precludes ploughing, other than stubble or shallow ploughing. The degree of modification required to ensure that all vehicles

TABLE 1

Vehicle characteristics required for various field operations

Operation	Draught[1]	Payload[1]	Speed[1]	Implement lifts	Work platform
Spraying or broadcasting	L	M/H	H	Yes	No
Planting	M	L	VL	Yes	Yes
Leaf vegetable harvesting	L	H	VL	Possibly	Yes
Root crop harvesting	M	H	L/M	Possibly	Possibly
Cereal harvesting	L	H	M/H	Yes	No
Cultivations (draught)	H	L/M	L/M	Yes	No
Cultivations (powered)	L	M/H	L	Yes	No
Sowing	L	M/H	L/H	Yes	No
Transport	L	H	H	No	No

[1]VL = very low; L = low; M = medium; H = high.

run in the same tracks, depends very largely on the range of crops being grown. For a cereals enterprise, for example, the most difficult machine to modify is the combine harvester and the most logical approach is to convert everything else to fit in with the track width of this vehicle. Some tractors may have this additional width as an optional setting, but others would require some modification, and most have a reduced load-carrying capacity at this setting.

The implement controls required for a tractor-based zero-traffic system do not need to differ from those currently provided. With gantries, the workplace, which is generally at one end of the vehicle, must allow the driver to face in each of four travel directions, and still have all the controls to hand.

MACHINE DESIGNS AND THEIR EFFECT ON COSTS

Gantries

Most gantries, because of the remote position of their engines in relation to the driven wheels, have a hydrostatic transmission. Although other methods of drive are possible, it is unlikely that presently available alternatives will appreciably affect the overall cost of the vehicle. Table 2 shows that the transmission forms a large proportion of the cost of a two-wheel drive gantry. If additional torque is required at the wheels, the price of the transmission will generally increase in direct proportion to the extra torque. Therefore, preferably, the tasks selected for a particular vehicle should each have a similar torque demand. For example, a vehicle designed for spraying is not compatible in draught requirement with one designed for cultivations (Table 1). Compared to the cost of the transmission, the capital cost of the wheels and the load carrying beam are only a small proportion of the component cost of the vehicle (Table 2). The cost of the beam can be expected to increase in direct proportion to its weight, i.e., as the square of the span.

Electro-hydraulic controls are more expensive than manually operated valves, but as they are increasingly being used on tractors, price differences between the machines from this source are likely to be small. It is not clear whether a roll-

TABLE 2

Likely component costs for building a two-wheel drive, 12-m span gantry and associated labour costs (source: Dowler Gantry Systems Ltd)

Proportion of total component cost (%)								Labour (% of total cost)
Engine	Trans-mission	Wheels	Centre beam	Frame & linkages	Hydr-aulics	Controls	Cabin	
10	25	5	6	20	15	14	5	41

over protection structure is required for the driver. Although the vehicle is in no danger of topling sideways, there is the possibility that certain designs of gantry could pitch forwards around their drive wheels. However, unlike a tractor, this could be prevented by a stabilising arm. The main reason for the higher price of gantries, compared with tractors (Table 8), is the low production number. Although the hydrostatic transmission will maintain a small additional cost factor compared with mechanical equivalents, volume production would probably reduce present prices by about 25%.

Tractors

The major additional cost to a tractor-based system is associated with modifying all the field vehicles concerned to have the same wheel track. If we assume that on a cereals farm the combine harvester wheel track determines this distance, then only the smaller tractors in the system may require their axles to be extended beyond the normally available settings. This cost may be in the order of 5% of the purchase price of the vehicle, and in many instances could be carried out in the farm workshop. Extension of the combine unloading auger, to span the width between traffic lanes, could be built into the initial specification, and the extra cost offset by specifying the narrowest tyres (appropriate to the load being carried) to minimise the width of the traffic lanes.

Extending trailer axles during manufacture would have little effect on the purchase price of a trailer. Where root crops are grown, the costs of adaptation may be higher, particularly for harvesters, which could require axle steering to keep them on the traffic lanes and lateral extension of both the axle and the crop unloading system. In these cases, costs could be as high as 20% of the new price of the machine.

OPERATING COSTS

Operating costs are governed by a wide range of factors, one of the most significant being that associated with the high draught and energy requirement of cultivations. Zero-traffic systems lower this cost by avoiding increases in soil strength created by wheel compaction (Lamers et al., 1986; Tullberg, 1988; Chamen et al., 1990a,b; Dickson and Campbell, 1990). Further savings in energy are possible because the machines run on compacted rather than cultivated soil (Burt et al., 1986), which entails a 30% lower coefficient of rolling resistance (CRR)(McAllister, 1983). Hydrostatic transmissions are generally less efficient than their mechanical equivalents and they therefore have higher running costs. Calculations by Tinker (1991) suggest that, whereas a constant mesh mechanical tractor transmission may have an efficiency of 88%, the hydrostatic equivalent may only manage 70%. However, this advantage of mechanical tractor transmissions is being eroded by "power shift" or torque magnifiers, for example, which increase the parasitic losses of the transmission (Tinker, 1991). The overall efficiency of tyre and track systems has tended to be very similar, because the greater CRR (= towing force/normal load) of tracks has been compensated by their improved coefficient of traction (= draught capability/normal load). However, tracks can get closer to their maximum efficiency over a wider range of soil conditions than tyres, and their CRR could be reduced by eliminating some of the drive losses (M.J. Dwyer, personal communication, 1990). Current developments in rubber track systems could offer advantages both in efficiency and cost per unit of traction force available (see Chapter 21).

Gozani and Zur (1989) suggest that the maintenance and repair costs of the Israeli gantry (Bar et al., 1988) were 10% of the capital cost of the vehicle, compared with 8% for an equivalent tractor, each over a life span of seven years. Factors which affect these costs include tyre or track wear, power train maintenance, working conditions, operator skill and maintenance procedures. Vidrine et al. (1979) found that the power shift transmissions of logging skidders, which are employed largely on draught operations, accounted for about 10% of maintenance downtime. Unfortunately, there are few reported data available for equivalent hydrostatic transmissions. The cumulative hourly tractor repair and maintenance costs (30% of the purchase price at 5000 h and 60% at 8000 h) listed by Morris (1988, p. 198), are likely to apply also to tractor-based zero-traffic systems. However, because work rates are higher with zero-traffic, tractor costs per hectare will be noticeably less.

WORK RATES

The operational efficiency of machinery systems in the field is governed by many factors. These include: (1) field shape; (2) incorrect over-lap of implement passes; (3) forward speed; (4) time for headland turning; (5) field access; (6) handling of produce or materials; (7) machinery maintenance; (8) flexibility in the number of field workers; (9) operator skill. These factors are, in turn, affected by the engineering design, the control systems and the methods of field operation. Tillett et al. (1988, p. 63) reported that the use of a gantry improved the flexibility of labour utilisation and field access when working with cauliflowers. Study of a gantry system during fertilizer application (Chamen et al., 1986), showed that the work rate during this operation was very similar to tractor-based systems. More recent work by LePori and Chamen (1989) suggested that gantry working width and operating speed have a far greater influence on the work rate than field size or working pattern. The loss in efficiency of conventional operations due to under- or over-lap of subsequent cultivation passes, is little documented, but it is anticipated that zero-traffic systems, because of their precise matching of adjacent passes, could reduce existing inefficiencies. Transport between fields also affects overall work rate. On the one hand, tractor systems are highly manoeuvrable but require that wide implements are folded for

Fig. 2. A 9-m commercially-built gantry for harvesting cauliflowers. The vehicle, running on 2.5-m long tracks, was based on the design of an experimental unit built at the Silsoe Research Institute. (Photo: J. Hornby, Westgate Farm, Preston, U.K.).

transport, whereas gantries, although perhaps less manoeuvrable, can simply raise their purpose-built implements within their road width, and travel lengthways.

If permanent soil-based traffic lanes are created in the field, Taylor (1983), Lamers et al. (1986) and Monroe and Taylor (1989) have all shown that timeliness of operations can be improved as a result of earlier field access following rainfall. With high-value salad or vegetable crops, the rapid unloading of produce is also important to overall efficiency because of the large number of workers whose harvesting work is likely to be interrupted. For these purposes a tracked gantry, which provides good field access and a high load carrying capacity, requires less frequent stops for unloading (Fig. 2). Another, less obvious, advantage with a tracked vehicle of this nature, is the reduced team size required for harvesting cauliflowers. Tillett et al. (1988, p. 63) reported on a commercial gantry, which, because it inflicted little damage on the crop, encouraged more selective harvesting, combined with more frequent passes and a smaller labour team. As a result, both crop quality and marketable yield were improved. Multiple harvesting passes are required for a wide range of fruit and vegetable crops.

THE EFFECT OF ZERO-TRAFFIC SYSTEMS ON CROP YIELD

When comparing the economics of zero- and conventional traffic systems, the loss in productive area of the traffic lanes is a major factor. However, because in many crops the row spacing allows limited wheel access, and many conventional systems use annual tramlines in cereal crops, the additional losses with zero-traffic may not be as great as expected. Limited data have suggested that the loss in yield of cereals due to tramlines, is equivalent to half the area lost (Ministry of Agriculture, Fisheries and Food, 1977; Austin and Blackwell, 1980; Darwinkel, 1984; Lamers et al., 1986). Janssens (1991), in a slightly different approach, suggested that investments above those for conventional traffic on a 60-ha farm, were only economically feasible if the total area losses were <6.5%.

The direct effect of soil compaction on crop yield has been widely investigated. In reports by Eriksson et al. (1974), Riley (1983), Canarache et al. (1984) and Lamers et al. (1986), the effect of conventional machinery on cereal crop yields was shown to be a reduction in the order of 5%. In reviews by Soane et al. (1982) and Håkansson et al. (1988), it was evident that responses of cereals to soil compaction were very variable. However, results from 15 experiments with repeated annual compaction over a period of 7 years, suggested that a yield reduction of 5% could be expected compared with zero-traffic. If high axle loads (e.g., 10 Mg, typical of a large combine harvester) were a regular occurrence, a reduction of >10% was recorded. On soils with a low clay content, the damaging effects were less and could often be remedied by a single ploughing operation, provided no further compaction was applied to the seedbed. Work by Marks and Soane (1987) over a period of seven years indicated that autumn-sown crops, grown on medium and heavy textured soils, did not respond to subsoil loosening in a conventional traffic regime.

Yield responses to zero-traffic of root crops and other more compaction-sensitive crops are an order of magnitude greater than with cereals (Kayombo and Lal, 1986; Polc, 1989; Dickson et al., 1992). Gantry systems also have the potential to reduce the amount of crop damage and Tillett and Audsley (1985) showed that a capital expenditure of US $ 360 ha[-1] could be justified for each percentage point of improvement in market value of a cauliflower crop and US $ 164 ha[-1] for a similar improvement in yield. Hilton and Mohr (1987) in their "desk-top" study showed that, with broccoli, cauliflower and lettuce, the annual cost of a gantry could be met easily by a 10% improvement in price as a result of improved quality. If a 5% increase in yield were achieved as well, the gain was about 2.5 times that required to justify the gantry. For carrots a similar potential was identified, but for lucerne, a tractor-based system was considered to be more economically viable than a gantry system. Dickson et al. (1992), in three years of trials with zero-traffic for potatoes, reported an average 17% increase in marketable yield compared with conventional practice.

WHOLE-FARM ECONOMICS

Having discussed a number of factors which effect the economics of zero-traffic systems, the remainder of this chapter considers the use of this information and results from elsewhere in a case study (Chamen and Audsley, 1993). A large farm, with an extensive crop rotation, typical of East Anglia, was modelled using a computer-based operational analysis (Audsley, 1981). This compared the profitability of a conventional and a number of zero-traffic systems. Tests were then made to identify the relative economic strengths and weaknesses of the systems.

Five machinery options, designed for use in a rotation of wheat, barley, beans and oilseed rape, were compared. Where different powered tractors were compared, the computer model was capable of choosing the least cost option. The systems were: (1) conventional practice (CT), using tractors of 56, 75 or 112 kW; (2) a 6-m tractor-based zero traffic system (T6), using 75- or 112-kW tractors; (3) a partial gantry system (PG), using both a low-powered, wide-working gantry (12 m, 70 kW, 5.8 kW m^{-1}) and conventional 56- or 75-kW tractors;

TABLE 3

Abbreviated names and description of the five machinery systems used in the study (after Chamen and Audsley, 1993)

Machinery system	Description	Tyre width (mm)
Conventional (CT)	Conventional tractor-based plough system with implement widths to suit the tractor power (56, 75 or 112 kW) and soil type.	458
6-m tractor (T6)	Tractors (75 or 112 kW) with axles extended to c. 2.4 m. All implements in multiples of 6-m width; double wheel tracks at 6-m centres. No ploughing.	532
Partial gantry (PG)	Ploughing with conventional tractors (56 or 75 kW) and harvesting with conventional equipment. All other operations with a 12-m wide, two-wheel drive gantry (70 kW, 5.8 kW m^{-1}).	488
Two gantries (G6/12)	Ploughing with a 6-m wide, four-wheel drive gantry (125 kW, 21 kW m^{-1}) and harvesting with a 6-m wide, two-wheel drive, 163-kW gantry. All other operations, including transport, with a 12-m wide, two-wheel drive gantry (70 kW, 5.8 kW m^{-1}).	488
Full gantry (G6)	All operations, other than harvesting, with the 6-m wide, 125 kW gantry (as for G6/12). Harvesting with the 163-kW gantry (as for G6/12).	488

(4) a two-width gantry system (G6/12), using a low-powered, wide-working gantry (12 m), a high-powered, narrow-working gantry (6 m, 125 kW, 21 kW m^{-1}) and a similar width 163-kW gantry-based harvester; (5) a 6-m full gantry system (G6), using only the high-powered, narrow-working gantries (see Table 3 for further details).

In the first instance the systems were compared on the basis of the following assumptions: (1) there is a 5% increase in yield for those systems where no seedbed compaction occurs between ploughing and harvest, and a 7% increase in yield where wheel compaction is avoided altogether. Yield loss from the permanent traffic lanes or annual tramlines is equivalent to half the area lost to them; (2) in all the systems the same operations are used to produce the crops. In line with recorded data, the zero- and reduced-traffic systems have lower cultivation energy inputs. Ploughing depth on non-trafficked soil is 0.15 m compared with 0.2 m for the conventional and reduced traffic systems.

The crop production systems

All systems, other than T6, use ploughing as the means of primary cultivation. For practical reasons, ploughing is not possible with the T6 system, and a combined disc and twisted share cultivator is used (straw incorporator). With this system, largely due to less effective weed control (Agricultural Development and

TABLE 4

Summary of machinery operations required to establish a rotation of four crops (winter wheat, winter and spring barley, winter and spring beans, oilseed rape) on medium (M) and heavy (H) soil (after Chamen and Audsley, 1993)

Operation	Crop	Machine system				
		CT	T6	PG	G6/12	G6
Spreading P and K	All	*	*	*[1]	*	*
Cultivation and then broadcasting beans	Beans	M	-	M	M	M
Ploughing[2]	All	*	-	*	*	*
Straw incorporation (2 passes)	All	-	*	-	-	-
Spring-tine cultivation	All	*	M	*	M	M
Rotary harrowing	All	H	H	H	H	H
Drilling	All	*[3]	*	*[3]	*[3]	*[3]
Rolling	All except beans	*	*	*	*	*

* M and H; - = not used in this system.
[1]After primary cultivation to mark out traffic lanes.
[2]Beans ploughed 0.15 m deep on medium soil.
[3]Except beans on M.

Advisory Service, 1988), the gross margins of wheat and barley are reduced by US $ 114 ha⁻¹. Table 4 provides a summary of the operations used to produce a rotation of four crops: wheat, barley, beans and oil seed rape.

Procedure for calculation of machinery workrates and costs

The aim of the analysis is to systematically calculate the appropriate work rate for a given system and operation on two soil types. The difference between the soils is based on the specific soil resistances when ploughing, which, at a forward speed of 5.5 km h⁻¹ and working depth of 0.2 m, are: (1) medium soil, 70 kPa; (2) heavy soil, 120 kPa. For draught operations it is assumed that the energy (kWh ha⁻¹) for a given operation is the same whatever tractor is used. Forward speed and coefficients of traction and rolling resistance are also assumed to be similar for all tractors. Therefore, once a basic work rate is determined for one size of tractor, the work rate for all other sizes can be calculated *pro rata* from the tractor power. The procedure adopted allows the workrates for other operations and systems to be calculated on a common basis. The parameters used in the calculations are shown in Table 5.

TABLE 5

Parameters used in the calculation of machinery workrates and costs

A	Available power (kW)
a	Application rate (kg ha⁻¹ or l ha⁻¹)
C	Hopper or tank capacity (kg or l)
c	Factor for type of draught cultivation (= 1 for ploughing)
D_a	Average draught (kN)
D_d	Dynamic draught (kN)
d	Depth of work (m)
E	Field efficiency (typically 75%)
f	Number of plough furrows
g	Time to fill hopper or tank (h 1000 kg⁻¹ or l⁻¹, typically 0.167h)
i	Soil type index (medium = 1.5; heavy = 2.5)
k	Spot work rate (ha h⁻¹)
P	DIN power of tractor/gantry (kW)
p,q	Constants in work rate equations
R	Overall time for operation (h ha⁻¹)
S	Specific draught (kPa)
s	Daily set-up time (h)
T	Overall combine harvester throughput (Mg h⁻¹)
t	Travel cycle time for reloading hopper (h) (e.g., in field = 0.5 km travel at 8 km h⁻¹ for field travel time + 0.12 h for mixing chemicals)
U	Useful power (kW)
v	Operating speed (km h⁻¹)
w	Implement width (m)
y	Crop yield (Mg ha⁻¹)

Relationships between parameters used in the calculations are as follows:

$$D_a = Swd \tag{1}$$

$$D_d = bD_a \tag{2}$$

where $b = 1.32$ for tractor sensing systems (Crolla, 1975); $b = 1.20$ for non-ploughing tasks.

$$A = mP \tag{3}$$

where $m = 0.95$ at the power-take-off (pto); $m = 0.88$ at the axle; $m = 0.70$ at the gantry wheel motors.

$$U = nA \tag{4}$$

where $n = 0.65$ for two-wheel drive (2-WD) tractors; $n = 0.72$ for four-wheel drive (4-WD) tractors and gantry; $n = 1.00$ for pto-driven implements.
From eqns. (1)-(4) the following may be calculated:

$$v = 3.6U/D_d \tag{5}$$

$$k = vw/10 \tag{6}$$

$$R = 1.5/k \text{ (to allow for turning, maintenance, etc.)} \tag{7}$$

Where the implement width is fixed, the following speed limits were applied to ensure effective operation. Roll: 7 km h^{-1}; spring-tine cultivator: 12 km h^{-1}; rotary harrow: 8 km h^{-1}; drill: 12 km h^{-1}. Labour was assumed to cost US \$ 20,000 year^{-1}.

Primary cultivation
The draught requirement for primary cultivation is a function of operation and soil type and can be expressed as $D_a = (\alpha i + \beta)cw$, where α and β depend on the type of traffic the soil has been subjected to. Re-arrangement of the above equations establishes a relationship of the form:

$$R = c(p+qi)/P \tag{8}$$

The values for c, p and q are derived from the following considerations and are shown in Table 6. On heavy soil, the plough resistance following the adoption of a partial gantry system is about 15% less than that following conventional traffic (Chamen and Leede, 1989), while following a zero-traffic system the correspon-

TABLE 6

Constants p and q and the coefficients for type of draught cultivation (c), used in eqn. (8) (after Chamen and Audsley, 1993)

Machine system	p	q	Cultivation	c
CT	-9.6	96.0	Plough	1.00
T6	47.6	37.9	Plough and press	1.15
PG	12.9	73.2	Straw incorporator	0.41
G6/12	64.1	43.2		
G6	72.5	34.7		

ding reduction in plough resistance is about 45% (Arndt, 1966; Lamers et al., 1986; Tullberg and Lahey, 1989; Chamen et al., 1990a,b; Dickson and Campbell, 1991). On medium soil, these draught reductions are assumed to be less by a factor proportional to the soil specific resistances. Thus, on medium soil, the zero-traffic system reduces plough resistance by 70/120 x 45 = 26%. Where two widths of gantry are used, the additional traffic lane (which is equivalent to 4% of the area) is removed during primary cultivation and is assumed to have the plough resistance associated with conventional practice. This reduces the work rate of the full gantry system by 10%. A furrow press increases the draught of the operation by 15%.

Chittey et al. (1986a,b) found that the engine power required by a tine and disc cultivator, was about 22 kW m^{-1} width at 8 km h^{-1}. In the zero-traffic systems, the actual width cultivated is reduced to allow for the width of the traffic lanes.

Secondary cultivation

Secondary cultivation work rates are particularly problematical to define, as there are a large number of alternatives which can be used, depending on factors such as soil condition and previous cultivation and crop. The machinery options for this study have been reduced to two: (1) a draught operation represented by an implement equipped with spring tines; (2) a powered operation represented by a rotary harrow. For secondary cultivation the differences between medium and heavy soil are less pronounced, because the main bonding of the soil has been broken. However, on the heavy soil, clod strength is greater and more energy is required to create a seedbed. Data from Cope and Patterson (1989) indicate that power-driven implements have a factor of 2.2 (medium to heavy soil) and non-driven implements a factor of 1.0. These factors suggest that, with power-driven implements, it is possible to increase the amount of energy imparted to the soil if required, whereas with passive machines little extra energy can be bestowed on a heavy soil compared with one of a medium texture. To create a seedbed on the medium soil, the spring-tine cultivator is used and field

TABLE 7

Examples of the time required (h 100 ha^{-1}) for tillage operations using the different systems on a medium soil (approx. 25% clay, 25% silt and 50% sand). The power of the selected tractor (kW) is as shown and the width of the gantry (m) is indicated by G6 or G12 (after Chamen and Audsley, 1993)

Operation	Machine system					
	CT	T6	PG		G6/12	G6
	75 kW	112 kW	75 kW	G12	G6 & G12	G6
Ploughing	162^{4f}	-	148^{4f}	-	71^{6+7t}	68^{7t}
Ploughing + pressing	187^{4f}	-	169^{4f}	-	-	-
Straw incorporation	67$^{3.5m}$	35	-	-	-	-
Spring-tine cultivation	41^{4m}	24	-	48^{6m}	48^{12+6m}	27^{6m}
Rolling	24^{12m}	16^{12m}	-	32^{12m}	32^{12+12m}	24^{6m}

^{t} = number of plough furrows used; ^{m} = implement width (m); $^{6+}$ = type of gantry and implement, e.g., $^{6+7t}$ = 6-m wide, 125-kW gantry with a 7-furrow plough; $^{12+12m}$ = 12-m wide, 70-kW gantry with 12-m wide implement (see Table 8 for other operations).

data (Chittey et al., 1986a,b; Cope and Patterson, 1989) indicate that 7 kW m^{-1} would be required by this implement at 7 km h^{-1}. This includes the force required to remove the compaction caused by the wheels immediately in front of the tines. Tullberg and Lahey (1989) suggest that this force, which of course is absent in a zero-traffic system, is equal to the rolling resistance of the tractor. The relative energy inputs for secondary cultivation were therefore adjusted to take this into account, depending on the machine system being used. Work rates and implement widths on medium soil are shown in Table 7.

On the heavy soil, a seedbed was created by a combination of drawn and powered implements, which required 10 and 69 kWh ha^{-1}, respectively. On both the medium and heavy non-trafficked soil, the energy input was assumed to be reduced by 26 and 45%, respectively, as for primary cultivation. For the full zero-traffic system, the required energy input was obtained on heavy soil by replacing the spring-tine cultivator and power harrow passes by one pass of the power harrow, and on medium soil by one pass of the spring-tine cultivator. The work rate for each power unit was then calculated by inserting the appropriate value of draught (D_a) in the above procedure. For rolling, an energy requirement of 4.4 kW m^{-1} at 7 km h^{-1} was assumed. Because the 6-m gantry system is limited to a 6-m roll, the output of this equipment is limited by forward speed.

Sowing and chemicals application
To calculate the overall work rate when drilling, spraying and fertilizer

spreading, a generalised formula was established:

$$R = 9(at/C + ag/1000 + 10/wv + 0.04/w + 0.04)/(9 - 9s)E \tag{9}$$

A "standard" field is defined as 500 x 500 m (i.e., 25 ha) and is located at 1 km from the farmstead. The turning time at each headland is $(0.12 + 0.12w)/60$ h for all systems and operations, thus the time taken for each field length is: $1/2v + (0.12 + 0.12w)/60$ h. The work rate when spreading is therefore $0.5w/10[1/2v + (0.12 + 0.12w)/60]$ ha h^{-1}. The time taken to fill the hopper or tank is: $a/C(t + gC/1000)$ h ha^{-1} and, assuming a 9-h day, the value for eqn. (9) can be derived. If a bowser or trailer is used to take materials to the field:

$$s = 0.25 + 2(1/15 + 1/8) + 20/60 \tag{10}$$

where 15 and 8 are the travel speeds (km h^{-1}) when the containers are empty and full, respectively, and the overall time for each operation is as follows.

Spraying from bowser (t = 0.146; s = 0.967; a = 200)

$$R = 0.11 + 204.2/C; \text{ (CT, v = 8.0)} \tag{11}$$

$$R = 0.11 + 196.6/C; \text{ (gantries, v = 8.7)} \tag{12}$$

Fertilizer spreading (t = 0.063; s = 0.967; w = 0.02C (tractors); w = 0.01C (gantries)

$$R = 0.060 + 0.00025a + 64.68/C + 0.094a/C; \text{ (CT, v = 12.1)} \tag{13}$$

$$R = 0.060 + 0.00025a + 75.52/C + 0.094a/C; \text{ (gantries, v = 13.0)} \tag{14}$$

Gantries are attributed with slightly higher forward speeds for these operations (authors' unpublished field data), which is possible because of their wider track and better stability.

Drilling (t = 0.063; s = 0.967; g = 0.463; 17.5% of total drilling time is taken to fill the hopper using 50-kg bags). For all operations/systems:

$$R = 0.060 + 0.00069a + 31.32/P + 0.060/w + 0.094a/C \tag{15}$$

The forward speed for drilling is calculated on the assumption that 8 kW m^{-1} is required at 8 km h^{-1}, i.e., $v = 0.4767P/w$. For the gantry systems it is assumed that only half the power is needed to achieve the same speed (i.e., P = 2P in eqn. (15)). For tractor systems, hopper sizes are related to the width of the drill:

TABLE 8

Examples of the times required for spraying, fertilising, drilling and harvesting a wheat crop with the different systems (after Chamen and Audsley, 1993)

Operation		Quantity (l ha^{-1}) or kg ha^{-1})	Time required (h 100 ha^{-1})	
			With tractor	With gantry
Spraying		200	21	20
Fertilizer	P and K (bulk)	420	19	19
spreading	N (bulk)	544	29	30
Drilling, 6-m drill		175	59 (75 kW)	42 (G6)
Harvesting	Yield (Mg ha^{-1})	7.50	54 (CT)	60 (G6)

typically $C = 900w/4$ and typically $w = 0.0667P$. The gantries are assumed to have a 1200-kg hopper for both fertilizer and seed. Sowing rates are 175 kg ha^{-1} (cereals), 215 kg ha^{-1} (beans) and 7 kg ha^{-1} (rape).

When spreading fertilizer, drilling and, in the spring, rolling, the tractor systems are restricted to using the 56-kW or 75-kW tractors. Table 8 shows some of the times required for carrying out the different operations on wheat.

Harvesting

The work rates for harvesting (Table 8) were calculated on the assumption that the gross output of a 6-m, 163-kW combine harvester is 20 Mg h^{-1} in wheat. The overall work rate is taken to be 70% of the gross value, and the work rates for other sizes of combine are calculated proportionally from their power. The gantry harvester (Fig. 3) was assumed to have a power equivalent to the 163-kW conventional combine. The time required for harvesting wheat is calculated from:

$$R = y/T \tag{16}$$

In barley, the work rate is 12.5% lower than in the same yield of wheat and hence $R = 1.125y/T$. Similarly, in beans, $R = 2.49y/T$ and in oilseed rape, $R = 4.05y/T$.

The yields are calculated from a standard for the conventional system, assuming a first-year crop (with tramlines at 24-m spacing), sown and harvested at the optimum time. For wheat grown on medium soil, this standard yield $= 6.0 + 1.0i$. Therefore, the nominal yield of wheat from the T6 system on medium soil ($i = 1.5$) is $y = 0.92(6.0 + 1.0 \times 1.5) = 6.9$ Mg ha^{-1}, where 0.92 accounts for the loss of crop from the traffic lanes.

With both the full and partial gantry systems, it is likely that only one gantry will be available for unloading crop from the harvester, whereas the tractor

Fig. 3. An experimental gantry-mounted harvesting system (gross output 6 Mg h^{-1}) in a crop of spring wheat.

systems use two tractors and trailers. From an analysis of the transport logistics, it was calculated that unloading with only one transport unit in wheat and barley would reduce the overall work rate from 14 to 12 Mg h^{-1}. The model was allowed to choose whichever of one or two unloading units was best in each circumstance.

Capital costs of machinery

The majority of the prices for the equipment were obtained from the Farm Management Pocketbook (Nix, 1990), but best estimates and some known prices of different gantries were also used (Table 9).

The price of some cultivation implements on the gantry were calculated from the cost per m width of a commercially-built spring-tine cultivator for a 12-m gantry. The ratio of this cost to the equivalent of a tractor-mounted machine was 0.64. Thus, the cost of a tractor-mounted spring-tine cultivator, at US $ 2000 m^{-1}, was reduced to US $ 1280 m^{-1}, giving a price of US $ 7680 for a 6-m unit. To maintain comparability it was assumed that tractors and gantries have the same operational life.

RESULTS AND DISCUSSION OF CASE STUDY

Basic scenarios

Table 10 shows the farm gross margin (profit before deduction of fixed costs, such as rent, office, etc.) predicted by the model on both the medium and heavy

TABLE 9

The new purchase price (US $) of the machinery in relation to the width or capacity used in the study[1] (after Chamen and Audsley, 1993)

Machine	Tractor system	Gantry systems	
2-WD tractor	6,474 + 474P	(70 kW)	124,000
4-WD tractor	-3,542 + 740P	(125 kW)	171,000
Plough + press	2,480 + 2,720f	-	2,680f
Straw incorporator	4,250w	-	-
Spring-tine cultivator	2,000w	(6 m)	7,680
Rotary harrow	3,500w	(6 m)	14,000
Roll	2,334w - 4,004	(12 m)	10,000
Fertilizer spreader (trailed)	5.84C - 1,500	(12 m, mounted)	11,000
Drill	3,500w - 2,000	(6 m)	8,000
		(12 m)	16,000
Sprayer	1,000 + 8.5C	(24 m)	16,000
Combine harvester	40,634 + 10,474T	(6 m)	190,000

[1]For explanation of symbols, see footnote to Table 7 and notation (Table 5).

soils, using the assumptions described. As the need for power harrowing on heavy land was severely restricting the performance of the PG system, it was decided to allow 56-kW and 75-kW tractors with suitable low ground pressure tyres, to carry out secondary cultivations. This additional system is denoted as PGT and the assumed yield increase was reduced to 2%. Results showed that the range in gross margins on medium soil was relatively small (about 9%), whereas on heavy

TABLE 10

Farm Gross Margins (FGM) for all systems and basic scenarios (after Chamen and Audsley, 1993)

Machine system	Wheeling (m)	Rate of harvesting (Mg h^{-1} overall)	FGM (US $ ha^{-1})	
			Medium soil	Heavy soil
CT	24[2]	14	622	550
T6	6[2]	14	586	600
PG	12[1]	14	600	446
G6/12	12[1]	12	642[3]	588[3]
G6	6[1]	12	624	618
PGT	12[1]	14	596	506

[1]One traffic lane at specified distance.
[2]Two traffic lanes at specified distance.
[3]Yield only increased by 5% because of the residual effects of the 6-m traffic lane.

TABLE 11

Number of machines and the labour needed by each system on a 250-ha farm on medium (M) and heavy (H) soil (after Chamen and Audsley, 1993)

Machine/Labour	Machine system and soil type											
	CT		T6		PG		G6/12		G6		PGT	
	M	H	M	H	M	H	M	H	M	H	M	H
56-kW tractor	0.7	0.7	-	-	0.7	0.0	-	-	-	-	0.2	0.0
75-kW tractor	0.3	0.8	0.4	0.5	0.4	1.5	-	-	-	-	0.9	2.2
112-kW tractor	0.4	1.3	0.5	1.2	-	-	-	-	-	-	-	-
70-kW 12-m gantry	-	-	-	-	0.7	1.6	0.7	1.4	-	-	0.4	0.7
125-kW 6-m gantry	-	-	-	-	-	-	0.4	0.6	0.8	1.5	-	-
14-Mg ha^{-1} combine	0.6	0.7	0.5	0.6	0.8	0.7	-	-	-	-	0.6	0.7
Combine gantry	-	-	-	-	-	-	0.6	0.7	0.6	0.7	-	-
Labour	1.5	2.7	1.0	1.8	1.8	2.9	1.2	2.2	1.1	1.9	1.5	2.8

land it approached 30%. The reasons for some of these differences are apparent from Tables 11 and 12. Table 11 shows the optimal machinery and labour requirements for each of the systems. For a specific farm size, adjustments to the size of machines would need to be made to ensure that approximately whole numbers of men and machinery are required. For example, a farm size requiring 0.2 112-kW tractors is not likely to be economically viable, but one requiring 0.7 is. Practice has shown that, providing the number of machines required is reasonably close to an integer value greater than zero, whole numbers do not generally have a major effect on comparisons. Table 12 shows how, on heavy soil,

TABLE 12

Effect on FGM of whole numbers of men and machinery relative to FGM of the conventional system (CT) on heavy soil (cf. Table 9) (after Chamen and Audsley, 1993)

System	Difference in FGM (US $ ha^{-1})	
	250 ha	400 ha
T6	-6	+8
PGT	0	-18
G6/12	-122	-18
G6	-48	-8

the comparisons are affected when FGM is calculated on this basis. In general, the comparisons are little affected, but it is clear that G6/12 and G6 are poorly matched to a 250-ha farm. In the case of G6/12, one 70-kW gantry is insufficient, while one 125-kW gantry is too many. On 400 ha, two 70-kW gantries and one 125-kW gantry are a good combination.

From Tables 11 and 12 it is also clear that the 6-m tractor system (T6) needs a minimum size of about 500 ha on both medium and heavy soil to be viable. However, on heavy soil the conventional system (CT) would be practical on about 250 ha and would require a labour force of three. The same applies for the PG and PGT systems. The G6/12 and G6 systems need only two people. On a 400-ha farm the labour requirements are all increased by one.

A major problem with gantries is their high cost compared with tractors of the same power. Thus, while the 112-kW tractor has an annual cost of US $ 20,000 the 70-kW gantry costs US $ 30,000 per annum. Consequently, with the conventional system the least-cost objective is to have ample tractors for as few operators as possible, whereas with the gantry systems, the same objective requires ample labour for as few gantries as possible.

Table 13 lists the optimal area of each crop grown, shown as a percentage of the total area of the farm. In all systems, oilseed rape is extremely profitable (accepting that the price for this crop is extremely variable) and, therefore, the maximum area permissible is grown. In most cases, about 50% of the area is cropped with winter wheat because it is US $ 8 ha^{-1} more profitable than winter barley. The largest difference is in the area of spring crops. The partial gantry system grows the largest area, with 30% of the total on heavy land, while the tractor systems both have <20% of the area devoted to spring crops. On heavy land, with the T6 system only 15% of the total area is sown in the spring.

TABLE 13

Effect of different systems on the optimum area of each crop grown (% of farm area, on medium (M) and heavy (H) soils) (after Chamen and Audsley, 1993)

Crop	Machine system and soil type											
	CT		T6		PG		G6/12		G6		PGT	
	M	H	M	H	M	H	M	H	M	H	M	H
Winter wheat	53	55	37	53	46	45	50	48	50	53	45	45
Winter barley	0	0	0	0	0	0	0	0	0	0	0	0
Winter beans	5	2	15	7	3	0	2	0	0	0	8	0
Spring barley	2	0	18	2	9	10	5	7	6	2	10	10
Spring beans	15	18	5	13	17	20	18	20	20	20	12	20
Oilseed rape	25	25	25	25	25	25	25	25	25	25	25	25

A major difficulty in the comparison of systems is to match like with like. To compare similar power levels, the 112-kW conventional tractor system must be compared with the 125-kW gantry systems (G6/12 and G6). However, the gantries, because of their zero-traffic operation and despite their reduced transmission efficiency, can do up to about 60% more work than the 112-kW tractor in conventional systems, and so farm sizes are not comparable. Similarly, the PG system is comparable to a farm with 56-kW tractors and is suitable for much smaller farms than the other gantry systems.

Comparison of the 6-m gantry and the conventional systems on heavy land shows that the power and labour costs are similar at US $ 394 ha^{-1} and US $ 378 ha^{-1}, respectively. However, the additional loss of crop due to one traffic lane every 6 m, compared with two every 24 m (annual tramlines), is 3% and is equivalent to US $ 50 ha^{-1}. Cereal harvesting work rate is reduced by transport limitations from 14 to 12 Mg ha^{-1} in the gantry system, and this costs US $ 14 ha^{-1}. Overall, the disadvantage to the 6-m gantry system (G6) totals US $ 66 ha^{-1}.

Sensitivity tests

Although well-documented research results were used as a basis for the assumptions made in the initial comparisons, the extent to which these would be experienced on a farm is uncertain, and other problems related to these changes might occur. Therefore, it is essential that the sensitivity of the FGM to a number of these assumptions, and to other changes, should be explored. The results of these tests are presented in Table 14 and discussed in more detail below.

Effect of assumptions related to zero-traffic operations
When ploughing, it was assumed that this could be shallower than with the conventional system because there was no need to ameliorate soil compaction caused by the wheels. However, ploughing to a depth of 0.15 m might be inadequate for effective weed control, and the negative effect on FGM of deeper ploughing is shown in Table 14.

In the U.K., subsoiling is often practised with conventional machinery systems as a means of counteracting yield losses due to compaction. The additional cost of this operation is shown in Table 14. On heavy soil, a yield increase of 3% would be required to cover this cost. An increase of about 5% would be necessary to bring this system in line with the best of the zero-traffic systems. However, as discussed earlier, there is little experimental evidence, particularly on heavy soils, to suggest that subsoiling followed by conventional traffic is an effective method of maintaining yield.

The variability of yield responses to zero-traffic would indicate that, on some soils, an increase in yield might be absent, perhaps because of a trace element deficiency initiated by the loose soil conditions. Table 14 shows what happens to the FGM if the assumed 2, 5 and 7% increases in yield do not materialise. In all

TABLE 14

The effect of machinery system changes and other variables on FGM (US $ ha^{-1}) (cf. Table 10) (after Chamen and Audsley, 1993)

Change to system	Machinery system	Medium soil	Heavy soil
Ploughing to 20 cm depth	G6/12	-18	-42
	G6	-22	-38
Subsoil 1 year in 4	CT	-16	-40
No increase in yield (2%)	PGT	-26	-28
No increase in yield (5%)	PG	-64	-70
	G6/12	-62	-70
No increase in yield (7%)	T6	-76	-86
	G6	-82	-96
Gantry price reduced to US $ 100,000	PG	+40	+42
Labour costs US $ 28,000 year^{-1}; effect relative to CT	T6	+12	+32
	PGT	-2	-4
	G6/12	+8	+20
	G6	+10	+26
Crop prices -20%; effect relative to CT	T6	+26	+30
	PGT	0	+8
	G6/12	+2	+8
	G6	+10	+12
Spreading width 12 m	CT	-34	-32
Harvesting work rate 10 Mg h^{-1}	CT	-38	-44
	PGT	-40	-44
Harvesting work rate 14 Mg h^{-1}	G6/12	+12	+16
	G6	+12	+14
Maximum tractor size 56 kW	CT	-38	-112
Straw incorporation	CT	-46	-26
Double pto power of gantry for power harrowing	PG	-	+50
	G6/12	-	+34
Combine harvester gantry used for cultivations from October onwards	G6	0	+20

cases, the FGM of the zero traffic systems is reduced to below that of CT. If the price of the gantry in the PG system is reduced by 20%, FGM is increased by 7% on medium soil and by 9% on heavy soil.

Effect of labour costs and crop returns
An increase in the cost of labour, relative to other costs and prices, shows the advantage of systems which have a low labour input. It is particularly favourable to T6 and the complete gantry systems, where for every US $ 2000 per annum increase in labour costs, there is a relative improvement in FGM of US $ 6-8 ha⁻¹.

If crop prices are reduced in real terms to address the problems of surpluses in Europe, the cost of subsidies and in response to the General Agreement on Tariffs and Trade (GATT), there would be a large reduction in farm profit. In practice there may also have to be some husbandry changes, such as reducing the amount of nitrogen applied. However, England (1986) showed that a 20% reduction in cereal prices only reduced the optimal use of nitrogen by about 3%. Reducing the crop price shows there will be a significant swing to spring crops and that oilseed rape is relatively less profitable. The FGM of the T6 system would, with a 20% drop in crop price, become closely competitive with the conventional system on medium soil (US $ 586 + 26; Tables 10 and 14) and would improve its standing still further on heavy land (US $ 600 + 30). Overall, results of this comparison are similar to those showing the effect of increased labour costs.

Effect of machinery choices
Choice of a particular machine can have a large effect on the profitability of the different systems even though every attempt has been made to make them comparable.

Conventional system (CT). A 12-m rather than a 24-m spreading width reduces FGM by about US $ 32 ha⁻¹, largely due to the increased area of land lost to traffic lanes. If harvesting work rate, which is affected both by the size of combine and by the transport system, is restricted by transport limitations to 10 Mg h⁻¹, there is a marked reduction in FGM. Conversely, the systems which restricted the 20-Mg h⁻¹ harvester to 12 Mg h⁻¹ (G6/12 and G6; see *"Basic scenarios"* and Table 10), are improved if the transport system allows them to operate at the full 14 Mg h⁻¹ possible.

A farm restricted to 56-kW tractors, which is comparable to the PG system, reduces FGM by US $ 112 ha⁻¹ on heavy soil. Using straw incorporation reduces the FGM of the CT system by US $ 46 ha⁻¹ on medium soil and by US $ 26 ha⁻¹ on heavy soil. This makes it the least profitable system.

6-m tractor system (T6). The enforced use of a straw incorporator in place of a plough, reduces farm profit to the same extent as in the conventional system.

Partial gantry system (PG). Allowing tractors to do secondary cultivations (PGT compared with PG; Table 10) increases FGM by US $ 60 ha^{-1} on heavy soil, despite the slight reduction in yield assumed. On medium soil this change has little effect on FGM, reflecting the easier working conditions. Doubling the pto power of the gantry at a cost of US $ 30,000 (to allow faster power harrowing), increases FGM on heavy soil by US $ 50 ha^{-1}.

Two gantry system (G6/12). Doubling the pto power of the 12-m gantry as above on heavy soil, increases FGM by US $ 34 ha^{-1} and makes this one of the most profitable systems.

6-m gantry system (G6). Allowing the harvesting gantry to be used for low draught operations from October onwards, increases FGM by US $ 20 ha^{-1} on heavy soil, but has no effect on the profitability of the medium soil farm.

CONCLUSIONS

(1) The design and cost of vehicles intended for zero-traffic operations is highly dependent upon the range of tasks which they must perform. With gantries, the highest proportion of the total component costs of these machines is in the hydrostatic transmission. Differences in the operational costs of zero- and conventional-traffic systems are difficult to specify, but the hydrostatic transmissions of existing gantries are about 18% less efficient than are their purely mechanical counterparts.

(2) The loss of crop yield due to traffic lanes in the field, whether these are of a temporary or permanent nature, is equivalent to about half the area lost. Where traffic is excluded from the cropping area, yields of cereals are likely to be increased by between 5 and 10%.

(3) On a medium soil in the U.K., farms with complete gantry systems growing wheat, barley, beans and rape would be as profitable as conventional practice. On heavy soil they are about US $ 50 ha^{-1} more profitable than the conventional system. A 6-m tractor-based zero-traffic system is US $ 36 ha^{-1} less profitable than conventional practice on medium soil, but US $ 50 ha^{-1} more profitable on heavy soil. The profitability of this system is restrained by the large area of land lost to traffic lanes (approx. 18%), and the reduced weed control resulting from the preclusion of ploughing.

(4) All the zero-traffic systems rely on some crop yield improvement to remain competitive. Where a mixture of gantry and conventional machines are employed on medium soil, the system is US $ 22 ha^{-1} less profitable than conventional practice. On heavy soil, FGM is reduced by US $ 104 ha^{-1} compared with conventional practice. This is the same as the reduction in profit caused by restricting the conventional system to the use of 56-kW tractors, which are equivalent in power to the 70-kW gantry used in this system. A 25% reduction

in the cost of a gantry used in this situation would increase farm gross margins by about US $ 40 ha^{-1}.

(5) With the exception of the partial gantry system, which can be fully utilised on a farm of 250 ha, farms of 400-500 ha are needed to justify use of the zero-traffic equipment considered here.

(6) The engine power of gantries should be relatively greater than their tractor counterparts by an amount equivalent to their reduced transmission efficiency. Gantry span should be the maximum practical for the range of operations which the vehicle must perform.

REFERENCES

Agricultural Development and Advisory Service, 1988. Gross Margin Budgets, 1988: Arable Crops. Agricultural Development and Advisory Service, London, U.K., 46 pp.

Arndt, W., 1966. Traffic compaction of soil and tillage requirements. IV. The effect of traffic compaction on a number of soil properties. J. Agric. Eng. Res., 11: 182-187.

Audsley, E., 1981. An arable farm model to evaluate the commercial viability of new machines or techniques. J. Agric. Eng. Res., 26: 135-149.

Austin, R.B. and Blackwell, R.D., 1980. Edge and neighbour effects in cereal yield trials. J. Agric. Sci., Camb., 94: 1-26.

Bar, Z., Gilboa, J. and Gozani, E., 1988. Field experiment with FPU. Am. Soc. Agric. Eng., St. Joseph, MI, U.S.A., Pap. 88-1081, 8 pp.

Burt, E.C., Taylor, J.H. and Wells, L.G., 1986. Traction characteristics of prepared traffic lanes. Trans. ASAE, 29: 393-397.

Canarache, A., Colibaş, I., Colibaş, M., Horobeanu, I., Patrú, V., Simota, C. and Trandafirescu, T., 1984. Effect of induced compaction by wheel traffic on soil physical properties and yield of maize in Romania. Soil Tillage Res., 4: 199-213.

Chamen, W.C.T. and Audsley, E., 1993. A study of the comparative economics of conventional and zero-traffic systems for arable crops. Soil Tillage Res., 25: 369-390.

Chamen, W.C.T. and Leede, P.R., 1989. Dowler field gantry project. Silsoe Res. Inst., Silsoe, U.K., Contract Rep. CR/349/89/8647, 26 pp.

Chamen, W.C.T., Chittey, E.T. and Armstrong, M.C., 1986. Development and assessment of a wide-span vehicle for arable crops. Silsoe Res. Inst., Silsoe, U.K., Div. Note DN 1357, 43 pp.

Chamen, W.C.T., Vermeulen, G.D., Campbell, D.J. and Sommer, C., 1990a. EEC Cooperative project on reduction of soil compaction. Am. Soc. Agric. Eng., St. Joseph, MI, U.S.A., ASAE Pap. 90-1073, 28 pp.

Chamen, W.C.T., Chittey, E.T., Leede, P.R., Goss, M.J. and Howse, K.R., 1990b. The effect of tyre/soil contact pressure and zero traffic on soil and crop responses when growing winter wheat. J. Agric. Eng. Res., 47: 1-21.

Chittey, E.T., Chamen, W.C.T. and Patterson, D.E., 1986a. Cultivation systems for straw incorporation on clay soils : 1983-84. Silsoe Res. Inst., Silsoe, U.K., Div. Note DN 1313, 20 pp.

Chittey, E.T., Chamen, W.C.T. and Patterson, D.E., 1986b. Cultivation systems for straw incorporation on clay soil, 1984-85. Silsoe Res. Inst., Silsoe, U.K., Div. Note DN 1318, 23 pp.

Cope, R.E. and Patterson, D.E., 1989. Mechanisms of soil aggregate reduction, 1987-1988. Silsoe Res. Inst., Silsoe, U.K., Div. Note DN 1544, 62 pp.

Crolla, D.A., 1975. The performance of off-road vehicles under fluctuating load conditions. Proc. Conf. Off-highway Vehicles, Tractors and Equipment, Inst. Mech. Engineers, London, U.K., pp. 91-99.

Darwinkel, A., 1984. Yield responses of winter wheat to plant removal and to wheelings. Neth. J. Agric. Sci., 32: 293-300.

Dickson, J.W. and Campbell, D.J., 1990. Soil and crop responses to zero- and conventional-traffic systems for winter barley in Scotland, 1982-1986. Soil Tillage Res., 18: 1-26.

Dickson, J.W., Campbell, D.J. and Ritchie, R.M., 1992. Zero and conventional traffic systems for potatoes in Scotland, 1987-1989. Soil Tillage Res., 24: 397-419.

England, R.A., 1986. Reducing the nitrogen input on arable farms. J. Agric. Econ., XXXVII: 13-24.

Eriksson, J., Håkansson, I. and Danfors, B., 1974. The effect of soil compaction on soil structure and crop yields. Swedish Inst. Agric. Eng., Uppsala, Sweden, Bull. 354, 101 pp.

Gozani, A. and Zur, Y., 1989. Comparison of the tillage costs with the FPU system and with regular tractors. Dep. Prod. Econ., Hakiria, Israel, 13 pp.

Håkansson, I., Voorhees, W.B. and Riley, H., 1988. Vehicle and wheel factors influencing soil compaction and crop response in different traffic regimes. Soil Tillage Res., 11: 239-282.

Hilton, D. and Mohr, G., 1987. Investigation of the commercial potential for agricultural gantry systems in Australia. Darling Downs Inst. Adv. Educ., Qld., Australia, Tech. Rep. NERDDP EG89/783, 142 pp.

Janssens, S.R.M., 1991. Rendabiliteit van een verminderde bodembelasting. (Profitability of reduced ground pressure). Res. Sta. Arable Farming Field Prod. Vegetables (PAGV), Lelystad, Netherlands, Rep. 127, 57 pp. (in Dutch).

Kayombo, B. and Lal, R., 1986. Influence of traffic-induced compaction on growth and yield of cassava (Manihot esculenta Crantz). J. Root Crops, 12: 19-23.

Lamers, J.G., Perdok, U.D., Lumkes, L.M. and Klooster, J.J., 1986. Controlled traffic farming systems in the Netherlands. Soil Tillage Res., 8: 65-76.

LePori, W.A. and Chamen, W.C.T., 1989. Analysis of gantry field operations. Am. Soc. Agric. Eng., St. Joseph, MI, U.S.A., ASAE Pap. 89-1620, 19 pp.

Marks, M.J. and Soane, G.C., 1987. Crop and soil response to subsoil loosening, deep incorporation of phosphorous and potassium fertiliser and subsequent soil management on a range of soil types. Part 1: Response of arable crops. Soil Use Manage., 3: 115-123.

McAllister, M., 1983. Reduction in the rolling resistance of tyres for trailed agricultural machinery. J. Agric. Eng. Res., 28: 127-137.

Ministry of Agriculture, Fisheries and Food, 1977. Boxworth Experimental Husbandry Farm, Cambridge, U.K., Ann. Rev., 58 pp.

Monroe, G.E, and Taylor, J.H., 1989. Traffic lanes for controlled-traffic cropping systems. J. Agric. Eng. Res., 44: 23-31.

Morris, J., 1988. Estimation of tractor repair and maintenance costs. J. Agric. Eng. Res., 41: 191-200.

Nix, J., 1990. Farm Management Pocketbook. Wye College, University of London, Ashford, U.K., 203 pp.

Polc, M., 1989. Vplyv utlačania pôdy na úrodu plodín. (The influence of soil compaction on crop yield). Mechanizace Zemedelstvi, 39: 175-177 (in Slovak).

Riley, H., 1983. Relationship between soil density and cereal yield. Forskning-og-Forsök-i-Landbruket, 34: 1-11.

Soane, B.D., Dickson, J.W. and Campbell, D.J., 1982. Compaction by agricultural vehicles: A review. III. Incidence and control of compaction in crop production. Soil Tillage Res., 2: 3-36.

Taylor, J.H., 1983. Benefits of permanent traffic lanes in a controlled traffic crop production system. Soil Tillage Res., 3: 385-395.

Taylor, J.H., 1989. Controlled traffic research: an international report. Proc. 11th Conf. Int. Comm. Agric. Eng., Dublin, Ireland, Vol. 3, pp. 1787-1794.

Tillett, N.D. and Audsley, E., 1985. The potential economic benefits of gantries to mechanise the production of leaf vegetables. Silsoe Res. Inst., Silsoe, U.K., Div. Note DN 1288, 21 pp.

Tillett, N.D., Holt, J.B., Chestney, A.A.W. and Reed, J.N., 1988. Experimental field gantry for leaf vegetable production: specification, design and evaluation. J. Agric. Eng. Res., 41: 53-64.

Tinker, D.B., 1991. Integration of tractor engine, transmission and implement depth controls, part 1: transmissions. Silsoe Res. Inst., Silsoe, U.K., Div. Note DN 1602, 48 pp.

Tullberg, J.N., 1988. Controlled traffic in sub-tropical grain production. Proc. 11th Conf. Int. Soil Tillage Res. Org. (ISTRO), Edinburgh, U.K., Vol. 1, pp. 323-327.

Tullberg, J.N. and Lahey, G.V., 1989. Energy cost of wheeltrack tillage. Agric. Eng. Austr., 18: 42-45.

Vidrine, C.G., Carothers, J.E. and Robbins J.W.D., 1979. "Downtime" in the use of four wheeled-drive rubber-tired logging skidders. Trans. ASAE, 22: 2-6.

Soil Compaction in Crop Production
B.D. Soane and C. van Ouwerkerk (Eds.)
©1994 Elsevier Science B.V. All rights reserved.

CHAPTER 25

Control and Avoidance of Soil Compaction in Practice

W.E. LARSON[1], Anna EYNARD[1], A. HADAS[2] and J. LIPIEC[3]

[1]University of Minnesota, Department of Soil Science, St. Paul, MN, U.S.A.
[2]Agricultural Research Organization, Volcani Center, Bet Dagan, Israel
[3]Polish Academy of Sciences, Institute of Agrophysics, Lublin, Poland

SUMMARY

On most soils, soil compaction can be prevented or alleviated. Land managers recognize the detrimental effects of soil compaction in a number of ways: decreased stand uniformity and plant growth, restricted root growth, reduced soil aeration, reduced water infiltration, reduced internal drainage and reduced effectiveness of underground drainage systems. Tilling compacted soils results in increased labor, machine and energy costs.

Compactability is related to soil water content and hence, soil water management plays a major role in management of compaction. Drainage and/or irrigation can aid in maintaining the soil water matric potential in the desirable range.

Matching machine operations to soil conditions for prevention or alleviation of compaction is a major management tool. Load on the soil surface is a determinant in soil compaction. Ground contact pressure can be manipulated by number, kind, size and inflation pressures of tires. Speed of machine operation, number of passes and soil water content are important in operation of machines. Choice of the machine is conditioned by the desired result, soil water content and available power. Controlled traffic may be a desirable option for cropping practices where traffic is frequent and soil conditions favor compaction.

Agronomic practices for prevention or alleviation of compaction include crop rotations which require a minimum of traffic, crops that produce organic materials for maintenance of soil organic matter, crops with penetrating and fibrous root systems, liming and animal grazing management.

INTRODUCTION

Experience throughout the world indicates that crop and forest producers are increasingly concerned with the detrimental effects of soil compaction. Land managers observe these effects in a number of ways: decreased crop stand and stand uniformity, reduced plant growth, restricted root growth, reduced soil aeration, reduced water infiltration, reduced internal drainage, reduced effectiveness of underground drainage systems, increased machine amortization, and increased labor and energy costs for tilling the compacted soil.

The purpose of this chapter is to discuss the management options available for both preventing and alleviating soil compaction. A specific practice for a given region has to be developed in conformity with the crop rotation, the soil, the climate and the available machinery.

PREVENTION OF COMPACTION

Soils vary greatly in their susceptibility to compaction. Models are available to predict the compactability of soils and the consequences of compacted conditions. The effects of soil and external conditions (e.g., soil water, machine load) on compaction effects have been discussed in this book (see Chapters 2-4). Practical strategies which land managers can use will be discussed in the following treatise.

Water management

Because compactability is related to soil water content, soil water management plays a major role in the management of compaction. Models relating water content to compaction have been studied on clay soil (Steinhardt, 1974). Mechanical manipulation of the soil with little or no structural damage is feasible within a limited range of water potentials, which defines the soil workability (Van Wijk and Buitendijk, 1988). The desirable range of soil water potentials depends on the required operation (e.g., subsoiling), method of operation and machinery type, and soil type. These factors define also the range of trafficability (i.e., the ability of bearing the traffic load without structural damage), but the range limits for different operations and soils are not the same. Spraying or harvesting may be feasible when tillage is not (Reeve and Fausey, 1974). On the other hand, the soil surface may be sufficiently dry for seedbed preparation but at the same time the subsoil may be too wet to bear traffic (Krause and Lorenz, 1984). Workability has a wider range in water content for plowing than for seedbed preparation (Krause and Lorenz, 1984). Timeliness of operations is also of primary importance in forestry management (Wert and Thomas, 1981), as well as for crops grown under critical climatic conditions (e.g., sowing in a short wet spring or harvesting prior to early autumn frosts).

Soil compaction results from forces acting on the soil and its resultant reaction, namely partial or complete failure of its matrix structure when its strength is overcome by stress. The degree of soil failure depends on soil strength characteristics (Schafer and Johnson, 1982; see Chapters 3 and 4) and the external and internal forces acting on the soil (see Chapters 1, 4, 17 and 18). Soil strength is commonly defined in the context of one of the following modes of failure: (1) shear and compression, which usually involve volume changes; (2) tension failure, when parts of the soil matrix are completely separated through

TABLE 1

Soil consistency, workability and trafficability as related to soil water content[1]

	Soil consistency at soil water matric potential (MPa)[2]			
	<-2.0 (SL)	-2.0/-0.5 (LPL)	-0.5/-0.3	>-0.03 (UPL)
Consistency	hard	friable	plastic	liquid
Resistance to tillage	high	low	medium	very low
Bearing capacity	high	high to moderate	low	very low
Resistance to compression	very high	high to moderate	low	high
Efficiency of tillage implements	chisel plow subsoiler land plane	moldboard plow rototiller rolling cage harrow	spade plow	-

[1]SL = shrinkage limit; LPL = lower plastic limit; UPL = upper plastic limit.
[2]The range of values relates to differences in soil texture.

shearing and cracking; (3) plastic flow, especially in wet clay soils.

Until now, stress-strain-volume relationships have not been adequately established for agricultural soils (Schafer and Johnson, 1982; Koolen and Kuipers, 1983). Tension failure is usually encountered in tillage or when soil volume is changed. There is no simple way to quantify plastic flow of agricultural soils under traffic or cultivation.

Since no simple means to define soil strength are available (see Chapters 1-4) on which farm managers may base their decisions concerning minimizing soil compaction practices, their only practical way is to refer to soil consistency as a guide to overcome the complex interrelationships between soil failure and soil strength. Soil consistency is a manifested soil physical reaction to stresses, e.g., its resistance to compression, shear and tension. Soil consistency varies with water content, clay type and clay fraction content, organic matter content and type of exchangeable cations (Baver et al., 1972; Yong and Warkentin, 1975).

Table 1 shows the relationships between soil water and soil consistency, resistance to tillage, bearing capacity, resistance to compression and efficiency of tillage tools. Consistency limits for different soil types are given in Table 2. Fig.

TABLE 2

Soil consistency limits (%, w/w) for different soil types (from Archer, 1975)

Soil type	Clay content (%, w/w)	Shrinkage limit	Lower plastic limit	Upper plastic limit
Sandy loam	12	14	16	21
Silty clay loam	23	18	25	40
Clay	51	13	36	83

1 shows the dynamic factors involved in tillage as related to soil water content. Smith et al. (1985) have established the following relationships:

$$\text{Plastic limit} = 2.077\, w_{hyg} + 15.242 \qquad\qquad r^2 = 0.67 \qquad (1)$$

$$= 0.258\, (\%\ \text{clay}) + 14.329 \qquad\qquad r^2 = 0.60 \qquad (2)$$

$$\text{Liquid limit} = 3.648\, w_{hyg} + 22.329 \qquad\qquad r^2 = 0.85 \qquad (3)$$

$$= 0.575\, (\%\ \text{clay}) + 15.345 \qquad\qquad r^2 = 0.77 \qquad (4)$$

where w_{hyg} = hygroscopic water content.

Since the strength of the soil and of the individual clods increases significantly with decreasing water content from the upper to the lower plastic limit and then decreases slowly as the water content is further reduced, the optimum water content for working the soil is just below the lower plastic limit (Godwin and Spoor, 1977). Gill (1967), Allmaras et al. (1969) and Ojeniyi and Dexter (1979) have pointed out that the optimum soil water content for tillage was about 0.9 LPL (lower plastic limit) and, if multipass tillage operations were performed, greater variability in attained porosity was observed for soil tilled at 1.3 LPL than at 0.65 LPL. These observations emphasize the importance of soil water management and timeliness of tillage operations. No such general observations were given for minimizing soil compaction.

Shaffer and Clapp (1987) found that soil strength was correlated with the soil matric potential (range of 0 to -70 kPa) and the soil bulk density (range of 1 to 2 Mg m^{-3}). In their work, soil strength refers to root development but not to compaction. Söhne (1953, 1958) suggested that his concentration factor should have values of 4, 5 and 6 for hard, firm, and soft soils, respectively. These values describe soil firmness or resistance to deformation (see Chapter 3). Ram (1984) found that the values of the concentration factor increased with an increase in soil water content at a constant bulk density and decreased with an increase in bulk density at the same water content.

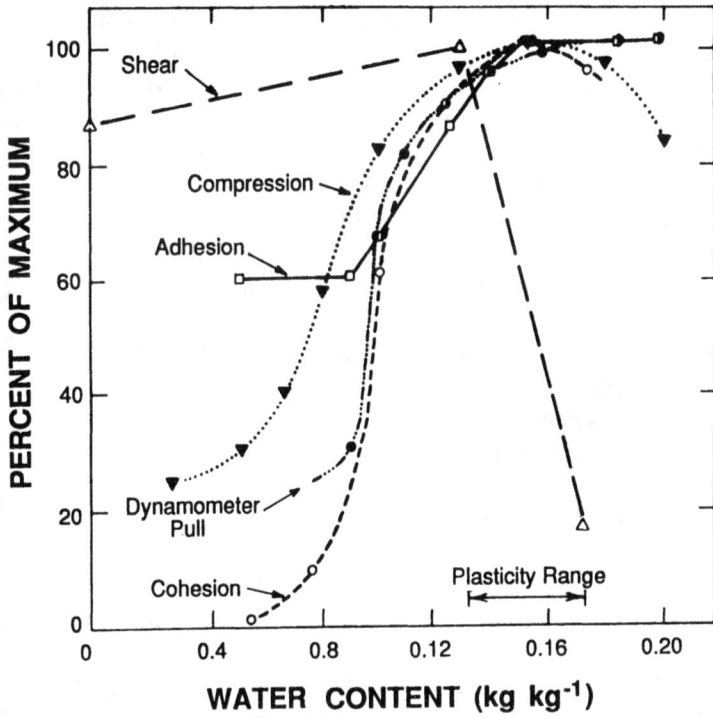

Fig. 1. Relationship of soil strength factors involved in tillage to soil water content, with special reference to the plasticity range. (The maximum value for each of these factors was taken as 100) (from Baver et al., 1972).

Drainage

Soil compactability is related to drainage through its effect on the water content (Cannell et al., 1978). Soils which are compacted in the field are often those that do not quickly drain to field capacity (Howard et al., 1981). Artificial drainage is used to remove excess water and it can help in creating more uniform water conditions when wet spots tend to persist (Reeve and Fausey, 1974).

When a high water table exists, drainage is essential to permit field operations with little or no soil compaction risks. A suggested water table depth to minimize compaction is >50 cm, depending on soil type and soil strength (Steinhardt and Trafford, 1974). Subsoiling, used to alleviate compaction damages, may also have an indirect effect by improving drainage (Eck et al., 1977; Trouse, 1983).

Irrigation

It is possible to alter the soil consistency by irrigation (Krause and Lorenz,

1984). The Mesopotamian farmers' almanac stated: "Before you till your fields, open the sluices of the irrigation ditches, but take care not to inundate the fields too much." (Lampe, 1984). However, irrigation has a limited power of controlling water content because soil factors affect water movement into and out of the soil (Letey, 1985). Different modes of irrigation lead to different patterns of wetting in the field, which in turn may help to avoid soil compaction and allow traffic immediately after irrigation (e.g., strip irrigation) or may prohibit traffic completely (e.g., sprinkler irrigation). By varying the amount, rate and duration of irrigation, it is possible to increase or decrease the soil bulk density, depending on soil swelling or soil structure collapse.

By maintaining compacted layers near field capacity it is possible to lower their mechanical impedance and to encourage loosening by root growth (Bowen, 1981), but the efficiency of crop root systems to loosen soil may be limited by aeration deficiencies arising from slow oxygen diffusion into wet, compacted layers.

MACHINERY MANAGEMENT

Matching machine operations to the conditions of the soil for prevention or alleviation of soil compaction is a major management tool. Potentially, great latitude exists in the choice of machinery, the method of operation, wheel management and traffic distribution on the soil. All of these factors need to be considered by a manager or a farmer in view of the soil's wetness, availability of implements and tractive machinery.

Soil compaction results from failure of a soil matrix under stress imposed at its surface or by implements at a given depth. Stress is commonly defined as load per unit of contact area. The theoretical models predict that the compactive stress decreases with the square root of depth, and a given stress will propagate deeper as the width of the contact area between the loaded unit (e.g., implement or tractive device) and the soil increases (see Chapter 4). Thus, for a given contact area, an increase in load will cause an increase in the depth of soil compaction.

For a given load, the greater the contact area, the lower the stress per unit area, but it will affect a greater soil volume to a greater depth (see Chapter 4). These theoretical predictions have been verified experimentally although some discrepancies were observed.

Load

There are many benefits to be gained from a limitation to the axle loads of agricultural vehicles (see Chapter 20). The load imposed on the soil surface is a primary factor in soil compaction, particularly at depths >30 cm (Porterfield and Carpenter, 1986). The higher the load, the deeper the compaction, unless the surface layer is firm (Taylor, 1987). Increasing the load was shown to result also in higher bulk density at shallow depths (Gameda et al., 1987b). In other experiments, stresses applied under wet conditions led to subsoil compaction (Voorhees et al., 1986; Gameda et al., 1987a).

The use of heavy machinery during tillage should be avoided on wet land. In reduced tillage, the load requirement is often less because the power need is reduced (Young et al., 1985). With active PTO-driven tools, less load of the tractor is required to contribute to traction (Cooper, 1971; Krause and Lorenz, 1984). Yet, such implements may lead to tillage pans which need to be broken down occasionally with heavier equipment. Draught animals do not usually compact the soil to any harmful extent (Krause and Lorenz, 1984). However, because of the load concentration under hooves, moldboard plowing and comb harrowing by water buffalo compacted the soil more than did the load of a 5-7-kW tiller (Kuether, 1977).

Reducing the load of machinery is one of the primary ways to reduce compaction in forest soils, although a pulling vehicle can compact the soil as much as a heavy lifting vehicle (Greacen and Sands, 1980). However, reducing loads leads to reduced traction, which at times becomes the limiting factor in deciding which operation should be performed. In humid regions, where farmers usually work on moist to wet soils, the question arises whether to use wheel tractors with low tire inflation pressure and lower soil surface stress, or crawler-type tractors. Practically, on friable or soft soils, there is no real difference in soil compaction between these two types of tractive devices, provided the total load is the same (Erbach et al., 1988).

Ground contact pressure

Wheels
For a wheeled vehicle of given weight, the compactive stress can be reduced by: (1) decreasing the ground contact pressure, e.g., by using larger wheels or tracks, (2) increasing the contact area, e.g. by using dual wheels; (3) reducing the inflation pressure in the tires (Soane et al., 1981). By lowering the inflation pressure, traction is increased (Cooper and Reaves, 1985) and, within limits, compaction is reduced (Krause and Lorenz, 1984; Söhne, 1985). Traction depends on ground contact pressure and shear strength of the soil. Reduced ground contact pressure may increase traction due to increased contact area. If the soil is either very wet, well aggregated or friable (soft), reduced ground contact pressure will result in a greater reduction in shear than an increase in contact area. The reduction of ground contact pressure by the use of dual wheels and low ground pressure tires is likely to be particularly important on seedbeds (Fig. 2).

Increasing the tire size does not always affect its performance (Cooper and Reaves, 1985). Increasing the width of tires increases the wheeled surface area (Soane and Pidgeon, 1975). For a given total load, an increase either in diameter or in width of tires, results in compressive stresses extending to deeper soil layers

Fig. 2. Techniques for the reduction of ground contact pressure on seedbeds. Top: dual wheels; bottom: low pressure tires.

than for smaller tires (Soane et al., 1981).

Use of tires with low inflation pressure on transport vehicles has been successful in reducing compaction in the main regions of the former U.S.S.R.

(except for some northwestern regions, Siberia and the Far East), where the soil water content does not exceed 0.9 field water capacity (Sadovnikov et al., 1988). This has resulted in decreasing power requirements and fuel consumption.

On soft, wet or tilled soils, cage wheels can be used instead of duals to reduce ground contact pressure (Davies et al., 1982) and slip (Soane et al., 1981). Advantages of cage wheels are: (1) they are inexpensive (Soane et al., 1982); (2) they can aid in reducing compaction (Krause and Lorenz, 1984); (3) they reduce rutting (Soane and Pidgeon, 1975). However, compaction from conventional cage wheels may be only slightly less than from rubber tires and a zone of limited compaction may form below the cage wheel (Soane, 1973).

Tracks

The increase of compactness below metal or rubber crawler tracks may not always be as large as below rubber tires (Soane, 1973; Wolf and Hadas, 1987; Erbach et al., 1988). Consequently, the use of tracklayers instead of tires (see Chapter 21) may effectively reduce compaction, especially in fine-textured soils (Soane et al., 1981).

Research in the former U.S.S.R. has indicated that the decrease in crop yields due to compaction can be minimized by using tracks instead of wheels (Evtenko et al., 1984; Puponin, 1984; Esheev and Kalashnikov, 1988; Yushin et al., 1988). The average ground contact pressure exerted by wheeled tractors weighing 7.5 Mg, was 1.6-2 times higher than that imposed by tracked tractors (Puponin, 1984). The grain yield of spring barley (*Hordeum vulgare*), peas (*Pisum sativum*), oats (*Avena sativa*) and winter wheat (*Triticum aestivum*), was lower by 8-20% when wheeled tractors were used instead of tracked ones for seedbed preparation (Puponin, 1984). The use of flexible pneumatic tracks (tractor "Ruslan") reduced ground contact pressure by up to 20-30 kPa, which resulted in a 10-20% lower yield depression (Prokopetz, 1980). Such an improvement was predicted by Taylor and Burt (1975), who compared tires with pneumatic tracks.

Tracked vehicles are especially recommended for spring operations when the soil is wet and for the transport of heavy loads from the field, especially of potatoes (*Solanum tuberosum*), vegetables and sugar beet (*Beta vulgaris*) (Bondarev et al., 1987).

Operation of machinery

Speed

When soil compaction is to be avoided or reduced, the main factors to be considered in the operation of machines and tractors are: (1) forward speed; (2) soil water content; (3) ground contact presure; (4) number of passes.

Traveling at high speed causes less compaction than traveling at low speed (Dexter and Tanner, 1974), but steerability may be impaired. Providing sufficient wheel loading and sufficient speed, adequate power at less slip will be developed, which may minimize compaction (Davies et al., 1973), but not in all cases (Hadas, 1987; Hadas et al., 1986, 1988). In some operations (e.g., plowing), an increase in speed requires an increase in draft, which necessitates additional weight or power or both.

The effect of forward speed of agricultural vehicles is dependent on the circumstances of the soil-machine interaction (Karczewski, 1978; Puponin, 1984; Agafonov et al., 1988). Karczewski (1978) showed in a soil bin experiment that at forward speeds of 1 and 12 km h^{-1}, the resulting maximum soil bulk densities were 1.65 and 1.49 Mg m^{-3}, respectively. This is in agreement with the results of Puponin (1984) who found a considerably smaller soil deformation with increasing forward speed. In field experiments on a rough soil surface in Sweden, an increase in the forward speed from 2 to 7 km h^{-1} resulted in a slightly lower degree of compactness. When the speed increased further to 12 km h^{-1}, a slightly higher compactness was obtained (Ljungar, 1977). This increase in compactness at higher speeds was ascribed to the tractor bouncing on a rough soil surface, resulting in greater stress transmission. In the case of dual wheels, an increase in the speed from 0 to 9.1-9.7 km h^{-1} caused a smaller contact area between the soil and the tire and, as a consequence, the ground contact pressure increased by 9.9-18.7%, depending on the wheel size (Agafonov et al., 1988).

Machinery designed for high speeds may be of great help in controlling compaction, whatever the cropping system. For conventional tillage, high-speed plowing can be performed with a variable-angle moldboard plow (Cooper and Reaves, 1985), but the tilth produced may differ from that produced at low speeds. In direct-drilling systems, high-speed vehicles should be used up to a speed limit specific for each operation, with regard to amount of application and accuracy in distribution and placement, not only for sowing but also for the application of fertilizers, herbicides, and pesticides (Elliott, 1980). However, controlling the speed of travel has only limited importance. Thus, in practice, even if mean rut depth and bulk density can be reduced by about 10% by increasing the speed from 0.5 to 10 km h^{-1}, there is still a high risk of complete structural damage when operating on wet soil (Horn et al., 1989).

Soil conditions

The optimum water content to minimize soil compaction varies with the kind of operation and the tool. Tillage implements which loosen from the surface downwards, such as those with tines working at different depths, can operate over a wide range of soil water conditions and depths (Spoor, 1980). The optimum water content for trenching is around the mid-plastic limit, whereas deep tillage by moldboard or disk plow should be done between the mid-plastic and the upper-plastic limits, in order to avoid the formation of large clods, smearing and adhesion of soil to implements, particularly in clay soils. Subsoiling clay pans is effective only if the clay pan is shattered, which is possible only at relatively low water contents, viz., the lower friable to hard range of soil consistency (Eck and

Unger, 1985). If the soil is too wet to be shattered by conventional subsoiling, deep loosening may be possible by double-digger subsoiling (Rowse and Stone, 1980). To achieve the best loosening, chiseling or ripping should be done in soil drier than the lower plastic limit (Eck and Unger, 1985).

Each tool has a critical working depth, which is the maximum useful depth below which tillage results in soil compaction rather than in loosening. The critical working depth reduces as soil water content and plasticity increase (Spoor and Godwin, 1978). In order to improve management procedures, the plastic limits, implement critical depth and tilth produced should be determined.

In the same field, significant differences in compactness are generally found between trafficked and non-trafficked zones. The differences are often largest in the topsoil but, deep in the profile, differences may still be found (Voorhees et al., 1978). Repeated passes may further increase the severity of compaction (Duval et al., 1989; Ohu and Folorunso, 1989). However, three-fourths of the increase in bulk density resulting from four passes of a tire running in the same rut, may occur at the first pass, which causes almost 90% of the sinkage (Cooper and Reaves, 1985).

By means of combined cultivation implements, which are equipped with active and passive tools, e.g., chisel and rotovator, it is possible to link operations, thus reducing the required weight and power of tractors and the number of passes (Cooper, 1971; Hadas et al., 1988).

Controlled traffic

One of the opportunities to minimize the detrimental effects of soil compaction is by confining wheel traffic to permanent lanes (see Chapter 22). The soil is wheel trafficked to pre-arranged positions only, in contrast to the generally random wheelmarks. The tracks may be permanent or temporary (Soane et al., 1982). In the case of permanent tracks, trenches of stones in stony soils (Chamen et al., 1980) and, experimentally, concrete tracks have been tested (Pollard and Elliott, 1978). Wide-frame tractors and gantry-type tractors are operated under these permanent lane conditions.

Advantages of controlled traffic are improved traction efficiency, timeliness and controlled soil conditions for optimized crop development, while reducing soil compaction (Puponin, 1984; Young et al., 1985; Taylor, 1987). However, if the soil is very dry, the effectiveness of traction on the compacted path can be severely reduced because the lugs cannot penetrate the soil layer and, thus, the soil contact area is limited to the lugs area (Cooper and Reaves, 1985).

During harvest, grain is often transferred to a wagon as the harvester moves over the field. Likewise, unthreshed cereals are sometimes stored in stacks in the field prior to threshing (Puponin, 1984). Where compaction is a problem, transference of the grain to wagons or trucks and storage of stacks, should be confined to the headlands to minimize compaction in the bulk of the field.

Implements

The kind of tool determines the compactness resulting from each operation. Maintenance of machinery in good operating condition is important in order to prevent compaction (Gill, 1971), directly because sharp blades and tips cut and shatter better whereas blunt tips compress, and indirectly because an increase in the operating efficiency may reduce the number of passes and loading time. For the same reason, tools or knives with an excessively large (blunt) cutting angle should be avoided (Krause and Lorenz, 1984). On the one hand, fragmented soil, especially when moist or wet, is more susceptible to compaction. On the other hand, fragmented peds are stronger along one or two of their axes than along the others (Hadas, 1990). Moreover, their spatial arrangement should be known (Hadas and Shmulewich, 1990). Thus, tillage implements should be designed for creating a geometrical arrangement of peds and fragments which will be least sensitive to recompaction.

Compaction during plowing can be decreased with winch-operated plows (Soane et al., 1982) or by out-of-furrow plowing. In the latter case, mechanical furrow-follower systems may allow self-steering at a set distance from the furrow so that the traffic load is concentrated on the uncultivated land (Hilton and Chestney, 1973). Recently, new plow bodies have been designed for high-speed plowing, such as the bottom plow (Kishida, 1985). Disc plows have been modified and there is a tendency toward plows with powered discs, which make use of both the PTO and the drawbar of tractors (Kishida, 1985).

Among the possibilities for effective control of compaction is the use of a universal machine with a large working width that performs seedbed preparation, fertilizing and seeding in one pass. Use of such a machine may diminish the number of passes by vehicles by 100-200% (Kushnarev, 1987). Well planned use of wide tools, with repeated passes in the same wheel tracks, reduced the normally trafficked area (without field transport) on fields of sugar beet, maize (*Zea mays*) and winter wheat, by 40, 48 and 63%, respectively (Kushnarev et al., 1987). This system increased the grain yield of spring barley grown on a Mollisol by about 9%, and reduced fuel consumption considerably (Puponin, 1984). It is important to select tractors and transport trailers with the same wheel spacing (Puponin, 1984).

Traffic damage may be reduced by operating combined cultivation implements and drills on pre-arranged wheel tracks (Soane and Pidgeon, 1975; Kishida, 1985). The operations can be carried out traveling at high speed and with fewer passes, but the soil water content range for optimum trafficability and workability is narrowed and larger tractors are often required, which leads to increased traffic load and ultimately to compaction and more traffic in the lanes. However, the traction requirement may be reduced by reducing the working width without reducing the speed and the compaction may be reduced also by a proper wheel choice (Krause and Lorenz, 1984). The use of heavy harvesters should be subject to strict precautions to minimize soil damage (Soane et al., 1982). Spraying with low-volume sprayers should be encouraged.

In some cases, changes are necessary to make conventional equipment suitable for wide-bed controlled-traffic systems (Williford, 1987). In all controlled traffic systems an automatic or operator-assisted steering control is important, in order to prevent the traffic lanes from becoming excessively wide (Young et al., 1985). Such automatic systems are now in use in some wide-frame tractors and gantries, for instance the system described by Hadas et al. (1990).

AGRONOMIC PRACTICES

Soil compaction can often be prevented or alleviated by proper choice of agronomic practices. Soil organic matter maintenance, liming and choice of cropping systems are important.

Organic matter maintenance

Preservation and addition of organic residues is a powerful means for minimizing or preventing compaction. In a recent review article Soane (1990) has pointed out that soil "compressibility" or "compactability" diminishes as organic matter content increases. Six different mechanisms were given by which soil "compactability" can be decreased by increased soil organic matter content: (1) improved internal and external binding of soil aggregates (improved cohesion); (2) increased soil elasticity and rebounding capabilities (relaxation mode); (3) dilution effect (reduced bulk density) due to mixing organic residues with the soil matrix; (4) temporary or permanent existence of root networks; (5) localized change in electrical charge of soil particle surfaces; (6) change in the soil internal friction (e.g., lubrication effect).

By proper tillage, crops and crop rotations, the organic matter content of soils may be preserved or increased, whatever the land use (agriculture, forestry, range). By putting arable land to grass, an increased return of organic material, a decreased rate of organic matter decomposition and an accumulation of soil organic matter will occur (Jenkinson, 1977). Cover crops can preserve the soil organic matter, especially if they are highly productive in root mass, as evidenced by ley farming in Australia (Krause and Lorenz, 1984). If all crop residues (straw, fodder) are returned to the soil, high yielding grain crops can maintain or increase the organic matter in the plow layer. Under permanent pasture, the organic matter content may reach high levels. It also increases in direct-drilled land if there is heavy stubble and it may become higher than under overgrazed pasture and conventionally tilled soil (Chan, 1989). Yet, traffic or concentrated burial of residues may lead to variable decomposition rates and thus increase field variability.

Burning of residues, particularly on sandy soils and in semi-arid regions, both on forest and agricultural soils, should be avoided (Greacen and Sands, 1980; Krause and Lorenz, 1984). The direct application of organic materials is beneficial in cropping systems for buffering the influence of other practices. In a soil amended with lignified redwood, the growth of grass roots was relatively little affected by irrigation and compaction, compared with unamended soil (Letey, 1985). Manure additions increased the organic matter content and decreased the bulk density, the decrease in bulk density depending more on the amount of manure than on the incorporation technique (Sommerfeldt and Chang, 1985). In semi-arid regions, residues left on the surface until the start of the next rainy season help retard decomposition (Krause and Lorenz, 1984).

In spite of the marked benefits listed above, field incorporation of bulk manure or crop residues often result in horizontal and vertical non-uniformity. The degree of non-uniformity and its direction (vertical and horizontal) depends on the physical properties of the buried material and the implement used (MacIntyre et al., 1987; Soane, 1990). These non-uniformities may lead to impaired water distribution in the field, especially on irrigated fields (Rawitz et al., 1990).

Liming

The addition of lime to the soil may affect its resistance to compaction, in addition to its effect on production of organic matter. In choosing the right doses, attention should be paid to the form of lime and the fineness of grinding. The presence of free lime is particularly effective in clay soils, making them less compactable than neutral or acid soils of similar texture (Cannell et al., 1978). $Ca(OH)_2$ is more effective than ground limestone ($CaCO_3$) (Bowen, 1981). The clay content and mineralogy are essential features to determine the effect of lime level on soil compaction. Liming increased pH and modified clay dispersibility in variable-charge soils (Roloff and Larson, 1989).

Liming may reduce the plasticity of soils. The plasticity will be decreased mainly in clays dominated by montmorillonite and will increase with moderate to high exchangeable sodium content (Pettry and Rich, 1971; Smith et al., 1985). In highly weathered soils, over-liming may reduce the aggregate size and the stability of microbially-formed aggregates, due to the stimulation of microbial activity or to the higher susceptibility of the basic form of binding agents to the attack by soil organisms (Kamprath, 1971).

The use of other soil conditioners, such as poly-electrolytes, may reduce the compactability for several years because of their aggregating action, but they may be too expensive for general field application (Bowen, 1981). Some polybutadiene polymers were able to improve soil conditions, but possible phyto- and zoo-toxicity may limit their use (Van Impe et al., 1988). Moreover, to be effective they need an optimum pH, a soil temperature of at least $12°$ C, an optimum soil water content and the addition of curing agents (e.g., glyoxal, iron sulfate, melanic acid).

Cropping system

A cropping system is a compromised set of procedures which complement each other or substitute for one another in case there is a change in climate, crop to be grown, machinery availability, labor or economic constraints.

Crop yields will respond to excessive soil compaction but soil compaction induced by traffic is avoidable in many cases where mechanization is unavoidable. Therefore, for an effective crop sequence practice, it is important to select those alternatives which will reduce or even prevent soil compaction on a long-term basis.

Compaction occurs from pressure created by tillage tools or from vehicular traffic. Without tillage (e.g., zero-tillage), soils reach an equilibrium bulk density, which is dependent on the inherent soil properties and the amount of traffic over the soil, interacting with the soil water and temperature regimes. An equilibrium may be attained during the first year at depths >21 cm and after about three years at shallower depths (Pidgeon and Soane, 1977). The natural processes leading to the change in soil bulk density are very complex and involve: (1) turnover of organic matter (decomposition, soil locally stressed by roots, localized desiccation, etc.); (2) wetting and drying, swelling and shrinkage, freezing and thawing; (3) slaking during dissolution of salts which cement soil particle contact points together and then, after re-deposition, may re-cement other contact points.

In monoculture systems, an annual increase in bulk density and in the depth at which the maximum bulk density occurs, often takes place (Gameda et al., 1987a), because annual tillage can alleviate this constraint only near the soil surface. When double cropping is adopted, compaction problems are particularly likely to arise (Kuether, 1977). Crop rotation, especially with grassland, may result in higher organic matter content in the topsoil, but the resulting level of compaction will depend on the intensity of traffic on the grassland (see Chapter 15). The absence of tillage causes a lower oxidation rate of the organic matter in the soil and the year-round cover limits erosion of the surface organic-matter enriched soil layers (Bauer and Black, 1981). The surface litter and crowns of grass plants protect the macropores, created by the root channels, from closure by traffic even when some surface compaction occurs (Meck et al., 1989).

Overgrazing, however, may compact the soil to a greater extent than conventional cultivation. Grazing may cause immediate compaction of cultivated fallow or direct-drilled fields (Burch et al., 1986). Grazing may also be an important factor in causing compaction of some forest soils (Greacen and Sands, 1980). However, well managed pastures in rotation, with reduced vehicle traffic and animal grazing, can be the best use of the land in order to control compaction. If not overgrazed, soil organic matter will not decrease and the soil will stay covered and protected from the impact of compactive stresses. Grazing acts also indirectly by modifying the vegetation. The variation in the flora and relative amount of species affect the compactability of the soil. Management should be directed to increase the species and varieties more adapted to grow in difficult conditions, reducing the bulk density and possibly breaking compacted layers. Furthermore, by proper selection of crop sequences or crop rotation, traffic load and number of tillage operations at improper times (e.g., on wet soils), one can manage to reduce hazards of soil compaction and the consequential need for alleviating tillage operations (Kay et al., 1988).

REMEDIAL ACTIONS

When compaction takes place to such an extent that plant growth is hindered and the productivity decreased, it becomes necessary to break the compacted layers and to loosen the soil. Excessive compactness can be reduced by tillage and by organic matter management, both related to the cropping system adopted.

Tillage

Clearing the land to start crop production is usually followed by primary tillage which loosens the soil. When care is not taken to prevent compaction during the cropping cycle and maintain the initial condition, tillage will be needed again and cropping cycles become compaction-loosening cycles.

The effect of tillage by any tool is a function not only of soil water content, texture and structure, but also of the surface roughness and the amount of residues and roots at the surface (Henriksson, 1989). The most effective method of tillage depends upon the soil conditions. In chiseled soil, pockets are created where the water concentrates so that highly compactable zones remain (Godwin and Spoor, 1977). However, in fine-textured or very compacted soils, even if moldboard plowing loosens the soil, unfractured clods may persist (Soane and Pidgeon, 1975), and several more tillage passes will be required (Hadas et al., 1978; Wolf and Hadas, 1987). In an African Oxisol derived under savannah, disking was the most successful tillage treatment for alleviating surface soil compaction (Onwualu and Anazodo, 1989). However, moldboard plowing was particularly effective in reducing the bulk density of fine-textured Alfisols of the semi-arid tropics (El-Swaify et al., 1985).

The benefits of tillage over no-tillage systems increase as soil compaction increases (Onwualu and Anazodo, 1989). While tillage is a fast cure for excessive compactness at the surface, it may cause damage deeper in the soil (Greacen and Sands, 1980; Duval et al., 1989). Reducing compaction from plowing is possible by out-of-furrow plowing with steering aids or by using shallow plows with large working widths. Fixed tines for breaking up the compacted layers below plow depth are useful (Soane and Pidgeon, 1975) and, sometimes, increasing the depth of primary tillage may be required. However, working on plastic soils below the critical depth may increase deep compaction rather than loosening (Spoor and Godwin, 1978) and thus it may become necessary to replace tillage by subsoiling.

Subsoiling

Effectiveness
The purpose of subsoiling is the disruption of deep compacted layers (Eck et al., 1977). Subsoiling is most effective when applied to air-dry or dry soil, which may rarely occur. As the water content increases, the effectiveness of the subsoiler decreases.

The effect of deep tillage on the soil water regime often determines its appropriateness. When compaction creates waterlogging, loosening of the compact layer is useful only if good drainage is achieved (Ide et al., 1984). By increasing the soil volume explorable by roots, subsoiling gives the most evident results in dry seasons and in arid and semi-arid regions, where it may be necessary to repeat subsoiling every 2-5 years (Lhotský et al., 1981; Krause and Lorenz, 1984; Loukotková, 1988). Subsoiling may not be required in the case of irrigated crops but it may allow flexibility in the irrigation schedule if deep compaction occurs (Buxton and Zalewski, 1983).

If the subsoil is acid and/or has a toxic level of aluminum, liming should be associated with subsoiling to improve soil conditions (Shiel and Rimmer, 1984; Eck and Unger, 1985).

The effect of subsoiling on crop response is markedly dependent on the soil conditions, plant species and weather conditions. Unger and Werner (1985) reported that subsoiling of a compact B_t horizon in an Alfisol resulted in an increase of 5-25% in root penetration and water uptake by sugar beet and ryegrass (*Lolium perenne*) from the 40-80-cm layer, despite rather favorable soil water conditions. Kushnarev et al. (1987) reported that, at various sites in the former U.S.S.R., subsoiling to 45 cm depth increased the yield of agricultural crops in dry years on the average by 4.4% in fine-textured soils and by 2.3% in sandy loam soils compared to plowing to 20 cm depth. On a sod-spodosolic soil of medium texture, subsoiling undertaken twice per rotation (over 7 years), increased the crop yield by 9.9%, and by 13.7% when a rotovator was used as well. The economic effect for a subsoiler plus a rotovator compared to a subsoiler alone was higher by 12-26.4% for potatoes, 19.9-25.2% for barley and 18.4-20.9% for a peas-oats mixture.

The effect of deep loosening on potato yield in Estonia was related to the increased volume of the ridges. This volume was 619 m^3 ha^{-1} in compacted soil and increased after plowing and loosening of the furrow bottom to 1535 m^3 ha^{-1} (Nugis, 1987). The persistence of a positive effect of subsoiling depends on soil conditions (Lhotský et al., 1981; Maslov, 1981; Loukotková, 1988; Pittelkow et al., 1988). On soils with a good structure, the effect may last up to 5-7 years (Maslov, 1981). The changes in soil properties due to subsoiling were more persistent on loess soil and rendzina than on sandy soil, where they disappeared almost entirely in the fourth year (Kęsik, 1980). However, the response of crop yields was most evident in sandy soil, which can be attributed to the fact that subsoiling disrupts the original positioning of non-elastic sand grains fixed with calcium carbonate and ferrous hydroxide, which restricts root penetration.

If subsoiling is followed by reduced traffic, the loosening effect may last for over 10 years (Duval et al., 1989) but it may disappear after two years of conventional tillage (Bishop and Grimes, 1978). In the case of controlled traffic, the effect of subsoiling was observed to persist for more than five years (Cooper and Reaves, 1985). The effect of subsoiling tends to last longer in clay soils where deep moldboard plowing may reduce compaction for at least four years (Eck and Unger, 1985). The lower bulk density created by trenching lasted for at least three years (Heilman and Gonzalez, 1973). Deep tillage by thoroughly mixing the profile may be effective for over 12 years in clay loams (Eck et al., 1977). No fragipan re-formation was observed 16 years after trenching (Bradford and Blanchar, 1980). Long-term persistence was obtained also by disk plowing in Texas, U.S.A. (Hauser and Taylor, 1964). In general, the persistence is longer when biological activity stabilizes the fissures (Spoor, 1980).

Subsoiling implements

Various tools can be used for loosening, shattering or inverting and mixing subsoils. The best shattering of plowpans is by chiseling relatively dry soil (Eck and Unger, 1985). A complete mixing of the profile can be obtained by trenching, which may be effective in breaking fragipans. The necessity of mixing against fragipan reformation, however, may vary with texture through the profile (Bradford and Blanchar, 1980).

In a stratified fine sandy loam, deep tillage by slip plowing appeared more effective than chiseling (Kaddah, 1976). A slip plow has a heavy deep chisel, about 3 m long, with a 30-cm flat plate, tilted upwards at the base of the chisel. Subsoiling with a rotovator, working to a depth of 45 cm and attached to a moldboard plow working to a depth of 30 cm, was successful in reducing wheeling compaction in England (Rowse and Stone, 1980). Bosse and Baur (1984) developed and tested various systems for plow pan loosening. The best loosening effects in terms of tractive force requirement were obtained with an implement fitted with 9-cm wide chisels at 70 cm spacing. Chisel and duck-foot sweeps of 12 cm width proved suitable for shallow loosening in the plowed furrow, whereas chisels of 9 cm width for loosening at 20 and 30 cm depths, were effective. Jóri (1988) reported that chisel plows with rake angles of 20-30° and 20-30-cm wide wings, mounted on shears, are favorable in terms of power requirement, size of cross section of soil disturbed, and loosening effect.

The soil bulk density may be modified in different patterns: (1) deep plowing results in deep loosening of all the soil; (2) deep chiseling results in deep-tilled channels and disturbance of the soil around them; (3) paraplowing combines fragmentation with limited mixing and shattering by a lifting and twisting action without mixing; (4) subsoiling by slit-tillage results in loosened slots, on either side of which the soil stays firm (Blackwell et al., 1989). Slit-tillage by a vibratory plow is less expensive and longer lasting than chiseling. Cutting narrow vertical slots through the plowpan beneath the row at planting was effective in the southern U.S.A. (Elkins et al., 1983) and it requires 12-40% less energy than conventional chiseling at the same depth (Cooper and Reaves, 1985). The deep-tilled area is narrow and, like other techniques of under-the-row subsoiling, recompaction due to wheel traffic is less likely than in conventional primary tillage or random subsoiling (Trouse, 1983).

Deep slit tillage

In Germany, it was found that in many cases loosening could be confined to channels or harmfully compacted areas (Unger and Steinert, 1983; Unger, 1984; Steinert, 1985; Stracke et al., 1985; Pittelkow et al., 1988; Unger et al., 1989). Vertically loosened channels (3 cm wide and 30 cm apart) permitted root penetration to depths similar to those in uncompacted soil and the use of soil water followed the same pattern as the root growth (Unger and Steinert, 1983; Unger, 1984; Steinert, 1985; Pittelkow et al., 1988). To alleviate the negative effect of a plowpan (upper subsoil layer), Unger et al. (1989) used a technique called "slit plowing" consisting of slits 10 cm wide, 20 cm deep and 30 cm apart. The slits serve as guiding zones for root growth and are also important for water drainage and air transport. The use of this technique resulted in 5-20% higher crop yields during the first five years.

To verify the loosening effect, Rogasik et al. (1984) used a computer-aided system for spatial scanning of penetration resistance. To aid the quantification of the primary tillage requirement, Petelkau et al. (1988) developed a technique for describing spatial variability of soil compaction, which is based on the relationship between soil mechanical resistance and bulk density.

Under-the-row subsoiling appeared successful in the southeast U.S.A. for corn (*Zea mays*), soybean (*Glycine max*), peanut (*Arachis hypogaea*), cotton (*Gossypium hirsutum*) and cucumber (*Cucumis sativus*) (Cooper and Reaves, 1985), as did chiseling in the center of the bed (precision tillage) for potatoes in California (Bishop and Grimes, 1978). Maximum reduction of compactness can be obtained if bedding follows subsoiling because, in addition to the physical disruption, the tilled slits are partially filled with the Ap material mounded up (Cassel et al., 1978). Under-the-row subsoiling, like deep loosening from tilling, has the advantage of lasting longer because most of the load is carried by the non-loosened areas (Hartge and Sommer, 1980). However, recompaction of the loosened soil tends to take place because the lateral effect of traffic extends as far as 30 cm (Cooper and Reaves, 1985) and even limiting the tillage to slots may not be sufficient to protect the soil from compaction. Cross-subsoiling, either by chisel or by slip plow, was more effective than subsoiling in only one direction (Kaddah, 1976). The effect of subsoiling tends to be reduced with increasing distance from the tilled channels (Batchelor and Keisling, 1982). Gypsum-enriched slots can reduce the soil compactability depending on: (1) the relative size of the slot and tires used (slot width should be 7-22.5 cm); (2) the unloosened soil strength; (3) the direction of passes (it should be transversal); (4) the load (a concentration of stress over the slot should be avoided) (Blackwell et al., 1989).

Rotations

A crop rotation including ley can remedy or alleviate soil compaction, although deep tillage is usually also desirable. Under pasture, the structure weakened by the previous cropping cycles can be restored because of increased organic matter content and biological activity. Both fast growing tap-rooted plants (creating large macropores) and fibrous root system plants (creating many smaller pores) are best suited for achieving the recovery of damaged soils by forming continuous vertical pores (Goss, 1987). Rotation with plants having desirable root systems, may save power and time, and last longer than loosening the soil by chiseling (Elkins and Van Sickle, 1984). Pan-loosening plants can be included in the rotation (Bowen, 1981), perhaps mixed with other species. Many years under pasture are required to restore the structure to that of a virgin sod (Russell, 1977). It is because of the long time required that the use of a crop rotation seems more suited for avoiding compaction than for relieving damaged soil (Marion and Merriam, 1985).

Kay et al. (1988) analyzed rates of changes in soil structure and their manifestations (e.g. soil compactability, tensile strength, stability to water), under different cropping systems. They found functional relationships between cropping sequence history and changes in soil structure and properties. While these findings are based on a narrow database, they introduce a logical way of analyzing databases related to soil susceptibility to compaction as time-dependent functions of tillage, traffic and cropping practices.

ALLEVIATION MEASURES

When compaction is not too severe, it is possible, over a short or medium term, to alleviate or overcome the detrimental effects without any specific treatment. By management (e.g., controlling soil water content, fertilization, selection of species) it is possible to alleviate to some extent the effect of poor soil physical conditions on plant growth (Letey, 1985). Drainage can compensate for the more rapid reaching of saturation which may occur in the topsoil in minimum cultivation systems (Cannell et al., 1978).

When sufficient irrigation is supplied, it may not be necessary to break plow pans or natural hard layers by subsoiling (Hauser and Taylor, 1964; Ibrahim and Miller, 1989). Deep tillage is more desirable where irrigation is less effective, as sometimes occurs in the lowest parts of the field (Musick et al., 1981). Without loosening deep compaction, irrigation should be more frequent (Eck et al., 1977;

Buxton and Zalewski, 1983). Lower rates and longer duration of water application may counter the lower infiltration rates which tend to occur in compacted soils (Chancellor, 1976). Fertilization may provide the nutrients, whose availability can be reduced because of the reduced volume of soil explored by roots in compacted soils. Nutrient application, in particular nitrogen application, needs careful management because nutrient uptake can interact with the effects of compaction, markedly affecting plant growth (Sills and Carrow, 1983). Large applications of fertilizer, frequent irrigations and installation of otherwise unneeded drainage systems, while helpful, are expensive and should only be considered as temporary approaches to compaction problems.

Planting crops that are less sensitive to water stress, water-logging and mechanical impedance, may alleviate to some extent the deleterious effects of compaction. For example, sorghum (*Sorghum bicolor*) is more draught tolerant than maize, and fall-planted cereals may mature before late season water shortages occur. Smucker and Erickson (1989) reported that tolerant dry bean genotypes yielded better on a compacted Aeric Haplaquept than did non-tolerant genotypes. Compensatory changes in root growth, ion and water uptake and respiration, appear to vary among cultivars (Schumacher and Smucker, 1987). Roots of some species can develop internal aerenchyma pore spaces, tolerate or eliminate anaerobic metabolites, and avoid stress by developing a compensatory root system away from the stress area (Smucker and Erickson, 1989).

Some plants with strong tap root systems appear to be able to penetrate compacted layers better than others. Alfalfa (*Medicago sativa*), sweet clover (*Melilotus alba*), and kudzu (*Pueraria lobata*) have been reported to penetrate compacted layers and show a benefit for several years because of the macropores they create.

The use of different crop species or cultivars to alleviate the detrimental effects of soil compaction has not received much attention by researchers. While crop selection can reduce the problem, the more satisfactory method is to eliminate the compacted soil condition so that the full genetic potential of crops can be realized.

CONCLUSIONS

(1) The more intensive the use of the land, the higher the risk of compaction. As management intensity increases to make it more productive, more care should be taken to avoid soil degradation.

(2) Systems are available for maintaining soil productivity at the highest level, year after year. Effective means may vary, but the best prevention is to fit the cropping, tillage and harvesting system to the soil.

(3) Compaction cannot be controlled by only one means and compaction tends to arise even when attempts have been made to avoid it. Remedial actions may then be necessary to restore the optimum compactness in the soil.

(4) Machine operations should be matched with the conditions of the soil for prevention or alleviation of compaction.

(5) Load on the soil surface from tractors, tillage and harvesting equipment, is a major determinant in soil compaction and can be manipulated by appropriate selection of the kind, size and inflation pressures of pneumatic tires.

(6) Speed of machine operation, number of passes and soil water content, are important considerations in the operation of machines.

(7) The choice of machines should be based on the desired result, soil conditions and available power. Controlled traffic may be a desirable option where traffic is frequent and soil conditions favor compaction.

(8) Agronomic practices for prevention or alleviation of compaction include crop rotations requiring a minimum of traffic, crops that produce organic materials for maintenance of soil organic matter, selection of most adapted crops, frequent irrigation, high nutrient levels and liming.

(9) Emphasis should be given to prevention of undue compaction rather than to costly remedial or alleviation practices.

REFERENCES

Agafonov, V.I., Sedov, M.V., Belkovskiy, V.N. and Pachev, V.P., 1988. Izmeneniye ploshchadey kontakta pnevmashin v zavisimosti ot rezhima ikh kacheniya. (Changes in contact area of pneumatic wheels with soil as affected by rolling conditions). Sbornik Nauchnykh Trudov, VIM, Moscow, U.S.S.R., 118: 164-169 (in Russian).

Allmaras, R.R., Burwell, R.E. and Holt, R.F., 1969. Plowlayer porosity and surface roughness from tillage as affected by initial porosity and soil moisture at tillage time. Soil Sci. Soc. Am. Proc., 31: 550-556.

Archer, J.R., 1975. Soil consistency. In: Soil Physical Conditions and Crop Production. Ministry Agric. Fish. Food, London, U.K., Tech. Bull. 29, pp. 289-297.

Batchelor, J.T. and Keisling, T.C., 1982. Soybean growth over and between subsoil channels on two loamy sands. Agron. J., 74: 926-927.

Bauer, A. and Black, A.L., 1981. Soil carbon, nitrogen, and bulk density comparisons in two cropland tillage systems after 25 years and in virgin grassland. Soil Sci. Soc. Am. J., 45: 1166-1170.

Baver, L.D., Gardner, W.H. and Gardner, W.R., 1972. Soil Physics. Wiley, New York, NY, U.S.A., 498 pp.

Bishop, J.C. and Grimes, D.W., 1978. Precision tillage effects on potato root and tuber production. Am. Potato J., 55: 65-72.

Blackwell, P.S., Jayawardane, N.S., Blackwell, J., White, R. and Horn, R., 1989. Evaluation of soil recompaction by transverse wheeling of tillage slots. Soil Sci. Soc. Am. J., 53: 11-15.

Bondarev, A.G., Rusanov, V.A. and Medvedev, V.W., 1987. Zaklucheniye. (Summary). In: V.A. Kovda (Editor), Pereuplotneniye Pakhotnykh Pochv. (Compaction of Arable Soils). Akad. Nauk U.S.S.R., "Nauka", Moscow, U.S.S.R., pp. 205-209 (in Russian).

Bosse, O. and Baur, A., 1984. Untersuchungen zu technischen Lösungen für die Krumenbasislockerung. (Studies in technical solutions for plough pan loosening). Akad. Landwirtsch.-Wiss. D.D.R., Berlin, Germany, Tag.-Ber. 227: 193-198 (in German).

Bowen, H.D., 1981. Alleviating mechanical impedance. In: G.F. Arkin and H.M. Taylor (Editors), Modifying the Root Environment to Reduce Crop Stress. Am. Soc. Agric. Eng.,

St. Joseph, MI, U.S.A., ASAE Mono. 4, pp. 19-57.

Bradford, J.M. and Blanchar, R.W., 1980. The effect of profile modification of a Fragiudalf on water extraction and growth by grain sorghum. Soil Sci. Soc. Am. J., 44: 374-378.

Burch, G.J., Mason, I.B., Fisher, R.A. and Moore, I.D., 1986. Tillage effects on soils: physical hydraulic responses to direct drilling at Lockhart, N.S.W. Aust. J. Soil Res., 24: 377-391.

Buxton, D.R. and Zalewski, J.C., 1983. Tillage and cultural management of irrigated potatoes. Agron. J., 75: 219-225.

Cannell, R.Q., Davies, D.B., Mackney, D. and Pidgeon, J.D., 1978. The suitability of soils for sequential direct drilling of combine harvested crops in Britain; a provisional classification. Outlook Agric., 9: 306-316.

Cassel, D.K., Bowen, H.D. and Nelson, L.A., 1978. An evaluation of mechanical impedance for three tillage treatments on Norfolk sandy loam. Soil Sci. Soc. Am. J., 42: 116-120.

Chamen, W.T.C., Collins, T.S., Hoxey, R.P. and Knight, A.C., 1980. Mechanisation opportunities likely to be provided by engineering in the 21st century. J. Proc. Inst. Agric. Eng., 35: 63-70.

Chan, K.Y., 1989. Friability of a hardsetting soil under different tillage and land use practices. Soil Tillage Res., 13: 287-298.

Chancellor, W.J., 1976. Compaction of soil by agricultural equipment. Univ. California, Div. Agric. Sci., Bull. 1881, 53 pp.

Cooper, A.W., 1971. Effects of tillage on soil compaction. In: K.K. Barnes, W.M. Carleton, H.M. Taylor, R.I. Throckmorton and G.E. Vanden Berg (Editors), Compaction of Agricultural Soils. Am. Soc. Agric. Eng., St. Joseph, MI, U.S.A., ASAE Mono. 1, pp. 315-364.

Cooper, A.W. and Reaves, C.A., 1985. Selected research accomplishments of the National Tillage Machinery Laboratory. Proc. Int. Conf. Soil Dynamics, Auburn, AL, U.S.A., Vol. 1, pp. 57-97.

Davies, D.B., Finney, J.B. and Richardson, J.B., 1973. Relative effects of tractor weight and wheel slip in causing soil compaction. J. Soil Sci., 24: 399-408.

Davies, D.B., Eagle, D.J. and Finney, J.B., 1982. Soil Management. Farming Press, Ipswich, U.K., 254 pp.

Dexter, A.R. and Tanner, D.W., 1974. Time dependence of compressibility for remoulded and undisturbed soils. J. Soil Sci., 25: 153-164.

Duval, J., Raghavan, G.S.V., Mehuys, G.R. and Gameda, S., 1989. Residual effects of compaction and tillage on the soil profile characteristics of a clay-textured soil. Can. J. Soil Sci., 69: 417-425.

Eck, H.V. and Unger, P.W., 1985. Soil profile modification for increasing crop production. Adv. Soil Sci., 1: 65-100.

Eck, H.V., Martinez, T. and Wilson, G.C., 1977. Alfalfa production on a profile-modified slowly permeable soil. Soil Sci. Soc. Am. J., 41: 1181-1186.

Elkins, C.B. and Van Sickle, K., 1984. Punching holes in plow pans. Solutions, 28: 38-41.

Elkins, C.B., Thurlow, D.L. and Hendrick, J.G., 1983. Conservation tillage for long-term amelioration of plow pan soils. J. Soil Water Conserv., 38: 305-307.

Elliott, J.G., 1980. Use of lightweight high-speed vehicles in direct drilling. J. Sci. Food Agric., 31: 417-418.

El-Swaify, S.A., Pathak, P., Rego, T.J. and Singh, S., 1985. Soil management for optimized productivity under rainfed conditions in the semi-arid tropics. Adv. Soil Sci., 1: 1-64.

Erbach, D.C., Melsin, S.W. and Cruse, R.M., 1988. Effects of tractor tracks during secondary tillage on corn production. Am. Soc. Agric. Eng., St. Joseph, MI, U.S.A., ASAE Pap. 88-1614.

Esheev, S.B. and Kalashnikov, S.F., 1988. Vliyaniye khodovykh sistem traktorov na plodorodiye kashtanovykh pochv Buryti. (Influence of tractive systems of tractors on productivity of chestnut soils of Buryti). Sbornik Nauchnykh Trudov, VIM, Moscow, U.S.S.R., 118: 126-130 (in Russian).

Evtenko, V.G., Yushin, A.A. and Blagodatnyi, J.N., 1984. Bewertung der Wirkung bodenschonender Fahrwerkssysteme an landwirtschaftlichen Maschinen und Geräten auf den Boden und die Effektivität ihres Einsatzes. (Influence of soil-conserving carriage systems of farm machines and implements on the soil and effectiveness of their use). Akad. Landwirtsch.-Wiss. D.D.R., Berlin, Germany, Tag.-Ber., 227: 49-55 (in German).

Gameda, S., Raghavan, G.S.V., McKyes, E. and Theriault, R., 1987a. Subsoil compaction in a clay soil. I. Cumulative effects. Soil Tillage Res., 10: 113-122.

Gameda, S., Raghavan, G.S.V., McKyes, E. and Theriault, R., 1987b. Subsoil compaction in a clay soil. II. Natural alleviation. Soil Tillage Res., 10: 123-130.

Gill, W.R., 1967. Soil implement relations. In: Proc. Conf. Tillage for Greater Crop Production. Am. Soc. Agric. Eng., St. Joseph, MI, U.S.A., pp. 32-36, 43.

Gill, W.R., 1971. Economic assessment of soil compaction. In: K.K. Barnes, W.M. Carleton, H.M. Taylor, R.I. Throckmorton and G.E. Vanden Berg (Editors), Compaction of Agricultural Soils. Am. Soc. Agric. Eng., St. Joseph, MI, U.S.A., ASAE Mono. 1, pp. 431-458.

Godwin, R.J. and Spoor, G., 1977. Soil factors influencing work days. J. Proc. Inst. Agric. Eng., 32: 87-90.

Goss, M.J., 1987. The specific effects of roots on the regeneration of soil structure. In: G. Monnier and M.J. Goss (Editors), Soil Compaction and Regeneration. Balkema, Rotterdam, Netherlands, pp. 145-155.

Greacen, E.L. and Sands, R., 1980. Compaction of forest soils: a review. Aust. J. Soil Res., 18: 163-189.

Hadas, A., 1987. Soil compaction under quasi-static and impact stress loading. Soil Tillage Res., 9: 181-186.

Hadas, A., 1990. Directional strength in aggregates as affected by aggregate volume and by wetting and drying cycle. J. Soil Sci., 41: 85-93.

Hadas, A. and Shmulewich, I., 1990. Spectral analysis of cone penetrometer data for detecting spatial arrangement of soil clods. Soil Tillage Res., 18: 47-62.

Hadas, A., Wolf, D. and Meirson, I., 1978. Tillage implements-soil structure relationships and their effect on crop stands. Soil Sci. Soc. Am. J., 42: 632-637.

Hadas, A., Wolf, D. and Rawitz, E., 1986. Evaluation of draft requirements-soil compaction relations in tilling moist soil. Soil Tillage Res., 8: 51-64.

Hadas, A., Larson, W.E. and Allmaras, R.R., 1988. Advances in modeling machine-soil-plant interactions. Soil Tillage Res., 11: 349-372.

Hadas, A., Shmulewich, I., Hadas, O. and Wolf, D., 1990. Forage wheat yields as affected by compaction and conventional vs. wide-frame tractor traffic patterns. Trans. ASAE, 33: 79-85.

Hartge, K.H. and Sommer, C., 1980. The effect of geometric patterns of soil structure on compressibility. Soil Sci., 130: 180-185.

Hauser, V.L. and Taylor, H.M., 1964. Evaluation of deep tillage treatments on a slowly permeable soil. Trans. ASAE, 7: 134-136, 141.

Heilman, M.D. and Gonzalez, C.L., 1973. Effect of narrow trenching in Harlingen clay soil on plant growth, rooting depth and salinity. Agron. J., 65: 816-819.

Henriksson, L., 1989. Effects of PTO-driven harrows on seedbed qualities and yields of cereals. In: W.E. Larson, G.R. Blake, R.R. Allmaras, W.B. Voorhees and S.C. Gupta (Editors), Mechanics and Related Processes in Structured Agricultural Soils. Kluwer, Dordrecht, Netherlands, p. 243.

Hilton, D.J. and Chestney, A.S.W., 1973. Low-cost self-steering devices for out-of-furrow ploughing. J. Proc. Inst. Agric. Eng., 28: 102-106.

Horn, R., Blackwell, P.S. and White, R., 1989. The effect of speed of wheeling on soil stresses, rut depth and soil physical properties in an ameliorated transitional red-brown earth. Soil Tillage Res., 13: 353-364.

Howard, R.F., Singer, M.J. and Frantz, G.A., 1981. Effects of soil properties, water content, and compactive effort on the compaction of selected California forest and range soils. Soil Sci. Soc. Am. J., 45: 231-236, 1006.

Ibrahim, B.A. and Miller, D.E., 1989. Effect of subsoiling on yield and quality of corn and potato at two irrigation frequencies. Soil Sci. Soc. Am. J., 53: 247-251.

Ide, G., Hofman, G., Ossermerct, C. and Van Ruymbeke, M., 1984. Root-growth response of winter barley to subsoiling. Soil Tillage Res., 4: 419-431.

Jenkinson, D.S., 1977. Studies on the decomposition of plant material in soil. V. The effects of plant cover and soil type on the loss of carbon from 14C labelled ryegrass decomposing under field conditions. J. Soil Sci., 28: 424-434.

Jóri, J.I., 1988. Chisel plough and the primary tillage in Hungary. Hungarian Agric. Eng., 1: 17-19.

Kaddah, M.T., 1976. Subsoil chiseling and slip plowing effects on soil properties and wheat grown on a stratified fine sandy soil. Agron. J., 68: 36-39.

Kamprath, E.J., 1971. Potential detrimental effects from liming highly weathered soils to neutrality. Soil Crop Sci. Soc. Flo. Proc., 31: 200-203.

Karczewski, K., 1978. Wpływ prędkości przejazdu na zmiany zagęszczenia gleby przez koła maszyn rolniczych. (The influence of passing speed of agricultural machinery wheels on soil compaction). Zesz. Probl. Post. Nauk Roln., 201: 69-78 (in Polish).

Kay, B.D., Anger, D.A., Groenevelt, P.H. and Baldock, J.A., 1988. Quantifying the influence of cropping history on soil structure. Can. J. Soil Sci., 68: 359-368.

Kęsik, T., 1980. The influence of deep loosening on some physical properties of soil and plant crops. Proc. 2nd Int. Conf. Phys. Prop. Agric. Mat., Gödöllő, Hungary, Vol. 2, pp. 45: 1-7.

Kishida, Y.Y., 1985. World trends and needs in agricultural and tillage machinery systems as we approach the 21st century. Proc. Int. Conf. Soil Dynamics, Auburn, AL, U.S.A., Vol. 1, pp. 129-140.

Koolen, A.J. and Kuipers, H., 1983. Agricultural Soil Mechanics. Springer, Berlin, Germany, 241 pp.

Krause, R. and Lorenz, F., 1984. Soil Tillage in the Tropics and Subtropics. Hoehl-Druck, Bad Hersfeld, Germany, Schriftenreihe GTZ 150, 310 pp.

Kuether, D.O., 1977. Soil compaction and wetland rice tillage systems. Am. Soc. Agric. Eng., St. Joseph, MI, U.S.A., ASAE Pap 77-1021, 14 pp.

Kushnarev, A.S., 1987. Rol kombinirovannykh i shirokozakhvatnykh mashin i agregatov v umensheni uplotnyayushchevo vozdeistviya na pochvy. (Importance of combined and wide machines in diminishing of compactive effect on soil). In: V.A. Kovda (Editor), Pereuplotneniye Pakhotnykh Pochv. (Compaction of Arable Soils). Akad. Nauk U.S.S.R., "Nauka", Moscow, U.S.S.R., pp. 144-149 (in Russian).

Kushnarev, A.S., Puponin, A.N. and Matyuk, N.S., 1987. Agrotechnitzeskiye priyemy razuplotneniya pochv. (Agrotechnical measures for soil loosening). In: V.A. Kovda (Editor). Pereuplotneniye Pakhotnykh Pochv. (Compaction of Arable Soils). Akad. Nauk U.S.S.R., "Nauka", Moscow, U.S.S.R., pp. 158-165 (in Russian).

Lampe, K.J., 1984. Foreword. In: R. Krause and F. Lorenz, Soil Tillage in the Tropics and Subtropics. Hoehl-Druck, Bad Hersfeld, Germany, Schriftenreihe GTZ 150, p. 10.

Letey, J., 1985. Relationship between soil physical properties and crop production. Adv. Soil Sci., 1: 277-294.

Lhotský, J., Vachal, J. and Ehrlich, P., 1981. Hodnocení účinnosti a životnosti hloubkového melioračního kypření a vylehčování těžkých a zhutnělých půd. (Evaluation of the effectiveness and durability of deep ploughing and loosening of heavy textured compacted soils). Sbor. UVTIZ-Meliorace, 17: 81-94 (in Czech).

Ljungar, A., 1977. Olika faktorers betydelse for traktorernas jordpackningsverkan. Matningar 1974-1976. (Importance of different factors on soil compaction by tractors. Measurements in 1974-1976). Agric. College Sweden, Div. Soil Manage., Rep. 52, 43 pp. (in Swedish).

Loukotková, V., 1988. Ovlivnění fyzikálního stavu oglejené půdy na spraši různými agromelioračnimi technologiemi. (Effect of various agromeliorative technologies on physical conditions of gleyed loess soils). Vědecké Práce VUZZP, 5: 99-107 (in Czech).

MacIntyre, D., Sharp, M.J. and Gray, A.G., 1987. Rotary implement (the "Sturplow"). The Agric. Eng., 42: 120-123.

Marion, J.L. and Merriam, L.C., 1985. Predictability of recreational impact on soils. Soil Sci. Soc. Am. J., 49: 751-753.

Maslov, B.S., 1981. Opyt osusheniya tyazhelykh pochv i zadachi nauki. II. Osusheniye tyazhelykh pochv. (Problem of drainage in heavy soils and duties of science. II. Drainage of heavy soils). Kolos, Moscow, U.S.S.R., pp. 5-20 (in Russian).

Meck, B.D., Rechel, E.A., Carter, L.M. and DeTar, W.R., 1989. Changes in infiltration under alfalfa as influenced by time and wheel traffic. Soil Sci. Soc. Am. J., 53: 238-241.

Musick, J.T., Dusek, D.A. and Schneider, A.D., 1981. Deep tillage of irrigated Pullman clay loam - a long-term evaluation. Trans. ASAE, 24: 1515-1519.

Nugis, E., 1987. Dernovo-podzolistyye pochvy Estonii. (Sod podzol soils in Estonia). In: V.A. Kovda (Editor), Pereuplotneniye Pakhotnykh Pochv. (Compaction of Arable Soils). Akad. Nauk U.S.S.R., "Nauka", Moscow, U.S.S.R., pp. 35-45 (in Russian).

Ohu, J.O. and Folorunso, O.A., 1989. The effect of machinery traffic on the physical properties of a sandy loam soil and on the yield of sorghum in north-eastern Nigeria. Soil Tillage Res., 13: 399-405.

Ojeniyi, S.O. and Dexter, A.R., 1979. Soil factors affecting the macro-structure produced by tillage. Trans. ASAE, 22: 339-343.

Onwualu, A.P. and Anazodo, U.G.N., 1989. Soil compaction effects on maize production under various tillage methods in a derived savannah zone of Nigeria. Soil Tillage Res., 14: 99-114.

Petelkau, H., Gätke, C.R., Dannowski, M., Seidel, K. and Augustin, J., 1988. Bodenphysikalische Grundlagen für die Steuerung der Grundbodenbearbeitung. (Soil physical backgrounds for control of primary tillage). In: Erhöhung der Bodenfruchtbarkeit und der Erträge durch wissenschaftlichen Fortschritt. Forschungszentrum Bodenfruchtbarkeit, Müncheberg, Germany, pp. 362-379 (in German).

Pettry, D.E. and Rich, C.I., 1971. Modification of certain soils by calcium hydroxide stabilization. Soil Sci. Soc. Am. J., 35: 834-838.

Pidgeon, J.D. and Soane, B.D., 1977. Effects of tillage and direct drilling on soil properties during the growing season on a long-term barley monoculture system. J. Agric. Sci., Camb., 88: 431-442.

Pittelkow, U., John, K. and Körbs, P., 1988. Über die Auswirkungen von Brückenzonen in Krumenbasisverdichtungen auf Durchwurzelung und Wasserentzug bei differenzierter Bodenfeuchte und -dichte. Ergebnisse aus der Jenaer Bodenmodellanlage. (The effects of bridging zones in compacted upper subsoil on root penetration and water uptake at different soil water contents and bulk densities. Results from the Jena soil model installation). Arch. Acker- Pflanzenb. Bodenkd., 6: 379-387 (in German).

Pollard, F. and Elliott, J.G., 1978. The effect of soil compaction and method of fertiliser placement on the growth of barley using a concrete track technique. J. Agric. Eng. Res.,

23: 203-216.

Porterfield, J.W. and Carpenter, T.G., 1986. Soil compaction: an index of potential compaction for agricultural tires. Trans. ASAE, 29: 917-922.

Prokopetz, Yu.A., 1980. Vplyv ushilneniya gruntu khodovymi sistemami traktorov na urozhaynost selskogospodarskykh kultur. (Effect of soil compaction by tractive systems of tractors on crop yields). Vestnik Selskogospodarskoy Nauki., 4: 41-42 (in Russian).

Puponin, A.I., 1984. Obrabotka pochvy v intensivonom zemledeliy nechernozemnoy zony. (Soil tillage in intensive agriculture of non-chernozem zone). Kolos, Moscow, U.S.S.R., pp. 3-184 (in Russian).

Ram, R.B., 1984. Pressure measurement in the soil under the load. Soil Tillage Res., 4: 137-145.

Rawitz, E., Lior, H. and Rimon, M., 1990. The effect of drip-line placement and residue incorporation on the growth of drip-irrigation cotton. Soil Tillage Res., 16: 227-232.

Reeve, R.C. and Fausey, N.R., 1974. Drainage and timeliness of farming operations. In: J. van Schilfgaarde (Editor), Drainage for Agriculture. Am. Soc. Agron., Madison, WI, U.S.A., Agron. Mono. 17, pp. 55-66.

Rogasik, H., Weirauch, M., Böttcher, B. and Morstein, K.H., 1984. Nachweis der Lockerungswirkung bei Massnahmen der Krumenbasisbearbeitung mittels horizontalsondierung. (Using horizontal scanning to verify the loosening effect of plough pan tillage). Tag.-Ber., Akad. Landwirtsch.-Wiss. D.D.R., Berlin, Germany, 227: 199-203 (in German).

Roloff, G. and Larson, W.E., 1989. Strength of low and variable charge soils. In: W.E. Larson, G.R. Blake, R.R. Allmaras, W.B. Voorhees, and S.C. Gupta (Editors), Mechanics and Related Processes in Structured Agricultural Soils. Kluwer, Dordrecht, Netherlands, pp. 59-72.

Rowse, H.R. and Stone, D.A., 1980. Deep cultivation of a sandy clay loam. I. Effects on growth, yield and nutrient content of potatoes, broad beans, summer cabbage and red beet in 1977. Soil Tillage Res., 1: 57-68.

Russell, E.W., 1977. The role of organic matter in soil fertility. Phil. Trans. R. Soc. Lond. B, 281: 209-219.

Sadovnikov, A.N., Nebogin, I.S., Ilchenko, I.R. and Yushkov, E.S. 1988. Otzenka effektivnosti snizheniya davleniya na Chernozemnuyu pochvu dvizhitelya pritzepa 2 PTS - 4 M. (Evaluation of the effectiveness of diminishing of compactive pressure in chernozem soil exerted by the trailer 2 PTS -4M). Sbornik Nauchnykh Trudov, VIM, Moscow, U.S.S.R., 118: 149-158 (in Russian).

Schafer, R.L. and Johnson, C.E., 1982. Changing soil conditions - the soil dynamics of tillage. In: D.M. Kral, B. Hawkins, P.W. Unger and D.W. Van Doren (Editors), Predicting Tillage Effects on Soil Physical Properties and Processes. Am. Soc. Agron., Madison, WI, Spec. Publ. 44, pp. 13-28.

Schumacher, T.E. and Smucker, A.J.M., 1987. Ion uptake and respiration by dry bean roots. Plant Soil, 99: 411-422.

Shaffer, M.J. and Clapp, C.E., 1987. Cotton root submodel. In: M.J. Shaffer and W.E. Larson, (Editors), NTRM Soil-Crop Simulation Model for Nitrogen, Tillage and Crop Residue Management. USDA-ARS, Conserv. Res. Rep. 34-1, pp. 63-72.

Shiel, R.S. and Rimmer, D.L., 1984. Changes in soil structure and biological activity on some meadow hay plots at Cockle Park, Northumberland. Plant Soil, 76: 343-356.

Sills, M.J. and Carrow, R.N., 1983. Turfgrass growth, N use, and water use under soil compaction and N fertilization. Agron. J., 75: 488-492.

Smith, C.W., Hadas, A., Dan, J. and Koyumdjisky, H., 1985. Shrinkage and Atterberg limits in relation to other properties of principal soil types in Israel. Geoderma, 35: 47-65.

Smucker, A.J.M. and Erickson, A.E., 1989. Tillage and compactive modification of gaseous flow and aeration. In: W.E. Larson, G.R. Blake, R.R. Allmaras, W.B. Voorhees and S.C. Gupta (Editors), Mechanics and Related Processes in Structured Agricultural Soils. Kluwer, Dordrecht, Netherlands, pp. 205-221.

Soane, B.D., 1973. Techniques for measuring changes in the packing state and cone resistance of soil after the passage of wheels and tracks. J. Soil Sci., 24: 311-323.

Soane, B.D., 1990. The role of organic matter in soil compactibility: a review of some practical aspects. Soil Tillage Res., 16: 179-202.

Soane, B.D. and Pidgeon, J.D., 1975. Tillage requirements in relation to soil physical properties. Soil Sci., 119: 376-384.

Soane, B.D., Blackwell, P.S., Dickson, J.W. and Painter, D.J., 1981. Compaction by agricultural vehicles: a review. II. Compaction under tyres and other running gear. Soil Tillage Res., 1: 373-400.

Soane, B.D., Dickson, J.W. and Campbell, D.J., 1982. Compaction by agricultural vehicles: a review. III. Incidence and control of compaction in crop production. Soil Tillage Res., 2: 3-36.

Söhne, W.H., 1953. Druckverteilung im Boden und Bodenverformung unter Schlepperreifen. (Pressure distribution in the soil and soil deformation under tractor tires). Grundl. Landtech., 5: 49-63 (in German).

Söhne, W.H., 1958. Fundamentals of pressure distribution and soil compaction under tractor tires. Agric. Eng., 39: 276-281, 290.

Söhne, W.H., 1985. Soil dynamic research at the National Tillage Machinery Laboratory: an international perspective. Proc. Int. Conf. Soil Dynamics, Auburn, AL, U.S.A., Vol. 1, pp. 98-128.

Sommerfeldt, T.G. and Chang, C., 1985. Changes in soil properties under annual applications of feedlot manure and different tillage practices. Soil Sci. Soc. Am. J., 49: 983-987.

Spoor, G., 1980. Subsoiling and deep cultivation. J. Sci. Food Agric., 31: 418-419.

Spoor, G. and Godwin, R.J., 1978. An experimental investigation into the deep loosening of soil by rigid tines. J. Agric. Eng. Res., 23: 243-258.

Steinert, P., 1985. Modellversuchsergebnisse zur Funktionsprüfung von Leitbahnen in Verdichtungsschichten. (Results from model experiments for testing guide channels in compacted soil layers). Tag.-Ber., Akad. Landwirtsch.-Wiss. D.D.R., Berlin, Germany, 231: 199-207 (in German).

Steinhardt, R., 1974. Evaluating penetration resistance and wheel sinkage response to soil water suction changes in a draining clay soil. Soil Sci. Soc. Am. J., 38: 518-522.

Steinhardt, R. and Trafford, B.D., 1974. Some effects of sub-surface drainage and ploughing on the structure and compactability of a clay soil. J. Soil Sci., 25: 138-152.

Stracke, W., Reich, J., Werner, D. and Mäusezahl, D., 1985. Technisch-technologische und bodenkundliche Ergebnisse zu bodenmeliorativen Mechanisierungslösungen für den krumennahen Unterboden. (Technical, technological and pedological results of solutions for mechanizing amelioration work in the upper subsoil layer). Tag.-Ber., Akad. Landwirtsch. D.D.R., Berlin, Germany, 231: 305-315 (in German).

Taylor, J.H., 1987. A rationale for controlled traffic research. Acta Hortic., 210: 9-18.

Taylor, J.H. and Burt, E.C., 1975. Track and tire performance in agricultural soils. Trans. ASAE, 18: 3-6.

Trouse, A.C., Jr., 1983. Observations on under-the-row subsoiling after conventional tillage. Soil Tillage Res., 3: 67-81.

Unger, H., 1984. Über die Funktion von Schlitzen als Wurzelleitbahnen in verdichteten Schichten der Krumenbasis bindiger Substrate. (The function of slots as root channels in compacted layers of the plough pan of cohesive substrata). Tag.-Ber., Akad. Landwirtsch.-

Wiss. D.D.R., Berlin, Germany, 227: 185-191 (in German).

Unger, H. and Steinert, P., 1983. Erste Ergebnisse aus Modelluntersuchungen zu einer neuen Prinziplösung für die Überbrückung von Schadverdichtungen an der Krumenbasis schwerer Böden. (Preliminary results from model experiments on a new solution for overcoming detrimental compactions in the lower part of the topsoil of heavy soils). Tag.-Ber., Akad. Landwirtsch.-Wiss. D.D.R., Berlin, Germany, 215: 221-232 (in German).

Unger, H. and Werner, D., 1985. Erhöhung der Nutzbarkeit des Bodenwasservorrates durch Beeinflussung der Zugriffsfähigkeit und Zugriffsmöglichkeit der Pflanzenwurzel. (More efficient use of soil water reserves by influencing the capability and possibility of access of plant roots). Tag.-Ber., Akad. Landwirtsch.-Wiss. D.D.R., Berlin, Germany, 231: 161-177 (in German).

Unger, H., Werner, D., Pittelkow, U. and Reich, J., 1989. Ergebnisse des Schachtpflügens bei der Strukturmelioration krumenbasisverdichteter Lö- und V-Standorte. (Results of slit ploughing in the amelioration of the structure of the compacted lower part of the topsoil of loess soils). In: Vörtrage wissensch. Tagung FZB, 26-29 Juni 1989, Müncheberg, Germany, pp. 218-226 (in German).

Van Impe, W.F., De Boodt, M. and Meyus, I., 1988. Improving the bearing capacity of topsoil layers by means of a polymer mixture grout. In: J. Drescher, R. Horn and M. de Boodt (Editors), Impact of Water and External Forces on Soil Structure. Catena, Cremlingen, Germany, Catena Suppl. 11, pp. 1-14.

Van Wijk, A.L.M. and Buitendijk, J., 1988. A method to predict workability of arable soils and its influence on crop yield. In: J. Drescher, R. Horn and M. de Boodt (Editors), Impact of Water and External Forces on Soil Structure. Catena, Cremlingen, Germany, Catena Suppl. 11, pp. 129-140.

Voorhees, W.B., Senst, C.G. and Nelson, W.W., 1978. Compaction and soil structure modification by wheel traffic in the northern Corn Belt. Soil Sci. Soc. Am. J., 42: 344-349.

Voorhees, W.B., Nelson, W.W. and Randall, G.W., 1986. Extent and persistence of subsoil compaction caused by heavy axle loads. Soil Sci. Soc. Am. J., 50: 428-433.

Wert, S. and Thomas, B.R., 1981. Effects of skid roads on diameter, height and volume growth in Douglas-fir. Soil Sci. Soc. Am. J., 45: 629-632.

Williford, J.R., 1987. Controlled traffic research with modified production equipment. Acta Hortic., 210: 19-24.

Wolf, D. and Hadas, A., 1987. Determining efficiency of various moldboard ploughs in fragmenting and cutting air-dry soils. Soil Tillage Res., 10: 181-186.

Yong, R.N. and Warkentin, B.P., 1975. Soil Properties and Behaviour. Elsevier, Amsterdam, Netherlands, Developments in Geotechnical Engineering 5, 449 pp.

Young, S.C., Grisso, R.D., Dumas, W.T. and Johnson, C.E., 1985. Long term tillage requirements with controlled traffic. Proc. Int. Conf. Soil Dynamics, Auburn, AL, U.S.A., Vol. 5, pp. 1139-1151.

Yushin, A.A., Evtenko, V.G. and Blagodatnyi, Yu.N., 1988. Effektivnost primeneniya khodovykh sistem so snizhennym urovnem vozdeystviya na pochvu. (Effectiveness of tractive systems with lowered effect on soil). Sbornik Nauchnykh Trudov, VIM, Moscow, U.S.S.R., 118: 174-181 (in Russian).

PART G
==============================

CONCLUSION

Soil Compaction in Crop Production
B.D. Soane and C. van Ouwerkerk (Eds.)
627

CHAPTER 26

Conclusions and Recommendations for Further Research on Soil Compaction in Crop Production

C. van OUWERKERK[1] and B.D. SOANE[2]

[1]Institute for Soil Fertility Research (IB-DLO), Haren Gn, Netherlands
[2]Scottish Centre of Agricultural Engineering, SAC, Penicuik, U.K.

SUMMARY

Soil compaction has been estimated to be responsible for the degradation of an area of 83 million ha world-wide, of which 33 million ha lies in Europe and 18 million ha in Africa. Soil compaction problems are of economic importance in the production of a wide range of crops throughout the world. Current research on this subject is leading to a greater understanding of the soil/machine/crop/weather interactions, which strongly influence the incidence and severity of these problems. Modifications to vehicles, their running gear and their management systems, have been shown to offer opportunities to reduce greatly the incidence of compaction problems. However, as yet the commercial uptake of these techniques is at a rudimentary level. There is a pressing need to adopt a more rigorous approach to the conduct of basic research and to extend the use and validation of models to predict soil, crop, environmental and economic responses to different systems of machinery management for a range of crops, soils and weather conditions.

INTRODUCTION

Globally, human-induced soil degradation is a serious problem, which occurs on an area of nearly 2 billion ha (Oldeman et al., 1991). Table 1 shows that water and wind erosion arc by far the most important types of soil degradation; they account for 55.6 and 27.9%, respectively, of the total area affected by human-induced soil degradation. Chemical and physical soil degradation cover 12.2 and 4.2%, respectively, of the total degraded area.

Physical deterioration consists of soil compaction, waterlogging and subsidence of organic soils. On a world scale, physical soil degradation has only minor significance (83.3 million ha). However, in Europe it accounts for no less than 36.4 million ha or 16.6% of the total degraded area, which covers 218.9 million ha. Also in Africa, although accounting for only 4% of the total degraded area,

TABLE 1

World area (million ha) affected by human-induced soil degradation (after Oldeman et al., 1991)

Type of degradation	Degree of degradation				Total
	Light[1]	Moderate[2]	Strong[3]	Extreme[4]	
Water	343.0	526.8	217.2	6.6	1093.6
Wind	268.6	253.5	24.3	1.9	548.3
Chemical	92.9	103.4	42.1	0.8	239.2
Physical	44.2	26.8	12.3	-	83.3
Total	748.7	910.5	295.9	9.3	1964.4

[1]Slightly reduced productivity but manageable in local farming systems.
[2]Greatly reduced productivity; major ameliorations required.
[3]Not any more reclaimable at farm level; major engineering works required.
[4]Unreclaimable and beyond restoration; virtually lost for crop production.

the physically degraded area is impressive: 18.7 million ha.

By far the most important subtype of physical soil degradation is soil compaction. World-wide it covers 68.3 million ha (i.e., 81.9% of the total physically degraded area), of which 33.0 million ha are in Europe and 18.2 million

TABLE 2

Area (million ha) degraded by soil compaction (after Oldeman et al., 1991)

Region	Degree of degradation[1]				Total
	Light	Moderate	Strong	Extreme	
Asia[2]	4.6	5.0	0.2	-	9.8
Africa	1.4	8.0	8.8	-	18.2
South America	2.9	0.8	0.3	-	4.0
Central America	-	0.1	-	-	0.1
North America	0.5	0.4	-	-	0.9
Europe[3]	24.8	7.8	0.4	-	33.0
Australasia	0.7	-	1.6	-	2.3
Total	34.9	22.1	11.3	-	68.3

[1]For explanation, see Table 1.
[2]Asia includes the Asian part of the CIS countries.
[3]Europe includes the European part of the CIS countries.

ha in Africa (Table 2), while waterlogging and subsidence of organic soils occur on 10.5 and 4.6 million ha, respectively, which account for 12.6 and 5.5% of the total physically degraded area, respectively. A light degree of compaction occurs

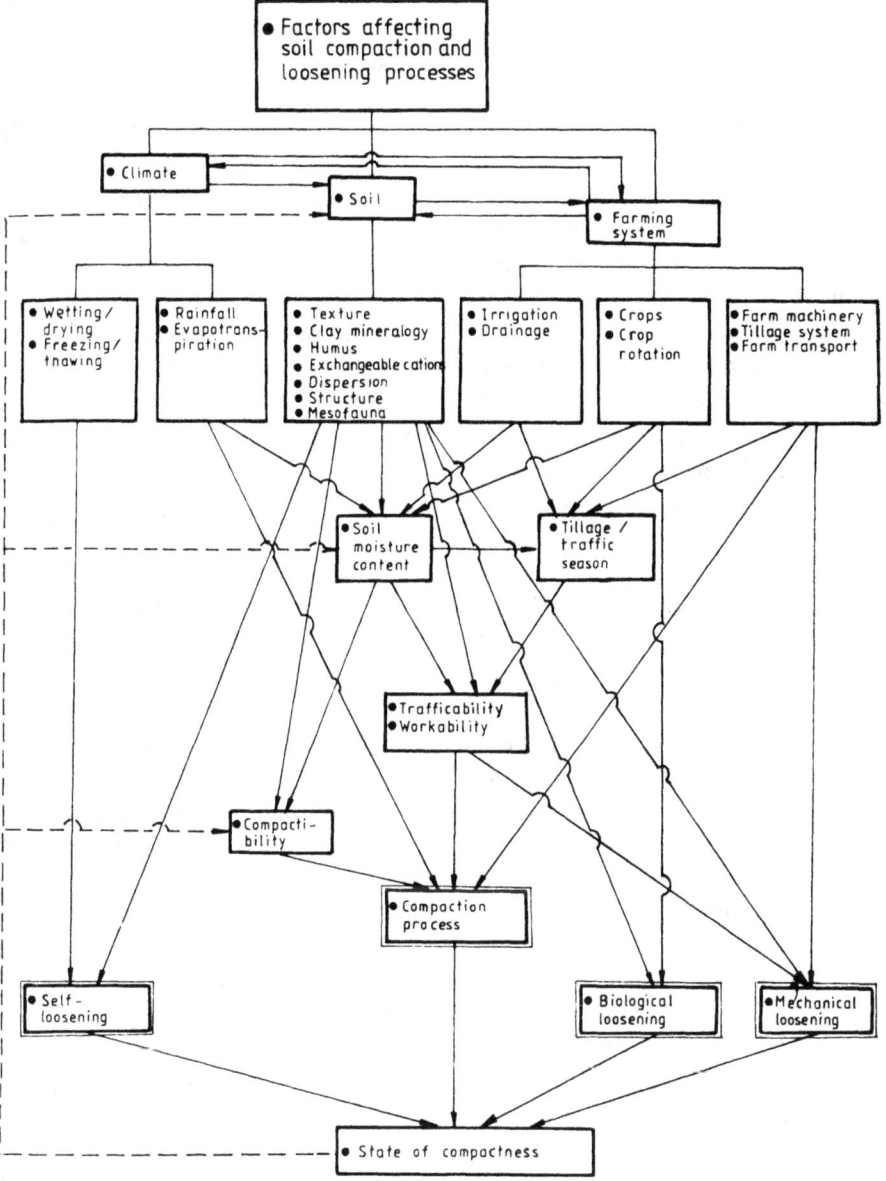

Fig. 1. Conceptual model of factors and processes affecting the state of soil compactness (from Canarache, 1991).

on 34.9 million ha (51.1% of the total area degraded by soil compaction), moderate compaction covers 22.1 million ha (32.4%) and strong compaction is found on 11.3 million ha (16.5%).

In Europe, soil compaction accounts for 91% of all physical degradation (Van Lynden, 1994). Likewise, in Asia, Africa, North America and Australasia, although the proportion of the area affected by physical deterioration is relatively small (1-4% of the total degraded area), soil compaction accounts for 80-100% of all physical deterioration. Only in Central and South America is the importance of soil compaction reduced (2 and 51% of the total physically degraded area, respectively).

The information reported above on the geographical distribution and the degree of human-induced soil degradation must be regarded as tentative owing to the very limited reliable information available in many parts of the world.

Soil compaction is a very complex problem as many interacting factors are involved (Fig. 1). In crop production, soil compaction problems are essentially of a practical and commercial nature. Therefore, we attach considerable importance to the need to stimulate closer contact between research scientists, extension workers and crop producers. Although compaction problems are widely recognised, the interpretation and translation of research findings into commercially-acceptable systems has been rarely attempted. Hence, few opportunities have arisen for the evaluation of farm-scale crop production systems which incorporate effective compaction controls. It seems likely that the economics of compaction control may be more favourable than is generally believed. However, as yet the observational data required to confirm this viewpoint is largely lacking although computer simulation models are increasingly employed to provide predictions of economic responses.

The global impact of compaction may be far greater and more serious than was previously suspected from the results of studies concerned only with the impact on crop production. In a 15-year global assessment of soil degradation, funded by the UN Environmental Programme, soil compaction was reported to be an extensively distributed component in the degradation of cultivated soils in all continents (Hammond, 1992). However, research on the environmental impact of soil compaction is still at a rudimentary level and, as yet, there has been little opportunity for the scientists concerned to review progress and to draw up plans for international cooperative research, which is essential for a subject of this calibre and global character.

Much information has been presented in the foregoing chapters concerning the incidence, mechanisms and control of soil compaction problems. The objective of this chapter is to assemble conclusions which we consider to be of general relevance and to make recommendations concerning priorities for further research.

CONCLUSIONS CONCERNING THE INCIDENCE AND CONTROL OF SOIL COMPACTION IN CROP PRODUCTION

Present-day compaction research

Numerous research projects related to the incidence and control of soil compaction in crop production are in progress throughout the world, with particularly active programmes in North America and Europe. These programmes are generally sponsored and controlled by national research organizations. However, the resources of nationally-funded programmes are usually inadequate to cover all the relevant aspects. Increased international cooperation is being stimulated by funding from international agencies such as FAO, EU, etc., and is seen as an important route to greater achievement within an acceptably short timescale. This is especially true because of the significance of soil/machine/crop/weather interactions which, extending beyond national boundaries, may not be evaluated effectively within the confines of any one country.

The success of future research into compaction problems, and the subsequent application of the results in practice, will be enhanced by a wider public awareness of the current programmes being undertaken worldwide on agricultural, forest and horticultural crops.

Field traffic

There is still much variation in the level of mechanization in different countries (Fig. 2), and even within individual countries. Much of this variation is closely related to the size of holdings and the size of machinery employed. The influence on compaction hazards of machinery size, usually expressed in terms of tractor power, is still not fully evaluated.

The steady increase in the size, power and weight of field machinery and tractors appears likely to continue in North America and Western Europe, in association with the steady decline in the farm labour force. However, in Eastern Europe, where very large tractors have been a common feature in the past, changes in the structure of farm management may give rise to a reduction in the average size of field machinery. In developing countries, where there is a steady, if slow, transition from hand to animal and ultimately to tractor power, it seems likely that mechanization will gradually introduce wheeled vehicles into areas where presently they are found to only a very limited extent.

Soil strength and responses to the passage of vehicles

A knowledge of the mechanical behaviour of soils when subjected to applied loads confirms the complexity of the compaction process, especially in structured

Fig. 2. Contrasts within Europe in the level of mechanization for harvesting grass for conservation in 1989, with low compaction risk on dry soils in central Europe (top) and with high compaction risk on moist soils in northwestern Europe (bottom).

soils. The strength of structured soils may greatly exceed that of the same soil in an homogenized state. Soil behaviour under load may be of non-deforming, hardening or flow type and in the field all three types may be of relevance under different conditions.

Stress-strain relations can be quantified using both uniaxial and triaxial approaches, but the relative merits of each approach have not been fully evaluated. Uniaxial tests have the merit of simplicity and speed, whereas, in contrast, triaxial tests tend to be slow but allow more realistic stress regimes to be applied.

Soil responses in the field to the passage of vehicles need to be expressed in terms of both horizontal and vertical distribution, especially where traffic is irregularly distributed. Microscopic studies can reveal how and when soil responses to applied loads occur and, in particular, can be used to investigate the kinds and quantity of voids and their interconnections, which have such a dominant effect on soil behaviour.

While the measurement of changes in bulk density and its related properties provides essential information, the relevance of this information is much enhanced if the data is converted to a relative basis by comparison with the bulk density of the same soil under a standard load. A number of such relative relationships have been proposed, of which the "degree of compactness" is perhaps the most frequently used.

Soil behavioural properties, such as hydraulic, aeration and strength characteristics, are all of major importance in influencing the subsequent "quality" of the soil after compaction. However, their relative importance in crop production will depend strongly on the prevailing circumstances (soil, crop, weather, etc.).

Compaction has a dominant effect on infiltration and water distribution in the profile. Numerical models to predict water flow in layered media are available but, in structured soils, there is a need to distinguish between matrix and macropore flow. Apart from the direct effects of the soil water status on crop growth, the indirect effects on the dynamics of nutrients and pesticides are also important.

At high soil water contents, compaction may readily induce deficient aeration. The evaluation of soil aeration *in situ* is a complex problem and parameters such as air-filled porosity, oxygen diffusion rate, soil gas composition and air permeability, may all have relevant applications.

Soil strength parameters are relevant to wheel/soil, weather/soil and soil/plant interactions. However, the heterogeneity of field soils leads to difficulties in measuring and interpreting soil strength. It is unlikely that a single method will be applicable throughout the whole range of soil water status. The volume of soil employed in a strength measurement technique must relate to the intended application of the results in view of the scale effects associated with soil structure.

Compaction in the subsoil, which may persist for decades, is now acknowledged to be primarily the result of high axle loads. However, the response of the subsoil to such loads will depend markedly on its initial strength and structural condition, about which as yet little knowledge exists. Where the subsoil has sufficient strength to resist imposed stresses without appreciable decrease in hydraulic conductivity, the influence of high axle loads will be less marked than where the subsoil is readily compactable.

Crop responses to soil compaction

Experiments conducted in small plots may give misleading information on the effects of compaction on crop yield and quality because of the tendency to conduct such experiments under "ideal" management conditions. This is particularly important with respect to the influence on timeliness of field operations during periods of adverse weather conditions.

Virtually all crops show sensitivity to high soil compactness in the root zone. While most of the research undertaken to date has been concerned with arable crops, other groups of crops are now subject to research programmes. Forest trees are found to show growth responses to soil compaction, especially on sandy soils and often for decades after the mechanical thinning or harvesting operations, which usually are responsible for the soil damage. Perennial forage crops subject to repetitive traffic have been found to be adversely affected by the resulting compacted soil conditions. Opportunities to use low ground contact pressure tyres as a means of reducing the incidence of compaction problems, appear to be even more promising on such crops than on arable crops.

Compaction problems in the tropics

Particularly serious compaction problems have been identified in the tropics. These problems may often be related more to the influence of adverse land clearing techniques, soil structural instability and highly erosive rainfall, than to the type and extent of mechanization in crop production. However, there is a continuing trend for the introduction of wheeled vehicles into tropical crop production and this is likely to intensify compaction problems. Although the importance of correct soil management procedures in land clearing and subsequent crop production are recognised in principle, the economic restrictions in developing countries may make it extremely difficult to achieve these desirable procedures in practice. There is a need for more long-term, field-scale projects, to examine compaction problems on the very wide range of soils and climates in the tropics. It is particularly important that the results of such experiments are disseminated as widely as possible and greater attention should be paid to multilingual publication opportunities.

Tillage requirements and compaction

The requirements for primary and secondary tillage and remedial deep tillage are closely related to the previous incidence of compaction. Where reductions in the depth or frequency of tillage are practised, such as in minimum tillage or no-till systems, the need for periodic aggressive loosening to relieve compaction in the topsoil is now generally recognised. The ineffectiveness of deep tillage to relieve compaction in the subsoil has often been demonstrated but is still not fully appreciated by many farmers.

Weather effects on soil compaction

It is now well known that weather conditions have a highly important influence on compaction problems, particularly as a result of the influence of soil water content on: (1) soil compactability; (2) water supply to the crop; (3) aeration status of compacted soils. As yet the quantification of these effects is at an elementary stage. In the relatively short frost-free growing seasons of northern U.S.A., Canada and northern Europe, soils may have high water contents at planting and harvest and, therefore, tend to be highly vulnerable to compaction. In contrast, the compaction of dry soils at planting time in Central and Eastern Europe is generally found to give beneficial results in crop establishment and early crop development.

Environmental implications of soil compaction

Soil compaction is now recognised as a component of world-wide soil degradation, which may have implications in the environment far removed from the immediate area of crop production (ISTRO, 1993). In particular the unfavourable physical, chemical and biological conditions which occur in compacted soils, may contribute to increases in the amount of erosion and the loss of nutrients, especially nitrogen, into the atmosphere and to the ground water. These effects will attract a social cost in addition to the economic losses facing the individual farmer. In the past, compaction from traffic has been regarded as an inherent feature of crop production methods and has not been evaluated as a contributor to environmental pollution processes. However, these environmental implications now assume greater significance and are likely to demand separate evaluation, in addition to research oriented to the effects of soil compaction on crop production.

Compaction control

The primary opportunities for the reduction of compaction under vehicles are, in principle, straightforward. Axle loads, ground contact pressure and traffic intensity should all be reduced, especially when soils are wet. The need to reduce the distribution of traffic is particularly important in view of the much greater influence of the initial wheel pass than of subsequent passes. In practice, these objectives are often difficult or impossible to achieve within the present circumstances of low labour/high power crop production (Fig. 3).

Until now, attempts to reduce the incidence of compaction problems in commercial crop production have not been particularly effective. In certain countries, standards have been advocated which specify the maximum axle loads

Fig. 3. Sugar beet harvesting in the Netherlands is a low labour/high power operation in which many heavily loaded wheels pass over the field.

or ground contact pressures recommended for use on agricultural vehicles. However, the responses to these recommendations from commercial enterprises, particularly from the multi-national agricultural vehicle manufacturers, have not yet been evaluated. To have general applicability, such specifications should be related to the soil water status at the time of traffic. Other recommendations, such as those advocating the avoidance of vehicle traffic on wet soils, are difficult to observe in practice, particularly under the management systems of very large farms when the scheduling of field operations tends to be specified without reference to weather conditions (Fig. 4). Effective action to control compaction in practice is likely to involve a complex of different techniques rather than being dependent on only one method. Such a multiple treatment approach is, however, not generally included in field compaction experiments.

There may be opportunities to exploit entirely novel mechanisms and types of running gear for agricultural vehicles. For instance, manual or automatic control of tyre inflation pressure, which, although expensive, would provide a unique method of maintaining tyre deflection at an acceptable level on hard surfaces, while permitting deflation and a considerable increase of contact area on soft surfaces.

New types of wheels and tracks which are specifically designed to minimize

Fig. 4. Evidence of severe soil damage resulting from the transport of slurry at times of high soil water content.

compaction should also continue to be investigated, even when their use may appear to diverge widely from current practice. Currently, there is considerable interest in the exploitation of rubber-belt tracks to achieve low ground contact pressures without the restrictions on road travel associated with steel tracks.

Research and development programmes in a number of countries have shown that zero traffic, employing either standard vehicles with extended axles or purpose-built gantries, can overcome many of the more severe problems arising from excess soil compactness and can reduce the power requirements for cultivations. However, among farmers the cost of gantries is usually perceived as being excessive, compared to the benefits which can be realised from their use. Consequently, commercial uptake, even for high-value crops, is still very limited.

During recent years, the international market in agricultural machinery has experienced a marked concentration of manufacturing capacity, leading to a smaller number of larger companies. These companies are likely to be increasingly receptive to proposals for the solution of compaction problems if they can be shown to have a worldwide significance on market requirements. However, to achieve this it will be necessary for scientists to adopt a greater awareness of the problems facing machinery manufacturers and to make greater efforts to express the results of field experiments in economic terms.

Economic aspects of soil compaction

The economic acceptability of alternative types of vehicles and their running gear must be assured before they are likely to be adopted in commercial practice. Economic studies in this subject are still at a fairly rudimentary stage. However, recent studies have demonstrated the dominant role of farm size, soil type, weather and cultivation systems on the predicted economic responses to changes in machinery selection and management.

The benefit-cost relations of new types of machinery tend to be poorly understood, owing to the lack of operational experience. Gantries are expensive, partly because of low volume production and partly because the hydrostatic transmission systems usually employed, tend to be inherently less efficient than their mechanical counterparts used in conventional tractors. Model studies on farms of 400-500 ha in the U.K. indicate that the use of a complete gantry system would probably be equally profitable as the use of conventional machinery on medium textured soil and perhaps more profitable on heavy soil. However, as yet commercial uptake has been limited.

RECOMMENDATIONS FOR FURTHER RESEARCH ON SOIL COMPACTION IN CROP PRODUCTION

General

(1) There is a need to achieve a greater relevance and application of R & D resources to soil compaction problems by re-assessing the priorities for research and extension activities.

(2) Positive efforts should be made to achieve greater uniformity in usage with respect to the terminology, units and symbols related to soil compaction problems. Concepts, such as ground contact pressure and stresses within soils, need a more rigorous and unified terminology, which should be relevant to vehicles of all types. A new radical approach should be made rather than the continued adoption of usage which has evolved haphazardly in different disciplines in the past.

(3) It is now recognised that the extent and complexity of wheel/soil/crop/ weather interactions require a pooling of resources on a worldwide, multi-disciplinary and multi-crop basis if a rigorous scientific understanding of compaction problems is to be achieved. This can then be translated, with the aid of engineers, economists and manufacturers, into machinery systems which have been evaluated and found capable of keeping soil compaction within acceptable limits and are technically and commercially acceptable to farmers, fruit and vegetable growers and foresters.

(4) Greater assistance should be provided to promote research and development in the developing countries, particularly in the tropics, where

changes in agricultural technology are introducing compaction problems which are likely to have serious implications on food production programmes. In particular, greater attention should be given to: (1) the responses of a wider range of tropical crops to compaction; (2) the influence of soil and weather effects; (3) the role of plant species and cultivars having root systems capable of penetrating compact and semi-rigid soils; (4) the role of tillage implements to control compaction; (5) the role of different cropping/tillage systems, such as zero-tillage with mulch cover, as a means of overcoming compaction problems.

(5) Mathematical models are likely to be a highly effective way of predicting: (1) the effect of vehicle running gear on soil properties under a range of traffic intensities in the field; (2) the effect of these soil properties on the growth, yield and quality of crops under a range of weather conditions. While some progress has been made in this subject, considerably more data will have to be collected before such models can be fully developed and subsequently validated for general application.

(6) The adoption or rejection of new mechanization systems depends largely on factors other than scientific merit. Therefore, research workers should pay greater attention to feedback of information from commercial farmers concerning compaction problems and the various ways in which they try to solve these problems.

(7) There is a need among research workers for greater awareness of the economic factors which influence the uptake of new systems by farmers. Thorough analyses of benefit-cost aspects of possible solutions to compaction problems are required. Alternative machinery and soil management systems should be evaluated in field-scale and whole-farm trials in order to obtain realistic data on handling and operating characteristics, which may well have a decisive influence on their acceptability to farmers. The economics of timeliness of field operations in relation to soil compaction are of paramount importance but they are rarely taken into account in field compaction experiments.

(8) There is a need to strengthen and coordinate investigations into the environmental implications of compaction.

Field machinery

(1) Further studies are needed on the effects of mass and power of tractors and other vehicles on the incidence of compaction, with particular attention to the traffic distribution in farm-scale systems.

(2) Attempts to reduce the incidence of compaction in the subsoil through restrictions in the commercial use of vehicles with high axle loads, require further investigation over a greater geographical area and should be related to economic studies to evaluate the effects of such controls on whole-farm profitablity.

(3) Studies should be made on the opportunities for systems of compaction control in the field, involving a number of different procedures rather than

reliance on only a single treatment.

(4) Opportunities for achieving a greater degree of control of traffic distribution with conventional machinery should be explored.

(5) Evaluation, with respect to compaction and other aspects of performance, is needed of rubber-belt tracked tractors in long-term, field-scale experiments in comparison to conventional tractors.

(6) More appropriate and readily standardised methods are required for measuring and expressing the compaction capability of running gear.

(7) Further information is needed on the actual distribution and intensity of traffic in cropping systems with different types of machinery on farm-scale commercial production units.

Soil responses to the passage of vehicles

(1) Greater attention should be given to the influence of soil structure and heterogeneity on the behaviour of both topsoils and subsoils under compactive loading and the subsequent responses of crop plants.

(2) Soil compaction models should be further developed to take into account the complex distribution of loading forces in the field, e.g., the actual distribution of wheel traffic and the presence of lugs on tyres. Alternative types of soil reaction, e.g., plastic flow, should also be included in model studies. The role of horizontal forces under running gear on soil responses needs further evaluation. Models should be extended to include three-dimensional analysis of soil responses to loading forces.

(3) The dynamic distribution of normal and shear stresses over time under field conditions should be evaluated.

(4) Studies are needed on the effects of soil compactness on fertilizer use efficiency, particularly the influence on denitrification and losses of nitrate to the ground water, and the opportunities to reduce fertilizer applications without loss of yield.

(5) Models on the influence of soil compaction on soil hydraulic, aeration and strength properties, should be extended to take account of the indirect effects on plant rooting, mechanical resistance and the availability of nutrients.

(6) There is a need for further work on the *in situ* measurement of aeration in layered soils.

(7) Greater use should be made of visual examination of compacted soils so that observations within compaction experiments may be more easily and quickly compared with soil conditions on commercial farms.

(8) The advantages of different criteria for relative compactness of the soil should be tested and compared for a wide range of soils, crops and climates.

(9) On forest soils, further work is needed to evaluate the economic advantages of a layer of slash on the surface and the spatial and temporal control of vehicle traffic to reduce soil damage from vehicles. The use of deep ripping as a cure for

compaction of forest soils needs evaluation in view of the limited advantages of this technique in arable crops.

(10) The opportunities of reducing soil compactability through management practices leading to an increase of soil organic matter content and improved stability of soil structure, require further examination.

Crop responses to soil compaction

(1) Further studies are needed to investigate the relationships between crop responses and measured soil conditions in the field, especially to establish simple and statistically effective relationships which have relevance, both at the start and during the growing season.

(2) Further work is needed to define under what site conditions negative or positive crop responses may be obtained from increases in soil compactness.

(3) Crop growth models should be applied to compacted soils under a variety of simulated weather conditions and subsequently subjected to rigoruos validation.

(4) Further work is needed on those aspects of crop quality, disease resistance and other aspects of economic importance which are influenced by soil compactness.

(5) Greater attention should be paid to the effect of compaction on root distribution, especially through the use of simple visual observations, to augment soil measurements in experiments and to examine the distribution of roots of commercially-grown crops.

(6) The ability of certain deep-rooted plants and macrofauna to penetrate compacted soils, and thus to ameliorate adverse conditions biologically, should be further examined. The role of organic matter added to the soil surface may also be relevant to the action of soil fauna in stabilizing soil structure and in loosening compacted soils, especially in the tropics.

(7) Strong regional differences have been found in the responses of crops to soil compaction and greater attention should be paid to extend the understanding of these effects on a wider scale.

(8) Increased research is urgently needed to investigate the role of compaction in the production of tropical crops and to promote the adoption of improved management systems to reduce the problems observed.

(9) Vehicle traffic on perennial forage crops should be further studied, with particular emphasis on the role of ancillary traffic, e.g., slurry application, and on the opportunities to reduce fertilizer rates in the context of reduced soil compactness.

Role of weather effects in compaction studies

(1) More studies are required on the effect of various weather parameters on

crop responses to compaction in a wider range of climates.

(2) The implications of changes in climate, particularly the increase or decrease of rainfall, on the incidence of compaction problems should be studied.

(3) A statistical approach should be used to assess the probabilities of variation in seasonal weather and the resulting soil conditions and crop responses to compaction.

Economic studies

(1) More exact information is required on the economic advantages of timeliness arising from the use of traffic systems which involve zero or low ground contact pressure.

(2) Further work is needed to relate the economic consequences of extrapolating the results of small-scale experiments to whole-farm situations.

(3) Additional information is wanted to quantify the direct and indirect effects of changes in machinery management systems on farm profitablity.

(4) There is a need to validate economic models of compaction and to compare the results obtained with different models, particularly with respect to regional differences.

REFERENCES

Canarache, A., 1991. Factors and indices regarding excessive compactness of agricultural soils. Soil Tillage Res., 19: 145-164.

Hammond, A.L. (Editor), 1992. World Resources 1992-93. Oxford University Press, New York, NY, U.S.A., 385 pp.

ISTRO, 1993. Extended abstracts ISTRO Workshop Soil Compaction and the Environment. Proc. Int. Conf. Protection of the Soil Environment by Avoidance of Compaction and Proper Soil Tillage, 23-27 August 1993, Melitopol, Ukraine. Melitopol Institute of Agricultural Mechanization, Melitopol, Ukraine, Vol. 2, 44 pp.

Oldeman, L.R., Hakkeling, R.T.A. and Sombroek, W.G., 1991. World Map of the Status of Human-Induced Soil Degradation, An Explanatory Note. ISRIC, Wageningen, Netherlands/UNEP, Nairobi, Kenya, 34 pp.

Van Lynden, G.W.J., 1994. The European Soil Resource. ISRIC, Wageningen, Netherlands/Council of Europe, Strasbourg, France (in press).

INDEX

Notes: (1) Soil properties are listed under the name of the property, not under "Soil". (2) Types of soils are listed under "Soil type".

Abies lasiocarpa, see Fir, sub-alpine,
Aeration, in soil, *see also* Gas, in soil, *and* Oxygen, in soil, 167-85
- air permeability, 6, 62, 168, 172, 177-9, 180, 228
- anaerobic conditions, 223, 227
- critical limits, 169, 179-80, 257-8
- diffusion processes, 167-8
- effect of compaction, 177, 247, 349
- effect of organic matter, 183, 223
- effect of plough pan, 180
- effect of soil water status, 86, 168, 171, 179, 257-60
- effect of tillage, 180, 185
- effect of weather conditions, 168
- gaseous components, 173-4
- in layered soils, 179
- indicators, 168-9
- mass flow, 172
- measurement techniques, 176-7
- properties, 168-76
- requirements of plants, 180, 241
- requirements of roots, 185, 253-4
- requirements of soil organisms, 169
- stress on emerging crop, 241
Aggregate(s), soil, *see also* Clod(s), 93, 193, 208
- angle of internal friction, 60
- bulk density, 56
- compressive strength, 201
- geometric mean diameter, 273
- pores, 46, 361
- size distribution, 46-7, 202, 205, 458, 515
- stability, 52, 57, 103, 191, 294, 449
- strength, 56, 59-60, 194-5, 204
- tensile strength, 56, 201, 206
Agric horizon, 103
Agricultural systems, 288
Agro-ecosystem(s), 224, 229
Agroforestry, 230
Air, in soil, *see* Aeration, in soil,
Air permeability, *see* Aeration, in soil,
Air pycnometer, 176

Alder, red, 319
Alfalfa *see* Lucerne,
Alley-cropping, 230, 304
- effect on soil properties, 230
Alnus rubra, see Alder, red,
Anaerobic conditions, *see* Aeration, in soil,
Animal(s), draught, 4, 428, 437
- compaction under hoofs, 4-5, 423, 427, 526, 603
- ground pressure under hoofs, 4, 427
Anisotropy, Index of, 97, 107-8
Ant(s), 224, 229
Aporrectodea caliginosa, 225
- *A. longa*, 225
- *A. rosea*, 225
Arachis hypogaea, see Peanut,
Atterberg test,
- liquid limit, 53, 600
- plastic limit, 53, 600
- plasticity index, 53
Avena sativa, see Oats,
Axle(s),
- extension problems, 573
- number of, 473, 559
- steerable, 495
- tandem/twin, 472, 491, 559
- triple, 473
Axle load, 467, 602
- effect of high, 485-7, 576
- effect on subsoil, 367, 483-96
- restriction/limitation, 397, 491, 493, 635-6
- upper limits on roads, 491

Bacteria, in soil, 219-23
Balloon, rubber,
- to measure bulk density, 120-1, 129
Barley, 16, 204, 228, 267, 270, 370, 376, 381, 491, 525, 527, 543, 547, 560-2, 577-92, 613
- as nurse crop, 375
- spring, 269, 369, 373, 376-7, 379, 380, 381,

423, 544, 605, 608
- winter, 177, 179, 376, 525, 527
Bean, 302, 433-5, 577-92, 617
Bearing capacity, soil, 191, 202, 331, 333
Bed(s), wide, 524
Beetle(s), 224, 330
- dung, 224-25
Bermuda grass, 306
- aeration requirements, 183
Berry, northern red, 375
Beta vulgaris, see Sugar beet,
Biological activity, in soil, 215-30
Biomass, production, 2
Biopore(s), *see* Pores, biopores,
Biota, soil, 46-7, 215-30
Bonds, in soil,
- effect on soil strength, 192
- inter-aggregate, 206
- particle-to-particle, 192, 198
- role of organic matter, 198
Boussinesq theory, 62, 76, 80, 82, 501
Bracharia spp., 306
Brassica napus, see Oil seed rape,
Brazilian test, *see* Strength, compressive,
  soil,
Bridge, *see* Gantry,
Bromegrass, smooth, 268, 344, 361
Bromus inermis, see Bromegrass,
Brush blade, 335-6
Bulk density, *see also* Compactness, 113-35
- as a measure of compaction, 6, 113, 266,
  273, 290, 324, 349, 367
- balloon method, 120-1, 129
- clod method, 121-3, 129
- converted to void ratio, 115
- core sampling method, 117-8, 128-9
- critical, 329, 381
- effect on crop yields, 217, 266-79, 373
- effect on hydraulic properties, 144
- equilibrium, 116, 611
- frame sampling method, 118-9, 129
- influenced by stresses, 454
- maximum, 148
- measurement methods, 116-29
- measurement problems, 127-9, 134-5
- optimum, 269-70, 297, 308, 370, 376-7
- presentation of results, 130-3
- radiation methods, 123-129
- related properties, 114, 299
- relative expressions, 115-6

- sand replacement method, 119-20, 129
- selection of measuring method, 127-9

Cable systems,
- in forests, 325, 327, 333
- of traction, 4, 228
Cajanus cajan, see Pigeon pea,
Calcium, in soil, 610
Capillary forces, 192, 199
- rise equation, 143
Carbon dioxide, in soil air, 173-4
- concentration, 173-4, 182-3
Carrot, 374, 380, 576
Cart, *see* Trailer,
Casagrande apparatus,
- liquid limit, 53
- shear box, 201
Cassava, 288, 292-3
Cauliflower, 530, 574-6
Cedar,
- Japanese, 322, 336
- western red, 319
Cementation, 294
- compounds, 294, 366
- effect of clay, 192
- effect of organic matter, 192
Central tyre inflation system, (CTIS)
  468, 472
Centrosema spp., 306
Cereals, 228, 307, 370, 373, 378, 380,
  423-4, 431, 526, 543, 556-8, 572, 576,
  617
Chisel plough, *see* Plough, chisel,
Clay mineral type, 55, 192, 267, 270, 307
Climate, *see* Weather,
Clod(s), *see also* Aggregate(s), 193, 208, 2:
- bulk density, 121-2, 273
- cloddiness, 192
- crushing strength, 200
- flotation method, 121
- fragmentation, 200
- in potato ridges, 205
- produced by tillage, 153, 291
- size distribution, 202
- strength, 204-5, 206
Clover, 350, 375
- red, 344, 346, 354, 375, 381
- sweet, 617
- white, 344
Clover/grass mixture, 354

Cocksfoot, 344, 352
Coherence, soil,
- disintegration of, 103
Cohesion, soil, 39, 55, 60, 65, 192, 199
Compactability, soil, 6, 10, 11, 48, 329, 332-3, 366, 418, 611-2, 616
- effect of organic matter, 5, 198, 331
- effect of water content, 297, 378
- measurement of, 202
Compacted soils,
- areas affected, 628-30
- layers affected, 202-4
Compaction, soil, 266, 423
- agro-economic consequences, 276, 539-63
- alleviation of, 282, 330, 616-7
- amelioration, 272, 305, 336, 357-61
- amelioration by roots, 298
- avoidance of, 357-61
- beneficial effects, 246, 293, 308, 375
- capability of wheels/vehicles, 11, 391, 406, 411-2, 418, 464
- conferences on, 8, 529
- control by agronomic practices, 609-12
- control by machinery management, 602-9, 635-7
- control by traffic management, 300
- control by water management, 598-602
- costs of, 540-5
- definition of, 2, 46, 329,
- distribution with depth, 130-3, 155-6, 482-5
- during harvesting, 304
- during ploughing, 4, 271
- during tillage, 153
- early studies, 5, 265, 270-2
- economic aspects, 276, 539-63
- effect of growing season, 328
- effect of inflation pressure, 452
- effect of lime, 610
- effect of load, 277-8
- effect of organic matter, 228, 380, 609
- effect of shear stress, 106
- effect of soil conditioners, 610
- effect of soil type, 294-5, 375-7
- effect of vehicle speed, 605-6
- effect of vibration, 366
- effect of water content, 38, 105, 148, 277-8, 328-9, 366, 562, 598-603, 606
- effect on aeration properties, 177-85, 260
- effect on crop diseases, 544
- effect on crop responses, 267, 576
- effect on hydraulic properties, 141-59, 218, 240, 242
- effect on nitrogen availability, 152, 544
- effect on pore size distribution, 151, 247
- effect on redox potential, 183-5
- effect on soil respiration, 181
- effect on soil strength, 191-209
- effect on solute transfer, 152
- effect on water retention, 147
- environmental aspects, 13, 216-8, 630
- extent of, 270-3
- geographic distribution, 3, 628-30
- global approach to problems, 3, 628-30
- horizontal distribution, 155-6
- in crop nutrition, 254, 544
- in forest soils, 317-37
- incidence of, 11, 628
- influence on nitrate leaching, 152
- methods of measuring, 48, 482
- models, 39-40, 72-85, 629-30
- multi-dimensional, 155-6
- multi-disciplinary aspects, 10
- multi-wheel pass effects, 273, 274, 293, 454-6, 484, 561
- natural, 153, 203, 288-90, 294, 307, 611
- optimum level, 269, 293, 297, 308, 370, 376-7, 513
- persistence, 270-3, 277, 484-5
- prediction, 192
- prevention, 229, 598-602
- publications on, 7, 71
- relative, 11, 115, 268-70
- remedial measures, 612-6
- residual, 271, 273, 298
- resistance, 37, 39
- subsoil, 5, 6, 103, 276-79, 367, 376, 481-9
- tests, 6, 48, 53
- under draft animals, 4-5
- under grazing animals, 303
- under row crops, 6, 268
- workshops on, 8
Compaction research programmes,
- present-day, 631
- recommendations, 638-42
Compaction Risk Factor, 421, 425-7
Compactness, soil, 10, 113-6, 455, 606
- definition, 2
- degree of, 11, 115, 377
- effect on water relations, 141-59

- optimum level for plant growth, 12, 18, 177, 260-1, 269, 308, 452, 465, 513
- relative expression, 115-6, 455
- tolerance of species/varieties, 298, 373, 375, 617
Compost, 223
Compressibility, soil, 10, 48, 55, 332
Compression, soil, 48, 77
- confined, 53-4
- effect of water content, 148, 150
- index, 268
- unconfined, 53
- uniaxial, 148, 202, 219
- virgin compression line, 77, 448
Concentration factor, 41, 64, 81, 84
- effect of aggregation, 63, 83
- effect of depth, 64, 82, 85
- effect of pre-compression, 64
- effect of texture, 62-3, 82
- effect of time, 83
- effect of water, 63, 82, 481
- estimation, 548
- soil conditions, 82, 600
Cone index, see also Penetration resistance, 548
Cone penetrometer, see Penetrometer, soil, cone,
Consistency, soil, 599
Consolidation, soil, 48
Contact area, ground, 63, 399-400, 403, 460
- of hoofprints, 427
- of lugs, 76
- of tracks, 402-3, 411, 510, 517
- of tyres, 73-4, 427, 555, 603
Contact pressure/stress, see Pressure, ground,
Controlled traffic, see also Zero-traffic, 521-34, 607
- benefits, 302-3, 532-4, 607
- crop losses in wheel lanes, 359, 525, 575-6, 591
- crop responses to, 303, 359
- effect on cotton yield, 524, 526, 530
- effect on forage crops, 526
- effect on nitrogen uptake, 526
- effect on potato yield, 525, 529
- effect on soil conditions, 281, 522-7
- effect on tillage energy, 302, 525, 544
- effect on tillage requirement, 303, 523-6,

533, 589
- effect on timeliness, 527, 533
- effect on trafficability, 525
- effect on vegetable crops, 526, 530, 532
- effect on winter barley yield, 525
- in forests, 527
- in tropical crops, 302-3
- in wide-bed system, 524, 526, 530
- mapping of fields, 528, 533
- rationale, 522-4
- relation to subsoiling, 302
- whole-farm economics, 533
- with gantries, see Gantry,
- with modified tractors, 525, 561-2, 571-3
Core samples, 117-8, 128-9
- effect of height/diameter, 48, 202
- effect of wall thickness, 117
- from auger-type samplers, 118
- from drive-type samplers, 117
- methods of sampling, 117-8
- problems in using, 117, 134
Corn, see Maize,
Cotton, 87, 293, 433-7, 530-1, 615
- growth model, 87
- root growth, 87
- yield, 87, 524, 526
Coulter, drill,
- smearing during sowing, 241
Cowpea, 13, 288, 291, 298
- roots, 291, 305
Cracks, soil, 46, 60, 98, 100-1, 157, 185, 19□ 197, 219
Crop(s),
- arable, 423-33, 463
- bulb, 526
- citrus, 5
- cover, 298, 302, 306-7, 609
- flower, 534
- forage, see Forage crops,
- forest, see Forest crops,
- fruit, see Fruit crops,
- horticultural, 534
- irrigated, 2
- leguminous, see Legumes,
- root, see Root crops,
- row, 268, 281, 293, 303
- vegetable, 378, 433, 475, 526, 530, 534, 562, 575-6, 605
Crop growth, 11, 85-7, 237-61, 267
- deep rooting, 227

- dry matter production, 274
- effect of soil structure, 257, 260
- emergence, 191, 238, 241, 270, 274, 368-9, 373, 513
- germination, 238, 241, 319
- growing period, 238, 240, 533
- maturity, 276
- potential rate, 237-8
- species tolerance of compactness, 298, 373, 375, 617
- varietal tolerance of compactness, 617
Crop production, 2, 266
Crop quality, 576
- influenced by soil compactness, 16, 355, 374, 525, 543
Crop residue(s), 198, 223
- burial/incorporation, 223, 609
- burning, 609-10
- on surface, 229-300, 610-1
Crop rotation, 13, 268
Crop yields, 12, 238
- economic value, 550-3
- effect of high compactness, 13, 271, 273-9, 367-81
- effect of low compactness, 13
- effect of soil structure, 548
- effect of subsoil compaction, 373, 485-7, 545
- effect of subsoil loosening, 615
- effect of weather, 274-9
- influenced by nutrient supply, 273-4
- influenced by soil compactness, 267, 273-9, 297-8, 369-80, 634
- optimum soil compactness, 12, 260-1, 297-8, 376-7, 513, 542
- responses of species to compactness, 13, 278, 298, 373, 375, 617
- responses of varieties to compactness, 373, 375, 617
Cropping system, 268, 611-2
- double-cropping, 611
- effect on compaction, 268, 611, 616
- green manuring, 103
- intercropping, 103
- monoculture, 611
- multi-, 533
- rotation, 611, 616
Crust, soil,
- depositional, 103
- effect on aeration, 180, 241

- erosional, 103, 295
- formation, 103, 241, 303
- influence on emergence, 241
- influence on water flow, 153, 229, 295, 297, 303
Cucumber, 615
Cucumbis sativus, see Cucumber,
Cultivation, *see* Tillage,
Currant, black, 375
Cynodon dactylon, see Bermuda grass,

Dactylis glomerata, see Cocksfoot,
Daucus carota, see Carrot,
Deforestation, 229-30
Deformation, soil, 29, 34, 36-7, 59
- elastic, 59, 84
- flow type, 42, 62
- hardening type, 41
- homogeneous, 35, 36
- non-deforming type, 41
- paths, 33-7
- plastic, 59, 84-5, 448, 464
- streamlines, 34
- types, 37, 41
Degradation, soil, 3, 8, 290, 295, 300, 627, 630
- biological, 3, 215-30
- chemical, 3, 627
- physical, 3, 627-30
Denitrification, 218, 224
- effect of anaerobic conditions, 168, 223, 250
- effect of compaction, 183, 255, 526
Densification, *see* Compaction,
Density, relative, 115
Dicotyledons, 219, 225, 298
Diplopod(s), 224
Direct drilling, *see* No tillage,
Disc harrow, 273
Disc plough, 612
- powered, 608
- soil reaction to, 614
Disk, *see* Disc,
Disk plow, *see* Disc plough,
Draft, *see* Draught requirements,
Drainage, 247, 378, 601, 616
- in compact soil, 240, 601
- influence on operations, 601
Draught requirements, 192
- for tillage, 544, 572

Drop-cone, *see* Penetrometer, soil, drop-cone,
Drying, of soil, 155, 185, 199, 204, 206
Duripan, 330

Earthworm(s), 224, 229-30
- activity, 218, 228, 230, 330, 361
- anecic, 224, 228
- burrows (channels), 46, 224-6, 228, 301
- casts, 225, 230
- deep-burrowing, 225, 230
- effect of no-till, 228
- effect of tillage, 228
- effect of traffic, 228
- endogeic, 224-5
- epigeic, 224, 228
- shallow-working, 228, 230
Economic assessment, of soil
- compaction problems, 15-6, 276, 474-5, 491, 539-63, 569-92, 630, 638
- analysis techniques, 546-9
- benefit-/cost relations, 540, 556-8
- costs, 281, 491, 516-7, 533, 540-5, 549, 552-6
- examples, 549-63, 569-92
- profitability, 540
- trade-offs, 281
- whole field/farm aspects, 550, 552, 576-92
Enchytraeid(s), 224
Energy requirements, 475
- for tillage, 466, 544, 573
Environment, 216, 223, 630
- atmospheric pollution, 216
- effects of compaction, 13, 288, 360, 489-90, 630, 635
- micro-, 217
- pesticide residues, 216
- water pollution, 216, 361
Erosion, 191, 295, 297, 303
- effect of ploughing, 290
- influence of compaction, 544-5
- water, 103, 300, 367, 627
- wind, 300, 627
Ethane, in soil air, 174
Ethylene, in soil air, 168, 267
Evaporation, of water, 241
- effect on soil temperature, 241
Evapotranspiration, 257
- effect on water potential, 245

Eudrilus eugeniae, 230

Fabric, soil, *see* Structure, soil,
- microfabric, *see* Microstructure, soil,
Farm, size, 267, 346, 424-7, 557-8, 587
Farming systems, 216, 267-8
Fauna, soil, 217, 224
- burrowing, 224-5
- faunal channels, 255-6
- macro-, 217, 224, 227
- meso-, 99, 217, 224, 227
- micro-, 227
Fermentation, in soil, 218
Fertilizers, 223, 281-2
- effect on compacted soils, 335-7, 360, 38 475, 544, 617
- spreading machinery, 344
Fescue, 344
Festuca spp., *see* Fescue,
Field,
- length, 495
- time, 431-8
Field Load Index (FLI), 422, 425-7, 434-8 441
Fir,
- Douglas, 319, 324
- sub-alpine, 321
Flax, 525
Flow, soil, *see also* Deformation, soil, 33, 40
Fly larvae, in soil, 224
Footprint, *see also* Contact area, ground,
- of hoofs, 423
- of tracks/tyres, 510, 524
Forage crop(s), 343-61, 375
- compaction effects, 345-61, 370, 375, 37
- cutting, *see* Mowing,
- harvester, 345
- harvesting, 345, 350, 359, 493
- irrigated, 344
- management, 344-5, 361
- protein content, 351, 355
- vehicles, 345, 357
Forces, on soil, 46, 57
Forest crops, *see also* Tree(s),
- effects of compaction, 317-37
- productivity, 318
- regeneration, 320-1
Forest operations, 318, 331-7
- cable logging, 325, 327, 333

- effect of soil water, 331-3
- extraction, 324-5
- harvesting, 319, 321, 325-31
- logging, 7, 319-20, 322, 327, 332, 334-5
- management, 331-7
- on frozen soils, 331
- skidding, 322, 325, 327-8, 331
- thinning, 318, 322-4, 327
Forest residues,
- incorporation, 335
- slash, 331, 333
- surface litter, 322, 331
Forest soils,
- amelioration, 335-7
- compactability, 329
- compaction, 7, 527
- organic matter, 331
- recovery, 330-1
- root mat, 331
- texture, 329-30
- water content, 332-3
Forest vehicles, 318, 481
- ground contact area, 318
- ground pressure, 327-9, 333-5
- forwarder, 325, 328, 333
- skidder, 333
Fragipan, 614
Frame sampling, 118-9, 129
Freezing and thawing, 58, 156, 185, 206, 271-3, 279, 484-5
- depth in soil, 267, 271
- influence on soil compactness, 270, 330, 361, 379-80, 484-5
- in subsoil, 267, 271, 485
Friction, soil, 192
- angle of internal, 39, 60-1, 65
Fröhlich's concentration factor, see Concentration factor,
Fruit crops, 375, 562, 575
- orchard fruits, 534
Fuel costs, for tillage, 302, 544

Gamma rays, to measure soil density, 123-9, 134-5
- absorption, 124
- backscatter method, 124, 126, 128-9
- calibration, 124-5
- double energy method, 126
- scattering, 124
- statistical aspects, 126-7

- transmission gauge, 124, 128-9, 134-5, 202
Gantry, 268, 521-34, 569-92
- area of crops, 588
- construction cost, 572
- design criteria, 571-3
- Dowler, 529-30
- draught requirements, 572
- driver controls, 534
- electrically powered, 531
- farm gross margin, 585-92
- Field Power Unit (F.P.U.), 530-1
- for chemicals application, 527, 574, 582-4
- for management of compaction, 523
- for primary cultivation, 578, 580-1
- for secondary cultivation, 578, 581-2
- for sowing, 582-4
- gantry agriculture, 529
- harvesting with, 522, 530, 534, 575, 584-5
- in glasshouses/greenhouses, 532, 534
- labour costs, 575, 588, 591
- MERHAV, 530
- "Monotrail", 529
- movement in fields, 534
- operating costs, 573-4, 579-85
- power/energy requirements, 577-92
- purchase price, 561, 573, 585, 588
- road travel, 575
- spanning beam, 571
- steam powered, 522, 528
- steering, 303, 534, 609
- tracks, 530, 575
- transmission mechanism, 530, 532, 572-3
- transport, 534
- width aspects, 577-92
- work rate, 574-5, 579-85
Gas, in soil, see also Aeration, soil, and Oxygen, in soil,
- composition, 168, 173-4, 177, 181
- diffusion coefficient, 168-71, 176, 178-9, 180
- diffusion processes, 167
- flow, 185
Glasshouse, see Greenhouse,
Glycine hispida, see Soybean,
Glycine max, see Soybean,
Gossipium hirsutum, see Cotton,
Grass(es), 272, 306, 373, 453
- compaction effects, 351-4, 360
- for silage, 544, 556
- perennial, 343-4, 463

Grassland, 306, 346-7, 359, 375, 433-4, 438, 453, 456
- loosening, 359-61
- slurry application, 345-6, 353, 358, 438
- use of low ground pressure, 349, 352, 357-9, 463-7, 470
Gravel, 288
- in subsoil, 289, 291, 301
- shallow horizons, 294
Grazing, 304
- effect on compaction, 611
- effect on runoff, 303
- mechanical, 344
- over-, 611
- zero/controlled, 303-4, 344
Greenhouse, 267
- trials with tree seedlings, 318-20
Ground contact pressure, see Pressure, ground,
Ground pressure, see Pressure, ground,
Groundmass, 93, 105-6
- void types in, 93
Grubs, 330
Gypsum, incorporation, 615

Harrow, disc, 335-6
Harrowing,
- effect on soil conditions, 301
Harvester, 282, 473-4
- combine, 183, 268, 277, 365, 429, 473, 481, 556-8
- forage, 473-4
- sugar beet, 473-4, 481
- weight, 473-4
- wheel spacing problems, 471
Harvesting, 575
- damage to soil structure, 154, 183, 240, 304, 324-31, 491
- effect of soil water, 378
- high axle loads, 154
- manual, 289
- traffic, 424
Hay/Hayland, 155, 268, 344
- tame, 547
- yield, 352, 360, 375
Headland, field, 442
- compaction on, 290, 304
- restriction of traffic to, 282, 607
Hemlock, 321
- western, 319, 324

Herbicide, application, 302
Hoe, 307
Homogenized soil, 48, 53, 55-7, 61
Hordeum vulgare, see Barley,
Horse, see Animal, draught,
Humification, 224
Hydraulic conductivity, saturated, 143
- effect of compaction, 148, 150, 242
- estimated, 154-5
- measurement, 145-7
Hydraulic conductivity, unsaturated, 143, 147
- effect of compaction, 147, 242, 381
- in relation to pore size, 144, 154
- in relation to water retention, 143-4
- measurement, 145-7
Hydrogen, in soil air, 174
Hydrogen sulphide, in soil air, 174, 223
Hydromica, 376
Hyperiodrilus africanus, 230

Ice-lens, 58, 273
Illuviation, after tillage, 103
Implements, field,
- mass, 429-31
Infiltration of water, into soil, 156
- effect of compaction, 141, 150-2, 223
- effect of crust, 297, 302-3
- effect of tillage, 150-1, 302
- into compact soil, 240
- measurement, 146
- ponded infiltrometers, 146
- tension infiltrometers, 145, 151
Inflation pressure, tyre, 17, 41, 72-3
- automatic control, 561, 636
- minimum, 17
Internal friction, see Friction, soil,
Irrigation, 5, 367, 601-2, 616
- compacting effect, 367, 602
- influence on soil properties, 290, 602
- requirement, 308

Jurin's equation, 143

Kaolinite, 307
Kneading, of soil, 65
K_s, see Hydraulic conductivity, saturated,
Kudzu, 306, 617

Land clearing, 288, 295-7, 304-5

- hand methods, 296, 304
- mechanical, 289, 291, 296-7, 306
- shear blade, 304
Larch, Japanese, 322
Laterite, in soil, 288, 294
Leaching,
- effect of compaction, 360, 526
- of nutrients, 216, 489
Legume, 272, 306-7, 343-4
Leucaena leucocephala, 230, 304
Lime/Liming, 610
- effect on compactability, 610
- in subsoils, 613
Liquid limit, 53, 600
Load index, 7, 395, 420
Lolium multiflorum, see Ryegrass, Italian,
L. perenne, see Ryegrass, perennial,
Loosening, *see* Soil loosening,
Lotus corniculatus, see Trefoil, birdsfoot,
Low/Reduced ground pressure (LGP) system, 456-75
- crop responses, 456, 462-5
- definition, 467
- economic effects, 474-5
- for cereals, 556-558
- for ryegrass, 556
- for wheat, 457-60, 464-5
- soil responses, 458-65, 525
Lower plastic limit, *see* Plastic limit,
Lucerne, 268, 271, 344, 375, 576
- compaction effects, 350-1, 355
- effect of soil type, 344, 355
- root system, 219, 271, 344, 355, 376, 380, 617
- traffic, 359, 423, 526
Lumbricus terrestris, 224-5
Lupin, 380
- white, 376
Lupinus albus, see Lupin, white,

Machinery, *see also* Vehicle(s),
- changes in mass with time, 265, 268, 271, 276, 479-81
- costs, 541, 549, 551-2
- to reduce soil compaction, 283, 549
- mass reduction effects, 280
Maize, 155, 180, 229-30, 259, 268, 281, 288, 289-90, 301, 302, 379, 433-5, 511, 525, 543, 547-8, 608, 615, 617

- grain yield, 266, 271, 273-9, 289, 302, 304, 306, 371-2, 376, 513, 547, 551-3, 559, 560, 562
- nutrient uptake, 255, 267
- planting systems, 281
- roots, 13, 205-6, 242, 255, 289
- silage yield, 283, 547, 551-3
Manganese, deficiency, 460, 465, 525
Manihot esculenta, see Cassava,
Manure, *see* Organic waste,
Mass/power ratio, *see* Tractor, wheeled, mass/power ratio,
Mechanical impedance, of roots, 218-9, 244, 254, 368
- critical limits, 219, 260
Mechanization, 288
- degree, 421
- effect on soil compactness, 365, 631
- effect on traffic, 423, 427
Medicago sativa, see Lucerne,
Melilot, 381
Melilotus alba, see Clover, sweet,
- *M. officinalis, see* Melilot,
Mercury displacement, in bulk density measurement, 121-2
Methane, in soil air, 174
Microbe, *see* Micro-organism(s),
Microbial activity, 215-8, 223
- effect of aeration, 167
Microclimate, 230
Microelectrode, 177
- platinum, 175
Micro-flora, 227
Micro-organism(s), soil, 215-30
Microstructure, soil, 92-108, 154
- analytical methods, 94
- back-scattered electron method, 95-6
- collapse, 106
- correlation with physical properties, 98
- digitized images, 94
- domain segmentation, 97, 107
- drying of samples, 95
- effect of compaction, 98, 107
- effect of loading, 106
- effect of tillage, 98
- energy dispersive X-ray analysis, 94
- image analysis, 96-7
- impregnation of samples, 95
- microporosity, 96, 98-105
- microscopic techniques, 94

- microstructural elements, 94
- of clays, 106-7
- of plough pan, 101
- orientation of particles, 97, 106
- realignment of particles, 106-8
- sample preparation, 94
- scanning electron microscopy, 94, 97
- thin sections, 95
- transmission electron microscopy, 94
Minimum tillage, see Tillage, minimum,
Mite(s), 224
Mobility Index, 420
Model(s), 18, 208, 227
- aeration, 86, 260
- analytical, 80-5
- bulk density, 422
- catchment wetness, 332
- compaction, 39, 45, 72-85, 391, 454-5, 481, 511, 548, 629
- cotton growth, 87
- crop growth, 85-7, 157, 257
- crop response to compaction, 45, 276, 371, 373
- crop yield loss, 422, 547-8
- economic, 546-8, 577
- finite element, 78-80, 157
- gantry operation, 577
- GOSSYM, 87, 157
- heat/water flow, 157
- hydraulic conductivity of surface crust, 153
- hydraulic storage/runoff, 154
- mechanical impedance, 260
- NTRM, 157
- root growth, 85-7, 157
- soil biota, 227
- soil layer(s), 153
- soil strength, 86
- soil structure, 227
- stress distribution in soil, 77-85, 208
- SWACRO, 257
- water flow, 153-4, 156-9
- water retention, 154
Mohr-Coulomb failure, 39, 79, 202
- failure line, 61
Moisture, soil, see Water, soil,
Monocotyledons, 219, 225, 298
Montmorillonite, 376, 610
Mowing, of forage,
- traffic effects, 344-5, 359

Mucuna utilis, 306
Mulch, 230, 300-1, 336
- bark residue, 337
- crop residue, 229, 300
- effect on soil temperature, 157
- effect on water flow, 157, 229

Nematode(s), 219, 222
Nitrate, in soil, 254
- influenced by compaction, 250
- movement in earthworm channels, 225-6
- supply rate, 250
Nitrification, influenced by compaction, 25
Nitrogen, in soil,
- as a fertilizer, 274, 617
- as a nutrient, 360
- availability to plants, 240, 255, 267, 462, 465, 544
- fixation, 337
Nitrous oxide, in soil air, 174
Nodule/Nodulation,
- of soybean, 267
No tillage/No-till, 289, 301-2, 456, 525, 606, 611
- compaction problems, 150, 180, 282, 289, 380
- crop responses, 289-90, 381
- effect on biological activities, 217, 228-9
- effect on hydraulic properties, 150-2, 229
- effect on soil aeration, 180, 223, 559
- effect on soil bulk density, 273, 302, 611
- effect on soil strength, 58
- effect on soil temperature, 301
- soil recovery under, 15, 185, 282, 298, 301
- soil responses, 611
Nuclear radiation, see Gamma rays, to measure soil density,
Nutrients, 13, 465
- availability, 65, 223
- leaching, 216, 489
- micro-, 460
- mobility, 65
- supply, 238, 249-50, 465
- uptake, 238, 254-5, 381, 489, 617

Oats, 268, 369, 370, 373, 376, 463, 491,

527, 547, 605, 613
Oil displacement,
- for clod bulk density, 121
- for hole volume, 120
Oil seed radish, 375-6
Oil seed rape, 374, 376, 379, 577-92
Onion,
- response to LGP, 457, 475
- root distribution, 14
Onobrychis viciifolia, see Sainfoin,
Orchardgrass, *see* Cocksfoot,
Organic matter, in soil, 6, 366, 609-10
- accumulation, 609, 611
- application, 610
- decomposition, 223-4, 229, 609
- effect on compaction, 228, 331, 380, 609
- effect on earthworms, 228
- effect on respiration, 183
- effect on soil strength, 55, 192, 198, 206, 331
- maintenance, 609
- surface litter, 611
Organic waste, 223
- manure, 223, 380, 610
Organisms, soil, 216
Oryza sativa, see Rice,
Oxygen, in plant,
- stress, 249
- transport via aerenchyma, 248, 254
Oxygen, in soil, 173-4, 238, 241, 247-9, 253-9
- biological demand, 253
- concentration, 174, 181-2, 257-8
- consumption rate, 241, 257-8
- diffusion coefficient, 247, 257-8
- diffusion rate (ODR), 168, 174-5, 177, 183, 247
- effect of surface sealing, 241, 253
- in relation to soil water, 241
- lower critical aeration limit, 180, 258
- supply, 238, 241, 247-9
- transport in soil, 255
- upper critical aeration limit, 180, 258
- uptake by roots, 238, 255

Packing density, soil, 76-7
Paraplow, 359, 614
Particle(s), soil, 46-7
- density, 134
- shape, 97
- specific gravity, 114

Particle orientation, soil, 97, 108
- effect of stress, 107-8
- of clay minerals, 106
Particle-size distribution, soil, 97
Paspalum spp., 306
Pastures, 343-61
- compaction under, 303-4
- faunal channels, 101
- restoration of structure, 101, 359-61, 616
- structure of soils under, 347-51
Pea(s), 275, 379, 527, 550-1, 560-1, 605, 613
- canning, 491
- field, 373
Peanuts, 307, 615
Ped(s), *see also* Aggregate(s), soil, 93
Pedality, *see also* Aggregate(s), soil, 93, 154
Penetration resistance, *see also* Cone Index, 12, 200, 269, 273
- effect of tillage, 203, 615
- effect of traffic, 203-4, 423
- effect on root growth, 86, 219
- to measure soil strength, 200
Penetrometer, soil, 201
- cone, 201
- drop-cone, 117, 201
Permeability, soil, air, *see* Aeration, air permeability,
- intrinsic, 219
- water, 142
Pesticide, 216, 226
Phalaris arundinacea, see Reed, canary grass,
Phaseolus spp., *see* Bean,
Phleum pratense, see Timothy,
Phosphorus/Phosphate, 381
- potential supply rate, 250
- uptake kinetics, 255
Photosynthesis,
- effect on shoot growth, 257
Phragmites australis, 228
Picea abies, see Spruce, Norway,
- *P. englemanii, see* Spruce, Englemann,
- *P. glauca, see* Spruce, white,
- *P. sitchensis, see* Spruce, Sitka,
Piezoelectric-type pressure sensor, *see* Pressure sensors,
Pigeon pea, 305

Pine,
- loblolly, 321
- lodgepole, 319, 321
- ponderosa, 321-2, 324
- radiata, 319, 321-2, 325, 333, 335
- Scots, 319, 324
- white, 325
Pinus contorta, see Pine, lodgepole,
- *P. ponderosa, see* Pine, ponderosa,
- *P. radiata, see* Pine, radiata,
- *P. sylvestris, see* Pine, Scots,
Pisum sativum, see Pea(s),
Plant(s), *see* Crop(s),
Planting,
- machinery, 268
- manual, 289
- planter width, maize, 281
Planting date, effect on profitability, 544
Plastic limit, 53, 600
Plasticity,
- effect of lime, 610
- index, 53
Plate sinkage tests, 202
Plinthite, *see* Laterite, in soil,
Plough/Ploughing, 16, 303, 427, 430, 436,
    442, 614
- chisel, 150-2, 185, 273, 306-7, 612, 614-5
- compaction in furrow, 271, 282, 301, 483,
    496, 608, 612
- disc, *see* Disc plough,
- draught in compact soils, 544, 580-1, 612
- effect on degradation, 290, 301-2
- effect on water storage, 301
- high speed, 606, 608
- moldboard, *see* mouldboard,
- mouldboard, 153, 155, 204, 273, 278-9,
    282, 550, 612
- out-of-furrow, 608, 612
- "segment", 488-9
- shallow, 612
- slip, 614-5
- to reduce soil compactness, 486, 496
- tyres for, 470-1, 562
Plough pan, 180, 615
- effect on roots, 218-9
- formation, 101, 267, 270, 302
- properties, 101, 205
Plow, *see* Plough,
Poisson's ratio, 79-80
Pollution, 216

- air, 216
Polyelectrolyte, 610
Pore(s), soil, *see also* Void(s), soil,
- accessible, 220-3
- biopores, 100, 154, 217, 219, 268
- continuity of pores, 98, 143, 171, 180,
    381
- diameter, 172, 219
- faunal, 229
- habitable, 220, 222
- inaccessible, 220
- inter-aggregate, 46
- intra-aggregate, 46
- macropores, 218, 227, 239, 269-70, 301,
    356
- shape of pores, 143
- size of pores, 143
- stability, 98, 219
- "structural", 196
- "textural", 195
- tortuosity, 143, 171, 180
- vertical, 58, 219
Pore-size distribution, 46-7, 144, 242
- determined by water retention, 143
- effect of compaction, 100, 381
- effect of tillage, 99-100
- related to hydraulic conductivity, 101
Porosity, soil, 143
- air-filled, 114, 168-9, 176, 179-80, 185,
    219, 223, 241, 247, 253, 266, 349,
    458-9, 548
- bulk density conversion, 115
- definition, 114
- effect of compaction, 448
- inter-aggregate, 143, 268, 361
- intra-aggregate, 143
- macro-, 100, 185, 239
- optimum, 260-1, 298
- prediction, 77
- related to fluid conductivity, 98, 143
- related to soil management, 98-100
- total, 100, 114-5, 143, 204
Potassium, 254
- supply rate, 250
Potato, 6, 205, 246, 366, 373-4, 378, 424-7,
    433, 435-7, 441, 457-8, 471, 475, 527,
    529, 544, 576, 605, 613, 615
- leaf injury, 267
- marketable yield, 247, 266-7, 374
- root growth, 205, 376

- tuber quality, 267
- tubers, 246, 254, 376-7
Potential, matric, water, *see* Water, soil,
 matric potential,
Power/weight ratio, *see* Tractor, wheeled,
 mass/power ratio,
Precipitation, *see* Rainfall,
Pre-compression stress, 49, 58-9, 64
Pressure, ground (contact), 178
- average, 75, 402, 405-6, 427, 449-56
- cumulative, 274, 349, 550
- defined by inflation pressure, 75, 406
- distribution, 73-4
- effect of soil condition, 75
- effect on compactness, 350
- effect on crop yield, 274-5, 370
- estimation, 72-3, 75, 406
- maximum permissible, 328-9, 366, 374,
 378, 465, 492-3, 501
- on deformable surface, 406
- on hard surface, 405-6, 452
- reduced, 357-9, 460-3, 545
- standard, 451-2
- under forest vehicles, 327-9
- under tracks, 365, 402-3, 409-11, 510-2,
 605
- under wheels, 365, 603-5
Pressure, pore air, 46
Pressure, pore water, 46
Pressure sensors, in soil, 50-1, 453
Proctor test, 6, 53
- comparison with field tests, 332
- maximum bulk density, 115, 269
Protozoa, 219-20, 222-3
Pseudotsuga menziessi, see Fir, Douglas,
Psophocarpus spp., 306
Puddling, of soils, 329
- effect of texture, 330
- effect on structure, 65
- susceptibility, 333
Pueraria lobata, see Kudzu,
Pycnometer, air, *see* Air pycnometer,

Quality, soil, 24, 42-3, 270, 633

Radiation, nuclear, *see* Nuclear radiation,
Radish, oil, 375
Rainfall, 15, 240, 267, 295
- effect on plant growth, 274-6
- effect on soil aeration, 248

- effect on soil strength, 295
- intensity, 288, 295
Rainstorm, 295
Raphanus clinensis, see Oil seed radish,
R. sativus, see Oil seed radish,
Raspberry, 375
Recompaction, 306, 487, 533
- during ploughing, 204
- in subsoil, 283, 454, 475, 487, 615
Redox potential, soil, 177
- as indicator of aeration, 168, 175-6
- critical limits, 185
- effect of compaction, 183
- effect of flooding, 175-6
- effect of oxygen diffusion rate, 176
- relation to denitrification, 185
- variation with depth, 183
- variation with time, 183-4
Reed, 228
- canary grass, 344
Remoulding, soil, *see also* Homogenized
 soil, 195, 201
Respiration, in roots, 167, 252-4
- effect of oxygen supply, 223, 253
Respiration, in soil, 181, 253
- effect of pore size, 223
Respiratory quotient, 168
Ribes nigrum, see Currant, black,
- *R. schlechtendalli, see* Berry, northern red,
Rhizobia spp., cells, relation to pore size,
 220-2
Rice, 288-9, 502-3
- paddy, 289, 308, 531
Ridge(s),
- crops on, 281
- soil, 157-9, 458, 613
Ripper,
- rock, 335
- winged, 335
Ripping, 303, 306, 335-7, 607
Rolling resistance, 191-2, 465, 509
- coefficient of, 465-6
- effect of soil condition, 466, 573
Root(s)/Root system,
- aerenchyma, 617
- channels, 46, 60, 101, 219, 228
- clustering, 244, 248, 250
- contact with soil, 242, 247-8
- deep, 227, 246, 301
- density, 12

- depth, 243, 248
- diseases, 544
- distribution, 12, 244, 380
- effect of compactness, 13, 380
- effect on gas movement, 185
- effect on soil strength, 198
- effect on soil structure, 218, 239-40
- endodermis, 251
- grafting, in trees, 318
- growth rate, 238, 244
- hairs, 227, 253
- impedance, 380
- in gravel layers, 289
- in layered soil, 205
- in subsoil, 381
- in wheel tracks, 244
- injury, 323, 359-60, 380
- length, 253-5
- mass, 380
- mat, 331, 354
- morphology, 380
- penetration ability, 13, 218, 298, 301, 380, 381, 616-7
- pressure, 218
- resistance, internal, 251-3
- respiration, 167, 223
- size, 227, 238, 253, 381
- taproot(s), 219, 227, 306
- tip, 251, 253
- uptake of oxygen, 253
- water uptake efficiency, 244
Root crops, 254, 373, 378, 424, 459, 465, 526, 576
- harvesting problems, 148, 240, 458, 471
Root environment, 308
- influenced by compact soil, 266, 289
- rootability, 243, 250
- root-soil contact, 242, 247-8
Root growth, 85, 191, 615
- in compact soil, 86-7, 185, 205, 218, 246, 260, 289-90, 292-3, 305-6, 376, 380, 616
- in subsoil, 243, 247-8
- species differences, 13, 298, 617
- to exploit soil water, 245-7, 250, 289
- varietal differences, 617
Rotovator, 613-4
Rubus idaeus, see Raspberry,
Running gear, *see also* Tyre(s), Track(s), 12, 17, 391-412

Runoff, 154
- from compact soils, 242, 303
- from desiccated soils, 297
Rut(s), 130-2, 134, 152, 321, 329, 332, 427, 438, 544
- depth, 334, 349, 557
- effect of ground pressure, 453, 464, 504
- length, 419, 423, 427
- maximum allowable depth, 333, 493
- soil conditions under, 423, 455
Rye, 369-70
- winter, 373, 376, 529
Ryegrass, 344, 356, 381
- Italian, 354, 375
- perennial, 352, 359, 460-2, 526, 613

Sachruum spp. *see* Sugar cane,
Sainfoin, 344
Sand replacement method, for bulk density, 119-20, 129
Saturated flow, *see* Hydraulic conductivity, saturated,
Saturation, degree of, 46-7, 114
Secale cereale, see Rye,
Settlement, soil, 203
Shear tests,
- direct, 55, 200-1
- torsion, 201
- vane, 201
Shearing resistance, 55, 61
Shrinkage, soil, 157, 199, 329, 376
Sieving, wet, 52
Silage, 344, 360
- machinery, 358-99
- traffic, 349, 358
- trailer, 345, 352, 357-8
Skid trail, 324, 331-2, 336
- compaction under, 322, 329, 330-1
- roots under, 330
Skidder, 328, 332
- crawler/tracked, 327
- dual wheels, 333
- torsion suspension, 333
- wheeled, 327
- wide tyres, 326, 333
Slaking, soil, 103, 153, 241, 295, 301
Slash-and-burn, 304
Slitting, soil, 615
Slurry, 345, 438

- distribution, 346, 491
- injection, 472
- tankers, 345, 353, 358, 472, 481
Small grains, see Cereals, Wheat, Barley,
Sodium, in soil, 206, 610
Soil conditioners, 198, 610
Soil loosening,
- by implements, see Tillage
- by plant roots/tubers, 291, 293, 306, 361
Soil mechanics, 6, 10, 23-43, 192, 208
- Critical State, Theory of, 76
Soil qualities, see Quality, soil,
Soil type,
- Aeric Haplaquept, 617
- Alfisol, 230, 289, 292, 293-4, 298, 301,
   306-7, 612
- alluvial, 376
- brown, 376
- chernozem, 370, 373, 376, 379-80
- chestnut, 367
- duplex, 332
- gley-podzolic, 376
- Latosol, 291
- loess, 181, 183, 355, 376, 381, 613
- Luvisol, 63
- Mollisol, 157
- moraine, 376
- Oxisol, 294, 302, 306, 612
- peat, 322, 359
- podzolic, 336
- polder, 227
- pumice, 322, 335
- red, 376
- red-brown, 526
- rendzina, 370, 613
- salt-affected, 367
- sod-podzolic, 376
- sod-spodosolic, 613
- Typic Hapludoll, 83
- Udic Haploboroll, 83
- Ultisol, 294, 306
- Vertisol, 291
- Xanthozem, 320
Solanum tuberosum, see Potato,
Solute(s), in soil, 152
Sorghum, 293, 617
- yield, 293, 307
Sorghum bicolor, see Sorghum,
Soybean, 267-8, 273, 281, 288, 291-2,
   373, 544, 615

- grain yield, 275-8, 291-2, 302, 306, 376, 380
- reduced nodulation, 267, 291
- roots, 13, 252, 291
Spanner, see Gantry,
Specific gravity, soil, 114
Specific volume, soil, 114
Speed, of vehicles, see Vehicle(s),
Spraying machinery, 277
Spreading/Spreaders,
- fertilizer, 277
- manure, 277, 346
- slurry, see Slurry,
Springtail(s), 224
Spruce, 323
- Englemann, 321
- Norway, 319, 324-5
- Sitka, 319
- white, 319
Steam engine, 4
- compaction under, 4, 522
- mass, 4
- running gear, 5
Steering,
- automatic, 534
- furrow follower systems, 612
- multi-axial, 471, 474, 495, 573
Stizolobium spp., 306
Stones, see also Gravel, 117, 127, 294
- problems in soil measurements, 134
Storage, of water, 153
Strain,
- homogeneous, 30, 36
- normal, 31, 48
- octahedral normal, 32
- octahedral shear, 32
- principal, 31
- shear, 31-2, 48
- theory, 29-37, 48
- vertical, 35
Straw,
- effect on soil strength, 198
Stream tube, 34, 36
Streamlines, 34
Strength, compressive, soil, 201
- brazilian test, 201
- effect of compression speed, 37-8
- uniaxial test, see Uniaxial compression
   test,
Strength, shear, soil, 59
- direct shear test, 55, 200-1

- effect of aggregates, 60-1
- effect of shear speed, 54
- ring shear test, 201
- torsional shear test (vane), 201, 204, 462
Strength, soil, 191-209, 631-3
- by penetration tests, 12, 200
- by torsion shear tests, 201
- by vane shear tests, 201
- cone penetration tests, *see* Penetrometer, soil, cone,
- effect of aggregates, 59-60, 193-5
- effect of clay content, 60
- effect of clay mineral, 192
- effect of colloid fraction, 197
- effect of compactness, 60, 194, 205-6
- effect of cracks, 194, 197
- effect of freezing/thawing, 58, 60, 206
- effect of mineral content, 195
- effect of organic matter, 60, 192, 195, 198
- effect of root channels, 219
- effect of soil volume, 194, 197
- effect of texture, 57, 192, 195
- effect of tillage, 58
- effect of water content, 199
- effect of water potential, 57, 60, 199
- effect of wetting/drying, 60, 199, 206
- factors, 24, 55
- hard-setting, 218
- influence on emergence, 191
- measurements, 49-55, 191, 199-202, 209
- structural, 448-9, 465
- variation within field, 204
- variation within profile, 193, 202, 204
Strength, tensile, soil, 268
- direct tension test, 201
- effect of clay content, 197
- effect of water content, 206
- textural, 197, 199
Stress, within soil, 46
- at/below track-soil contact, 409-11
- at/below tyre-soil contact, 29, 63, 81, 403-9
- deviatoric, 79
- distribution with depth, 62-4, 77-85, 450-1, 481-2
- dynamic, 65
- effect of inflation pressure, 453-4
- effect of load, 29-30, 41, 63, 81, 481-3

- effect of tyre/track width, 512
- effect of vehicle speed, 63
- effective, 46
- horizontal normal, 81
- lateral distribution, 404, 407-9, 449
- measurement, *see* Pressure sensors,
- neutral, 46
- normal, 26, 46, 84, 391
- octahedral normal, 29, 84
- octahedral shear, 29
- pre-compaction, 37, 448
- pre-compression, 49, 58-9, 64
- pre-consolidation, 37
- principal, 28, 76, 83
- radial, 391, 403-4, 407
- shear, 24, 46, 81, 106, 108, 391
- tangential, 75-6, 391, 405, 409
- tensor, 25-6
- theory, 24-9
- under lugs, 407-9
- under tracks, 511
- under wheels, 481-4
- uniaxial, 448
- vertical normal, 81, 83, 492
- vibrational, 391
Stress-strain relationships, 42-3, 49, 57, 77 85, 191, 632
- measurement, 37-41
- time dependence, 203
Structure, soil, 46, 92, 216, 366
- definition, 92
- effect on crop growth, 46, 257, 290
- effect of crop root systems, 298
- effect on water movement, 103, 154-5
- influenced by tillage, 92, 108
- microstructure, *see* Microstructure, soil,
- morphological evaluation, 202
- regeneration, 185, 361
- stability, 92, 219, 268, 291, 294, 366
- variability, 239
Stylosanthes guianesis, 298, 306
Subsoil,
- anaerobic, 227, 248
- compaction, 6, 103, 154, 218-9, 246, 276 367, 373, 376, 453, 481-9
- gravel, 301
- hydraulic properties, 154-5
- loosening, 101, 487-9
- microstructure, 101
- pedal properties, 154-5

- recompaction, 204, 359, 454
Subsoiler, *see also* Ripper, Chisel
    plough, 614-6
- draught, 544
- mini, 359
- plough-mounted, 103, 488-9
- rotary, 614
- soil breakup, 383, 613-4
- winged, 335
Subsoiling, 16, 103, 283, 302, 359, 475,
    487-9, 558, 613-5
- chisel plough, 307
- crop responses, 306, 360, 576, 589, 613-5
- cross-, 615
- effect of soil type, 283, 613
- effect of subsequent traffic, 283
- effect of water content, 283, 359, 606-7,
    613
- effect of weather conditions, 283, 360,
    613
- effect on soil conditions, 613
- effectiveness, 283, 306, 613-4
- implements, 614-5
- persistence, 283, 613-4
- under-the-row, 615
- with gypsum, 615
- with lime, 613
Suction, *see* Water, soil, potential,
Sugar beet, 254, 265-6, 368, 373, 378-9,
    423-7, 431, 433-7, 441, 457, 471, 473-5,
    481, 493, 529, 605, 608, 613
- effect of controlled traffic, 525, 527
- leaves, 223
- quality, 527
- sugar content, 373
- use of LGP, 471
Sugar cane, 304, 502
Sunflowers, 526
Surface energy, 59
Surface, soil, 157-9
- roughness, 153, 458, 606, 612
- water storage, 154, 240-1, 367
Surface tension, water, 61
Sweeps, duck foot, 614
Swelling/Shrinkage effects, 55, 208, 376, 379

Tanker, slurry, 471-2, 481
Temperature, soil,
- effect of compactness, 241, 252, 257, 349
- effect on plant growth, 238

- influence of evaporation, 241
- reduction with no-tillage, 301
Termite(s), 224, 229, 301
Texture, soil,
- effect on compaction, 55, 366, 376
Thuja plicata, see Cedar, western red,
Tillage, *see also* No tillage,
- ameliorating compaction, 272, 612
- compaction during, 153, 216, 335
- conservation, 217-8, 229, 282
- critical depth, 335
- deep, 290, 292, 295, 306
- effect of soil water, 153, 185
- effect on hydraulic properties, 152-3, 156
- effect on soil biota, 216
- in fall, 272
- in forest soils, 335-6
- inversion, 216
- minimum, 58
- non-inverting, 228, 359
- over-intensive, 300
- primary, 272
- reduced, 217-8, 282
- requirement, 16, 634
- pan, 154-5, 205, 290, 603
- ridge, 155, 157
- rotary, 613
- shallow, 289, 307
- slit/slot, 360, 614-5
- strip catchment, 302
- to control compaction, 306-7
- traffic interaction, 15
- winch drawn, *see* Cable systems,
- zero, *see* No tillage,
- zonal, 302
Tillage tool(s),
- sharpness, 608
- slant-leg, 359
- to remove compaction, 359
- vibrating/vibratory, 614
- wear, 544
- weight, 429-431
Timeliness, field operations, 277, 280,
    475, 490, 523, 527, 533, 544, 552,
    562, 575, 598
Timothy, 344, 352
- roots, 347
Tine, *see also* Plough, chisel, 612
Tire(s), *see* Tyre(s),
Tomato, 526, 531

Track(s), *see also* Tractor, tracked/crawler,
  402-3, 501-18, 605
- compaction capability, 412, 510
- compaction under, 132-3, 483, 510-2
- construction, 503, 504-6
- flexible, 402, 410-1, 503, 505-6
- ground pressure, 365, 370, 402-3, 483,
  510-1
- grouser, 505-6, 508-9
- half-, 403, 503, 509
- mean maximum pressure, 411
- nominal contact pressure, 402
- on harvesters, 502-3, 507
- on steam engines, 5
- on wagons, 507
- pneumatic, 507, 518, 605
- rigid, 402, 410, 503, 505-6
- roadwheel, 402, 411, 503, 506, 517
- rolling resistance, 509
- rubber-belt, 403, 411, 506, 508, 511-2,
  518, 560
- stress distribution under, 409-11, 509-11,
  517
- suspension, 402-3, 411, 517
- tension control, 403, 411, 506, 517
- width, 411, 510-2
Traction, 437, 465
- coefficient, 466
- effect of ground pressure, 603
- effect of inflation pressure, 603
- under tracks, 508, 574
Tractor, tracked/crawler, *see also*
  Track(s), 402-3, 502-3, 508-18, 550, 560,
  605
- ballasting, 516
- cost, 516-7
- crop response, 512-5, 605
- disadvantages, 503, 515-8, 560
- maize responses, 513
- mass, 370
- operator comfort, 515
- turning characteristics, 516
- use in forestry, 331, 333
- use in land clearing, 297
- use in seedbed preparation, 502
- use on roads, 506, 515-6
- vibration under, 366, 511
- weight, *see* mass,
Tractor, wheeled,
- cost, 560

- development, 5, 7, 365
- four-wheel drive, 268, 429, 469, 501, 550,
  559-60
- ground pressure, 365, 370
- internal combustion engine, 5
- mass, 7, 268, 365, 370, 423, 479-80, 501,
  549-50
- mass/power ratio, 5, 423, 428-9, 550
- optimum power, 552
- power/size, 268, 541, 549-53
- power trends, 7, 365, 423, 479-80, 501
- tyre selection, 469-70
- weight, *see* mass,
- weight/power ratio, *see* mass/power ratio
- wide wheel track, 525, 561-2, 571-3
Traffic, 418-42, 631
- area/coverage of wheelings, 327, 345, 35,
  419, 423-4, 529
- controlled, *see* Controlled traffic,
- distribution, 204, 279, 418-9, 438, 442
- effects of, 43, 152, 179, 183, 216, 228, 29,
- forage crops, 344-5
- harvest, 154, 183, 282, 304, 325-31, 350,
  435, 442
- hazard, 424-5
- high ground pressure, 148
- in forests, 317-37
- in plough furrow, 155
- index, 356
- intensity, *see also* Traffic Intensity, 345,
  359, 487, 490, 495
- inter-row, 272, 274, 281-2, 374
- lanes, *see* Wheel lanes,
- loading events, 434-42
- loading time, 428
- low ground pressure, 148, 349, 447-75
- mapping, 438-41
- patterns in fields, 438-42
- post-harvest, 436
- pre-plant, 177, 180, 183, 293, 368, 434
- quantification, 418-23
- reduced, 300
- restriction to headlands, 282
- zero, *see* Zero-traffic,
Traffic Intensity, 419, 421-7
Trafficability, 192, 208, 227, 238, 598
- effect of compaction on, 240
- under tracks, 509-10
Trailer, 471
- grain, 268, 277, 479

- sugar beet, 268
Tramline(s), 16, 268, 463
- permanent, see Wheel lane(s),
- temporary, 561, 576
Transpiration, 253
Tree(s),
- deep-rooting, 230
- establishment, 318-20
- form, 325
- germination, 319-20
- growth, 319, 321, 324-5
- response to compaction, 317-37, 324,
    329
- response to fertilizers, 335-7
- response to tillage, 335-6
- root grafting, 318
- root growth, 319, 321, 328-9, 331
- root mat, 331
- root pruning, 323-4
- root rot, 325
- root system, 322-3
- seedlings, 319-22
- volume, 321-2
Trefoil, birdsfoot, 344
Triaxial test, 38-9, 40, 42, 76, 148, 202
- consolidated drained, 54
- consolidated undrained, 54
- ram speed, 40
- size/shape of samples, 38
- unconsolidated undrained, 54
Trifolium pratense, see Clover, red
- T. repens, see Clover, white,
- T. spp., see Clover,
Triticum aestivum, see Wheat,
Tropics, 229-30, 287-309, 634
- semiarid, 290, 293, 302, 307, 612
Tsuga heterophylla, see Hemlock, western,
Tyre(s), see also Wheel(s), 392-401
- carcass stiffness, 24, 63, 392, 394, 399, 406
- code markings, 392-6
- compaction under, 132-3, 411
- contact area, 76, 399-400, 555
- contact coefficient, 555-6, 558
- cost, 554-6
- cross-ply, 392, 397
- deflection, 397-401
- diameter, 73
- dual, 333, 348, 359, 374, 456, 468, 474,
    482, 558, 603, 606
- dynamic behaviour, 391, 401, 466

- flotation, 393, 472-3
- front, 559
- ground pressure, 427, 449-56
- high ground pressure, 148
- inflation pressure, 17, 41, 72-3, 76, 358,
    395-7, 399, 406, 438, 468, 472, 561
- load capacity, 396-7
- low ground pressure, 148, 333, 349, 357,
    359, 374, 394, 447-75, 554-8, 562, 604-5
- lug(s), 74-6, 401, 408, 449
- markings, 394-6
- ply-rating, 73, 394, 555
- radial, 392, 397
- size, 392-6
- smooth, 74
- speed relations, 396, 467, 470
- standards, 392-7
- tread pattern, 392
- triple, 559
- wear, 544
- wide, see low ground pressure,
- width problems, 470-1
Tyre and Rim Association, 392
- Scandinavian, 394

Uniaxial compression test, 37-8, 42, 76, 82-3,
    148, 150, 178
- confined, 53-4, 202, 448
- unconfined, 53
Upper plastic limit, see Liquid limit,

Vane shear tests, see Shear tests, vane,
Vehicle(s), see also Machinery,
- changes in mass with time, 265, 268,
    271, 276, 479-81
- compaction capability, 411-2, 418
- design, 17
- forestry, 481, 493
- high axle load, 277-8, 485-7
- high ground pressure, 455
- light weight, 17, 493, 545
- low ground pressure (LPG), 455
- mass, 17, 345, 365, 493
- military, 402
- operation criteria, 571
- reduced mass benefits, 489-93
- speed, 560-1, 605-6
- transport, 471-3, 556-8
- weight, see mass,
Vibration, 65

- under tracks, 366, 511
- under tyres, 366
Vigna unguiculata, see Cowpea,
Viscosity, soil, 40
Void(s), soil, *see also* Pore(s), soil,
- planar, 154
- tillage, 93, 98-100, 103, 153-4, 229
- tubular, 154
Void ratio, 114-5, 148
- structural, 196, 206
Void-size distribution, *see* Pore-size
    distribution,
Volumenometer, 121

Wagon, *see* Trailer,
Waste, organic, *see* Organic waste,
Water, plant,
- stomatal resistance, 252-3
- stress, 257
- transpiration, 253, 257
- uptake, 255
Water, soil,
- available to plants, 245
- capillary flux from groundwater, 243, 246
- content, 114, 368
- diffusivity, 144
- effect on aeration, 257
- effect on soil strength, 599
- infiltration, *see* Infiltration of water, into
    soil,
- matric potential, 143, 155-6, 330, 600
- optimum for compaction, 330
- potential, 105, 179, 257, 368
- retention, 141, 143-4, 147, 156
- sorptivity, 146-9, 152, 157
- storage, 153
- suction, *see* Water, soil, potential,
- supply to roots, 13, 238, 241-7
- surface, storage, 154
- uptake by roots, 238, 244, 251-3, 381
Water table, 243, 246-7, 253, 601
Waterlogging, 613, 627, 629
- transient, 178, 525
Weather, 15, 239, 378-80, 562-3, 635
- crop responses to, 238, 274-6, 379, 551-3
- rainfall, 15, 295, 559, 562
Weight, transfer, 472
Weight/power ratio, *see* Tractor, wheeled,
    mass/power ratio,
Wetting and drying, soil, 156, 185, 203, 206,
    248, 270, 330, 484, 510

Wheat, 16, 228-9, 266-7, 270, 275-6, 288,
    290-1, 302, 306, 370-1, 381, 458, 462,
    491, 525-7, 577-92
- roots, 290, 380
- winter, 156, 183, 269, 373, 376-7, 379-80,
    424-7, 433, 435-7, 441, 457, 459, 460,
    463-5, 527, 544, 547, 605, 608
Wheel(s), *see also* Tyre(s),
- cage, 132-3, 605
- compaction capability, 11, 391, 406, 411-2,
    418, 464
- contact area, 399-400
- dual, 333, 348, 374, 456, 468, 471, 474,
    482, 558, 603, 606
- dynamic load, 73
- load, 41, 396-7, 406, 438
- multiple front, 559
- packing, 271
- rigid, 76
- slip, 65, 401, 466, 544
- spacing problems, 281-2, 561, 571
- triple, 559
Wheel lane(s), 525
- permanent, 525, 527, 532, 534, 575,
    607
- temporary, 607
- use of stones, 607
Wheel roads, *see* Wheel lane(s),
Wheel paths, *see* Wheel lane(s),
Wheel track, 204-5
Wheel way, *see* Wheel lane(s),
Wheeling(s),
- area, 204, 366, 484
- multiple pass effects, 484
- number, 487
Wood, production, 317, 325
Work rate, in field, 574-5, 579-85
Workability, soil, 208, 240, 598

X-ray transmission tomography, 126

Young's modulus, 79-80

Zea mays, see Maize,
Zero-tillage, *see* No tillage,
Zero-traffic, *see also* Controlled traffic, 456,
    458-65
- crop responses, 15, 179, 359, 456, 526-7,
    576
- soil responses, 183, 349